PROGRESS IN COLLOID & POLYMER SCIENCE

Editors: F. Kremer (Leipzig) and G. Lagaly (Kiel)

Volume 100 (1996)

Trends in Colloid and Interface Science X

Guest Editors:

C. Solans, M. R. Infante
and M. J. García-Celma (Barcelona)

 Springer

STEINKOPFF
DARMSTADT

IV

ISBN 978-3-662-16064-0
ISSN 0340-255 X

Die Deutsche Bibliothek –
CIP-Einheitsaufnahme

Trends in colloid and interface science.

Früher begrenztes Werk
in verschiedenen Ausg.

10 (1996)
(Progress in colloid & polymer science ;
Vol. 100)
ISBN 978-3-662-16064-0
ISBN 978-3-7985-1665-6 (eBook)
DOI 10.1007/978-3-7985-1665-6
NE: GT

© 1996 by Springer-Verlag Berlin Heidelberg
Originally published by Dr. Dietrich Steinkopff Verlag GmbH & Co. KG, Darmstadt in 1996
Softcover reprint of the hardcover 1st edition 1996

Chemistry editor: Dr. Maria Magdalene Nabbe; English editor: James C. Willis; Production: Holger Frey, Bärbel Flauaus.

Type-Setting: Macmillan Ltd., Bangalore, India

Progr Colloid Polym Sci (1996) V
© Steinkopff Verlag 1996

PREFACE

The IXth European Colloid and Interface Society (ECIS) Conference was held in Barcelona, 17 – 22 September, 1995. It was attended by over 320 participants from 30 different countries. The scientific program was composed of 301 contributions (70 lectures and 231 poster communications). It covered the theoretical, experimental and technical aspects of almost all fields of modern Colloid and Interface Science.

This volume contains a selection of the contributions presented at the Conference. It is divided into the following sections:
– Surfactant aggregates: Micelles, vesicles and liquid crystals
– Colloidal particles: Interaction, structure and aggregation
– Emulsions and concentrated systems
– Microemulsions
– Mixed colloidal systems
– Rheology
– Biocolloids
– Interfaces, films and membranes.

On behalf of the Organizing Committee, we would like to thank all participants for their scientific contributions which resulted in a successful Conference. We are especially grateful to the members of the Scientific Committee, M. Corti, H. Hoffmann, D. Langevin, H. Lekkerkerker, B. Lindman, M. A. López Quintela, C. Mans, P. Schurtenberger and Th. F. Tadros. They helped us in the difficult task of selecting the contributions for oral presentations from a rather large number of high quality Abstracts.

We gratefully acknowledge the support of the Faculty of Chemistry of the University of Barcelona. They kindly allowed us to use their facilities in which to hold the Conference.

We also gratefully acknowledge the financial support from the European Colloid and Interface Society and from the following Spanish Organizations: Generalitat de Catalunya, Ministerio de Educación y Ciencia (MEC), Universitat de Barcelona (UB), Real Sociedad Española de Química (RSEQ), Consejo Superior de Investigaciones Científicas (CSIC), Mettler Toledo S.A.E., Petroquímica Española S.A., ICI Española S.A. and Optilas Ibérica S.A.

We sincerely thank the staff of the "Department de Tecnologia de Tensioactius" (CSIC) and the "Department d'Enginyeria Química i Metallurgia" (UB) for their enthusiastic and efficient collaboration in the organization of the Conference. We are also indebted to Mr. and Mrs. Guruswamy for their valuable and generous help.

On behalf of the Organizing Committee,

C. Solans, M. R. Infante, M. J. García-Celma

Progr Colloid Polym Sci (1996) VII
© Steinkopff Verlag 1996

CONTENTS

Surfactant aggregates: Micelles, vesicles and liquid crystals

H. Kunieda, K. Shigeta,   Formation and structure of reverse vesicles 1
K. Nakamura, T. Imae:

H. Edlund, M. Bydén,   Phase diagram of the 1-dodecyl pyridinium bromide-dodecane-water sur-
B. Lindström, A. Kahn: factant system 6

K. Berger, K. Hiltrop: Characterization of structural transitions in the SLS/n-alcohol/water
system 9

T. Kato: Surfactant self-diffusion and networks of wormlike micelles in concentrated solutions on nonionic surfactants 15

T. Lin, Y. Hu, W.-J. Liu, J. Samseth,   Thermodynamic theory and experimental studies on mixed short-chain
K. Mortensen: lecithin micelles 19

V. Babić-Ivančić, D. Škrtic,   Phase behavior in sodium cholate/calcium chloride mixtures 24
N. Filipović-Vinceković:

J. R. Chantres, M. A. Elorza,   The effect of sub-solubilizing concentrations of sodium deoxycholate on
B. Elorza, P. Rodado: the order of acyl chains in dipalmitroylphosphatidylcholine (DPPC).
Study of the fluorescence anisotropy of 1,6-diphenyl-1,3-5-hexatriene
(DPH) 29

D. Težak, M. Martinis, S. Punčec,   Multifractality of lyotropic liquid crystal formation 36
I. Fischer-Paković, S. Popović:

Y. A. Shchipunov, E. V. Shumilina: Surfactant aggregation and response of ion-selective electrode 39

R. Bikanga, P. Bault, P. Godé,   3-deoxy-s-alkyl-d-glucose derivative micelle formation, CMC and
G. Ronco, P. Villa: thermodynamics 43

A. K. Van Helden: Computer simulation studies of surfactant systems 48

Colloidal particles: Interaction, structure and aggregation

V. Reus, L. Belloni, T. Zemb,   Equation of state of a colloidal crystal: An USAXS and osmotic pressure
N. Lutterbach, H. Versmold: study 54

R. H. Ottewill, A. R. Rennie: Interaction behaviour in a binary mixture of polymer particles 60

V. Peikov, Ts. Radeva, S. P. Stoylov,   Electric light scattering from polytetrafluorethylene suspensions. II. In-
H. Hoffmann: fluence of dialysis 64

M. Pailette, S. Brasselet, I. Ledoux,   Two photon Rayleigh scattering in micellar and microemulsion systems . 68
J. Zyss:

A. Fernández-Barbero,   Dynamic scaling in colloidal fractal aggregation: Influence of particle sur-
M. Cabrerizo-Vílchez,   face charge density 73
R. Martínez-García:

H. Verduin, J. K. G. Dhont: Effects of shear flow on a critical colloidal dispersion: a light scattering
study 81

R. M. Santos, J. Forcada: Synthesis and characterization of latex particles with acetal functionality . 87

B. Gerharz, H. J. Butt, B. Momper: Morphology of heterogenous latex particles investigated by atomic force
microscopy 91

M. C. Miguel, J. M. Rubí: Dynamic properties of magnetic colloidal particles: Theory and experiments 96

A. Cebers: Two-dimensional concentration domain patterns in magnetic suspensions: Energetical and kinetic approach 101

O. Valls, M. L. Garcia, X. Pagés, J. Valero, M. A. Egea, M. A. Salgueiro, R. Valls: Adsorption-desorption process of Sodium Diclofenac in Polyalkylcyano-acrylate nanoparticles 107

P. Verbeiren, F. Dumont, C. Buess-Herman: Determination of the complex refractive index of bulk tellurium and its use in particle size determination 112

J. Bonet Avalos, A. N. Semenov, A. Johner, J. F. Joanny, C. C. van der Linden: Structure of adsorbed polymer layers: Loops and tails 117

Emulsions and concentrated systems

J. Müller, T. Palberg: Probing slow fluctuations in nonergodic systems: Interleaved sampling technique 121

F. Renth, E. Bartsch, A. Kasper, S. Kirsch, S. Stölken, H. Sillescu, W. Köhler, R. Schäfer: The effect of the internal architecture of polymer micronetwork colloids on the dynamics in highly concentrated dispersions 127

R. Pons, G. Calderó, M. J. García, N. Azemar, I. Carrera, C. Solans: Transport properties of W/O highly concentrated emulsions (gel-emulsions) 132

J.-L. Salager: Quantifying the concept of physico-chemical formulation in surfactant-oil-water systems – State of the art 137

G. Marion, K. Benabdeljalil, J. Lachaise: Interbubble gas transfer in persistent foams resulting from surfactant mixtures 143

Yu. V. Shulepov, S. Yu. Shulepov: Equilibrium states and structure factor of concentrated colloidal dispersions in optimized random phase approximation 148

A. Kasper, S. Kirsch, F. Renth, E. Bartsch, H. Sillescu: Development of core-shell colloids to study self-diffusion in highly concentrated dispersions 151

M. Olteanu, S. Pertz, V. Raicu, O. Cinteza, V. D. Branda: Concentrated graphite suspensions in aqueous polymer solutions 156

Microemulsions

M. Gradzielski, D. Langevin, B. Farago: Experimental investigation of the structure of nonionic microemulsions and their relation to the bending elasticity of the amphiphilic film 162

F. Bordi, C. Cametti, P. Codastefano, F. Sciortino, P. Tartaglia, J. Rouch: Effect of salinity on the electrical conductivity of a water-in-oil micro-emulsion 170

M. Camardo, M. D'Angelo, D. Fioretto, G. Onori, L. Palmieri, A. Santucci: Dielectric relaxation of microemulsions 177

M. G. Giri, M. Carlà, C. M. C. Gambi, D. Senatra, A. Chittofrati, A. Sanguineti: Percolation in fluorinated microemulsions: A dielectric study 182

S. Amokrane, P. Bobola, C. Regnaut: Adhesive spheres mixture model of water-in-oil microemulsions 186

C. Vázquez-Vázquez, J. Mahía, M. A. López-Quintela, J. Mira, J. Rivas: Preparation of Gd_2CuO_4 via sol-gel in microemulsions 191

S. M. Andrade, S. M. B. Costa: Fluorescence studies of the drug *Piroxicam* in reverse micelles of AOT and microemulsions of Triton X-100 195

Mixed colloidal systems

V. Degiorgio, R. Piazza, G. Di Pietro: Depletion interaction and phase separation in mixtures of colloidal particles and nonionic micelles 201

K. Kostarelos, Th. F. Tadros, P. F. Luckham: The effect of monovalent and divalent cations on sterically stabilized phospholipid vesicles (liposomes) 206

N. Fauconnier, A. Bee, J. Roger, J. N. Pons: Adsorption of gluconic and citric acids on maghemite particles in aqueous medium 212

D. Bastos-Gonzáles, R. Hidalgo-Alvarez, F. J. de las Nieves: Influence of heat treatment on the surface properties of functionalized polymer colloids 217

N. Stubičar, K. Banić, M. Stubičar: Kinetics of crystal growth of α-PbF$_2$ and micellization of non-ionic surfactant Triton X-100 at steady-state condition 221

O. Lopez, A. de la Maza, L. Coderch, J. L. Parra: Selective solubilization of the stratum corneum components using surfactants 230

V. Sovilj, P. Dokic, M. Sovilj, A. Erdeljan: Influence of surfactant-gelatin interaction on microcapsule characteristics 235

Rheology

S. von Hünerbein, M. Würth, T. Palberg: Microscopic mechanisms of non-linear rheology of crystalline colloidal dispersions 241

A. Guerrero, P. Partal, M. Berjano, C. Gallegos: Linear viscoelasticity of O/W sucrose-palmitate emulsions 246

J. Llorens, E. Rudé, C. Mans: Structural models to describe thixotropic behavior 252

J. Castle, A. Farid, L. V. Woodcock: The effect of surface friction on the rheology of hard-sphere colloids . . . 259

A. Cerpa, M. T. Garcia-González, P. Tartaj, J. Requena, L. R. Garcell, C. J. Serna: Rheological properties of concentrated lateritic suspensions 266

Biocolloids

N. L. Burns, K. Holmberg: Surface charge characterization and protein adsorption at biomaterials surfaces 271

R. Bru, J. M. López-Nicolás, A. Sánchez-Ferrer, F. García-Carmona: Cyclodextrins as molecular tools to investigate the surface properties of potato 5-lipoxygenase 276

K. Holmberg, S.-G. Oh, J. Kizling: Microemulsions as reaction medium for a substitution reaction 281

S. Avramiotis, A. Xenakis, P. Lianos: Lecithin W/O microemulsions as a host for trypsin. Enzyme activity and luminescence decay studies 286

L. Molina, A. Perani, M.-R. Infante, M.-A. Manresa, M. Maugras, M.-J. Stébé, C. Selve: Synthesis and properties of bioactive surfactants containing β-lactam ring 290

C. Otero, L. Robledo, M. I. del Val: Two alternatives: Lipase and/or microcapsule engineering to improve the activity and stability of *Pseudomonas* sp. and *Candida rugosa* lipases in anionic micelles 296

T. Hianik, R. Krivánek, D. F. Sargent, L. Sokolikova, K. Vinceová: A study of the interaction of adrenocorticotropin-(1-24)-tetracosapeptide with BLM and liposomes 301

M. De Cuyper: Impact of the surface charge of magnetoproteoliposomes on the enzymatic oxidation of cytochrome c 306

Interfaces, films and membranes

J. R. Lu, J. A. K. Blondel, D. J. Cooke, R. K. Thomas, J. Penfold: The direct measurement of the interfacial composition of surfactant/ polymer mixed layers at the air-water interface using neutron reflection . 311

R. Miller, V. B. Fainerman, P. Joos: Dynamics of soluble adsorption layer studied by a maximum bubble pressure method in the µs and ms range of time 316

A. Prins, A. M. P. Jochems, H. K. A. I. van Kalsbeek, J. F. G. Boerboom, M. E. Wijnen, A. Williams: Skin formation on liquid surfaces under non-equilibrium conditions . . . 321

N. N. Kochurova, A. I. Rusanov: Dynamic surface properties of aqueous solutions 328

G. Bähr, P. Grigoriev, M. Mutz, E. John, M. Winterhalter: Electric potential differences across lipid mono- and bilayers 330

V. I. Gordeliy, V. G. Cherezov, A. V. Anikin, M. V. Anikin, V. V. Chupin, J. Teixeira: Evidence of entropic contribution to "hydration" forces between membranes 338

A. Arbuzova, G. Schwarz: Pore kinetics of mastoparan peptides in large unilamellar liquid vesicles . 345

G. Brezesinski, K. de Meijere, E. Scalas, W. G. Bouwman, K. Kjaer, H. Möhwald: Head-group variations and monolayer structures of diol derivatives . . . 351

I. Porcar, R. M. Garcia, C. M. Gómez, V. Soria, A. Campos: Macromolecules in ordered media III. A fluorescence study on the association of poly-2-vinylpyridine with a phospholipid bilayer 356

P. Jauregi, J. Varley: Lysozyme separation by colloidal gas aphrons 362

Author Index 368

Subject Index 370

Progr Colloid Polym Sci (1996) 100:1–5
© Steinkopff Verlag 1996

H. Kunieda
K. Shigeta
K. Nakamura
T. Imae

Formation and structure of reverse vesicles

Dr. H. Kunieda (✉) · K. Shigeta
Department of Physical Chemistry
Division of Materials Science
and Chemical Engineering
Faculty of Engineering
Yokohama National University
Tokiwadai 156, Hodogaya-ku
Yokohama 240, Japan

K. Nakamura
Department of Living Science
Faculty of Education
Niigata University
Igarashi Nino-cho 8050
Niigata 950-21, Japan

T. Imae
Department of Chemistry
Faculty of Science
Nagoya University Chikusa-ku
Nagoya 464, Japan

Abstract Both normal and reverse vesicles can be formed in the mixture of sucrose dodecanoates with different hydrocarbon chain number. The conditions to produce both types of vesicles are discussed using the geometrical packing model. The conditions are roughly the same for both types of vesicles. The formation of reverse vesicles is confirmed by means of video-enhanced microscopy (VEM) and cryo-transmission electron microscopy (Cryo-TEM).

Key words Reverse vesicles – sucrose alkanoates – cryo-electron microscopy – video enhanced microscopy

Introduction

Contrary to normal vesicles formed in water, recently it was found that closed bimolecular layers can be also formed in nonpolar media [1–11]. The self-organizing structure was named reverse vesicles. The orientation of amphiphilic molecules in bilayers of reverse vesicles is opposite to that in normal vesicles. Reverse vesicles were found in various mixed surfactant systems in which hydrophilic and lipophilic surfactants were combined. Correlation between the surfactant molecular shape and resulting self-organizing structure can be explained by geometrical packing model [12]. Since reverse vesicles are new self-organizing structures the packing condition of surfactant has not been reported.

Sucrose alkanoates are unique and biocompatible surfactants which possess a strong hydrophilic sucrose ring. Ordinary polyoxyethylene-type nonionic surfactants are in general completely miscible with hydrocarbons, but do not self-associate in the absence of water [13]. On the other hand, the monomeric solubilities of sucrose alkanoates in nonpolar solvents are rather low and they tend to self-associate from self-organizing structure in nonpolar media.

In this paper, the condition to produce reverse vesicles is discussed and the formation of reverse vesicles in sucrose alkanoate systems is reported. The morphology of reverse vesicles is shown by means of video enhanced microscopy (VEM) and cryo-transmission electron microscopy (Cryo-TEM).

Experimental

Materials

Sucrose dodecanoates, L-1695 and L-595 were kindly supplied by Mitsubishi Chemical Corp. L-1695 consists of 83.6 wt% sucrose monododecanoate, 15.2 wt% sucrose didodecanoate, and 1.2 wt% sucrose tridodecanoate. L-595 consists of 30.3 wt% sucrose monododecanoate, 39.3 wt% sucrose didodecanoate, and 30.4 wt% sucrose tridodecanoate. L-1695 is water-soluble whereas L-595 forms milky dispersion in water. By mixing L-1695 and L-595, we can prepare sucrose dodecanoates with different number of hydrophobic chains from 1.13 to 1.84.

Methods

Sample preparation

L-1695 and L-595 were dissolved in methanol in order to obtain a homogenized mixture. After evaporation of methanol in vaccuum, isooctane was added to the dried mixture. The sample was sonicated by means of an ultrasonicator (Shimadzu, USP-50) at 9 W for 5 min. The vesicular dispersion was twice extruded using a 0.6 μm Millipore filter. Branched hydrocarbon, isooctane is used in order to avoid the crystallization of solvent when vitrifying.

Optical, and electron microscopy

VEM: A differential-interference-phase-contrast (Nomarski-type) microscope (Nikon, X2F-NTF-21) equipped with an image processor (Hamamatsu Photonics Co., Argus 10) was used for VEM observation.

Cryo-transmission electron microscopy: A droplet of vesicular dispersion was placed on a TEM grid and quickly vitrified in liquid nitrogen at its freezing point. The frozen sample was observed at X20 000 magnification on a transmission electron microscope, Hitachi H-800, operating at 100 kV.

Results and discussion

Condition to produce reverse vesicles

For normal vesicles in water, when the surfactant bilayer forms a closed structure, the interfacial part of the outer layer is the most compressed part as is shown in Fig. 1(a).

Fig. 1 Schematic representation of normal and reverse vesicles. A_s is an interfacial area per one hydrophilic group (sucrose ring) and l is hydrocarbon chain length

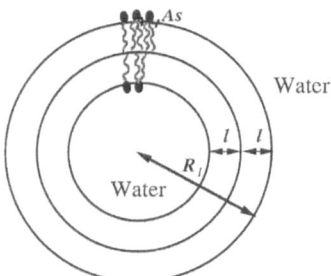

Normal vesicle

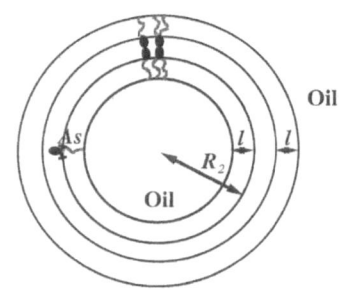

Reverse vesicle

Table 1 Packing parameter and formation of vesicles

SE1	SE2	Average number of hydrocarbon chains	Normal vesicles in water			Reverse vesicles in oil		
			$v/A_s \cdot l$	R (nm)	Formation	$v/A_s \cdot l$	R (nm)	Formation
100 wt%	0 wt%	1.13	0.49	—	M	0.55	3.4	RV
90	10	1.19	0.52	—	M	0.58	3.7	RV
80	20	1.25	0.55	—	M	0.61	4.0	RV
70	30	1.31	0.57	—	M	0.64	4.3	RV
60	40	1.38	0.60	3.9	NV	0.67	4.7	RV
50	50	1.45	0.63	4.2	NV	0.71	5.2	RV
40	60	1.52	0.66	4.6	NV	0.74	5.9	RV
30	70	1.59	0.70	5.2	NV	0.78	7.0	RV
20	80	1.67	0.73	5.7	NV	0.81	8.1	RV
10	90	1.75	0.77	6.6	NV	0.86	11	RV
0	100	1.84	0.80	7.8	NV	0.90	15	RV

Progr Colloid Polym Sci (1996) 100:1–5
© Steinkopff Verlag 1996

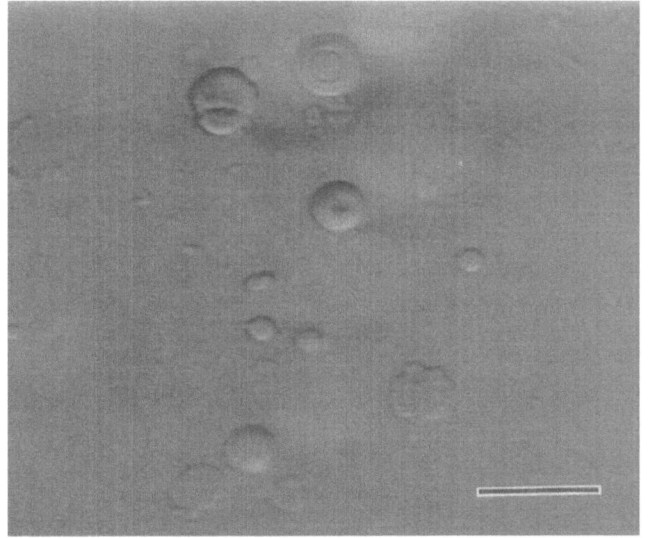

(a)

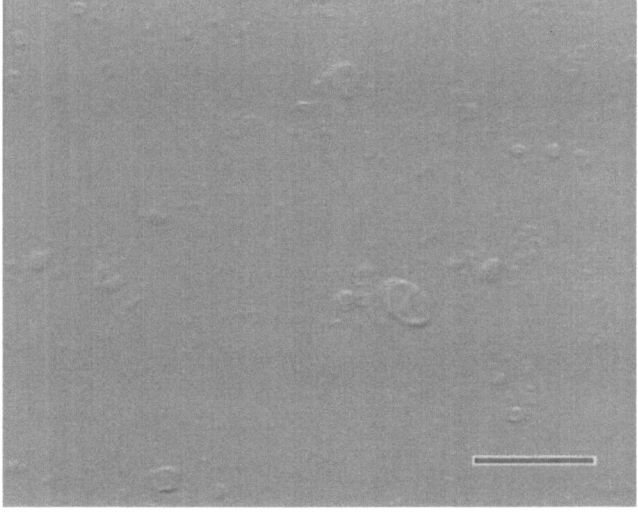

(b)

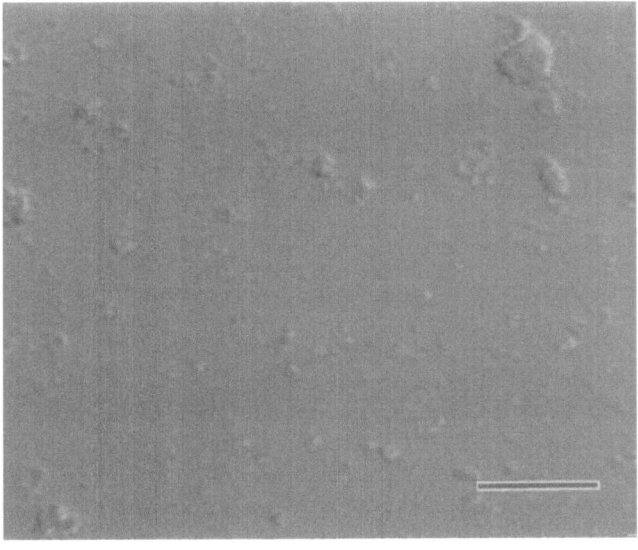

(c)

Fig. 2 VEM pictures of normal (a) and reverse (b and c) vesicles. (a) 3 wt% L-595, 97 wt% water. (b) 1 wt% L-1695, 99 wt% isooctane. (c) 1 wt% of equal weight mixture of L-1695 and L-595, 99 wt% isoctane. The bar indicates 25 μm

By using this consideration, Israelachvili proposed the following equation to determine the minimum possible radius of normal vesicles.

$$R_1 = \frac{l}{1 - v/A_s l},\qquad(1)$$

where l is the hydrophobic chain length of surfactant, v is the volume of hydrophobic chain and A_s is the interfacial area per surfactant molecule. Thus, in order to form normal vesicles in water, $v/A_s l < 1$ is the condition for surfactant. If this value is less than $1/2$, it is suggested that normal micelles form [12].

In the case of reverse vesicles, the orientation of amphiphilic molecules in the bilayer is opposite and the hydrophilic group is oriented toward the inside of the bilayer as is shown in Fig. 1(b). When the bilayer forms the closed structure, the interfacial part of the inner layer must be compressed. Therefore, as well as normal vesicles, the condition for the minimum size of reverse vesicles can be represented by

$$R_2 = \frac{l}{1 - v/A_s l}.\qquad(2)$$

Namely, the conditions to produce normal and reverse vesicles are similar. If the $v/A_s l$ exceeds unity, reverse micelles tend to form [16]. However, the lower limiting value for reverse vesicles has not been known yet.

Sucrose dodecanoate system

Monomeric solubilities of sucrose alkanoates in both water and oil are very small. Hence, it is considered that most of the surfactant molecules form self-organizing structure if the surfactant concentration is well above the CMC. Sucrose alkanoates have the same hydrophilic group and sucrose rings, and only the number of fatty acids attached to it is changed. By using Eqs. (1) and (2), we calculated the possible minimum radius of normal (R_1) and reverse (R_2) vesicles in the sucrose dodecanoate systems and the result are shown in Table 1. The calculated packing parameters are also shown.

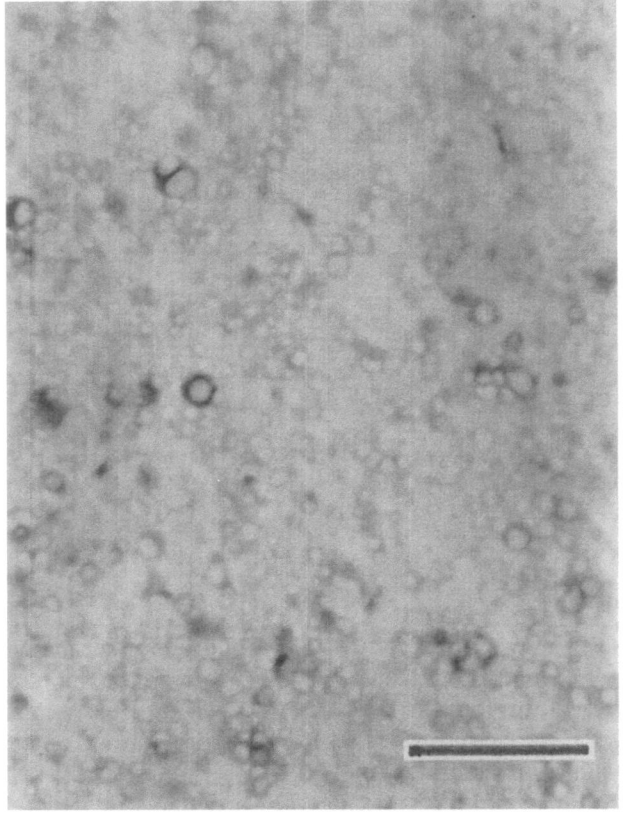

(a)

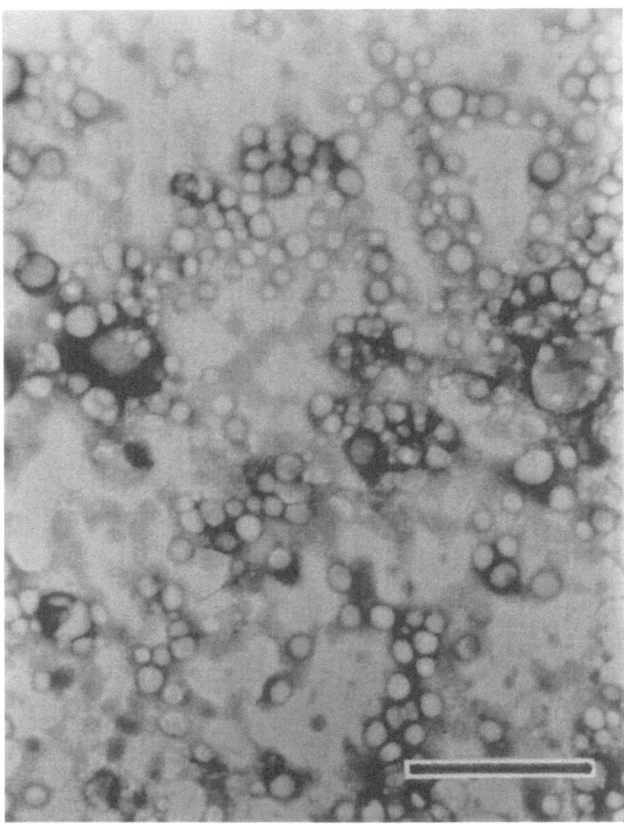

(b)

Fig. 3 Cryo-TEM picture of reverse vesicles. (a) 0.5 wt% L-1695, 99.5 wt% isooctane. (b) 0.5 wt% of equal weight mixture of L-1695 and L-595, 99.5 wt% isooctane. The bar indicates 1 μm

The volume (0.323 nm³ per one hydrocarbon chain of dodecanoate) and length (1.54 nm) of hydrophobic chain were calculated by Tanford equation [17] and the interfacial area ($A_s = 0.48$ nm² for normal vesicles $A_s = 0.43$ nm² for reverse vesicles) was obtained from the previously measured SAXS data for lamellar liquid crystal [10]. It is clear from Table 1 that the minimum radius of vesicles decreases with decreasing the average hydrocarbon chain number. It is suggested from the present rough calculation that both normal and reverse vesicles can be obtained in the present system.

In the above discussion, we assume that the mixing ratio of surfactants is always unchanged in vesicles. It is possible that distribution of single- or double-chain surfactants in the inner surfactant layer is different from that in the outer layer.

VEM and Cryo-TEM pictures of vesicles

Phase behavior of sucrose dodecanoate mixtures in water or isooctane was observed and the result is also indicated

in Table 1. In the case of water system, normal vesicles are changed to micelles as the hydrocarbon chain number decreases. On the other hand, formation of reverse vesicles was observed at any sucrose dodecanoate mixing ratio as is shown in Table 1. Figure 2 shows the VEM pictures for normal and reverse vesicles.

In Figs. 2(a)–(c), we can observe reverse vesicles whose size distribution is quite large. In order to verify the close structure of small particles, we observed the solutions by means of Cryo-TEM and the result is shown in Fig. 3.

Although the preparation method is exactly the same, the diameter of most of the reverse vesicles is approximately 0.1 μm or less in both L-1695 and L-1695 + L-595 mixture systems. As predicted from Table 1, the possible minimum sizes are similar in both systems though the surfactant mixing ratios are largely different.

Acknowledgment The authors wish to thank Ms. N. Kanei for VEM measurement. The authors also wish to thank Mitsubishi Chemical Corp. for supplying sucrose dodecanoates.

5

References

1. Kunieda H, Nakamura K, Evans DF (1991) J Am Chem Soc 113:1051
2. Kunieda H, Nakamura K, Davis HT, Evans DF (1991) Langmuir 7:191
3. Nakamura K, Machiyama Y, Kunieda H (1992) J Jpn Oil Chem Soc (YUKAGAKU) 41:480
4. Kunieda H, Yamagata M (1992) J Colloid Interface Sci 150:277
5. Kunieda H, Makino S, Ushio N (1991) J Colloid Interface Sci 147:286
6. Kunieda H, Nakamura K, Infante MR, Solans C (1992) Adv Mater 4:291
7. Kunieda H, Akimaru M, Ushio N, Nakamura K (1993) J Colloid Interface Sci 156:446
8. Kunieda H, Nakamura K, Olsson U, Lindman B (1993) J Phys Chem 97:9925
9. Ushio N, Solans C, Azemer N, Kunieda H (1993) J Jpn Oil Chem Soc (YUKAGAKU) 42:915
10. Kunieda H, Kanei N, Uemoto A, Tobita I (1994) Langmuir 10:4006
11. Nakamura K, Uemoto A, Imae T, Solans C, Kunieda H (1995) J Colloid Interface Sci 170:367
12. Israelachvili JN (1985) Intermolecular and Surface Forces Academic Press, Inc., London, Part III
13. Olsson U, Jonstromer M, Nagai K, Soderrman O, Wennerstrom H, Klose G (1988) Prog Colloid Polym Sci 76:75
14. Kunieda H, Ushio N, Nakano A, Miura M (1993) J Colloid Interface Sci 159:37
15. Rosen MJ (1989) Surfactants and Interfacial Phenomena. John Wiley & Sons. New York, Chapter 3
16. Kunieda H, Kanei N, Tobita I, Kihara K, Yuki A (1995) Colloid Polym Sci 273:584
17. Tanford C (1972) J Phys Chem 76:3020

Progr Colloid Polym Sci (1996) 100:6–8
© Steinkopff Verlag 1996

H. Edlund
M. Bydén
B. Lindström
A. Khan

Phase diagram of the 1-dodecyl pyridinium bromide–dodecane–water surfactant system

H. Edlund · M. Bydén
Dr. B. Lindström (✉)
Department of Chemistry and
Process Technology
Chemistry, Mid Sweden University
851 70 Sundsvall, Sweden

A. Khan
Physical Chemistry 1
Chemical Centre
Box 124
221 00 Lund, Sweden

Abstract The isothermal ternary phase diagram for the 1-dodecyl pyridinium bromide–water–dodecane surfactant system has been determined at 40 °C by ^{2}H NMR and polarizing microscopy methods. For the binary surfactant–water system, a normal hexagonal liquid crystalline phase and an isotropic normal micellar solution phase are identified.

On addition of dodecane, a new isotropic cubic liquid crystalline phase is formed between the micellar and the hexagonal phases.

Key words Dodecyl pyridinium bromide – surfactant – phase diagram – liquid crystal

Introduction

Knowledge of phase diagrams is of vital importance for understanding the surfactant self-association phenomenon. Electrostatic forces and geometrical constraints are shown to play a dominant role in the surfactant self-association process and in determining the phase equilibria and range of stability of phases in binary and ternary systems containing ionic surfactants [1]. The influence of the charge distribution in a large headgroup [2, 3] can also be expected to influence both the phase stabilities and structure of aggregates. We have undertaken a systematic study of the phase behavior of dodecyl pyridinium surfactant in water where the alkyl chain is attached to different positions of the pyridine ring [3]. In this work, we present the ternary phase diagram for the 1-dodecyl pyridinium bromide (1-DPB)–dodecane–water surfactant system.

Experimental section

Materials

Unless otherwise stated, the chemicals were of analytical grade and used as received from commercial suppliers. The surfactant, 1-dodecyl pyridinium hydrobromide, was prepared using methods previously described [2–4].

Sample preparation

The samples were prepared by weighing appropriate amounts of substances into glass tubes which were flame-sealed. The samples were equilibrated by repeated centrifugation during a few days and after that they were kept at the appropriate temperature. NMR measurements were carried out several times until no noticeable change was detected in the spectra.

Methods

The phase diagram of the surfactant system was constructed on the basis of the experimental results obtained by polarizing microscopy and ^{2}H NMR methods.

Polarizing microscopy

All samples were first examined against crossed polaroids for sample homogeneity and birefringence. Characteristic

Progr Colloid Polym Sci (1996) 100: 6–8
© Steinkopff Verlag 1996

textures of the hexagonal liquid crystals were detected easily under the polarizing microscope [5].

Water deuteron NMR

The water deuteron NMR method has been used widely to determine the phase diagrams of complex surfactant systems [6–9] and the method has been described in detail [6]. The boundary lines of single phases determined by ^{2}H NMR and polarizing microscopy methods are correct to about 1–2 wt % in the ternary phase diagram. ^{2}H NMR spectra were obtained at a resonance frequency of 41.47 MHz or a JEOL EX 270 spectrometer with a superconducting magnet of 6.34 T. The quadrupolar splitting, Δ, was measured as the peak-to-peak distance and given in frequency units.

Results and discussion

^{2}H NMR spectra of a large number of samples in the whole composition range were recorded. These spectra provide information on the characterization of phases if two or more anisotropic phases appear in the system [6, 7]. For the heterogeneous regions one gets information about the phase equilibria without achieving a macroscopic separation of the phases. The phase diagram is shown in Fig. 1.

Phase diagram of the 1-DPB system

1-dodecyl pyridinium bromide is easily soluble in water at room temperature [3] (50 wt % surfactant) yielding a clear isotropic solution phase, L_1. The normal hexagonal liquid crystalline phase E forms in the binary surfactant–water system and it extends between 53 and 86 wt % of 1-DPB [3]. A clear, isotropic and very viscous cubic liquid crystalline phase, I, (43–63 wt % surfactant) was found when dodecane (3 wt % up to 11 wt %) was added to the system. This phase was rather insensitive to composition and variation in temperature around the ambient (40 °C) temperature. In the high water region, the hexagonal phase was in equilibrium with the I phase (E + I region) while on the water-poor side E phase was in equilibrium with the hydrated surfactant (E + G region). A maximum of about 8–9 wt % dodecane could be solubilized in the L_1 phase, whereas the I as well as the E phase could solubilize about 11 wt % of dodecane.

The consistency of the E phase was rather soft and the samples showed a hexagonal texture in the polarizing microscope [5]. This phase also produced single splittings

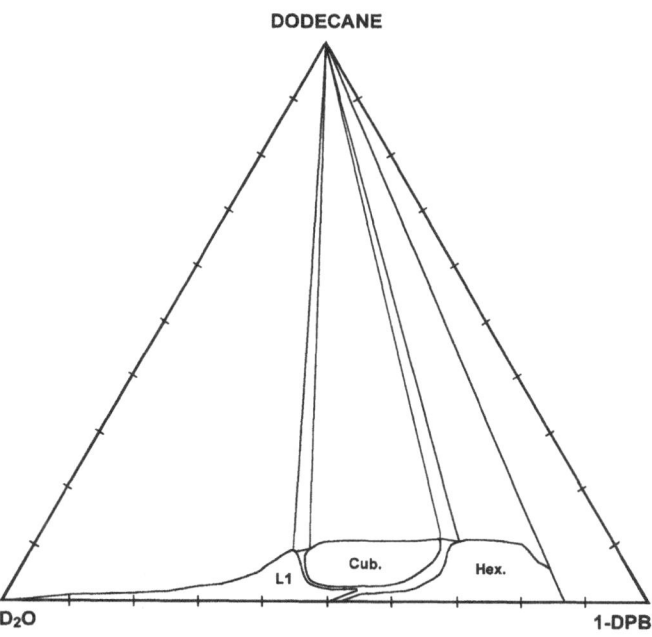

Fig. 1 Isothermal ternary phase diagram for the 1-dodecyl pyridinium bromide–dodecane–water surfactant system at 40 °C

Table 1 Representative water deuteron splittings one-, two- and three phase regions in the ternary phase diagram shown in Fig. 1

Sample comp. (wt %)		Dodecane	Quadrupolar splittings (Hz)	Phase/s
D_2O	1-DPB			
26	65	9	615	Hex.
15	76	9	760	Hex.
15	80	5	930	Hex.
47	53	–	400	Hex.
27	73	–	940	Hex.
19	81	–	1100	Hex.
48	52	–	400	Hex. + L_1
41	57	2	460	Hex. + Cub.
30	65	5	610	Hex. + Cub.
25	59	16	620	Hex. + Cub. + oil
7	20	83	680	Hex. + oil
20	65	15	720	Hex. + oil

in the ^{2}H NMR spectra. The splitting values (Δ) were rather small and the values were found to increase with the decrease of water content (Table 1).

Preliminary x-ray experiments indicate that rods in the hexagonal phase showed a pronounced swelling from about 20 to about 30 Å when 11 wt % dodecane was added. No lamellar liquid crystalline phase was found.

The heterogeneous regions

All heterogeneous regions (two- and three-phase regions) as required to satisfy the phase rule, were identified, but their boundaries were not determined accurately (Fig. 1).

Acknowledgments We are grateful to Professor M.C. Holmes and Dr. M.S. Leaver at the University of Central Lancashire, England for providing us with the SAXS facility. Financial support from the Mid Sweden University is also acknowledged.

References

1. Jönsson B, Wennerström H (1987) J Phys Chem 91:338
2. Jacobs PT, Geer RD, Anacker EW (1972) Colloid Interface Sci 39:611
3. Edlund H, Lindholm A, Carlsson I, Lindström B, Hedenström E, Khan A (1994) Prog Colloid Polym Sci 97:134
4. Ames DE, Bowman RE (1952) J Chem Soc 44:1057
5. Rosevear FB (1968) J Soc Cosmet Chem 19:581
6. Khan A, Fontell K, Lindblom G, Lindman B (1982) J Phys Chem 86:4266
7. Fontell K, Khan A, Lindström B, Maciejewska D, Puang-Ngern S (1991) Colloid Polym Sci 269:727
8. Khan A, Jönsson B, Wennerström H (1985) J Phys Chem 89:5180
9. Ulmius J, Wennerström H, Lindblom G, Arvidson G (1977) Biochemistry 16:5742

Progr Colloid Polym Sci (1996) 100:9–14
© Steinkopff Verlag 1996

K. Berger
K. Hiltrop

Characterization of structural transitions in the SLS/*n*-alcohol/ water system

K. Berger · Dr. K. Hiltrop (✉)
Universität - GH - Paderborn
Fachbereich 13
Physikalische Chemie
33095 Paderborn, FRG

Abstract Ternary systems sodium dodecylsulphate (SLS)/*n*-alcohol/ water with alcohols of different chainlengths and a constant water content of 55% have been investigated by polarized optical microscopy and small-angle x-ray scattering. Upon addition of cosurfactant a hexagonal phase with constant lattice parameter transforms via a two-phase region into a lamellar phase. If surfactant and cosurfactant have similar chainlengths only a small two-phase region is observed and at low cosurfactant content a temperature-sensitive lamellar structure is formed. This structure can be described as a defective one with water-filled pores. At higher alcohol content the well-known behavior of a "classical" lamellar phase with bilayers of great lateral extension is observed. The maximum mole fraction of cosurfactant the lamellar phase of our system can incorporate is 0.77.

Key words Lyotropic – liquid crystals – lamellar phase – SLS – x-ray diffraction

Introduction

In common amphiphile/water systems the two-dimensional hexagonal phase (H) of long rodlike micelles and the one-dimensional lamellar phase (L_α) composed of alternating water and surfactant sheets of great lateral extension have a wide range of stability with respect to concentration and temperature [1–4]. Recently in the intermediate region defective lamellar structures of bilayers with finite lateral extension pierced by circular or elongated water pores have been observed as well in binary [5–8] as in ternary surfactant systems [9–12].

Our aim was to get further information about the stability of this new defective one-dimensional ordered structure by changing the chainlength of the cosurfactant. Thus we have studied the transition from the curved interface of the hexagonal phase into the planar one of the classical lamellar phase for the model system sodium dodecylsulphate/*n*-alcohol/water in great detail. The mea- surements have been performed by varying the amount of cosurfactant and surfactant, but leaving the water content constant at 55 wt%.

Experimental

Materials and sample preparation

SLS (Merck, purity >99%, for biochemical and surfactant investigations) and *n*-alcohols (p.a. quality) were of commercial origin. The H_2O was double distilled and millipore filtered.

The samples were prepared by weighing the appropriate amounts of SLS, *n*-alcohol and water into glass tubes which were then sealed. They were vigorously mixed for several days in a Staudinger-wheel below 40 °C and then allowed to equilibrate at the desired temperature during at least 3 weeks. Just before performing the measurements the mass of the glass tubes was controlled and the composition was calculated assuming that the losses (< 1%) were caused by evaporation of water.

Microscopy

The samples were examined by polarized microscopy between glass plates in orthoscopic geometry. The characterization of the phases was done by comparison to standard textures [13]. Temperature control was achieved by a home-built stage which allowed regulation of temperature with an accuracy of better than 0.1 °C. Especially at the extreme phase boundaries, where even small changes in temperature cause phase transitions, the samples were additionally examined in sealed flat microslides (Camlab).

For fluid samples two-phase regions could easily be detected by macroscopic phase separation after centrifugation. To ensure phase boundaries additional ^{2}H-NMR measurements, which are not presented here, were performed on selected samples.

X-ray scattering

The main tool in our investigation was small-angle x-ray scattering using a Kratky-camera with monochromatic Cu-K$_\alpha$ radiation ($\lambda = 1.542 \cdot 10^{-10}$ m). The scattering patterns were detected by a position-sensitive proportional counter equipped with a metallized quartz wire (Braun) to obtain a high resolution. Thus an uncertainty in $\vec{q}$ ($|\vec{q}| = 4\pi \sin\theta/\lambda$) of $\Delta|\vec{q}| = 2.5 \cdot 10^{-2}$ nm^{-1} was achieved. The uncertainty was approximately $0.5 \cdot 10^{-10}$ m for a lattice parameter of $35 \cdot 10^{-10}$ m and $2.5 \cdot 10^{-10}$ m for $80 \cdot 10^{-10}$ m.

For each system at least two samples were prepared in Lindemann capillaries of 1 mm diameter which were flame-sealed immediately after filling. The samples were investigated and then the tightness of the capillaries was tested in vacuum. A few days later the measurements were repeated. Thus every lattice parameter is the average value of at least four measurements.

The experiments were performed at 30 °C. In solutions of the longer cosurfactants dodecanol and tetradecanol the measurements were performed at 35° and 40°C to prevent the mixtures from crystallizing.

Results

Phase diagrams

Phase transitions in ionic surfactants are usually induced by changes in concentration and not by variations of

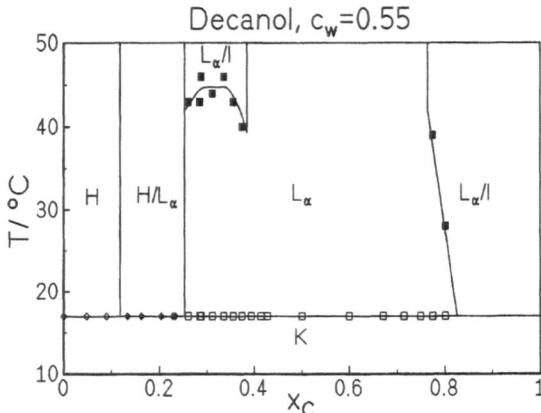

Fig. 1 Phase sequence for the cosurfactant decanol and a water content of $c_W = 0.55$

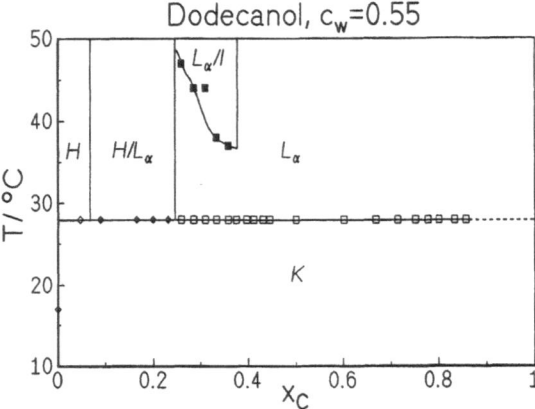

Fig. 2 Phase sequence for the cosurfactant dodecanol and a water content of $c_W = 0.55$

temperature because the electrostatic interaction exceeds the thermal energy by far. Hence our results are presented as a function of the mole fraction of cosurfactant in the aggregate x_C which is calculated on the assumption that neither surfactant nor cosurfactant are water soluble.

Figures 1 and 2 show the phase sequences obtained by polarized microscopy and ^{2}H-NMR-spectroscopy for the cosurfactants decanol and dodecanol. Due to hydrolysis of sodium dodecylsulphate at elevated temperatures the samples have only been examined up to 50 °C.

In the binary system surfactant/water a hexagonal phase is formed at $c_W = 0.55$. If we replace ionic surfactant by neutral cosurfactant a hexagonal phase transforms into a lamellar phase by crossing an intermediate two-phase region hexagonal/lamellar. The lamellar structure at low cosurfactant content is temperature sensitive and can easily be destroyed by slight heating. At elevated alcohol

Table 1 Phase transitions for the mixtures with different cosurfactants and water content of $c_W = 0.55$ as a function of the mole fraction of alcohol in the hydrophobic part ($\pm$ uncertainty of the last digit). H = hexagonal phase; L_α = lamellar phase; I = isotropic phase

Cosurfactant		Isothermal phase transitions $f(x_C)$						
Hexanol	H	$\xrightarrow{0.34(2)}$	H/L_α	$\xrightarrow{0.49(2)}$	L_α	$\xrightarrow{0.77(3)}$	L_α/I
Octanol	H	$\xrightarrow{0.13(4)}$	$H/L_\alpha/I$	$\xrightarrow{0.39(2)}$	L_α	$\xrightarrow{0.76(3)}$	L_α/I
Decanol	H	$\xrightarrow{0.11(3)}$	H/L_α	$\xrightarrow{0.25(2)}$	L_α	$\xrightarrow{0.77(2)}$	L_α/I
Dodecanol	H	$\xrightarrow{0.07(3)}$	H/L_α	$\xrightarrow{0.25(2)}$	L_α	$\xrightarrow{>0.86}$	L_α/I
Tetradecanol	H	$\xrightarrow{0.05(4)}$	H/L_α	$\xrightarrow{0.31(2)}$	L_α	$\xrightarrow{>0.65}$	L_α/I

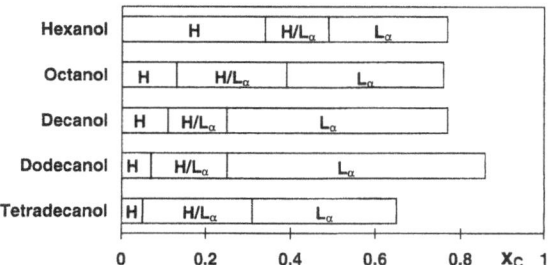

Fig. 3 Isothermal phase transitions for different cosurfactants and a water content of $c_W = 0.55$

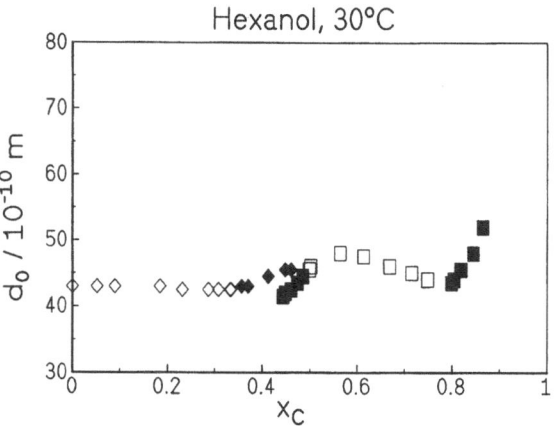

Fig. 4 Interplanar distance for the cosurfactant hexanol and a water content of $c_W = 0.55$; $\diamond$ = hexagonal region; $\blacklozenge$ = multiphasic region, hexagonal parameter; $\blacksquare$ = multiphasic region, lamellar parameter; $\square$ = lamellar region

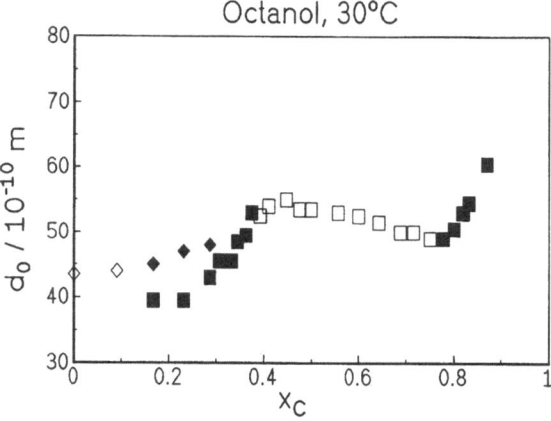

Fig. 5 Interplanar distance for the cosurfactant octanol and a water content of $c_W = 0.55$; $\diamond$ = hexagonal region; $\blacklozenge$ = multiphasic region, hexagonal parameter; $\blacksquare$ = multiphasic region, lamellar parameter; $\square$ = lamellar region

concentrations the lamellar phase becomes more stable with respect to variations of temperature. Due to the increased chainlength of the cosurfactant the crystallizing temperature for dodecanol is higher than for decanol.

For the other surfactant/cosurfactant systems no temperature induced transformations were observed. Thus the isothermal phase transitions at the corresponding temperatures are summarized in Table 1 and Fig. 3.

According to geometric packing considerations [14] the maximum alcohol content a hexagonal phase can incorporate is shifted towards lower mole fractions with increasing cosurfactant length. The dramatic change from hexanol with $x_C = 0.34$ to octanol with $x_C = 0.13$ is only partially caused by the enhanced water-solubility of hexanol.

The longer cosurfactants decanol and dodecanol induce a transition into the lamellar phase at low alcohol concentrations of $x_C = 0.25$. In comparison higher concentrations of the shorter molecules hexanol and octanol and the longer alcohol tetradecanol are necessary to stabilize a lamellar phase. Thus the width of the two-phase region hexagonal/lamellar is minimal in mixtures with the cosurfactant decanol.

Except in solutions of the cosurfactant dodecanol the transition from the lamellar phase into the two-phase region lamellar/isotropic occurs at a $x_C = 0.77$. Apparently the repulsive electrostatic force is too weak to stabilize a lamellar phase furthermore.

Position of the first diffraction maxima

In Figs. 4–8 the positions of the first diffraction maxima, which are associated with the so-called interplanar distance d_0, are plotted versus x_C for the different surfactant/cosurfactant systems. In the case of the hexagonal phase the interplanar distance d_0 is connected to the lattice parameter d by $d = 2/\sqrt{3}d_0$.

In the hexagonal phase the interplanar distance d_0 remains nearly constant with varying cosurfactant concentration. In the lamellar phase two different regions can be distinguished.

• At low concentrations of alcohol the lattice parameter depends strongly on composition. A defective

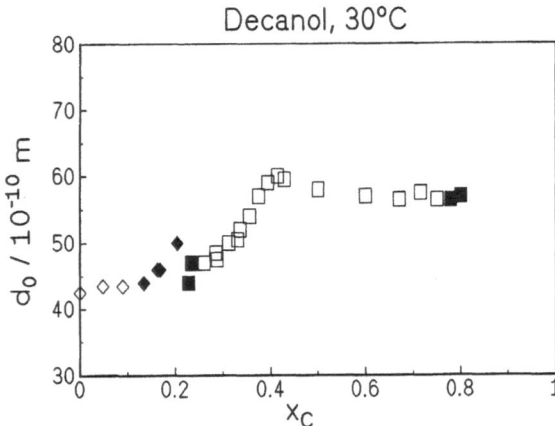

Fig. 6 Interplanar distance for the cosurfactant decanol and a water content of $c_W = 0.55$; $\diamond$ = hexagonal region; $\blacklozenge$ = multiphasic region, hexagonal parameter; $\blacksquare$ = multiphasic region, lamellar parameter; $\square$ = lamellar region

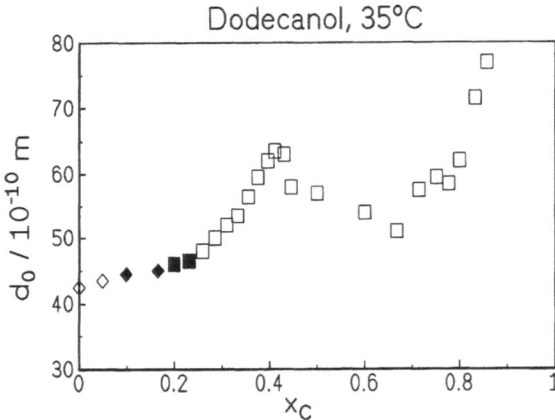

Fig. 7 Interplanar distance for the cosurfactant dodecanol and a water content of $c_W = 0.55$; $\diamond$ = hexagonal region; $\blacklozenge$ = multiphasic region, hexagonal parameter; $\blacksquare$ = multiphasic region, lamellar parameter; $\square$ = lamellar region

lamellar phase with varying amounts of water piercing the bilayer is formed (see the next section).

• At medium mole fractions of cosurfactant the interplanar distance changes only slightly with x_C. Apparently the lamellar structure consists of water-free bilayers of great lateral extension (see the next section).

The transition from the two-phase region hexagonal/lamellar to the monophasic lamellar domain as detected by polarized microscopy is not accompanied by a discontinuity in the interplanar distance. This shows that x-ray diffraction is not sufficient to obtain phase diagrams.

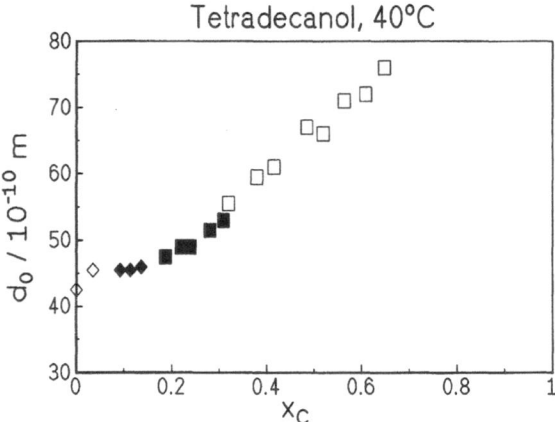

Fig. 8 Interplanar distance for the cosurfactant tetradecanol and a water content of $c_W = 0.55$; $\diamond$ = hexagonal region; $\blacklozenge$ = multiphasic region, hexagonal parameter; $\blacksquare$ = multiphasic region, lamellar parameter; $\square$ = lamellar region

Bilayer thickness

The hexagonal phase

The bilayer thickness of the amphiphilic aggregates d_u can be calculated from the x-ray data and the volume fraction of the hydrocarbon chains ϕ_u using the group volumes in [15].

According to [16] the diameter of the cyclindrical aggregates in the hexagonal phase as a function of the interplanar distance d_0 is:

$$d_u = d_0 \left(\frac{8}{\sqrt{3\pi}} \right)^{1/2} \phi_u^{1/2} . \tag{1}$$

The x-ray data reveal that the interplanar distance in the hexagonal phase and consequently d_u remain constant on addition of cosurfactant and seem to increase only in the subsequent two-phase region.

Table 2 shows that the bilayer thickness of the cyclindrical micelles is $\approx 31.5 \cdot 10^{-10}$ m irrespective of the applied cosurfactant. The value in solutions with tetradecanol is increased. The calculated diameter is only slightly smaller than the length of two dodecylchains in all-trans-conformation of $33.4 \cdot 10^{-10}$ m indicating that the thickness of the aggregate is limited by the longer surfactant molecule.

The classical lamellar phase

If the lamellar phase at medium cosurfactant content consists of bilayers of great lateral extension, we expect a linear dependence on composition. At $x_C = 0$ a pure

Table 2 Micelle diameter of the hexagonal phase at a water content of $c_W = 0.55$

Cosurfactant $d_u/10^{-10}$ m	Hexanol 31.1	Octanol 31.6	Decanol 31.8	Dodecanol 31.7	Tetradecanol 33.2

Table 3 Parameters of the classical lamellar phase ($d_{c,u} = a + bx_C$). a = bilayer thickness for $x_C = 0$; b = slope of the bilayer thickness

Cosurfactant	Experimental values		Limiting values	
	$a/10^{-10}$ m	$b/10^{-10}$ m	$a/10^{-10}$ m	$b/10^{-10}$ m
Hexanol	23.1	−7.5	33.4	−15.2
Octanol	21.5	−0.8	33.4	−10.1
Decanol	21.7	3.7	33.4	−5.1
Dodecanol	28.7	−10.2	33.4	0

Table 4 Parameters of the non-classical lamellar phase ($d_u = a + bx_C$). a = apparent bilayer thickness for $x_C = 0$; b = slope at the apparent bilayer thickness; r = correlation coefficient

Cosurfactant	$a/10^{-10}$ m	$b/10^{-10}$ m	r
Hexanol	−1.2	38.2	0.97
Octanol	8.5	29.5	0.96
Decanol	6.3	41.5	0.99
Dodecanol	6.9	43.8	0.98
Tetradecanol	11.2	33.2	0.99

Table 5 Maximum volume fractions of defects $\varphi_{w,max}$ a lamellar phase can incorporate

Cosurfactant	Hexanol	Octanol	Decanol	Dodecanol
$\varphi_{w,max}$	0.06	0.04	0.22	0.27

surfactant bilayer would be formed; its thickness is limited by two dodecylchains in all-trans-conformation. The maximal thickness of a cosurfactant lamella would be given by two alcohol chains.

In Table 3 the fits to the experimental values are compared with the expected limiting values. No data are given for mixtures with tetradecanol because in the entire region of the lamellar phase the lattice period increased strongly with composition.

The y-interception a of about $22 \cdot 10^{-10}$ m is independent of the used cosurfactant evidencing that the alkylchains are not in a "crystalline" but in a "melted" state and thus significantly shorter than the limiting value. As expected, the slope b increases with the cosurfactant length. Only the data for dodecanol with $a = 28.7 \cdot 10^{-10}$ m and $b = -10.2 \cdot 10^{-10}$ m are strange and cannot be explained so far.

The value of the slope is about $8 \cdot 10^{-10}$ m larger than expected. This bilayer thickening can be explained by denser packing of the headgroups due to weakened electrostatic repulsion on addition of cosurfactant. The reduced flexibility of the hydrocarbon chains has already been detected in other ternary surfactant systems [17].

In summary, the lamellar structure at medium alcohol concentrations resembles a classical lamellar phase of bilayers with great lateral extension. Even small changes in chemical composition can accurately be detected by x-ray diffraction.

The non-classical lamellar phase

At the boundary to the hexagonal phase curved interfaces in the lamellar phase may occur. Water can pierce the bilayer and a defective structure is formed.

In a first approximation we assumed a linear dependence of the apparent bilayer thickness on the cosurfactant content. The results are summarized in Table 4.

The correlation coefficients of $r > 0.95$ indicate that this assumption is a good approximation. The y-interceptions a are small or even negative and cannot be identified with the real thickness of a pure surfactant bilayer. The slopes b are between 30 and $40 \cdot 10^{-10}$ m but are not correlated to the cosurfactant length. Especially it cannot be distinguished if the increase of d_0 occurs in the two-phase region hexagonal/lamellar as for hexanol and octanol or in the monophasic lamellar region as for the longer cosurfactants.

By comparing the minimal apparent bilayer thickness to the expected bilayer thickness the maximum volume fraction of water that enters the bilayer as defects can be calculated [12]. From Table 5 it is seen that in solutions of the shorter cosurfactants nearly no defects are observed whereas the longer cosurfactants can form defective lamellar structures with up to 25% defect water in the bilayer.

Discussion

Experiments have already shown that surfactant and cosurfactant are not homogeneously distributed within ribbon-shaped aggregates [18]. The uncharged alcohol prefers the planar interfaces whereas the charged surfactant tries to occupy a greater area in the edges.

Our study reveals that the microsegregation of these molecules within one aggregate will only be possible if surfactant and cosurfactant have similar chainlengths.

Thus in these systems a defective lamellar phase at the boundary to the hexagonal phase is observed. If both molecules have different lengths curved and planar interfaces cannot be realized in the same but only in different shaped micelles. Then in the region between the hexagonal and the classical lamellar phase no temperature sensitive defective structure but extensive two-phase regions occur.

Furthermore, we have no evidence of a distortion of the cylindrical micelles of the hexagonal phase in the SLS/n-alcohol/water system at higher alcohol content.

One explanation for our results is that the short cosurfactants are not able to fill the hydrophobic interior of an amphiphilic aggregate. The thickness of the curved regions where the surfactant is enriched exceeds the diameter of the flat regions by far. Energetically unfavorable interactions between water and the hydrophobic tails would occur. Hence the mixture separates into a surfactant rich hexagonal phase and a surfactant poor lamellar phase.

Acknowledgment We would particularly like to thank G. Juennemann for technical assistance.

References

1. Ekwall P (1975) Composition, properties and Structures of ... In: Brown GH (ed) Advances in Liquid Crystals. Academic Press, New York, 1
2. Luzzati V (1968) X-Ray Diffraction Studies of Lipid Water Systems In: Chapman D (ed) Biological Membranes. Academic Press, New York, 71
3. Tiddy GJT (1980) Physics Report 57:1
4. Hiltrop K (1994) Lyotropic Liquid Crystals In: Stegemeyer H (ed) Liquid Crystals. Steinkopff, Darmstadt, p 143
5. Funari SS, Holmes MC, Tiddy GJT (1994) J Phys Chem 98:3015
6. Funari SS, Holmes MC, Tiddy GJT (1992) J Phys Chem 96:11029
7. Leaver MS, Holmes MC (1993) J Phys II France 3:105
8. Kekicheff P, Cabane B, Rawiso M (1984) J Phys Lett Paris 45:813
9. Hendrikx Y, Charvolin J (1992) Liq Cryst 11:677
10. Quist PO, Halle B (1993) Phys Rev E 47:3374
11. Quist PO, Fontell K, Halle B (1994) Liq Cryst 16:235
12. Berger K, Hiltrop K (accepted for publication) Colloid Polym Sci
13. Rosevear FB (1954) J Am Oil Chem Soc 31:629
14. Israelachvili JA (1985) Thermodynamic and Geometric Aspects of Amphiphilic Aggregation. In: Degiorgio V, Corti M (eds) Physics of Amphiphiles: Micelles, Vesicles and Microemulsions. North-Holland, Amsterdam, p 24
15. Tanford C (1980) The Hydrophobic effect 2nd ed. New York, Wiley 52
16. Husson F, Mustacchi H, Luzzati V (1960) Acta Cryst 13:660
17. Klason T, Henriksson U (1982) The Motional State of Hydrocarbon Chains in the Ternary System ... In: Mittal KL, Fendler EJ (eds), Solution Behaviour of Surfactants. Plenum, New York, p 417
18. Alpérine S, Hendrikx Y, Charvolin J (1985) J Phys Lett Paris 46:27

Progr Colloid Polym Sci (1996) 100:15–18
© Steinkopff Verlag 1996

T. Kato

Surfactant self-diffusion and networks of wormlike micelles in concentrated solutions of nonionic surfactants

Dr. T. Kato (✉)
Department of Chemistry
Faculty of Science
Tokyo Metropolitan University
Minamiohsawa, Hachioji
Tokyo 192-03, Japan

Abstract Phase behavior, surfactant self-diffusion coefficients (D) and small-angle x-ray scattering (SAXS) have been measured on concentrated solutions of a nonionic surfactant ($C_{16}E_7$) where entanglement of wormlike micelles has been suggested in our previous light scattering study. In the lower temperature range, the self-diffusion coefficient/activation energy for self-diffusion processes (E_D) are much smaller/larger than those in the liquid crystal (LC) phases where the lateral diffusion is dominant, which confirms the validity of our diffusion model previously reported where intermicellar migration of surfactant molecules at the entanglement point is considered. As the temperature is raised above about 45 °C, however, E_D increases rapidly towards the value for the LC phases. Above about 60 °C, the D value coincides with that expected from the temperature dependence of D in the cubic phase. From these results and the structure of the cubic phase, it is inferred that crosslinking of wormlike micelles occurs above about 45 °C and that its extent increases with increasing temperature. SAXS results are consistent with such a change in the network structures. Relations with phase behaviors are also discussed.

Key words Self-diffusion – micelles – nonionic surfactant – concentrated solutions – lyotropic liquid crystal

Introduction

In recent years, much attention has been paid to the network structure of wormlike micelles including cross-links formed by fusion of micelles [1–7]. In previous studies [8, 9], we have measured the surfactant self-diffusion coefficient (D) on semidilute solutions of a nonionic surfactants $C_{16}E_7$, $C_{14}E_6$, and $C_{14}E_7$ (C_nE_m) represents a chemical formula $C_nH_{2n+1}(OC_2H_4)_mOH$) by using the pulsed-gradient spin echo (PGSE) method. In the concentrated region (> 10 wt%), the self-diffusion coefficient increases with increasing concentration (c) and follows the power law ($D \sim c^{2/3}$) which can be explained by a simple model taking into account the intermicellar migration of surfactant molecules at the entanglement point. If the crosslinking occurs, surfactant molecules can diffuse over a large distance without "intermicellar" migration. Then the observed diffusion coefficient is dominated by the lateral diffusion alone. Such a situation is really encountered in liquid crystal phases. Especially, the bicontinuous cubic phase resembles such a crosslinking picture (or "multiconnected" network [3]) although the regularity disappears in the micellar phase.

In the present work, we have measured temperature dependence of the self-diffusion coefficient of $C_{16}E_7$ in more concentrated region including the cubic and hexagonal phases. In addition, small-angle x-ray scattering (SAXS) has been measured in order to discuss the self-diffusion results from the structural viewpoints.

Experimental section

Materials

$C_{16}E_7$ was purchased from Nikko Chemicals, Inc. in crystalline forms and used without further purification. Deuterium oxide purchased from ISOTEC, Inc. (99.9%) was used after being degassed by bubbling of nitrogen to avoid oxidation of the ethylene oxide group of surfactants.

Methods

Phase diagram of the $C_{16}E_7$–D_2O system was determined by using polarizing microscopy and SAXS. Measurements of SAXS were made by using an apparatus (MAC Science) constructed from an x-ray generator (SRA, MXP18, 18 kW), incident monochrometer (W/Si multilayer crystal), Kratky slit, and imaging plate (DIP 200).

PGSE was measured on protons at 399.6 MHz on a JEOL JNM-EX400 Fourier transform NMR spectrometer. The magnitude of gradient was kept constant at about 2.2 Tm^{-1}. Details of the measurements have already been reported [8, 9]. The measurements were made above about $T_c - 15$ K (T_c, the lower critical solution temperature) where entanglement of wormlike micelles is expected from the light scattering studies [8, 10].

Results and discussion

Surfactant self-diffusion coefficient and activation energy for self-diffusion processes

Figure 1 illustrates a partial phase diagram of the $C_{16}E_7$–D_2O system. Although the phase diagram of this system has not been reported even for a H_2O solution, main features of it are similar to those of homologous systems.

In Fig. 2, logarithm of the self-diffusion coefficient of $C_{16}E_7$ (D) is plotted against the reciprocal temperature. It should be noted that in the hexagonal phase the direction of each rod may be random. It can be seen from the figure that in the lower temperature range the self-diffusion coefficient in the micellar phase is much lower than those in the liquid crystal phases. As the temperature is increased, the self-diffusion coefficient in the micellar phase rapidly increases.

From these plots, the activation energy for the self-diffusion processes (E_D) is obtained by using the equation

$$E_D = -R\frac{\partial \ln D}{\partial(1/T)} \tag{1}$$

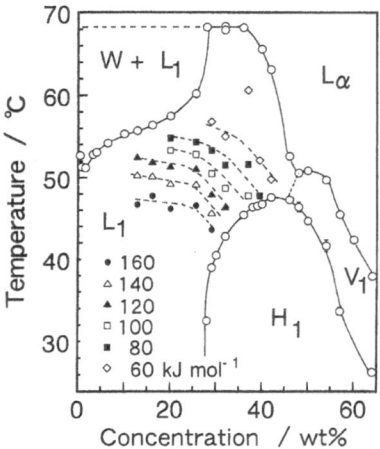

Fig. 1 Partial phase diagram of $C_{16}E_7$–D_2O system. L_1, micellar solution; $W + L_1$, co-existing liquid phases; H_1, hexagonal phase; V_1, cubic phase; L_α, lamellar phase. Dashed lines represent constant activation energy contours obtained from the results in Fig. 3

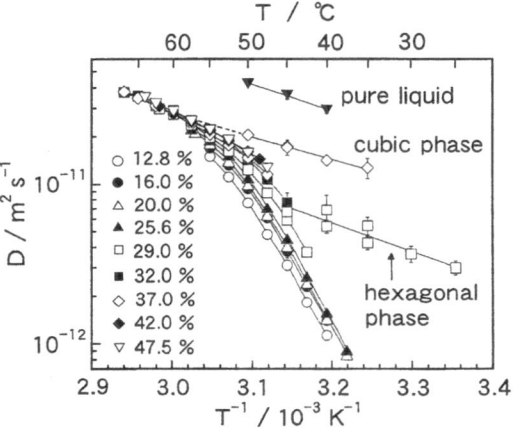

Fig. 2 Logarithm of surfactant self-diffusion coefficient vs. reciprocal temperature. The concentrations for the micellar solutions (in wt%) are indicated in the figure

where R is the gas constant. The results are shown in Fig. 3. This figure demonstrates that the activation energy in the micellar phase is much larger than that in the cubic phase below about 45 °C. It should be noted that the activation energies in the hexagonal phase and in the pure liquid are nearly equal to that in the cubic phase as can be seen from Fig. 2.

If wormlike micelles are crosslinked, the observed diffusion coefficient is dominated by the lateral diffusion. Then, the activation energy in the micellar phase, E_D (micelle), is expected to be close to those in the liquid crystal phases, E_D (LC), where the lateral diffusion is

Progr Colloid Polym Sci (1996) 100:15–18
© Steinkopff Verlag 1996

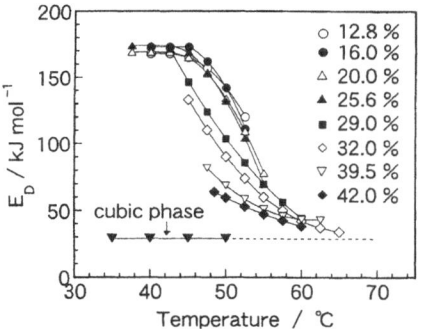

Fig. 3 Temperature dependence of activation energy for self-diffusion processes obtained from the results in Fig. 2

dominant. Therefore, the above result suggests that the crosslinking does not occur below about 45 °C.

In previous papers [8, 9], we proposed a diffusion model taking into account the intermicellar migration of surfactant molecules. In this model, it is assumed that i) a surfactant molecule diffuses in a micelle along its contour during the time τ_{mig} and then migrates to adjacent micelles at one of the entanglement points, ii) τ_{mig} satisfies the condition $R_g^2/D_L \ll \tau_{mig} \ll \Delta$ where R_g is the radius of gyration of wormlike micelles, D_L the intramicellar (lateral) diffusion coefficient, and Δ the diffusion time, and iii) the lifetime of micelles is longer than τ_{mig}. Then the surfactant self-diffusion coefficient can be expressed as

$$D = D_M + \langle d \rangle^2 / (6\tau_{mig})$$
$$= D_M + Ac^{2/3}, \tag{2}$$

where D_M is the self-diffusion coefficient of micelles, $\langle d \rangle$ the mean distance between the centers of mass of adjacent micelles, and A a constant depending on the temperature. In our previous paper [9], we showed that Eq. (2) can explain the observed concentration dependence of D in the $C_{16}E_7$, $C_{14}E_6$, and $C_{14}E_7$ systems in the range from $T_c - 15$ K to $T_c - 5$ K (35–45 °C in the $C_{16}E_7$ system).

According to the above model, the activation energy for the self-diffusion processes in the concentrated region (where D_M is much smaller than D) can be divided into two terms. One is the temperature dependence of the mean aggregation number of micelles and the other is the activation energy for the lateral diffusion and intermicellar migration processes. In the previous paper [9], we showed that the anomalously large value of the observed activation energy (in the range 35–45 °C for the $C_{16}E_7$ system) can be explained by these two terms.

As the temperature is raised above about 45 °C, however, E_D (micelle) decreases rapidly and approaches the value of E_D (LC). Above about 60 °C, moreover, the self-diffusion coefficient depends on concentration only slightly and falls into the line obtained by extrapolating the plot

in the cubic phase to higher temperatures (see the dashed line in Fig. 2). Taking into account the structure of the cubic phase in the homologous system (space group Ia3d for $C_{12}E_6$ [11]), we may infer that the crosslinking of wormlike micelles occurs above about 45 °C and its extent increases with increasing temperature.

Relations with phase behavior

Figure 1 includes constant activation energy contours which may be useful for discussing change in the structure of networks with concentration and temperature if the activation energy can be regarded as the measure of the extent of crosslinking. According to the interpretation described before, crosslinking does not occur below the line corresponding to $E_D = 160$ kJ mol^{-1}. Above this line, the extent of crosslinking increases with increasing temperature or concentration. Near the cubic and lamellar phases, the crosslinking may be almost completed.

In order to confirm these conclusions, we have measured SAXS in the similar concentration and temperature range. In the micellar phase, a broad peak has been observed at the position close to those of main peaks in the liquid crystal phases (corresponding to the (10) plane in the hexagonal phase, the (211) plane in the cubic phase, and the (100) plane in the lamellar phase). This suggests that the structure of micellar phase is correlated with those of liquid crystal phases. It should be noted that the SAXS form factor of the $C_{16}E_7$ micelles may also give a peak. However, our preliminary measurements of the neutron scattering indicate that the peak position in the concentration range studied is dominated mainly by the structure factor.

Figure 4 shows the concentration dependence of the repeat distance obtained from the peak position in the micellar phase. Although the cubic phase does not exist in such a low concentration range, we have tried to calculate the distance between the (211) planes of the cubic structure, d_{211}, by using the following equations.

$$d_{211} = a/\sqrt{6} = (2/\sqrt{3})l$$
$$\phi_{hc} = (3\sqrt{2}/4)(r_{hc}/l)^2[1 - 0.491(r_{hc}/l)] \tag{3}$$

where a is the cell constant, l the rod length (including the junctions), ϕ_{hc} and r_{hc} the volume fraction and the radius of the hydrocarbon core of the rod, respectively. The ϕ_{hc} value was calculated from the average specific volume of hydrocarbons (1.25 cm^{3}g^{-1}). The solid line in Fig. 4 represents calculated d_{211} for $r_{hc} = 18$ Å which corresponds to the radius of the hydrocarbon core of the rod in the hexagonal phase calculated from the observed repeat distance assuming infinite rod length. It can be seen from

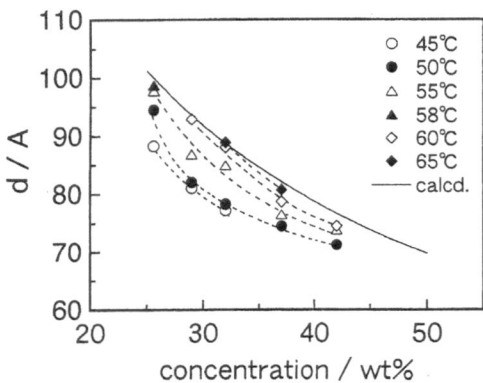

Fig. 4 Concentration dependence of repeat distance obtained from the position of the SAXS peak in the micellar phase. The solid lines represent the repeat distance between the (2 1 1) planes of the cubic phase (space group Ia3d) calculated from Eq. (3)

the figure that the repeat distance increases with increasing temperature and approaches the calculated line near the phase boundary (see Fig. 1). This suggests that the struc-

ture of the network begins to resemble that of the cubic phase in the higher temperatures, which is consistent with the self-diffusion results.

Conclusions

The present results may be summarized as follows.

1) In the lower temperature range (below 45 °C), the self-diffusion results can be explained by our model previously reported where intermicellar migration of surfactant molecules at the entanglement point is taken into account.

2) Above about 45 °C, the crosslinking of wormlike micelles occurs (our diffusion model fails). The extent of the crosslinking increases with increasing temperature. Near the cubic and lamellar phases, crosslinking may be almost completed.

3) The SAXS results are consistent with the self-diffusion results.

References

1. Hoffmann H (1994) Proceedings of the ACS Symposium: American Chemical Society, Washington DC. References therein
2. Appell J, Porte G, Khatory A, Kern F, Candau SJ (1992) J Phys II France 2:1045–1052
3. Khatory A, Kern F, Lequeux F, Appell J, Porte G, Morie N, Ott A, Urbach W (1993) Langmuir 9:933–939
4. Khatory A, Lequeux F, Kern F, Candau SJ (1993) Langmuir 9:1456–1464
5. Drye TJ, Cates ME (1992) J Chem Phys 96:1367–1375
6. Monduzzi M, Olsson U, Soderman O (1993) Langmuir 9:2914–2920
7. Shikata T, Hirata H, Kotaka T (1987) Langmuir 4:354–359
8. Kato T, Terao T, Tsukada M, Seimiya T (1993) J Phys Chem 97:3910–3917
9. Kato T, Terao T, Seimiya T (1994) Langmuir 10:4468–4474
10. Kato T, Anzai S, Seimiya T (1990) J Phys Chem 94:7255–7259
11. Rancon Y, Charvolin J (1987) J Physique 48:1067–1073

Progr Colloid Polym Sci (1996) 100:19–23
© Steinkopff Verlag 1996

SURFACTANT AGGREGATES

Thermodynamic theory and experimental studies on mixed short-chain lecithin micelles

T. Lin
Y. Hu
W.-J. Liu
J. Samseth
K. Mortensen

Prof. Dr. T.-L. Lin (✉) · Y. Hu · W.-J. Liu
Department of Nuclear Engineering and
Engineering Physics
National Tsing-Hua University
Hsin-Chu 30043, Taiwan

J. Samseth
Institutt for Energiteknikk
P.O. Box 40
2007 Kjeller, Norway

K. Mortensen
Risø National Laboratory
4000 Roskilde, Denmark

Abstract A thermodynamic theory for ideally mixed rodlike micelles is used to predict the size distribution and mean aggregation number of mixed dihexanoylphosphatidyl-choline (diC$_6$PC) and diheptanoyl-phosphatidylcholine (diC$_7$PC) micelles. The mixed diC$_6$PC and diC$_7$PC micelles are also studied by small-angle neutron scattering. The measured neutron scattering spectra are analyzed to give the size distribution and mean aggregation number of the mixed micelles by using indirect transform method. The experimentally determined mean aggregation number of the mixed micelles is in general in good agreement with that predicted by the thermodynamic theory. The addition of small amounts of diC$_6$PC to a 10 mM diC$_7$PC solution is found to make the micelles become significantly smaller and the addition of small amounts of diC$_7$PC to a 10 mM diC$_6$PC solution is found to make the micelles grow slightly larger.

Key words Thermodynamic theory – mixed micelles – small-angle neutron scattering – short-chain lecithins

Introduction

Amphiphilic molecules (surfactants) form micellar aggregates in aqueous solutions. In practical applications, we often mix different surfactants or co-surfactants to give the desired properties. Micelles formed by mixing different surfactants are of great interest for practical use and for scientific research. Theoretical modeling of mixed micelles is essential for a better understanding of mixed micelles. There have been some thermodynamic modeling and analyses of mixed surfactant-alcohol aggregates, mixed nonionic micelles, and mixed ionic micelles [1–3]. Some mixed micellar systems have also been widely studied by experiments [4–8]. Here, we are interested in studying the mixed dihexanoylphosphatidylcholine (diC$_6$PC)

and diheptanoylphosphatidylcholine (diC$_7$PC) micelles. diC$_6$PC and diC$_7$PC are synthetic zwitter-ionic surfactants. diC$_6$PC is known to form globular micelles while diC$_7$PC forms polydisperse rodlike micelles in aqueous solutions. The structures and thermodynamic properties of pure one-component diC$_6$PC or diC$_7$PC micelles were well studied by small-angle neutron scattering (SANS) [9–11]. Since the properties of diC$_6$PC and diC$_7$PC were well characterized, the mixed diC$_6$PC and diC$_7$PC micellar system is an ideal model system for studying mixed micelles. SANS was used here to study the mixed diC$_6$PC and diC$_7$PC micelles at various mixing ratios. The measured SANS spectra were analyzed by using indirect transform method to determine the size distribution of mixed micelles. It was shown previously that the formation and growth of pure diC$_7$PC rodlike micelles can well be described by the "ladder" thermodynamic theory. A thermodynamic theory for ideally mixed two-component rodlike micelles can be derived in close analogy to the one-component theory. The SANS results will be compared with that predicted by this thermodynamic theory.

Thermodynamic theory of ideally mixed rodlike micelles

Here, we will describe briefly the thermodynamic theory for ideally mixed two-component rodlike micelles. A detail description of this theory has been published elsewhere [12]. A thermodynamic theory similar to the one for one-component rodlike micelles can be developed for two-component rodlike micelles. The mixed rodlike micelles can be modeled as composed of two end caps and one cylindrical middle section. The mixed micelles are formed by surfactant 1 and surfactant 2 molecules. For a dilute sample, the concentration (in mole fraction) of a mixed rodlike micelles X_i, formed by N_1^s and N_2^s surfactant molecules in the two end caps and N_1^c and N_2^c surfactant molecules in the cylindrical section, is given by

$$X_i = \beta_1^{N_1} \beta_2^{N_2} e^{-(N_1^s \Delta_1 + N_2^s \Delta_2 + \varepsilon_{mix}^s (N_1^s + N_2^s) + \varepsilon_{mix}^c (N_1^c + N_2^c))/kT}. \quad (1)$$

Here, we have $\beta_1 \equiv X_1^f e^{-\delta_1^s/kT}$, $\beta_2 \equiv X_2^f e^{-\delta_2^s/kT}$, and

$$\varepsilon_{mix}^s (N_1^s + N_2^s) = -kT \ln \frac{(N_1^s + N_2^s)!}{N_1^s! N_2^s!}. \quad (2)$$

$$\varepsilon_{mix}^c (N_1^c + N_2^c) = -kT \ln \frac{(N_1^c + N_2^c)!}{N_1^c! N_2^c!}. \quad (3)$$

Here, δ_1^s and δ_2^s are, respectively, the change in free energy per molecule for surfactants 1 and 2 to move into the end caps of the mixed rodlike micelles. δ_1^c and δ_2^c denote, respectively, the change in free energy per molecule for surfactants 1 and 2 to move into the cylindrical section of the mixed rodlike micelles. X_1^f and X_2^f are, respectively, the concentration in mole fraction of the free monomers of surfactants 1 and 2 in the solution. Δ_1 and Δ_2 are defined by $\Delta_1 \equiv \delta_1^s - \delta_1^c$ and $\Delta_2 \equiv \delta_2^s - \delta_2^c$. $\varepsilon_{mix}^s (N_1^s + N_2^s)$ is the free energy of mixing the surfactants 1 and 2 in the end caps of the rodlike micelle and $\varepsilon_{mix}^c (N_1^c + N_2^c)$ is the free energy of mixing the surfactants 1 and 2 in the cylindrical section of the rodlike micelle. The constraints are the total concentrations of surfactant molecules added in the solution, which can be expressed as

$$X_1 = X_1^f + \sum_i (N_1^s + N_1^c) X_i. \quad (4)$$

$$X_2 = X_2^f + \sum_i (N_2^s + N_2^c) X_i. \quad (5)$$

Here, X_1 and X_2 are, respectively, the total concentration in mole fraction of surfactant 1 and 2 added in the solu-

tion. The summation is over all allowable combinations of N_1^s, N_2^s, N_1^c, and N_2^c. The total number of surfactant molecules forming the two end caps, N^s, should be subject to some constraints due to packing limitations. A linear model is proposed for determining N_1^s and N_2^s as a function of N^s $N_{0,1}$ and $N_{0,2}$:

$$N^s = N_{0,1} \frac{N_1^s}{N_1^s + N_2^s} + N_{0,2} \frac{N_2^s}{N_1^s + N_2^s}. \quad (6)$$

$N_{0,1}$ and $N_{0,2}$ are, respectively, the number of surfactant molecules forming the two end caps of the one-component rodlike micelles for surfactant 1 and 2. δ_1^s, δ_2^s, δ_1^c and δ_2^c are assumed to be constants in the present model. For a given set of δ_1^s, δ_2^s, Δ_1, Δ_2, $N_{0,1}$ and $N_{0,2}$, one can determine β_1 and β_2 at any given mixing concentrations of X_1 and X_2 by using Eqs. (4) and (5). Once β_1 and β_2 are determined, one can compute the size distribution and other useful information about the mixed micelles. For computing the size distribution of mixed diC_6PC and diC_7PC micelles, the values of δ_1^s, δ_2^s, Δ_1, Δ_2, $N_{0,1}$ and $N_{0,2}$ are, respectively, set equal to $-7.70\,kT$, $-9.87\,kT$, $0.438\,kT$, $0.610\,kT$, 16, and 27 [11].

SANS experiments and the indirect transform method

diC_6PC and diC_7PC were obtained from Avanti Polar Lipids, Inc. (USA). They were in powder form and used without further purification. The samples were prepared in D_2O solution. The diC_7PC to diC_6PC concentration ratios of these samples are respectively equal to 10 to 0, 10 to 2, 10 to 4, 10 to 6, 2 to 10, 4 to 10, 6 to 10, and 8 to 10, in units of mM. The small-angle neutron scattering measurements were done at the Institutt for Energiteknikk, Kjeller Norway, by using the SANS spectrometer located at its research reactor. The samples were kept at 25 °C during measurements. The measured neutron scattering intensities were corrected and normalized to give the normalized scattering intensity per unit sample volume, $I(Q)$, in units of $1/cm$. Here, Q is the scattering vector and it is equal to $4\pi \sin(\theta/2)/\lambda$, where θ is the scattering angle.

Here, we will show briefly the method of indirect transform to determine the size distribution of polydisperse rodlike micelles. The scattering intensities from polydisperse mixed rodlike micelles formed by surfactant 1 and surfactant 2 molecules can be given as

$$I(Q) = \sum_N C_N \frac{1}{N} (N_1 \Delta \rho_1 V_1 + N_2 \Delta \rho_2 V_2)^2 P_N'(Q, R, L). \quad (7)$$

For simplicity of calculation, the mixed micelles are assumed to be uniformly mixed and the mixing ratio within each micelle is the same. When surfactant 1 and surfactant

Progr Colloid Polym Sci (1996) 100:19–23
© Steinkopff Verlag 1996

2 molecules have close similarities in their structures and scattering length densities, these assumptions should be applicable. C_N is the concentration of surfactants forming micelles with aggregation number N, and $N = N_1 + N_2$. $P'(Q, R, L)$ is the normalized form factor of cylindrical rods with radius R and length L. $\Delta\rho$ is the scattering length density difference between the surfactant molecule and the solution. V is the volume of one surfactant molecule. C_N can be expanded into n cubic B-spline functions $\Phi_v(N)$:

$$C_N = \sum_{v=1}^{n} C_v \Phi_v(N) . \tag{8}$$

Then, $I(Q)$ can be written as

$$I(Q) = \sum_{v=1}^{n} C_v \Psi_v(Q) , \tag{9}$$

where

$$\Psi_v(Q) = \sum_N \Phi_v(N) \frac{1}{N} (N_1 \Delta\rho_1 V_1 + N_2 \Delta\rho_2 V_2)^2 P'(Q, R, L) . \tag{10}$$

The coefficients C_v can be determined by minimizing [13]

$$\sum_{i=1}^{m} \left[I(Q_i) - \sum_{v=1}^{n} C_v \Psi_v(Q_i) \right]^2 \frac{1}{\sigma^2(Q_i)} + \lambda \sum_{v=1}^{n-1} (C_{v+1} - C_v)^2 . \tag{11}$$

Here, $\sigma(Q_i)$ is the statistical uncertainty associated with each scattering intensity $I(Q_i)$. λ is a stabilizing parameter. The ratio of aggregation number N to its rod length L, denoted as α, and R_c can be determined by plotting $\ln(I(Q)Q^2)$ versus Q^2. The radius of the rodlike micelles R can be related to R_c by $R = R_c\sqrt{2}$. R and L are needed in computing $P'(Q, R, L)$.

Results and discussion

Figure 1 shows the measured neutron scattering spectra as a function of scattering vector for mixed micellar solutions containing 10 mM diC$_7$PC and 0 to 6 mM diC$_6$PC. The solid lines are the fitted scattering spectra calculated from the size distribution obtained by indirect transform method. The fitted scattering spectra are in good agreement with the measured data. For dilute samples, the scattering amplitude, the scattering intensity at $Q = 0$, should be proportional to the total surfactant concentration forming micelles and the weight weighted mean aggregation number. As shown in Fig. 1, the decrease of the scattering amplitude with increasing diC$_6$PC concentration indicates that the mixed micelles becomes smaller with addition of diC$_6$PC. The original pure diC$_7$PC rod-

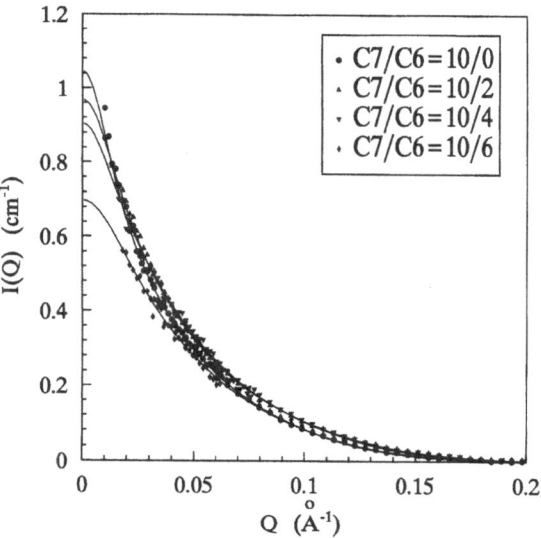

Fig. 1 The measured neutron scattering intensities for mixed diC$_6$PC and diC$_7$PC micelles at 10 mM diC$_7$PC and 0 to 6 mM diC$_6$PC. The solid lines are the fitted scattering spectra by using the indirect transform method

like micelles will become shorter rods upon addition of diC$_6$PC. It can be shown that the mixed diC$_7$PC and diC$_6$PC micelles are still rodlike micelles by plotting $\ln(I(Q)Q)$ versus Q^2. For particles with cylindrical structures, the middle section of the scattering data in such plots will fall on a straight line.

The measured neutron scattering spectra for mixed micellar solutions containing 10 mM diC$_6$PC and 2 to 8 mM diC$_7$PC are shown in Fig. 2. The scattering intensities are found to increase with the addition of diC$_7$PC. Careful analysis is required to tell whether the increase of the scattering intensity is due solely to the increase of the total surfactant concentration in the solution or also due to the increase of the size of mixed micelles with the addition of diC$_7$PC. When the scattering intensities are normalized by the total surfactant concentration for each sample, it is found that the normalized scattering intensities are higher for cases with higher diC$_7$PC concentration. This means that the micelles become larger upon the increase of diC$_7$PC concentration while the diC$_6$PC concentration remains at 10 mM. The measured neutron scattering spectra were analyzed by using the indirect transform method to recover the size distribution of the mixed micelles. The solid lines shown in Fig. 2 are the fitted results by the indirect transform method. The fitted scattering spectra are in good agreement with the experimental data.

As an example, the size distributions of the mixed micelles for the sample containing 10 mM diC$_7$PC and

22

T. Lin et al.
Theory and experiments on mixed micelles

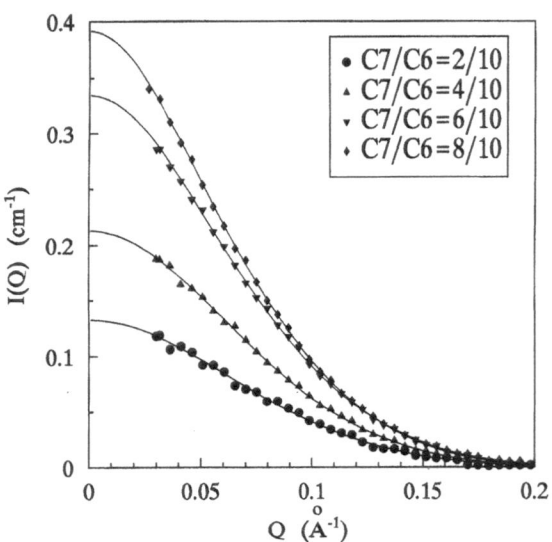

Fig. 2 The measured neutron scattering intensities for mixed diC6PC and diC7PC micelles at 10 mM diC6PC and 2 to 8 mM diC7PC. The solid lines are the fitted scattering spectra by using the indirect transform method

[diC7PC]/[diC6PC] = 10/6

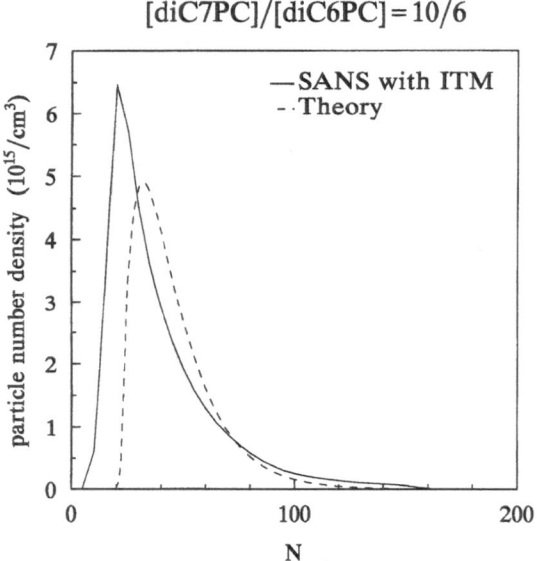

Fig. 3 The size distributions of the mixed micelles for the sample containing 10 mM diC7PC and 6 mM diC6PC. The solid line is the size distribution obtained by the analysis of measured neutron scattering data by using the indirect transform method and the dashed line is the size distribution predicted by the thermodynamic theory

6 mM diC6PC are shown in Fig. 3. The solid line is the size distribution obtained by the analysis of measured neutron scattering data by using the indirect transform method.

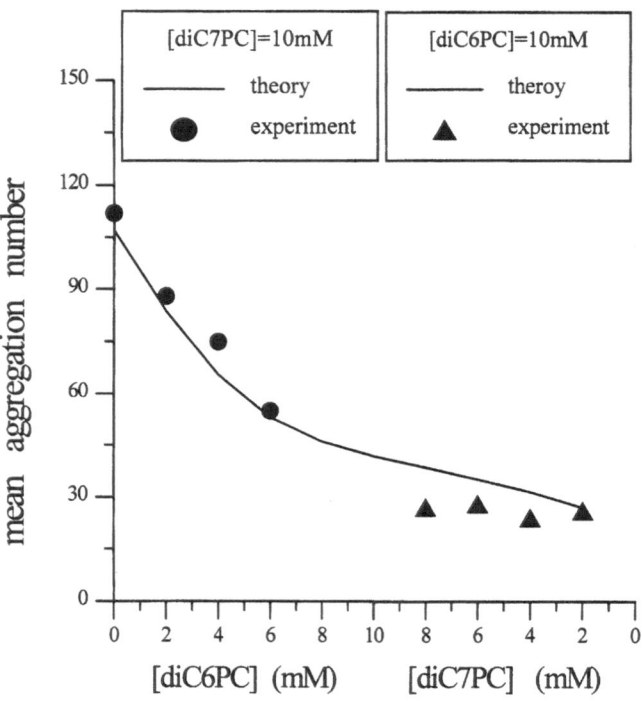

Fig. 4 The weighted mean aggregation number as obtained by the analysis of the measured scattering data by indirect transform method and that predicted by the thermodynamic theory

The dashed line is the size distribution predicted by the thermodynamic theory. These two size distributions have similar distribution profiles and they also roughly agree with each other in magnitude. The indirect transform method does not use any prescribed size distribution model in fitting the measure scattering data. It is remarkable that the size distribution calculated completely by theory is in close agreement with the experimentally determined one. The major discrepancy is in the lowest size region. Compared with that predicted by the thermodynamic theory, the indirect transform method overestimates the number of micelles in the region with N below about 30; in compensation, it overestimates the number of micelles in the region with N between 30 to 70. In the analysis by indirect transform method, we did not put a lowest size limit for the mixed micelles. In fact, the proportion of micelles with an aggregation number below 16 (the estimated value of $N_{0,1}$, for diC6PC) should be very small. In the thermodynamic theory for mixed rodlike micelles described in this paper, there is a lowest limit for the mixed micelles, which is set equal to $N_{0,1}$ or $N_{0,2}$ (whichever is smaller). It is believed that better agreement can be obtained if a lowest size limit for mixed micelles is imposed during the analysis of the measured scattering data by indirect transform method.

Although there are small discrepancies in the obtained size distributions between the theory and the analysis by indirect transform method, there should be only small effects on the obtained weighted mean aggregation numbers. However, the effects will be greater for samples with small size micelles, such as the sample with 10 mM diC$_6$PC and small amounts of diC$_7$PC. The weighted mean aggregation number as obtained by the analysis of the measured scattering data by indirect transform method and that predicted by the thermodynamic theory are all plotted in Fig. 4. As stated in this section, the agreement between the theory and the experiment is much better for samples with large mixed micelles (sample with 10 mM diC$_7$PC and lower amounts of diC$_6$PC) and the analysis of the experimental data by indirect transform method underestimates the mean size for samples containing small micelles (samples with 10 mM diC$_6$PC and lower amounts of diC$_7$PC). The addition of small amounts of diC$_6$PC to a 10 m diC$_7$PC solution is found to make the micelles significantly become smaller, and the addition of small amounts of diC$_7$PC to a 10 mM diC$_6$PC solution is found to make the micelles grow only slightly larger. In general, the mixed micelles become smaller as the ratio of diC$_6$PC to diC$_7$PC increases.

Conclusions

It is shown in this study that the thermodynamic theory for ideally mixed rodlike micelles developed in the paper can be successfully used in studying the mixed rodlike micelles formed by mixing diC$_6$PC and diC$_7$PC. The indirect transform method can be very useful for recovering the size distribution of polydisperse rodlike micelles. The size distribution obtained by the analysis of the experimental data by the indirect transform method is, in general, in good agreement with that predicted by the thermodynamic theory. There is also good agreement between the experimentally determined weighted mean aggregation number of the mixed micelles and that predicted by the thermodynamic theory.

Acknowledgments The authors are grateful for the use of small-angle neutron scattering spectrometers at Institut for Energiteknikk, and are thankful for support by the National Science Council, ROC, grants NSC84-2113-M-007-005 and NSC85-2113-M-007-033.

References

1. Ben-Shaul A, Rorman DH, Hartland GV, Gelbart WM (1986) J Phys Chem 90:5277–5286
2. Bergstrom M, Eriksson JC (1992) Langmuir 8:36–42
3. Puvvada S, Blankschtein D (1992) J Phys Chem 96:5579–5592
4. Pilsl H, Hoffmann H, Hoffmann S, Kalus J, Kencono AW, Linder P, Ulbricht W (1993) J Phys Chem 97:2745–2754
5. Long MA, Kaler EW, Lee SP, Wignall GD (1994) J Phys Chem 98:4402–4410
6. Cantu L, Corti M, Degiorgio V (1990) J Phys Chem 94:793–795
7. Abe M, Uchiyama H, Yamaguchi T, Suzuki T, Ogino K, Scamehorn JF, Christian SD (1992) Langmuir 8: 2147–2151
8. Alami E, Kamenka N, Raharimihamina A, Zana R (1993) Colloid Interface Sci 158:342–350
9. Lin T-L, Chen SH, Gabriel NE, Roberts MF (1986) J Am Chem Soc 108: 3499–3507
10. Lin T-L, Chen SH, Gabriel NE, Roberts MF (1987) J Phys Chem 91:406–413
11. Lin T-L, Chen SH, Roberts MF (1987) J Am Chem Soc 109:2321–2328
12. Lin T-L, Hu Y, Liu WJ (1996) Accepted for publication in Langmuir
13. Glatter O (1977) J Appl Cryst 10: 415–421

Progr Colloid Polym Sci (1996) 100:24–28
© Steinkopff Verlag 1996

V. Babić-Ivančić
D. Škrtić
N. Filipović-Vinceković

Phase behavior in sodium cholate/calcium chloride mixtures

V. Babić-Ivančić · D. Škrtić
Dr. N. Filipović-Vinceković (✉)
Department of Chemistry
Ruđer Bošković Institute
Bijenička c. 54
10000 Zagreb, Croatia

Abstract Phase behavior of sodium cholate/calcium chloride mixture strongly depends on both molar ratio and actual concentration of each component. Phase diagram exhibits precipitation regions where mixture of several kinds of solid crystalline and liquid crystalline phase coexist and regions where micelles are formed. Considerable variation of properties is caused by different arrangements of cholate anions which are able to self-organize in molecular structure and supramolecular aggregates due to their amphiphillic character. Solubility, hydrophilicy and aggregation behavior are regulated by the carboxylic group via protonation/deprotonation processes that result in the formation of insoluble and soluble salts.

Key words Bile salt – precipitation – liquid crystals

Introduction

The most important biological anionic surfactants, bile salts, are shown to be responsible for most surfactant activity of urine [1] and have a potential inhibitory role in urinary stone development [2]. Local regulation of calcium concentration by bile salts may be responsible for the extraordinary fine control of kidney stone process formation.

In order to better understand the complex interaction in biological systems at the molecular level, we have focused our investigation on the interactions in mixtures of sodium cholate and calcium chloride. Such investigations cover equimolar soluble complex formation, precipitation, micellization and liquid crystalline phase formation [3]. Two different equilibria are considered responsible for the sequence of phases observed with increasing reactant concentration: i) an equilibrium resulting in the precipitation and ii) another equilibrium resulting in micellization and liquid crystalline phase formation.

In this contribution the precipitation equilibria in sodium cholate/calcium chloride systems have been investi-gated by light microscopy Micellization itself has not been widely discussed although certain reference to it has been made.

Experimental

Many methods were used previously to examine different facets of sodium cholate/calcium chloride interactions. Detailed description of the experimental approaches utilized may be found in ref. [3].

Experimental systems were prepared either with a constant concentration of calcium chloride and increasing sodium cholate (NaC) concentration, or vice versa. The appearance of solid phases, their morphological and textural changes are detected by light microscopy (Leitz, Ortophan). Precipitation boundaries were assessed as described in detail earlier [4]. pH was measured using a combined glass–calomel electrode connected to an Ionalyzer (Orion Research, Model 901). All measurements were performed at 298 K.

Progr Colloid Polym Sci (1996) 100:24–28
© Steinkopff Verlag 1996

Results and discussions

As shown in Fig. 1, phase diagram for NaC/CaCl$_2$ consisted of several areas, the regions where solid crystalline (SC) and/or liquid crystalline (LC) phase precipitate and the regions characterized by continuous transition from micellar (M) and gel-like (G) phase as the concentrations of both components rises [3].

In the region of clear solutions (CS), no crystals are seen under the microscope. In this area ions, monomers and soluble ion-pairs (CaC$^+$) coexist [3]. Within the precipitation region the separate areas corresponding to a single crystalline phase and mixtures of solid and liquid crystalline coexisting phases are indicated. In the region of non micellar solutions three boundaries (I, II and III) are observed depending on actual concentration and molar ratio of calcium and cholate ions. Addition of calcium chloride to NaC solutions lowered the final pH value of the mixtures. Changes in morphology of SC and textures of LC phases followed the changes of pH values accordingly. The region of a single SC phase (region I in Fig. 1) with pH < 6.5 and thin plate-like solid crystals of different dimensions (Fig. 2a) was not broad. Region with pH between 6.5 and 7.0 (region II) where SC and LC coexist (Figs. 2b, c) was, in contrast, broad. In the vicinity of micellar region (M) pH values increased above 7.0.

In the systems prepared with high excess of NaC large nongeometric solid crystals (Fig. 2d) coexisted with micellar phase (region IV in Fig. 1). With increasing NaC concentration, the crystals became smaller and their number decreased, indicating gradual dissolution. The pH values increased from 7.2 to 8.1 paralleling the NaC concentration increase. The reduction of crystallite size was accompanied by the appearance of coacervate (Fig. 3a). Increase in calcium chloride concentration at NaC concentrations close to the micellar region caused the transition from SC to LC phase. The transition from region III to IV was characterized by coexistence of both phases (Fig. 3b). At the highest concentration of both components transparent and homogeneous gel is formed (region V in Fig. 1). Micrographs from this region exhibited typical dispersion of liquid crystalline particles with Maltese crosses (Figs. 3c, d). The size distribution was quite broad with particle diameters ranging up to 100 μm.

Theoretically, Na$^+$, H$^+$ and Ca^{2+} ions could cause cholate anions to precipitate. Due to the high salinity tolerance and low Krafft point of NaC (5), it is assumed that H$^+$ and Ca^{2+} are the only cations involved in the solid phase formation. Precipitation reactions can therefore be presented by the following equations:

$$H^+ + C^- = HC \tag{1}$$

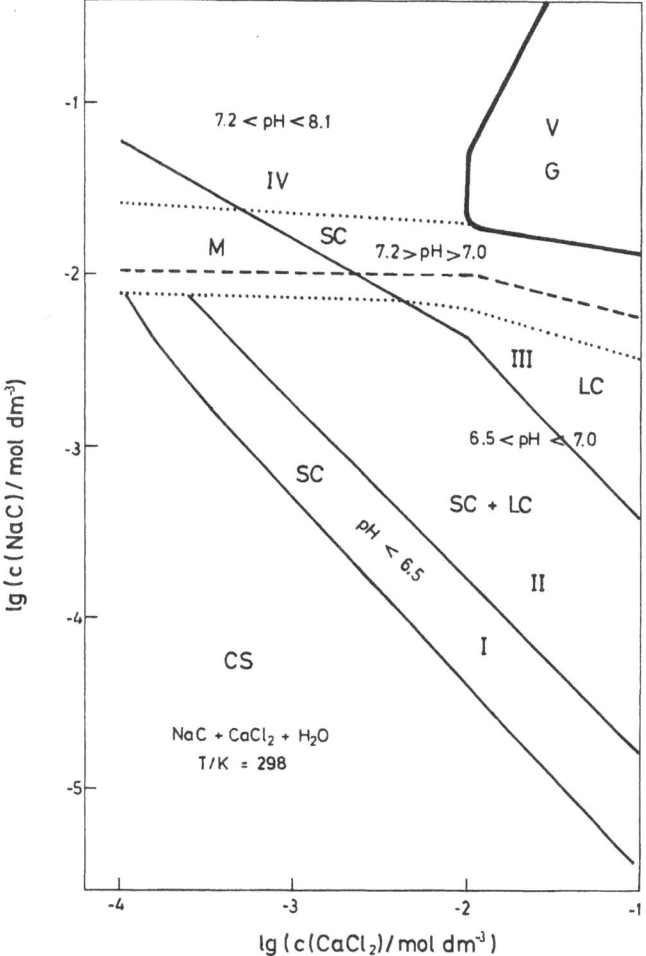

Fig. 1 Phase diagram of the sodium cholate/calcium chloride system: CS – ions and/or ion-pairs; SC – solid crystalline phase; LC – liquid crystalline phase; M – micelles; G – gel like phase. Full lines correspond to the precipitation boundaries; dotted lines denote different pH areas; dashed line denotes micellar region; full heavy line denotes gel-like region

and

$$Ca^{2+} + 2C^- = CaC_2 \tag{2}$$

where C$^-$ denotes cholate anion, HC is cholic acid and CaC$_2$ is calcium cholate. Solubility equilibrium is described by the solubility product, K_{sp}. Corresponding equilibria are:

$$K_{sp} = a(H^+)a(C^-) = f^2 c(H^+)c(C^-) \tag{3}$$

and

$$K_{sp} = a(Ca^{2+})a(C^-)^2 = f^6 c(Ca^{2+})c(C^-)^2 \tag{4}$$

where a and c stand for activities and concentrations of solutes. f is the activity coefficient for monovalent ions

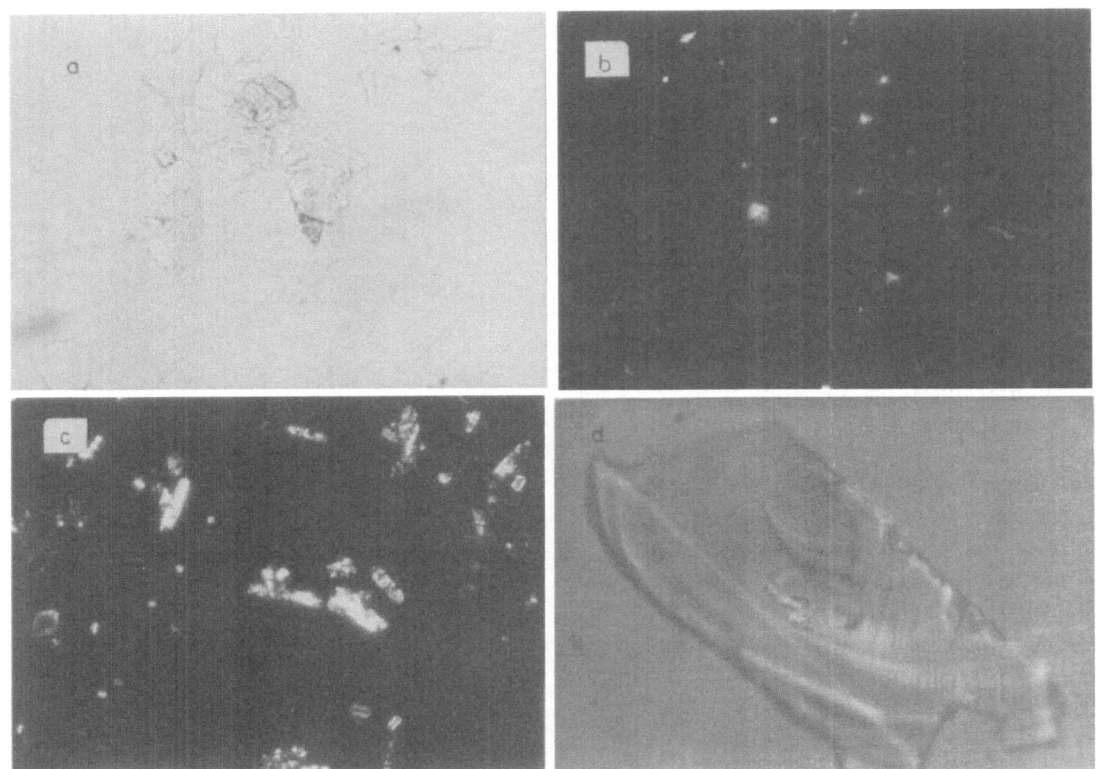

Fig. 2 Microphotographs (magnification 440 X) of phases from sodium cholate/calcium chloride systems: a) thin plate-like solid crystals from the system: $c(NaC) = 0.00006$ mol dm^{-3} + $c(CaCl_2) = 0.03$ mol dm^{-3}; b) liquid crystalline phase from the system: $c(NaC) = 0.004$ mol dm^{-3} + $c(CaCl_2) = 0.005$ mol dm^{-3}; c) mixture of thin plate-like crystals (not shown) with liquid crystalline phase from the system: $c(NaC) = 0.0006$ mol dm^{-3} + $c(CaCl_2) = 0.005$ mol dm^{-3}; d) non-geometric large crystals from the system: $c(NaC) = 0.03$ mol dm^{-3} + $c(CaCl_2) = 0.023$ mol dm^{-3}

given by the Debye–Hückel limiting law

$$-\log f = Al^{1/2}/(1 + l^{1/2}) , \tag{5}$$

where A is temperature dependent constant and l is the ionic strength.

Indirect estimates of the composition of the precipitate may be obtained from the slope of the precipitation/solubility line. The valencies of precipitation components determine the slope of straight line indicating the equilibria of the solubility product. The values of the slopes determined for regions I to IV are listed in Table 1. Except for the region IV the ratio of cholate to calcium ions in crystals was higher than the expected ratio of 0.5 corresponding to CaC_2. Therefore, the crystals observed in regions I to III most probably had a structural composition different from that of CaC_2.

The sparingly soluble cholic acid precipitates from solution when pH is < 6.5 and a solid crystalline precipitate formed in region I was most probably HC. As pH values increased, the appearance of solid-plate like crystals was reduced and the amount of LC phase increased. This

Table 1 Molar ratio of calcium to cholate, Ca/C, determined from the slope of precipitation/solubility lines

Precipitation region	Ca/C
I	0.9
II	1.0
III	1.0
IV	0.5

finding is in accord with decreasing possibility of HC precipitation with the increase of pH above 6.5.

Possibly the composition of precipitate formed in regions II and III may involve hydrogen, calcium, cholate and other anions present (chloride and hydroxide). Gu et al. [6] suggested that OH$^-$ supplied the charge neutrality in precipitates with cholate to calcium ratio being from 1 to 3. At pH values below 7 and higher calcium chloride concentrations the possibility of OH$^-$ involvement in precipitate is diminished.

For a qualitative explanation of the results obtained, one may consider the incorporation of chloride ions into

Progr Colloid Polym Sci (1996) 100:24–28
© Steinkopff Verlag 1996

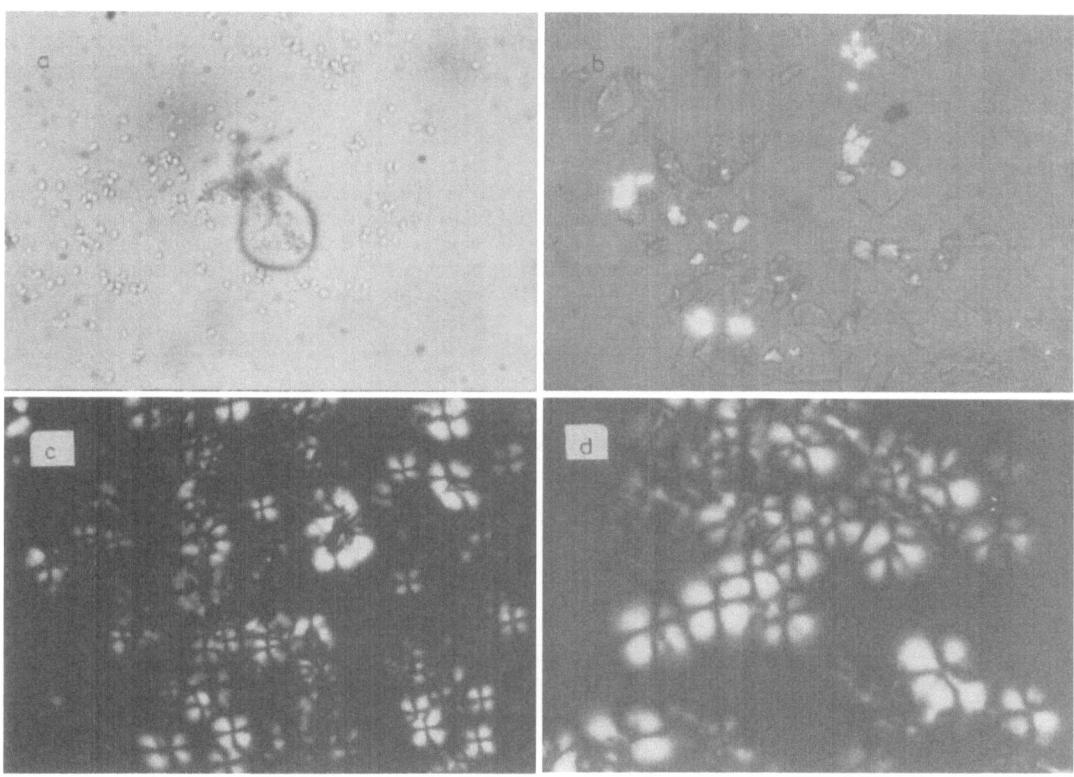

Fig. 3 Microphotographs of phases from sodium cholate/calcium chloride systems: a) coacervate formed from the system: $c(NaC) = 0.3$ mol dm^{-3} + $c(CaCl_2) = 0.023$ mol dm; magnification 660 X; b) mixture of solid crystalline and liquid crystalline phase from the system: $c(NaC) = 0.01$ mol dm^{-3} + $c(CaCl_2) = 0.01$ mol dm^{-3}; magnification 440 X; c) lamellar phase from the system: $c(NaC) = 0.4$ mol dm^{-3} + $c(CaCl_2) = 0.03$ mol dm^{-3}; magnification 180 X; d) lamellar phase from the system: $c(NaC) = 0.04$ mol dm^{-3} + $c(CaCl_2) = 0.1$ mol dm^{-3}; magnification 180 X

the precipitate. Precipitation of calcium cholate chloride (CaCCl) from mixture of sodium cholate and calcium chloride at slightly acidic condition was described by Hogan [7]. The bilayer arrangement of cholate moieties in CaCCl may cause the birefringence indicating the formation of liquid crystalline mesophases.

The ratio of calcium/cholate ions in the precipitate as determined from the line of region IV (Ca/C = 1:2) indicated CaC_2 formation and the mean solubility product for CaC_2 was calculated for each experimental point. Calculated value (log $K_{sp} = -7.1$) was lower than that obtained by Gu et al. [6] CaC_2 (log $K_{sp} = -6.2$). The difference may be explained by taking into consideration micelle formation and the finding that there is only a narrow concentration region where only CaC_2 is formed.

At higher NaC concentration, the presence of Na$^+$ affected the precipitation by favoring micellization; micelles and precipitate coexist (region IV in Fig. 1). Increase in micelle concentration generally redissolved CaC_2 precipitate and coacervate phase before gelation occurred. Since the concentration of NaC at which coacervate is

formed is lower than would be found in the absence of $CaCl_2$, one can conclude that the presence of calcium ions causes a shift of the NaC mesophase equilibria to a lower concentration. In the gelation region (region V in Fig. 1), solid crystalline calcium cholate and liquid crystalline sodium cholate coexisted. The composition and structure of these mixtures were determined by elementary and x-ray analysis [3]. The observed mesophases are identified as the smectic-A type.

It is apparent that in solutions containing both free and micellar cholate anion the addition of Ca^{2+} could result in several simultaneous reactions. The addition of $CaCl_2$ lowers the pH value and allows the terminal carboxylic group to be protonated (the terminal carboxylic group regulates hydrophilicity, solubility and aggregation behavior via protonation/deprotonation processes). At higher pH, carboxylic acid forms precipitate which composition is not clear at present and needs further investigation. Further increase in both pH and Na$^+$ resulted in the formation of water soluble salt and the existence of micelles which incorporate part of cholate ions and bind sodium and calcium ions in the mixed counterion layer at the micelle/solution interface [3].

In conclusion, precipitation of bile acid is mainly controlled by pH, whereas precipitation of insoluble calcium salt is controlled by the activity of Ca^{2+} and by the concentration of cholate anion. Cholate anion may exist in several states: as an insoluble protonated acid, as simply dissolved acid and/or its anion, a simple micelle or an insoluble calcium salts. Above described results, although obtained in a simplified model system, may certainly contribute to better understanding of the complex interactions between calcium ions and bile salts that occur in the human body under both normal and pathological conditions.

Acknowledgments The financial support of the Ministry of Science, Technology and Informatics of the Republic Croatia (Grant 1-07-189) and the Commission of the European Community, Directorate-General XII for Science, Research and Development (Grant No. C11-CT88-345) is gratefully acknowledged.

References

1. Mills CO, Ellias E, Martin GH, Wan MTC, Winder AF (1988) J Clin Chem Clin Biochem 26:187–194
2. Škrtić D, Filipović-Vinceković N, Babić-Ivančić V, Tušek Božić Lj (1993) J (Crystal Growth 60:103–111
3. Filipović-Vinceković N, Babić-Ivančić V, Šmit (1996) J Colloid Interface Sci 177: 646–657
4. Božić J, Krznarić I, Kallay N (1979) 257:201–207
5. Carey MC, Small DM (1972) Arch Inter Med 130:506–527
6. Gu JJ, Hofmann AF, Ton-Nu HT, Schteingart CD (1992) 33:635–646
7. Hogan A, Ealick SE, Bugg CE, Barnes S (1984) J Lipid Res 25:791–798

Progr Colloid Polym Sci (1996) 100:29–35
© Steinkopff Verlag 1996

J.R. Chantres
M.A. Elorza
B. Elorza
P. Rodado

The effect of sub-solubilizing concentrations of sodium deoxycholate on the order of acyl chains in dipalmitoylphosphatidylcholine (DPPC). Study of the fluorescence anisotropy of 1,6-diphenyl-1,3,5-hexatriene (DPH)

Dr. J.R. Chantres (✉) · M.A. Elorza
B. Elorza · P. Rodado
Department of Physical Chemistry II
Faculty of Pharmacy
Complutensian University
28024 Madrid, Spain

Abstract The destabilization of phosphatidylcholine bilayer membranes by the bile salt sodium deoxycholate was studied from steady-state fluorescence anisotropy measurements. FATVET liposomes of DPPC composition were prepared by sequential extrusion through polycarbonate membranes and characterized for their overall inner volume, average size and size distribution, and lamellarity. Interactions between acyl chains in the lipid matrix, which reflect in rotational diffusion motion in the DPH molecules, are perturbed by the presence of the bile salt in the medium (even at low concentrations) below and above the main transition phase temperature of pure DPPC bilayers. Its effects on the lipid matrix are clearly reflected in the DPH steady state fluorescence anisotropy (r_s) measurements. The resolution of r_s into its static (r_∞) and dynamic component (r_d) shows that DOC affects both the amplitude and the velocity of DPH movements.

However, at temperatures below the gel ↔ liquid crystal phase transition point, the static component (that reflects chain order) is more markedly affected than is the dynamic component (which reflects bilayer fluidity). Thus, at 31 °C, angle θ, a measure of the amplitude of DPH oscillations, rises from 23° to 52° over the DOC concentration range from zero to 2.0 mM (equivalent to a effective molar ratio in bilayer of $R_e^{25°} = 0.12$); at 45 °C, however, it varies from 62° to 77°. These changes in the bilayer packing status may be responsible for the alteration in the retention ability of liposomal formulations of the cytostatic agent 5-fluorouracil at sub-solubilizing concentrations of deoxycholate [Elorza et al., *J. Pharm. Sci.* (submitted)]

Key words Deoxycholate – surfactant–liposome interactions – drug delivery systems – 5-fluorouracil – extrusion (prep. meth.) – fluorescence anisotropy – DPH

Abbreviations EYL, egg-yolk lecithin; DPPC, L-α-dipalmitoyl phosphatidylcholine; DOC, sodium deoxycholate; 5-FU, 5-fluorouracil; DPH, 1,6-diphenyl-1,3,5-hexatriene; 5(6)-CF, 5(6)-carboxyfluorescein; Tris, tri(hydroxymethyl)aminomethane; MLVs, multilamellar vesicles; FATVETs, vesicles obtained by cyclic freezing-thawing and extrusion; V_i, overall internal volume of the liposomal suspension; PCS, photon correlation spectroscopy; $\langle d_h \rangle$, mean hydrodynamic diameter; CMC, critical micellar concentration; D_t, total detergent concentration in the medium; D_b, detergent concentration in the aqueous medium; D_h, detergent concentration in bilayer; R_e, effective [detergent]/[lipid] molar ratio; R_e^{sol}, R_e at complete solubilization; R_e^{sat}, R_e at the onset of the lamellar to micelle transition; SD, standard deviation.

30

J.R. Chantres et al.
Effect of sub-solubilizing concentrations of DOC on acyl chains in DPPC

Introduction

The use of liposomes for transport and delivery of drugs provides a means for increasing the therapeutic index of some. Liposome–surfactant (detergent) interactions play a central role in liposome-mediated drug delivery mechanisms, which have been comprehensively reviewed [1–6]. 5-Fluorouracil, a potent cytostatic agent used for 20 years in the treatment of solid tumors and dermal diseases, is a firm candidate for liposomal encapsulation with a view to its parenteral or topic administration in order to avoid some of its undesirable side-effects (particularly those affecting the gastrointestinal duct and bone marrow) [7]. While *in vitro* and *in vivo* experiments have provided encouraging results, there remain a number of problems to be solved in this respect, such as those arising from the instability of liposomes in the biophase (*viz.* retention disturbances caused by some plasma components). In fact, some amphiphilic molecules in biological fluids can induce its release, so liposome–surfactant interactions play a crucial role in the mechanism by which the drug is released by lipid vesicles.

The aim of this work was to derive information on the destabilizing action of deoxycholate (DOC) ion on phospholipid bilayers. For this purpose, we studied its effects on the packing state of hydrocarbon chains of FATVETs liposomes of dipalmitoylphosphatidyl choline composition by monitoring the steady state fluorescence anisotropy of DPH molecules imbibed in the bilayer and analyzing the results in terms of its static and dynamic components, according with the "wobbling-in-cone" model by Kinosita et al. [8–10].

Materials and methods

Lipids and other reagents

Lyophilized L-α-dipalmitoyl phosphatidylcholine (DPPC) 99% pure, sodium deoxycholate (DOC), 5-fluorouracil (5-FU), and the fluorescent probe 1,6-diphenyl-1,3,5-hexatriene (DPH) were supplied by Sigma. 5(6)-Carboxyfluorescein was purchased from Eastman Kodak.

The stated purity of the phospholipids was checked by the usual TLC methods [11]. They were stored as 2:1 v/v chloroform/methanol solutions in a nitrogen atmosphere at $-20\,^{\circ}$C. All were titrated according to Bartlett [12]. Deoxycholate was recrystallized twice from hot ethanol. Its CMC in the aqueous medium used was determined by using the fluorescent dye method; we have estimated a value of 2.7 mM. DPH was stored as a tetrahydrofuran (THF) solution in the dark. The polycarbonate filters used were purchased from Nucleopore. All other reagents were analytical-grade chemicals and bidistilled water was used throughout.

Preparation of vesicles

Multilamellar vesicles (MLVs) were formed from dried lipid films by resuspension in an aqueous buffer consisting of 5 mM Tris-HCl and 150 mM NaCl (pH 7.4, 290 mOsm) following rehydration in a nitrogen atmosphere for 2 h. The initial phospholipid concentration in the aqueous medium was $10\,\mu\text{mol}\,\text{mL}^{-1}$ in all preparations. The osmolality of the solutions was measured with a digital cryo-osmometer (Knauer). Freeze-thaw extrusion vesicles (FATVETs) were prepared by subjecting MLVs to five freeze-thaw cycles in liquid nitrogen and warm water ($60\,^{\circ}$C), followed by sequential cyclic extrusion through two stacked polycarbonate filters of $0.2\,\mu$m (5 cycles) and $0.1\,\mu$m pore size (10 cycles) at 55×10^5 Pa using an extruder (Lipex Biomembranes) jacked at a thermostating bath. Extrusion was performed at $55\,^{\circ}$C for FATVETs[DPPC] ($T_\text{m} = 41\,^{\circ}$C for MLVs).

Characterization of dispersions

Vesicle dispersions were characterized from their internal volume, V_i, average vesicle size and size distribution, and lamellarity.

The overall internal volume of the dispersions was measured by using 5(6)-carboxyfluorescein [5(6)-CF] as label. The average V_i value (expressed in liters of encapsulated water per mole of phospholipid) and standard deviation obtained in six individual experiments were $1.70 \pm 0.08\,\text{L}\,\text{mol}^{-1}$. The z-average hydrodynamic diameter $\langle d_\text{h} \rangle$ of the vesicles and the size distribution of their suspensions were determined by photon correlation spectroscopy (PCS) using a Malvern Zetamaster S submicron particle analyzer. FATVETs formulations were found to be monodisperse (polydispersity index = 0.06), with $\langle d_\text{h} \rangle = 92 \pm 7$ nm. Lamellarity was estimated by ^{31}P-NMR. Measurements were performed at $25\,^{\circ}$C using a Bruker AMX-500 spectrometer. From the integrated peak areas found before and after Mn^{2+} addition, the percentage relative loss of signal was calculated. The loss of signal is caused by the extensive line broadening that results from the interaction between the paramagnetic ions and the outward-facing head groups of the phospholipids in the vesicles. The average from three measures shows that percentage of external surface area is $48 \pm 3\%$; so FATVETs[DPPC] used in this study are predominantly unilamellar.

Fluorescence anisotropy measurements

The variation of the steady-state anisotropy of DPH imbided in lipid bilayers as a function of temperature and the effect of DOC on it were monitored on a Perkin–Elmer MPF-44A spectrofluorimeter as described elsewhere [13]. Samples were excited with vertically polarized light of $\lambda_{ex} = 360$ nm and emitted light analyzed at $\lambda_{em} = 430$ nm. A bandpass of 10 nm was used for both the excitation and emission beam, in addition to a 390 nm cut-off filter. Steady-state anisotropy values were determined from the following equation:

$$r_s = \frac{I_{VV} - GI_{VH}}{I_{VV} + 2GI_{VH}},$$

(1)

where I_{VV} and I_{VH} are the parallel and normal component, respectively, of the fluorescence emission relative to the vertical polarization plane of the excitation beam and $G = (I_{VH}/I_{HH})$ is the Azumi–McGlyn factor [14]. All fluorescence emission measurements were corrected for light scattering (less than 1% in every case).

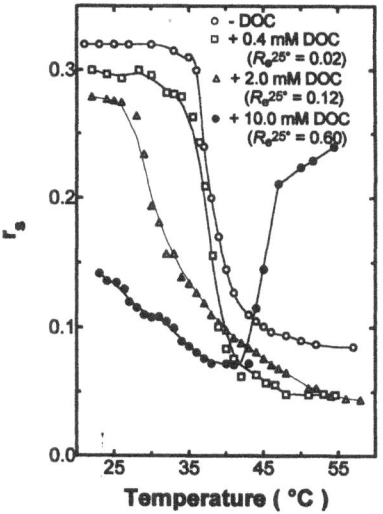

Fig. 1 Steady-state fluorescence anisotropy, r_s, of DPH in FATVETs[DPPC] vs. temperature as function of the total concentration of deoxycholate in the medium. (Lipid concentration was 0.5 μmol mL^{-1} (*ca.* 0.4 mg mL^{-1}); molar ratio DPH/DPPC: 1/500; $\lambda_{ex} = 360$ nm; $\lambda_{em} = 430$ nm)

Results and discussion

DPH is a rod-like molecule which absorption and emission dipoles are almost parallel and lie along the principal molecular axis that is placed perpendicular to bilayer plane. The mean position of probe into bilayers is equivalent to segment $C_{10} - C_{11}$ of the hydrocarbon chains of phospholipids. Rotational motions of DPH molecules reflect the movement of the chains and so, any modification on them (i.e., induced by the incorporation of a foreign molecule from aqueous phase) will result in changes on the fluorescence anisotropy of DPH. Figure 1 shows the influence of temperature on the steady-state fluorescence anisotropy of DPH in absence of DOC and some examples of its alteration in presence of DOC in the medium.

The effect of surfactant on bilayer integrity depends on the "effective" molar ratio detergent/phospholipid in the bilayer (R_e) and not on the "total" molar ratio (R_t), R_e being defined as $R_e = D_b/PL$ and R_t as $R_t = D_t/PL$ with $D_t = D_b + D_w$. A partition coefficient is also defined as $K = R_e/D_w$. In a previous study [13], we have found that the R_e value corresponding to the onset of bilayer solubilization R_e^{sat}, (i.e., bilayers being saturated by deoxycholate) of FATVETs[DPPC] is 0.3 at 25 °C and 50 °C. In turn, R_e value corresponding to completion of bilayers solubilization, R_e^{sol}, lie between 1.2 and 1.4 at both temperatures. Furthermore, the same results were obtained with FATVETs[EYL] at 25 °C. This mean that R_e^{sat} and R_e^{sol} are independent of the physical state of the bilayers.

However, this factor certainly affects the partition coefficient K that for FATVETs[DPPC] at 25 °C is 0.07 mM^{-1} and at 50 °C is 0.16 mM^{-1}, and for FATVETs[EYL] at 25 °C is 0.21 mM^{-1}. According to the definition of partition coefficient K, this only has meaning in the range of sub-solubilizing DOC concentrations. R_e can be estimated, in that range, for any liposome sample by means of the relation:

$$R_e = \frac{D_t}{((1/K) + PL)},$$

(2)

as can be concluded from previous expressions. DOC concentration corresponding to results depicted in Fig. 1 is equivalent to an effective molar ratio $R_e^{25°}$ of 0.02, 0.12 and 0.6, respectively. As can be seen, the thermotropic gel ↔ liquid crystal phase transition of DPPC molecules in the bilayer structure still took place at a DOC/phospholipid effective mole ratio of 0.12, as derived from the r_s vs T profile. The transition vanished by a DOC/phospholipid effective mole ratio of 0.6, beyond which the r_s vs T profile departed from the usual shape for a lamellar structure (a smooth decay in r_s); in fact, no trace of the transition remained above a DOC/phospholipid effective mole ratio of 0.3 (results not shown).

According to the "wobbling-in-cone" model proposed by Kinosita et al. [8] motions of DPH are restricted by the bilayer components. This leads to resolve the steady-state fluorescence anisotropy in two components, one static (r_∞)

and another dynamic (r_d), being: $r_s = r_\infty + r_d$, a relation exists between r_∞ and the opening angle θ^1 of the cone around of the perpendicular to the bilayer plane, whose expression is:

$$\left(\frac{r_\infty}{r_0}\right) = \left[\left(\frac{1}{2}\right)\cos\theta(1 + \cos\theta)\right]^2, \qquad (3)$$

when r_0 is initial limiting anisotropy, in this study we have adopted for DPH the value $r_0 = 0.4$. Later than Kinosita et al. theory, Heyn y Jähnig [15, 16] established the relation between the order parameter S y r_∞:

$$\left(\frac{r_\infty}{r_0}\right) = S^2; \quad 0 \le S \le 1, \qquad (4)$$

thus, r_∞ are related with the order degree of hydrocarbon chains into bilayer. van Blitterswijk et al. [17], Pottel et al. [18], and van der Meer et al. [19] have demonstrated the possibility to derive the r_∞ value from the measured steady-state data r_s by an empirical relationship applicable in a variety of artificial and biological membranes labeled with various probes, that relation is:

$$r_0 r_s^2 / (r_0 r_s + (r_0 - r_s)^2/m) \qquad (5)$$

the parameter m, which has a positive value, expresses the difference between the rotational diffusión of the probe in the bilayer and that in an isotropic reference oil. For DPH van der Meer et al. [19] found the value $m = 1.7$ for DPPC bilayers between $5°-60°C$. This value was used in our study. Figure 2 show the variation of static r_∞ (Fig. 2a) and dynamic r_d (Fig. 2b) components of r_s in function of temperature, in absence and presence of 2.0 mM DOC ($R_e = 0.12$ at $25°C$). As r_∞ as r_d are affected. Before the gel $\leftrightarrow$ liquid crystal occurs, the static component drops, while the dynamic component increases. No variations in r_∞ and r_d were observed between $25°C$ and the onset transition temperature, in each one; this can mean that the partition coefficient K does not change significantly into that temperature range. In the temperature range at which the transition occurs, presence of DOC causes diminution of r_∞ and increases r_d. When the transition has been completed r_d values falls below those obtained in absence of surfactant, while r_∞ ones continue being less than those obtained in its absence. Also, it is interesting to observe that when the transition has taken place, in absence of DOC, r_∞ and r_d are almost constants, while in its presence both values keep their tendency to go down, at least up to $55°C$. From results shown in Fig. 2 it is deduced that the incorporation of deoxycholate into DPPC bilayers causes

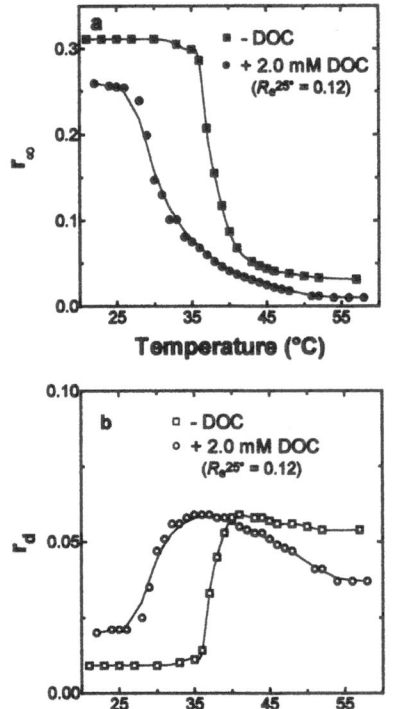

Fig. 2 a) Static, r_∞, and b) dynamic, r_d components of r_s as function of temperature in absence and presence of 2.0 mM DOC, equivalent to an "effective" molar ratio (surfactant/phospholipid) at $25°C$, of $R_e^{25°} = 0.12$. Bars symbolized standard deviation of at least six measurements

a disturbance in interactions among chains of which turn out: a) the temperature of the gel$\leftrightarrow$liquid crystal transition descends, b) the temperature range in which the transition occurs increases. As a consequence a increment in the amplitude and rate of the rotational motions of DPH molecules takes place.

Figure 3 depicts the variation of r_s (Fig. 3a), r_∞ (Fig. 3b) and r_d (Fig. 3c) as function of the total concentration of DOC in the medium. Results correspond to temperatures of $31°$ and $45°C$, to which DPPC bilayers, in absence of DOC, are in gel phase and liquid crystal phase, respectively. At $31°C$ and surfactant concentrations below 1.0 mM (i.e., $R_e^{25°} = 0.06$) the action of DOC on r_s components is small. From 2.0 mM (i.e., $R_e^{25°} = 0.12$) both components undergo important changes which end up by being equal to that found at $45°C$. This is a consequence that from 2.0 mM DOC the transition start to $T \le 28°C$ ($T_m \le 33°C$). It is likely that more "fluid" regions in bilayers arise that support an increase in surfactant incorporation at them. We have found [13] that the partition coefficient driving deoxycholate distribution between lipidic phase and aqueous phase depends especially on

---

^{1} θ is defined as the angle at which the membrane probe encounters an infinite energy barrier.

Progr Colloid Polym Sci (1996) 100:29–35
© Steinkopff Verlag 1996

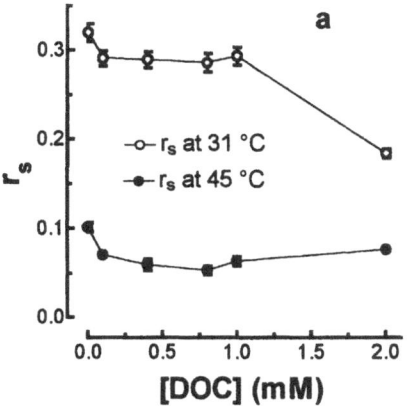

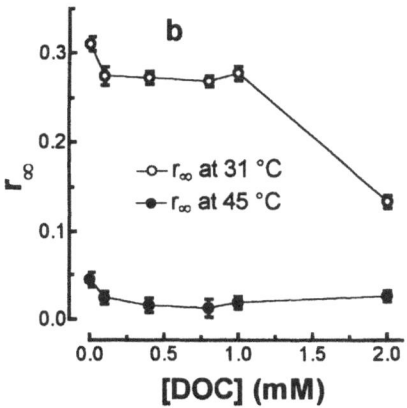

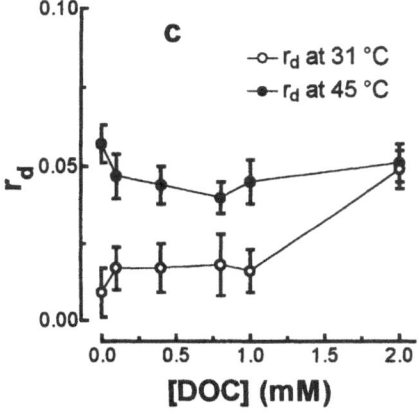

Fig. 3 Variation of steady-state fluorescence anisotropy r_s(a), and its static r_∞(b) and dynamic r_d(c) components as function of deoxycholate concentration in the medium. Results at 31° and 45 °C, temperatures below and above, respectively, the temperature of the gel ↔ liquid crystalline phase transition of DPPC bilayers. Bars symbolize standard deviation of at least six measurements. (FAT-VETs of DPPC showed a $T_m = 38\,°C$ [13] against $T_m = 41\,°C$ displayed by MLVs, because the smaller curvature radio of the unilamellar vesicles)

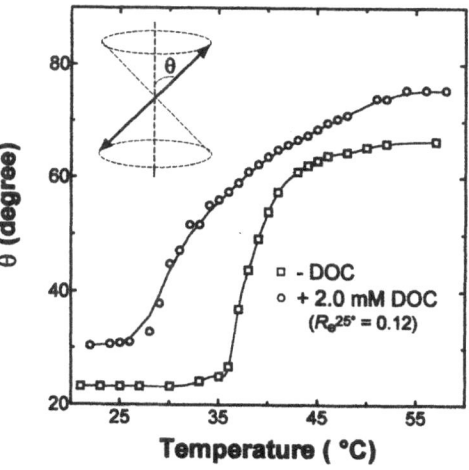

Fig. 4 Variation of "wobbling-in-cone" angle θ of DPH molecules imbided in DPPC bilayers as function of temperature, in absence and presence of 2.0 mM DOC ($R_e^{25°} = 0.12$)

the physical state of bilayer. At 45 °C, with the FATVETs[DPPC] in liquid crystalline state, DOC presence does not seem to modify r_∞ and r_d diminishes a little, i.e., a "condensant" effect similar to the induced by cholesterol does not seem to take place [20].

Figure 4 depicts the variation of the "wobbling-in-cone" angle θ of DPH molecules imbibed in FATVETs[DPPC] bilayers as function of temperature, in absence and presence of 2.0 mM DOC. In absence of surfactant θ remains constant with a value of $\theta \approx 23°$ at $T < (T_m - 2°)$ and with a value of $\theta \approx 67°$ at $T > (T_m - 2°)$; all change takes place in the small range of $T_m \pm 2°$. In presence of 2.0 mM DOC (i.e., $R_e^{25°} = 0.12$) θ remains constant ($\theta \approx 30°$) until $T \approx 28\,°C$; starting from here θ increases progressively more (sharply at the beginning) until $T \approx 55\,°C$ reaches the value of $\theta \approx 77°$.

Figure 5 shows the influence of total surfactant concentration on θ values at 31° and 45 °C, in the range equivalent to $0 < R_e^{25°} \le 0.3$. While bilayers are in the ordinate gel phase, DOC presence is assumed by a small increase in θ that, however, is not very sensitive to increasing surfactant concentrations in the medium, which is likely due to the difficulties its molecules encounter by becoming incorporated into bilayers in those conditions. When the transition is triggered as consequence of the surfactant-destabilizing action, θ increases at the same time that increasingly more DOC molecules become incorporated into bilayers, which in turn promotes a larger chain motion freedom that is reflected in a progressive increase in θ. At 45 °C, with bilayers in the mesophase liquid crystalline and an easer accessibility for the incorporation of DOC molecules (as judged by the partition coefficient K value at 50 °C), the parameter θ appears more sensitive

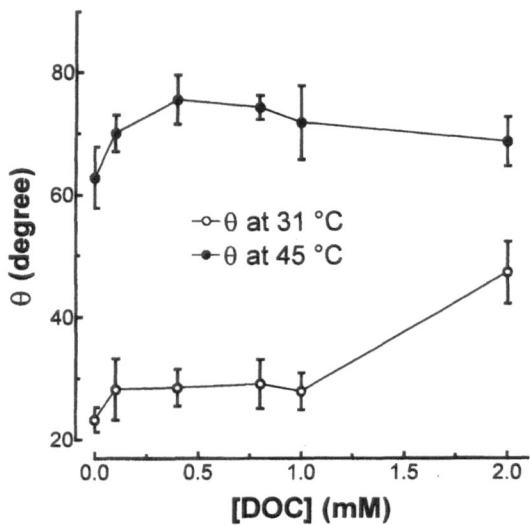

Fig. 5 Variation of "wobbling-in-cone" angle θ as function of deoxycholate concentration in the medium. Results at 31° and 45 °C, temperatures below and above, respectively, of the temperature of the gel $\leftrightarrow$ liquid crystalline phase transition of DPPC bilayers. Bars symbolized standard deviation of at least six measurements

to the changes in surfactant concentration, although the variation range is restricted at 10°. When the DOC total concentration is 5.0 mM (i.e., $R_e^{25°} = 0.30$) lipid phase becomes saturated by the bile salt and θ values obtained at 31° and 45 °C agree well in practice.

Conclusions

a) As sense by DPH molecule, even at low [DOC]$_{total}$, perturbating action of surfactant on fatty acyl chains is

obvious. The tendency of DOC to be imbibed normal to the surface of bilayer membranes, becomes intercalated among phospholipid molecules [21], collaborate to it. Its effects on lipid matrices are clearly reflected in steady-state fluorescence anisotropy measurements of DPH.

b) From our study, the bilayer perturbation induced by DOC at sub-lytic levels on leakage of 5-FU is believed to result from the packing and structural changes undergone prior to solubilization of the bilayers.

The phospholipid bilayer is an anisotropic medium and the incorporation of a foreing molecule into its structure from the aqueous phase cannot be contemplated as a macroscopic partition phenomenon. Therefore, the amount of surfactant that a phospholipid bilayer can accept depends on the packing state of the phospholipid molecules (either acyl chains as head polar groups), the chemical structure of surfactant, and the surfactant–phospholipid interactions within the bilayer. Disposition of deoxycholate molecules may involve introduction of polar hydroxyl groups in positions 3 and 12 into bilayers' hydrophobic core, which will have a destabilizing effect on DOC-phospholipid mixed vesicles.

In summary, the stability of liposomal formulations against amphiphiles present in physiological fluids can be improved by ensuring that acyl chains in bilayers are tightly packed; this can be accomplished by using phospholipids of a high T_m and/or including cholesterol in the formulation. In any case, this should not compromise the drug-releasing ability of liposomes, so bilayer stability and permeability should always be carefully balanced.

Acknowledgments The authors wish to express their gratitude to financial support in the form of a grant from the Programa Nacional de I + D (SAF 93-0249).

References

1. Helenius A, McCaslin DR, Fries R, Tanford C (1979) Methods Enzymol 56:734–740
2. Helenius A, Simons K (1975) Biochim Biophys Acta 415:29–79
3. Lichtenberg D, Robson RJ, Dennis EA (1983) Biochim Biophys Acta 737:285–304
4. Lasch J, Schubert R (1993) in Liposome Technology. Vol II (Gregoriadis G ed) pp 233–260, CRC Press, Boca Raton, FL (USA)
5. Goñi FM, Urbaneja MA, Alonso A (1993) in Liposome Technology, Vol. II (Gregoriadis G ed) pp 261–273, CRC Press, Boca Raton, FL
6. Lasch J (1995) Biochim Biophys Acta 1241:269–292
7. Chabner BA (1982) In: Chabner BA (ed) Pharmacologic Principles of Cancer Treatment. Saunders, Philadelphia, pp 132–142
8. Kinosita JrK, Kawato S, Ikegami A (1977) Biophys J 20:289–305
9. Kawato S, Kinosita JrK, Ikegami A (1977) Biochemistry 16:2319–2324
10. Kinosita JrK, Ikegami A (1984) Biochim Biophys Acta 769:523–527
11. Kovács L, Zalka A, Dobó R, Pucsok J (1986) J Chromatogr B Biomed Appl 382:308–313
12. Bartlett GR (1959) J Biol Chem 234:466–468
13. Elorza MA, Elorza B, Rodado P, Chantres JR (1996) J Pharm Sci (submitted)

14. Azumi T, McGlynn S (1962) J Phys Chem 37:2413–2420
15. Heyn MP (1979) FEBS Lett 108:359–364
16. Jähnig F (1979) Proc Natl Acad Sci USA 76:6361–6365
17. van Blitterswijk WJ, van Hoeven RP, van der Meer BW (1981) Biochim Biophys Acta 644:323–332
18. Pottel H, van der Meer W, Herreman W (1983) Biochim Biophys Acta 730:181–186
19. van der Meer BW, van Hoeven RP, van Blitterswijk WJ (1986) Biochim Biophys Acta 854:38–44
20. Kawato S, Kinosita JrK, Ikegami A (1978) Biochemistry 17:5026–5031
21. Saito H, Sugimoto Y, Tabeta R, Suzuki S, Izumi G, Kodama M, Toyoshima S, Nagata C (1983) J Biochem (Tokyo) 94:1877–1885

Progr Colloid Polym Sci (1996) 100:36–38
© Steinkopff Verlag 1996

Multifractality of lyotropic liquid crystal formation

Đ. Težak
M. Martinis
S. Punčec
I. Fischer-Palković
S. Popović

Dr. Đ. Težak (✉) · S. Puncec
I. Fischer-Palković
University of Zagreb
Faculty of Science
Department of Chemistry
Marulicev trg 19
P.O. Box 163
10000 Zagreb, Croatia

M. Martinis · S. Popović
Rugjer Boskovic Institute
Bijenicka cesta 54
10000 Zagreb, Croatia

Abstract The molecular organization in lamellar mesophases was proposed on the basis of dynamic and static light scattering, polarizing and transmission electron microscopy, and x-ray diffraction measurements. The basic symmetries of the dynamics of formation and structures of the lamellar bilayers and multilayers of the surfactants of alkylbenzenesulphonate series were found to be consistent from the beginning of the precipitation process from a homogeneous solution until the aggregation to the macroscopic aggregates, due to the self-similarity.

Key words Aggregation – lamellar mesophase – multiscaling – self-similarity

Introduction

The linear alkylbenzenesulphonate surfactants are used for industrial and household detergents; therefore, the precipitation reactions with metal ions which could be found in aqueous eco-systems, are of interest. In the previous paper [1, 2] the reactions of different metal ions in water and in sea-water were described. A model of multiscaling for the calculation of fractal dimensions in the formation of the liquid crystaline phases was proposed [3]. This theoretical and experimental treatment allows us to consider the process of the formation of lamellar phases as self-similar.

Experimental

Materials

Dodecylbenzenesulphonic acid (HDBS) from "Ventron", Germany (an isomeric mixture of linear and branched compounds containing approximately 80% of C_{11} and C_{12}, and 18% of C_9, C_{10}, C_{13} and C_{14} isomers), was used as surfactant which forms lamellar phases. Analytically pure magnesium and aluminum nitrates from "Kemika", Zagreb, dissolved in water, were standardized complexometrically, and used as counterions in the reactions with surfactants. Doubly distilled water was used in all experiments.

Methods

Dynamic (DLS) and static (SLS) light scattering measurements were made over a range of scattering angles ($\theta = 5$–$145°$) using an Otsuka photometer. X-ray diffraction patterns were obtained using a standard Siemens x-ray diffractometer with a counter and Si-crystal monochromatized CuK_α radiation. The photomicrographs were obtained by using a Leitz Wetzlar optical microscope with polarizing equipment. The samples for transmission electron microscopy (TEM) using the freeze etching method, as well as electron micrographs, were made courtesy of the University of Braunschweig, Germany.

Results and discussion

The lamellar phases were found to be formed in surfactant/water [4] as well as in surfactant/electrolyte systems [1].

The basic symmetry of the dynamics of formation and structure of liquid crystals was found to be consistent, i.e., it is self-similar, but multifractal in terms of time [3]. The particle sizes presented in Fig. 1 (R is a radius presenting the maximum particle size of the Gaussian distribution of polydispersed aggregates) showed lower values for high surfactant concentrations due to a faster nucleation than for low concentrations. The assumption that the aggregates of amphiphile molecules can be dynamically treated as multifractal objects [3], can be improved by the tem-

poral dependence of particle radii that follows two time-periods, the first of them belonging to the short-range interactions of growth, and the later period of consecutive aggregation, while the curves in Fig. 2 reach the constant values.

The first step of mixing the reacting components to form colloidal (lyotropic liquid crystalline) systems is the most important: the initial complexation from a

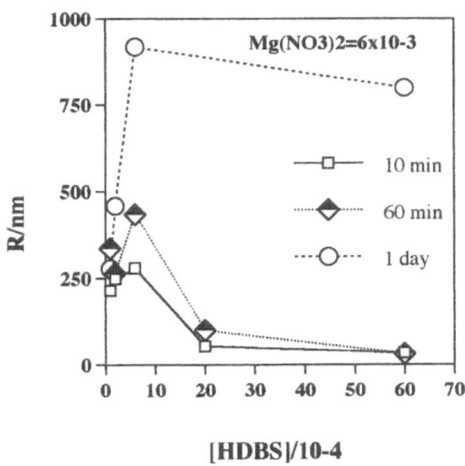

Fig. 1 Average cluster radii of magnesium dodecylbenzenesulphonate liquid crystalline phase depending on the concentration of HDBS. All concentrations are expressed in $mol\,dm^{-3}$

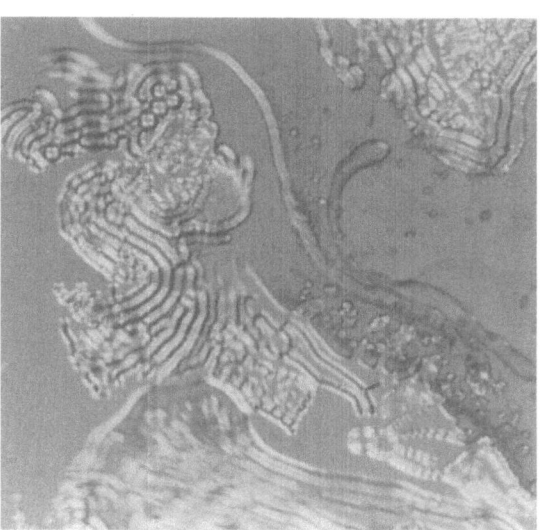

Fig. 3 Photomicrograph of the recently prepared aluminum dodecylbenzenesulphonate liquid crystalline phase. Crossed polarizers, 1λ-plate; magnification 80 X

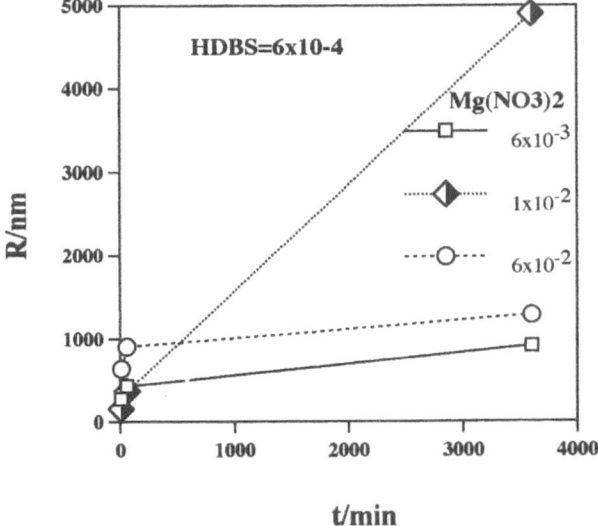

Fig. 2 Time-dependence of the average cluster radii of the magnesium dodecylbenzenesulphonate. All concentrations are expressed in $mol\,dm^{-3}$

Fig. 4 TEM micrograph of the aged sample of the magnesium dodecylbenzenesulphonate showing the flat and bent lamellar multilayers. Magnification 50 000 X

38

Đ. Težak et al.
Lyotropic liquid crystal formation

homogeneous phase usually causes the formation of bi-layer cylinders which very quickly perturb to lamellar droplets (primary particles) (Fig. 3). Some cylinders show only the light traces due to their faster ordering into the droplets than it is the exposure time. After a longer period these droplets form larger spherulites (aggregated secondary particles) showing maltese crosses. The formation of flat lamellar multilayers, and folded lamellar bilayers like a primary spherulite can be seen in the TEM micrograph (Fig. 4).

The basic interplanar distances, amounting to (32.4 ± 0.03) Å for alumnum, (33.2 ± 0.6) Å for magnesium, and similar values for other bi- and trivalent metal ions [5], showed the same values for different sizes of spherulites, and for any time of sampling.

It can be considered that the association of surfactant molecules is a self-similar process, regarding from the first step of the formation of a new phase from the homogeneous solution throughout the consecutive aggregation processes of the formation of lamellar structure.

References

1. Težak Đ, Strajnar F, Milat O, Stubičar M (1984) Progr Colloid Polym Sci 69:100–105
2. Težak Đ, Babačić O, Đerek V, Galešić M, Heimer S, Hrust V, Ivezić Z, Jurković D, Rupčić S, Zelović V (1994) Colloids Surfaces A: Physicochem Eng Aspects 90:261–270
3. Težak Đ, Martinis M, Punčec S, Fischer-Palković I, Strajnar F (1995) Liquid Crystals 19:159–167
4. Težak Đ, Hertel G, Hoffmann H (1991) Liquid Crystals 10:15–27
5. Težak Đ, Popović S, Heimer S, Strajnar F (1989) Progr Colloid Polym Sci 79:293–296

Progr Colloid Polym Sci (1996) 100:39–42
© Steinkopff Verlag 1996

Surfactant aggregation and response of ion-selective electrode

Y.A. Shchipunov
E.V. Shumilina

Dr. Y.A. Shchipunov (✉) · E.V. Shumilina
Institute of Chemistry
Far East Department
Russian Academy of Sciences
690022 Vladivostok, Russia

Abstract The report is concerned with the functioning of ion-selective electrodes responding to surface-active agents in solutions of aggregated surfactants. Contrary to the current mechanisms destined for inorganic sensors, it has been suggested that adsorbed species and surface aggregates (hemimicelles and admicelles) on the membrane surface are essential factors in the response and functioning of surfactant electrodes.

Key words Ion-selective electrode – surfactant – association – micellization – adsorption – surface aggregation

Introduction

It is usually recognized that in diluted solutions surfactants begin with self-organizing into micelles, consisting of tens of molecules, at a certain critical micelle concentration (CMC). On the other hand, there are rather numerous observations indicative of aggregation of few surfactant molecules into associates below the CMC. The reality of the associates is evidenced by different physico-chemical techniques, but not by surfactant electrodes (for details, see refs. [1–3]). The reasons for their insensitivity to the substance association and the mechanisms of their functioning in the presence of aggregates are the subjects of this report.

Experimental

Wire-coated electrodes responsive to inorganic ions, nitrate anions, and to organic ones, tetraphenylboron (TPB) and dodecylsulfate (DS) anions, were prepared by dip-coated method detailed in [3]. An ion-selective membrane consisted of 50% w/w PVC, 47% w/w dibutyl phthalate and 3% w/w ion exchanger. Methyltridodecylammonium nitrate was used as an ion exchanger for a nitrate elec-

trode, high-molecular ammonium salts of DS and TPB or salt of TPB and tetraphenylphosphonium for DS and TPB electrodes, respectively.

The response of an ion-selective electrode was examined in the following electrochemical cell:

Reference electrode	Aqueous solution with tested anions	Ion-selective electrode examined.

The reference silver/silver chloride electrode LAB-REF with two junctions and a porous Teflon stopper were a gift from Innovative Sensors (Anaheim Cal., USA). The emf's (potentials) in an electrochemical cell were measured with a pH-voltmeter PH-673M or a voltmeter S-4313 with high input.

Results and discussion

The features of the examined electrodes are shown from calibration graphs, i.e., potential measured in the electrochemical cell vs. the logarithm of concentration of tested salts in aqueous solution, in Fig. 1. Five important points should be noted.

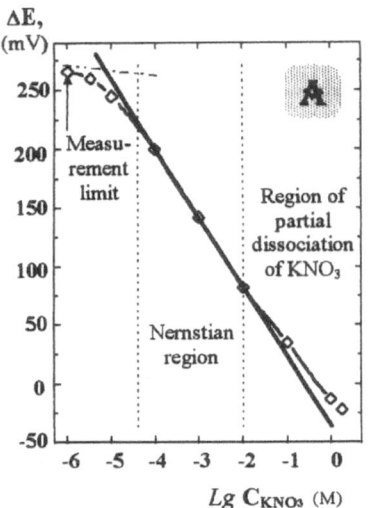
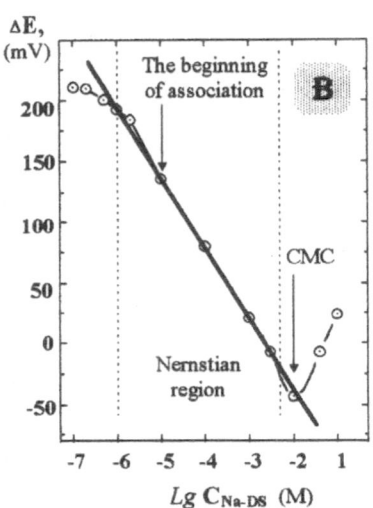
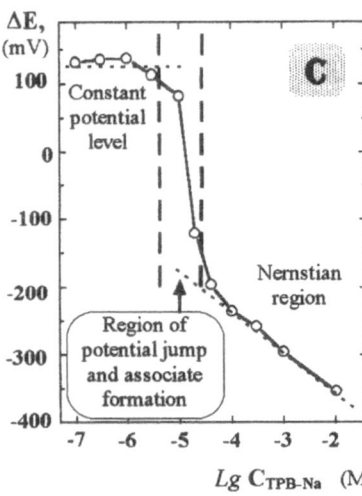

Fig. 1 Potential measured in an electrochemical cell with nitrate (A), dodecyl sulphate (B), and tetraphenylboron (C) electrode as a function of the logarithm of concentration of potassium nitrate (A), sodium dodecyl sulphate (B), and sodium tetraphenylborate (C) in aqueous solution

Fig. 2 Calibration graphs for tetraphenylboron electrode. A) Ion exchanger content in membrane: 1) 1.5; 2) 3; and 3) 12% w/w. B) Sodium chloride concentration in aqueous solution: 1) 0; 2) 0.01; and 3) 1 M

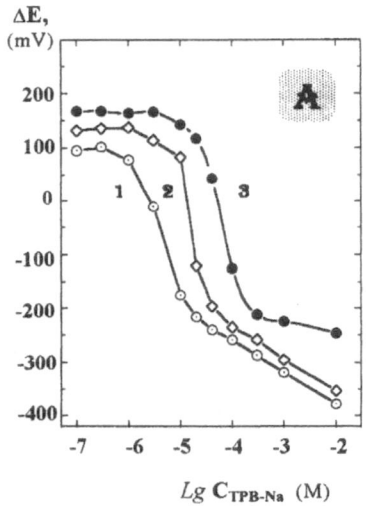
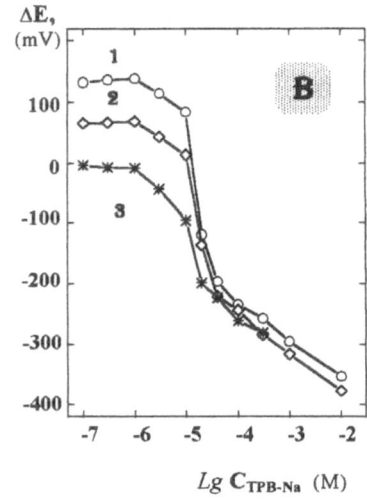

i) The nitrate electrode (Fig. 1A) gives proof that the ion-selective electrodes are every bit as sensitive to the dissociation-association of salts in bulk solutions as the other physicochemical techniques. This follows from the deviation from the Nernstian response at concentrations greater than 10^{-2} M, caused by partial dissociation of KNO_3.

ii) The DS electrode exhibits a typical behavior peculiar to surfactant sensors (Fig. 1B); there is a kink on the curve in the vicinity of the CMC, but the electrode demonstrates a linear Nernstian response in the presence of associates.

iii) The TPB electrode is distinct in its intricate behavior from the other ones examined here. Three regions are resolved on a calibration graph (Fig. 1C). A sharp drop in the potential correlates well with the beginning of TPB association in bulk solutions. A subsequent linear Nernstian response occurs in the presence of TPB associates.

iv) A sigmoidal concentration dependence of the potential measured with TPB electrode resembles those observed in the course of potentiometric titration of one substance with another. This has been previously noted by Morf et al. [4], who observed a closely similar response

Progr Colloid Polym Sci (1996) 100:39–42
© Steinkopff Verlag 1996

Fig. 3 Filling by potential-determining ions of the membrane/water interface where charged centers are formed owing to the dissocia-tion of ion exchanger in diluted aqueous solution (A). (B → C) Formation of hemimicelles with increasing concentration of tetraphenylboron anions in water. (D → E) Monolayer and bimolecular filling of interface by dodecyl sulphate anions, resulting in the admicelle formation

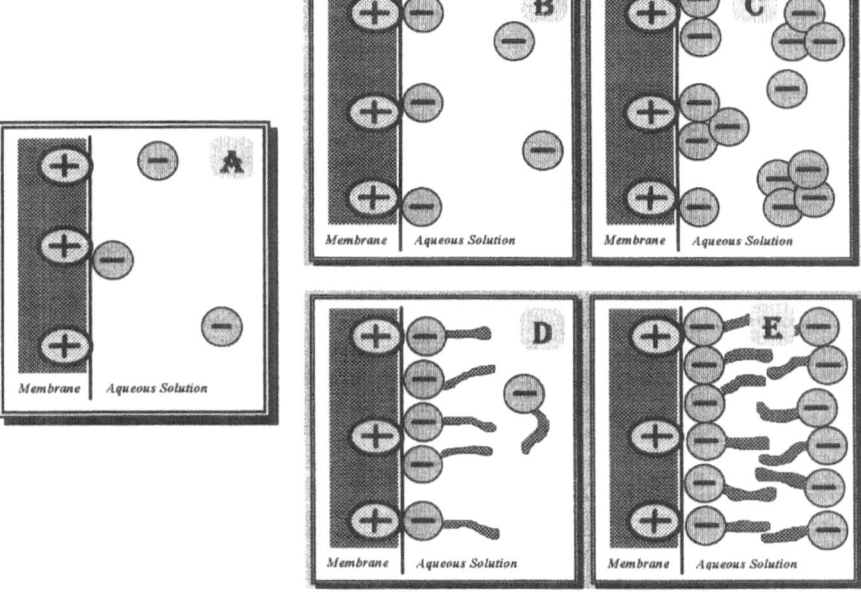

of ion-selective electrodes made up of sparingly soluble inorganic salts.

v) A response of the TPB electrode depends on the ion exchanger content in the membrane matrix and the ionic strength of an aqueous solution. This is illustrated by Fig. 2. It should be mentioned that the increase of ion exchanger content results primarily in a shift in the position of the potential jump along the horizontal coordinate, whereas the value of constant potential in diluted solutions decreases with growing ionic strength.

The above-mentioned features of the examined electrodes help to elucidate the lack of their sensitivity to the surfactant association and to suggest a mechanism for their functioning. This is illustrated by schematic drawings in Fig. 3.

The experimental results allow the conclusion that for the electrode response the surfactant adsorption and adsorbed species on the membrane surface are important. The adsorption in diluted solutions must occur at charged centers (Fig. 3A) formed by an ion exchanger owing to its dissociation in accordance with the equation:

$$C^+ A^- (m) \rightleftharpoons C^+ (m) + A^- (s) ,$$

where C^+ is the cation, A^- is the anion, m is the membrane, s is the solution. The process of adsorption at this stage can be considered a titration of surface charges with potential-determining ions (Fig. 3B). The potential change

in an electrochemical cell may be described by the following equation:

$$\Delta E = \Delta E^\circ - RT/F \ln (C_{A^-} - C_m) ,$$

where ΔE° is the standard electrode potential, C_{A^-} is the anion concentration in a bulk solution, C_m is the concentration of charged centers at the membrane surface. The jump-like potential change takes place at $C_{A^-} \approx C_m$.

The following adsorbed species tend to concentrate also in the vicinity of oppositely charged centers owing to the electrostatic attraction (Fig. 3C). This implies a formation of associates on the surface, named hemimicelles. It should be stressed that the surface aggregation goes on in parallel to the bulk association (for evidence, see ref. [3]). As a consequence, the compositions of the adsorption layer and the bulk solution are changed in a like manner. This may be quite responsible for the lack of electrode sensitivity to the surfactant association and for the Nernstian response.

With increasing surfactant concentration, the complete filling of membrane surface can be attained. When this occurs, the concentration of adsorbed species will be held constant which will result in an independence of an electrode potential on the bulk concentration of potential-determining ions. According to current views, such surfactants as SDS are prone to form first monomolecular film (Fig. 3D) and then bimolecular adsorption layer (Fig. 3E) at the solid/liquid interfaces, called admicelles. Complete filling is often observed in the neighborhood of the CMC

[5–7]. This could be a novel approach to the interpretation of the concentration independence of electrode potential beyond the CMC. The current concepts are based on the constant concentration of monomers in bulk solutions.

References

1. Shchipunov YuA (1984) J Colloid Interface Sci 102:36–45
2. Somasundaran P, Ananthapadmanabhan KP, Ivanov IB (1984) J Colloid Interface Sci 99:128–135
3. Shchipunov YuA, Shumilina EV (1995) J Colloid Interface Sci 173:192–201
4. Morf WE, Kahr G, Simon W (1974) Anal Chem 46:1538–1543
5. Harwell JH, Hoskins JC, Schechter RS, Wade WH (1985) Langmuir 1:251–262
6. Besio GJ, Prud'homme RK, Benziger JB (1988) Langmuir 4:140–144
7. Rupprecht H, Gu T (1991) Colloid Polym Sci 269:506–522

Progr Colloid Polym Sci (1996) 100:43–47
© Steinkopff Verlag 1996

SURFACTANT AGGREGATES

R. Bikanga
P. Bault
P. Godé
G. Ronco
P. Villa

3-deoxy-s-alkyl-d-glucose derivative micelle formation, CMC and thermodynamics

Dr. R. Bikanga (✉) · P. Bault · P. Godé
G. Ronco · P. Villa
Laboratoire de Chimie Organique
et Cinétique
Université de Picardie Jules Verne
33 Rue Saint Leu
80039 Amiens Cedex 1, France

Abstract We have studied micelle formation of two series of D-glucose amphiphilic derivatives: 3-deoxy-S-alkyl-1,2-O-isopropylidene-α-D-glucofuranose, marked MAG-S_n and 3-deoxy-S-alkyl-D-glucopyranose, marked GLU-S_n, where n is the carbon atom number of the alkyl chain R in the alkyl group ($R = n$-C_nH_{2n+1}; $n = 8, 12, 16, 18$). CMC in water and corresponding surface tension γ have been determined at 25°, 30°, 40° and 50 °C by the Wilhelmy plate method. Results show that surface tension is not much altered, either by alkyl chain length or by free hydroxyl group number (2 for MAG-S_n series; 4 GLU-S_n series), whereas these parameters have a great influence on CMC values. Experimental determination of free energy, enthalpy and entropy of micellization

shows that micelle formation is an entropic phenomenon ($T\Delta S_m > \Delta H_m$); the monomer association is easier for GLU-S_n type than for MAG-S_n one ($\Delta H_{m(MAG-Sn)} > \Delta H_{m(GLU-Sn)}$); the interactions between hydrophilic heads and water molecules are greater for GLU-S_n compounds than for MAG-S_n ones ($\Delta S_{m(MAG-Sn)} > \Delta S_{m(GLU-Sn)}$). Quantitative influences of both the alkyl chain length and the free hydroxyl group number, on CMC correlations and on thermodynamics data have been studied. The results are very close to those reported for the corresponding 3-O-alkyl D-glucose and 3-O-acyl-D-glucose derivatives.

Key words D-glucose derivatives – nonionic surfactants – micelle – CMC – thermodynamic data

Introduction

We recently described the synthesis of a large range of 3-deoxy-S-alkyl-D-glucose derivatives [1–3], either as potential drugs (n-butyrates as antitumorals [1,4]; valproates as antiepilectics [3,5]) or non-ionic surfactants (deoxy-S-alkyl = S-n-C_nH_{2n+1}; $n \geq 8$). The preliminary results on surfactant properties incited us to work out their study. In the present paper, we expose, critical micellar concentration (CMC) and thermodynamics data of 3-deoxy-S-alkyl-1,2-O-isopropylidene-α-D-glucofuranose

series, marked MAG-S_n and 3-deoxy-S-alkyl-D-gluco-pyranose series, marked GLU-S_n.

Results are compared to those obtained with the analogous 3-O-alkyl-D-glucose derivatives [6] and 3-O-acyl-D-glucose derivatives [7].

Materials and methods

Samples

The following 3-deoxy-S-alkyl-D-glucose derivatives, prepared and purified as previously indicated [2], were studied:

MAG-S$_n$ GLU-S$_n$

R = n-C$_n$H$_{2n+1}$; n = 8, 12, 16 and 18

The above compounds are compared to the corresponding 3-O-alkyl-D-glucose derivatives [6]:

MAG-O$_n$ GLU-O$_n$

R = n-C$_n$H$_{2n+1}$; n = 8, 12, 16 and 18

and 3-O-acyl-D-glucose derivatives [7]:

MAG-C$_n$ GLU-C$_n$

R = n-C$_n$H$_{2n+1}$; n = 7, 11, 15 and 17

Water solubility and CMC determination

The solubility S_0 in distilled and filtered (Millipore 0.45 μm) water has been determined for each sample at 25 °C.

In order to obtain CMC values, a solution scale 3 S_0/4, S_0/2, S_0/4, S_0/10, S_0/20, S_0/50, S_0/100 and S_0/200, was prepared for each compound, by successive dilutions of a primary solution S_0. Surface tension γ was measured for each solution at 25°, 30°, 40° and 50 °C, by the Wilhelmy plate method (Prolabo TD 2000 Tensiometer). The critical

Table 1 Solubility S_0, in water, at 25 °C of MAG-S$_n$ and GLU-S$_n$ compounds

MAG-S$_n$	S_0 (10^{-4} Mol·L^{-1})	GLU-S$_n$	S_0 (10^{-4} Mol·L^{-1})
MAG-S$_8$	8.50	GLU-S$_8$	12.2
MAG-S$_{12}$	0.45	GLU-S$_{12}$	9.9
MAG-S$_{16}$	0.34	GLU-S$_{16}$	6.9
MAG-S$_{18}$	0.28	GLU-S$_{18}$	5.7

micellar concentrations are obtained from the classical diagrams $\gamma = f(\log C)$ where C is the concentration expressed in Mol·L^{-1}.

Results and discussion

Critical micellar concentration (CMC)

The experimental results with MAG-S$_n$ and GLU-S$_n$ series, show that:

- the solubility S_0 (Table 1) is largely higher than the corresponding CMC value (Table 2) at the same temperature;
- CMC values decrease either when lipophilicity increases (by increasing alkyl chain length) or when hydrophilicity decreases (by changing the free hydroxyl group number from 4 to 2);
- neither the alkyl chain length nor the free hydroxyl group number have a significant influence on surface tension γ.

The same structural influences were observed with the corresponding 3-O-alkyl derivatives (MAG-O$_n$; GLU-O$_n$)[6] and 3-O-acyl derivatives (MAG-C$_n$; GLU-C$_n$) which are much more soluble.

Critical micellar concentration correlations

Quantitative influence of both alkyl chain length and temperature on CMC has been studied.

Alkyl chain influence was examined using Ueno's relation [8]

$$\log(CMC) = an + b , \tag{1}$$

where n is the carbon atom number of alkyl chain. The experimental a and b constants were determined at different temperatures and compared to those obtained with the corresponding ethers (MAG-O$_n$, GLU-O$_n$)[6] and esters (MAG-C$_n$; GLU-C$_n$)[7].

Table 2 CMC and corresponding γ values of MAG-S$_n$ and GLU-S$_n$ compounds

$T\,(^\circ C)$	n	MAG-S$_n$		GLU-S$_n$	
		CMC (10^{-4} Mol·L^{-1})	γ (mN·m^{-1})	CMC (10^{-4} Mol·L^{-1})	γ (mN·m^{-1})
25	8	6.8	34.8	6.8	32.0
	12	0.09	31.8	1.7	29.2
	16	0.02	32.8	0.65	31.6
	18	0.02	37.8	0.36	38.1
30	8	6.3	33.8	6.0	31.6
	12	0.09	31.5	1.4	29.0
	16	0.02	32.4	0.62	31.3
	18	0.01	37.4	0.32	37.7
40	8	4.8	33.0	5.3	30.2
	12	0.07	30.6	1.0	26.5
	16	0.01	31.7	0.56	31.2
	18	0.009	36.5	0.28	35.0
50	8	3.8	31.9	5.0	27.0
	12	0.07	29.8	0.95	25.4
	16	0.01	31.1	0.41	31.1
	18	0.005	35.3	0.25	34.5

Table 3 Experimental a and b constants of the Ueno's relation[8]

$T\,(^\circ C)$	MAG-S$_n$		GLU-S$_n$	
	a	b	a	b
25	-0.26 ± 0.06	-1.5 ± 0.9	-0.13 ± 0.01	-2.2 ± 0.09
30	-0.27 ± 0.06	-1.3 ± 0.8	-0.12 ± 0.01	-2.3 ± 0.15
40	-0.27 ± 0.05	-1.4 ± 0.8	-0.12 ± 0.02	-2.4 ± 0.2
50	-0.28 ± 0.05	-1.4 ± 0.6	-0.13 ± 0.02	-2.4 ± 0.2
	MAG-O$_n^6$		GLU-O$_n^6$	
	a	b	a	b
25	-0.21 ± 0.05	-1.6 ± 0.7	-0.11 ± 0.02	-2.2 ± 0.3
30	-0.23 ± 0.04	-1.4 ± 0.6	-0.11 ± 0.02	-2.3 ± 0.2
40	-0.22 ± 0.05	-1.6 ± 0.6	-0.10 ± 0.01	-2.4 ± 0.2
50	-0.22 ± 0.05	-1.8 ± 0.7	-0.10 ± 0.01	-2.6 ± 0.2
	MAG-C$_n^7$		GLU-C$_n^7$	
	a	b	a	b
25	-0.22 ± 0.04	-1.4 ± 0.6	-0.09 ± 0.03	-2.3 ± 0.5
30	-0.22 ± 0.04	-1.4 ± 0.6	-0.09 ± 0.04	-2.3 ± 0.5
40	-0.23 ± 0.05	-1.4 ± 0.7	-0.10 ± 0.04	-2.3 ± 0.6
50	-0.23 ± 0.05	-1.4 ± 0.7	-0.10 ± 0.05	-2.3 ± 0.6

Table 4 Experimental A and B constants of the Meguro's relation [8]: $\text{Ln(CMC)} = A(1/T) + B$

n	MAG-S$_n$		GLU-S$_n$	
	A	B	A	B
8	2380 ± 90	-15.3 ± 0.3	1190 ± 200	-11.3 ± 0.6
12	1190 ± 310	-15.6 ± 1.0	2380 ± 450	-16.7 ± 1.5
16	3270 ± 960	-24.1 ± 2.8	1780 ± 310	-15.6 ± 1.0
18	4830 ± 1020	-29.5 ± 3.3	1430 ± 110	-15.1 ± 0.4
n	MAG-O$_n^6$		GLU-O$_n^6$	
	A	B	A	B
8	2630 ± 80	-15.7 ± 0.3	2610 ± 290	-15.7 ± 0.9
12	2580 ± 180	-19.0 ± 0.6	1380 ± 400	-13.1 ± 1.3
16	2740 ± 680	-20.9 ± 2.2	760 ± 220	-11.9 ± 0.7
18	3190 ± 1150	-22.6 ± 3.7	1380 ± 350	-14.1 ± 1.1
n	MAG-C$_n^7$		GLU-C$_n^7$	
	A	B	A	B
7	340 ± 40	-8.0 ± 0.1	260 ± 40	-7.3 ± 0.1
11	1880 ± 440	-16.5 ± 1.4	11010 ± 140	-12.31 ± 0.5
15	2000 ± 1	-18.1 ± 0.1	1540 ± 200	-14.3 ± 0.7
17	1360 ± 330	-16.6 ± 1.1	1120 ± 20	-12.3 ± 0.1

Table 3 shows that Ueno's correlation can be applied to the 24 glucose derivatives herein examined as for the alkyloctaethylene glycol compounds [8]. We also observe that:

– a and b constants are both independent of the temperature and, in a least case, of the nature of atom or atom group (S, O or OCO) linking the lipophilic chain R to the hydrophilic part.

– a and b constant values express a large hydrophilicity influence since the slope a and the origin ordinate b vary in the ratio 1/2 and 3/2 respectively when the free hydroxyle group number varies from 2 to 4.

Temperature influence on CMC values was expressed by Meguro's relation [9]:

$$\text{Ln(CMC)} = A(1/T) + B \,. \tag{2}$$

Table 5 Micellization free energy, enthalpy and entropy of $MAG-S_n$, $GLU-S_n$, $MAG-O_n$, $GLU-O_n$, $MAG-C_n$ and $GLU-C_n$ componds at 25 °C*

	n	$MAG-S_n$	$GLU-S_n$	$MAG-O_n^6$	$GLU-O_n^6$	n	$MAG-C_n^7$	$GLU-C_n^7$
ΔH_m	8	19.8 ± 0.7	9.9 ± 1.7	21.9 ± 0.7	21.8 ± 2.4	7	2.8 ± 0.3	2.2 ± 0.3
	12	9.9 ± 2.6	19.8 ± 3.7	21.5 ± 1.5	11.5 ± 3.3	11	15.6 ± 3.7	9.2 ± 1.2
	16	27.2 ± 7.2	14.8 ± 2.6	22.8 ± 5.7	6.3 ± 1.8	15	16.6 ± 0.8	12.8 ± 1.7
	18	40.2 ± 8.5	11.9 ± 0.9	26.5 ± 8.5	11.5 ± 2.9	17	11.3 ± 2.7	9.3 ± 0.2
ΔS_m	8	127 ± 3	94 ± 5	131 ± 3	130 ± 8	7	66 ± 1	61 ± 1
	12	130 ± 8	139 ± 13	158 ± 5	109 ± 11	11	137 ± 12	101 ± 4
	16	200 ± 23	123 ± 8	174 ± 18	99 ± 6	15	151 ± 1	116 ± 6
	18	245 ± 28	125 ± 3	188 ± 31	117 ± 9	17	138 ± 9	102 ± 1
$T\Delta S_m$	8	37.8 ± 0.7	28.0 ± 1.5	39.0 ± 0.7	38.7 ± 2.2	7	19.7 ± 0.2	18.2 ± 0.2
	12	38.7 ± 2.5	41.4 ± 3.7	47.1 ± 1.5	32.5 ± 3.3	11	40.8 ± 3.5	30.1 ± 1.3
	16	59.6 ± 6.9	36.6 ± 2.5	51.8 ± 11.2	29.5 ± 1.7	15	45.0 ± 0.3	34.6 ± 1.7
	18	73.0 ± 8.2	37.2 ± 1.0	56.0 ± 9.2	34.9 ± 2.7	17	41.1 ± 2.7	30.4 ± 0.2
ΔG_m**	8	-18.0 ± 1.4	-18.1 ± 3.2	-17.2 ± 1.4	-17.0 ± 4.6	7	-16.8 ± 0.5	-16.0 ± 0.5
	12	-28.9 ± 5.1	-21.6 ± 7.4	-25.6 ± 3.0	-21.0 ± 6.6	11	-25.2 ± 7.2	-20.9 ± 2.5
	16	-32.3 ± 14.1	-21.8 ± 10.9	-29.0 ± 11.2	-23.2 ± 3.5	15	-28.4 ± 0.3	-21.7 ± 3.4
	18	-32.8 ± 16.7	-25.3 ± 4.2	-29.6 ± 17.7	-23.4 ± 5.6	17	-29.8 ± 5.4	-21.1 ± 0.4
ΔG_m**	8	-18.1	-18.1	-17.1	-17.0	7	-17.0	-16.0
	12	-28.8	-21.5	-25.6	-20.8	11	-25.2	-20.8
	16	-32.5	-21.9	-28.8	-23.3	15	-28.3	-22.0
	18	-32.5	-25.4	-29.1	-21.4	17	-29.8	-21.1

* ΔH_m, $T\Delta S_m$, ΔG_m in $kJ \cdot Mol^{-1}$; ΔS_m in $J \cdot K^{-1} \cdot Mol^{-1}$.
** from relation (5).
*** from relation (3): $298R Ln(CMC)$.

The experimental A and B constants were determined for the type $MAG-S_n$ and $GLU-S_n$ compounds in order to compare (Table 4) their values to those of the corresponding ethers ($MAG-O_n$, $GLU-O_n$)[6] and esters ($MAG-C_n$; $GLU-C_n$)[7].

Relation (2) combined to the classical ones [10, 11]:

$$\Delta G_m = RT \, Ln(CMC) \tag{3}$$

and

$$\Delta G_m = \Delta H_m - T\Delta S_m \tag{4}$$

allows to express the micellization thermodynamics ΔG_m, ΔH_m and ΔS_m from experimental A and B values:

$$\Delta G_m = RA + RTB \tag{5}$$

$$\Delta H_m = RA \tag{6}$$

$$\Delta S_m = -RB \tag{7}$$

Table 5 reports corresponding ΔH_m and ΔS_m values as well as ΔG_m values obtained from relations (2) and (5).

These results suggest the following comments for all the ethers, esters and thioethers herein studied:

1) Micellization free energy (ΔG_m) values indicate that:
• micelle stability increases by increasing alkyl chain length;
• micelle stability decreases by increasing hydrophilicity (by changing free hydroxyl group number from 2 to 4). This difference in micelle stability increases progressively by increasing alkyl chain length.

2) According to the relation (4), the micelle formation is an entropic phenomenon, since ΔH_m part is lower than $T\Delta S_m$ part is free energy of micellization.

3) Micelle formation seems easier from the more hydrophilic series ($GLU-S_n$, $GLU-O_n$ and $GLU-C_n$) than from monoacetal ones ($MAG-S_n$, $MAG-O_n$ and $MAG-C_n$) since: $\Delta H_{m(GLU-S_n)} < \Delta H_{m(MAG-S_n)}$, $\Delta H_{m(GLU-C_n)} < \Delta H_{m(MAG-C_n)}$ and $\Delta H_{m(GLU-O_n)} < \Delta H_{m(MAG-O_n)}$.

4) Stronger interactions between hydrophobic heads of $GLU-S_n$ compounds and water molecules can explain that $\Delta S_{m(GLU-S_n)}$ values are lower than $\Delta S_{m(MAG-S_n)}$ ones.

5) Both micelle stability (ΔG_m) and micelle formation (ΔH_m) seem to be little dependent on the nature of the atom or atom group (S, O or OCO) linking the alkyl chain to the hydrophilic part, however, ΔH_m comparison indicates that thioether series seems slightly more stable than the two other series.

Conclusion

The 24 studied amphiphilic D-glucose thioether, ether and ester derivatives analyzed in this study constitute a range of non-ionic surfactants in which CMC values at 25 °C

Progr Colloid Polym Sci (1996) 100:43–47
© Steinkopff Verlag 1996

vary from $1.6\,10^{-3}$ to $5\,10^{-7}$ $Mol \cdot L^{-1}$ by changing carbon atom number in the alkyl chain from 7 to 18 and free hydroxyl group number from 4 to 2.

Among the different presented results, we can also emphasize that atom or atom group linking the alkyl chain to the hydrophilic part seems to have a negligible influence on:

- Ueno's correlation (similar constants a and b in the relation $\log(CMC) = an + b$),

- micelle stability (ΔG_m) and micelle formation (ΔH_m). In a forthcoming work, we shall study, with new amphiphilic D-glucose derivative series, an eventual influence of other alkyl chain linkages on the parameters examined in this paper.

Acknowledgment The authors are grateful to the Centre de Valorisation des Glucides, the Biopôle de Picardie and to the Générale Sucrière for financial support.

References

1. Pieri F, Ronco G, Segard E, Villa P, Brev (1987) FR 87/13294; PCT FR 88 00470; WO 89 02895; Brev. USA/07/501080 (1990)
2. a) Postel D, Ronco G, Villa P, Brev. FR 88/05230, Université de Picardie (1988), publié Brev. FR n° 26300445 (1989)
   b) Postel D, These, Amiens (1990)
3. Armand V, Louvel J, Postel D, Pumain R, Ronco G, Villa P, Brev. FR 94/08542 (11/07/94)
4. a) Planchon P, Raux H, Magnien V, Ronco G, Villa P, Crepin M, Brouty-Boye D (1991) Int J Cancer 48:443
   b) Pouillart P, Cerutti I, Ronco G, Villa P, Chany C (1992) Int J Cancer 51:596
   c) Pouillart P, Ronco G, Cerutti I, Trouvin JH, Pieri F, Villa P, J Pharm Sciences 81(3), 241 (1992)
5. Armand V (1994) Thesis, Amiens
6. Bikanga R, Godé P, Ronco G, Van Roekeghem P, Villa P (1994) Jorn Com Esp Deterg 25:607
7. Goueth PY, Gogalis P, Bikanga R, Godé P, Postel D, Ronco G, Villa P (1994) J Carbohydr Chem 13(2):249
8. Ueno M, Takasawa Y, Tabata Y, Sawamura T, Kawahashi N, Meguro K, Yukagatu (1981) CA 94:90032 e, 30:421
9. Meguro K, Muto K, Ueno M (1980) Nippon Kagatu Kaishi, CA 92:169876 a, 394
10. Sharma B, Rakshit AK (1989) J Colloid Interface Sci 129(1):139
11. Zourab SM, Saber VM, Abo-El Dahab H (1991) J Dispersion Sci Technology 12(1):25–36

Progr Colloid Polym Sci (1996) 100:48–53
© Steinkopff Verlag 1996

A.K. Van Helden

Computer simulation studies of surfactant systems

Dr. A.K. Van Helden (✉)
CTSOL/2 Kon. SHELL Laboratorium
Amsterdam
P.O. Box 38 000
1030 BN Amsterdam, The Netherlands

Abstract Traditionally, surfactant research in industry has relied almost exclusively on experiments. Lately, we have attempted in our laboratory to use computer simulations to complement this experimental effort. Molecular dynamics and Monte Carlo techniques have now become an integral part of surfactants and fluids research, not only offering new insight (primarily of interest to academia), but also enhancing the researcher's intuition through visualisation of complex systems and providing guidance to the synthesis of novel products, important to industry.

Surfactant assembly and oil solubilization in micellar solutions have been simulated as well as the adsorption of surfactant monolayers at air–water interfaces. These results have lead to novel surfactants with special performance characteristics relevant to laundry detergents.

Simulation of two monolayers showed compression-induced phase transitions and configurational energy hysteresis, which has also been observed experimentally during loading/unloading of solid surfaces. This work provides a good basis to explore novel surfactants with desired frictional characteristics, relevant to, for example, thin film lubrication.

Key words Simulation – surfactant – micelle – monolayer – detergents – lubricants

Introduction

The technology of soap and surfactants has a prominent and longstanding place in society. The origin of soap (in the chemical sense of saponified fats and oils) can with certainty be put as early as 500 AD. We know how the technology gradually evolved, switching from hard- to soft soap in the 12th century; improving product purity by glycerine separation in the 17th century; employing synthetic soda instead of caustic potash as a feedstock, since Muspratt and Gamble commercialized the LeBlanc soda process in the 1820's [1]. More recently, in the 1940's, the pace of technological development quickened considerably, when synthetic surfactants were introduced by the (petro)chemical industry which gradually displaced soap in certain applications, Fig. 1. Today a wide spectrum of commodity- and speciality surfactants are used in a variety of consumer products and industrial applications, see e.g. [2]. They compete with each other on the basis of cost/performance, formulation flexibility and environmental compatibility. In the 1990s, the world-wide consumption level of soap and surfactants has reached the 14 million tons per annum mark.

The technological advances of this century would have been impossible without the effort and attention given by many scientists to surfactant and interface properties. Experimental and theoretical work as early as 1774 by

Benjamin Franklin, followed by many others since Langmuir's work on monolayers has provided a good understanding of how surfactants adsorb at and modify the properties of interfaces. Since the early 1980s computer simulations are successfully applied to solid–liquid and liquid–gas interfaces and have become a third route, complementary to experiments and theory in the academic world. Quite recently, computer simulations of surfactant systems have become feasible, see for instance [3].

The subject of this paper is to summarize simulation work on surfactant systems conducted in our laboratory, i.e., in an industrial research environment. The main message is that computer simulations have become an integral part of surfactant and fluid research. They offer insight and new knowledge (primarily of interest to the academic world). But in the area of surfactants and fluids, simulation can now also contribute to technological advances by the industry, enhancing the researcher's intuition by visualizing complex systems and providing guidance to the synthesis of promising new products. Two examples will be given to illustrate the point, both from applications relevant to the petrochemical industry, one relating to laundry detergents and the other to lubricants. The first example employs a coarse grained molecular representation and a system with a large number of particles

(30 000). Those calculations were done using a transputer network of 400 processors. The second example uses a detailed molecular model, less particles in the system (5000) and results were obtained with a high-performance workstation.

Simulation methods

Molecular simulations are not without limitations. For example, one is not capable to simulate a large number of molecules for an unlimited amount of time, using a detailed molecular model and still get quantitative agreement with experiment. There is a competition between the required accuracy of the simulation data, and the minimum amount of phase space, minimum number of molecules and a minimum detail of molecular representation that are sufficient to access phenomena of interest in the investigation.

A variety of techniques can be used, including molecular mechanics (MM), quantum chemistry (QC), molecular dynamics (MD), Brownian dynamics (BD), Monte Carlo (MC) calculations and others [4]. Figure 2 summarises my coworker's view over which method can be used to calculate certain properties. We have mainly used MD and MC, because in almost all of our applications, one needs to calculate thermodynamic properties. In order to calculate

Fig. 1 SOAP = alkali salts of fatty acids, derived from fats and oils; BABS = branched alkyl benzene sulphonate (sodium salt); LABS = linear alkyl benzene sulphonate (sodium salt); PS = sodium alkane sulphonate; AEO = alcohol ethoxylate; AEOS = alcohol ethoxy sulphate (alkali salts); APEO = alkyl phenol ethoxylate

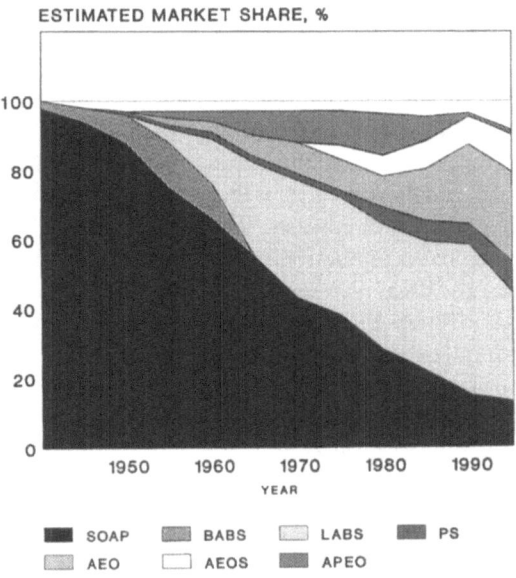

SURFACTANT CONSUMPTION
W.EUROPEAN CONSUMER PRODUCTS

ESTIMATED MARKET SHARE, %

Fig. 2 Single molecule properties include indicators such as end-to-end distance, electronic structure, chain conformation. Structural properties are mostly radial distribution functions and structure factors. Thermodynamic properties such as pressure and surface tension. Thermal properties include all free energy terms such as Helmholtz and Gibbs free energies. Phase equilibria includes temperatures, pressures and chemical potentials. Chemical reactions properties include electronic structures, bond strength and product distributions. Performance properties are usually very complex indicators such as cleaning, solvent power or smell, etc.

Simulation Methods of Choice for Selected Properties

		Simulation Method					
		MMM QC	CA BD	MD	MC	CP QMC	QSPR QSAR
P	single-molecule	●			●	●	
r	structural		●	●	●	●	
o	thermodynamic			●			
p	thermal				●		
e	phase equil.				●		
r	transport		●	●			
t	chem. reaction	●				●	
y	performance						●

these properties a molecular modelling technique is needed in which the free energy is minimized. MD and MC simulations appear to be the ideal methods to do so.

Although commercial packages are available, we developed home-made software to ensure a good appreciation of its capabilities and limitations as well as state-of-the-art programming and flexibility. (The downside, of course of such inhouse software development is the reduced emphasis on user-friendliness, driven by the desire to constantly update these programs with the latest physical models and computational aspects.)

Laundry detergents

Surfactants are key ingredients in laundry detergents. Their role in the laundering process is to remove certain soils from textile fibers and suspend them in the washing solution. This is achieved by surfactant adsorption at textile- and soil surfaces, thus weakening the adhesion of soil to the fabric. Also surfactant aggregates or micelles in solution promote the removal of soil into the washing liquor. Studies on model systems had shown that the best detergency performance is usually obtained at surfactant concentrations above the critical micelle concentration (CMC) see for instance [5, 6]. In domestic laundry approximately 100–200 g laundry detergent, of which 20–30 g surfactant is currently used per wash, to remove roughly 0.1–1 g of soil from (5 kg) fabric. Since the mid 1980s there is a trend in Western Europe to reduce the detergent dosage per wash for environmental, detergent manufacture and cost reasons. As a result the concentration of detergent ingredients has dropped and (anionic) surfactants are currently used at concentrations close to or even below their CMC. This trend renewed our interest in the role of surfactants and micelles in cleaning. In the early 1990s, experimental and simulation work started in our laboratory in an effort to identify surfactants that could work well at much lower concentrations. Using the earlier work of Carroll [7], Van Os developed an experimental technique to measure the rate of removal of oil drops from a fiber into an aqueous surfactant solution. Dramatic effects of surfactant structure were found. Conventional alcohol ethoxylate surfactants solubilize oil much faster than regular alkylbenzene sulphonate surfactants. We also found that certain alkyl xylene sulphonates solubilized oil at a spectacular rate, a hundred times faster even than the nonionics tested [8]. Simulation results offered a key to generic surfactant structures that may even be more active than these aklyl xylene sulphonates.

With the MD technique applied to a system of 30 000 particles and relatively simple molecular models, Smit et al. were able to calculate surfactant aggregation and monolayer adsorption in oil/water/surfactant systems [9]. Essentially all the features of micellization, oil solubilization and interface adsorption could be investigated. As expected from geometric considerations, surfactant aggregation in solution depends on surfactant structure. Surfactants with a small hydrophilic headgroup form bilayers. Bulky headgroups yield spherical micelles. Chain like hydrophilic groups lead to a wide size distribution of spheres [10].

Subsequently, Karaborni found that when an oil droplet is immersed into such a micellar solution, it solubilizes via three mechanisms [11]. The first is a direct effect of the finite solubility of oil molecules in water. Dissolved oil molecules are trapped by micelles in their immediate vicinity. In the second mechanism, oil molecules are transferred through oil droplet-micelle collision during which exchange of surfactant and oil occurs. The third mechanism is due to the low surface energy of the oil/water interface: oil molecules exit the droplet collectively, desorbing together with a large number of surfactants. Interestingly, the rates of these processes appear to be dependent on the nature of the oil and the surfactant molecules.

Meanwhile Kern and coworkers had found that so-called gemini surfactants exhibit curious visco-elastic effects in aqueous solutions; a six orders of magnitude increase of viscosity at low surfactants concentrations, while the viscosity dropped again at intermediate surfactant concentrations [12]. Simulations with gemini surfactants in our laboratory showed that the viscosity-concentration curve may well be explained by the formation of rather different aggregation morphology compared to conventional single chain surfactants [13], Fig. 3. Gemini molecules form threadlike micelles instead of spherical micelles. At intermediate concentrations these thread like micelles are branched, thus reducing the end-to-end distance compared to lower concentrations. These results provide a good explanation for the viscosity behavior found by Kern. Moreover, simulation results indicated that gemini's have very low CMC's. This suggested that selective oil solubilization might be a possibility on the basis of the dominance of mechanism one (see above). This effect would be much stronger with geminis containing very short (hydrophobic) spacers. Van Os et al. verified these predictions experimentally, using model quaternary alkyl ammonium bromide gemini surfactants. Indeed, these materials display the visco-elastic effect also found by Kern and as expected, they also showed selective solubilization. Toluene is solubilized in preference over n-hexane. And again as predicted the visco-elastic effect as well as the selective solubilization are much more pronounced with a 2 carbon spacer than with 6 or 10 carbon spacers [14].

This work illustrates potential benefits of simulation not only for academic, but also for industrial research: new

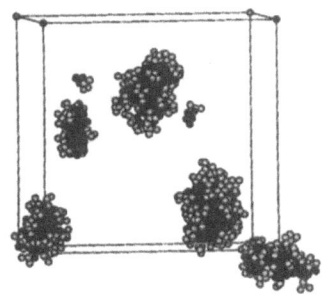

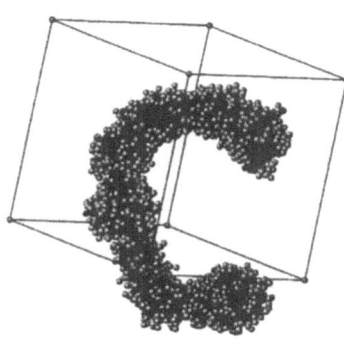

Fig. 3 Instantaneous configuration of one aggregate formed by single chain surfactants and gemini surfactants with a spacer of one oil-like particle between the head groups. Surfactant weight fraction is 28%. For reasons of clarity water molecules are not shown, the particles that belong to the headgroups are displayed in light color, and those belonging to the surfactant tails in dark color

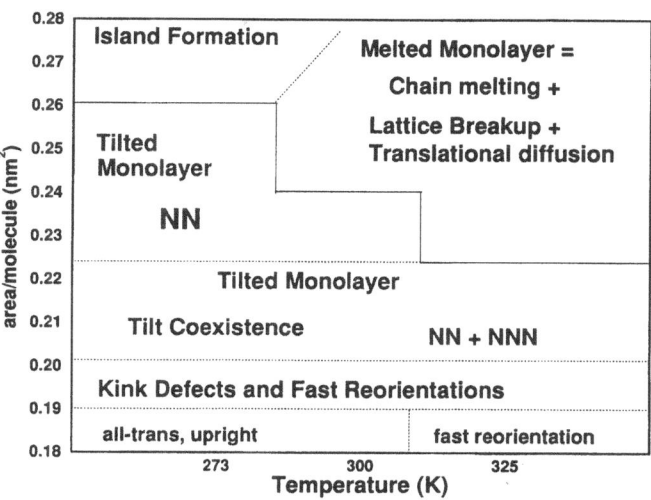

Fig. 4 The two-dimensional phase diagram of alkyl carboxylate with 19 methylene units in the surfactant tail. Tilt angle is Next Nearest Neighbor (NNN) instead or Nearest Neighbor (NN)

insight in solubilization mechanisms; a qualitative explanation of the viscosity behavior of gemini solutions; and guidance toward gemini surfactants with short hydrophobic spacers displaying selective oil solubilization and strong visco-elastic behavior in water.

Lubrication

Thin film lubrication is a crucial process in the safe and reliable operation of many devices and machines, from computer disks to internal combustion engines. The desired wear and friction properties are usually conferred by amphiphilic monolayers adsorbed at each of the sliding surfaces [15].

From laboratory experiments it is well known that loading-unloading cycles can lead to hysteresis. Israellach-villi and coworkers contend that such effects are due to entanglements between surfactant molecules from both monolayers [16]. They argue that since entanglements are much easier than distanglements in the amorphous state,

the required energy to separate the monolayers is much more than that during loading. On the other hand, one could argue that while entanglements are perfectly feasible with polymer molecules, they seem much less probable with molecules comprising only 20 carbon atoms.

MD and MC work was able to shed light on this issue and discredited the entanglement hypothesis. In this case a very detailed molecular model has been used; 19 segments having the same size and energy characteristics as CH_2 groups, a head group representing the carboxylate functionality. The system contained approximately 5000 particles. Further details of the model and the results are given in [17]. Subtle phase changes could be observed in the simulation results, which had to my knowledge never been detected experimentally. Siepmann further developed the MC technique with the so-called "configurational bias" method [18] and was able to simulate (two-dimensional) vapor–liquid phase equilibria of long chain molecules. As a result the complete phase behavior of surfactant monolayers could be modeled at a wide range of surface densities and temperatures [19], showing many phases and even the co-existence of two-dimensional liquid and vapor phases, Fig. 4.

Subsequently, in an effort to simulate thin film lubrication and colloid stability phenomena, Karaborni [20] investigated two parallel monolayers with 25 A2/molecule packing density, compressed at 25 °C and a rate of 7 ms^{-1} (low enough to allow for equilibration of the chain confrontation-trans fraction between two successive compressions). During compression and relaxation, molecular conformations and their rate of reorientation were clculated. These parameters indicated a transition of

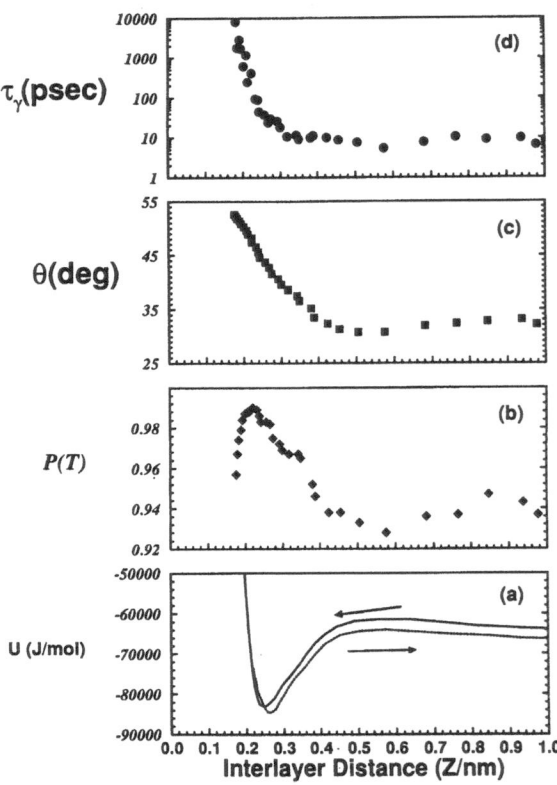

Fig. 5 Average properties during approach as function of the interlayer distance Z. (a) Configurational energy of the total system; (b) average tilt angle in degrees; (c) average trans-function; (d) relaxation time of an autocorrelation function indicating the molecular rate of reorientation

a semi-crystalline liquid phase to a solid phase of the monolayers upon compression, Fig. 5. The ordering phase transition is due to excluded volume effects. In other words, decreasing volume available for the molecules induces a better packing. The simulated configurational energy of the whole system during approach and separation showed a clear hysteresis. The analysis strongly suggests that the energy hysteresis measured experimentally, is due to subtle phase transitions rather than chain entanglements.

Further simulations explored the effect of adsorption strength, conformational freedom and surfactant chain length on the total configurational energy and monolayer phase behavior. These results clearly indicate that hysteresis and therefore the (stick-slip) frictional forces in thin lubrication films can be related to monolayer and surfactant structure, suggesting novel generic surfactant structures for desired friction characteristics.

Conclusions

State-of-the-art Monte Carlo and molecular dynamics simulations have become powerful tools in surfactant and fluid research. They can aid the interpretation of experimental and theoretical results and provide new insight and mechanistic knowledge, primarily of interest to the academic world. But they are now also used in industrial research to increase the scientists' intuition through visualization of complex systems, and most importantly to predict structure/property relations of novel (generic) surfactants. The latter may considerably reduce the effort to synthesize and evaluate new products on a trial and error basis.

Two examples illustrate these points; one in the area of laundry detergents, in particular aqueous solutions of gemini surfactants; the other related to thin film lubrication.

In the first, gemini surfactants exhibit selective solubilization as predicted by simulation. Experimental verification showed that toluene is more strongly solubilized than *n*-hexane by cationic gemini solutions. Simulation also correctly suggested that both selective oil-solubilization and viscoelasticity would be most prominent for short spacer geminis. The second example showed how the energy hysteresis between monolayers during loading-unloading cycles, relate to subtle phase transitions within these surfactant layers and how this insight was used to explore novel surfactants to obtain desired stick-slip friction characteristics.

Acknowledgment I am grateful to Sami Karaborni, Nico van Os and Arie Vreugdenhil for their helpful discussions in preparing this paper.

References

1. Derry TK (1960) A Short History of Technology. Oxford University Press, Oxford
2. Falbe J (1987) Surfactants in Consumer Products. Springer Verlag, Berlin
3. Van Os NM (ed) (1993) Tenside, Surfactants, Detergents. Carl Hanser Verlag, München
4. Allen MP, Tildesley DJ (1987) Computer Simulation of Liquids. Oxford Science Publications
5. Kurzendorfer CP, Lange H (1969) Fette-Seife 71:561
6. Gocho H, Sato M (1989) Daigaku 20:29
7. Carroll BJ (1981) J Colloid Interface Science 79:126
8. Van Os NM (1994) presented at the ACS Colloids and Surfaces Meeting, California

9. Smit B, Hilbers PAJ, Esselink K, Rupert LAM, Van Os NM, Schlijper AG (1990) Nature 348:624
10. Esselink K, Hilbers PAJ, Van Os NM, Smit B, Karaborni S (1994) Colloids and Surfaces A91:155
11. Karaborni S, Van Os NM, Esselink K, Hilbers PAJ (1993) Langmuir 9:1175
12. Kern F, Lequeux R, Zana R, Canau SJ (1994) Langmuir 10:1714
13. Karaborni S, Esselink K, Hilbers PAJ, Smit B, Karthaeuser J, Van Os NM, Zana R (1994) Science 226:254
14. Dam Th. submitted to Colloids & Surfaces
15. Yoshizawa H (1993) Science 259:1305
16. Israelachvilli JN (1992) Vac Sci Technology A10:2961
17. Karaborni S (1993) Langmuir 9:1334
18. Siepmann JI, Karaborni S, Klein ML (1994) J Phys Chemistry 98:6675
19. Karaborni S (1994) presented at the Royal Soc Chem meeting, Bristol
20. Karaborni S (1994) Phys Rev Letters 73:1668

Progr Colloid Polym Sci (1996) 100:54–59
© Steinkopff Verlag 1996

COLLOIDAL PARTICLES

V. Reus
L. Belloni
T. Zemb
N. Lutterbach
H. Versmold

Equation of state of a colloidal crystal: An USAXS and osmotic pressure study

V. Reus (✉) · L. Belloni · T. Zemb
Service de Chimie Moléculaire
(DRECAM/DSM)
Bât. 125, CEA SACLAY
91191 Gif sur vette, France

N. Lutterbach · H. Versmold
Lehrstuhl für Physikalische Chemie II
RWTH Aachen
Templergraben 59
52056 Aachen, FRG

Abstract We investigate the structure and the osmotic pressure versus concentration of a colloidal crystal of charged bromopolystyrene particles of diameter 100 nm.

In a concentration range between 2 and 12% volume fraction and in the presence of ion exchange resin which fixes a very low (micromolar) salt concentration, a colloidal crystal is obtained. The packing structure evidenced via Ultra-Small-Angle X-ray scattering (USAXS) are either fcc or fcc in equilibrium with bcc. A strong diffuse band before the first Bragg peak is observed. The Q-values ratio of this diffuse pre-peak to the first Bragg peak is 1.44.

Osmotic pressures of these colloidal crystals are measured with different salt contents. Particle structural charge and diameter are known.

Assuming screened electrostatic potential is the main repulsive interaction stabilising the system, all measurements can be rationalized using a simple electrostatic model.

The melting of the fcc structure occurs with an average displacement from equilibrium position, deduced from the screened electrostatic potential, which is 10% larger than for molecular solids.

This type of study of the colloidal crystal osmotic pressure versus distance (in the range 300 nm) is equivalent to an ultraprecise atomic force measurement, since it allows measurements of forces smaller than 10^{-12} N.

Key words Colloidal crystal – USAXS – osmotic pressure – bromopolystyrene particles

Introduction

Electrostatically stabilised colloidal crystals occur at high dilution (a few percent volume fraction) of highly charged latex particles [1]. These samples give an interesting opportunity to detect with a very good sensitivity long-range attractive forces, like those reviewed in a recent paper by Ise [2]. This detection of long-range forces at the same range as lattice parameter (300 nm) requires simultaneous measurement of the crystallographic structure and osmotic pressure of the samples. This study was pioneered by Ottewill et al. [3] more than 20 years ago in the osmotic pressure range of a few atmospheres for high latex concentration. We explore in this paper the range of lower concentrations.

Material and methods

The polystyrene particles, labeled with Bromine were prepared as previously described [4]. The USAXS camera, using Cu Kα radiation, was described previously [5]. Semi-linear collimation effects were desmeared by means of the program kindly made accessible to us by Strobl [6]. Quasi-elastic light scattering allows precise determination of the average radius of the particles to be 50 nm. For low concentrations, USAXS was replaced by light scattering as described in [7].

Osmotic pressure determination was made by measurement of water height difference between a reservoir and the sample. Corrections taking into account surface tension in the capillaries were neglected. Mixed anionic cationic resins were used for deionization, leaving a residual ionic salt concentration of the order of one micromole/l [4].

Conductivity and pH of the sample were checked in closed circuit, which included a quartz capillary (1 mm diameter) used as the sample holder for the USAXS measurements.

Crystallographic structure of the latex crystallites

Figure 1 shows the USAXS scattering spectra of a sample containing 3.5% volume fraction of particles in the absence of salt. The experiment is made in powder average: the crystallites of the colloidal crystal are much smaller than the cross-section of the sample cell. A strong peak appears at low q, i.e., at $1.5 \cdot 10^{-3}$ Å^{-1}, followed by 5 Bragg peaks allowing identification and precise determination of the lattice constant. Shearing in the capillary via pumping with a peristaltic pump induces immediate melting of the sample towards a liquid structure. The broad peak characteristic of a liquid order is always superposed on the first Bragg peak. Assuming a fcc packing, the lattice

constant calculated from the volume fraction ($\phi_V = 3.5\%$) is about 3500 Å. A strong diffuse peak occurs inside the basic cubic cell: this is the proof of a correlation occurring at larger distances than distance between nearest neighbors.

This observation is similar to the one described in the case of sterically stabilized latex particles by Pusey et al. [8]. However, in this experiment, the diffuse "pre-peak" occurred at larger distances than the first Bragg peak of the hcp random packing occurring in their samples. The diluted charged latex system under investigation in our case induces a "pre-peak". We noticed that this prepeak

Fig. 2 Powder scattering with monochromatic beams: (a) light scattering obtained at very low volume fraction ($\phi_V = 0.07\%$), in reduced coordinates compared to the indexing of the centered cubic lattice bcc; (b) USAXS of a perfect colloidal crystal without diffuse peak at higher volume fraction ($\phi_V = 4\%$), compared to the face centered cubic lattice fcc

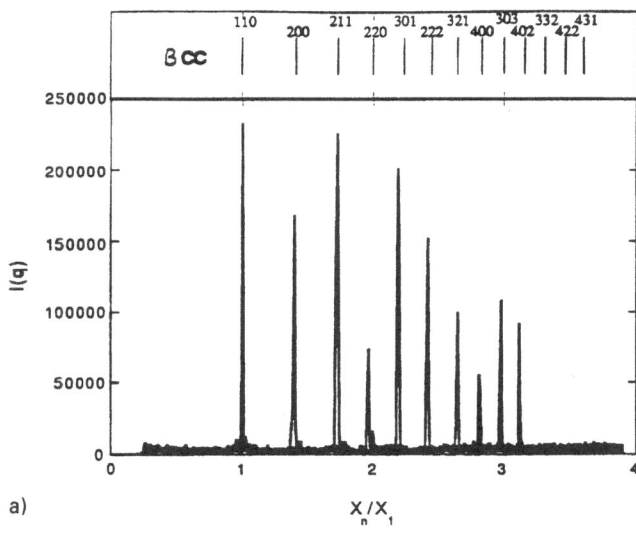

a)

Fig. 1 (---) USAXS spectrum of a deionised latex sample at a volume fraction 3.5%: the strongest peak at low angle is the diffuse scattering, (——) spectrum obtained when the sample is under shear in the capillary

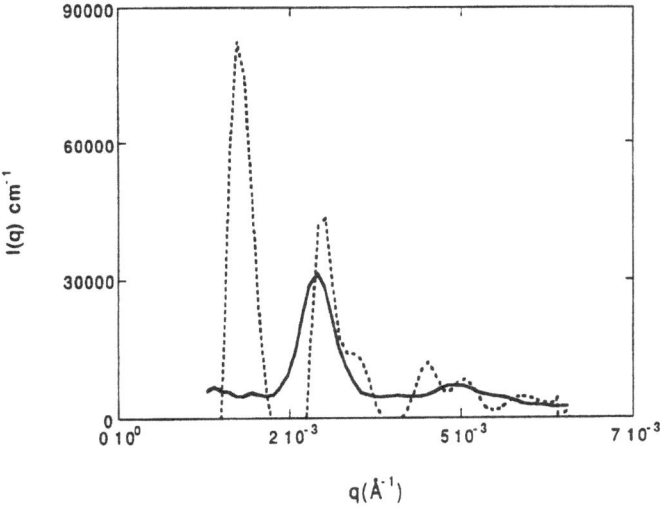

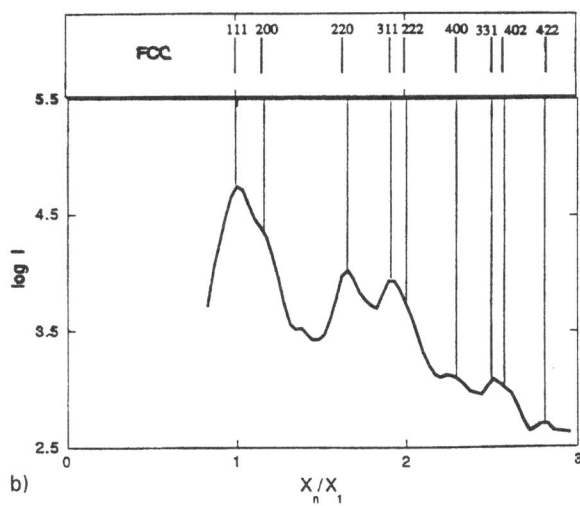

b)

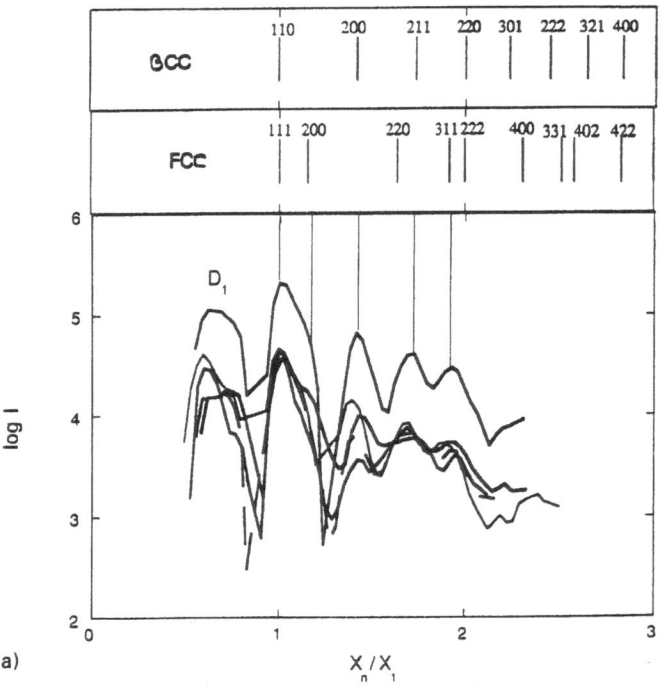

a)

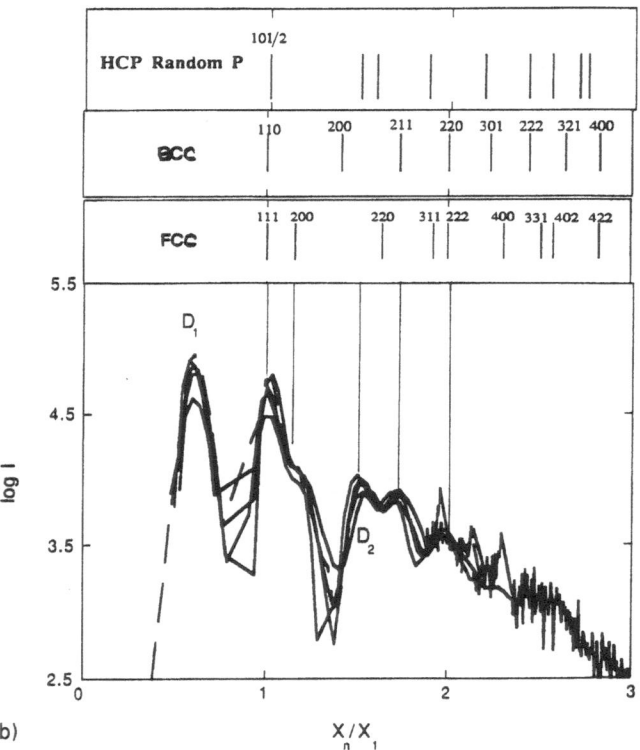

b)

Fig. 3 USAXS scattering obtained in the capillary with fast growing crystallites (a) diffuse peak D_1 with fcc lattice; (b) two diffuse peaks (D_1 and D_2) compared to fcc and bcc and hexagonal random packing indexing

was strongest when the growing of the crystallites was fast. Long ripening of the order of a few hours produced series of Bragg peaks without the strong diffuse "pre-peak".

Bragg peak positions move – as expected – inversely proportionally to the third power of volume fraction. At low concentrations (between 0.04 and 0.3% in volume fraction), a bcc lattice is observed, as shown in Fig. 2a: ten Bragg peaks were measured by light scattering at 0.07% in volume fraction. At higher concentrations, fcc structures are observed. An USAXS spectrum is shown together with the expected positions of peaks of a fcc lattice in Fig. 2b. There is no extinction due to the form factor, i.e., the scattering of an isolated latex particle: the first zero of the scattering for monodisperse homogeneous spheres is located at $q = 10^{-2}$ Å^{-1}, in a region not visible on Fig. 2b.

Figures 3a and b show spectra in reduced coordinates ($X_1 = Q$ value corresponding to the first Bragg peak), superposed for different concentration. In no case, did we observe lattice constants smaller than the one expected due to voids attributed to long-range attractive forces such as those detected by Ise [2]. One strong and broad diffuse band at $X_n/X_1 = 0.69$ was always obtained. A coexistence of fcc with small parts of the sample in bcc packing is a possible interpretation. Figure 3b shows results where a second diffuse peak was detected. In this latter case, this second diffuse band is not located at the same position as the 200 reflection of the bcc lattice. In no case could the observed spectra be indexed in the hexagonal random packing structure.

Osmotic pressure results

For each sample, USAXS allows the determination of the lattice constant with a precision of the order of 1%. In order to establish the equation of state, i.e., the relation of pressure versus volume fraction in the crystalline state, we need to have a simultaneous measure of the osmotic pressure. We used the same equipment as the one used to investigate the colloidal sol in the liquid state.

The pressure versus volume fraction is shown in Fig. 4a. Since the osmotic pressures correspond to water level difference of the order of 1 cm, the osmotic pressure was determined using the method described by Fuoss and Mead [9]. True equilibrium was checked by determining the osmotic pressure during dilution and reconcentration of the sample: no hysteresis was detected. Therefore, the osmotic pressure corresponds to thermodynamic equilibrium. The same data are plotted in reduced coordinates in Fig. 4b. The observed pressure is about 400 times larger than the pressure of a perfect gas with the same particle density. The effective charge of the particles is of the order

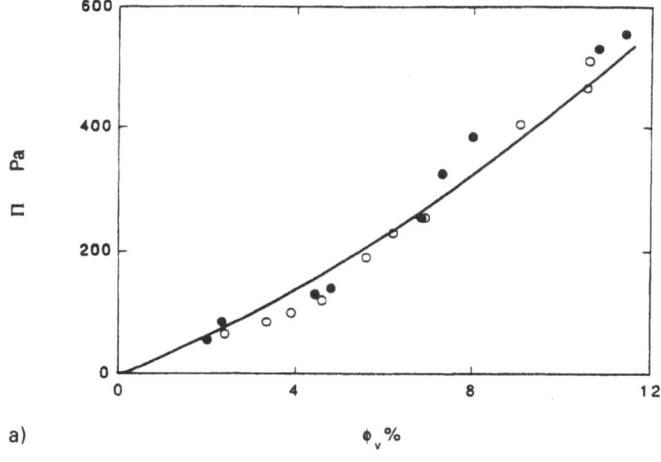

a)

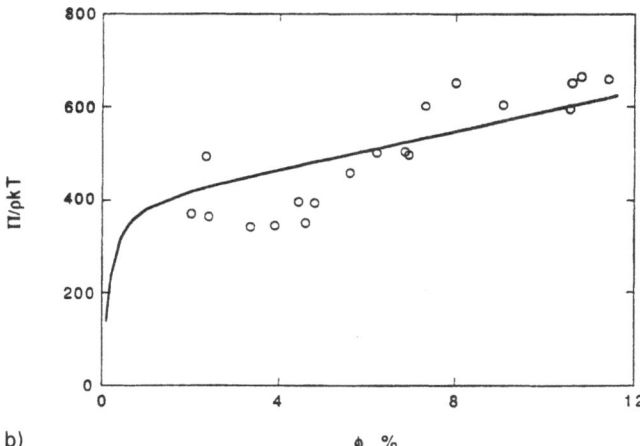

b)

Fig. 4 Variation of the osmotic pressure versus volume fraction in latex. Open dots are obtained during dilution while black dots are obtained during reconcentration of the sample. (a) pressure in Pascal versus volume fraction; (b) in reduced units compared to the perfect gas pressure

Discussion: comparison with theoretical phase diagrams

We can now compare our results to the theoretical phase diagram calculated by Robbins [10]. The two physical constants involved are $\lambda = \kappa l$ (κ is the inverse of the Debye screening length and l the interparticle distance) and the ratio of thermal energy to electrostatic energy at the interparticle spacing l:

$$\frac{kT}{U(l)} = \frac{l}{Z_{\text{eff}}^2 L_B \exp(-\kappa l)}.$$

There are three regions in the theoretical phase diagram: bcc, fcc and fluid state. Observed bcc and fcc are plotted in the same figure (Fig. 5a) that the theoretical lines obtained

Fig. 5 Dilution experiment: (a) colloidal crystals obtained without salt compared to theoretical phase diagram, including the liquid and the two colloidal crystal domain. Abcissa is $\lambda = \kappa l$; ordinate is the interaction energy at the distance between nearest neighbors in units of kT (b) α is the ratio of the average displacement from equilibrium position to the lattice constant as function of λ. Lindeman's criterion is given by the full line

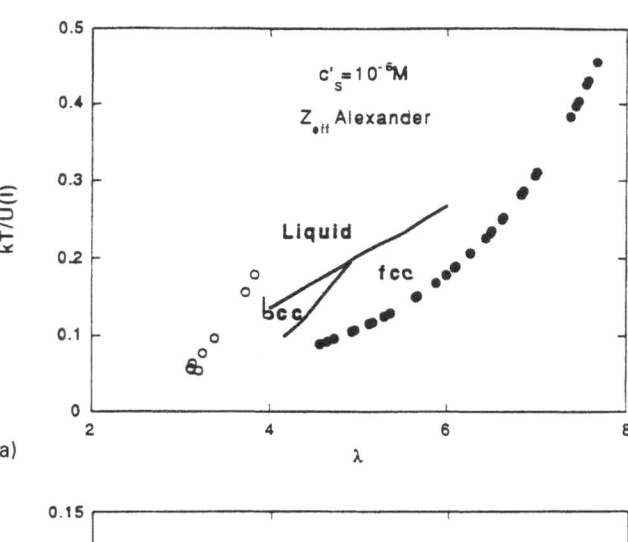

a)

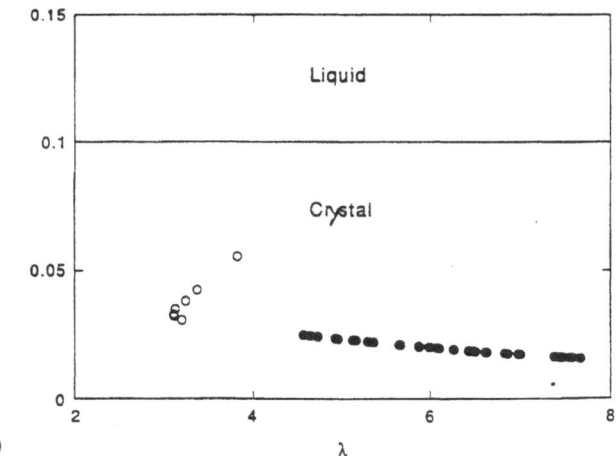

b)

of 400, much less than the structural charge (17 700 charge per particle). To our knowledge, this is the most direct method to obtain an experimental value of the effective charge [4]. Full lines in Figs. 4a and b are the predictions of the Poisson–Boltzmann Cell (PBC) model, assuming that the residual monovalent salt concentration in the reservoir is 10^{-6} M, the radius 510 Å and electrostatic interaction is the only one responsible for the observed pressure.

The agreement between prediction of this theory and observed values is excellent: we have here a series of samples with known lattice constants as well as osmotic pressure, mainly imposed by counter-ions.

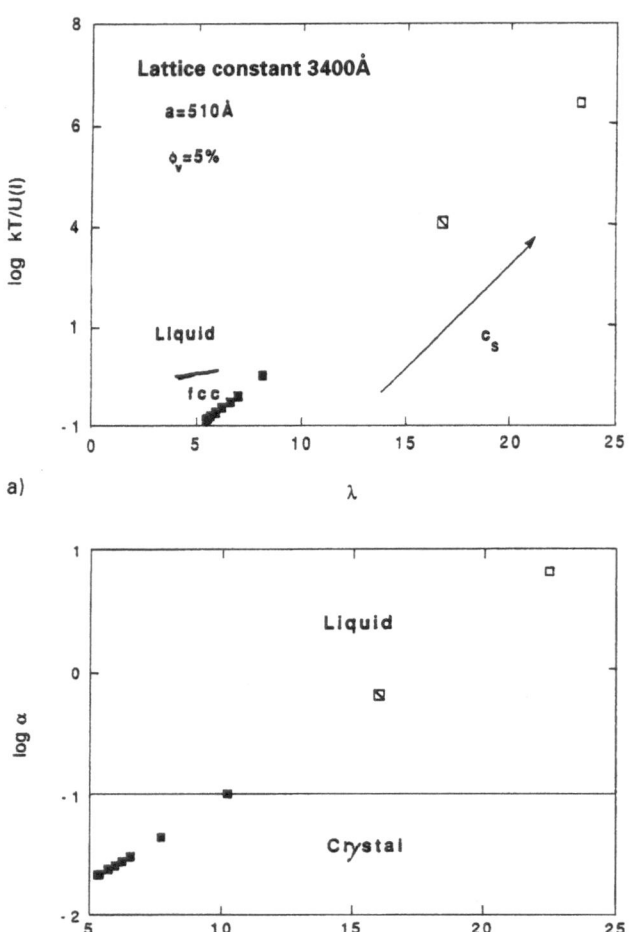

a)

b)

Fig. 6 Melting of the colloidal crystal (a) same coordinates as Fig. 5a. Squares are the experimental points, showing agreement with the expectation for melting (b) same coordinates than figure 5b; black squares are colloidal crystals; open squares is fluid. State was difficult to determine from scattering for the intermediate point

Table 1 Melting of colloidal crystal using salt for a volume fraction $\Phi_V = 5\%$, $d = 3400$ Å

Salinity in the reserv. c'_S(M)	Salinity in the solution c_S(M)	Πth PBC in cm water	Π exp in cm water	Conduct reserv. in μS/cm	State
10^{-6}	$9\cdot10^{-9}$	1.8	2.1		crystal
10^{-5}	$8.9\cdot10^{-7}$	1.64	1		crystal
$2\cdot10^{-5}$	$3.5\cdot10^{-6}$	1.4	1.4	5	crystal
$4\cdot10^{-5}$	$1.3\cdot10^{-5}$	0.77	0.5	10.3	crystal
10^{-4}	$5.7\cdot10^{-5}$	0.2	0.3	18	crystal
$5\cdot10^{-4}$	$4.2\cdot10^{-4}$	0.0048			?
10^{-3}	$8.8\cdot10^{-4}$	0	0		liquid

from the equilibrium position is calculated by adding the potentials of two neighbors (one-dimensional reduction of the problem) with an excentric PBC model. This square displacement, in reduced units of lattice spacing, is α. Lindeman's criterion for atomic crystals is that melting occurs when α is of the order of 0.1 [11]. Without added salt, all our samples are in a crystalline phase (Fig. 5b). In the case of electrostatically stabilized colloidal crystals, we observed melting by means of USAXS for $\alpha > 0.1$ (Fig. 6b). We think that this surprisingly large value for melting is due to the long range of a electrostatic interaction.

Finally, we consider that each particle with its counterions is surrounded by a unit cell called Wigner–Seitz cell. The surface S of this cell in the case of a fcc lattice is $6 d^2$ where d is the lattice constant. The osmotic pressure Π can be converted in force per particle F, by a simple geometric relation:

$$F = \frac{\Pi S}{N},$$

where N is the number of nearest neighbors.

Repulsive electrostatic forces per particle measured in this study are of the order of 10^{-11}–10^{-12} N per latex particle, comparable to the performance of the best AFM devices available. Theoretical evaluation of the force via PBC model are included as a thick line in Fig. 7a.

If one considers the force exerted via the screened electrostatic forces between neighboring particles through space available for the screening the counter-ions, the force per particle is given by the repulsive term of the DLVO potential for the same lattice constants:

$$F(r) = \frac{Z_{eff}^2 L_B \exp(2\kappa a)}{(1 + \kappa a)^2}\left(\frac{(1 + \kappa r)\exp(-\kappa r)}{r^2}\right).$$

This repulsive force per particle, indirectly detected in our experiment, is shown in Fig. 7b.

by molecular dynamic simulations, showing an excellent agreement between the theoretical prediction and the observed structures.

Electrostatic energy in reduced units as well as screening can be varied by adding salt. This has been done for a 5% volume fraction sample ($d = 3400$ Å) and salt concentration in the reservoir in the range of 10^{-6} to 10^{-3} M (Fig. 6a). We observe a melting for salinities larger than predicted by the theory. Table 1 gives compositions of these samples with added salt.

Repulsive force per particle

The electrostatic potential between particles is known. Therefore, the average square displacement of one particle

Progr Colloid Polym Sci (1996) 100:54–59
© Steinkopff Verlag 1996

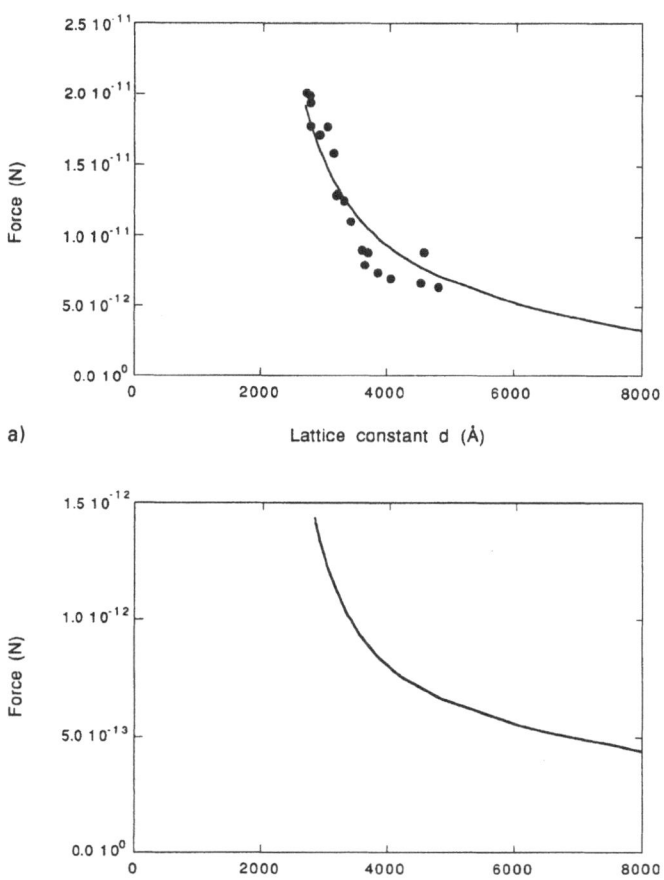

Fig. 7 (a) Force between particles obtained by dividing the osmotic pressure by the one of the faces of the Wigner–Seitz cell's surface as a function of the lattice constant d; the full line is the Poisson–Boltzmann cell prediction (b) Repulsive part of the DLVO interaction between two latex polyions as a function of the lattice constant d

Conclusion

Using simultaneous osmotic pressure and USAXS experiments, we are able to measure the pressure versus distance relation for monodisperse charged latex colloidal crystals with lattice constants of the order of 3000 to 4000 Å. No anomaly of the density or effect of long-range attraction was detected: all observed values are consistent with a strong prevalence of electrostatic repulsion versus all other type of interaction. We have now a device equivalent to an ultra-precise AFM arrangement, allowing detection of forces of the order of a few pN per particle. Adding very low contents of polymers, multivalent co-ions or other type of perturbing of the long-range forces has dramatic effects on the pressure: this is a new type of investigation method for possible long-range interactions.

The most surprising result is the observation of a strong pre-peak, occurring for fast growing crystallites. This prepeak is not observed when the sample grows during a few days in a closed capillary including a small bit of resin: deionization assured by diffusion of the ions slows down the crystallite growth in these conditions. The ratio of the diffuse peak to the first Bragg peak is constant: the only possible explanation at the present stage is that during rapid growth, some sites of the growing lattice are not filled by the available particles. During rapid growth, sites without particles are included in the lattice as isolated "voids" in the crystal. These voids are not the large holes filled with solvent detected by Ise [2], but isolated sites in the fcc lattice without the corresponding particle. The probability that voids are produced on adjacent sites is low when growth occurs rapidly. This induces the effective strong repulsion between voids, responsible for the broad diffuse band observed inside the first cell.

References

1. Sirota EB, Ou-Yang HD, Sinha SK, Chaikin PM, Axe JD, Fujii Y (1989) Physical Review Letters 62, 13, 1524
2. Dosho S, Ise N, Ito K, Iwai S, Kitano H, Matsuoka H, Nakamura H, Okumura H, Ono T, Sogami IS, Ueno Y, Yoshida H, Yoshiyama T (1993) 9:394; Yoshida H, Ise N, Hashimoto T (1995) Langmuir 11:2853
3. Barclay L, Harrington A, Ottewill RH (1972) Kolloid Z u Z Polym 250:655
4. Reus V, Belloni L, Zemb Th, Lutterbach N, Versmold H (1995) Journal de Chimie Physique 92:1233
5. Lambard J, Lesieur P, Zemb Th (1992) J Phys I, 2, 5, 1191
6. Strobl GR (1970) Acta Cryst, A26, 367
7. Segschneider C, Versmold H (1990) J Chem Education 67:967
8. Pusey PN, van Megen W, Bartlett P, Ackerson BJ, Rarity JG, Underwood SM (1989) Physical Review Letters 63:25
9. Fuoss RM, Mead DJ (1943) J Phys Chem 47:59
10. Robbins MO, Kremer K, Grest GG (1988) J Chem Phys 88:3286
11. Lindemann FA (1910) Zeitschrift für Physik 11:609

Progr Colloid Polym Sci (1996) 100:60–63
© Steinkopff Verlag 1996

COLLOIDAL PARTICLES

R.H. Ottewill
A.R. Rennie

Interaction behaviour in a binary mixture of polymer particles

Prof. Dr. R.H. Ottewill (✉)
School of Chemistry
University of Bristol
Cantock's Close
Bristol BS8 1TS, United Kingdom

A.R. Rennie
Polymers and Colloids Group
Cavendish Laboratory
University of Cambridge
Cambridge CB3 OHE, United Kingdom

Abstract The behaviour of small charge-stabilised particles (A, radius 168 Å) in the presence of larger particles (B, radius 510 Å) in 6×10^{-5} mol dm^{-3} sodium chloride solution has been investigated using small-angle neutron-scattering. The number concentration ratio, N_A/N_B, used was 15. The results indicated that the particles remained colloidally stable in the mixture but that some clustering of the smaller particles occurred in a "fluid-like" arrangement of the larger particles.

Key words Polystyrene latices – neutron scattering – structure – complex fluids

Introduction

Despite the technological importance of multimodal dispersions, for example, in processes such as film formation, only a relatively small amount of research has been reported on the structure of mixtures of colloidal particles, where each type of particle has a well-defined particle-size distribution [1–6]. The present work was started with the objective of mixing together charge-stabilised polystyrene particles (latices) of different sizes in various number ratios, to form a binary-particle mixture under well-defined conditions.

A schematic illustration of some of the possible behaviours in a binary mixture is shown in Fig. 1; possible effects include a) formation of colloidal crystals, b) segregation of the two phases, c) heterocoagulation. Of these possibilities mixed-colloidal crystal formation has been well illustrated using sterically-stabilised nearly-hard sphere particles at high volume fractions and crystals corresponding to stoicheometry of both AB_2 and AB_{13} (A = large and B = small particle) were observed by scanning electron

microscopy and light diffraction [5]. Segregation and heterocoagulation have also been observed in previous studies [7, 8]. A point of interest is whether other structures are possible in either sterically-stabilised or charge-stabilised systems.

The present work was carried out using polystyrene particles of spherical shape in an aqueous environment with a low electrolyte concentration, 6×10^{-5} mol dm^{-3} sodium chloride. The latter electrolyte concentration gives a condition where long-range electrostatic interaction can occur between the particles, even at low volume fractions, since the Debye reciprocal length is ca. 400 Å.

The technique used to study the structure of the binary mixtures was small-angle neutron scattering, SANS. This method was chosen since it allowed, by using one set of particles in the form of deuterated polystyrene and the other set in the hydrogenated form and by choosing an appropriate mixture of H_2O and D_2O, that each set of particles could be put into a condition of essentially zero-scattering. Hence, the structure of each set could be examined in the presence of the other.

Progr Colloid Polym Sci (1996) 100:60–63
© Steinkopff Verlag 1996

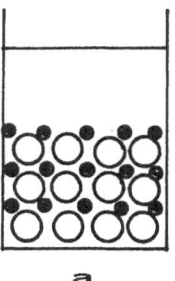 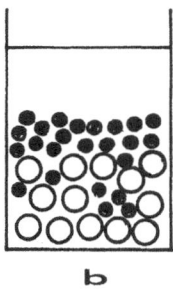 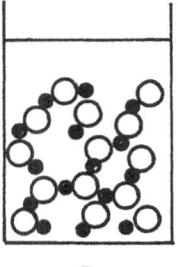

a **b** **c**

Fig. 1 Schematic illustration of some possible structures in binary mixtures, a) crystals, b) particle segregation, c) heterocoagulation

Table 1 Latex, particle size

Latex	Method	Radius/Å
A	SANS[a]	168 ± 5
	Dynamic light scattering[b]	190 ± 20
B	SANS[a]	510 ± 10
	Electron microscopy[c]	490 ± 20

a: weight average   b: ~z-average   c: number average

Table 2 Coherent neutron scattering lengths

Materials	$\rho/10^{10} \, cm^{-2}$
Latex A, h_8-polystyrene	1.41
Latex B, d_8-polystyrene	6.35[†]
H_2O	−0.56
75% D_2O in H_2O	4.62
D_2O	6.35

[†] experimentally determined

Experimental

Materials

The small particles (latex A) were prepared by emulsion polymerisation using h_8-styrene and the larger particles (latex B) using d_8-styrene. The particles sizes obtained are listed in Table 1.

The weight fractions of latices A and B were determined from their dry weights. For the binary mixtures known volumes of each latex were taken and the amounts checked by weighing so that both the weight fraction and the volume fraction were known accurately.

Small-angle neutron scattering

The SANS experiments were carried out at the National Institute of Standards and Technology (NIST), Cold Neutron Research Facility, Gaithersburg, MD, using the NG7 30 m spectrometer. This was configured to measure over a scattering vector range, Q, from 0.002 to 0.02 Å^{-1} using an incident neutron wavelength of 10 Å. A sample-detector distance of 15.34 m was used. NIST calibration standards were used to convert the measured intensities into absolute units.

Theory

The intensity of scattering, $I(Q)$, from a volume fraction, ϕ_p, of noninteracting particles each of volume V_p can be written in the form [9],

$$I(Q) = \phi_p V_p [\rho_p - \rho_m]^2 P(Q) \tag{1}$$

where, for elastic scattering, Q can be taken as $4\pi \sin(\theta/2)/\lambda$, with θ = the scattering angle and λ = the wavelength of the incident beam; ρ_p and ρ_m are respec-

tively the coherent neutron scattering lengths of the particle and the medium respectively. The values used in the present work are listed in Table 2.

$P(Q)$ defines a particle shape factor which for a particle of radius R can be written as,

$$P(Q) = \left(\frac{3(\sin QR - QR \cos QR)}{(QR)^3} \right)^2 . \tag{2}$$

Thus by using this expression in Eq. (1) and comparing calculated and measured intensities over a range of Q values the particle size, R, can be determined. A particle-size distribution function [10] was also included in the analysis.

When interaction occurs between the particles an additional function needs to be included; this is termed the structure factor, $S(Q)$, when the equation for scattered intensity becomes,

$$I(Q) = \phi_p V_p [\rho_p - \rho_m]^2 P(Q) S(Q) . \tag{3}$$

Analytical expressions for $S(Q)$, which is related to the Fourier transform of the radial distribution parameters, have been given by Hayter et al. [11, 12] based on the interaction potential for sphere-sphere interaction [13], in terms of particle radius R, surface potential, ψ_s, and salt concentration. Thus in Eq. (3), $P(Q)$ represents the intraparticle scattering and $S(Q)$ the interparticle scattering.

In the case of a binary mixture the scattering equation becomes more complex and has the form,

$$\begin{aligned} I(Q) = &\, \phi_A V_A [\rho_A - \rho_m]^2 P(Q)_A S(Q)_{AA} \\ &+ \phi_B V_B (\rho_B - \rho_m)^2 P(Q)_B S(Q)_{BB} \\ &+ 2(\rho_A - \rho_m)(\rho_B - \rho_m) \\ &\times [\phi_A V_A \phi_B V_B P(Q)_A P(Q)_B]^{\ddagger} S(Q)_{AB} . \end{aligned} \tag{4}$$

The three terms $S(Q)_{AA}$, $S(Q)_{BB}$ and $S(Q)_{AB}$ represent the interparticle interactions between the A particles in the presence of the B particles, the B particles in the presence of the A particles and between the A and the B particles.

Since there are three unknowns in Eq. (4) then three data points are needed at each Q in order to solve for, $S(Q)_{AA}$, $S(Q)_{BB}$ and $S(Q)_{AB}$. Therefore experiments were carried out with H_2O, 75% D_2O and D_2O as the dispersion media. In the first the A particles were close to a contrast match point and the scattering was dominated by the deuterated B particles; in D_2O the B particles were matched. 75% D_2O, in which neither the A nor the B particles were matched gave the third data set.

Results and discussion

Figure 2 shows the experimental results for a mixture of A and B with a number ratio, N_A/N_B, of 15 and a radius ratio, R_A/R_B, of 0.33. From the results shown in Figure 2 and the compositions given in Table 3 the partial structure factors $S(Q)_{AA}$, $S(Q)_{BB}$ and $S(Q)_{AB}$ were extracted as a function of Q. These are shown in Fig. 3.

Our interpretation of the partial structure factors at the present stage can be summarised in the following way:

$S(Q)_{BB}$: The form of the curve Fig. 3b indicates that the particles are colloidally stable and strongly interacting by electrostatic repulsion as expected for particles with a surface potential of 15 mV in an electrolyte concentration of 6×10^{-5} mol dm^{-3} sodium chloride. The position

Table 3 Compositions of binary mixtures

Medium	ϕ_A	ϕ_B	ϕ_A/ϕ_B
H_2O	0.0104	0.0194	0.537
75% D_2O	0.0112	0.0209	0.537
D_2O	0.0115	0.0214	0.537

of the main peak indicates an average correlation distance between the particles of ca. 2000 Å, indicative of short-range order. At the higher Q values the peaks are damped as expected for a "fluid-like" structure of the particles and as observed in earlier experiments with monodisperse latices [14]. In the absence of the A particles the main peak occurred at essentially the same Q value. The indication from these results is that there is very little perturbation of the structural arrangement of the B particles occurring as a consequence of the presence of the A particles.

$S(Q)_{AA}$: In the absence of B particles the form of $S(Q)_A$ against Q was very similar in shape to that shown for $S(Q)_{BB}$ but scaled according to particle size, as expected for monodisperse electrostatically interacting spheres. In the mixture there is a profound change in the behaviour of the A particles as indicated by the $S(Q)_{AA}$ against Q curve, Fig. 3a. The upturn at low Q indicates that clustering of the small particles is not entirely random in the "fluid-like" structure of the larger particles. This behaviour is a consequence of the much stronger electrostatic repulsion which occurs between the layer B particles. However, the peak at a Q of ca. 0.010 Å^{-1} indicates that there is still some correlation between the A particles on a distance scale of the order of 700 Å.

Fig. 2 Ln $I(Q)$ against Q for a mixture of latices A and B in 6×10^{-5} mol dm^{-3} sodium chloride: $\circ$, H_2O: $\triangle$, 75% D_2O: $\bullet$, D_2O

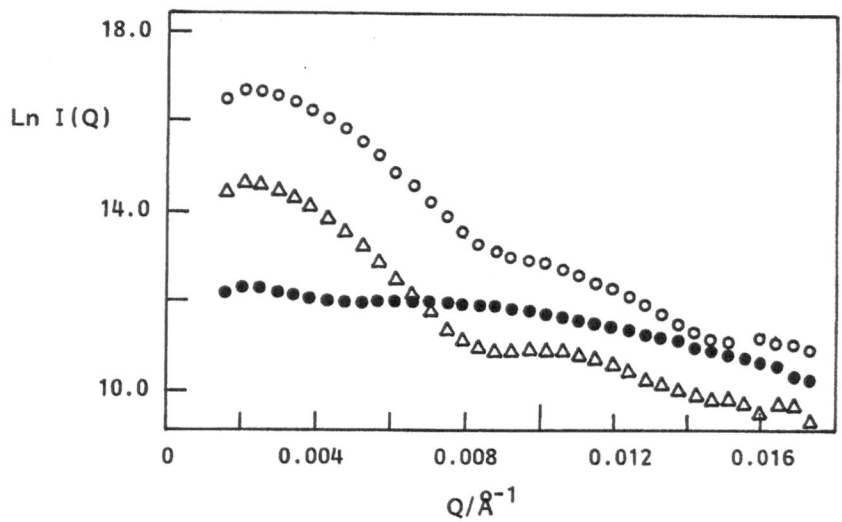

Progr Colloid Polym Sci (1996) 100:60–63
© Steinkopff Verlag 1996

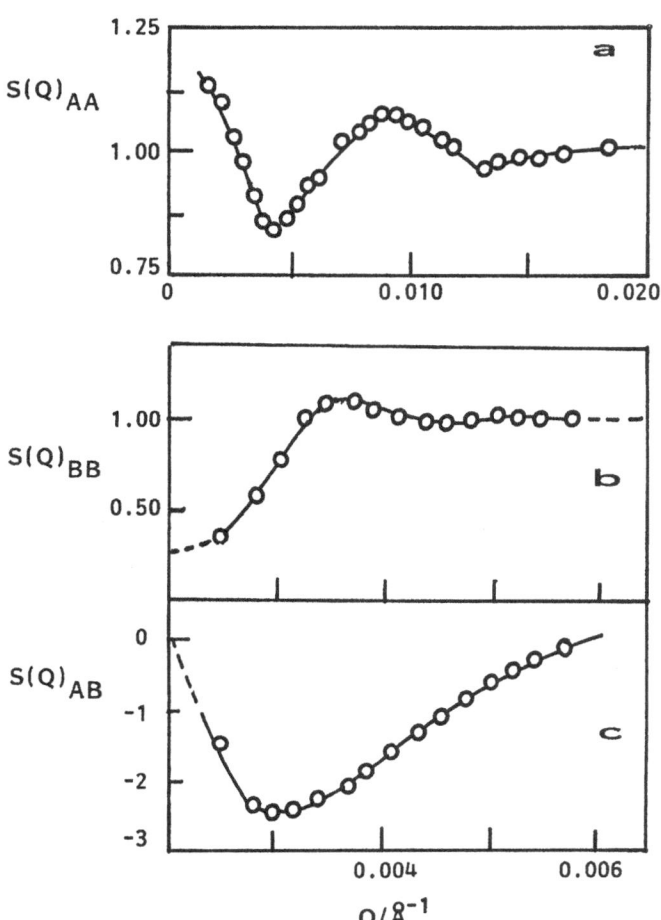

Fig. 3 Partial structure factors against Q for $N_A/N_B = 15.0$ a) $S(Q)_{AA}$, b) $S(Q)_{BB}$, c) $S(Q)_{AB}$

$S(Q)_{AB}$: The form of $S(Q)_{AB}$, Fig. 3c, indicates a negative correlation between the A and the B particles suggesting some separation of the two species in the overall structure which are uncorrelated with the structures formed by the A–A and B–B interactions. In essence, the presence of the B particles and their excluded volume as a consequence of the electrical double layer means that the A particles are excluded from these regions. In addition, calculation of the potential energies of interaction for the pair potentials between A–A, B–B and A–B particle pairs indicates that the force of repulsion has a nonlinear dependence on the size of the particles, that is, the force of A–B repulsion is not the mean of the A–A and B–B repulsive forces; the B–B repulsive force is the strongest.

The form of the scattering curves did not show any evidence of particle coagulation, that is, both types of particles remained colloidally stable in the binary mixture. Since the B particles were not completely ordered and remained in a "fluid-like" state the most likely explanation of the results is that a higher number concentration of A particles were repulsed into the larger void spaces between the B particles so that overall the distribution in the system was non-uniform.

Acknowledgments We acknowledge with thanks discussion with Drs. H.J.M. Hanley and G.C. Straty during the course of this work at the Cold Neutron Facility, NIST, Gaithersburg, MD. We also thank John Barker and the staff of the Facility for their help and use of equipment. RHO and ARR also wish to thank the DTI Colloid Technology Project and the Science and Engineering Council (UK) for support.

References

1. Hanley HJM, Pieper J, Straty GC, Hjelm R, Seeger PA (1990) Faraday Discuss Chem Soc 90:91
2. Bartlett P, Ottewill RH, Pusey PN (1990) J Chem Phys 93:1299
3. Hanley HJM, Straty GC, Lindner P (1991) Physica 174:60
4. Bartlett P, Ottewill RH (1992) J Chem Phys 96:3306
5. Bartlett P, Ottewill RH, Pusey PN (1992) Phys Rev Let 68:3801
6. Ottewill RH, Hanley HJM, Rennie AR, Straty GC (1995) Langmuir 11:3757
7. Waller R, B.Sc. thesis, 1975, University of Bristol
8. Goodwin JW, Ottewill RH (1978) Faraday Discuss Chem Soc 65:338
9. Guinier A, Fournet G (1955) Small Angle Scattering of X-Rays, Wiley, New York
10. Espenschied WF, Kerker M, Matijević E (1964) J Phys Chem 68:3093
11. Hayter JB, Penfold J (1981) Mol Phys 42:109
12. Hansen JP, Hayter JB (1982) Mol Phys 42:651
13. Verwey EJW, Overbeek JThG (1948) Theory of Stability of Lyophobic Colloids, Elsevier, Amsterdam
14. Cebula DJ, Goodwin JW, Jeffrey GC, Ottewill RH, Parentich A, Richardson RA (1983) Faraday Discuss Chem Soc 76:37

Progr Colloid Polym Sci (1996) 100:64–67
© Steinkopff Verlag 1996

COLLOIDAL PARTICLES

Electric light scattering from polytetrafluorethylene suspensions. II. Influence of dialysis

V. Peikov
Ts. Radeva
S.P. Stoylov
H. Hoffmann

Dr. V. Peikov (✉) · Ts. Radeva
S.P. Stoylov
Institute of Physical Chemistry
Bulgarian Academy of Sciences
Sofia 1113, Bulgaria

H. Hoffmann
Department of Physical Chemistry I
University of Bayreuth
95440 Bayreuth, FRG

Abstract Aqueous suspensions of Polytetrafluorethylene (PTFE) rod-like particles have been studied by electric light scattering at different particle concentrations (0.05–4% w/w) and degrees of dialysis. For particle concentrations 1 and 4% w/w the electric light scattering effect changes its sign from positive to negative when going to lower frequencies at low electric field strengths for both undialyzed and dialyzed samples, which could be due to the interparticle interactions. Dialysis decreases the amount of the bound ionic surfactant and consequently, the surface charge of the particles. As a result, the interparticle electrostatic interactions are much stronger in the undialyzed samples in comparison to the dialyzed ones. The higher surface charge could also explain the observed increase of the relaxation frequency of the surface electric polarizability in the undialyzed suspensions.

Key words Electric light scattering – PTFE particles – dialysis of suspension – change of surface charge

Introduction

Polytetrafluorethylene (PTFE) particles dispersed in aqueous media have been widely used as a model system for investigation of electro-optic phenomena at high particle concentrations [1, 2].

The most attractive property of PTFE particles as model particles in electro-optic experiments is the good optical matching in aqueous media. Consequently, there are almost no optical complications (multiple light scattering) at high concentrations. Also the electro-optic effects are relatively high because of the optical anisotropy of the particles. On the other hand, the surface electric properties of the PTFE particles, dispersed in aqueous media, are not yet completely understood. For example, the origin of the surface charge as well as its change upon dialysis or addition of different additives (surfactants, polymers, electrolytes, etc.) is still under consideration.

At increasing PTFE concentration a change of the sign of the electro-optic effect from positive at electric field frequencies above 1 kHz to negative at lower frequencies has been observed by electric birefringence [3, 4] and electric light scattering [5]. Similar "anomalous" electro-optic behavior has been often observed in many aqueous dispersions, but still is not sufficiently explained. Different explanations could be found in the literature for every particular case – permanent dipole moment of the particles, electrostatic and hydrodynamic interparticle interactions, optical effects, etc. The change of PTFE particle orientation from longitudinal at high frequencies to transversal one at low frequencies has been confirmed experimentally by an independent study of SANS in an electric field [6]. Consequently, the causes for the observed electro-optic behavior are not of purely optical origin. Currently, it is believed that electrostatic and/or hydrodynamic interparticle interactions cause the transversal orientation of PTFE particles. The exact mechanism of that effect is studied intensively [7–9].

In the latest electric birefringence study of PTFE ellipsoidal particles of Degiorgio et al. [10], one interesting possibility for changing the value of the particle surface charge is suggested. The authors add a nonionic surfactant to the PTFE dispersions which displaces the ionic surfactant bound to the particle surface. The changes in the surface charge depend on the amount of the added nonionic surfactant. The authors suggested PTFE particles as a perspective model for studying the properties of the electric double layer of colloidal particles.

In the present work, results of electric light scattering of PTFE rod-like particles are discussed. The influence of the ionic strength and surfactant concentration on the low frequency "anomaly" at high particle concentration is considered and the possible explanations of this effect are discussed.

Materials and methods

The basic suspension of PTFE rod-like particles (density $2.31 \, \text{g cm}^{-3}$, average refraction index $n = 1.377$, number average length 210 nm, average diameter 30 nm) is obtained from the Department of Physical Chemistry I, Bayreuth and is described in detail in ref. [4]. It is prepared by emulsion polymerization of tetrafluorethylene in the presence of perfluorinated surfactants $C_8F_{17}COOH/C_8F_{17}COONH_4$ (75:25) and stocked at 10% w/w concentration [4]. The dialyzed sample is prepared by dialysis for 60 days against deionized water up to a surface tension of $71.7 \, \text{mN m}^{-1}$. The surface charge of the particles is due to the bound surfactant, left after dialysis (about 95–99% from it is left adsorbed on the particles) and to the persulfate initiators used for the polymerization.

The electric light scattering effect α is defined as:

$$\alpha = \frac{I_E - I_0}{I_0},$$

where I_E and I_0 are the intensities of the scattered light, when an electric field of strength E is applied on the suspension and without field.

The electro-optic effect is measured under a 90° angle of observation and a 90° angle between the applied electric field and the plane of observation, with white incident light. More details for the method and the apparatus used are given in ref. [7].

Results and discussion

The frequency dependence of the stationary component of the electric light scattering effect for different particle concentrations prepared by dilution with bidistilled water of the stock (undialyzed) PTFE suspension is given in Fig. 1. With the rise of concentration, the value of the electro-optic effect in the kHz frequency region α_{kHz} decreases and a large negative electro-optic effect appears at lower frequencies. The observed transversal orientation of PTFE particles corresponds to the results for dialyzed PTFE suspensions [3–5]. Moreover, the frequencies at which the sign of the electro-optic effect changes from negative to positive are close to those found previously [3–6]. The hertz electro-optic effect α_{Hz}, which equals the difference between the plateau value of the kHz electro-optic effect and the plateau value in the low frequency region, increases with the rise of particle concentration. This electro-optic effect corresponds to orientation of the particle symmetric axis perpendicular to the applied electric field. The increase of the value of α_{Hz} in concentrated suspension could be due to electrostatic and hydrodynamic interparticle interactions [9, 11].

The high frequency relaxation of the electro-optic effects in undialyzed suspension moves to much higher frequencies than that observed in dialyzed samples, especially in dilute suspensions (Fig. 2). One of the reasons for this is the higher ionic strength of the undialyzed solutions which leads to an increase of the relaxation frequency of the surface electric polarizability in comparison to the dialyzed samples. Our previous experiments show that

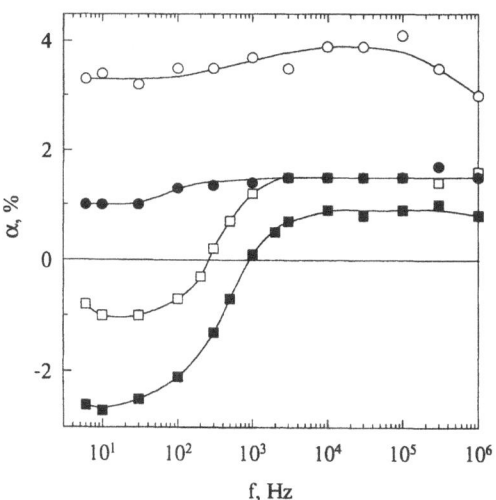

Fig. 1 Dependence of the electro-optic effect measured in undialyzed suspensions on the electric field frequency at constant field strength intensity of 77 V cm^{-1} for different PTFE concentrations: ○ – 0.05; ● – 0.1; □ – 1 and ■ – 4% w/w. Conductivity: $3.2 \times 10^{-5} \, \text{S cm}^{-1}$; $3.2 \times 10^{-5} \, \text{S cm}^{-1}$; $5.6 \times 10^{-5} \, \text{S cm}^{-1}$ and $4.5 \times 10^{-4} \, \text{S cm}^{-1}$, respectively

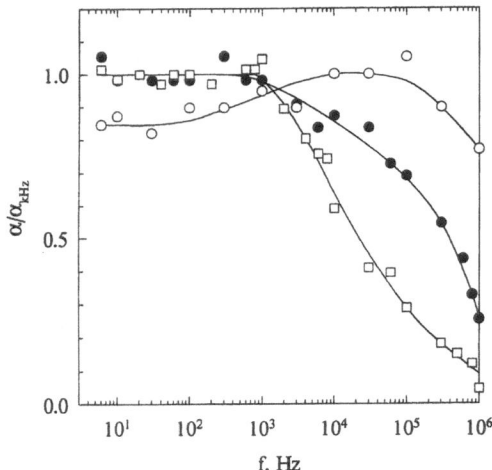

Fig. 2 Dependence of the normalized electro-optic effect α/α_{kHz} on the electric field frequency for 0.05% w/w PTFE suspension measured at constant field strength intensity of 77 V cm^{-1}: ○ – undialyzed, conductivity – 3.2×10^{-5} S cm^{-1}; □ – dialyzed, conductivity – 0.6×10^{-5} S cm^{-1}; ■ – dialyzed, conductivity – 2.6×10^{-5} S cm^{-1}

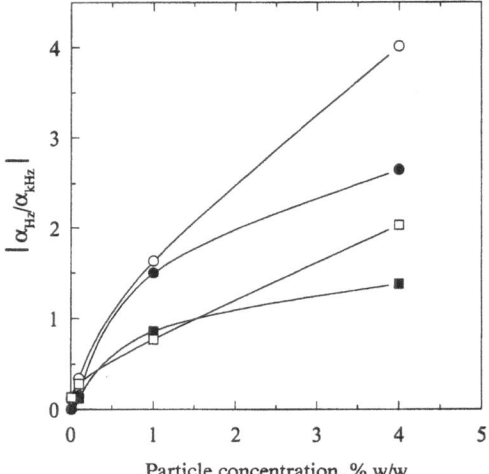

Fig. 3 Dependence of the ratio of the electro-optic effects $|\alpha_{Hz}/\alpha_{kHz}|$ on the particle concentration for dialyzed (●, ■) and undialyzed (○, □) suspensions measured at 77 V cm^{-1} (●, ○) and 174 V cm^{-1} (■, □)

such effect exists, but it is not strong enough to explain the observed difference in the relaxation frequency of 0.05% w/w dialyzed and undialyzed samples. The interparticle interactions are negligible at that concentration and their contribution to the displacement of the relaxation frequency could also be neglected.

The concentration dependence of the ratio $|\alpha_{Hz}/\alpha_{kHz}|$ measured at low (77 V cm^{-1}) field strength is given in Fig. 3 for undialyzed and dialyzed suspensions. This ratio strongly increases with the rise of particle concentration and reaches about 4 for undialysed 4% w/w sample. The value of the ratio for dialyzed 4% w/w suspension is about two times smaller than that for undialyzed one. The connection between particle interactions and the value of α_{Hz} is demonstrated from the strong decrease of α_{Hz} at decreasing the particle concentration. At low concentration (0.05% w/w) there is no interparticle interactions and the value of $|\alpha_{Hz}/\alpha_{kHz}|$ is about 0.12 for both undialyzed and dialyzed suspensions. This value could be attributed to the transversal polarization (or permanent dipole moment) of the individual PTFE particle. The higher value of the ratio $|\alpha_{Hz}/\alpha_{kHz}|$ measured for the most concentrated undialyzed sample corresponds to higher degree of particle interaction in comparison to the dialyzed samples. The stronger particle interaction is allowed independently from the higher ionic strength of the concentrated undialyzed sample (theoretically and experimentally is shown that the rise of ionic strength strongly decreases the electrostatic interactions [12]). This rule is not observed in the present study. The reason might be the decrease of the particle surface charge

upon dialysis. It is known that the surface charge of PTFE particles is due to the bound surfactant, and to the persulfate initiators used in the process of polymerization. Hence, the dialysis is one possibility to change the amount of the bound ionic surfactant and, consequently, the surface charge of the particles. In our case the efficiency of the dialysis is low – only 1–5% of the bound surfactant is removed after dialysis against deionized water for 60 days [4]. Even so, the particle ζ-potential decreases from ~ 50 mV before dialysis to ~ 40 mV after dialysis. This 20% decrease is due to decreasing surface charge of the particles. On the other hand, the ζ-potential of PTFE particles is independent from the ionic concentration and pH of the media in broad limits (from $\sim 5 \times 10^{-5}$ to $\sim 10^{-2}$ M NaCl and pH from 1.5 to 9.7). Our results show that the effect of the increased ionic strength on the interparticle interactions is overcompensated by the higher surface charge of PTFE particles in undialyzed samples. As a result, the interparticle electrostatic interactions, which are usually believed to be due to overlapping of the particle double layers, are much stronger in the undialyzed samples in comparison to the dialyzed ones. The higher surface charge could also explain the increased relaxation frequency of the surface electric polarizability in the undialyzed suspensions.

Acknowledgment This work was financially supported by the Commission of European Communities in the frames of COST Project 244. Thanks are due to Dr. J. Widmaier for the electrophoretic measurement and to Mr. K.H. Lauterbach for the help in the sample preparation.

Progr Colloid Polym Sci (1996) 100:64–67
© Steinkopff Verlag 1996

References

1. Batchelor P, Meeten GH, Maitland GC (1987) J Colloid Interface Sci 117: 360–365
2. Bellini T, Piazza R, Sozzi C, Degiorgio V (1988) Europhys Lett 7:561–565
3. Angel M, Hoffmann H, Huber G, Rehage H (1988) Ber Bunsenges Phys Chem 92:10–16
4. Angel S (1991) Preparation and characterizing of Polytetrafluorethylene-fibrilles-dispersions, Thesis, Bayreuth
5. Radeva Ts, Peikov V, Stoylov SP, Hoffmann H (1995) Colloids and Surfaces, in press
6. Baumann J, Klaus J, Neubauer G, Hoffmann H, Ibel K (1989) Ber Bunsenges Phys Chem 93:874–878
7. Stoylov SP (1991) Colloid Electrooptics. Academic Press, London
8. Hoffmann H, Kramer U, Thurn H (1990) J Phys Chem 94:2027–2033
9. Cates ME (1992) J Phys II France 2:1109–1119; Schoot P, Cates ME (1994) J Chem Phys 101:5040–5046
10. Bellini T, Degiorgio V, Mantegazza F, Marsan FA, Scarnecchia C (1995) J Chem Phys 103:8228–8232
11. Stoylov SP (1988) Ferroelectrics 86: 245–253
12. Hagenbuchle M, Weyerich B, Deggelmann M, Graf C, Krause R, Maier EE, Schulz SF, Klein R, Weber R (1990) Physica A 169:532–538

Progr Colloid Polym Sci (1996) 100:68–72
© Steinkopff Verlag 1996

M. Paillette
S. Brasselet
I. Ledoux
J. Zyss

Two photon Rayleigh scattering in micellar and microemulsion systems

Dr. M. Paillette (✉)
Groupe de Physique des Solides
(CNRS URA 017)
Université Paris VII
Tour 23, 2, Place Jussieu
75251 Paris, Cedex 05, France

S. Brasselet · I. Ledoux · J. Zyss
Molecular Quantum Electronic Department
(CNRS URA 250)
Centre National des Telecommunications
France Telecom
196, Avenue Henri Ravera
92220 Bagneux, France

Abstract We report the first preliminary two photon Rayleigh scattering measurements done at 1.064 μm in micellar and microemulsion systems as a function of the molar water-to-surfactant ratio value (w_0) and volume fraction (ϕ) of the droplets. They provide the β values of the static quadratic hyperpolarizability of BHDC surfactant molecule dissolved in water and heavy water as well as those of individual droplets in BHDC/Benzene/Water and BHDC/Benzene/Heavy Water microemulsion systems.

Key words Nonlinear scattering – microemulsions – quadratic hyperpolarizability coefficients

Introduction

In a fluid, when nothing breaks off the isotropy, all the χ_{ijk} coefficients of the tensor of the macroscopic second order dielectric susceptibility vanish although individual molecules may imply a nonzero microscopic quadratic hyperpolarizability β_{ijk}.

Second harmonic can however be scattered from the quadratically induced dipoles $\beta_{ijk}\vec{E}(\omega)\vec{E}(\omega)$ where $\vec{E}(\omega)$ is the electric field associated to an intense light beam from a Q-switched laser. The second harmonic scattered intensity, at exact double frequency (2ω) of the excitation (ω) is proportional to the product:

$$\langle \beta_{ijk}(-2\omega, \omega, \omega)\beta_{lmn}(-2\omega, \omega, \omega)\rangle \vec{E}(\omega) \cdot \vec{E}(\omega)$$

where $\langle \rangle$ implies the average over all the molecule orientations, whatever their symmetry.

This incoherent second harmonic light scattering in dense medium was first observed by Terhune and coworkers [1] followed by the spectral analysis by Maker [2]. This diffusion is called two photon Rayleigh scattering or elastic harmonic light scattering (EHLS).

This light-scattering technique nearly disappeared due to its experimental difficulties until its revival by Persoons et al. [3] followed soon afterwards by Zyss and coworkers [4–6] who evidenced the octupolar contribution in centrosymmetric molecules.

In this paper, we present the first preliminary measurements and results in micelles and reversed micellar systems (water-in oil (w/o) microemulsion) using the EHLS technique. The surfactant molecule under investigation is the benzyldimethyl-n-hexadecyl ammonium chloride (BHDC) either dissolved in water (H_2O) or in heavy water (D_2O) in presence of benzene as oil in microemulsions or not in micelles.

Experimental

Material

The three-component microemulsion consisted of BHDC dissolved in water or in heavy water and benzene. The BHDC surfactant, supplied by Sigma without further purification is a benzalkonium chloride with a benzene structure bound to the polar headgroup. The water was double distilled and desionized (resistivity 18 MΩ·cm) and benzene RP used as supplied.

Three series of microemulsions were prepared with different values of the water-to-surfactant molar ratio $w_0 = 7, 20, 22$ respectively. This choice is based on the measured complex Kerr constant $B^*(2\omega)$ behaviors in this series. For $w_0 = 7$ and $w_0 = 22$, the Kerr constants are respectively negative and positive in all the explored volume fraction ϕ range [7]. For $w_0 = 20$, the Kerr constant is successively negative for ϕ values below $\phi_0 = 1.7\%$ and positive above [8, 9].

Different volume fractions ϕ ranged from 0.06% to 10%, measured in this study, were prepared using the dilution procedure described in ref. [10].

The hydrodynamic radii R_h of the droplets measured from dynamic light scattering [11] were respectively 36, 55 and 58 Å. The mean water core radii obtained from time-resolved fluorescence quenching [12] were respectively 21.5, 46 and 49 Å corresponding to w_0 values 7, 20, 22.

Experimental setup

The experimental setup and the procedure, similar in its principle to that used by Maker in ref. [2], are described in full detail in earlier publications [5, 6, 13].

We use a transverse and longitudinal single mode N_d^{3+}:YAG laser as the fundamental source ($\lambda = 1.064\ \mu m$) of 10 Mw peak power and 10 ns pulse duration (repetition rate 10 khz). A beam splitter permits the coherent generation of a 0.532 μm reference signal $I_{ref}(2\omega)$ proportional to the square of the incident signal $I(\omega)$.

The elastically scattered radiation $I(2\omega)$ twice the laser frequency, due to the fluctuations of the randomly induced molecular dipoles, is collected at a right angle to the direction of the incident laser beam focused inside the cell containing the liquid.

The total intensity of the second harmonic scattered light $I(2\omega)$, ($\lambda = 0.532\ \mu m$) may be expressed as follows:

$$I(2\omega) = G' \cdot N \cdot \Sigma_{ijklmn} \langle \beta_{ijk}(-2\omega, \omega, \omega) \\ \times \beta_{lmn}(-2\omega, \omega, \omega) \rangle \cdot I_{ref}(2\omega) \quad (1)$$

where N is the number density of molecules with static quadratic hyperpolarizabilities β averaged over all molecular orientations $\langle \rangle$. The factor G' depends upon the scattering geometry, the incident wavelength the Lorenz–Lorentz-type local field correction and a corrective factor accounting for the conversion efficiency of the reference signal.

Dust or residual impurities are carefully removed from the samples through 0.5 μm. Millipore filters in order to avoid the breakdown of the focused I.R. laser beam inside the sample cell. The weakness of the signals did not permit the use of a monochromator but interference filters centered at 5320 Å prevent possible parasite scattered radiations.

Relative calibration of the experimental setup is performed from carbon tetrachloride measurement as solvent. From Terhune et al. measurements [1] using a ruby laser ($\lambda = 0.6943\ \mu m$), in the limits of the mode quality of the incident beam and the uncertainties in the geometrical parameter estimates, we have choosen as reference the value:

$$\sqrt{\langle \beta^2 \rangle} = 0.35 \times 10^{-30}\ esu\,(cm^{9/2} \cdot erg^{-1/2}).$$

Figure 1 shows the dependence of the harmonic scattered intensity $I(2\omega)$ (in arbitrary units) for benzene as oil and for $w_0 = 7$ microemulsion system (volume fraction $\phi = 4.83\%$). The full lines correspond to the mean

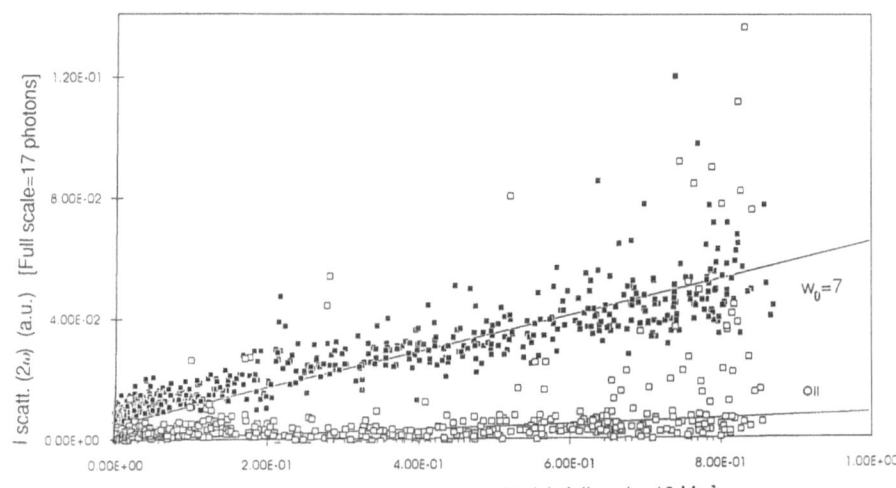

Fig. 1 Experimental dependence and expected linear (full lines) variation of the scattered second harmonic intensity $I(2\omega)$ as a function of the reference signal $I_{ref}(2\omega)$ (proportional to $I(\omega)^2$) for benzene(oil) and for the microemulsion system: BHDC surfactant/benzene/water ($w_0 = 7$; $\phi = 4.83\%$)

least-squares fit with the expected linear behavior. Unavoidable localized scattered emissions due to intensity fluctuations of the excitation laser beam are observed.

Data analysis

Parameter G' calibration

Pure carbon tetrachloride (CCl_4), benzene-(C_6H_6), and water (H_2O) samples, previously measured by EHLS technique, were both used to calibrate the parameter G'. We find $G' = 2.35 \times 10^{37}$ (esu units). Before and after each series of measurements, we repeat this calibration.

EHLS measurement of pure heavy water liquid (D_2O)

Up to now, the value of the static quadratic hyperpolarizability $\langle \beta^2 \rangle$ of this liquid remained unknown. We yield $\sqrt{\langle \beta^2 \rangle} = 0.135 \pm 0.02 \times 10^{-30}$ esu to be compared to the value for water (0.097×10^{-30} esu). Although H_2O and D_2O are chemically similar and their dielectric constants nearly identical, there may be differences in hydrogen (vs. deuterium) bondings to explain the result [14].

Investigations in micellar systems

From EHLS the micellar phase of BHDC in aqueous (H_2O) and heavy water (D_2O) solutions, at concentrations well below the CMC (5×10^{-3} M · (15)), were investigated.

In a first stage, we have considered these binary electrolyte solutions as mixtures of two components: BHDC surfactant (subscript s) and H_2O (resp. D_2O) (subscript w). Then Eq. (1) may be expressed as follows:

$$I(2\omega) = k \cdot I_{ref}(2\omega) = G' \cdot (N_s \cdot \langle \beta_s^2 \rangle + N_w \langle \beta_w^2 \rangle) \cdot I_{ref}(2\omega)$$

or

$$k = b \cdot N_s + a$$

where $b = G' \cdot \langle \beta_s^2 \rangle$ and $a = G' \cdot N_w \cdot \langle \beta_w^2 \rangle$.

We have to consider the mole fraction of the dissolved surfactant molecule only, the concentration of the solvent being assumed constant owing to the weak concentration of the dissolved molecules.

Because of the manifestation of associations, this procedure leads to erratic values for $\langle \beta_s^2 \rangle$. So from the experimental k_i values for pure solvent ($k_i = 1.15 \times 10^{-2}(H_2O)$, $= 2.1 \times 10^{-2}(D_2O)$), one gets $\langle \beta_s^2 \rangle$ from the expression: $\langle \beta_s^2 \rangle = (k - k_i)/G' \cdot N_s$, applied to each experimental value.

This method finally leads to:

$$\sqrt{\langle \beta^2(BHDC) \rangle} \text{ in } H_2O = 27 \pm 3 \times 10^{-30} \text{ esu},$$

and

$$\sqrt{\langle \beta^2(BHDC) \rangle} \text{ in } D_2O = 56 \pm 4 \times 10^{-30} \text{ esu}.$$

As aforementioned, these results confirm the difference in the hydrogen (vs. deuterium) bondings with the surfactant molecules.

For comparison, these large values of the static quadratic hyperpolarizability corresponds to those of para-nitroaniline-like molecules: $32 \pm 5 \times 10^{-30}$ esu obtained from electric field-induced second harmonic generation technique (EFISH) [16].

Measurements in BHDC reversed micelles in presence of water or heavy water

EHLS measurements were performed successively from BHDC/benzene/water and BHDC/benzene/heavy water microemulsion systems for different values of w_0 and variable values of the volume fraction ϕ occupied by the droplets.

For β measurements of droplet phase, we have to consider the number density of droplets N_d, defined as: $N_d = \phi/(4\pi/3)R_h^3$ the concentration of the oil N_o (benzene) being assumed constant owing to the weak concentration of droplets. In the case of a mixture of two different species, droplets (subscript d) and benzene molecules (subscript o), Eq. (1) has the form:

$$I(2\omega) = k \cdot I_{ref}(2\omega) = G' \cdot (N_d \cdot \langle \beta_d^2 \rangle + N_o \cdot \langle \beta_o^2 \rangle) \cdot I_{ref}(2\omega)$$
$$= k \cdot I_{ref}(2\omega).$$

The k factor depends linearly on the concentration of droplets N_d:

$$k = b \cdot N_d + a$$

where: $b = G' \cdot \langle \beta_d^2 \rangle$ and $a = G' \cdot N_o \cdot \langle \beta_o^2 \rangle$.

The dependence of the k coefficient as a function of the concentration N_d in benzene for the BHDC/benzene/H_2O system and w_0 values 7, 20, 22, is plotted in Fig. 2. The full lines represent a mean least-squares fit to the data with the expected linear behaviour. We also found the same behavior for the BHDC/benzene/D_2O system.

From the least-squares analysis of the experimental points, the slope b leads to $\langle \beta_d^2 \rangle$ and the intercept a to $\langle \beta_o^2 \rangle$. The results for the two microemulsion systems are listed in Table 1. The overall relative error is of the order of 20%.

For benzene molecule, the reference value of $\sqrt{\langle \beta^2 \rangle}$ is 0.23×10^{-30} esu, the mean-fitted value is $0.22 \pm 0.10 \times 10^{-30}$ esu.

Progr Colloid Polym Sci (1996) 100:68–72
© Steinkopff Verlag 1996

Fig. 2 Plot of the experimental k values as a function of the number density of droplets N_d for the three series: $w_0 = 7 (\times)$; $= 20 (\circ)$; $= 22 (\bullet)$ respectively of the reversed micellar system: BHDC/benzene/water. The different lines are the mean least-squares fit with the expected linear behavior: $k = bN_d + a$ ($w_0 = 7 (\times)$; $w_0 = 20 (\circ)$; $w_0 = 22 (\bullet)$)

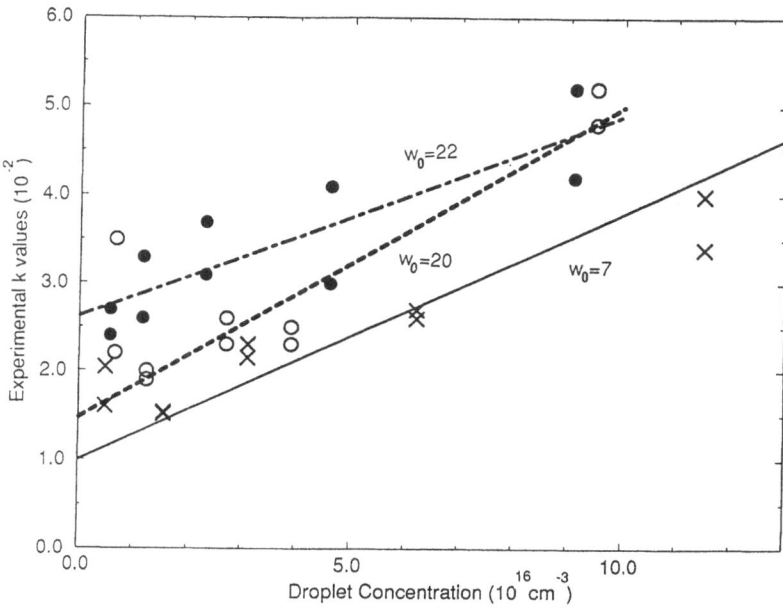

Table 1 Experimental values of the intercepts a, the slopes b deduced from the mean least-squares fit $k(N_d)$ and the quadratic hyperpolarizabilities $\sqrt{\langle \beta^2 \rangle}$ of benzene molecule (oil) and $\sqrt{\langle \beta^2 \rangle}$ of individual droplets for the different series ($w_0 = 7,20,22$) corresponding to the two reversed micellar systems explored: BHDC/Benzene/Water(H_2O) and BHDC/Benzene/Heavy Water(D_2O)

Systems	BHDC/C_6H_6/H_2O			BHDC/C_6H_6/D_2O	
w_0	7	20	22	7	22
$a \times 10^2$	1.07	1.47	2.25	1.3	2.1
$\sqrt{\langle \beta^2 \rangle} \times 10^{30}$ esu	0.12	0.29	0.38	0.13	0.16
$b \times 10^{19}$	2.49	3.5	3.18	12.5	7.9 − 12.8
$\sqrt{\langle \beta^2 \rangle} \times 10^{30}$ esu	103 ± 10	122 ± 12	116 ± 10	230 ± 30	183 ± 24, 234 ± 30

The static quadratic hyperpolarizability of droplets, whatever the system (H_2O or D_2O), seems to be independent of the size of the droplets. For identical w_0 values, the $\sqrt{\langle \beta^2 \rangle}$ values of droplets in the deuterated system are two times higher than of those in the aqueous system within the hypothesis that the size of the droplets is identical. We find again the same discrepancy previously observed in micellar systems. Electrical resistance and dielectric relaxation measurements from similar system show that the surface charge density of the droplets in the H_2O systems are three times higher than those in D_2O system at identical w_0 values [17].

Conclusion

From these preliminary results, we are yet unable, at this stage, to propose a physical mechanism to explain them.

Further experiments are planned, in particular EFISH measurements to yield the manifestation of the $\langle \beta \rangle^2$ coefficient due to the orientation of a possible permanent dipole moment of the droplets by a static electric field. It seems possible that the shape fluctuations of droplets contribute to this hyperpolarizability contribution [18]: a theoretical analysis is currently being made.

Already, the substitution of H_2O by D_2O reveals the effect of electrostatic interactions, the observed scattering should be obscured by interparticle interaction effects which could decrease the magnitude of the coefficients.

The EHLS experiment should be very promising to revisit the electronic manifestations in micelles and inverted micellar systems. Furthermore, the large possibilities of these systems and the magnitude of the quadratic hyperpolarizability coefficients should offer new molecular engineering schemes.

References

1. Terhune RW, Maker PD, Savage CM (1965) Phys Rev Lett 14:681
2. Maker PD (1970) Phys Rev A 1:923
3. Clays K, Persoons A (1991) Phys Rev Lett 66:2980
4. Zyss J, Chau Van T, Dhenaut C, Ledoux I (1993) Chem Phys 177:281
5. Zyss J, Dhenaut C, Chau Van T, Ledoux I (1993) Chem Phys Lett 206:409
6. Dhenaut C, Ledoux I, Samuel IDW, Zyss J, Bourgault M, Le Bozec H (1995) Nature 374:339
7. Paillette M (1991) Progr Colloid Polym Sci 84:144
8. Paillette M (1993) Nonlinear Optics 5:449
9. Koper GJM, Roman Vas C, van der Linden E (1994) J Phys II (France) 4:163
10. Gracias A, Lachaise J, Martinez A, Bourrel M, Chambu C (1976) CR Acad Sci (Paris) B 282:547
11. Chatenay D, Urbach W, Cazabat AM, Langevin D (1985) Phys Rev Lett 54:2253
12. Jada A, Lang J, Zana R, Makhloufi R, Hirsch E, Candau SJ (1990) J Phys Chem 94:387
13. Dhenaut C (1995) Thesis (University Orsay) unpublished
14. Chou SI, Shah DO (1981) J Colloid Interface Sci 80:49
15. Porter RM (1994) Handbook of Surfactants (Chapman and Hall-London)
16. Oudar JL, Le Person H (1975) Opt Comm 15:258
17. Chou SI, Shah DO (1981) J Phys Chem 85:1480
18. Borkovec M, Eicke HF (1988) Chem Phys Lett 147:195

Progr Colloid Polym Sci (1996) 100:73–80
© Steinkopff Verlag 1996

COLLOIDAL PARTICLES

A. Fernández-Barbero
M. Cabrerizo-Vílchez
R. Martínez-García

Dynamic scaling in colloidal fractal aggregation: Influence of particle surface charge density

Dr. A. Fernández-Barbero (✉)
M. Cabrerizo-Vílchez · R. Martínez-García
Grupo de Física de Fluidos y Biocoloides
Departamento de Física Aplicada
Universidad de Granada
Campus de Fuentenueva
18071 Granada, Spain

Abstract We have studied the influence of the surface charge density of polymer colloids on aggregation processes induced at a high salt concentration. In this way, the effect of the residual interaction between the particles on the aggregation mechanism was studied. The time dependence of the detailed cluster-size distribution and the time-independent scaling distribution were obtained by single particle light scattering. We found a change in the aggregation mechanism, expressed as a variation of the μ exponent value, when the surface charge is modified.

Key words Colloidal aggregation – dynamic scaling – fractal structure – single particle detection – light scattering

Introduction

The aggregation of colloidal prticles is a good model for describing cluster growing [1]. These processes are usually described by the time evolution of the cluster-size distribution and thus, its knowledge is an essential task. Due to the difficulties to detect and classify single clusters, accurate measurements of the detailed cluster-size distribution are seldom and by now, the number of theoretical predictions and computer simulations far exceeds the number of experimental results.

Both theory and experiments have shown a universal behavior, independent of the colloid, when the aggregation of clusters is diffusion limited (DLCA) or reaction limited (RLCA) [2–9]. DLCA occurs when every collision between diffusing clusters results in the formation of a bond. In this regime, the rate of aggregation is limited by the time it takes clusters to diffuse towards one another. RLCA occurs when only a small fraction of collisions between clusters leads to the formation of bigger clusters. In this case, the rate of aggregation depends not only on the diffusion of the clusters, but also on the time it takes

clusters to form a bond. The two most prominent features of the universal behavior are reaction kinetics and cluster morphology. The former is described using Smoluchowski's coagulation equation [10] and the latter by considering clusters as fractal structures [11, 12].

The aggregation mechanism depends on the interaction between clusters. When an aggregation process is induced at a high salt concentration the repulsive potential is screened and highly localized around the particles. The electrostatic interaction is reduced at large distances, but keeps a large value for the short-range residual interaction. The basic question we will address is how does this residual interaction affect the aggregation kinetics. In order to control the residual potential, it is necessary to act directly on the particle surface charge density. This may be performed by modifying the degree of ionization of the surface ionic groups by changing the pH of the solvent.

In this paper we analyze the influence of the particle surface charge density on the aggregation kinetics in processes induced at high salt concentration. Thereby, the effect of the residual interaction on the aggregation mechanism is studied. The detailed cluster-size distribution of aggregating colloids of polystyrene microspheres was measured by single particle light scattering [13–20]. We carried out a series of measurements for different surface charge densities observing dynamic scaling in all the cases. From the scaling functions, parameters λ and μ of the Van Dongen and Ernst homogeneous kernels [21] were determined and features of the aggregation mechanisms were deduced.

Theory

We interpret our results using the well known Smoluchowsk's equation [10] which expresses, the time evolution of the cluster-size distribution $N_n(t)$ for diluted systems, in terms of the reactions kernels k_{ij}. The structure of this equation is simple, although it is difficult to know the form of the reaction kernel for a given system. Most coagulation kernels used in the literature are homogeneous functions of i and j [22–24]. Van Dongen and Ernst [21] introduced a classification scheme for homogeneous kernels, based on the relative probabilities of large clusters sticking to large clusters, and small clusters sticking to large clusters. In this theory it is assumed that, in the scaling limit, i.e., long times and large clusters, the cluster-size distribution has the form [21, 25, 26]:

$$N_n(t) \sim s^{-2} \Phi(n/s) \tag{1}$$

where $s(t)$ is related to the average cluster size and $\Phi(x)$ is a time-independent scaling distribution ($x \equiv n/s(t)$). Dependence of homogeneous kernels on i and j at large i and/or j, may be characterized by two exponents defined as follows:

$$k_{ai,aj} \sim a^\lambda k_{ij} \quad (\lambda \le 2) \tag{2}$$

$$k_{i,j} \sim i^\mu j^\nu \quad (\nu \le 1) \; i \ll j \; \lambda = \mu + \nu \,. \tag{3}$$

Kernels with either $\lambda > 2$ or $\nu > 1$ are unphysical, since the reactivity cannot increase faster than the cluster mass. No restrictions are imposed on μ.

The homogeneity parameter λ describes the tendency of a large cluster to join up with another large cluster and governs the overall rate of aggregation. λ takes the value 0 in DLCA, and 1 in the case of RLCA. For $\lambda > 1$ an infinite-size cluster forms in finite time and a gelation process occurs.

The exponent μ establishes the rate at which large clusters bind to small clusters and its sign determines the shape of the size distribution. For $\mu < 0$ the large cluster-small cluster unions are favored and large clusters gobble up small ones resulting in a bell-shaped size distribution. Kernels with negative μ appears to be a good description

for DLCA. When $\mu > 0$ large-large interaction dominates and size distributions resulting from such kernels tend to be polydisperse since small clusters may still exist even in the presence of large ones and the size distribution decays monotonously with increasing n.

Both $s(t)$ and $\Phi(x)$ in (1) contain the exponents λ and μ of the kernels [21, 27]. When $\lambda < 1$ the function $s(t)$ grows as a power law of the time, $s(t) \sim t^{1/(1-\lambda)}$ and when $\lambda = 1$ it increases exponentially in time. The function $s(t)$ is related to the number-average mean cluster size $\langle n_n \rangle \equiv M_1/M_0$ (where $M_i = \sum n^i N_n$) and the relationship depends on the sign of μ. For kernels with $\lambda < 1$ [27, 28]:

$$s(t) \sim \begin{cases} \langle n_n \rangle & \mu < 0 \\ \langle n_n \rangle^{\frac{1}{(2-\tau)}} & \tau < 1 + \lambda \quad \mu = 0 \\ \langle n_n \rangle^{\frac{1}{(1-\lambda)}} & \mu > 0 \end{cases} \tag{4}$$

The analytical form of $\Phi(x)$ is known for large and small x [21, 27]:

$$\Phi(x \gg 1) \sim x^{-\lambda} \exp(-\beta x) \tag{5}$$

$$\Phi(x \ll 1) \sim \begin{cases} \exp(-x^{-|\mu|}) & \mu < 0 \\ x^{-\tau} & \tau < \lambda + 1 \quad \mu = 0 \\ x^{-(1+\lambda)} & \mu > 0 \end{cases} \tag{6}$$

For large x, $\Phi(x)$ decreases and is related only to the exponent λ, where β is a fitting parameter. For small x, $\Phi(x)$ depends on the sign of μ. Therefore, when $\mu < 0$, $\Phi(x \ll 1)$ increases with x, while for $\mu \ge 0$, $\Phi(x \ll 1)$ decays monotonously. Thus, the shape of the function $\Phi(x)$ depends critically on the sign of μ. For kernels with $\mu < 0$, $\Phi(x)$ is a bell-shaped function, while for $\mu \ge 0$, it decreases monotonously [21].

The above formulae (4) and (6) imply the following predictions for the long time behavior of the cluster-size distribution $N_n(t)$ (1). Thus, when $\lambda < 1$:

$$\begin{cases} \dfrac{N_n(t)}{N_1(t)} \to \infty & \mu < 0 \\ N_n(t) \sim t^{-w} n^{-\tau} \quad w > 1 \; \tau < 1 + \lambda & \mu = 0 \\ N_n(t) \sim t^{-1} n^{-(1+\lambda)} & \mu > 0 \end{cases} \tag{7}$$

Material and methods

The single particle light scattering instrument

A single particle optical sizer was built in our laboratory [29, 30] based on the Pelssers et al. device [15]. In this technique, single clusters, insulated by hydrodynamic focusing of a colloidal dispersion, are forced to flow across a focused laser beam. A measurement of the cluster-size distribution is performed by analyzing the light intensity scattered by single clusters at low angle, where intensity is monotonously related to the square cluster's volume: $I_n(\theta)/I_1(\theta) = n^2 \sim V^2$. The instrument is basically a flow ultramicroscope in which pulses of light from single particles are detected. Light from a laser goes through an entrance optical system in order to create a homogeneously illuminated zone at the center of the flow cell, where cluster separation is performed. Single particles cross this illuminated zone, scattering pulses of light which are detected at low angle (between 2.7° and 3.2°) and focused it onto a photomultiplier. The output signal is converted into voltage and digitalized. A computer controls an analog to digital converter board, and recognizes, classifies and counts the pulses by running an algorithm on-line. Size distribution is obtained by representing counts versus pulse intensity.

Figure 1 shows frequency histogram of an aggregating latex dispersion (Sekisui-NA1A19E). Peaks correspond to different cluster-sizes from singlets to hexaplets. The relationship $I \sim V^{(0.85 \pm 0.04)}$, between scattered light intensity and the volume of the clusters was found, which is close to Rayleigh's prediction, $I \sim V^2$. The cluster-size distribution is determined by integrating the peaks appearing in the histograms, and the time evolution is obtained by analyzing a consecutive series of the latter. The zero moment of the cluster-size distribution, $M_0 = \sum N_n$, is measured by adding the total number of counts in a histogram, including off-scale pulses.

Experimental system and experiment details

The experimental system was a polystyrene latex (Sekisui-N1A19E) with negative surface charge due to sulphate groups. The radius of the spherical microspheres, 292 ± 19 nm, was determined by transmission electron microscopy. The latex particles were cleaned by centrifugation and then by ion-exchange over a mixed bed. Buffers were used to set the pH of the samples: acetate at pH 4 and borate at pH 9. All chemicals used were of A. R. quality and twice-distilled water was purified using Millipore equipment. The critical coagulation concentration (c.c.c) was determined directly by dilution of latex particles in solutions of different salt concentration. The electrophoretic mobility of the particles, μ_e, was measured using a Zeta–Sizer IIc device (Malvern Instruments, U.K.) to control the modification in charge produced by pH. μ_e was determined as a function of pH at 0.015 M of ionic concentration. In every case it was found that $\mu_e < 0$, which confirmed the negative sign of the surface charge. Mobility was converted into ζ-potential using Smoluchowski's theory [31–32]. The electrokinetic surface charge, σ_{ek}, was calculated using the Gouy–Chapman

Fig. 1 Typical frequency histogram of an aggregating sample. Clusters from monomers up to hexaplets are shown. The last channel (not shown) records the total number of off-scale pulses, enabling M_0 to be determined. The variation of the scattered light intensity with aggregation number (cluster's volume) is plotted at the upper right hand corner. The best fit leads to $I \sim V^{(1.85 \pm 0.04)}$

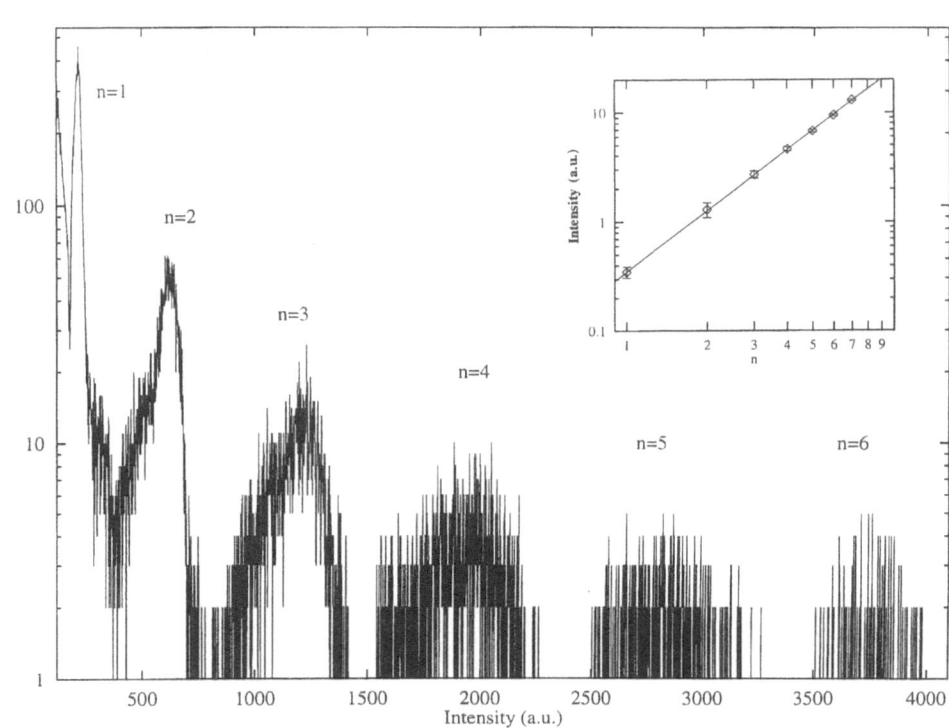

Table 1 Electrophoretic mobility, electrokinetic surface charge density and stability of the colloidal particles

pH	$-\mu_e/10^{-8}$ $(m^2V^{-1}s^{-1})$	$-\sigma_{ek}/10^{-2}$ (C/m^2)	c.c.c (M)	W
4.2 ± 0.1	2.1 ± 0.2	0.8 ± 0.2	≈ 0.10	1.9 ± 0.2
9.3 ± 0.1	3.7 ± 0.2	1.5 ± 0.2	≈ 0.35	3.2 ± 0.4

model for the electric double layer (e.d.l) [31]. Table 1 shows the two experimental cases studied. At pH = 4.2 the electrokinetic charge is 0.8×10^{-2} C m^{-2} and at pH = 9.3 the charge is larger by a further factor of 2. Table 1 also shows the critical KCl concentration of the suspensions and the stability factor ($W = k^{Brown}/k^{Exp}$) obtained using the single particle instrument.

Prior to undertaking our studies, fresh suspensions of microspheres were sonicated for 30 min to break-up any initial clusters. Aggregation was initiated by mixing the microspheres suspended in saltfree water and the aggregation agent. Aggregations were induced at high ionic concentration; 0.5 M of KCl, higher than the c.c.c. Immediately afterwards, the timer was started.

The aggregation experiments were carried out in a reaction vessel at 20 ± 1 °C and an initial particle concentration of $N_0 = 4.0 \times 10^8$ cm^{-3}. Precautions were taken so that nondestructive size distribution analysis could be

peformed. Control measurements under different stress conditions were carried out to test the cluster break-up in the focusing cell. From these we conclude that break-up does not affect our experiments. Small portions of aggregating colloid were slowly taken from the suspensions through a wide aperture pipette. Samples were diluted in the same solvent used for the dispersions in order to stop the aggregation during the measurements. Moreover, the particle concentration was optimized to ensure single particle detection ($\approx 10^7$ cm^{-3}).

Results and discussion

We analyze our data using dynamic scaling formalism. The number-average mean cluster size $\langle n_n \rangle = N_0/M_0(t)$ was determined. This is related to $s(t)$ and, together with the distribution $N_n(t)$, allows the scaling function $\Phi(x)$ to be determined.

Figure 2 shows the time dependence of $\langle n_n \rangle$. For the two pH $\langle n_n \rangle$ increases linearly in time, although for pH = 9.3 the kinetics is slower than at pH = 4.2 which agrees with the higher stability of the system (see Table 1).

The relationship between $\langle n_n \rangle$ and the scaling function $s(t)$ depends on the sign of μ (4). In order to determine this sign we plotted N_n vs n (Fig. 3), observing for pH = 4.2

Fig. 2 Time evolution of the number-average mean cluster size. In both cases, the mean cluster size grows linearly at long times. For pH = 9.3 the growth is slower than for pH = 9.3, which agrees with the stability measurements

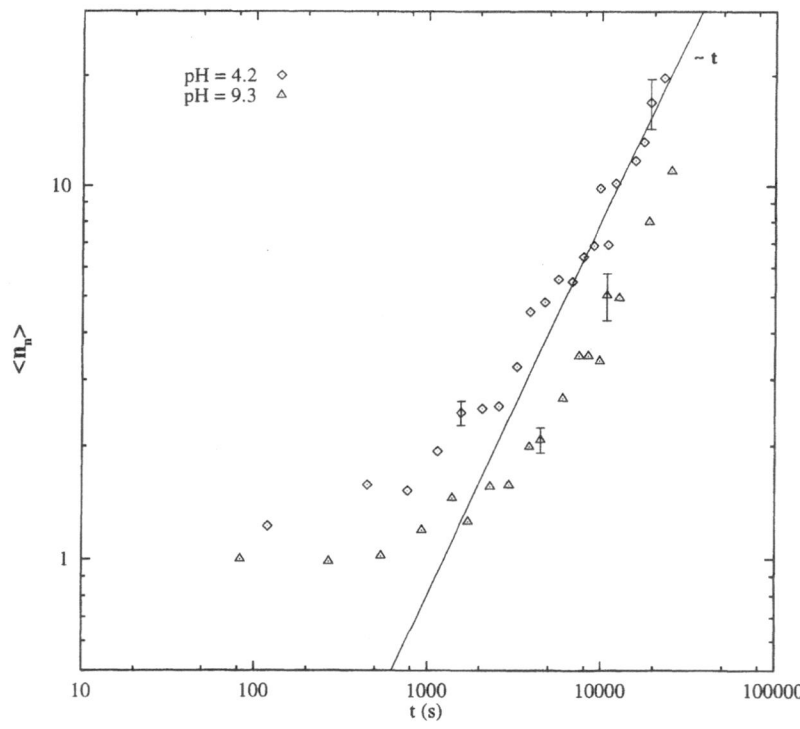

Fig. 3 Plot of $N_n(t)$ versus n for different times. For pH = 4.2 a peak is developed at long times, which indicates $\mu < 0$. For pH = 9.3 a power-law decay is shown, which suggests that $\mu \geq 0$

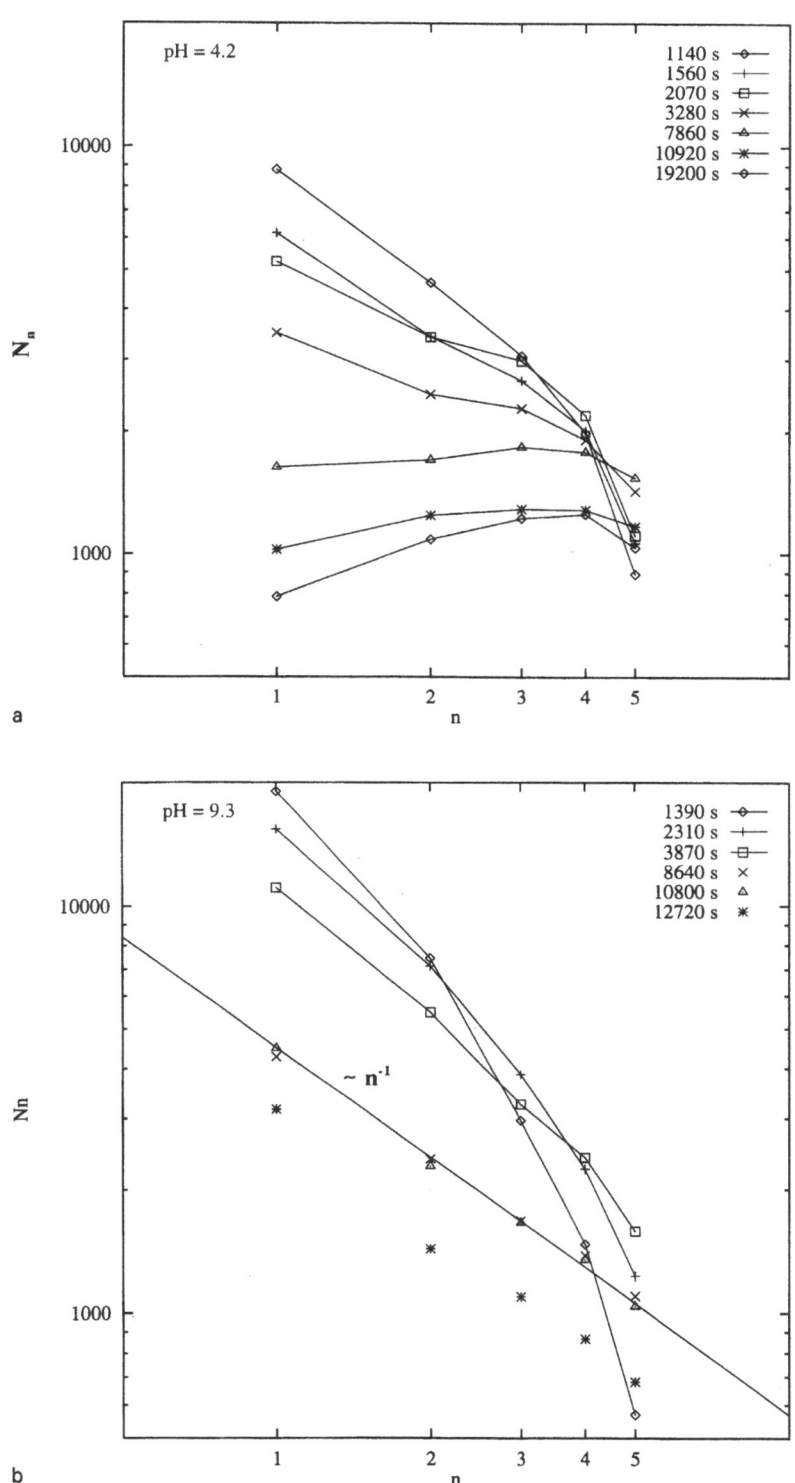

a

b

a bell-shaped curve when the aggregation time increases and then $\mu < 0$ (7). Thus, the relationship $s(t) \sim \langle n_n \rangle$ is satisfied (4). Depletion of small clusters is a characteristic of DLCA. Nevertheless, at pH = 9.3 (Fig. 3b), the distribution exhibits a power-law decay in n at long times, given by $N_n \sim n^{-1}$, which suggest $\mu \geq 0$ (7). The power-law decay is a characteristic of RLCA [33]. In order to determine whether μ is greater than or equal to zero, the long time

78

A. Fernández-Barbero et al.
Dynamic scaling in colloidal fractal aggregation

dependence of the cluster-size distribution was determined to be $N_n(t) \sim t^{-(0.8 \pm 0.1)}$. For kernels with $\mu = 0$ theory predicts $N_n(t) \sim t^{-w}$ with $w > 1$ (7). Thus, the exponent (0.8 ± 0.1), close to one, suggests that $\mu > 0$. For kernels with $\mu > 0$, the coagulation equation also predicts $N_n \sim n^{-(1 + \lambda)}$ (7) and thus, from the experimental exponent of n we find $\lambda \approx 0$. Consequently, the relationship $s(t) \sim \langle n_n \rangle$ is also satisfied at pH = 9.3 when the value

Fig. 4 Experimental $\Phi(x)$ distributions up to $t \approx 12t_{agg}$ for pH = 4.2 and $t \approx 14t_{agg}$ for pH = 9.3. The distributions are aligned for different times, showing dynamic scaling. For pH = 4.2 the bell-shaped curve confirms the negative sign of μ. For pH = 9.3 $\Phi(x)$ decays monotonously, demonstrating that the sign of μ has changed to $\mu \geq 0$. For $x > 1$, a decreasing exponential behavior is shown in both cases (solid lines), which suggests $\lambda \approx 0$

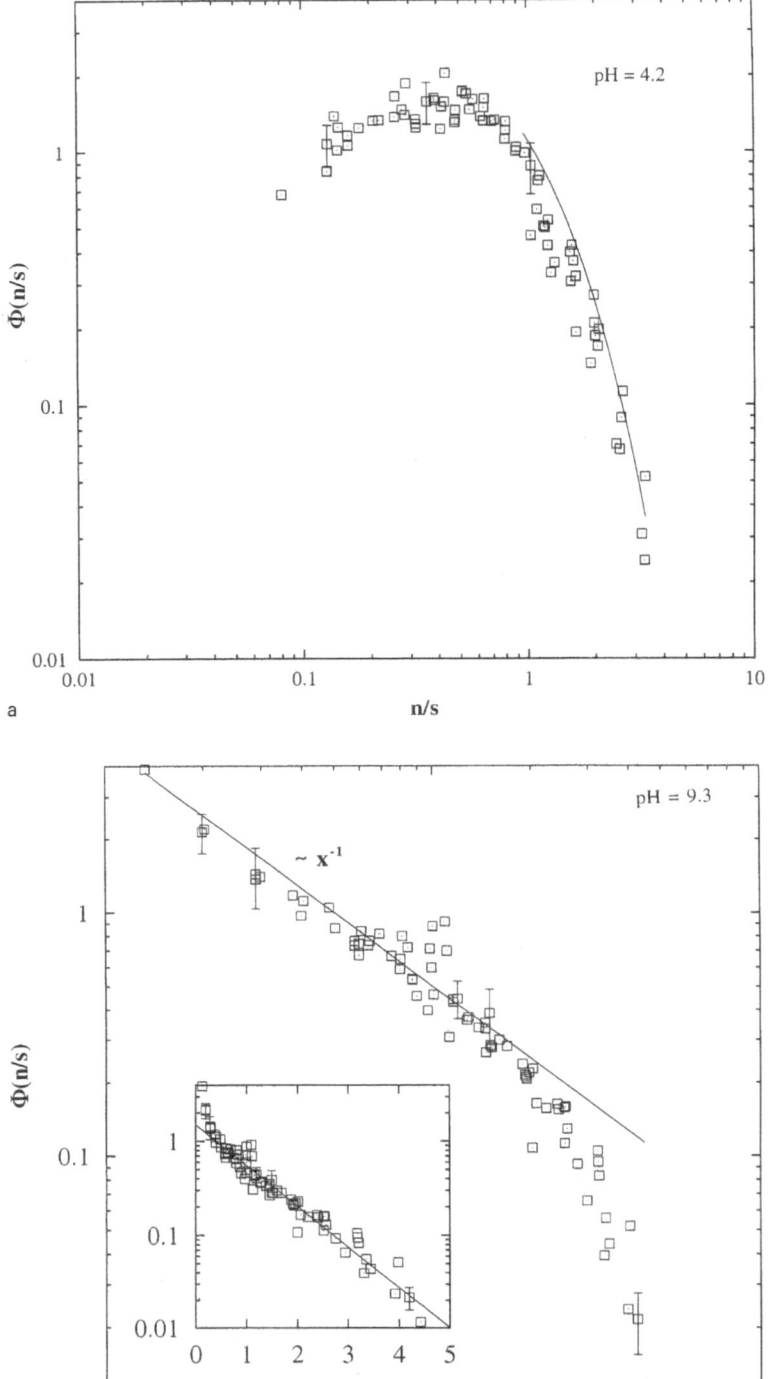

$\lambda = 0$ is used in Eq. (4). Therefore, for every case, $s(t)$ shall be identified with $\langle n_n \rangle$, and thus, the relationship $\langle n_n \rangle \sim t^{1/(1-\lambda)}$, when $\lambda = 0$, reproduces the experimental results, shown in Fig. 2.

$\Phi(x)$ was determined by plotting $s^2(t) N_n(t)/N_0$ versus $x \equiv n/s(t)$ for different times. For aggregation times longer than the characteristic aggregation time, $t_{agg} \equiv 2/N_0 k_s^{exp}$, the data collapses along a single master curve (Fig. 4). These results demonstrate that the scaling approach is appropriate for the experimental data. It is somewhat surprising since the data correponding to $t \approx t_{agg}$ does not represent the scaling limes.

For pH = 4.2 a bell-shaped curve appears, which confirms the negative sign of μ (6). At pH = 9.3, $\Phi(x)$ decays monotonously which confirms that $\mu \geq 0$ (6), and from the time dependence of the cluster-size distribution, we conclude that $\mu > 0$. Figure 4b shows that $\Phi(x < 1) \sim x^{-1}$. For kernels with $\mu > 0$, the theory predicts, $\Phi(x \ll 1) \sim x^{-(1+\lambda)}$ (6). Thus, our result leads to $\lambda \approx 0$, which confirms the value directly obtained from the cluster-size distribution (Fig. 3b).

A decreasing exponential behavior for $x > 1$ was found for all cases. Theory predicts for $x \gg 1$ a μ-independent behavior given by (5). Several values of λ were tried, concluding that the best fitting leads to $\lambda \approx 0$, which is in line with the value obtained above.

To sum up, our results indicate that in all the processes, induced at high salt concentrations, the homogeneity parameter has a value of zero. This result is in line with DLCA experiments [3, 5, 6], [34–40]. However, for the lower surface charge (pH = 4.2), the exponent μ is negative. This result is also in line with DLCA experiments [27, 34]. When the charge increases the sign of μ changes to positive which is in line with measurements of slow aggregation [6]. Table 2 shows the stability data and the kinetics characteristic for the two cases.

The original Smoluchowski Brownian kernel [10] which has $\lambda = 0$ and negative μ, can be used for describing our experimental results for pH = 4.2. Recently, Thorn et al. [41] have used a stochastic simulation method to recover the scaling behavior of microsphere colloids, finding for the DLCA simulation a bell-shaped $\Phi(x)$ curve which agrees with out experimental results.

For pH = 9.3, μ is positive and the Brownian kernel fails to fit the experimental results. The positive sign of μ is interpreted by a greater reactivity between the larger-sized clusters when the surface charge density increases. The power-law decay in the cluster-size distributions ($\mu \geq 0$) (Figs. 3b and 4b), the asymptotic time dependence ($\mu > 0$)

Table 2 Summary of stability data and long time kinetics

pH	W	$\langle n_n \rangle(t)$	$N_n(t)$	$\Phi(x)$	λ	μ	Regime
4.2 ± 0.1	1.9 ± 0.2	$\sim t$	peaks	peaks	0	<0	DLCA
9.3 ± 0.1	3.2 ± 0.4	$\sim t$	decays	decays	0	>0	Intermediate

and the stability data suggest that in this case the process is not DLCA. On the other hand, the value of λ, close to zero, that we have found does not indicate slow aggregation, and it is far from the exponent $\lambda = 1$ predicted analytically for RLCA kernels [33]. Probably, in this case, we are observing an intermediate regime between DLCA and RLCA. The stability data and the kinetics of the process indicate that in this case the residual interaction between particles controls the aggregation mechanism.

Conclusions

The temporal evolution of cluster-size distributions has been measured using a single particle light scattering instrument. Measurements were carried out at a high salt concentration and different surface charge densities. In all cases, we found that distributions exhibit dynamic scaling for times longer than t_{agg}.

For particles with low surface charge density, the scaled size-distribution is bell-shaped. In this case, the kernel is characterized by the parameters $\lambda \approx 0$ and $\mu < 0$, which agrees with the value of the exponents for the Brownian kernel and with results obtained by simulation using this kernel. Thus, for low surface charge densities, our results are in accord with DLCA. Large clusters bind preferentially to small clusters.

When the particle surface charge density increases, we find that the distribution exhibits a power-law decay, which demonstrates that the aggregation mechanism changes. In this case we found $\lambda \approx 0$ and $\mu > 0$. Thus, the residual interaction controls the aggregation, and when this interaction increases, large clusters bind preferentially to large clusters. This result suggests that the residual electrostatic interaction could control the reactivity between large and small clusters.

Acknowledgments We thank M. Cohen Stuart for helpful discussions when the single particle detection instrument was constructed. This work was supported by CICYT (Spain), Project MAT 94-0560.

80

A. Fernández-Barbero et al.
Dynamic scaling in colloidal fractal aggregation

References

1. Family F, Landau DP (eds) (1984) Kinetics of Aggregation and Gelation. North-Holland
2. Martin JE, Wilcoxon JP, Schaefer D, Odinek J (1990) Phys Rev A 41:4379
3. Bolle G, Cametti C, Codasefano P, Tartaglia P (1987) Phys Rev A 35:837
4. Lin HY, Lindsay HM, Weitz DA, Ball RC, Klein R, Meakin P (1990) Phys Rev A 41:2005
5. Asnaghi D, Carpineti M, Giglio M, Sozzi M (1992) Phys Rev A 45:1018
6. Broide ML, Cohen RJ (1990) Phys Rev Lett 64:2026
7. Carpineti M, Giglio M (1993) Adv Colloid Interface Sci 46:73
8. Stankiewicz J, Cabrerizo-Vílchez M, Hidalgo-Alvarez R (1993) Phys Rev E 47:2663
9. Jullien R (1987) Comments Cond Mat Phys 13:177
10. Von Smoluchowski M (1917) Z Phys Chem 92:129
11. Vicsek T (1992) Fractal Growth Phenomena, Word Scientific
12. Mandelbrot BB (1982) The Fractal Geometry of Nature, Freeman WH
13. Cahill J, Cummins PG, Staples EJ, Thompson LG (1987) J Colloid Interface Sci 117:406
14. McFadyen P, Smith AL (1973) J Colloid Interface Sci 45:573
15. Pelssers EGM, Stuart MCA, Fleer GJ (1990) J Colloid Interface Sci 137: 350, 362
16. Walsh DJ, Anderson J, Parker A, Dix MJ (1981) Colloid Polym Sci 259:1003
17. Buske N, Gedan H, Lichtenfeld H, Katz W, Sonntag H (1980) Colloid Polym Sci 258:1303
18. Beyer GL (1987) J Colloid Interface Sci 118:137
19. Bowen MS, Broide ML, Cohen RJ (1985) J Colloid Interface Sci 105: 605, 617
20. Bartholdi M, Salzman GC, Hielbert RD, Kerker M (1980) Appl Opt 19:1573
21. Van Dongen PGJ, Ernst MH (1985) Phys Rev Lett 54:1396
22. Leyvraz F, Tschudi HR (1982) J Phys A 15:1951
23. Ziff RM, Ernst MH, Hendriks EM (1983) J Phys A 16:2293
24. Hendriks EM, Ernst MH, Ziff RM (1983) J Stat Phys 31:519
25. Lishnikov AA (1973) J Colloid Interface Sci 45:549
26. Ernst MH (1986) In: Pietronero L, Tosatti E (eds) Fractals in Physics. North-Holland
27. Broide ML, Cohen RJ (1992) J Colloid Interface Sci 153:493
28. Taylor TW, Sorensen CM (1987) Phys Rev A 36:5415
29. Fernández Barbero A (1994) PhD Thesis, Universidad de Granada
30. Fernández-Barbero A, Schmitt A, Cabrerizo-Vílchez M, Martínez-García R, Physica A (to be published)
31. See for example, Hidalgo-Álvarez R (1992) Adv Colloid Interface Sci 34:217
32. Fernández-Barbero A, Martín-Rodríguez A, Callejas-Fernández J, Hidalgo-Álvarez R (1994) J Colloid Interface Sci 162:257
33. Ball RC, Weitz DA, Witten TA, Leyvraz F (1987) Phys Rev Lett 58:274
34. Pefferkorn E, Varoqui R (1989) J Chem Phys 91:5679
35. Weitz DA, Lin MY (1986) Phys Rev Lett 57:2037
36. Aubert C, Cannel DS (1986) Phys Rev Lett 56:738
37. Lin HY, Lindsay HM, Weitz DA, Ball RC, Klein R, Meakin P (1989) Nature 339:360
38. Zhou Z, Chu B (1991) J Colloid Interface Sci 143:356
39. Carpineti M, Ferri F, Giglio M, Paganini E, Perini U (1990) Phys Rev A 42:7347
40. Lin HY, Lindsay HM, Weitz DA, Klein R, Ball RC, Meakin P (1990) J Phys Condens Matter 2:3093
41. Thorn M, Seesselberg M (1994) Phys Rev Lett 72:3622

Progr Colloid Polym Sci (1996) 100:81–86
© Steinkopff Verlag 1996

COLLOIDAL PARTICLES

Effects of shear flow on a critical colloidal dispersion: a light scattering study

H. Verduin
J.K.G. Dhont

H. Verduin · J.K.G. Dhont (✉)
Van't Hoff Laboratory
Utrecht University
Padualaan 8
3584 CH Utrecht, The Netherlands

Abstract Spherical colloidal silica particles coated with stearyl alcohol and dissolved in benzene show a temperature dependent attraction which gives rise to a gas–liquid critical-point. The attraction is due to the fact that benzene is a marginal solvent for the stearyl alcohol.

Close to the critical-point, spatially extended structures exist, which are characterized by the correlation length. A shear flow severely distorts these extended structures and, as a result, the microstructure of the sheared dispersion will be highly anisotropic. This will affect light scattering properties, such as for instance the turbidity.

Theory predicts that in the mean field region the effect of shear is characterized by a dimensionless group λ that is proportional to $\dot{\gamma}\xi^4$ ($\dot{\gamma}$ is the shear rate and ξ is the correlation length of the quiescent dispersion at the given temperature). Furthermore, the turbidity in the sheared system minus the turbidity in the quiescent system can be described with a master curve which is only dependent on the parameter λ.

We report on light scattering results which are used to test the above-mentioned theory. The experiments include small-angle light scattering, in order to obtain the temperature dependent equilibrium correlation length and turbidity measurements as function of the temperature and shear rate. The turbidity results confirm the theoretically predicted λ-dependence and are satisfactorily described by the λ-dependent mastercurve, although there is some discrepancy between an experimentally obtained pre-factor and its theoretically expected value.

Key words Colloidal dispersion – shear flow – critical behavior – small-angle light scattering – turbidity

Introduction

Two well separated particles attain a large relative velocity when subjected to shear flow. With increasing distance (in the gradient direction), a smaller shear rate is sufficient to sustain a given relative velocity. For these large separations, diffusion is never fast enough to restore the shear induced displacements. As a result, extended structures are severely affected by shear flow, even for small shear rates.

In particular, close to the gas–liquid critical point, where microstructures exist with linear dimensions of the (ultimately diverging) correlation length, strong influence of weak shear flows is to be expected. This effect is even enhanced, due to a decrease of the diffusion coefficient, commonly referred to as critical slowing down. The shear rate dependence of the turbidity reflects, in an integrated form, the effect of shear flow on the long-range microstructure of a suspension. This shear rate dependence is singular at the critical point. That is, an *infinitesimally* small shear rate is then sufficient to induce a *finite* effect on the turbidity, due to the *infinite* extent of microstructures.

Theory [1] predicts that the effects of temperature and shear flow (characterized by the shear rate) on the turbidity can be described with a function which is dependent on a single dimensionless group $\lambda \sim \dot{\gamma}\xi^4$ ($\dot{\gamma}$ is the shear rate and ξ is the correlation length in the equilibrium system, which in turn is a function of the temperature difference from the critical point). The theory is valid in the mean field region. Theories for molecular systems beyond mean field [2, 3] predict a $\lambda \sim \dot{\gamma}\xi^3$ dependence.

In this paper we present shear rate and temperature dependent turbidity measurements and compare these to the theoretically derived relation (see also ref. [4]). As a colloidal model system we use stearyl alcohol coated silica spheres, dispersed in benzene which is a marginal solvent for the stearyl alcohol chains. This dispersion shows an upper critical point. We are not aware of shear flow experiments performed on critical colloidal dispersions. Experiments of this kind are performed on critical fluids [5] and indicate the same λ-dependence as for colloidal systems.

This paper is organized as follows. In the theoretical section we will summarize the theoretical approach to obtain the expression for the shear rate dependent turbidity and discuss how the correlation length can be derived from the equilibrium structure factor. The experimental section contains a description of the colloidal system and the set up used for the measurement of the small-angle critical part of the structure factor and of the shear rate dependence of the turbidity. In the subsequent section the results are presented and discussed. We close with conclusions.

Theory

Effect of a stationary shear flow on microstructure

The influence of a shear flow on the microstructure of a critical colloidal system can be calculated from the equation of motion for the pair-correlation function. This equation is obtained from the N-particle Smoluchowski equation by integration over all, except for two position coordinates [1]. Disregarding hydrodynamic interactions, one obtains the following (stationary) equation of motion for the pair-correlation function g,

$$\frac{\partial}{\partial t} g(\mathbf{R}|\dot{\gamma}) = 0 = 2D_0 \left[\nabla^2 g(\mathbf{R}|\dot{\gamma}) + \beta \nabla \cdot g(\mathbf{R}|\dot{\gamma}) \left\{ \nabla V(R) \right. \right.$$
$$\left. \left. + \bar{\rho} \int d\mathbf{r} [\nabla_r V(r)] \frac{g_3(\mathbf{r}, \mathbf{R}|\dot{\gamma})}{g(\mathbf{R}|\dot{\gamma})} \right\} \right]$$
$$- \nabla \cdot (g(\mathbf{R}|\dot{\gamma}) \Gamma \mathbf{R}), \qquad (1)$$

where D_0 is the Stokes–Einstein diffusion coefficient, $\beta = 1/k_B T$, k_B is Boltzmann's constant and T the temperature, V is the pair-potential, $\mathbf{R}$ denotes the separation between two particles, $\bar{\rho}$ is the number density N/V and Γ is the velocity gradient tensor, for which we will take the following form,

$$\Gamma = \dot{\gamma} \begin{pmatrix} 0 & 1 & 0 \\ 0 & 0 & 0 \\ 0 & 0 & 0 \end{pmatrix}, \qquad (2)$$

with $\dot{\gamma}$ the shear rate. This choice corresponds to a flow in the x-direction with its gradient in the y-direction.

A detailed discussion on the analysis of Eq. (1) is beyond the scope of this paper and can be found in ref. [1]. We will only sketch several steps for the derivation of an expression for the turbidity from Eq. (1). The first step is to use an appropriate expression for the three-particle correlation function g_3 which is essential for the description of long-range correlations. We used an improved version of the superposition approximation, which results in a divergent correlation length as the critical point is reached, as it should. Secondly the asymptotic solution of Eq. (1) for large distances is obtained by linearization with respect to the total correlation function $h = g - 1$. Fourier transformation then results in an expression for the structure factor which in turn is used to calculate the turbidity by integration over all possible angles. It turns out that the turbidity in a sheared critical colloidal dispersion minus the turbidity in the equilibrium system can be described by a single scaling function $T(\lambda)$,

$$\tau(\dot{\gamma}) - \tau^{eq} = \frac{C}{(k_0 R_V)^2} \frac{1}{(\beta \Sigma / R_V^2)} T(\lambda), \qquad (3)$$

with the dimensionless parameter λ defined as follows,

$$\lambda = \frac{Pe^0(\dot{\gamma})}{(\xi^{-1} R_V)^4 (\beta \Sigma / R_V^2)}. \qquad (4)$$

In Eqs. (3) and (4) $k_0 = 2\pi/\lambda_s$ with λ_s the wavelength of the light in the solvent, R_V is the range of the pair potential, Σ is a well behaved function of the density and the temperature and is related to the Cahn–Hilliard square gradient coefficient, Pe^0 equals $\dot{\gamma} R_V^2 / 2D_0$ and ξ is the correlation length in the equilibrium system.

The function $T(\lambda)$ can be calculated from the structure factor by numerical integration and is plotted in Fig. 1. It shows that an increase in the parameter λ results in a larger effect on the turbidity. An important feature is the proportionality of λ with $\dot{\gamma}\xi^4$. This means that close to the critical point, where the correlation length diverges, a small shear rate is sufficient to cause a large effect. The constant C appearing in Eq. (3) is related to the optical

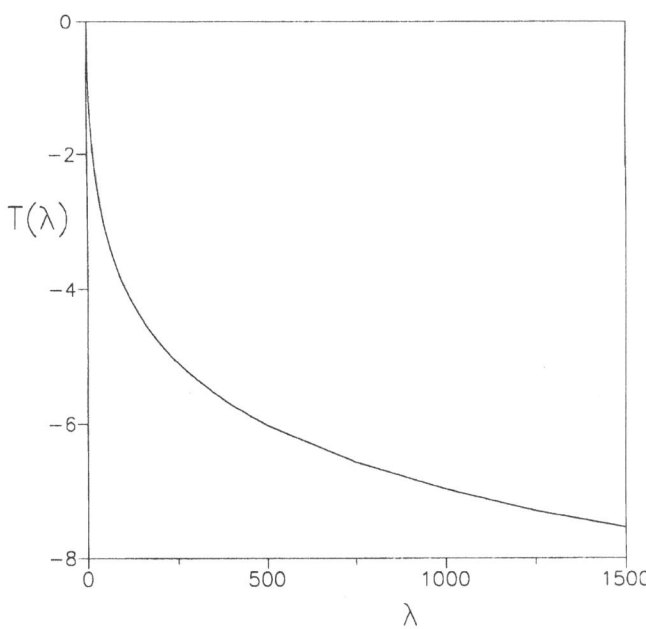

Fig. 1 The turbidity scaling function

contrast between the particles and the solvent,

$$C = \frac{4\pi^4 k_0^4}{(2\pi)^6 \varepsilon_s^2} \bar{\rho} \left[\int_{V_d} dr (\varepsilon(r) - \varepsilon_s) \right]^2, \tag{5}$$

where V_d is the volume of a colloidal particle and $\varepsilon(r)$ and ε_s are the dielectric constants of the particle and the solvent, respectively.

Equation (3, 4) are the relations we wish to verify experimentally. Measurements of $\tau(\dot{\gamma}) - \tau^{eq}$ plotted as a function of $\dot{\gamma}\xi^4$ should all collapse on a single curve. Notice that identical numerical values of $\dot{\gamma}\xi^4$ may correspond to different shear rates and temperatures. Moreover, when proportionality factors relating $\tau(\dot{\gamma}) - \tau^{eq}$ to $T(\lambda)$ in Eq. (3), and relating $\dot{\gamma}\xi^4$ to λ in Eq. (4) are known, all the data points should coincide with the master curve plotted in Fig. 1. To perform this mapping we must determine the temperature dependence of the correlation length ξ in the equilibrium system and a value for Σ. In the next we will discuss how to obtain these values.

Determination of the correlation length

The correlation length and the value for Σ can be obtained from the equilibrium structure factor which can be calculated from Eq. (1) and has the well known Ornstein–Zernike form,

$$S^{eq}(k) = \frac{(\beta\Sigma)^{-1} + \xi^{-2}}{k^2 + \xi^{-2}}. \tag{6}$$

The measured scattered intensity is directly proportional to $S^{eq}(k)$. Our interest is in the critical short wavevector contribution to the scattered intensity. To correct for non-critical contributions to the small wavevector scattered intensity, resulting from short-ranged interactions, the non-critical zero-wavevector contribution to the structure factor, $S_{nc}^{eq}(0)$, is subtracted from Eq. (6). In an experiment this is done in approximation, by subtraction of the high-temperature intensities from the measured low-temperature intensities, yielding a net intensity. The reciprocal net intensity is given by,

$$I_{net}^{-1} \propto \frac{k^2 + \xi^{-2}}{(\beta\Sigma)^{-1} - S_{nc}^{eq}(0)\xi^{-2} - (S_{nc}^{eq}(0) - 1)k^2}. \tag{7}$$

For small wavevectors, $(k\xi^{-2})^2 \ll 1$, the quotient of the intercept and slope of a plot of the reciprocal net intensity versus k^2 yields ξ^{-2}. Moreover, from a plot of the reciprocal slopes versus ξ^{-2}, a value for $(\beta\Sigma) \cdot S_{nc}^{eq}(0)$ can be extracted. Estimating of $S_{nc}^{eq}(0) \approx 1$ then yields an estimate for $\beta\Sigma$, which quantity is directly proportional to the Cahn–Hilliard square gradient coefficient.

The temperature dependent correlation lengths are fitted to the following equation,

$$\xi = \xi_0 \left(\frac{T - T_C}{T_C} \right)^{-\nu}$$

with ξ_0 a prefactor, T_C the critical temperature and ν the critical exponent for the correlation length. The critical exponent is 1/2 in the mean field region and 0.625 beyond mean field.

Experimental

The colloidal system

The colloidal system consists of spherical silica particles coated with stearyl alcohol [6]. The radius determined in cyclohexane at low concentration with dynamic and static light scattering is 39 ± 1 nm. The size polydispersity is about 12%. The specific volume (q) was determined from the relative viscosity, η_r, at low mass concentration. Using Einstein's law, which states that $\eta_r = 1 + 2.5\phi$, with ϕ the volume fraction ($= q \cdot c$, with c the mass concentration), we obtained $q = 0.69$ cm³/g. This value was used to calculate volume fractions from known mass concentrations. Mass concentrations were determined by drying a known volume of the silica dispersion for several hours until no change in the mass was recorded.

The interaction between the particles depends on the quality of the solvent. Dissolved in benzene, which is a poor solvent for the stearyl alcohol coating, an attractive interaction is induced which increases with decreasing temperature. Due to the short length of the stearyl alcohol chains (about 2 nm.) the attractive part of the pair potential is of short range. When the temperature is lowered significantly, the attractive forces induce phase separation.

The determination of the phase diagram of this colloidal system is extensively described in ref. [7]. Three different types of phase transition lines are found: the binodal, spinodal and the gelline. For volume fractions smaller than 0.19, the gel transition is located below the binodal and spinodal. At volume fractions larger than 0.19, the gel line masks the binodal and spinodal. We observed an upper-critical point at a volume fraction of 0.19. All experiments described here were performed at this volume fraction. The critical temperature obtained from cloud point measurements is 17.95 °C.

Structure factor measurements

The equilibrium structure factor is obtained from small-angle light scattering experiments. For detection we used a diode camera consisting of 512 diodes with a dynamic resolution of 10^{14} and an accuracy of 5% (EG&G/PARC model 1452A). To improve on the accuracy, the intensities of 10 adjacent diodes were averaged. The camera's diode chip covered an angle range from 2.9° to 7.4°. As a light source we used a He–Ne laser with a wavelength of 632.8 nm. The experimental ka-range is thus 0.029 to 0.075 (k is the wavevector and a is the particle radius). The camera is interfaced to a microcomputer for data control.

To suppress effects of multiple scattering, the sample was sealed in a thin flat 0.2 mm cuvette. The samples were made dust free by centrifugation for 10 min at a speed of 2000 rpm. The cuvette was immersed in a thermostated toluene bath with an optical cylindric glass wall, the temperature of which was measured with a Pt-100 element. Scattering angle dependent intensities were measured at 20 different temperatures ranging from 19° to 18°C.

Shear rate dependence of the turbidity

A temperature-controlled Couette type optical shear cell [8] was used for the turbidity measurements. A He–Ne laser beam (632.8 nm) was directed through the center of the inner cylinder perpendicular to the flow and vorticity direction. On both sides of the cell the laser beam passed a pinhole to ensure that no scattered light at very small angles is detected. The intensity of the laser beam was regulated by means of two polarization filters. A photo cell was used for detection.

At each different shear rate the transmittance for various temperatures was measured. In the low shear rate regime more data were taken, as large effects are expected for small shear rates close to the critical point. From the solvent corrected transmittance data, the turbidity for each shear rate and temperature was then calculated from Lambert–Beers law.

The temperature dependent zero shear turbidity was obtained by slowly cooling down the dispersion at a rate of 0.25 °C/h, starting at a temperature of 21° down to 17 °C. We verified that the rate of cooling does not affect the measured temperature dependence of the turbidity.

Results

Equilibrium correlation length

In Fig. 2 the experimental reciprocal net intensities are plotted as a function of k^2, for ten different temperatures. The solid lines in the figure represent linear fits. A few temperatures are omitted from this plot for clarity. According to Eq. (7), for small k-values, the quotient of the slope and the intersect equals the squared inverse correlation length, ξ^{-2}. The values for the correlation lengths thus obtained are plotted in Fig. 3 as function of the temperature on a double logarithmic scale. The drawn line is a linear fit according to Eq. (8) with a critical temperature of 17.95 °C. We obtain for the critical exponent, $v = 0.522 \pm 0.023$, which corresponds to the expected mean-field value of 1/2. For the prefactor ξ_0 we obtain a value of 190 ± 10 nm, which seems quite plausible in view of the diameter of the particles, which is 78 nm. The value for $\beta\Sigma$, scaled on R_V^2, is found to be equal to

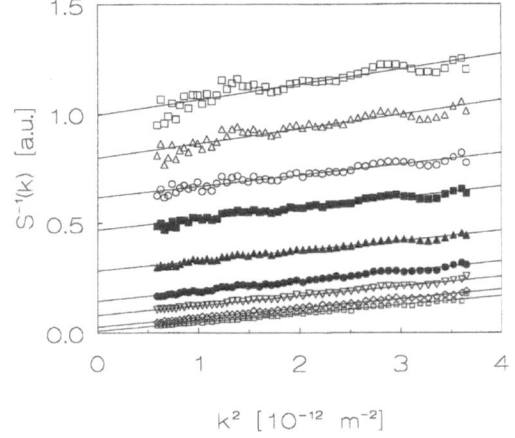

Fig. 2 Reciprocal structure factors plotted as a function of k^2. The quotient of the intercept and slope gives ξ^{-2}. The different temperatures are (from bottom to top): 17.98°, 18.05°, 18.10°, 18.14°, 18.24°, 18.34°, 18.45°, 18.56°, and 18.66 °C

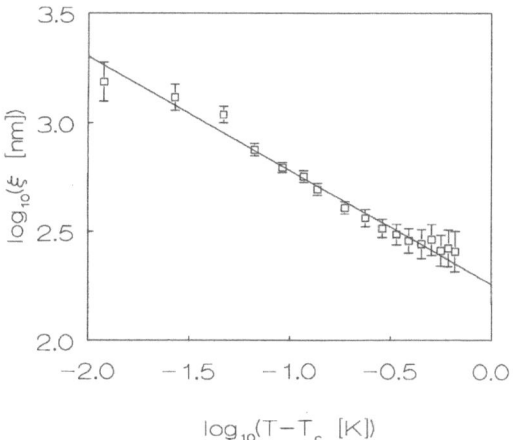

Fig. 3 Correlation lengths obtained from the reciprocal structure factor (see Fig. 2). The solid curve corresponds to a critical exponent of 0.522

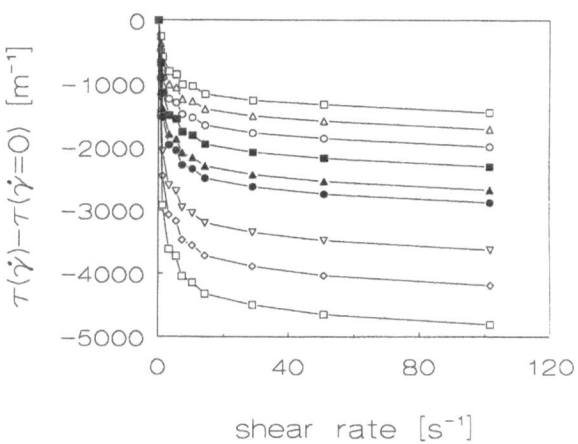

Fig. 4 The turbidity in the system minus the turbidity in the equilibrium state plotted as function of the shear rate. The different temperatures equal (from top to bottom): 18.41°, 18.35°, 18.30°, 18.25°, 18.21°, 18.18°, 18.10°, 18.06°, and 18.01 °C

3.60 ± 1.10. The range R_V of the pair potential is taken equal to the particle diameter. This estimated value for $\beta \Sigma / R_V^2$ is larger than the theoretical order of magnitude estimate of 0.1. This is not a bad correspondence, in view of the fact that both the experimental and theoretical values are crude estimates.

Shear rate dependence of the turbidity

In Fig. 4 the difference of the turbidity in the sheared and quiescent system is plotted as a function of the shear rate for several temperatures. The effect of shear flow is small at the higher temperatures but increases dramatically on approach of the critical point. In that case only a very small shear rate is needed to cause a significant decrease of the turbidity.

In Fig. 5 the data of Fig. 4 are scaled on the turbidity scaling function (Fig. 1). To this end, all the turbidity data are plotted as a function of $\dot{\gamma} \xi^4$. The two proportional constants which relate $\tau(\dot{\gamma}) - \tau^{eq}$ to $T(\lambda)$ and $\dot{\gamma} \xi^4$ to λ (see Eqs. (3) and (4)) are chosen in such a way that the theoretical curve and the data points have the most overlap. First of all, this figure shows that the $\dot{\gamma} \xi^4$-scaling is satisfied to within experimental error, since all data points given in Fig. 5 collapse onto the same curve. Secondly, the two proportionality constants mentioned above can be chosen such that all data match with the theoretical curve. This verifies the predicted functional dependence of $\tau(\dot{\gamma}) - \tau^{eq}$ on $\dot{\gamma} \xi^4$.

We can go a step further, and compare the two proportionality constants between $\tau(\dot{\gamma}) - \tau^{eq}$ and $T(\lambda)$, and between $\dot{\gamma} \xi^4$ and λ, with the theoretical values in Eqs. (4)

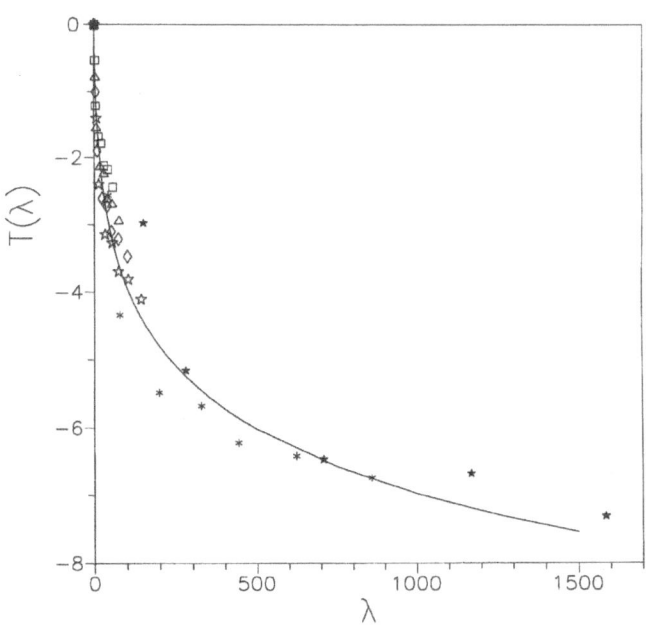

Fig. 5 Data of Fig. 4, plotted against $\dot{\gamma} \xi^4$. The proportionality constants for the λ-axis and T-axis are chosen in such a way that the data points are best mapped on the theoretical turbidity scaling function

and (5). For the $\dot{\gamma} \xi^4$-axis we used a proportionality constant equal to $5.0 \cdot 10^{26}$ s·m^{-4}. This constant should equal $1/[2D_0(\beta \Sigma / R_V^2)R_V^2]$, with $D_0 = k_B T/6\pi \eta a$, $\eta = 0.652 \cdot 10^{-3}$ Pa·s (viscosity of benzene at a temperature of 18 °C), $k_B T = 4.02 \cdot 10^{-21}$ Nm, $a = 39$ nm, $R_V = 78$ nm and $\beta \Sigma / R_V^2 = 3.60$. The constant calculated on the basis of these values is $2.72 \cdot 10^{24}\,\mathrm{s} \cdot \mathrm{m}^{-4}$. This indicates that the measured values are too large. Three reasons may account for this difference. Firstly, the experimentally obtained value for $\beta \Sigma / R_{\mathrm{v}}^2$ of 3.60 ± 1.10 is nothing but a crude estimate, since we assumed $S_{\mathrm{nc}}^{\mathrm{eq}}(0) \approx 1$. Secondly hydrodynamic interactions are not taken into account in our theory. Hydrodynamic interactions are probably accounted for by replacing D_0 by an "effective diffusion coefficient", which may be much smaller than D_0. Thirdly, the improved closure relation that is used to express the three-particle correlation function g_3 in Eq. (1) in terms of the pair-correlation function g, is an approximation. Although this closure relation accounts for the relevant physical phenomena, it may not be correct quantitatively, and contributes to the discrepancy between the measured and the calculated proportionality constant.

The prefactor used for the turbidity-axis is $2.10 \cdot 10^{-3}\,\mathrm{m}$. This value should resemble $[(k_0 R_{\mathrm{v}})^2 (\beta \Sigma / R_{\mathrm{v}}^2)]/C$. Using a value of 1.42 for the refractive index of the particles and 1.50 for the solvent, the factor is $1.81 \cdot 10^{-3}\,\mathrm{m}$,

which is satisfactory in view of the uncertainty in $(\beta \Sigma / R_{\mathrm{v}}^2)$.

Conclusions

The shear rate and temperature dependence of the turbidity is found to scale with $\dot{\gamma} \xi^4$ ($\dot{\gamma}$ is the shear rate and ξ the temperature dependent correlation length of the quiescent suspension), in accord with the mean-field theoretical prediction. The correlation length is found to diverge with the mean field exponent 1/2 for $T - T_{\mathrm{c}} > 0.05\,^{\circ}\mathrm{C}$. The functional dependence of the turbidity on $\dot{\gamma} \xi^4$ is also verified, although there is some discrepancy concerning the proportionality constant, relating the turbidity to the theoretical scaling function. This discrepancy is due to i) an inaccurate experimental value for the Cahn–Hilliard square-gradient coefficient, ii) the neglect of hydrodynamic interactions in the theory and, iii) the closure relation that is used to express the three-particle correlation function g_3, in terms of the pair-correlation function g, is an approximation.

References

1. Dhont JKG, Verduin H (1994) J Chem Phys 101:6193
2. Onuki A, Kawasaki K (1979) Ann Phys 121:456
3. Oxtoby DW (1975) J Chem Phys 62:1463
4. Verduin H, Dhont JKG (1995) Phys Rev E 52:1811
5. Beysens D, Gbadamassi M (1980) Phys Rev A 22:2250
6. van Helden AK, Jansen JW, Vrij A (1981) J Colloid Interface Sci 81:352
7. Verduin H, Dhont JKG (1995) J Colloid Interface Sci 172:425
8. Johnson SJ, de Kruif CG, May RP (1988) J Chem Phys 89:5909

Progr Colloid Polym Sci 100:87–90 (1996)
© Steinkopff Verlag 1996

COLLOIDAL PARTICLES

R.M. Santos
J. Forcada

Synthesis and characterization of latex particles with acetal functionality

R.M. Santos · Dr. J. Forcada (✉)
Grupo de Ingeniería Química
Departmento de Química Aplicada
Facultad de Ciencias Química
Universidad del País Vasco/EHU
Apdo. 1072
20080 Donostia-San Sebastían, Spain

Abstract Core-shell latex particles with surface-acetal functionality were synthesized for use in immuno-diagnostic testing. By acidification of the acetal groups to aldehydes, covalent bonding with the amino groups of biomolecules is possible. Acetal functionality was chosen due to the chemical instability of the aldehyde group. The synthesis of this kind of particles was carried out by a two-step emulsion polymeriza-tion in a batch reactor. Firstly, a monodisperse core of polystyrene (PS) was obtained; and in a second step, a shell of styrene (St), methacrylic acid (MAA) and methacrylamidoacetaldehyde di(n-methyl) acetal (MAAMA) was formed around the polystyrene cores. The shell has both acid and acetal functionalities. In order to analyze the effect of the pH of the reaction medium on the surface groups, reactions at acid and neutral pH were carried out.

The latex characterization consisted of the determination of the particle size distributions and the amount of functionalized surface groups.

The particles synthesized in neutral medium were covalently bonded with IgG a-CRP rabbit antibody. The complex latex-protein was immunologically active against the CRP antigen.

Key words Emulsion polymeriza-tion – acetal group – immunoassays

Introduction

Currently, the use of polymer latex particles in biomedical applications, mainly in developing new immunoassays, is of ever-increasing interest.

A new generation of functionalized particles have been synthetized in order to bond covalently the amino groups of biomolecules [1–3].

The covalent coupling of protein to the functionalized groups situated on polymer surfaces, prevents the physical desorption of the proteins and keeps the native chemical conformation of the biomolecules.

In this work, core-shell polymer latex particle were synthesized. The core is a monodisperse seed of polysty-rene, and the shell is a terpolymer with acid and acetal functionalities. By acidification of these acetal groups to aldehydes, covalent bonding to biomolecules is possible [1].

Due to the chemical instability of the aldehyde groups, the acetal functionality was chosen to mask the reactivity of the particles.

The synthesized particles with the acetal groups on the surface were covalently coupled to an IgG-a CRP (IgG anti-C-reactive protein), rabbit polyclonal antibody. The immunoreactivity of the particles sensitized with IgG a-CRP was demonstrated by carring out immunoassays at different concentrations of CRP antigen.

Materials and methods

Synthesis of the functionalized latexes

The functionalized core-shell particles were synthesized by means of a two-step emulsion polymerization in a batch reactor. The core was a seed of polystyrene (PS) and the shells were obtained by terpolymerization of styrene (St), methacrylic acid (MAA) and methacrylamidoacetaldehyde di(n-methyl) acetal (MAAMA), using the seed obtained previously.

The monomer St was distilled under reduced pressure and all other materials were used as received. Potassium persulfate, sodium dihydrogen phosphate and sodium hydrogen phosphate were used as initiator and buffer, respectively. The surfactants were aerosol MA-80 (sodium dihexyl sulfosuccinate) for the core and sodium lauryl sulfate for the shells. Deionized water was used throughout. The acetal monomer MAAMA is not a commercial one and its synthesis was carried out using aminoacetaldehyde di(n-methyl) acetal and methacryloyl chloride (50/50 molar) and potassium carbonate as buffer, the reaction medium being chloroform, previously dried with $CaCl_2$. The reaction temperature was $0\,°C$ and the reaction time was 1 h. After purification by washing with water, the synthesized monomer was characterized by NMR H^1 and C^{13}.

Polymerizations were carried out in a 2-liter glass reactor fitted with a reflux condenser, stainless-steel stirrer, sampling device, nitrogen inlet and feed inlet tubes. The seed was prepared at $90\,°C$ by means of batch emulsion homopolymerization of St, and after polymerization, the seed was kept overnight at $90\,°C$ to decompose the initiator. The seed was dialyzed and cleaned by means of a serum replacement cell.

The shell was synthesized at $70\,°C$ by means of a batch emulsion terpolymerization of St, MAA and MAAMA. The weight ratio of the termonomers was 4/12/3 (MAAMA/St/MAA). The initiator concentration was 3.2 wt % of the total monomer, and the emulsifier concentration was that calculated to cover 35% of the particle surface. To study the effect of the pH of the reaction medium on the surface groups, two polymerizations were carried out, the first one (LKMB1 latex) under acid conditions and the second one (LKM1 latex), under neutral conditions. In both reactions 40 g of the seed and 752 g of water were used.

The charge sequence of the reactor was: the seed, the emulsifier and buffer solutions (in the case of LKM1 latex),

the MAA monomer (neutralized with NaOH in the LKM1 synthesis), the styrene and finally the MAAMA. The reaction mixture was kept for 1 h at room temperature and under stirring in order to swell the seed particles. Once the reaction temperature ($70\,°C$) was reached, the aqueous initiator solution was charged into the reactor to start the polymerization. The stirring rate was 200 rpm and the reaction time 5 h.

Latex characterization

The overall conversion at the end of polymerizations were determined gravimetrically. The final pHs were measured. The particle size distributions of the seed and the final latexes were obtained by Transmision Electron Microscopy (TEM).

The amount of surface groups was determined by conductimetric titration. The latexes were cleaned by means of a serum replacement cell before the titration. The COOH surface groups were titrated with NaOH, and the amounts of the CHO surface groups were determined after acid cleavage of the acetals and titration with hydroxylamine [2].

Latex–protein complexes, Immunoassays

The covalent coupling of the purified IgG a-CRP rabbit polyclonal antibody to the LKM1 latex surface was carried out at pH 5, with cleaned latex having a surface area of $0.4\,m^2$ and 2.4 mg of antibody ($6\,mg/m^2$ latex). In the covalent bonding procedure [4], the acetal latex was activated by adding HCl to reach pH 2. The aldehyde groups react with the amino groups of the antibody leading to an unstable imine double bound that is reduced with sodium borohydride.

The amount of IgG covalently bound to the latex surface by the aldehyde groups was determined by measuring the optical absorbance at 280 nm, in the presence of Tween 20, nonionic surfactant, to remove the protein physically adsorbed during the covalent coupling.

The immunoreactivity of the latex-protein complex was measured by the changes in the turbidity of the dispersion containing latex-protein complex mixed with different concentrations of CRP antigen. The dispersion was stirred and after 5 min the increments in the optical absorbance were measured at 570 nm.

Results and discussion

The overall conversion of the polystyrene seed was 100% within the experimental error. Table 1 shows the overall

Table 1 Overall conversion of the shell and final pH of the reaction

Latex	%X_T	pH
LKM1	60	7.0
LKMB1	73	3.1

Table 2 Number average diameters and polydispersity index of the latexes

Latex	Dp (nm)	PDI
PS	103	1.017
LKM1	112	1.020
LKMB1	112	1.018

Table 3 Functional surface groups of the core-shell latexes

Latex	mEq COOH/g pol	mEq CHO/g pol
LKM1	0.055	0.120
LKMB1	0.380	0.069

conversion of the shell and final pH of the reactions. The results showed that when the reaction medium was acid, the conversion was higher than in the neutral medium. This result is due to the higher MAA monomer reactivity at acid pH than at neutral pH.

Table 2 shows the number average diameter and polydispersity index (PDI) of the latex particles. Both latexes have the same size and the monodispersity in accordance with the requirements for the application of this kind of particles in immunoassays.

In the Table 3 the amounts of functionalized surface groups (mEq/g polymer) obtained by means of conductimetric titration are shown. The content of carboxyl surface groups, provided by the MAA monomer when the reaction medium is acid (LKMB1 latex) is higher than in neutral medium. This result is due to the fact that when the reaction medium is acid, the carboxyl groups are protonated, whereas, in neutral medium there are negatively charged, and the MAA monomer remains in the aqueous phase. On the other hand, the amount of aldehyde groups obtained in neutral medium (LKM1 latex) is higher than in acid medium. In the latter medium, the acetal groups become aldehydes that are highly reactive and can be oxidized to carboxylic acid. These newly formed acid groups increase the amount of carboxyl groups in the LKMB1 latex, decreasing the aldehyde surface group content.

In the covalent coupling of the IgG a-CRP to the LKM1 latex, it was found that from the 6 mg/m² of the antibody added, 3.01 mg/m² were bonded covalently to the latex particles surface.

Figure 1 shows the changes in the optical absorbance versus the CRP concentration in solution to detect the immunological response of the complex LKM1 latex-IgG a-CRP obtained by covalent coupling of the IgG (3.01 mg/m²). The changes in absorbance were obtained after 5 min of reaction with the CRP and the typical bell curve of immunoprecipitin was observed.

Fig. 1 Immunoassays: absorbance versus the CRP concentration

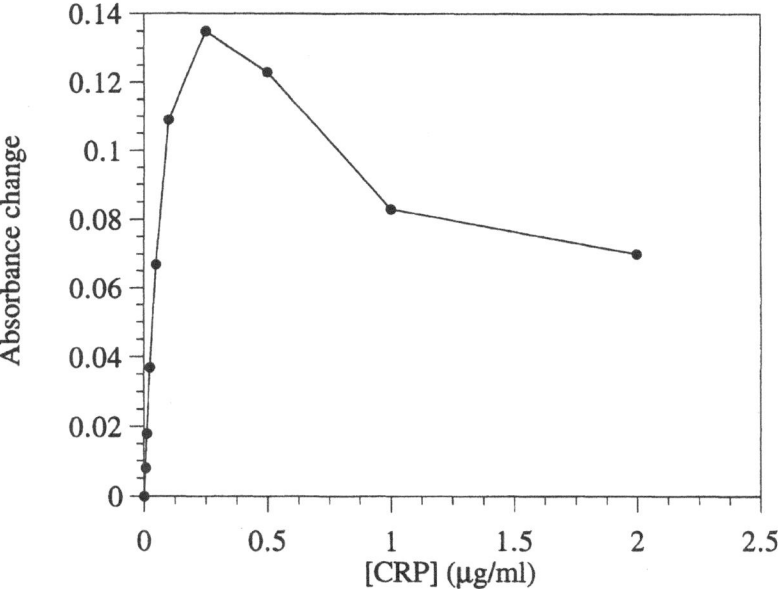

Conclusions

By changing the reaction medium pH, functionalized latex particles with different amounts of functionalized surface groups can be obtained.

The synthesized latex particles were found to be useful in the covalent bonding to IgG protein leading to latex-protein complexes with immunological response.

Acknowledgments This work is supported by the Comisión Interministerial de Ciencia y Tecnología (CICYT), project MAT 93-0530-CO2-02, the Universidad del País Vasco/EHU (UPV 221.215-EA155/93) and project PI 94/6 from the Eusko Jaurlaritza. The authors thank Grupo de Física de Fluidos y Biocoloides, Universidad de Granada (Granada, Spain) for giving R.M. Santos access to their laboratories for the immunoreactivity experiments.

References

1. Kapmeyer WH, Pauly H, Tuengler P (1988) J of Clin Lab Anal 2:76–83
2. Yan C, Zhang X, Sun Z, Kitano H, Ise N (1990) J Appl Polym Sci 40:89–98
3. Litchfield WJ, Craig AR, Frey WA, Leflar ChC, Looney CE, Luddy MA (1984) Clin Chem 30/9:1489–1493
4. Peula JM, Santos R, Forcada J, Hidalgo-Alvarez R, de las Nieves FJ (1995) J of Materials Science: Materials in Medicine 6:779–785

Progr Colloid Polym Sci (1996) 100:91–95
© Steinkopff Verlag 1996

COLLOIDAL PARTICLES

Morphology of heterogeneous latex particles investigated by atomic force microscopy

B. Gerharz
H.J. Butt
B. Momper

Dr. B. Gerharz (✉) · B. Momper
Hoechst AG
Abteilung F + E PM I
Gebäude D 581
65926 Frankfurt, FRG

H. J. Butt
Max-Planck-Institut für Biophysik
Kennedyalle 70
60596 Frankfurt, FRG

Abstract To investigate the morphology of individual latex particles with the atomic force microscope the particles were adsorbed onto mica and then dried. During adsorption and subsequent drying the packing of the particles is probably mainly determined by capillary forces, their deformation is caused by van der Waals forces. The shape of the particles on the mica surface depends on their original morphology and their viscoelastic properties. Chemically and also morphologically heterogeneous latex particles with therefore different viscoelastic properties were examined in comparison to the corresponding homogeneous latexes.

In addition, films formed with the latex particles were imaged to correlate the structure of individual particles with that of the film surface.

With the AFM it is possible to examine the morphology, size and viscoelastic inhomogenities of latex particles and thereby deduce their behavior during film formation. The particles examined here were on the monomer basis of n-butylacrylate and methylmethacrylate.

To compare the morphology of single particles with the surface structure of the corresponding films, we have chosen two different heterogeneous particles: For sample COM 1, we assumed a core-shell morphology, for sample COM 2 a hemispherical particle form. The morphology assumptions were based on studies which compare the minimum film formation temperatures with the glass transition temperatures of the different particle phases and their mass fractions. The results observed by the heterogeneous particles were compared to the corresponding homogeneous ones. Both composite particles exhibit the speculated morphology and, consequently, the expected properties after film formation.

Key words Emulsion polymerization – heterogeneous latexes – morphology – AFM-MFFT

Introduction

Latex dispersions are commonly used in paint, paper, adhesive and coating industries where the properties of the film are of great importance. The film formation depends on the viscoelastic properties and the morphology of the latex particles.

The morphology of latex particles can be examined by small-angle neutron or x-ray scattering, transmission electron microscopy, nuclear magnetic resonance spectroscopy and atomic force microscopy (AFM). The AFM yields topographic images of surfaces with high resolution. Imaging can be done in air without taking special care for dust-free environment. No special chemical modifications are necessary and the samples have not to be exposed to vacuum.

92

B. Gerharz et al.
Morphology of latex particles by AFM

The aim of this study was to correlate the morphology of composite particles with the topography of the film and the shape of single particles. In addition, from the shape of individual particles adsorbed to mica information about the behavior of the particles during film formation might be obtained. The driving forces for film formation and for the deformation of individual particles on a smooth surface are capillary and interfacial forces. In both cases the driving forces have to overcome particle rigidity. Capillary and interfacial forces and the viscoelastic properties of the latex determine the final structure of the film, and the shape of individual particles on a substrate.

Two classes of latex particles were investigated: Homogeneous particles, which were formed in a one-stage process, and composite latex particles, which were formed in two stages. Homogeneous particles were made of different amounts of methyl methacrylate (MMA) and n-butyl acrylate (BuA) yielding particles of different glass transition temperature, T_g. The morphology of composite latex particles is dictated by thermodynamic, and kinetic parameters, as well as the polymerization mechanism [1]. Two sorts of composite particles were prepared: one with a core-shell morphology, and one with hemispherical shape.

Materials and methods

The latex particles were composed of different proportions of n-butyl acrylate and methyl methacrylate (Table 1). The polymer glass-transition temperatures were measured with a differential scanning calorimeter with a heating rate of 20 K/min. The T_g of BuA was $-43\,°C$, that of MMA was $+122\,°C$. The minimum film formation temperature (MFFT) [2] was measured as described in [3]. The latexes were cast on a temperature gradient bar with a linear temperature range of 0 to 44 °C. After 6 h the temperature at which the film became optically clear was recorded as the MFFT. Mean particle diameters, d_L, were measured with dynamic light scattering [4]. The dried weight of all preparations was between 45.1 and 46.0%.

Homogeneous latexes were prepared in one step (HOM1-HOM4) and composite particles were prepared in two steps (COM1 and COM2) by semi-continuous emulsion polymerization (Table 1). The synthesis is described elsewhere [3] [BuGe95]. Latexes COM1 and COM2 were both composed of a relatively soft first phase and a hard second phase. In total, preparation COM1 yielded harder particles than COM2. While in COM1 the first phase was a 40:60 mixture of MMA and BuA, in COM2 the first phase was 100% BuA. The second phase was the hard MMA in COM1 and 60:40 mixture MMA:BuA in COM2. Since the shell in COM1 particles constitutes

Table 1 contains the chemical compositions of the particles investigated, their glass-transition temperature, T_g, their minimum film formation temperature, MFT, and their diameter, d_L, determined by dynamic light scattering. For composite particles the chemical compostions are listed separately for emulsions I and II. Due to their complex morphology they also have two T_g. Also the parameters measured with the AFM are summarized. The height was measured on single particles. The nearest-neighbor distance, NND, the peak-to-valley distance, Δz, and the roughness, rms, were measured on films

	HOM1	HOM2	HOM3	HOM4	COM1	COM2
Emul.I wt%	100	100	100	100	50	50
MMA wt%	100	60	40	0	40	0
BuA wt%	0	40	60	100	60	100
Emul.II wt %	–	–	–	–	50	50
MMA wt%	–	–	–	–	100	60
BuA wt%	–	–	–	–	0	40
T_g °C	122	37	6	−43	5, 112	−43, 33
MFFT °C	> 44	27	<0	< 0	>44	0
d_L nm	110	140	130	115	110	130
Height nm	94	45	26	9	45	36
NND nm	84	115	–	–	83	135
Δz nm	18	7	–	–	17	7
rms nm	8	6	3	–	14	12

about 50 wt %, its thickness was roughly 12 nm, assuming a total particle radius of 55 nm.

To image individual latex particles the emulsions were diluted $1:10^5$ in distiled water. For all sorts of particles the same concentration was used. A drop of 20 μl was placed on freshly cleaved mica and allowed to dry. Films were prepared by casting the latex onto fresh microscope slides with a wet film thickness of 100 μm, leading to a thickness of the dry film of about 50 μm. After 3 h to 2 days the samples were imaged in air with a Nanoscope[R] III (Digital Instr., Santa Barbara, California) in contact, constant force mode. We used silicon nitride cantilevers of 100 μm length (spring constant 0.09 ~ N/m) with integrated sharpened tips (Olympus, Tokyo, Japan). Typically the force between tip and sample was 10–20 nN. To describe the shape of single particles, we report the height and the diameter d_{AFM} measured at half maximal height. To characterize films the roughness (mean square deviation from the mean plane), the nearest-neighbor-distance, NND, and the peak-to-valley distance, Δz (difference between the height measured on top of a particle and the height measured between two neighboring particles), was measured [5].

Results and discussion

Homogeneous latex particles

To investigate the morphology of single composite particles as a comparison four homogeneous particles with

Progr Colloid Polym Sci (1996) 100:91–95
© Steinkopff Verlag 1996

different T_g were imaged (Table 1). Latex particles made of 100% MMA had a spherical shape when dried on mica. It was impossible to image isolated, single particles; only aggregates of several particles were observed (Fig. 1 top, left). This agrees with observations of Granier et al. [6] [7] who could also not image single latex particles when the T_g was well above the imaging (or a possible annealing) temperature. The mean height of particles HOM1 was 94 nm, the mean diameter d_{AFM} was 128 nm, which agrees with diameters determined by dynamic light scattering [4].

When the BuA content of the particles was increased to 40% (HOM2) the observed height of single particles decreased to roughly 45 nm (Fig. 1, top right). Hence, the particles were significantly deformed although the T_g to 37 °C was above imaging temperature (22 °C). The measured diameter was about 120 nm. Usually single particles lay isolated on the mica surface.

Often the top part of a particle was scraped off the surface leaving a donut-shaped ring of material behind. The height of such a ring was 8 nm. This ring could be composed of surfactant which was stripped off the particle surface when, during the drying process, the water–air interface moved across the particle surface.

Individual latex particles composed of 40% MMA and 60% BuA (HOM3) were significantly flattened (Fig. 1, bottom left); their average height and diameter were 26 nm and 160 nm, respectively. No aggregates, only isolated particles were observed. The particles were strongly bound to the mica and we could not scrape them off. The same spreading of the latex particles was observed by Granier et al. [7] when imaging single particles with $T_g = 5$ °C at room temperature, or when imaging particles which were annealed above their T_g.

Pure BuA particles (HOM4) were flat (9 nm height) and had a large diameter of 240 nm (Fig. 1, bottom, right). When scanning sometimes part of the particle material was scraped off the surface. Often the flat particles were irreversibly deformed or even destroyed after scanning them typically more than 10 times.

Composite latex particles

The morphology of the composite particles was deduced from results of MFFT measurements. MFFT measurements were done with composite particles of different compositions (the ratio of soft to hard phase was varied between 0.2 and 0.8, the mixtures of the constituting polymers were varied in the same range). The film formation of heterogeneous latices differs from the film formation of the corresponding mixtures. Only in the case of hemispherical particles is the film formation equal to the corresponding mixtures. According to [8, 9] hemispherical particles or mixtures of particles need to contain at least 30 wt% of the soft phase to reduce the MFFT (Fig. 2, Fig. 3). In mixtures of homogeneous particles with different composition an MFFT < 0 °C is available at nearly 30% soft latex with an MFFT < 0 °C (Fig. 4). In contrast, the film formation in composite particles is much more influenced by the morphology. Only by varying the phase composition can one generate core-shell and hemispherical particles with MFFTs varying from > 44 °C to < 0 °C (Fig. 5).

Fig. 1 Surface plots of typical homogeneous latex particles HOM1-HOM4 dried on mica. The z-scale was 110, 100, 100, and 50 nm/div for HOM1-4, respectively. Please note that the lateral scaling is different from the z scale so that particles might appear higher

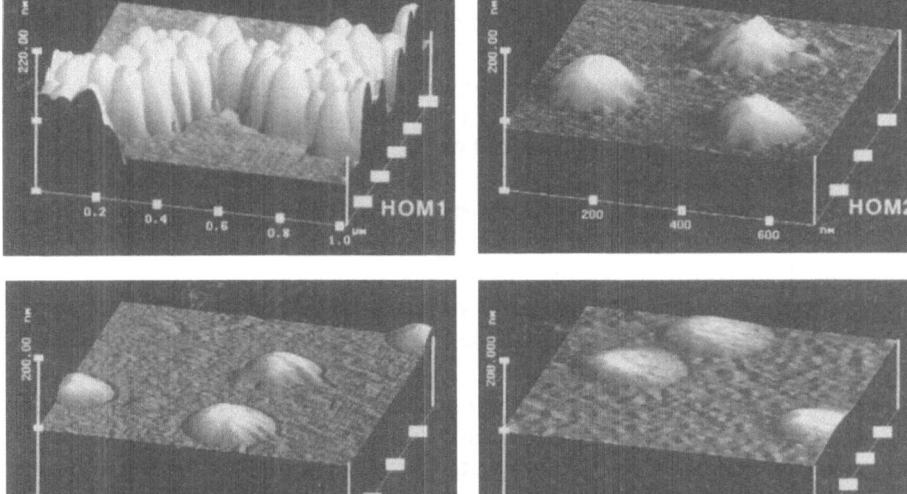

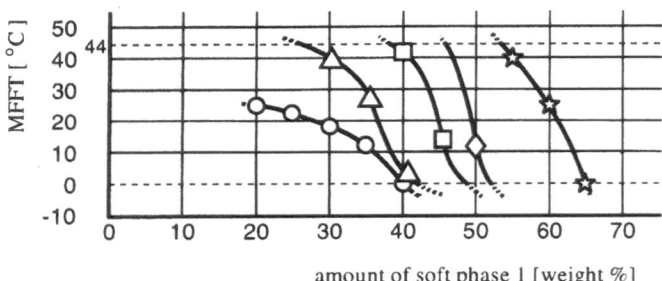

Fig. 2 MFFT-measurements of heterogeneous latices with a soft first phase (MMA/BuA = 20:80, constant ratio) and varied composition of the hard phase: ☆ 100% MMA, ◇ 90 MMA/10 BuA, □ 80 MMA/20 BuA, △ 70 MMA/30 BuA, ○ 60 MMA/40 BuA. The amount of soft phase in the latex particle is varied in the range given on the x-axis. Plotted is the MFFT [°C] in the measured range from 0 °C to 44 °C versus amount of soft phase I [wt %]

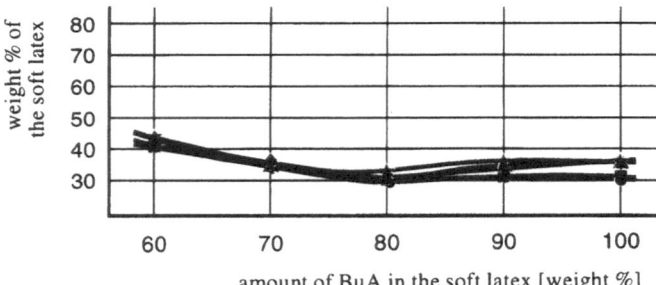

Fig. 4 Plotted is the amount of soft latex in the mixtures versus the ratio of BuA in this soft latex and selected are those mixtures which have the blended composition necessary to give a MFFT = 0 °C. The composition of the hard latex varies: ★ 100% MMA, ◆ 90 MMA/10 BuA, ■ 80 MMA/20 BuA, ▲ 70 MMA/30 BuA, ● 60 MMA/40 BuA

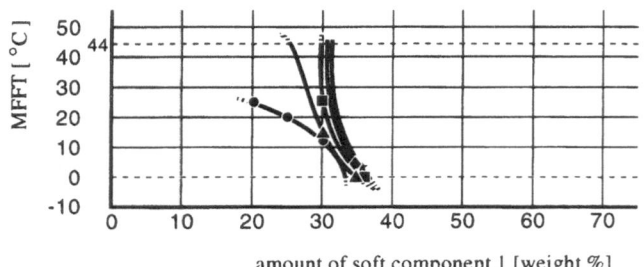

Fig. 3 MFFT-measurements of the corresponding mixtures with a soft component 1 (MMA/BuA = 20:80, constant ratio) and varied component 2 of the same composition as phase 2 in Fig. 2: ★ 100% MMA, ◆ 90 MMA/10 BuA, ■ 80 MMA/20 BuA, ▲ 70 MMA/30 BuA, ● 60 MMA/40 BuA. The amount of soft latex in the mixture is varied in the range given on the x-axis. Plotted is the MFFT [°C] in the measured range from 0° to 44 °C vs. amount of soft latex [wt %]

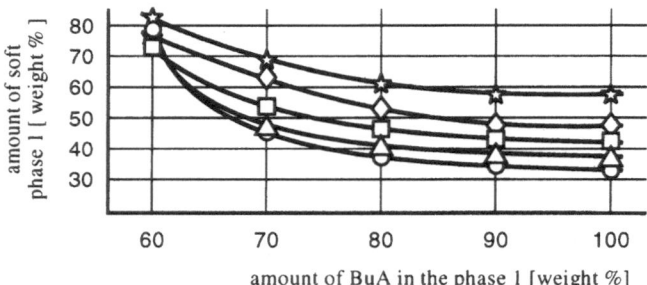

Fig. 5 Plotted is the amount of soft phase I versus the ratio of BuA in this soft phase and selected are those heterogeneous particles which have the composition necessary to give a MFFT = 0 °C. The composition of the hard phase II is given as follows: ☆ 100% MMA, ◇ 90 MMA/10 BuA, □ 80 MMA/20 BuA, △ 70 MMA/30 BuA, ○ 60 MMA/40 BuA

Modified COM2 particles, which only contained 30 wt % soft phase, did indeed form a film when cast onto a substrate. Hence, a hemispherical morphology is more likely. In comparision, COM1 behaved like a pure MMA latex although the soft phase constituted 50 wt %. Hence, we assumed a core-shell morphology.

To verify the assumed morphologies single particles and films of composite particles were imaged with the AFM. Single particles of the composite sample COM1 (Fig. 6, top left) had a shape similar to particles HOM2. Also, particles could be scraped of the surface leaving a donut-shaped ring of material behind. The particle had a height of 45 nm and a diameter of 150 nm. As observed by Sommer et al. [10], COM1 particles had a "raspberry"-like surface morphology. Normally, isolated particles were observed. In a few samples we also saw aggregates. In contrast to HOM2, particle shapes were often ellipsoid and the size varied considerably more.

For composite particles also the surface topography of layers is shown. Layers made from preparation COM1 (Fig. 6, bottom left) were relatively rough (14 nm). In some experiments we could not avoid scratching particles out of the film. Individual particles could be identified. These particles were relatively inhomogeneous in size and shape. They were strongly deformed. The nearest-neighbor-distance of 83 nm was much lower than the particle diameter measured by light scattering $d_L = 110$ nm. This was surprising since the MFFT was above 44 °C.

Single particles COM2 on mica were composed of a flat, broad layer (thickness: 7 nm, typical width: 300 nm) and a higher part (higher: 35 nm, diameter: 155 nm) usually in the center of the flat layer (Fig. 1, top right). The flat region was relatively soft and could be scraped off by repeated scanning with a high force. The high part did not significantly deform under the influence of the tip. The flat region was probably composed of the soft BuA phase.

Progr Colloid Polym Sci (1996) 100:91–95
© Steinkopff Verlag 1996

Fig. 6 Surface plots of typical individual heterogeneous latex particles dried on mica (top) and of films or layers formed from the particles (bottom). The z-scale was 90 and 100 nm/div for COM1 and COM2, respectively

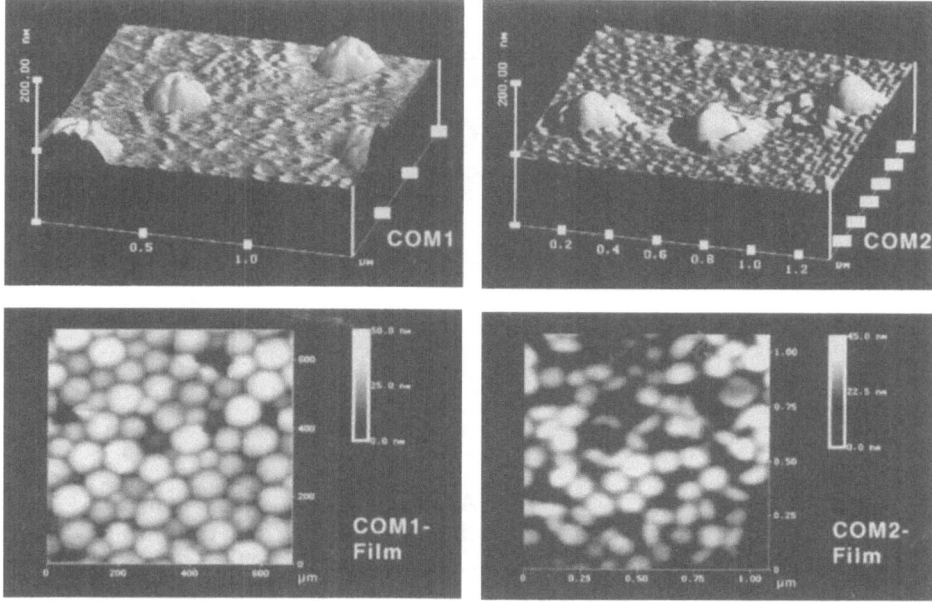

After adsorption of the particles to the mica, part of the soft phase might have flowed down the particle surface onto the mica. In a similar way, Okubo et al. [11] explained their results obtained with the TEM on composite particles made of butyl acrylate and styrene.

In films the roughness slightly decreased to 12 nm. Individual particles could be identified. Even more than in preparation of COM1 the shape of these particles varied. Often, flattened or hemispherical particles were observed. COM2 was the only sample where the nearest-neighbor distance of 130 nm in the film was similar to the diameter d_L determined by dynamic light scattering-although the MFFT was below 0 °C (Fig. 6, bottom right). Viscoelastic criteria for film formation [12–15] do not describe this process in soft composite films. These results agree, however, with the observation of Goh et al. [16], who found that in films formed from poly(butyl methacrylate) latex particles well above their MFFT the nearest-neighbor distance was equal to the particle diameter.

Conclusion

By imaging individual latex particles on mica it is possible to obtain information about the original morphology and viscoelastic properties of the particles. Film formation of heterogeneous latices is directly correlated to the particle morphology. The film formation behavior of the composite particles was directly compared to those of mixtures composed from the corresponding homogeneous latexes. Comparison of the MFFTs leads to the conclusion that one can at least deduce two extreme morphologies: core-shell and hemisphere. AFM results indicated that both assumed morphologies in the chosen heterogeneous particles exist. It also supports that core-shell and hemispherical particles can be generated by simply varying the phase composition. With the aim of AFM, one can detect inhomogenieties caused by different viscoelastic properties. This visualizes the morphology of heterogeneous latex particles.

References

1. Shen S, El-Aasser MS, Dimonie VL, Vanderhoff JW, Sudol ED (1991) J Polymer Sci A 29:857–867
2. Protzman TF, Brown GL (1960) J Appl Polymer Sci 4:81–85
3. Eckersley ST, Rudin A (1990) J Coatings Technol 62:89–100
4. Fischer JP, Nölken E (1988) Progr Colloid Polym Sci 77:180–194
5. Butt HJ, Gerharz B (1995) Langmuir 11:4735–4741
6. Granier V, Sartre A, Joanicot M (1993) J Adhesion 42:255–263
7. Granier V, Sartre A, Joanicot M (1994) Tappi Journal 77:220–229
8. Devon MJ, Gardon JL, Roberts G, Rudin A (1990) J Appl Polymer Sci 39:2119–2128
9. Tangyu C, Yongshen X, Yungdeng S, Yuhong H (1990) J Appl Polymer Sci 41:1965–1972
10. Sommer F, Duc TM, Pirri R, Meunier G, Quet C (1995) Langmuir 11:440–448
11. Okubo M, Katsuta Y, Matsumoto T (1982) J Polymer Sci: Polymer Lett 20:45–51
12. Brown GL (1956) J Polymer Sci 22:423–434
13. Vanderhoff JW, Tarkowski HL, Jenkins MC, Bradford EB (1966) J Macromol Chem 1:361–397
14. Mason G (1973) Brit Polymer J 5:101–108
15. Lamprecht J (1980) Colloid Polym Sci 258:960–967
16. Goh MC, Juhué D, Leung OM, Wang Y, Winnik MA (1993) Langmuir 9:1319–1322

Progr Colloid Polym Sci (1996) 100:96–100
© Steinkopff Verlag 1996

Dynamic properties of magnetic colloidal particles: Theory and experiments

M.C. Miguel
J.M. Rubí

M.C. Miguel · Prof. Dr. J.M. Rubí (✉)
Departament de Física Fonamental
Facultat de Física
Universitat de Barcelona
Diagonal 647
08028 Barcelona, Spain

Abstract We provide a general theory to study the dynamics of a ferromagnetic colloidal particle and the transport coefficients of a dilute dispersion under the influence of a magnetic field. Different measurements carried out for magnetic liquids show that there is finite coupling between the orientation of the magnetic moment of a ferromagnetic monodomain and the orientation of the particle itself (characterized by the orientation of its crystallographic axes). Because of this coupling, the relaxation of the magnetic moments takes place in two different ways that proceed simultaneously: rotation within the particle, and together with the particle with respect to the carrier liquid. We compute the rotational viscosity from a Green–Kubo formula and give an expression for different relaxation times. These characteristic times come from the dynamic equations for the correlation functions which, in the framework of the linear response theory, are involved in the calculation of some of the physical properties of the material we are interested in (optical, magnetic, ...). Our results agree quite well with experiments performed with different samples of ferromagnetic particles, which allow us to distinguish the different relaxation regimes occurring when the size and the nature of the magnetic material of the grains are freely modified.

Key words Magnetic particles – ferrofluids – colloidal systems – relaxation dynamics – non-equilibrium phenomena

Introduction

Systems of single-domain ferromagnetic particles immersed in a liquid phase exhibit a number of interesting relaxation phenomena which have been the subject of many experimental and theoretical analyses [1–4]. These phenomena are essential in the study of the dynamics of these particles and, particularly, have a clear influence when determining the effective viscosity, the dynamic birefringence, and the magnetic susceptibility. One of the main peculiarities of these systems is that their properties are greatly influenced by the presence of an external magnetic field. It is this fact which has been the basis of many practical applications [1].

The rotational dynamics of a ferromagnetic particle embedded in a liquid phase is the result of the competition of three orientational mechanisms related to the external field, the axis of easy magnetization, and rotational Brownian motion. That is, whereas the magnetic moment of the ferromagnetic particle relaxes towards the direction of the magnetic field, the axis of easy magnetization tends to be aligned with the magnetic moment, thus giving rise to different coupled relaxation phenomena. Until recently, the most frequent case studied in the literature deals with rigid-dipoles [4–6], for which the anisotropy energy is dominant due to the large value of the anisotropy constant, and because the radius of the particle usually exceeds a critical value. What is more, when looking at the relaxation phenomena described by a rigid dipole, one disregards the precessional motion of the magnetic moment, and, consequently, the associated dissipation. Under these conditions, we can no longer talk about the relaxation of the axis of easy magnetization towards the magnetic moment; instead both vectors relax together. However, there are materials for which the anisotropy energy may be comparable to the energy associated with the interaction with the magnetic field, or even smaller. Therefore, a general theory encompassing such a wide variety of situations and accounting for experimental results should be developed. The presence of different relaxation mechanisms affects the form of the effective viscosity of the system, which exhibits significant corrections when compared to the viscosity of a suspension of non-magnetic particles of the same shape. Another point of interest is the appearance of relaxation times which are usually involved in the characterization of certain physical properties, and which can be measured by different experimental techniques.

The purpose of this paper is to present a theory capable of giving expressions for the relevant transport coefficients of the system and of the corresponding characteristic relaxation times determining for instance, the effective viscosity, the dynamic birefringence, and the magnetic susceptibility of the suspension. The formalism we have developed is based on the linear response theory where the correlation dynamics comes from a Smoluchowski equation. As we will show in one of the sections, our result for the relaxation time of the rotation of the particle is compared to experimental data and agrees quite well with birefringence experiments.

We have distributed the paper in the following way: in the second section, we establish basic equations describing the dynamics of the degree of freedom. In the third section, we deal with the calculation of the rotational viscosity using a Green–Kubo equation proposed from the linear response theory. We also compute the relaxation time of the birefringence of the suspension, which is induced by the dispersed particles, when considering the different orientational mechanisms. We compare our results to experiments performed for two samples of very common ferromagnetic particles for which the size and the nature of the magnetic material clearly establish different values of the magnetic energy and the energy of anisotropy, and we obtain good agreement in both situations. Finally, in the last section we summarize our main results.

Dynamics of the degrees of freedom

The energy of a spherical single-domain ferromagnetic particle under the action of an external magnetic field is the sum of two contributions. These contributions originate from the externally imposed magnetic field and the presence of an axis of easy magnetization (for uniaxial crystals). Its expression is given by

$$U = -\vec{m} \cdot \vec{H} - K_a V_m (\hat{n} \cdot \hat{R})^2, \tag{1}$$

where $\vec{m} = m\hat{R}$ is the magnetic moment of the particles, $\vec{H}$ is the external magnetic field, K_a is an effective anisotropy constant (assumed positive), V_m is the magnetic volume of one of these spheres, and $\hat{n}$ is the unit vector along the direction of the axis of easy magnetization for uniaxial magnetic crystals. It is clear from Eq. (1) that in the general case where both contributions may take arbitrary values, the relaxation mechanisms of the degrees of freedom, $\hat{R}$ and $\hat{n}$, of the ferromagnetic spheres in suspension are coupled. It is also worth mentioning at this point that there will be essentially two relevant dimensionless parameters in the analysis. These parameters are $\sigma = K_a V_m / k_B T$, comparing anisotropy and thermal energies, and the parameter $\mu = mH/k_B T$, comparing the Zeeman and thermal energies. The derivative of this magnetic energy with respect to $\vec{m}$ determines the value and orientation of the effective magnetic field

$$\vec{H}_{eff} = -\frac{\partial U}{\partial \vec{m}} = \vec{H} + \frac{2K_a V_m}{m} \hat{n}(\hat{n} \cdot \hat{R}), \tag{2}$$

which includes the external field $\vec{H}$ and the anisotropy field $\vec{H}_a$ directed along the anisotropy axis. In the absence of the external field, the magnetic moment is just under the action of the anisotropy field and there are two equivalent equilibrium orientations $\hat{R} = \hat{n}$ and $\hat{R} = -\hat{n}$ between which the relaxation can take place. Now, the equilibrium condition is given by the absence of magnetic torques acting upon the particle, so that in the absence of thermal fluctuations the magnetic moment is parallel to $\vec{H}_{eff}$.

When this is the case there are two different orientational relaxation processes of the magnetic moment of the particle relative to its crystallographic axes. The *intrinsic motion* of the magnetic moment consists of a regular *precession* around the effective field and of *chaotic reorientations* due to thermal fluctuations. The regular motion relative to the crystalline axes of the particle is described by the classical Landau–Gilbert equation. This equation represents the precession of the magnetic moment with the Larmor frequency ω_L as well as the decay of this motion due to collisions or magnetoelastic interaction. Associated with this decay there is a characteristic time $\tau_0 = \alpha\omega_L$, where α is a dimensionless damping constant. Another characteristic time $\tau_D = (2D_m)^{-1}$ is connected with the rotational diffusion of the magnetic moment inside the particle with the diffusion coefficient $D_m = k_B T h$, where h plays the role of a rotational mobility of the magnetic moment.

Besides the internal motion relative to the crystallographic axis of the particle, the magnetic moment also undergoes an external rotation as a consequence of the motion of the magnetic particles in the liquid in which they are suspended. The ferromagnetic particles suspended in a nonmagnetic fluid are affected by the action of the carrier liquid through viscous friction and Brownian motion. The dynamics of $\hat{n}$ follows from the kinematic relation

$$\frac{d\hat{n}}{dt} = \vec{\Omega} \times \hat{n} , \tag{3}$$

where the angular velocity Ω can be obtained from the total angular momentum conservation equation.

As both the magnetic moment and the ferromagnetic particle itself undergo a systematic damping and random thermal fluctuations, the study of its dynamics can be accomplished by using Langevin and Fokker–Planck of Smoluchowski diffusion equations for the probability density $\psi(\hat{R}, \hat{n}, t)$. This equation can be obtained from the continuity equation in the space spanned by the degrees of freedom $\vec{m}$ and $\hat{n}$, with a Fick equation for the diffusion current [7]. In this Smoluchowski equation there appear two diffusion coefficients, namely $\tau_D = (2D_m)^{-1}$ related to the chaotic reorientations of $\vec{m}$ inside the particle due to thermal fluctuations, and the Brownian time $\tau_B = (2D_r)^{-1}$, where $D_r \equiv \frac{k_B T}{\zeta_r}$ is the Brownian rotational diffusion coefficient, and $D_m = k_B T h$. The Smoluchowski equation agrees with that obtained in ref. [8] by using a model similar to the itinerant oscillator model.

Dynamic properties of the dispersion

The structure of the Smoluchowski equation is rather complex and it does not seem feasible to solve it directly. However, from the diffusion equation one can easily obtain an infinite set of coupled equations for the moments of the distribution, or for the dynamic correlation functions. In order to solve the set we propose a simple procedure which appears to provide good results not only for the rigid dipole model but in any arbitrary situation including a moving suspension, the case of finite anisotropy energy, or for time dependent external fields. The method is based on the *decoupling approximation* of some of the quantities involved in the analysis, to be precise, of the quantities which vanish at equilibrium, when averaged, and those which are different from zero [9]. In particular, from the

dynamic equations for the correlation functions we are able to provide approximate expressions for their characteristic relaxation times, and they constitute the starting point to determine the transport coefficients using, for instance, the Green–Kubo formulas.

Rotational viscosity

The rotational viscosity of the suspension can be obtained from the corresponding Green–Kubo formula. This formula gives this transport coefficient in terms of the correlation function of the axial vector, $\vec{\Pi}_p^{(a)}$, related to the antisymmetric part of the contribution of the particles to the pressure tensor [10],

$$\eta_r = \frac{1}{V k_B T} \int_0^\infty dt \langle \Pi_{p,y}^{(a)}(t) \, \Pi_{p,y}^{(a)}(0) \rangle \tag{4}$$

where V is the volume of the system.

The rotational viscosity is a function of the parameters μ and σ, and of the ratio D_m/D_r. This last dependence comes from the fact that any departure of the magnetic moment of the particle from the equilibrium orientation is accompanied by a precession of the vector $\hat{R}$ with the corresponding dissipation of energy, as well as the dissipation due to the rotation of the particle in the viscous fluid. Previously, we have indicated that the parameter $D_m = k_B T h$ could be interpreted as a diffusion coefficient of the magnetic moment inside each particle. Thus, for a given value of the Brownian rotational diffusion coefficient D_r, the smaller the value of D_m the greater is the dissipation of energy and consequently the higher the rotational viscosity. This is exactly what we observe in Fig. 1. The initial slope of the curves depends on the ratio D_m/D_r, in such a way that the greater the ratio D_m/D_r the greater the value of μ to attain the saturation limit corresponding to a given value of σ. Moreover, the rotational viscosity tends to zero as a second power of the magnetic field strength, and reaches different saturation values for the different values of σ. The magnetic torque opposes the free rotation of the particles, giving rise to this viscosity, which characterize the macroscopic behavior of the magnetic fluid. Thus, for high values of σ, as the magnetic moment is more attached to the particle, the rotational viscosity increases. The maximum value is reached for a suspension of rigid dipoles.

Relaxation of the birefringence. Comparison with experiments

The relaxation of the optical birefringence of a medium, in which this magnitude is induced by a dispersion of

Progr Colloid Polym Sci (1996) 100:96–100
© Steinkopff Verlag 1996

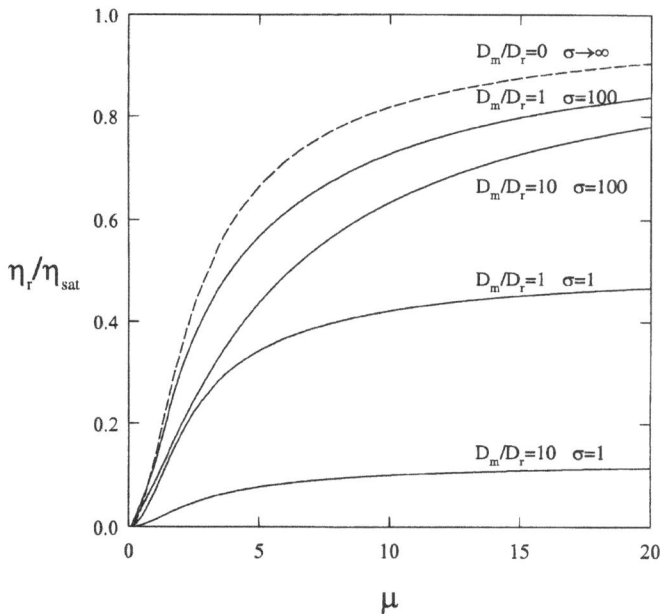

Fig. 1 Reduced rotational viscosity versus the parameter μ for different values of the ratio D_m/D_r and of the anisotropy parameter, as indicated in the plot. The dashed line corresponds to the rigid dipole limit

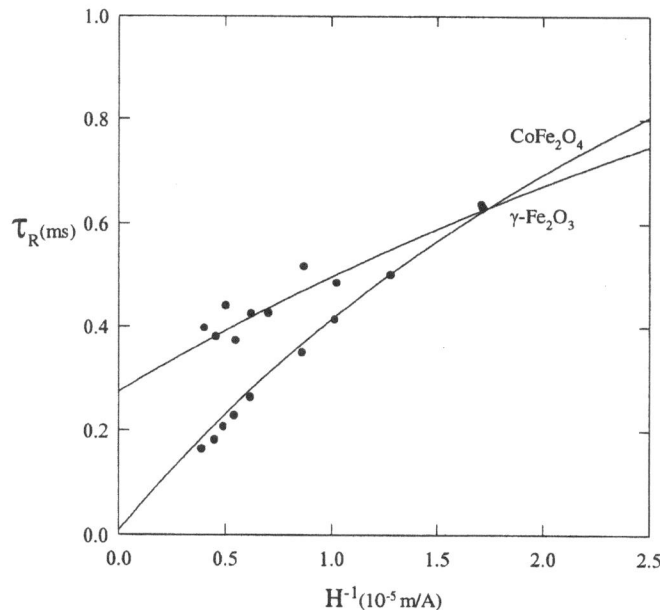

Fig. 2 Relaxation time of the particles as a function of H^{-1} for the Co-ferrite and maghemite samples. Experimental data of ref. [11] correspond to the dots. The dashed line corresponds to the limit $\mu \gg 1$ for the Co-ferrite

magnetic particles, is among the simplest experimental tools available for rheological studies of viscoelastic solutions. The main limitation of the usual transient birefringence devices is the polydispersity of the magnetic particles. Usually, a log-normal distribution of the diameters of the spheres is suitable to describe the sample. However, in order to get an exponential relaxation with only one characteristic time one must use a sample for which the dispersion of this distribution is very small.

The relaxation time associated with the reorientation of the particles under the action of a large field, τ_R, has been recently measured by Bacri et al. [11]. This quantity comes from the relaxation time of the light intensity collected in a photocell after crossing the sample, and when applying additional perpendicular pulses of magnetic field to perturb the system. These experiments permit us to distinguish the different relaxation regimes occurring when the size and the nature of the magnetic material of the particles are modified, which are determined by the parameters μ and σ. In Fig. 2, we represent the analytical expression we found for this time and compare it with the experimental data. The data correspond to two samples of magnetic particles of the same mean size but made of different magnetic materials, namely $CoFe_2O_4$ and $\gamma - Fe_2O_3$. The Co-ferrite sample has an anisotropy constant $K = 2 \cdot 10^5$ J/m³, and the saturation value of the magnetization is $M_s \simeq 250$ KA/m. For the maghemite,

the anisotropy constant is $K = 4 \cdot 10^3$ J/m³ and $M_s \simeq 270$ KA/m. With these values, the Co-ferrite sample can be considered as a rigid dipole ($\sigma \gg \mu$). On the other hand, the maghemite particles are such that $\mu \geq \sigma$. Regarding the values of the anisotropy constant, σ, and the ratio $m/k_B T = \mu_0 M_s V_m/k_B T$, we have taken $\sigma \sim 15$, $m/k_B T \sim 2.8 \cdot 10^{-4}$ m/A for the maghemite and $\sigma \sim 565$, $m/k_B T \sim 1.8 \cdot 10^{-4}$ m/A for the Co-ferrite. For both samples, $\tau_B \sim 4.5$ ms and $D_m/D_r \sim 1$.

As observed in the experiments for the Co-ferrite sample, τ_R tends to zero when $H \to \infty$. Both $\hat{R}$ and $\hat{n}$ quickly relax towards the field direction due to the rigidity of the dipoles. For the maghemite sample τ_R tends to a fixed, non-vanishing value (~ 0.3) ms. Under these conditions, the magnetic moments rapidly relax towards the field direction, but due to the moderate value of σ, the relaxation of the easy axis of magnetization, $\hat{n}$, or in other words, the mechanical relaxation of the particles, takes place in a finite period of time. In Fig. 2, we have also represented the extrapolation of the relaxation time coming from rather simple considerations made in ref. [11] for the rigid dipole limit under the action of a very large external magnetic field. Our results agree with the asymptotic behavior in its validity range, but at the same time, they show the deviations at intermediate and low magnetic field. These simple arguments can also be proposed for the opposite case $\mu \gg \sigma$ reproducing the asymptotic value $\tau_R \sim 0.3$ ms for the maghemite sample, but they are not able to explain the μ-dependence of the relaxation time for this material.

Conclusions

We have presented a general formalism to study the relaxation dynamics of ferromagnetic particles with the main purpose of providing explicit expressions for the viscosity and relaxation times, which can be applied to different situations ranging from the rigid-dipole limit to the limit where the anisotropy energy is dominant. Additionally, the appropriate relaxation times enable us to characterize different properties of the material, such as the birefringence and the complex magnetic susceptibility. The later is also obtained by a similar procedure, although it has not been considered in this paper.

The Smoluchowski equation allows us to obtain a hierarchy of dynamic equations for the different correlation functions that can be closed using appropriate decoupling approximations. The dynamic correlation functions provide expressions for the characteristic relaxation times, and they constitute the starting point to determine the transport coefficients using the Green–Kubo formulas. In particular, we have seen that the rotational viscosity reaches a saturation limit and depends on both parameters μ and σ. It is interesting to emphasize that the theory we have developed encompasses a wide number of situations characterized by the values of the magnetic and anisotropy energies of the ferromagnetic particles.

To check the validity of our formalism, we have compared our results for the relaxation time of the particles to birefringence experiments carried out with two types of ferromagnetic material, for which the dimensionless parameters μ and σ satisfy: $\sigma \gg \mu$ and $\mu \geq \sigma$ or $\mu \geq \sigma$, respectively. The first case corresponds to a rigid dipole, whereas in the second, the magnetic moment relaxes towards the field independently of the particle axis, which also relaxes towards the magnetic moment at a longer time scale. In both cases, our results are largely in agreement with experimental data.

Acknowledgments This work has been supported by DGICYT (Spain) under grant PB92-0895. MCM acknowledges support from the Generalitat de Catalunya.

References

1. Cabuil V, Bacri J-C, Perzynski R (eds) (1993) Proceedings of the Sixth International Conference on Magnetic Fluids. North-Holland, Amsterdam
2. Coffey WT, Cregg PJ, Kalmykov YuP (1992) In: Prigogine I, Rice SA (eds) On the Theory of Debye and Neel Relaxation of Single Domain Ferromagnetic Particles, Adv in Chem Phys Vol 83. Wiley Interscience, New York. p 263, Waldron JT, Kalmykov YuP, Coffey WT (1994) Phys Rev E 49:3976
3. Fannin PC, Charles SW, J Phys D Appl Phys 22 (1991) 187 (1989); J Phys D Appl Phys 24:76
4. Bacri JC et al. (1995) Phys Rev Lett 75: 2128
5. Raikher YuL, Shliomis MI (1994) In: Coffey W (ed) Relaxation phenomena in Condensed Matter. Adv in Chem Phys, Vol 87. Wiley Interscience, New York, p 595
6. Bashtovoy VG, Berkovsky BM, Vislovich AN (1988) Introduction to Thermomechanics of Magnetic Fluids. Springer-Verlag, Berlin
7. Rubí JM, Pérez-Madrid A, Miguel MC (1994) J of Non-Crystalline Solids 172:495
8. Shliomis MI, Stepanov VI (1993) J Magn Magn Mater 122:196
9. Miguel MC, Rubí JM (1995) Phys Rev E 51 No. 3 2190
10. Miguel MC et al. (1993) Physica A 193:359
11. Bacri J-C et al. (1993) In: Garrido L (ed) Proceedings of the XII Sitges Conference. J Magn Magn Mater 123:67. Springer-Verlag, Berlin

Progr Colloid Polym Sci (1996) 100:101–106
© Steinkopff Verlag 1996

A. Cebers

Two-dimensional concentration domain patterns in magnetic suspensions: Energetical and kinetic approach

Dr. A. Cebers (✉)
Institute of Physics
Latvian Academy of Sciences
Salaspils 1
2169 Latvia

Abstract The patterns formed in flat layers of magnetic suspensions and emulsions due to the competing short-range attractions and long-range magnetic repulsion forces are considered. On the basis of calculating energy of the ordered hexagonal and stripe structures of the liquid magnetics it was shown that the stripe phase is energetically preferred at the symmetrical distribution of the particles between concentrated and diluted phases. If one phase prevails the formation of the array of the separate droplets is expected. Results of the numerical calculation of the energies of the stripe and hexagonal patterns are confirmed by the numerical simulation results of the

kinetics of the magnetic field induced phase separation. Numerical simulation by the pseudospectral technique of the dynamics of the concentration perturbation show that at nonsymmetrical distribution of the particles between phases the periodic array of the separated droplets arises. Long-range magnetic repulsion forces are crucial for the formation of ordered arrays of the stripes or droplets since if they are absent characteristic coarsening phenomena at late stages of the phase separation are observed.

Key words Long-range interactions – stripe phase – hexagonal phase – concentration domains – coarsening

Introduction

Intricate patterns due to short-range attractive forces and long-range magnetic repulsion are formed in plane layers of magnetic fluids, magnetic suspensions and emulsions [1–4]. Although there exists definite understanding concerning behavior of the single magnetic droplets [1, 5, 6] full description of the different issues concerning collective properties of the patterns observed is absent at present. Here, we can mention only the approach towards the description of the disordered labyrinthine patterns by magnetic smectic approach [7, 8]. Among the issues where definite experimental data exist at present but theoretical

understanding is in fact lacking, one can mention the observation of the transition from the stripe phase to the hexagonal phase of the droplets [3], observation of the different morphologies of the patterns formed by the magnetic emulsions in dependence on the volume fraction of the droplets [9, 10], etc. It should be mentioned also here that even more complex behavior connected with the branchings of the droplets of the concentrated phase are possible [11].

Here the problem concerning the pattern formed by the magnetic suspensions or emulsions in plane layers under the action of the normal field is considered from two points of view. In the approximation of the vanishing thickness of the transition layer between concentrated phase and diluted one the energy contributions arising from interface boundaries and long-range magnetic interactions are calculated numerically for the magnetic liquid stripe phase and the hexagonal lattice of the magnetic droplets. It is shown that at near to equal amounts of the concentrated and diluted phases in the layer energetically the stripe pattern is preferred.

For situations not far from the critical point for the study of the field induced phase separation Ginzburg–Landau equations for the conserved order parameter with long-range magnetic interactions included are considered [12]. It was shown by numerical simulations that at equal amounts of the concentrated and diluted phases at the field induced phase separation the stripe phase is arising. In the case of the nonsymmetrical distribution of the particles between the phases the formation of the periodical system of magnetic droplets is confirmed.

Energy of the hexagonal and stripe phases

Let us have in the layer with thickness h the hexagonal lattice of the magnetic droplets with the radius of the

$$\times \left(1 - \frac{(2k^2-1)E}{k^3} - \frac{(1-k^2)K}{k^3}\right) + \frac{M^2 N}{2\pi\sigma S} \sum_{n \neq 0} \int dS \int dS'$$

$$\times \left(\frac{1}{\sqrt{(\vec{\rho} - \vec{\rho}' + \vec{\rho}_n)^2}} - \frac{1}{\sqrt{(\vec{\rho} - \vec{\rho}' + \vec{\rho}_n)^2 + h^2}}\right) \quad (1)$$

The second term accounts for the demagnetizing field energy of the single droplet, but the third accounts for the interaction between different droplets.

Here the following parameters are introduced: $\varphi = N\pi R_0^2/S$-volume fraction of the droplets, $Bm = M^2 h/\sigma$-magnetic Bond number, $y = 2R_0/a$, $x = h/a$, $k^2 = (2R_0/h)^2/(1 + (2R_0/h)^2)$.

For the calculation of the sum in the relation (1) the Ewald technique is employed:

$$\frac{1}{R} = \frac{2}{\sqrt{\pi}} \int_0^{1/2\sqrt{\tau}} e^{-t^2 R^2} dt + \frac{1}{R} erfc(R/2\sqrt{\tau}),$$

where τ is the Ewald parameter, the sum is split into two parts, one fast converging in the direct space, the other one fast converging in the Fourier space. As result the following expression for the energy is obtained

$$\frac{E_d}{2\pi\sigma S} = Bm\varphi(1-\varphi) + Bm\varphi\frac{4y}{3\pi x}\left(1 - \frac{(2k^2-1)E}{k^3} - \frac{(1-k^2)E}{k^3}\right) + \frac{Bmxy^4}{32\sqrt{3}}S \quad (2)$$

$$S = \frac{8}{\pi x^2} \sum_{n \neq 0} \int_{|\vec{r}| \leq 1} d\vec{r} f(r) \left\{\begin{array}{l} \frac{1}{|y\vec{r} + \vec{\rho}_n|} erfc(|y\vec{r} + \vec{\rho}_n|/\sqrt{\tau}) \\ - \frac{1}{\sqrt{(y\vec{r} + \vec{\rho}_n)^2 + x^2}} erfc(\sqrt{(y\vec{r} + \vec{\rho}_n)^2 + x^2}/\sqrt{\tau}) \end{array}\right\} - \frac{6}{x^2}\int_0^1 dr r f(r)\left\{\frac{1}{yr} erf\left(\frac{yr}{\sqrt{\tau}}\right)\right.$$

$$\left. - \frac{1}{\sqrt{y^2 r^2 + x^2}} erf\left(\frac{\sqrt{y^2 r^2 + x^2}}{\sqrt{\tau}}\right)\right\} + \frac{8\pi^2}{\sqrt{3}x^2}\left(x erf(x/\sqrt{\tau}) - \sqrt{\tau/\pi}\left(1 - \exp(-x^2/\tau)\right)\right)$$

$$+ \frac{8\pi^2}{\sqrt{3}x^2 y^2}\sum_{n \neq 0}\frac{J_1^2(k_n y)}{k_n^3}\left(\begin{array}{l} 2 erfc(k_n\sqrt{\tau}) \\ - (\exp(2xk_n))erfc(k_n\sqrt{\tau} + x/\sqrt{\tau}) + \exp(-2xk_n)erfc(k_n\sqrt{\tau} - x/\sqrt{\tau})\end{array}\right),$$

crossection R_0. The position of the center of the each droplet then will be determined by radius-vector $\vec{\rho}_n = n_1\vec{a}_1 + n_2\vec{a}_2$, where $\vec{a}_i$ are basic vectors of the hexagonal lattice. Accounting for the deviation of the field strength in the layer $\delta\vec{H}$ from the mean demagnetizing field $\langle\vec{H}\rangle$ as perturbation [13] it is possible to deduce the following expression for the demagnetizing field of the pattern

$$\frac{1}{8\pi}\int(\delta\vec{H})^2 dV/(2\pi\sigma S) = Bm\varphi(1-\varphi) + Bm\varphi\frac{4y}{3\pi x}$$

where $k_n = 2\pi(n_1^2 + n_2^2 - n_1 n_2)^{1/2}/\sqrt{3}$.

Surface energy of the magnetic droplets is accounted for by the relation

$$\frac{E_s}{2\pi\sigma S} = \frac{xy}{\sqrt{3}}.$$

For the numerical calculation of the energy of the hexagonal lattice of the droplets the following algorithm is applied. Double sums are transformed to $\sum_{n_1 > 0}\sum_{n_2 > 0}$. Internal sum is calculated up to the term 10^{-10}, external until the internal sum is less than 10^{-8}. Angular integration in the third term of the relation (2) was made by Tschebyscheff at 50 points in the interval $[0, \pi]$. Integral along radial variable is accomplished by Simpsons method. For the calculation of the Bessel function and error function interpolating formula are employed. The value of the Ewald parameter has been chosen equal to 0.6. For the calculation of the minimum of the energy of hexagonal lattice "golden mean" method has been applied.

Magnetic energy of the system of parallel stripes is expressed by the relation [13]

$$\frac{E_m}{2\pi\sigma S} = -\frac{Bm}{\pi}\frac{h}{l}\int_0^1 (1-t)\ln\left(1 + \frac{\sin^2(\pi\varphi)}{sh^2(xt/2)}\right)dt$$

$$+ Bm\varphi(1-\varphi). \tag{3}$$

Surface energy is expressed as

$$\frac{E_s}{2\pi\sigma S} = \frac{h}{\pi l}. \tag{4}$$

Condition of the minimum of the energy of the stripe pattern $E_m + E_s$ with respect to the structure period allows to obtain the equation for the period of the equilibrium pattern

$$\frac{h^2\pi^2}{l^2} = Bm\int_0^{\pi h/l} y\ln\left(1 + \frac{\sin^2(\pi\varphi)}{sh^2 y}\right)dy. \tag{5}$$

For the numerical calculation of the period of equilibrium structure Eq. (5) is solved by fixed point method. After that, according to the relations (3) and (4), the energy of the equilibrium stripe pattern is calculated.

The results of the energy calculations along the way described above can be summarized as follows. For the small volume fraction of the droplets the hexagonal pattern is preferable from the energetical point of view. At the volume fractions $\varphi > 0.38$ (for $Bm \in [50, 100]$) the stripe pattern is energetically more advantageous. What is also interesting to note is that the structure period of the equilibrium hexagonal pattern in wide region of the volume fractions of the concentrated phase depends on it quite weakly. That corresponds to the results of the experimental observations [14].

Numerical simulation of the field induced phase separation

It is now quite well established [15] that the thermodynamic stability of the liquid magnetic decreases at increase of the magnetic field strength. This means that for the given temperature there exists the critical value of the

magnetic field strength H_c such that

$$\frac{\partial\varphi}{\partial n}(n_c, T, H_c) = \frac{\partial^2\varphi}{\partial n^2}(n_c, T, H_c) = 0.$$

For small deviations from the critical field strength the thermodynamical potential can be expanded with respect to the deviation δn of the concentration from the critical one n_c:

$$f = \frac{1}{2}\frac{\partial^2\varphi}{\partial n\partial T}(H - H_c)(\delta n)^2 + \frac{1}{24}\frac{\partial^3\varphi}{\partial n^3}(\delta n)^4 + f_0(H) \tag{6}$$

taking into the account for the thermodynamical relation

$$\frac{\partial\varphi}{\partial n} = -\frac{\partial M}{\partial n};$$

the expression for the thermodynamical potential gives

$$f = f_0(H) - \frac{1}{2}\alpha(\delta n)^2 + \frac{1}{4}\gamma(\delta n)^4, \tag{7}$$

where

$$\alpha = \frac{\partial^2 M}{\partial n^2}(n_c, T, H_c)(H - H_c); \qquad \gamma = \frac{1}{6}\frac{\partial^3\varphi}{\partial n^3}(n_c, T, H_c).$$

Equation (7) shows that at $H > H_c$ there exist two phases with the concentrations $n = n_c \pm \sqrt{\alpha/\gamma}$.

Essential feature of the field induced phase separation consists of the fact that in the developing nonhomogeneous pattern long-range magnetic interactions between the inclusions of the arising phases must be accounted for. For that the energy of the magnetic field perturbations is calculated as follows. Up to the terms of the second order in δH thermodynamic potential

$$\tilde{F} = \int f dV - \frac{1}{8\pi}\int \vec{H}^2 dV,$$

accounting for the equation of the magnetostatic field

$$\text{div}(\mu\delta\vec{H}) = -4\pi\,\text{div}\left(\frac{\partial\vec{M}}{\partial n}\delta n\right)$$

and boundary conditions,

$$[\delta H_t] = 0; \qquad \vec{v}\left[\mu\delta\vec{H} + 4\pi\frac{\partial\vec{M}}{\partial n}\delta n\right] = 0$$

can be written as

$$\tilde{F} = \text{const} - \frac{1}{2}\int\frac{\partial\vec{M}}{\partial n}\delta n\delta\vec{H}dV.$$

Expressing the perturbation of the magnetostatic field through concentration perturbations ($\mu = 1$) which are supposed to be constant across the layer

$$\delta\psi = -\frac{\partial M}{\partial n}\int dS' \,\delta n(\vec{\rho}')\left(\frac{1}{\sqrt{(\vec{\rho}-\vec{\rho}')^2+(z-h)^2}}\right.$$
$$\left. -\frac{1}{\sqrt{(\vec{\rho}-\vec{\rho}')^2+z^2}}\right),$$

one arrives at the following expression for the energy describing the long-range magnetic interactions between the inclusions of the arising phases

$$\Delta\tilde{F} = \left(\frac{\partial M}{\partial n}\right)^2 \int dS\,\delta n(\vec{\rho})\int dS'\,\delta n(\vec{\rho}')\left(\frac{1}{\sqrt{(\vec{\rho}-\vec{\rho}')^2}}\right.$$
$$\left. -\frac{1}{\sqrt{(\vec{\rho}-\vec{\rho}')^2+h^2}}\right).$$

Since the phase separation is connected with arising of the new interfaces, it is necessary also to account for the corresponding surface energy. In the case when the deviations from the critical state are small, this can be done in the Cahn-Hilliard approximation [16] by adding the term to the volume density of the fluid $1/2\beta(\nabla n)^2$. As result, the following expression for the thermodynamical potential describing the field induced phase separation is obtained

$$\tilde{F} = h\int\left(-\frac{\alpha}{2}(\delta n)^2 + \frac{\gamma}{4}(\delta n)^4 + \frac{1}{2}\beta(\nabla\delta n)^2\right)dS$$
$$+ \left(\frac{\partial M}{\partial n}\right)^2\int dS\,\delta n(\vec{\rho})\int dS'\,\delta n(\vec{\rho}')\left(\frac{1}{\sqrt{(\vec{\rho}-\vec{\rho}')^2}}\right.$$
$$\left. -\frac{1}{\sqrt{(\vec{\rho}-\vec{\rho}')^2+h^2}}\right)$$

Linear phenomenological law for the particle flux according to

$$\frac{d\tilde{F}}{dt} = -T\frac{dS_t}{dt}$$

$$\frac{\partial h\delta n}{\partial t} = -\operatorname{div}\vec{J}$$

allows to obtain the following Ginzburg–Landau equation for the conserved order parameter (δ-friction coefficient of the particles)

$$\frac{\partial\delta n}{\delta t} + \frac{n_c}{\delta}\Delta\left(\alpha\delta n - \gamma(\delta n)^3 + \beta\Delta\delta n - 2\left(\frac{\partial M}{\partial n}\right)^2\right.$$
$$\left.\times\frac{1}{h}\int dS'\,\delta n(\vec{\rho}')\left(\frac{1}{|\vec{\rho}-\vec{\rho}'|} - \frac{1}{\sqrt{(\vec{\rho}-\vec{\rho}')^2+h^2}}\right)\right) = 0.$$

The following scalings are introduced: concentration – $\sqrt{\alpha/\gamma}$, length – $l = \sqrt{\beta/\alpha}$, time – $\delta l^2/n_c\alpha$. Then if the magnetic Bond number $Bm = 2M^2 h/\sigma$, where $M = \partial M/\partial n\,\alpha^{1/2}\gamma^{-1/2}$, $\sigma = \alpha^{3/2}\beta^{1/2}/\gamma$ is introduced, we arrive at the following differential equation for the concentration [12]

$$\frac{\partial\varphi}{\partial t} + \Delta\left(\varphi - \varphi^3 + \Delta\varphi - \frac{Bm}{(h/l)^2}\int dS'\,\varphi(\vec{\rho}')\right.$$
$$\left.\times\left(\frac{1}{|\vec{\rho}-\vec{\rho}'|} - \frac{1}{\sqrt{(\vec{\rho}-\vec{\rho}')^2+h^2}}\right)\right) = 0. \tag{8}$$

According to Eq. (8) for the growth increment of the periodic concentration distribution with the wave number k, we have

$$\lambda(k) = k^2\left(1 - k^2 - \frac{2\pi Bm}{(h/l)^2}\frac{1}{k}(1 - \exp(-kh/l))\right). \tag{9}$$

In the case $h \gg l$ according to Eq. (9) there exist modes with positive growth increment if $Bm < Bm_c$, where

$$\frac{2\pi Bm_c}{(h/l)^2} = 2\left(\frac{\sqrt{3}}{3}\right)^3. \tag{10}$$

Critical wave number in this case is

$$k_* = \left(\frac{\pi Bm_c}{(h/l)^2}\right)^{1/3}. \tag{11}$$

Relations (10) and (11) correspond to the scalings for the critical magnetic field strength in the layer with the layer thickness $H - H_c \approx h^{-2/3}$ and the period of the arising structure $\lambda \approx h^{1/3}$ found in [12], [17].

Thus, according to Eq. (8) in the flat layer at magnetic field strength greater than the critical one occurs growth of the periodic concentration perturbations at later stages leading to the formation of the 2D structure of the phase separated liquid magnetic. Morphologies of the structures arising can be studied by the numerical simulation.

For that, assuming the periodic distribution of the concentration in the layer $\varphi(x + L_x, y + L_y, t) = \varphi(x, y, t)$, it is possible to obtain the equations for the Fourier amplitudes of the concentration field by pseudospectral technique [18]. For the numerical simulations in the present case relative thickness of the layer h/l has been chosen equal to 5, but magnetic Bond number $Bm = 1$. The calculations have been carried out on the mesh 128×128 with the distance between points $\Delta_x = 1$ in our undimensional units. Simulations of the magnetic field induced phase separation have been carried out starting form the initial random perturbation around the homogeneous state in the interval $[-0.05; 0.05]$. Results of the numerical simulation of the phase separation starting from the randomly perturbed initial concentration $\varphi = 0$ are shown in Fig. 1. As one can see in this case according to the results of the

Progr Colloid Polym Sci (1996) 100:101–106
© Steinkopff Verlag 1996

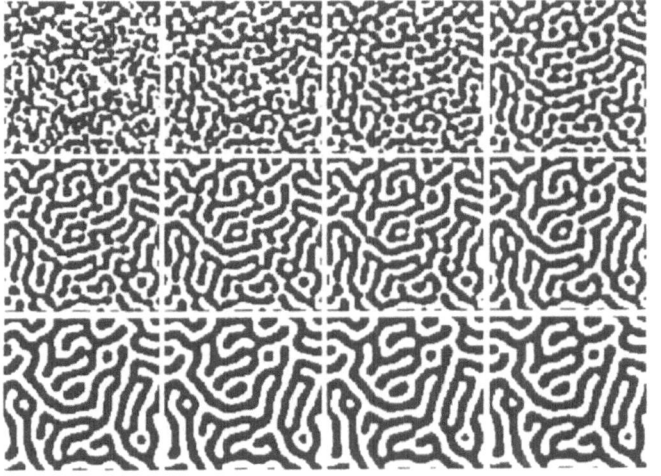

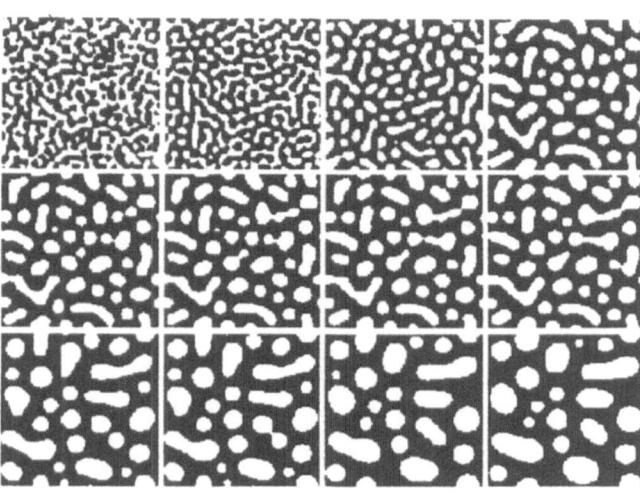

Fig. 1 Development of the stripe phase at symmetric distribution of the particles between concentrated and diluted phases. $h/1 = 5$; $Bm = 1$ which is below the critical value for initiating phase separation. Mesh 128×128. Sequential snapshots are made at following values of the undimensionalized time: 3.0; 15.1; 30.3; 75.7; 90.8; 121.1; 151.4; 302.8; 393.6; 484.5; 545.0; 787.2

Fig. 3 Coarsening at the phase separation far from critical point ($Bm = 0$). Mesh 128×128. In accordance with phase diagram [20] the formation of the ordered phases is not observed. Sequential snapshots are made at following values of the undimensionalized time: 3.6; 17.9; 71.4; 178.6; 214.3; 232.1; 250.0; 267.9; 607.1; 642.9; 928.6; 1107.1

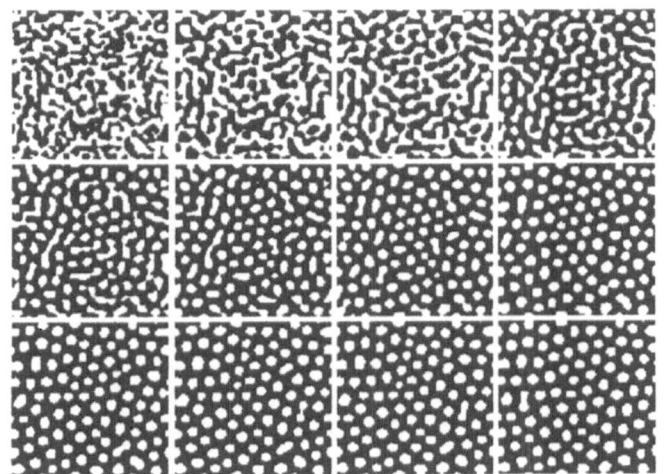

Fig. 2 Development of the hexagonal phase at nonsymmetric distribution of the particles between phases ($\langle\varphi\rangle = 0.2$). $h/1 = 5$; $Bm = 1$. Mesh 128X128. Sequential snapshots are made at following values of the undimensionalized time: 3.0; 15.1; 75.7; 90.8; 106.0; 121.1; 151.4; 302.8; 393.6; 469.3; 545.0

calculations presented in the first part of the paper, disordered stripe structure is formed. Results of the simulation in the case starting from randomly perturbed nonsymmetric initial concentration $\varphi = 0.2$ leads as expected to the formation of the disordered array of the separated droplets. It is interesting to note that the transition of 2D systems of the magnetic columns to glassy state has been experimentally studied in [19]. Kinetics of the pattern formation for that case is shown in Fig. 2. It is important to emphasize that for the formation of periodic concentration distributions the long-range magnetic repulsion forces are crucial. So in the case when magnetic forces are absent ($Bm = 0$) continuous coarsening of the pattern in time as it is shown on Fig. 3 occurs. That is also in agreement with the phase diagram of polarized amphiphile monolayers calculated in [20]; far from the critical point (at large magnetic field strength) ordered stripe and hexagonal phases disappear.

Acknowledgments This work is supported by ISF (long-term grant LJQ100) and Scientific Council of Latvia.

References

1. Cebers A, Maiorov MM (1980) Magnitnaya Gidrodinamika (In Russ) N1:27–35
2. Rosensweig RE, Zahn M, Shumovich R (1983) JMMM 39:127–132
3. Bacri JC, Perzynski R, Salin D (1987) Recherche 18:1150–1159
4. Langer SA, Goldstein RE, Jackson DP (1992) Phys Rev A46:4894–4904
5. Dickstein AJ et al. (1993) Science 261:1012–1015
6. Jackson DP, Goldstein RE, Cebers A (1994) Phys Rev E50:298–307

7. Cebers A (1994) Magnitnaya Gidrodinamika (In Russ.) 30:179–187
8. Bacri JC et al. (to be published)
9. Grasselli Y, Bossis G, Lemaire E (1994) J Phys II France 4:253–263
10. Jing Liu et al. (1993) In: 4th Er Fluids Conference Proceeding. Feldkirsh, Austria, pp 1–18
11. Hao Wang et al. (1994) Phys Rev Lett 72:1929–1932
12. Cebers A (1986) Magnitnaya Gidrodinamika (In Russ) 4:132–135
13. Cebers A (1990) Magnitnaya Gidrodinamika (In Russ) 3:49–54
14. Jing Liu et al. (1995) Phys Rev Lett 74:2828–2831
15. Cebers A (1982) Magnitnaya Gidrodinamika (In Russ) N2:42–48
16. Cahn JW, Hilliard JE (1958) J Chem Phys 28:258–267
17. Cebers A (1988) Magnitnaya Gidrodinamika (In Russ) N2:57–62
18. Rogers TM, Elder KR, Desai RC (1988) Phys Rev B37:9638–9649
19. Hwang YH, Wu X-L (1995) Phys Rev Lett 74:2284–2287
20. Andelman D, Brochard Fr, Joanny J Fr (1987) J Chem Phys 86:3673–3681

Progr Colloid Polym Sci (1996) 100:107–111
© Steinkopff Verlag 1996

COLLOIDAL PARTICLES

O. Valls
M.L. Garcia
X. Pagés
J. Valero
M.A. Egea
M.A. Salgueiro
R. Valls

Adsorption–desorption process of Sodium Diclofenac in Poly-alkylcyanoacrylate nanoparticles

Prof. Dr. O. Valls (✉) · M.L. Garcia
X. Pagés · J. Valero · M.A. Egea
M.A. Salgueiro · R. Valls
Unitat de Fisicoquímica
Facultat de Farmàcia
Universitat de Barcelona
Avda. Joan XXIII s.n.
08028 Barcelona, Spain

Abstract The adsorption–desorption process of Sodium Diclofenac in nanospheres of Polybutylcyanoacrylate (PBCA) and Poly-isobutyl-cyanoacrylate (PIBCA) was studied. The adsorption efficiency was obtained according to the type of polymer (PBCA or PIBCA), the molecular weight and concentration of the stabilizers of the polymerization process (dextranes), and the pH of the adsorption medium. The adsorption increases by dextrane addition at the polymerization medium (maximum for the Dextrane 70) and decreases when the pH of the adsorption medium increases. The adsorption is better in PIBCA than in PBCA.

A multilayer type BET's adsorption isotherm has been proposed. The desorption (release) of the drug, previously adsorbed in nanoparticles of PBCA and PIBCA with regards to the pH of de medium was also studied. The kinetics of drug release was a square root of time diffusion process, depending on the polymer (higher in the PBCA), and the pH of the medium (decreases when the pH increases). A variable amount (15–40%) of the drug remains loaded to the nanoparticles.

Key words Diclofenac – nanoparticles – alkylcyanoacrylates – polymer-adsorption – polymer-desorption

Introduction

Nanoparticles are solid colloidal nanometer-size carriers of polymeric nature in which the drug is dissolved, entrapped, encapsulated or adsorbed [1].

Despite that different kinds of polymers can be used in the design of nanoparticles as drug delivery systems, polyalkylcyanoacrylates are one of the most suitable due to their biocompatibility, biodegradability and ability to control drug distribution in the body [2].

It has also been show that the efficacy of nanoparticle formulations and the drug release profile from polymeric matrix depends on the way in which the active principle is incorporated. Usually, the drug is present during the polymerization process, so it is entrapped into the nanoparticle matrix. In these conditions a high amount of drug is incorporated into the particles.

The hydrophilic drugs (such as sodium diclofenac) are very difficult to incorporate into the lipophilic nanoparticle material. In this case, it is better to attempt to add the drug to the nanoparticles by an adsorption procedure.

The main purpose of this work is to study the association by adsorption of a hydrosoluble drug, Sodium Diclofenac, onto free nanospheres of Polybutylcyanoacrylate (PBCA) and Polyisobutylcyanoacrylate (PIBCA) prepared previously by the emulsion–polymerization procedure described by Couvreur et al. [3].

In order to improve the stability and the adsorption efficiency of the colloidal system several amounts of dextran of different molecular weights were added to the polymerization medium of free nanospheres. The effect of the dextrane on the drug adsorption has been evaluated. The effect of several other parameters able to modify the adsorption process, such as drug concentration or pH value of the adsorption medium, have also been evaluated.

Finally, factors affecting the drug desorption and the release profile from nanoparticles were also studied.

Materials and methods

Products

Sodium Diclofenac was purchased from Sigma (St. Louis, USA). Butyl-2-cyanoacrylate (Sicomet, BCA) was given by Sichel-Werke GmbH. Isobutyl-2-cyanoacrylate was purchased from Sigma (St. Louis, USA). Dextran 70 and 500 were supplied by Sigma and Dextran 10 and 100 were purchased from Pharmacia Fine Chemicals.

Preparation of nanoparticles

Drug-free polyalkylcyanocrylate nanoparticles with different dextrans, were prepared using the emulsion polymerization procedure proposed by Couvreur et al. [3]. After polymerization, the colloidal system was neutralized (pH 7) and centrifuged (2 h at 4000 rpm). The obtained sediments were resuspended with water solutions of Sodium Diclofenac at different pH values and stirred for 1 h.

Particle size analysis

Average particle size and polydispersity were evaluated by Photon Correlation Spectroscopy in a Malvern Autosizer II C (Malvern Instruments, Malvern, UK) [4].

Association efficiency

The adsorbed and free Diclofenac was measured by high performance liquid chomatography (HPLC), using a Hewlett Packard HP 1090 instrument with UV detector [5].

Determination of drug adsorbed to nanoparticles was carried out by separation of the free drug from drug loaded nanoparticles by centrifugation of the dispersion at 4000 rpm, for 2 h, in a Kontron ultracentrifugue. Sodium Diclofenac was determined in both clear liquid and sediment.

Drug desorption

The *in vitro* drug released from nanoparticles, along the time, at different pH values, was measured by a membrane diffusion technique [6]. The receptor medium was 250 ml of distilled water at 37 °C adjusted at different pH values with HCl 0.1 N.

Results

Frequently, during the preparation process of nanoparticles, dextran is added together with the monomer to the preparation medium. So this polysaccharide forms a covalent bond with the polymer giving a hydrophilic character to the nanoparticle surface. In this work the effect on the adsorption ability of the nanoparticles of the addition of variable amounts of dextrans of differents molecular weights was studied. Figure 1 shows the effect of the molecular weight of dextran, added at 0.5% concentration to the polymerization medium, on the adsorption efficiency (% of Sodium Diclofenac added to adsorption medium adsorbed on polymer surface). Figure 2 shows this adsorption efficiency as a function of dextrane 70 concentration in the adsorption medium.

The Sodium Diclofenac adsorption is related with the average nanoparticle size and polydispersity for PBCA and PIBCA, at two pH values, in Table 1. The effect of pH in the adsorption efficiency is shown in Fig. 3.

The main factor affecting the adsorption efficiency is the concentration of Diclofenac in the adsorption medium. This influence is shown in the Fig. 4.

The desorption (release) of Sodium Diclofenac from PBCA and PIBCA nanoparticles as a function of time was measured at two pH values (6.0 and 6.3). The release profiles are ploted in Fig. 5. The plots shows a quicker

Fig. 1 Average adsorption efficiency of Sodium Diclofenac on PBCA and PIBCA nanoparticles as a function of the molecular weight of dextrane added at the concentration of 0.5% to the polymerization medium. pH value of the adsorption medium 7.01

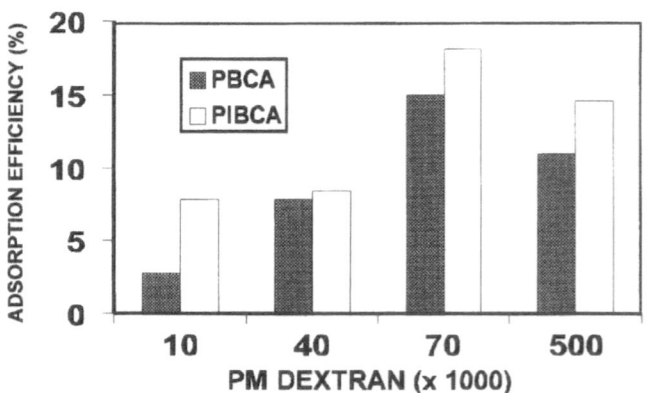

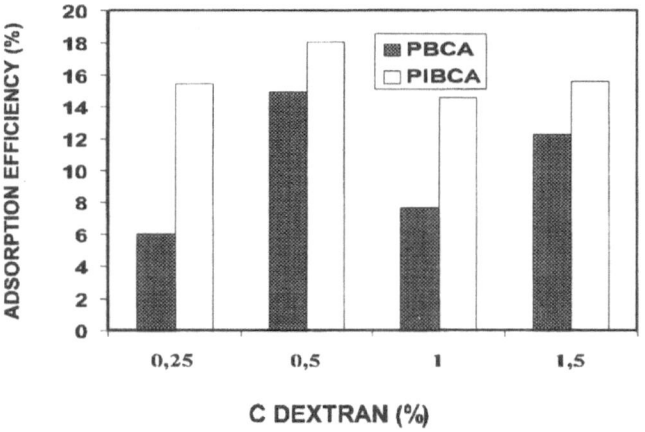

Fig. 2 Average adsorption efficiency of Sodium Diclofenac on PBCA and PIBCA nanoparticles as a function of the Dextrane 70 concentration added to the polymerization medium. pH value of the adsorption medium 7.01

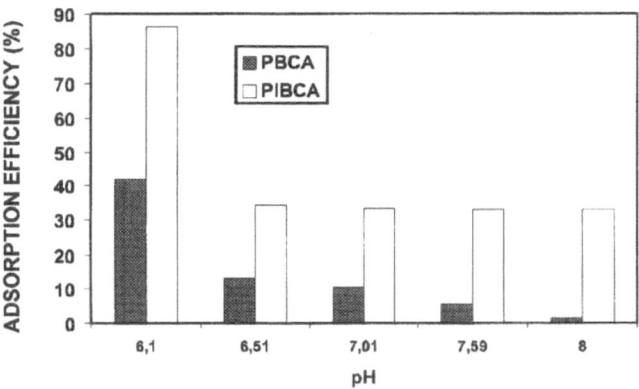

Fig. 3 Average adsorption efficiency of Sodium Diclofenac on PBCA and PIBCA nanoparticles, with Dextrane 70 at 0.5%, as a function of the adsorption medium pH value

Fig. 5 Release profiles of Sodium Diclofenac adsorbed on PBCA and PIBCA nanoparticles, made with 0.5% Dextrane 70, at two different pH values of the receptor medium

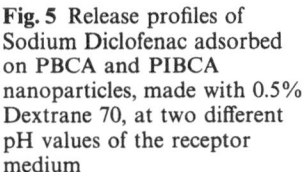

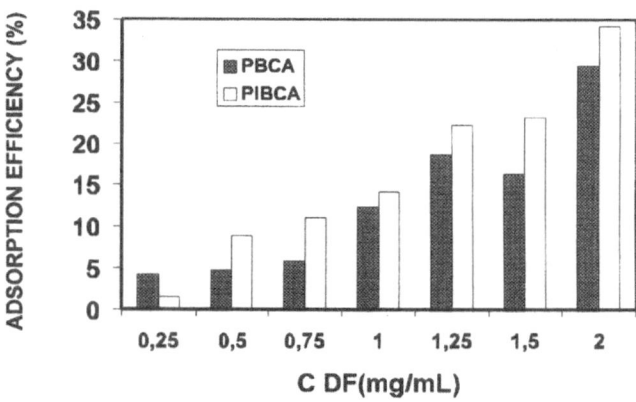

Fig. 4 Average adsorption efficiency of Sodium Diclofenac on PBCA and PIBCA nanoparticles, with Dextrane 70 at 0.5%, as a function of the concentration of the drug in the adsorption medium

release when the nanoparticles were made with PBCA and when the pH value was more acidic (pH 6.0).

Discussion

PIBCA nanoparticles show higher adsorption of Sodium Diclofenac than PBCA nanoparticles. This can be related to the lower size of PIBCA nanoparticles and to the fact that this polymer is branched out and consequently more porous than PBCA.

Dextran added to the polymerization medium contributes to stabilize the colloidal system. This polysacharide improves drug adsorption by increasing the number of the OH- groups in the nanoparticles surface. That fact facilitates the binding of hydrosoluble drugs, such as the Sodium Diclofenac.

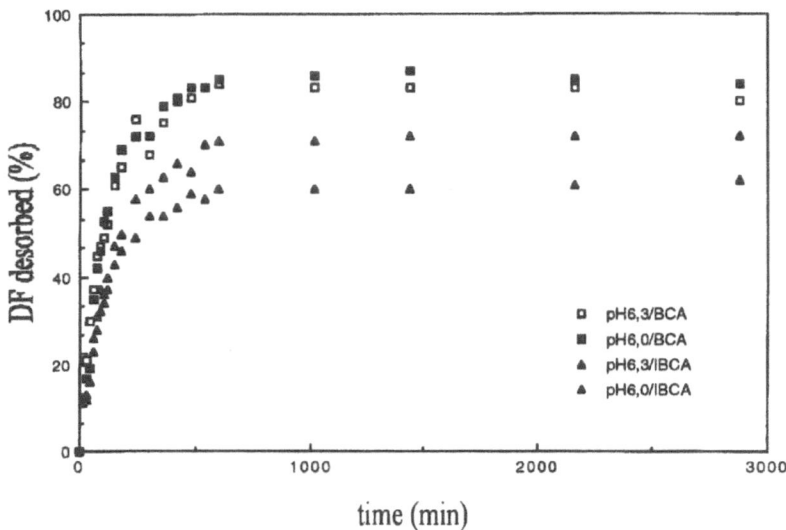

110

O. Valls et al.
Adsorption–desorption of Sodium Diclofenac in PBCA and PIBCA nanoparticles

	pH	PBCA Size (nm)	PBCA Polidispersity	PBCA A.D. (mg/mL)	PIBCA Size (nm)	PICBA Polidispersity	PICBA A.D. (mg/mL)
Table 1 Average adsorption efficiency (%) of Sodium Diclofenac related to the size and polydispersity of nanoparticles of PBCA and PIBCA, with Dextrane 70 (0.5%), at two pH values of the adsorption medium (A.D. Adsorbed Diclofenac)	6.0	260.6	0.094	0.450	236.1	0.099	0.523
	6.3	297.1	0.122	0.062	240.6	0.148	0.041

Molecular weight of dextrans is related to the adsorption efficiency, probably by the decrease of the particle size. The best adsorption efficiency was obtained with Dextrane 70 which reduces the size of the nanoparticles until 200–250 nm (see Table 1).

Sodium Diclofenac adsorption efficiency increases at low pH values of the adsorption medium because the lipophility is higher and thus its affinity to the polymer is better.

The adsorption isotherms obtained plotting the quantity of Sodium Diclofenac adsorbed on the free nanoparticles as a function of its concentration in the solution (Fig. 6) suggest a multilayer adsorption following a modified form of the BET's equation [7], proposed as:

$$\frac{C}{X(C_s - C)} = \frac{1}{X_m b} + \frac{b-1}{X_m b} * \frac{C}{C_s}, \tag{1}$$

where X is the adsorbed drug, C is concentration in the solution, C_s the solubility of the drug, X_m the adsorbed drug necessary to recover the nanoparticle forming a monomolecular film, and b a thermodynamic parameter related to adsorption enthalpy.

The Sodium Diclofenac desorption profile show a faster release for the PBCA nanoparticles and for high pH values.

The *in vitro* release efficiency of a drug from an inert matrix, like polymeric nanoparticles, usually follows the Higuchi equation [8], according to the square root relationship between the drug released and the time:

$$Q = k t^{1/2}, \tag{2}$$

where Q is the amount of drug released; t is the time; k is the liberation constant

$$k = ((De/\tau)(2A - eCs)Cs)^{1/2},$$

where D is the diffusion factor; e is the porosity factor of the matrix; τ is the tortuosity factor of the matrix; A the amount of drug in the matrix (weight/volume); Cs is the solubility of the drug. The above square relationship for the obtained results of this works led to the plot of Fig. 7. Two steps in the liberation profile are shown: first, during 7–8 h, there is a good correlation between the experimental data and the Higuchi equation, later the release efficiency decreases and a variable quantity of drug varying between 15 and 40% remains loaded to the polymer matrix.

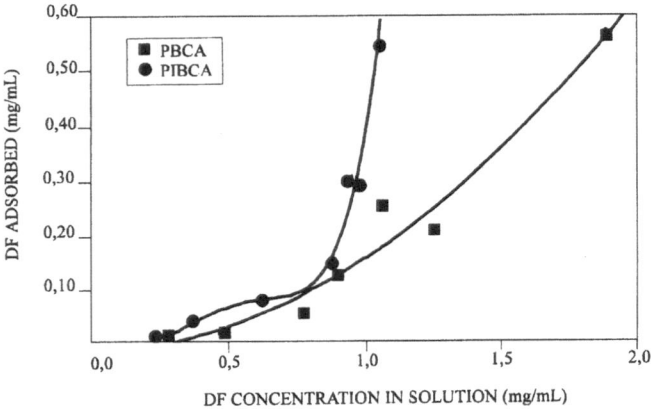

Fig. 6 Adsorption isotherms of Sodium Diclofenac on free PBCA and PIBCA nanoparticles made with 0.5% of Dextrane 70. Adsorption medium at pH 7.01 value

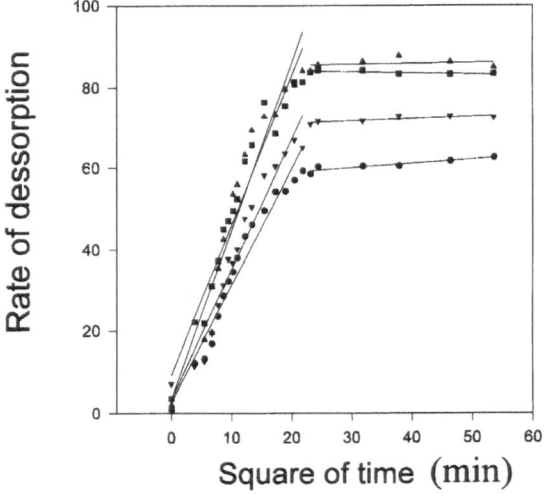

Fig. 7 Amount of Sodium Diclofenac released as a function of the square root of time ▲ PBCA pH = 6; ■ PBCA pH = 6.3; ● PIBCA pH = 6; ▼ PIBCA pH = 6.3

The Higuchi equation assumes that the solid-liquid interface has infinite size. After a determined time (t), the water has penetrated the polymer matrix to a determined depth. At the interface, the concentration (C) of dissolved drug is equal to the solubility of the drug Cs. In the dissolution medium outside the nanoparticles, the concentration is equal to zero. According to Fessi [9] there

Progr Colloid Polym Sci (1996) 100:107–111
© Steinkopff Verlag 1996

is a concentration gradient in the polymer, from $C = Cs$ at the interface to $C = 0$ on the outside of the nanoparticle. In practice, as long as there remains a part of the nanoparticle which has not been wetted by the desorption medium the Higuichi equation is valid. When the nanoparticle is completely wetted, C becomes lower than Cs, and the Higuchi equation can not longer be used. At this point a rapid change in the slope occurs, decreasing near to zero.

References

1. Kreuter J (1983) Pharm Acta Helv 58:196–208
2. Couvrer P (1988) CRC Crit Rev Ther Drug Carrier Systems 5:1–20
3. Couvreur P, Kante B, Roland P, Guiot P, Bauduin P, Speiser P (1979) J Pharm Pharmacol 31:331–332
4. Mc Connell ML (1981) Ann Rev Phys Chem 53:1007A–1018B
5. Keith KH Chan, Kunjbala H Vyas, Kenneth W (1982) Anal Letters 15:1649–1663
6. Washington C (1990) Int J Pharm 58:1–12
7. Brunnaver S, Emmet PH, Teller E (1938) J Amer Chem Soc 60:309–316
8. Higuchi T (1963) J Pharm Sci 52:1145
9. Fessi H, Puisieux JP, Marty JP, Cartensen JT (1980) Pharm Acta Helv 55:261

Progr Colloid Polym Sci (1996) 100:112–116
© Steinkopff Verlag 1996

COLLOIDAL PARTICLES

P. Verbeiren
F. Dumont
C. Buess-Herman

Determination of the complex refractive index of bulk tellurium and its use in particle size determination

Dr. P. Verbeiren (✉) · F. Dumont
C. Buess-Herman
Université Libre de Bruxelles
Service de Chimie Analytique
Faculté des Sciences
C.P. 255, 2, Bd. du Triomphe
1050 Brussels, Belgium

Abstract The complex refractive index of bulk tellurium as function of wavelength has been deduced from the measured optical properties of thin Te films. Films of different thicknesses were prepared by thermal evaporation of pure tellurium granules on fused silica substrates under high vacuum conditions. The thickness of these films was measured by Fizeau interferometry. Normal incidence reflectance measurements were achieved using a home-made reflectometer. This device was conceived to be adapted to an UVIKON 940 model spectrophotometer. The real and imaginary parts of the refractive index were calculated from reflectance and transmittance measurements using a method based on Koehler's work. Theoretical turbidity spectra of monodisperse and spherical Te hydrosols were calculated by the Mie theory using the obtained values of the refractive index.

Key words Tellurium – thin films – refractive index – spectroturbidimetry

Introduction

The knowledge of the wavelength dependence of the refractive index of materials is of major importance in colloid science. Its interest appears as well in the domain of the stability of colloidal suspensions, where it allows the computation of the Hamaker–Lifshitz constant, as in the particle size determination by the spectroturbidimetric method.

Tellurium is a poorly known system and the previously reported values of its refractive index are rare and not very reliable. In this paper new measurements of the refractive index of tellurium in the wavelength range 200–800 nm are presented and theoretical turbidity spectra of monodispersed Te hydrosols are calculated.

Calculation of the optical constants

The real and imaginary parts of the refractive index may be deduced from the measurements of the light transmitted and reflected by thin films of tellurium deposited on a definite substrate.

The principles of the method used below were described by Koehler [1].

Determination of the absorption coefficient k_1

Classical electromagnetic theory provides an equation which represents the dependence of the transmission coefficient T of a film with its thickness d_1 [2, 3].

If the absorption term, k_1, or the thickness of the film are sufficiently large, the transmission coefficient may then be written as

$$T = T_0 \exp\left(-4\pi \frac{k_1 d_1}{\lambda}\right), \tag{1}$$

where T_0 is a characteristic constant of the system, k_1 is the imaginary part of the refractive index of Te, d_1 is the thickness of the film and λ is the wavelength of the incident radiation.

Accounting for the presence of a substrate of finite thickness, Eq. (1) must be corrected for the multiple reflexions occurring in the substrate; this leads to

$$T_c = T_{0c} \exp\left(-4\pi \frac{k_1 d_1}{\lambda}\right) \tag{2}$$

or

$$\log \frac{1}{T_c} = \log \frac{1}{T_{0c}} + 5.457 \frac{k_1 d_1}{\lambda}, \tag{3}$$

where T_c is the corrected transmission coefficient.

$\log(1/T_c)$ is the usual definition of the optical density and is an experimental datum.

Thus, knowing the transmission optical densities of several films of different thicknesses at a given wavelength, Eq. (3) shows that a plot of $\log(1/T_c)$ versus the thickness d_1 gives a straight line the slope of which is $5.457\, k_1/\lambda$. The absorption term is deduced from this slope.

Determination of the real part of the refractive index n_1

k_1 being determined, the real part of the refractive index may be obtained using the simple equation for the normal incidence reflection coefficient R associated with the plane boundary between a semi-infinite dielectric and a semi-infinite absorbing medium (opaque film)

$$R = \frac{(n_0 - n_1)^2 + k_1^2}{(n_0 + n_1)^2 + k_1^2}, \tag{4}$$

where n_0 is the refractive index of the incidence medium (air).

In well-defined circumstances, an absorbing film deposited on a finite substrate may effectively have the same reflection coefficient as the boundary between two infinite media. If the film is sufficiently absorbing, the contribution of the fractions of light reflected at the boundary Te – substrate to the total reflected light will be negligible because of the absorption through the film.

Resolving Eq. (4) for n_1 leads to

$$n_1 = n_0 \left(\frac{1+R}{1-R}\right) \pm \sqrt{n_0^2 \left(\frac{1+R}{1-R}\right)^2 - (n_0^2 - k_1^2)}. \tag{5}$$

The choice of the sign to be adopted will be discussed below.

Preparation of the films

The films were prepared by thermal evaporation of pure tellurium granules under high vacuum conditions ($5\,10^{-6}$ Torr) on fused silica substrates. The granules were held in a tantalum boat. The films were deposited at a mean rate of 10 nm/s. The thickness of the films was measured by Fizeau interferometry.

Measurements

Transmission measurements were made using a double beam UVIKON 940 spectrophotometer. Normal incidence reflectance measurements were achieved using a home-made reflectometer. This device has been specially designed to be adapted to an UVIKON 940 spectrophotometer and allows an absolute measurement of the reflection coefficient at quasi-normal incidence angle of 2°.

The two measurement configurations of this reflectometer, the principle of which has been described by Strong [4], are represented in Fig. 1. These two configurations allow an absolute and accurate measurement of the reflection coefficients of the films.

Results

Figure 2 shows a typical transmission spectrum of a Te film. It has to be noted that Te presents an absorption

Fig. 1 Representation of the two measurement configurations of the reflectometer. M1 and M2 are two deflecting mirrors placed at an angle of 90°; L is a converging lens and S1, S2 and S3 are the different supports for the mirror M3 and the sample. The path of the measurement beam is represented by the arrows

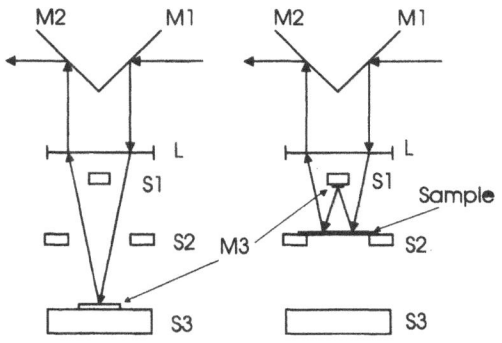

a Configuration   b Configuration

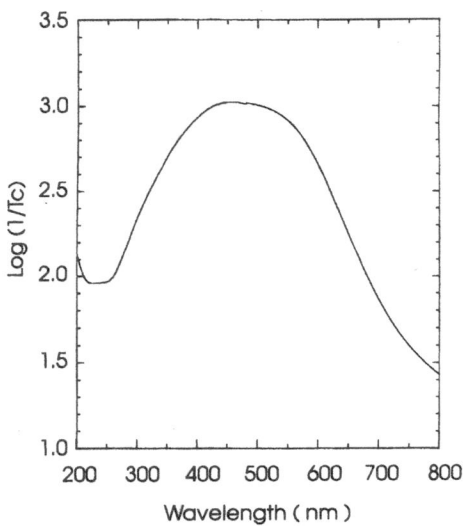

Fig. 2 Transmission spectrum of a Te film. ($d_1 = 74$ nm)

band in the wavelength range 200–800 nm. The values of the absorption term have been calculated as described previously. Plots of the transmission optical density versus thickness for different wavelengths are represented in Fig. 3. The best straight lines passing through the experimental points were obtained by linear regression.

The wavelength dependence of k_1 is presented in Fig. 4. As previously reported [5, 6], the absorption term exhibits a maximum at 550 nm.

The real term of the refractive index was deduced from the reflection measurements (Fig. 5). Two series of values for n_1 were obtained corresponding to the different solutions of Eq. (5). For $\lambda < 550$ nm, the lower values of n_1 were chosen while for $\lambda > 550$ nm the upper values were taken. This choice has been made in order to obtain the best agreement between our results and the previously reported values. However, additional measurements should be done to confirm these data. The spectral

Fig. 3 Transmission optical density versus thickness for different wavelengths: a) 200 nm, b) 375, c) 550 nm, d) 625 nm. Experimental points are represented by ∇ .

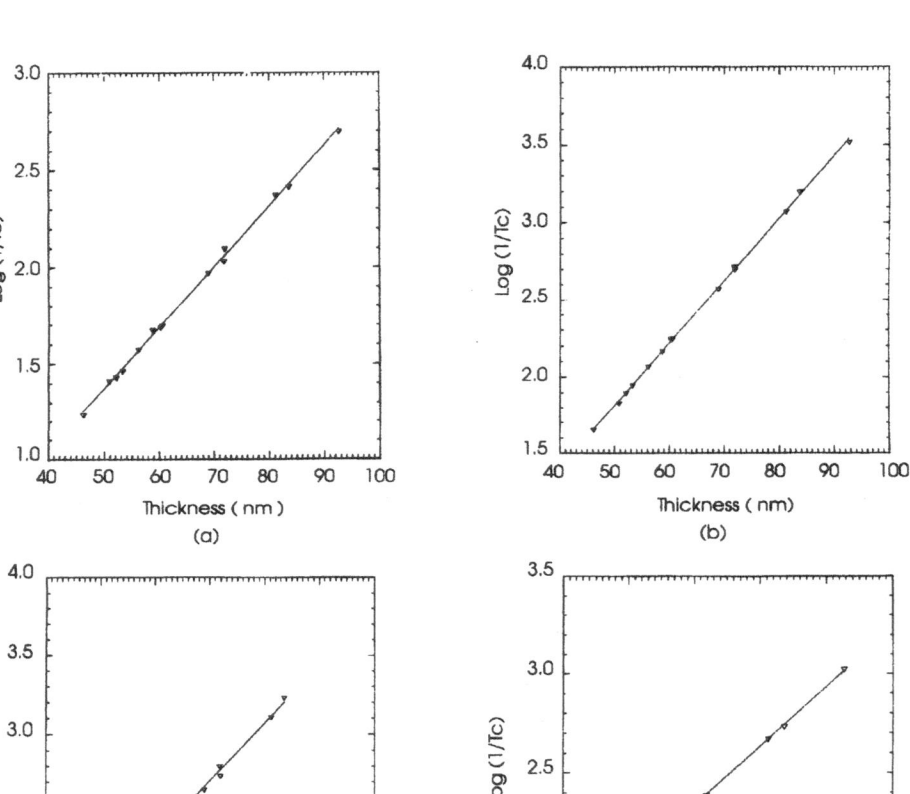

Progr Colloid Polym Sci (1996) 100:112–116
© Steinkopff Verlag 1996

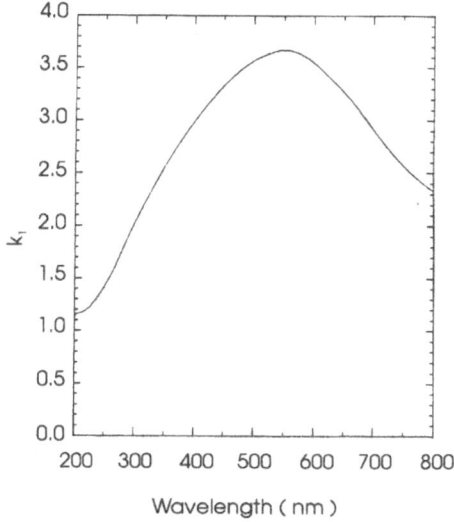

Fig. 4 Wavelength dependence of k_1

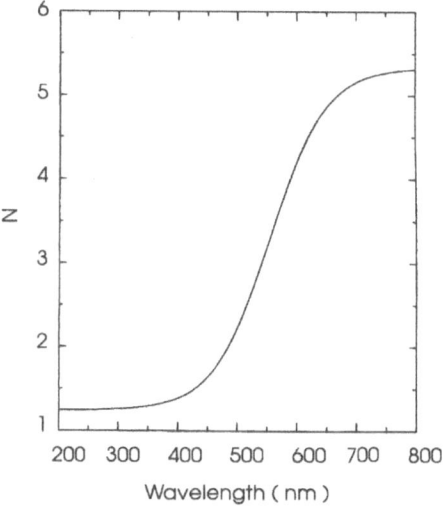

Fig. 6 Wavelength dependence of n_1

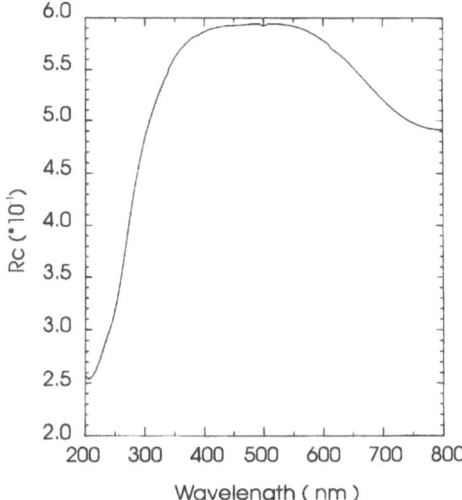

Fig. 5 Wavelength dependence of the reflection coefficient of a thick Te film

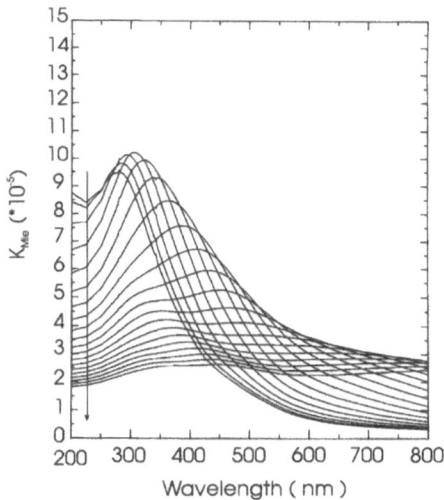

Fig. 7 Turbidity spectra of Te hydrosols. The arrow indicates the increase of the diameter from 10 to 200 nm by steps of 10 nm

dependence of n_1 is presented in Fig. 6. As predicted by dispersion theory, the real part of the refractive index strongly varies around the absorption band.

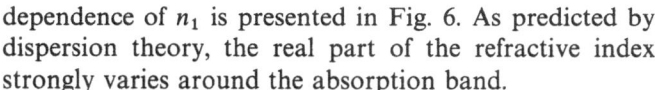

Theoretical turbidity spectra of Te hydrosols

Turbidity spectra of monodisperse Te hydrosols of different sizes were computed using the Mie theory with our values of the refractive index; they are presented in Fig. 7. It should be noted that the maximum shifts to higher wavelengths as the diameter increases. This peak also broadens with increasing size.

The position of the extremum of experimental spectra of Te hydrosols should be very useful for the determination of the particle size.

References

1. Koehler WF et al (1959) J Opt Soc Am 49:109
2. Vasicek A (1960) In: Optics of thin films. North Holland Pub Co, Amsterdam
3. Heavens OS (1965) In: Optical properties of thin solid films. Dover Pub Inc, New York
4. Strong J (1938) In: Procedures in experimental physics. Prentice Hall Inc, Englewood Cliffs, New-Jersey
5. Shklyarevskii IN et al (1974) Opt Spectrosc 36:457
6. Hodgson JN (1962) J Phys Chem Solids 23:1737

Progr Colloid Polym Sci (1996) 100:117–120
© Steinkopff Verlag 1996

J. Bonet Avalos
A.N. Semenov
A. Johner
J.F. Joanny
C.C. van der Linden

Structure of adsorbed polymer layers: Loops and tails

Dr. J. Bonet Avalos (✉)
Departament de Física Fonametal
Facultat de Física
Universitat de Barcelona
Diagonal 647
08028 Barcelona, Spain

A.N. Semenov
Physics Department
Moscow State University
Moscow 117234, Russia

A. Johner · J.F. Joanny
Institut Charles Sadron (UPR CNRS 022)
6, rue Boussingault
67083 Strasbourg, Cedex, France

C.C. van der Linden
Department of Physical and Colloid
Chemistry
Wageningen Agricultural University
Dreijenmplein 6
6703 HB Wageningen, The Netherlands

Abstract In this paper we discuss the elements of a recent mean-field theory developed to describe the structure of an adsorbed polymer layer. Recently it has been shown by Semenov and Joanny that the central part of an adsorbed polymer layer does not display a self-similar structure but the profiles depend on a characteristic length-scale z^*. This length separates the region dominated by the loops from the region where tails are dominant and it is only dependent on the size of the chain and on the dimension of the space. The results of the mean-field calculation are compared with the numerical solution of the full self-consistent field calculations.

Key words Polymer adsorption – scaling laws – loops and tails – interfaces

Introduction

Adsorbed layers of long flexible polymers may be much thicker than the monomer size a. This thick and diffuse polymer layer is responsible for important properties in colloid physico-chemistry as adhesion, lubrication, stabilization or controlled flocculation of colloidal dispersions [1].

In such a polymer layer one can roughly distinguish three main regions: *proximal* (dominated by wall effects, with a thickness of the order of the monomer size when the attractive potentials are of very short range); *distal* (the region at the border of the layer where the monomer concentration exponentially decays towards the bulk value) and a *central* region between both.

The structure of the proximal region needs a particular treatment [2] and may depend on monomer–monomer and monomer–wall interactions in real situations. For sufficiently large polymers (with polymerization index N of the order of 10^5, for instance) it is expected that the central and distal regions will show a rather *universal* behavior. We will focus on these regions where simple models may shed some light on the structure.

From the theoretical point of view, the structure of an adsorbed polymer layer has been studied in a *mean-field* picture by means of the so-called *ground-state dominance* [3], which essentially reduces the problem to the calculation of the order parameter $\psi_0(z)$, related to the monomer concentration by $\phi(z) \sim \psi_0^2(z)$. In such a procedure, one explicitly neglects the effects of the chain ends and considers that the layer is built up by loops, the chain ends being at infinity. In this way it is predicted that the monomer density profile decreases as $\phi(z) \sim z^{-2}$ all over the central region.

To incorporate the excluded volume correlations neglected in the mean-field picture, de Gennes [5] introduced a more elaborate procedure based on *Widom*'s [4] free energy functional, using the formal analogy between polymers with excluded volume interactions and critical phenomena [3]. In this approach, one still deals with a single-order parameter, which restricts its validity to regions where the effects of chain ends are again negligible. De Gennes predicted a *self-similar* structure of the adsorbed polymer layer without any characteristic length in the central region and a functional dependence of the density profile given by $\phi(z) \sim z^{-4/3}$. The mean-field result mentioned above is recovered if the analysis is performed in $d = 4$ dimensions, where excluded volume interactions are only marginally relevant.

The Dutch group of Wageningen [6] developed a procedure to numerically solve the complete self-consistent field problem, going beyond the validity of ground-state dominance results. They obtained a detailed view of the structure of the adsorbed polymer layer, which did not completely agree with the self-similar structure predicted by de Gennes.

Recently, by means of a scaling picture, Semenov and Joanny [7] showed that the self-similar structure of de Gennes predicted a density profile for monomers belonging to chain tails that monotonously increase with z. This indicated that at a certain distance of the wall, tails should dominate. They demonstrated that the layer has a characteristic length-scale $z^* \sim N^{1/(d-1)}$, where d is the dimensionality of the space, at which the density of monomers belonging to loops is of the same order of magnitude as that for monomers belonging to tails. For $z < z^*$ loops dominate while for $z > z^*$ tails dominate. Furthermore, when explicitly considering loops and tails in the structure, the authors found that in addition to the *Flory* exponent $\nu \simeq 3/5$, the so-called *susceptibility* exponent $\gamma \simeq 1.16$ was also involved.

We have developed a self-consistent field picture [8] beyond ground-state dominance explicitly taking into consideration the presence of tails in the adsorbed layer. We show that the central region is described by a set of two ordinary second order differential equations for *two* order parameters. Its simple structure permit to obtain all the asymptotic regimes for the concentration profiles as well as a far simpler numerical solution than the complete self-consistent field scheme. Although the excluded volume correlations are neglected, mean-field approaches serve to give a qualitative picture of physical situations on, which scaling analysis may be performed afterwards. In the same

spirit as in de Gennes approach in terms of Widom's free energy, some work is being done in the direction of constructing a free energy functional with two order parameter incorporating the excluded volume interactions and recovering the proper scaling behavior of the order parameters obtained in the scaling picture.

The elements of the mean-field theory and the comparison with the numerical results of the Dutch group are shown in what follows.

Loops and tails: two order parameters

In the mean-field picture the partition function of a chain section of n monomers, with an end at the space point z and the other elsewhere, satisfies the *Edwards* equation

$$\frac{\partial}{\partial n} \mathcal{Z}(n, z) = \frac{\partial^2}{\partial z^2} \mathcal{Z}(n, z) - \phi(z) \mathcal{Z}(n, z);$$

$$\text{with } \mathcal{Z}(0, z) = 1 , \tag{1}$$

where we have chosen a as the unit of length and $\phi(z) \equiv vc(z)$ stands for the effective volume fraction, v being the excluded volume parameter and $c(z)$ the number of monomers per unit volume. In this mean-field scheme, the total monomer concentration $\phi(z)$ is determined self-consistently. Therefore, the self-consistent field problem expressed by Eq. (1) involves the solution of a nonlinear integro-differential equation with partial derivatives.

However, due to the analogy between Edwards and Schrödinger equations, we can expand $\mathcal{Z}$ in terms of eigenfunctions $\psi_s(z)$ and eigenvalues E_s

$$\mathcal{Z}(n, z) = k_0 \psi_0(z) e^{\varepsilon n} + \sum_{s > 0} k_s \psi_s(z) e^{-E_s n}$$

$$\equiv \mathcal{Z}_a(n, z) + \mathcal{Z}_f(n, z) , \tag{2}$$

where $\varepsilon \equiv |E_0|$. Since the adsorbing potential wall is modeled by a δ-function and the wall is impenetrable, the spectrum of $\mathcal{Z}$ can be split into a *bound* state ($s = 0$) with $E_0 < 0$ and a continuum of *free* states ($s > 0$) and ($E_s > 0$). One should also realize that the partition function can be split into two depending on the boundary conditions at the wall. Effectively, we can write $\mathcal{Z} = \mathcal{Z}_a + \mathcal{Z}_f$ the first term on the right-hand side being finite at the wall and the second being zero. Both partition functions, however, are solutions of Edward's equation. In the limit of long chains and strong adsorption energy $N\varepsilon \gg 1$, we can approximately consider that $\mathcal{Z}_a$ can be identified with the bound state and the free states only contribute to $\mathcal{Z}_f$.

Such a distinction permits us to identify three *objects* in the layer: loops, tails and free chains, whose respective monomer densities are given by

$$\phi_l(z) = \frac{\phi_0}{N} \int_0^N dn \, \mathscr{L}_a(n,z) \mathscr{L}_a(N-n,z) \tag{3}$$

$$\phi_t(z) = \frac{\phi_0}{N} \int_0^N dn \, \mathscr{L}_a(n,z) \mathscr{L}_f(N-n,z) \tag{4}$$

$$\phi_f(z) = \frac{\phi_0}{N} \int_0^N dn \, \mathscr{L}_f(n,z) \mathscr{L}_f(N-n,z) \; . \tag{5}$$

To simplify the discussion, let us consider the very dilute case, in which the contribution of free chains can be neglected in the adsorbed layer, although calculations with different bulk solutions can also be carried out [8]. In view of Eqs. (2–5), note that while the monomer density due to loops is proportional to $\psi_0^2(z)$, the corresponding density due to tails is proportional to the product of $\psi_0(z)\,\varphi(z)$, the latter function being defined as

$$\varphi(z) \equiv \int_0^N dn \, e^{-\varepsilon n} \mathscr{L}_f(n,z) \; . \tag{6}$$

From Edwards equation (Eq. (1)), it is then possible to arrive at the set of differential equations for the two order parameters $\psi(z) \equiv \phi_0^{1/2} k_0 e^{\varepsilon N} \psi_0(z)$ and $\varphi(z)$

$$-\psi''(z) + (\phi(z)+\varepsilon)\psi(z) = 0 \tag{7}$$

$$-\varphi''(z) + (\phi(z)+\varepsilon)\varphi(z) = 1 \; , \tag{8}$$

where $\phi(z) = \phi_l(z) + \phi_t(z) = \psi^2(z) + B\psi(z)\varphi(z)$, is the total monomer density profile, determined self-consistently, and B a constant given by

$$B = \frac{\phi_0 k_0 e^{\varepsilon N}}{N} \; . \tag{9}$$

These two equations (7) and (8) can be obtained from the free energy functional

$$\mathscr{F} = \int_0^\infty dz \left(\psi'^2 + B\psi'\varphi' + \frac{1}{2}\phi^2 + \varepsilon\phi - \underline{B\psi} \right) \tag{10}$$

which, in addition to the usual contributions due to configurational entropy and excluded volume effects, explicitly incorporates an entropic term due to the presence of chain ends (underlined).

The characteristic length-scales in Eqs. (7) and (8) are, on one hand, that associated to $z^* \sim 1/B^{1/3}$ and, on the other, the cutoff length λ that determines the thickness of the adsorbed layer, related to ε according to $\lambda \sim 1/\varepsilon^{1/2}$.

Asymptotic behavior

The central regime is dominated by the existence of z^* and the self-similar profile is no longer valid. However, near the wall ($z/z^* \ll 1$) the total density profile is dominated by

loops while beyond z^* it is dominated by tails ($1 \ll z/z^* \ll \lambda/z^*$). This permits to find the asymptotic behaviors in these regions

$z/z^* \ll 1$	$1 \ll z/z^* \ll \lambda/z^*$
$\phi_l \sim 2/z^2$	$1800z^{*6}/z^8$
$\phi_t \sim 4z\ln(z/z^*)/z^{*3}$	$20/z^2$
$\phi \sim 2/z^2$	$20/z^2$

$$\tag{11}$$

Note the difference of a factor 10 in the prefactor for the total monomer density in both regimes. In the region $z \sim \lambda$, we have found an exponential decay for all densities

$1 \ll z/z^* \sim \lambda/z^*$
$\phi_l \sim z^{*6} e^{-2z/\lambda}/\lambda^8$
$\phi_t \sim e^{-z/\lambda}/\lambda^2$
$\phi \sim e^{-z/\lambda}/\lambda^2$

$$\tag{12}$$

In the asymptotic solution of Eqs. (7) and (8) in the region $z \ll z^*$ as shown in Eq. (11) we observe that between the wall and z^* the tail monomer density is an *increasing* function of the distance, all over the region where the contributions of the loops is dominant and decays as z^{-2}. In the region $z \gg z^*$, however, the contribution due to loops falls off very rapidly while the tail monomer density decays as z^{-2}.

Comparison with numerical full self-consistent field calculations

We now compare the numerical solution of Eqs. (7) and (8) with the full self-consistent field calculations of C.C. van der Linden by means of the Scheutjens-Fleer lattice method [9].

In Fig. 1 we plot in logarithmic scale the loop, tail and total monomer concentration versus distance, for $N = 10^5$, a very dilute bulk concentration $\phi_0 = 10^{-7}$. Straight lines indicate exponential decay. In this figure, the cutoff length λ is in fact smaller than z^*. The symbols correspond to van der Linden's lattice calculations while solid and dashed lines correspond to the numerical solution of Eqs. (7) and (8). R is chain's radius of gyration.

Figure 2 corresponds to another plot of the three concentrations but in this case for $N = 4 \times 10^4$, $\phi_0 = 10^{-7}$ and a stronger adsorption energy. Solid lines correspond to the present theory with a ratio $z^*/\lambda \simeq 1.13$. The disagreement at distances of the order of the radius of gyration is due to the penetration of free chains, neglected

120

J. Bonet Avalos et al.
Structure of adsorbed polymer layers: loops and tails

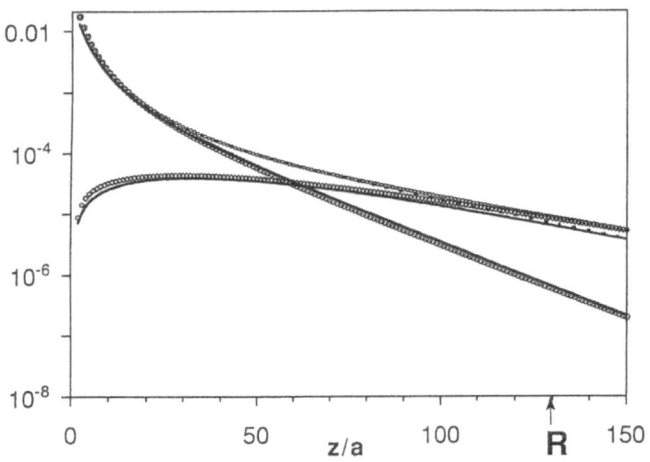

Fig. 1 Comparison in logarithmic scale between the solution of Eqs. (7) and (8) and numerical calculations of the full self-consistent field equations

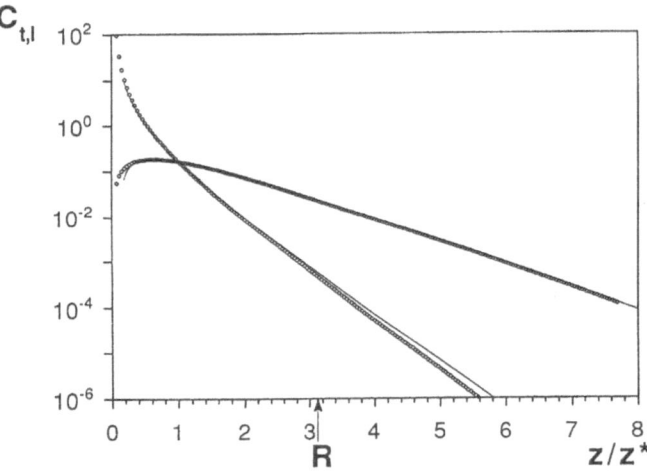

Fig. 2 Comparison in logarithmic scale between the solution of Eqs. (7) and (8) and numerical calculations of the full self-consistent field equations for a stronger adsorption energy

in Eqs. (7) and (8), as well as to the fact that the neglected term in the derivation of Eq. (7) is now of the same order of magnitude as the order parameters.

Conclusions

We have summarized the elements of a theory based on the distinction of loops and tails in the structure of an adsorbed polymer layer. The results of this theory compare well with the numerical solution of the full self-consistent field theory. Both clearly show that the structure of the layer is far more complex than the suggested self-similar profile introduced by de Gennes. Though in self-consistent field calculations the correlations due to excluded volume interactions are neglected, they serve to get a precise physical picture. Moreover, such a kind of calculation permits to also derive those prefactors that are disregarded in scaling theories. The knowledge of the prefactors are important when some effect depends on the competition of several terms that may be of the same order of magnitude as is the case, for instance, for the force between two adsorbing plates [10], which switches from repulsive at long distances to attractive at short distances.

A scaling theory may be built up to account for the exponents of some asymptotic behaviors, giving

$$
\begin{array}{ccc}
& z/z^* \ll 1 & 1 \ll z/z^* \ll \lambda/z^* \\
\hline
\phi_l \sim & z^{1/\nu-d} \sim z^{-4/3} & z^{-2d+(1-\gamma)/\nu} \sim z^{-6.27} \\
\phi_t \sim & z^{1\nu-1} - z^{2/3} & z^{1/\nu-d} \sim z^{-4/3} \\
\phi \sim & z^{-4/3} & z^{-4/3}
\end{array}
\tag{13}
$$

where the numerical values of the exponents are obtained by using $\nu = 3/5$, $d = 3$ and $\gamma = 1.16$, the latter being the susceptibility exponent. Using the values for these exponents in four dimensions $\nu = 1/2$, $d = 4$ and $\gamma = 1$, the exponents of the mean-field theory given in [11], are recovered.

References

1. Napper DH (1983) Polymeric Stabilization of Colloidal Dispersions. Academic Press (London)
2. Eiseriegler E, Kremer K, Binder K (1982) J Chem Phys 77:6296
3. de Genes PG (1979) Scaling Concepts in Polymer Physics. Cornell University Press Ithaca, New York, and references therein
4. Widom B, Phase Transitions and Critical Phenomena, Domb C, Green M, eds; Academic Press (New York 1972)
5. de Gennes PG (1981) Macromolecules 14:1637–1644
6. Fleer GJ, Cohen Stuart MA, Scheutjens JMHM, Cosgrove B, Vincent B, Polymers at Interfaces. Chapman and Hall (London 1993)
7. Semenov AN, Joanny JF (1995) Europhys Lett 29:279
8. Semenov AN, Avalos JB, Johner A, Joanny JF, Macromolecules (accepted)
9. Avalos JB, van der Linden CC, Semenov AN, Johner A, Joanny JF; Macromolecules (submitted)
10. Avalos JB, Semenov AN, Johner A, Jonnay JF; Europhys Lett (submitted)

Progr Colloid Polym Sci (1996) 100:121–126
© Steinkopff Verlag 1996

J. Müller
T. Palberg

Probing slow fluctuations in nonergodic systems: Interleaved sampling technique

Dr. J. Müller (✉) · T. Palberg
Institut für Physik
Universität Mainz
Staudingerweg 7
55099 Mainz, FRG

Abstract We present a new dynamic light scattering scheme to obtain ensemble-averaged correlation functions of slow fluctuations in nonergodic systems in an efficient way. On a rotating sample, a large set of *separate* correlation functions is measured in parallel, for each independent orientational component of the sample's density fluctuations. The ensemble-averaged correlation function spans a lag time range from 1 to 10^4 s. We describe our first implementation of this technique, discuss its statistical accuracy and show first results. Compared to plain ensemble averaging over a series of N measurements, the total measurement time is usually reduced by a factor N without significant degradation of statistical accuracy.

Key words Dynamic light scattering – photon correlation spectroscopy – nonergodic systems – ensemble averaging

Introduction

Photon correlation spectroscopy has been established as a convenient method for measuring dynami, structure factors at wave vectors in the visible range of the spectrum [1, 2]. In a standard photon correlation experiment, an estimate of scattered light intensity fluctuations is obtained by time averaging. While this procedure yields the desired ensemble averages in the case of ergodic systems, it is not applicable for nonergodic, partially frozen samples like gels or glasses. A number of experimental techniques have been proposed to construct the ensemble-averaged dynamic structure factor for this class of systems [3–5].

While these methods have been successfully applied to a large range of systems, we encountered difficulties in a particular experimental task: the investigation of very slow relaxation processes (time scales 1 to 10^4 s) on a background of even longer-lived fluctuations that may be considered frozen on the time scale of the experiment. This constellation is found, e.g., in the glassy state of hard-sphere colloids [6]. They show a weak, slow relaxation that persists beyond the glass transition and that has not been explained within the theoretical framework for the glass transition of ideal hard spheres. Our aim in this work is to establish an efficient dynamic light scattering method for probing these slow fluctuations. As previous studies on concentrated systems have shown that the suppression of multiple scattering by dual-detector, cross-correlation arrangements [7–9] would be beneficial, a method compatible with such special scattering geometries was sought.

Ensemble averaging procedures

We will briefly discuss available methods of ensemble averaging in order to introduce the basic concepts and terminology, and to point out the problems arising in the study of slow fluctuations. Throughout, the index T will denote time averages, while E refers to ensemble averages.

The obvious way to obtain a proper ensemble average is the plain, "brute force" method: Instead of a single one, many time-averaged correlation measurements are performed, but for slightly different sample positions or orientations. Each measurement yields a time-averaged estimator for the correlation functions, $\hat{g}_T^{(2)}(\mathbf{q}, \tau)$, at the scattering vector $\mathbf{q}$ and as a function of the lag time τ. An estimate for the ensemble-averaged correlation function, $\hat{g}_E^{(2)}(\mathbf{q}, \tau)$ is calculated from N of these measurements by weighting them with the respective time-averaged scattering intensities $\langle I \rangle_T$, and normalising with the total mean intensity $\langle I \rangle_E$ [5]:

$$\hat{g}_E^{(2)}(\mathbf{q}, \tau) = \frac{1}{N} \sum_{j=1}^{N} (\langle I \rangle_{T,j}^2 / \langle I \rangle_E^2) \hat{g}_{T,j}^{(2)}(\mathbf{q}, \tau) . \tag{1}$$

The desired dynamic structure factor $f(\mathbf{q}, \tau)$ can then be extracted in the usual way, as

$$g_E^{(2)}(\mathbf{q}, \tau) = 1 + \gamma f(\mathbf{q}, \tau)^2 \tag{2}$$

with an intercept γ defined by the spatial coherence of the detected scattering intensity.

The total measurement time for this scheme may, however, easily become impracticable: if slow fluctuations are of interest, each individual measurement requires a large observation time, and the number N of measurements needed to obtain proper ensemble averaging may well be of the order 100 to 1000.

A variant of this method, introduced by Chaikin et al. [4], uses steady translation or rotation of the sample while one continuous measurement progresses. The motion translates spatial sample fluctuations into temporal fluctuations. The resulting correlation function is directly equivalent to the result of "plain" averaging, but exhibits an additional relaxation to the baseline at a cutoff time, which reflects the spatial scanning of the speckles – the time it takes to move on to an adjacent scattering volume or independent angular position of the sample. As this cutoff time limits the accessible lag time range to typically 1 s [4], the method is most suited to investigations of faster relaxation processes. While a study of long-time fluctuations should be feasible using a suitably slow sample motion, the basic drawback of the plain approach remains – still a large number of consecutive observations of slow fluctuations would be needed.

A very elegant suggestion by Pusey and van Megen [5] avoids this time-consuming averaging process: These authors pointed out that – under some quite general conditions – the ensemble-averaged dynamic structure factor can be derived from just one time-averaged correlation measurement and an additional determination of the ensemble-averaged scattering intensity. The main experimental requirement is perfectly coherent detection, corresponding to an intercept $\gamma = 1$. This is readily achievable in suitable scattering geometries, although some special de-

signs (e.g., the cross-correlation scheme mentioned above) are ruled out. The more fundamental advantages and limitations of this method are recognised in a statistical analysis (see Schätzel [10] and below): While an excellent ensemble average is easily obtained, the statistical accuracy will then be dominated by the averaging over slow temporal sample fluctuations. As only one selected ensemble is observed, this will demand a prolonged measurement if slow processes are present.

Hence, the next logical step towards more efficient data collection is the parallel monitoring of several sample fluctuations. Bartsch et al. [11] have recently suggested and implemented an effective design relying on a CCD camera to observe about 50 speckles at once. The specific detector arrangement required in this approach is, of course, not compatible with special scattering geometries, like the cross-correlation experiment. Also, statistical considerations will show that a significantly larger number of parallel detectors is often desirable, but this becomes impractical in genuine parallel designs.

Experiment

We therefore developed a new ensemble averaging method that complements the technique of Chaikin [4]: While the shortest lag time observable in the correlation function is on the order of 1 s, the long time limit is essentially given by the duration of a single time-averaging measurement. A parallel ensemble average over a large number (on the order of 1000) of sample orientations or positions is performed during this measurement.

As in Chaikin's scheme, we rotate the sample while light scattering data are accumulated in a standard dynamic light scattering geometry. However, the observed intensity fluctuations are not fed into a single correlator (hence losing all correlation after the short cut-off time introduced by sample motion). Instead, we calculate separate correlation functions for each independent speckle that is observed during a rotation cycle. Each correlator is associated with one specific sample position, and consecutive intensity measurements are taken after every full rotation of the sample. Therefore, each correlator monitors the temporal evolution of one speckle, corresponding directly to a single, time-averaged measurement on a stationary sample. The time resolution is, of course, limited to the sample rotation period – typically 1 s in our experiment. An ensemble average, i.e., the sum of all correlation functions normalised by the total average intensity, can easily be calculated on-line.

The number of speckles appearing during one full rotation – i.e., the number of observable, independent Fourier components of the sample's refractive index

Fig. 1 Block diagram of the "interleaved sampling" experiment. Sampling and cuvette rotation are strictly synchronised via a common master clock; typical operating frequencies are indicated. PMT: photomultiplier tube, FIFO: first in/first out digital store. See text for further explanation

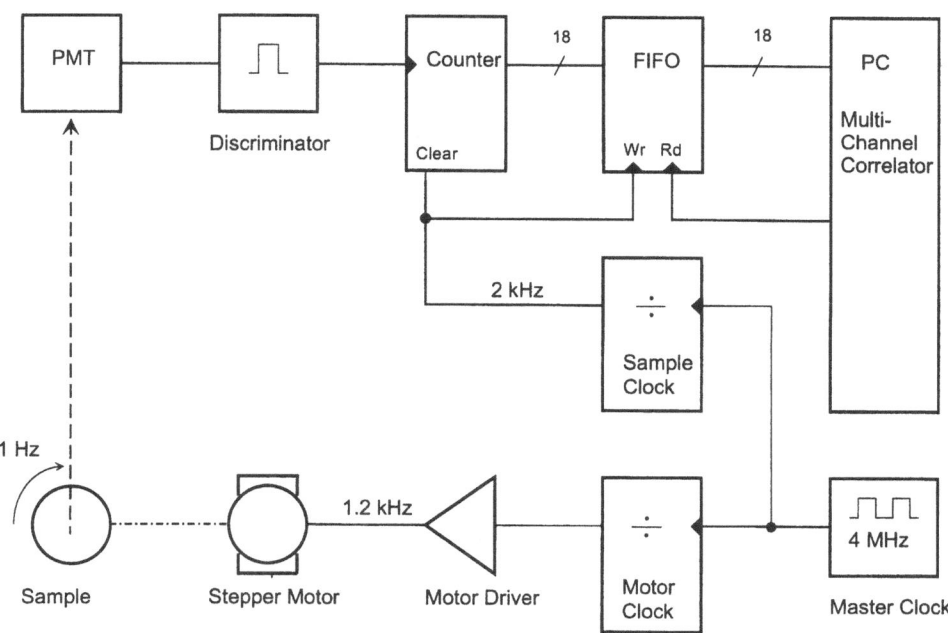

fluctuations – is determined by the diameter D of the measurement volume and the scattering vector q. A simple geometrical consideration shows that the sample must be rotated by an angle

$$\delta = 2\pi/qD \qquad (3)$$

to produce significant changes in the scattering phases; hence $N \approx qD$ independent speckles should be observed. For typical experimental geometries ($D = 100..200 \; \mu m$, $q = 5..30 \; \mu m^{-1}$), N will be in the range of approximately 500 to several thousand, providing an ensemble average of adequate statistical accuracy (see below).

Technical requirements for the implementation of our "interleaved sampling" scheme are quite modest. Of course, good reproducibility of the mechanical positioning of the sample cell, as well as perfect synchronisation between cell rotation and data sampling, are essential. We guarantee the latter by deriving both, the sampling and the motor driving clock signals, from a common master clock (Fig. 1). Programmable clock dividers provide for variable rotation and sampling frequencies. In our experimental setup, a simple stepper motor and gearbox are used to drive the sample. This arrangement yields a smooth, continuous movement at rotation frequencies up to 1.2 Hz; angular reproducibility was measured to be better than 0.1 mrad.

The multi-channel correlator was implemented in software on a standard 486 desktop PC; only the input unit consisting of one fast counter and a first in/first out store (to decouple sampling and data processing) had to be realised in hardware. Due to the slow sampling used in the individual correlation channels, up to approximately 4000 channels can be computed in parallel. The use of a Multiple Tau architecture with quasi-logarithmic lag time spacing [12] and symmetric normalisation [13] proved essential in order to obtain a large range of lag times and good statistical quality at lag times close to the total duration of the experiment.

Statistical accuracy

Compared to plain averaging, the "interleaved sampling" technique promises a dramatic reduction in measurement time for slow relaxations in nonergodic samples: Instead of a series of N time-averaged measurements, a single experiment of the same duration is expected to give equivalent information about both temporal and frozen fluctuations. Apparently, this massive reduction in total sampling time must come at a cost in terms of statistical accuracy. Specifically, data for the individual correlation functions are collected using "sparse sampling" – sample fluctuations are observed once, over a sampling time of typically 1 ms only, during a lag period τ_0 of approximately 1 s. To estimate the effect of sparse sampling, we follow Schätzel's considerations of photon correlation measurements on standard and nonergodic samples [10, 14], but will neglect the effects of multiple sampling times.

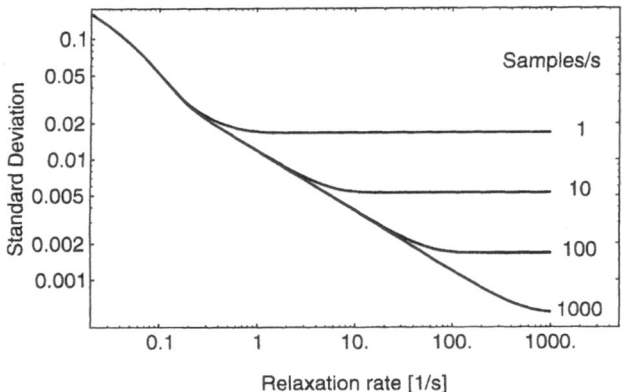

Fig. 2 Standard deviation of the time-averaged intensity correlation function $g^{(2)}(\tau)$ at lag time $\tau = 10$ s, for single-exponential relaxation, as a function of relaxation rate. A measurement duration of 1 h was assumed. Different curves show the effect of varying the number of (quasi-instantaneous) intensity samples taken per second

Three noise sources have to be considered: Photon detection noise, as well as classical averaging statistics over both temporal and frozen (ensemble) fluctuations of the sample. Photon noise can be safely neglected in most applications, as the number of photons detected even during the shortest sampling times is still quite large [14]. Noise contributions from temporal sample fluctuations depend on fluctuation times as well as sampling rates and require closer analysis: Basically, statistical accuracy will depend on the number of statistically independent states of the system that have been sampled. For fast fluctuations, with typical relaxation times $\tau_f \ll \tau_0$, we will indeed lose information if we do not watch their behaviour for most of the time. For the slow fluctuations of interest, however, taking one sample per lag time τ_0 is quite sufficient. More frequent measurements would simply sample the "same" statistically dependent state of the system again – the number of independent samples we can accumulate is not determined by the sampling rate, but the ratio of total measurement time T and characteristic fluctuation time τ_f.

Figure 2 gives a quantitative example for this point: It shows the standard deviation of the time-averaged intensity correlation function $g^{(2)}(\tau)$ at lag time $\tau = 10$ s. For simplicity, we consider single-exponential relaxation (as discussed by Schätzel [14]), and plot the standard deviation as a function relaxation rate. A total measurement duration of 1 h was assumed. The different curves show the effect of varying the number of (quasi-instantaneous) intensity samples taken per second. While the standard deviation contributed by fast fluctuations is reduced by increasing the sampling rate (roughly up to the relaxation rate), no change is found for slow fluctuations. Complex relaxation scenarios may be described by a spectrum of single-exponential relaxations. As long as the different

components are comparable in amplitude, the noise will always be dominated by the slowest components (note the logarithmic scale in Fig. 2) – hence, sparse sampling does not reduce the signal quality at all! Some caution is required, however, if a weak, slow component is to be measured in the presence of a faster relaxation with a strongly dominant amplitude. In this case, adequate sampling of the fast component may call for extra measurement time in a sparse sampling correlator.

In many cases, the total noise will actually be dominated by the ensemble averaging contribution. As pointed out by Pusey and van Megen [5], the frozen fluctuations obey Gaussian statistics. Therefore, their contribution to the standard deviation of the intensity correlator is simply proportional to $1/\sqrt{N}$. Approximating the noise from temporal fluctuations as $(\tau_f/T)^{1/2}$, Schätzel [10] estimated the total standard deviation as

$$\Delta \hat{g}_E^{(2)}(\mathbf{q}, \tau) = (\tau_f/T)^{1/2} [1 - f(\mathbf{q}, \infty)^2]$$

$$+ (2/\sqrt{N}) f(\mathbf{q}, \infty)^2, \qquad (4)$$

with the nonergodicity parameter $f(\mathbf{q}, \infty)$ as a weighting factor. Note that T is the *total* duration of a set of N "plain" measurements, or N times the actual duration of an interleaved sampling experiment. In both cases, the *individual* time-averaged measurements will usually have a duration larger than the fluctuation time τ_f (in order to observe this slow fluctuation). Hence, T/τ_f will be much larger than N, resulting in a dominance of ensemble averaging noise, except for systems with a very small nonergodicity parameter.

We conclude that noise in interleaved sampling experiments (as well as plain averaging experiments) will be dominated by ensemble fluctuations, as long as the system under study shows significant nonergodicity. Hence, the number of speckles observed by interleaved sampling should be maximised, which is possible – within the technical limits of the experiment – at no cost in measurement time. If temporal fluctuations determine the noise level, slow fluctuations will usually dominate. In both prevalent cases, one interleaved sampling experiment observing N speckles is equivalent to N plain measurements of the same duration.

Results

Interleaved sampling measurements are easily combined with short-time measurements, obtained by one of the standard methods discussed above, as both can be performed in the same experimental setup. This efficiently yields dynamic structure factors covering the full range of relaxation times from microseconds to about 10^4 s.

It should be noted that the intercept obtained in an interleaved sampling experiment may be reduced, in addition to the usual effect of spatial coherence, by averaging over the changing speckle pattern that is caused by sample rotation. Fitting the long-time data to a separate short-time measurement provides one way of normalising the data to account for this reduced intercept. We have to keep in mind, however, that in the overlapping time range the interleaved sampling data will usually be superior in statistical accuracy to the short-time measurement (due to better ensemble averaging). Hence, it would be both more efficient and more accurate to determine the intercept of the interleaved sampling measurement directly.

We can derive this intercept from the correlation function $g^{(2)}_{trans}(\mathbf{q}, \tau)$ of the artificial relaxation caused by sample translation. Two additional (quick) measurements are required to obtain this information, both basically following Chaikin's recipe of a measurement on a moving sample: Measurement A is performed while the sample rotates at the same speed as it did during the interleaved sampling experiment. The correlation function obtained will show a sharp decay at the typical life-time of a speckle – the result for our example is included as the dotted line in Fig. 3. Superimposed on this decay, however, is the initial relaxation of the sample itself. To separate this unwanted relaxation from the translation effects, we perform a second measurement B of the Chaikin type, but with the sample rotation slowed down by a factor of 10. This shifts the translation-induced decay to larger lag times, so only the intrinsic sample relaxation is visible over the relevant time scale; and we obtain the correlation function of the sample translation as $g^{(2)}_{trans}(\mathbf{q}, \tau) = g^{(2)}_A(\mathbf{q}, \tau)/g^{(2)}_B(\mathbf{q}, \tau)$. We note that both auxiliary measurements achieve the same ensemble averaging statistics as the main interleaved sampling measurement. The intercept reduction γ_{trans} due to sample translation is found by calculating the "triangular average" integral over this correlation function. This is analogous to the problem of correlogram distortion at early times due to averaging over classical sample fluctuations [14]:

$$\gamma_{trans} = \tau_S^{-2} \int_{-\tau_S}^{\tau_S} dt\, g^{(2)}_{trans}(\mathbf{q}, t)(\tau_S - |t|), \qquad (5)$$

for a sampling time τ_S.

Figure 3 shows the combined results of a conventional measurement and an interleaved sampling experiment on the same sample, a concentrated suspension of spherical polystyrene micro-network particles in the glassy state [15]. An initial, fast relaxation is followed by a massively broadened decay that stretches from the time domain covered by the conventional measurement into the inter-

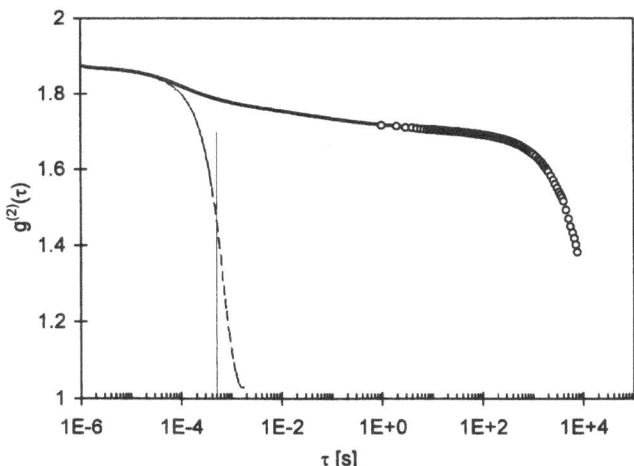

Fig. 3 Dynamic structure factor of a glassy suspension of colloidal micro-network spheres. Solid line: data obtained by plain averaging (100 measurements of 300 seconds each). Circles: data from an interleaved sampling experiment (duration 10^4 s, 2000 parallel samples, rotation period $\tau_0 = 1$ s). Dotted line: conventional correlation function measured on the rotating sample, showing the short-time relaxation of the sample, followed by a cut-off caused by the transition to the next scattering speckle. The dashed vertical line indicates the integration time used in the interleaved sampling measurement. This short-time information is used to normalise the interleaved sampling data (see text)

leaved sampling domain. This is followed by a pronounced decay on a typical time scale of 10^3 s. Note that the dynamic structure factor still shows strong nonergodicity on the time scale covered by the experiment.

Conclusions

We conclude that the "interleaved sampling" scheme provides a very efficient way to measure ensemble-averaged correlation functions over a time range of 1 to 10^4 s. It complements other ensemble averaging techniques, which often become impractical in this long time regime. Compared to the plain approach of taking many consecutive, time-averaged measurements, the total measurement time may be reduced by several orders of magnitude, without sacrificing statistical accuracy. The technique can be implemented by relatively simple additions to a standard light scattering setup. Hence, it is easily combined with standard, single-correlation methods to cover the full range of relaxation times.

Acknowledgments We thank Eckhard Bartsch of the Institut für Physikalische Chemie, Universität Mainz, for providing the micro-network samples.

References

1. Berne BJ, Pecora R (1976) Dynamic Light Scattering, Wiley, New York
2. Brown W (ed) (1993) Dynamic Light Scattering, Clarendon, Oxford
3. Joosten JGH, Geladé ETF, Pusey PN (1990) Phys Rev A 42:2161–2175
4. Xue J-Z, Pine DJ, Milner ST, Wu X-L, Chaikin PM (1992) Phys Rev 46: 6550–6563
5. Pusey PN, van Megen W (1989) Physica A 157:705–741
6. van Megen W, Underwood S (1993) Phys Rev Lett 70:2766–2769
7. Schätzel K, Drewel M, Ahrens J (1990) J Phys Condens Matter 2: SA 393–398
8. Segrè PN, van Megen W, Pusey PN, Schätzel K, Peters W (1995) J Mod Optics 42:1929–1952
9. Schätzel K (1991) J Mod Optics 38: 1849–1865
10. Schätzel K (1993) Appl Opt 32: 3880–3885
11. Frenz V, Bartsch E, to be published
12. Schätzel K (1985) Inst Phys Conf Ser 77 (Bristol: Institute of Physics): 175–184
13. Schätzel K, Drewel M, Stimac S (1988) J Mod Optics 35:711–718
14. Schätzel K (1990) Quantum Opt 2:287–305
15. Bartsch E, Frenz V, Möller S, Sillescu H (1993) Physica A 201:363–371

Progr Colloid Polym Sci (1996) 100:127–131
© Steinkopff Verlag 1996

F. Renth
E. Bartsch
A. Kasper
S. Kirsch
S. Stölken
H. Sillescu
W. Köhler
R. Schäfer

The effect of the internal architecture of polymer micronetwork colloids on the dynamics in highly concentrated dispersions

F. Renth · E. Bartsch (✉) · A. Kasper
S. Kirsch · S. Stölken · H. Sillescu
Institut für Physikalische Chemie
Universität Mainz
Jakob-Welder-Weg 15
55099 Mainz, FRG

W. Köhler · R. Schäfer
Max Planck-Institut für Polymerforschung
Postfach 4148
55021 Mainz, FRG

Abstract Motivated by the finding that colloidal dispersions of polymer micronetwork spheres with a crosslink density of 1:50 (inverse number of monomer units between crosslinks) show significant deviations from the dynamics of hard spheres in the colloid glass as seen by dynamic light scattering (DLS) (E. Bartsch, V. Frenz, H. Sillescu J. Non – Cryst. Solids 172–174 (1994), 88–97), we have undertaken a systematic study of the effect of the crosslink density on the dynamics at high concentrations. Long-time self-diffusion coefficients D_S^L and collective diffusion coefficients D_c were measured for colloids with crosslink densities of 1:10, 1:20 and 1:50 by forced Rayleigh scattering (FRS) and the newly developed thermal diffusion FRS (TDFRS) technique, respectively. Whereas no dependence of D_S^L on the crosslink density is found at low concentrations and the data coincide with theoretical results for hard spheres, strong effects of the internal architecture on self-diffusion are observed in the highly concentrated regime. Here, hard sphere behaviour is recovered only in case of the 1:10 particles, a glass transition being indicated at $\phi_g \sim 0.59$. Lowering the crosslink density leads to significantly higher ($\sim$ three decades) values of D_s^L at $\phi > \phi_g$. This may be due to an increased deformability of the spheres which could partially account for our DLS results. In contrast, D_c is more sensitive to variations of the degree of crosslinking. Here, a much faster increase of D_c with volume fraction as compared to hard spheres is observed already at low ϕ for 1:20 and 1:50 crosslinked particles, the effect being strongest for the lowest crosslink density. The results are tentatively interpreted in terms of a soft repulsive interaction potential, whose range increases on lowering the crosslink density.

Key words Micronetwork spheres – colloids – crosslink density – self-diffusion – forced Rayleigh scattering

Introduction

Spherical polymer micronetwork particles, swollen in a good solvent, can be considered as colloids which require no special stabilization to avoid aggregation. The interac-
tion potential is anticipated to depend on the degree of internal crosslinking, hard sphere behaviour being recovered in the limit of high crosslinking density (inverse number of monomer units between crosslinks). Studying the dynamics of highly concentrated micronetwork sphere dispersions with a crosslink density of 1:50 a glass

transition was discovered [1] which differs in several respects from the glass transition scenario observed for hard sphere colloids [2].

A better understanding of these findings was expected from studying the dependence of the structure and the dynamics of micronetwork colloids on the crosslink density up to extremely high concentrations. In measurements of the static structure factor $S(Q)$ and the radius of gyration R_g by small angle neutron scattering (SANS), we found that already in dilute dispersions 1:50 crosslinked spheres show clear deviations from hard sphere behaviour and a tendency to deswell [3]. No shrinking could be detected for the 1:20 crosslinked spheres. However, their $S(Q)$ started to deviate from hard sphere behaviour at volume fractions around $\phi \sim 0.47$. In contrast, a hard sphere $S(Q)$ could be obtained from dynamic light scattering (DLS) experiments with the 1:10 crosslinked micronetwork spheres up to high volume fractions [4].

Here we report on our first results upon the crosslink density dependence of the long time self-diffusion coefficient D_S^L, monitored by different tracer techniques (forced Rayleigh scattering (FRS), dynamic light scattering (DLS) and optical microscopy (OM)), and of the collective diffusion coefficient D_c, measured by a new variant of the FRS technique, named thermal diffusion FRS (TDFRS) [5].

Experimental details

Sample preparation

The polystyrene (PS) micronetwork spheres used in this study were mainly synthesized by an emulsion copolymerization of styrene with m-diisopropenylbenzene as a crosslinker, the emulsion being stabilized by an ionic surfactant [6]. In order to obtain tracer particles for the FRS experiments, a small amount of the colloids was first chloromethylated at the surface, and afterwards labelled with a photochromic nitro-stilbene-dye (ONS) [7].

Samples for the FRS self diffusion experiments were prepared in the following manner: Host and tracer polymers were dissolved in a good solvent. Then the solvent was removed to obtain homogeneous material containing about 1% of tracer particles. The appropriate amounts of refractive index matching solvent (2-ethylnaphthalene, which also provides a density match for PS) and the micronetwork spheres were then weighed into a small glass vessel and homogenized. Before filling the sample cells (standard rectangular UV cuvettes with 1 mm optical path length obtained from Hellma), the mixture was allowed to equilibrate for time intervals ranging from days to several weeks, depending on the respective volume fraction of the sample. In the TDFRS experiments the

same micronetwork spheres were studied that have been used in previous SANS experiments [3]. We also used the same solvent, toluene, as in the SANS measurements. Colloids and solvent, to which a small amount of the absorptive dye quinazirin was added, were weighed into a glass vessel. Due to the moderate viscosity of the sample material, it could be transferred into the sample cells (same as above) via a Millipore filter (0.4 μm) in order to remove dust particles.

The preparation of the host-tracer system for both dynamic light scattering (DLS) and optical microscopy (OM) measurements, consisting of poly-t-butylacrylate (P-(t)-BA) host colloids and core-shell tracer colloids (core: polystyrene (PS), shell: (P-(t)-BA), all crosslinked 1:10) is described in an accompanying paper [4].

The hydrodynamic radii R_H of the swollen and the unswollen (emulsified) spheres were determined via DLS, yielding the swelling ratio $S = R_H^3(\text{solvent})/R_H^3(\text{emulsion})$. For the larger spheres ($R_H > 150$ nm) the size polydispersity could be extracted from measurements of the particle form factor by static light scattering (SLS), for smaller colloids it was estimated from a cumulant analysis of DLS data [8]. The results of the characterization are collected in Table 1.

Experiments

A description of the DLS and OM measurements and the corresponding data evaluation can be found in ref. [4] and will be omitted here.

The FRS setup has been described elsewhere [7]. Here, we will only briefly recall the basic principle and state those details that are relevant to the study of colloidal

Table 1 Summary of the analytical data for the micronetwork colloids used to study the dynamics by forced Rayleigh scattering (FRS), thermal diffusion FRS (TDFRS), optical microscopy (OM) and dynamic light scattering (DLS). R_H: hydrodynamic radius[1,2]; σ: polydispersity[1]

crosslink density	material	R_H [nm]	σ	method
1:10	PS	81	~0.14	FRS
1:20	PS	56	~0.15	FRS
1:50	PS	108	~0.16	FRS
1:20	PS	20.4	~0.26	TDFRS
1:50	PS	25.0	~0.23	TDFRS
1:10	P-(t)-BuA	290	~0.07	DLS(host)
1:10	P-(t)-BuA shell PS core	255	~0.07	DLS(tracer)
1:10	P-(t)-BuA	480	~0.11	OM(host)
1:10	P-(t)-BuA shell PS core	455	~0.10	OM(tracer)

[1] determined by DLS. [2] in swollen state.

dispersions. The vertically polarized beam of an argon ion laser (Coherent Innova 90) equipped with an etalon and operating in the TEM_{00}-mode at wavelength $\lambda = 488$ nm is collimated to a beam diameter of about 0.8 mm and split into two beams of equal intensity, which are brought to interference within the sample. The resulting interference grating with a fringe spacing $d = \lambda/(2 \sin(\theta/2))$, θ being the intersection angle of the beams, is transformed via a photoreaction into a concentration grating of bleached and unbleached dye-molecules. A spatial grating of the absorption coefficient is thus created, which is read out after blocking one of the beams and attenuating the other by a factor of 10 000. Since the dye molecules are covalently bonded to the tracer colloids, the grating is destroyed by the diffusive motions of the latter and the time decay of the Bragg-scattered intensity, detected with a photomultiplier, monitors the self-diffusion of the colloids. The time dependence of the scattered intensity is given by:

$$I(Q, t) = [c_0 + c_1 \exp(-(t/\tau_d)^\beta)]^2 + c_2 . \tag{1}$$

In dilute dispersions $\beta = 1$ and the decay time τ_d of the FRS signal is related to the long-time self-diffusion coefficient D_S^L via

$$\tau_d = (D_S^L Q^2)^{-1}, \tag{2}$$

since FRS in our case works in the limit of $Q = 2\pi/d \to 0$ and $t \to \infty$. The coherent background c_0 is usually negligible and the incoherent background c_2 can be determined independently or treated as a fit parameter. For the more concentrated samples, a stretched-exponential decay ($\beta < 1$) of the scattered intensity could be observed. In those cases, the relaxation time τ_d in Eq. (2) was replaced by its average value $\langle \tau_d \rangle$

$$\langle \tau_d \rangle = \frac{\tau_d}{\beta} \Gamma\left(\frac{1}{\beta}\right), \tag{3}$$

where β and Γ denote the stretching exponent and the Gamma-function, respectively.

The basic idea of the thermal diffusion FRS (TDFRS) experiment is to create a spatially periodic temperature distribution by writing an optical interference grating into the sample via a slightly absorbing inert dye. The temperature gradients within the temperature grating induce thermal diffusion, giving rise to a concentration grating superimposed upon the thermal one. A phenomenological description of the TDFRS experiment is obtained from an extension of Fick's second law of diffusion [5]:

$$\frac{\partial c}{\partial t} = D_c \Delta c + D_T c(1 - c)\Delta T . \tag{4}$$

Here, c is the concentration in weight fractions. D_c and D_T denote the collective (or mutual) and the thermal diffusion coefficients, respectively. After writing the grating, the temperature equilibrates almost instantaneously and the

modulation depth of the concentration grating vanishes, leading to a decay of the Bragg scattered intensity according to:

$$I(Q, t) \propto \exp(-D_c Q^2 t) . \tag{5}$$

It should be emphasized that the collective diffusion coefficient is measured in TDFRS experiments, since a concentration gradient between the solvent and the solute molecules is built up. There is no bleaching of the dye and the dye is not attached to the polymer (its sole purpose being the absorption and thermalization of light). The TDFRS experiments are conducted at very small scattering angles of only a few degrees. Hence, inaccuracies as involved in the extrapolation to $Q = 0$ in order to obtain D_c via DLS are avoided.

The TDFRS setup has been described elsewhere [9]. An argon ion laser (488 nm) is used for writing and a helium–neon laser for reading of the grating. Polarization switching of one of the writing beams guarantees a constant average sample temperature. The optical density was adjusted to approximately 0.04 at 200 μm pathlength using quinizarin as inert dye.

Results and discussion

Figure 1 displays the results for the long time self-diffusion coefficient $D_S^L (\phi)$ for micronetwork colloids with crosslink densities of 1:10, 1:20 and 1:50 [10] as determined by FRS. For comparison, results obtained by DLS and OM for 1:10 crosslinked spheres are shown as well. The solid line shows the theoretical predictions for the long time self-diffusion coefficient of the hard sphere system according to Medina–Noyola [11]. There it was assumed that hydrodynamic interactions affect only the short time dynamics, i.e. are decoupled from the direct interactions, and D_S^L can be calculated as a product of the short time self-diffusion coefficient (which includes the hydrodynamic interactions) and a term containing essentially the static structure factor $S(Q)$ (representing the direct interactions). Due to the polydispersity of our samples, the micronetwork colloids do not crystallize. Thus, about $\phi \geq 0.49$ our experiments probe self-diffusion in a metastable state, which corresponds to the supercooled state of an atomic liquid close to the glass transition. As shown in Fig. 1, all the measured D_S^L/D_0 values coincide within experimental accuracy up to volume fractions ϕ of about 0.55, irrespective of the crosslink density, even though the values decrease by almost two decades in this range. Moreover, the data are in good agreement with the theoretical results of Medina–Noyola for hard spheres [11]. The systematic deviations of the DLS results for the highest three volume

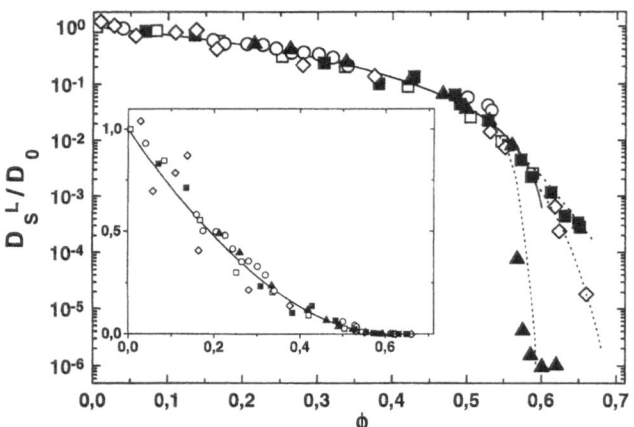

Fig. 1 Volume fraction dependence of D_S^L/D_0 for polystyrene micronetwork colloids with crosslink densities 1:10 (▲), 1:20 (■) and 1:50 (◇) as measured with FRS (data for the 1:50 spheres are from ref. [10]). For comparison data obtained by DLS (○) and optical microscopy (□) on poly-(t)-butylacrylate host-tracer colloidal dispersions (cf. ref. [4]) with a crosslink density of 1:10 are included as well. The solid line represents theoretical results obtained for hard spheres by Medina–Noyola [11]. The dashed lines are a guide for the eye. The insert depicts the same data on a linear scale

fractions reflect the difficulty of monitoring the true long time limit of self-diffusion at high densities with this technique. Further, the comparison with the Medina–Noyola prediction becomes doubtfully beyond $\phi = 0.5$ due to the known deficiencies of the hard sphere structure factor at high volume fractions, when calculated in the Percus–Yevick approximation [12].

For micronetwork colloids with a crosslink density of 1:10, the long time self-diffusion coefficient drops by four decades (from $D_S^L/D_0 = 10^{-2}$ to $D_S^L/D_0 = 10^{-6}$) within the narrow volume fraction range between $\phi = 0.55$ and $\phi = 0.59$. It seems to remain constant at higher volume fractions; however, this may be due to difficulties in determining the long-time limit of the FRS decay curves or insufficient equilibration of these highly concentrated suspensions. This behaviour compares well with theoretical predictions for hard sphere colloids at the glass transition [13]. There, D_S^L/D_0 drops from 10^{-2} at the relative distance $\varepsilon = (\phi - \phi_g)/\phi_g = 0.07$ to 10^{-6} at $\varepsilon = 0.005$ (cf. Fig. 1 of ref. [13]). This indicates that for the 1:10 crosslinked colloids a glass transition occurs at $\phi_g \sim 0.59$, in full agreement with hard sphere PMMA colloids, where $\phi_g = 0.58$ was found [2].

In contrast, the weaker crosslinked colloids show an increased mobility with respect to hard sphere behaviour. D_S^L monitors roughly the time domain of the dynamics which corresponds to the so-called structural relaxation of collective dynamics. Thus, the increased mobility is consistent with our previous observation, that 1:50 crosslinked spheres show long time density fluctuations at volume frac-

tions above ϕ_g [1], which are absent in hard sphere PMMA colloids. The latter were shown to undergo an ideal glass transition where structural relaxation is fully arrested on the time scale of the DLS experiments above ϕ_g [2].

Previously, polydispersity as well as deviations from the hard sphere potential due to the "softness" of the colloids were discussed as possible origin of those long time fluctuations. On the basis of the present data, the crosslink density effects on the interaction potential seem to be the dominant effect, as all FRS samples had roughly the same polydispersity. Even though the effect of the decreased crosslink density can be understood qualitatively as imparting additional mobility to the micronetwork spheres due to their increased deformability, it is difficult to interpret this feature quantitatively. It is at present not understood why 1:50 crosslinked spheres have a smaller D_S^L than the 1:20 ones (cf. Fig. 1). One possible explanation could be that with a crosslink density of 1:50 dangling ends and loops present at the sphere surface may be long enough for entanglement-like effects starting to become important. An increased friction between the spheres could result and partially offset the effect of the increased softness of the particles on the dynamics [14].

The data in Fig. 1 suggest, that deviations from hard sphere behaviour become apparent only close to the glass transition. This could either mean that the interactions change at about ϕ_g, or that D_S^L is an observable that depends not very sensitively on slight differences in particle interactions. The latter possibility is indicated by the fact that in SANS measurements of the static structure factor deviations from hard sphere behaviour were observed already below $\phi \sim 0.55$ for 1:20 and 1:50 crosslinked spheres [3]. In order to clarify this point, we studied the volume fraction dependence of the collective diffusion coefficient D_c of 1:20 and 1:50 crosslinked colloids, since this quantity is known to depend very sensitively on particle interactions [15]. Long-range soft repulsive forces, typical for charged latices at low ionic strength of the solvent, lead to a stronger increase of D_c with volume fraction as compared with hard spheres, while attractive forces (e.g. adhesive hard spheres) induce even a decrease of D_c [15]. The collective diffusion coefficients measured with the new TDFRS technique [5] are compared in Fig. 2 with literature data for hard spheres [16–19]. The deviations from hard sphere behaviour are obvious. The collective diffusion coefficients of our micronetwork colloids increase much faster with volume fraction, exceeding the hard sphere D_c by almost a factor of two at $\phi = 0.2$ in the case of the 1:50 crosslinked spheres. Since we are working in a nonpolar solvent, we expect electrostatic interactions to be irrelevant (sulfate groups from the initiator not being dissociated). Thus the "softness" of the spheres caused by decreasing crosslink density acts like a soft repulsive

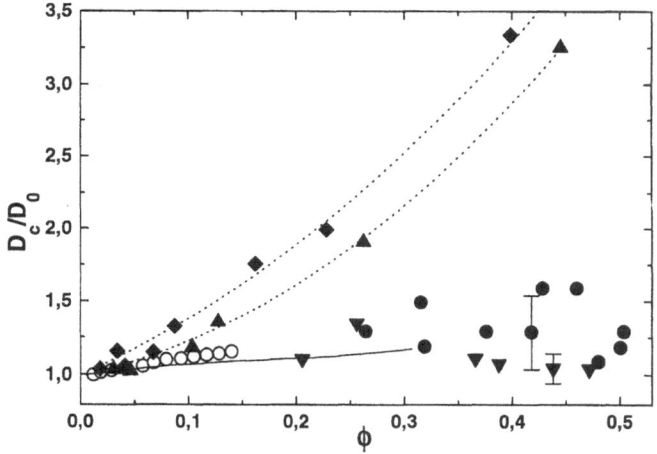

Fig. 2 Volume fraction dependence of D_c/D_0 for the 1:20 (▲) and 1:50 (♦) crosslinked PS micronetwork spheres as measured with TDFRS. For comparison, literature data for hard sphere colloids are given: experimental data from Kops–Werkhoven and Fijnaut (○: ref. [16] and ▼: ref. [17]), and van Megen et al. (●) [18]; theoretical results from Beenakker and Mazur [19] (solid line), which include many-particle hydrodynamic interactions. The dashed lines are guides for the eye

potential, whose interactions become more long ranged with decreasing crosslink density (D_c is larger for the 1:50 crosslinked spheres than for the 1:20 ones).

In summary, we have found indications that micronetwork spheres swollen in a good solvent interact with a soft repulsive potential, whose steepness and range decrease and increase, respectively, with decreasing crosslink density. Hard sphere behaviour is recovered only for crosslink densities as high as 1:10 in agreement with measurements of the static structure factor by SANS [3] and light scattering [4]. While the differences are well expressed in the collective diffusion coefficient already at low concentrations, they become significant for the long time self-diffusion only at high volume fractions. Further work is required to understand the effects of the crosslink density on collective and self motion in more detail.

Acknowledgments The authors thank A. Doerk for the preparation of the colloid particles. Financial support by the Sonderforschungsbereich 262 of the Deutsche Forschungsgemeinschaft is gratefully acknowledged. One of us (A.K.) acknowledges financial support by the Stipendienfonds der Chemischen Industrie.

References

1. Bartsch E, Frenz V, Sillescu H (1994) J Non Cryst Solids 172–174:88–97
2. van Megen W, Underwood SM (1993) Phys Rev E 49:4206
3. Stölken S, Bartsch E, Sillescu H, Lindner P (1995) Prog Colloid Polym Sci 98:155–159
4. Kasper A, Kirsch S, Renth F, Bartsch E, Sillescu H (1996) these proceedings
5. Köhler W (1993) J Chem Phys 98:660–668
6. Antonietti M, Bremser W, Müschenborn D, Rosenauer C, Schupp B, Schmidt M (1991) Macromolecules 24:6636–6643
7. Sillescu H, Ehlich D (1990) In: Fouassier JP, Rabek JB (eds) in Lasers in Polymer Science and Technology: Applications Vol III 211–226; Coutandin J, Ehlich D, Sillescu H, Wang CH (1985) Macromolecules 18:587–589
8. Koppel DE (1972) J Chem Phys 57:4814–4820
9. Köhler W, Rossmanith P (1995) J Phys Chem 99:5838
10. data for the 1:50 crosslinked micronetwork spheres were taken from Bartsch E, Frenz V, Möller S, Sillescu H (1993) Physica A 201:363–371
11. Medina-Noyola M (1988) Phys Rev Lett 60:2705–2708
12. Hansen JP, McDonald IR (1986) Theory of Simple Liquids, Academic Press (London)
13. Fuchs M (1995) In: Yip S (ed) Relaxation Kinetics in Supercooled Liquids – Mode Coupling Theory and its Experimentals Tests, Transp Theory Stat Phys 24:855–880
14. Since the sample preparation of the measurements reported in ref. [10] was somewhat different from the procedure used here, we cannot fully exclude the possibility that the surprising difference of the 1:20 and 1:50 results is due to experimental artefacts. Further experiments will be necessary in order to substantiate our results.
15. Pusey PN (1991) In: Levesque D, Hansen JP, Zinn-Justin (eds) Liquids, Freezing and the Glass Transition, Les Houches Session L1, Elsevier, Amsterdam, pp 763–942
16. Kops-Werkhoven MM, Fijnaut HM (1981) J Chem Phys 74:1618–1625
17. Kops-Werkhoven MM, Fijnaut HM (1982) J Chem Phys 77:2242–2253
18. van Megen W, Ottewill RH, Owens SM, Pusey PN (1985) J Chem Phys 82:508–515
19. Beenakker CWJ, Mazur P (1984) Physica A 126:349

Progr Colloid Polym Sci (1996) 100:132–136
© Steinkopff Verlag 1996

EMULSIONS AND CONCENTRATED SYSTEMS

R. Pons
G. Calderó
M.J. García
N. Azemar
I. Carrera
C. Solans

Transport properties of W/O highly concentrated emulsions (gel-emulsions)

Dr. R. Pons (✉) · G. Calderó
M.J. Carcía · N. Azemar
I. Carrera · C. Solans
Dep. Tecnologia de Tensioactius
C.I.D. (C.S.I.C.)
C/C7 Jordi Girona 18-26
08034 Barcelona, Spain

Abstract W/O highly concentrated emulsions formed in ternary water–nonionic surfactant–hydrocarbon systems consist of water droplets dispersed in a microemulsion phase. These emulsions show structural, rheological and optical properties that make them good candidates for the cosmetic and pharmaceutical industry as delivery systems. In addition, their high content of water and low concentration of surfactant and oil makes them attractive for economical, environmental and toxicological reasons. Our previous studies on their preparation, formation, structure and mechanical properties allow us to understand the basics underlying the transfer of added molecules. In order to study the transport properties of these systems, we have undertaken two types of experiment. In the first type, an emulsion with an added molecule is brought in contact with the same emulsion without additive. The concentration profile of the additive in both emulsions is determined after the system has evolved for a certain time. This allows for the determination of a global diffusion coefficient. In the other type of experiment the emulsion containing an added molecule is immersed in water. The release of the added compound from the emulsion to the receptor solution is monitored as a function of time. We model the diffusion behaviour with a microscopic model that takes into account the microstructure of the system. We present some results showing the modulation of the diffusion by changes of the composition variables.

Key words Diffusion – HIPRE – concentrated emulsion – release

Introduction

Water-in-oil highly concentrated emulsions have been studied in the last few years [1–7]. These emulsions have been shown to consist of an aqueous phase in a W/O swollen micelles matrix. These emulsions can be formulated with more than 99% dispersed phase and very low contents of oil and surfactant, i.e., less than 1% of the total. The surfactants used are low HLB nonionic surfactants and the oils are typical straight chain hydrocarbons. Properties like conditions of formation [1–3], preparation [5], structure [6] and rheology [7] are now well understood. Recently, we have been studying transport properties in these systems.

Studies of transport properties in complex systems have been undertaken to understand the phenomenon of drug release [8] or to study the structure (self diffusion NMR studies [9]). In the former studies a sample is usually brought in contact with a receptor solution through a membrane. These technique have the problem of accounting for the contribution of the membrane [8]. Self-diffusion NMR gives information about the self-diffusion coefficients of the participating molecules within the average environment surrounding them. This means that the water self-diffusion coefficient in these emulsions is a contribution of the coefficient for unbound water in the large water droplets (this coefficient is restricted at low times because the diffusion is restricted to the interior of the water droplets) plus a contribution of the water present within the swollen micelles [9]. In this paper we propose two methods to measure mutual diffusion coefficients in highly concentrated emulsions and give some results.

Experimental

Materials

All materials were used as received. Decane was Fluka, (Purum). The surfactant was Cremophor WO7 from BASF and used as received. This surfactant is hydrogenated castor oil condensed with polyethyleneoxide of 7 mol average chain length. Mandelic acid was Fluka (p.a.). Citric acid was Fluka Microselect. $CaCl_2$, NaCl and $NaSO_4$ were Merck. Doubly distilled water was used throughout.

Methods

The highly concentrated emulsions have been prepared by addition of water or aqueous solution to a mixture of hydrocarbon and surfactant with vigorous agitation. The analysis of the added molecule was performed by diluting a sample of emulsion in 10 times its volume of water. The content of added molecule was analyzed either by UV, HPLC or conductivity by comparison with calibration curves. HPLC was a setup from different brands, the detection was performed by UV. Conductivity was measured using a Crisson 525. UV spectra were recorded using a Shimadzu 265 FW.

Diffusion experiments

Experimental setup

Two types of experiment have been performed. In the first type (Exp. 1) an emulsion (which contains some added molecule to the aqueous phase) is put in contact with an emulsion of the same composition except for the added molecule. The system is let to evolve for a certain time and

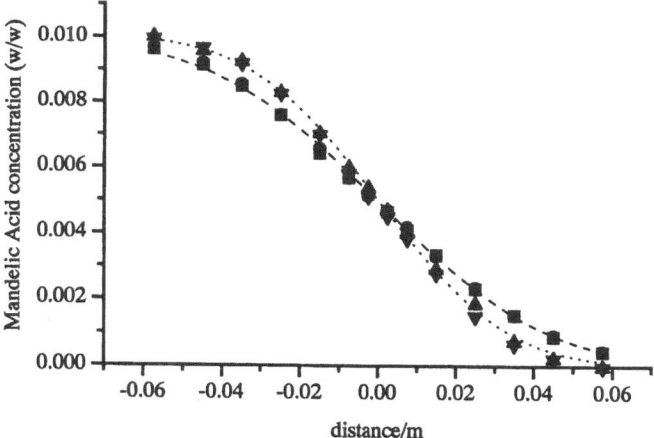

Fig. 1 Concentration as a function of distance obtained from experiment 1 for two different diffusion times and two replicates. ● and ■ 9 days of diffusion $D = 7.6$ and $7.5\,e^{-10}m^2s^{-1}$. ▲ and ▼ 5 days of diffusion $D = 7.7$ and $7.5\,e^{-10}m^2s^{-1}$. Analysis method is UV. The curves are the best fit of Eq. (1)

then the emulsion is cut (cuts parallel to the contact plane). These samples are well mixed and the content of added molecule is determined. From this experiment (with boundary conditions that imply that the system extends to infinity left and right of the dividing plane) a diffusion coefficient can be obtained by solving Fick's equations [10]. The concentration profile is:

$$C = \frac{C_0}{2}\left[1 - \mathrm{erf}\left(\frac{x}{\sqrt{4Dt}}\right)\right]. \tag{1}$$

In Fig. 1 experimental results of replicas after 5 and 9 days of diffusion are shown together with the fit of Eq. (1). The equation well fits the experimental results and the diffusion coefficients are independent of the total diffusion time. The reproducibility of the two measurements is very good. Replicas typically give values with a dispersion of 2% or less. The analyses of some samples have been performed, in several cases, by different methods (namely HPLC and UV), the results agree within a 5% dispersion.

In the second diffusion experiment (Exp. 2) the emulsion fills a cylindrical recipient. This recipient opened at the upper part is submerged in water. The system is kept at constant temperature and the aqueous solution is stirred at constant speed. This speed must be sufficient to ensure a good homogeneity, but not be so strong as to disturb the emulsion. This experiment can be done with these emulsions because the continuous phase is insoluble in water and the emulsion is stiff. Samples of the solution are removed for analysis as a function of time. In this case the diffusion coefficient cannot be obtained by solution of Fick's equations because the boundary conditions do not

134

R. Pons et al.
Transport properties of W/O highly concentrated emulsions

allow for the analytical solution of the equations. To obtain diffusion coefficients from these experiments we have performed a simulation that allows its estimation (see later for details of the simulation and how the diffusion coefficients are obtained).

Diffusion coefficient determination

As mentioned above, experiment 1 has an analytical solution of Fick's equation. In this solution there is the implicit assumption that the system has dimension infinity left and right of the dividing plane. This assumption is never met. However, the approximation is very good if at the extremes of the system the concentration of added molecule has not changed considerably (less than 10%). In experiment 2 there exists an analytical solution that assumes that the concentration of added molecule in the receptor solution is constant and the diffusing part has dimension infinity. These boundary conditions are valid only at very short diffusion time. Experimentally, at very short diffusion time the experimental errors are larger than at later times (when preparing the cylinder for the experiment some water droplets may be broken in the process and diffuse almost instantaneously due to the agitation; i.e., error in the timing of sampling affects the measure more strongly, etc). These problems and the possibility of implementing the nature of the microstructure of the system compelled us to simulate the system.

The simulation is performed taking a vector whose elements represent the concentration and the index their position. At time zero we set the initial concentrations in such a way that the first n indices correspond to C_0 and the remaining $n + 1$ to $2n$ correspond to zero concentration. At every time-step the new concentration at a certain position is calculated as:

$$C(x,t) = C(x,t-1) + d(C(x,t-1) - C(x+1,t-1))$$
$$- d(C(x-1,t-1) - C(x,t-1)) \, . \qquad (2)$$

This is an integral form of the first Fick's law. In this equation d is proportional to the diffusion coefficient. To simulate Exp. 2 the concentration of the cells $n + 1$ to $2n$ is averaged after every time step to simulate the homogeneity of the receptor solution. To simulate one and other experiments two different outputs are needed: for Exp. 1 the vector of concentration with its indices, for Exp. 2 the value of concentration in the cells $n + 1$ to $2n$ as a function of time-step.

From the simulation of the first experiment we make sure that the simulation works correctly and we can test for the accuracy of the approximation that we make when we calculate the diffusion coefficient from the fit of Eq. (1).

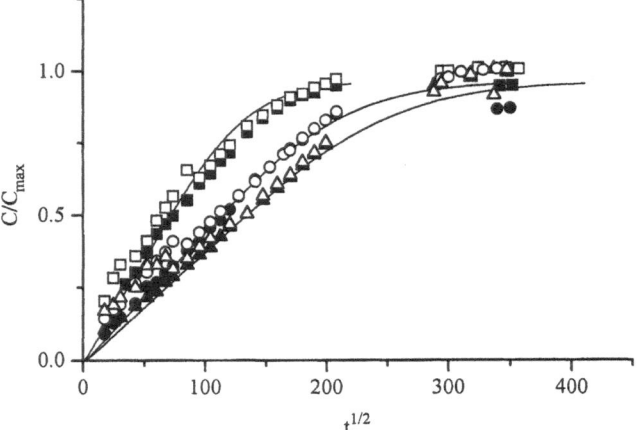

Fig. 2 Concentration as a function of time obtained from experiment 2 for three different volume fractions. ■ $\phi = 0.99$, ● $\phi = 0.95$ and ▲ $\phi = 0.90$. Analysis method is HPLC for the filled symbols and UV for the open symbols. The curves correspond to the simulated concentration profiles

To find the diffusion coefficient from the simulation we must superimpose the result of a simulation to a set of experimental results. We can do this by multiplying the simulation distances by a factor that brings both curves together. Superimposition of both curves means that the reduced coordinate $x/(Dt)^{1/2}$ takes the same values. From this proportionality we can obtain the experimental diffusion coefficients. The values obtained from the fit or from the simulation coincide if the system has not evolved for a very long time, this means that the concentration in the extreme points of the system has not changed more than 10%. If the system has evolved for longer times, the fit of Eq. (1) gives values of diffusion coefficients that are larger than the ones obtained by the simulation.

The determination of diffusion coefficients from experiment 2 is parallel to the determination from the simulation in experiment 1. The proportionality of the reduced coordinate $x/(Dt)^{1/2}$ is maintained in this case for the emulsion part of this experiment. In this case our variable is the time and we use a factor to superimpose experimental and simulated data that multiplies the simulation time. In this case x is fixed and corresponds to the depth of the cylinder filled with emulsion. In Fig. 2 the curves represent the results of the simulation.

Results and discussion

We find a fair agreement between the diffusion coefficients obtained from both experiments. In Table 1 the experimental diffusion coefficients for several systems are shown. Both types of experiment give reasonable values, however,

Table 1 Diffusion coefficients obtained from experiment 1 using several methods of analysis. D_1 and D_2 obtained from the fit of Eq. (3)

ϕ	Mandelic acid, HPLC	Mandelic acid, UV	Mandelic acid, HPLC	Citric acid	NaCl	Fructose
0.99	$7.25e^{-10}m^2s^{-1}$	$7.60e^{-10}m^2s^{-1}$	$7.0e^{-10}m^2s^{-1}$			
0.98	$6.30e^{-10}m^2s^{-1}$	$6.05e^{-10}m^2s^{-1}$				
0.95	$4.23e^{-10}m^2s^{-1}$	$4.95e^{-10}m^2s^{-1}$	$2.8e^{-10}m^2s^{-1}$			$3e^{-12}m^2s^{-1}$
0.90	$2.64e^{-10}m^2s^{-1}$	$2.95e^{-10}m^2s^{-1}$	$2.0e^{-10}m^2s^{-1}$	$3e^{-12}m^2s^{-1}$	$1e^{-11}m^2s^{-1}$	
0.80	$1.80e^{-10}m^2s^{-1}$	$2.40e^{-10}m^2s^{-1}$				
D_1	$8.0e^{-10}m^2s^{-1}$	$7.4e^{-10}m^2s^{-1}$	$8.1e^{-10}m^2s^{-1}$			
D_2	$0.4e^{-10}m^2s^{-1}$	$0.6e^{-10}m^2s^{-1}$	$0.3e^{-10}m^2s^{-1}$			

their applicability is different. The first experiment can be performed with any kind of sample provided its consistency is high enough to prevent any convection in the system. In the second experiment this requirement is more critical because the sample is brought in contact with a solution. In both experiments the system should be quite stable and should not show any changes within the experimental time. The experimental time ranges from a few hours for fast diffusion coefficients to a few days for lower diffusion coefficients. The first method can be applied to any W/O or O/W emulsion without any change in the experimental set up. The second method implies the use of a receptor solution that does not swell the emulsion.

The reproducibility of the diffusion coefficient measured by method 1 is much higher than the reproducibility obtained from the measurements by method 2. Method 1 gives reproducibility of around 2% while method 2 often gives a dispersion of values in the range of 50%.

The simulation allows us to check for the influence of the structure of the emulsions tested. For instance we can check for the effect of having two differently diffusing kinds of cells with different proportion. This simulates the two phases present in our systems and the effect of volume fraction. From these simulations we have seen that the following equation applies:

$$\frac{1}{D} = \frac{\phi}{D_1} + \frac{(1-\phi)}{D_2}, \qquad (3)$$

where D_1 corresponds to the diffusion coefficient of the phase that is present as ϕ, normally the dispersed phase and D_2 corresponds to the diffusion coefficient of the phase that is present as $1 - \phi$. This is equivalent to say that the total resistance of the system corresponds to the sum of the partial resistances with the appropriate weights (this would correspond, in the language of electric circuits, to the resistance of a system composed by two resistances in parallel). In Fig. 3 we represent values of diffusion coefficients obtained by both methods on equivalent emulsions with mandelic acid as added molecule. Equation (3) represents adequately all sets of data. D_1 and D_2 obtained from

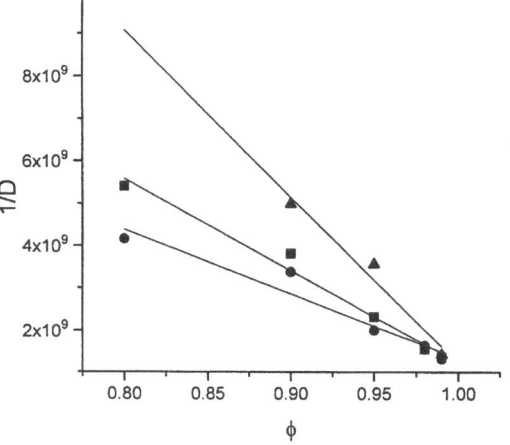

Fig. 3 Reciprocal diffusion coefficient as a function of volume fraction. ■ from experiment 1 (analyzed by HPLC), ● from experiment 1 (analyzed by UV), ▲ from experiment 2 (UV and HPLC analysis). The straight line corresponds to the best fit of Eq. (3)

these fits are shown in Table 1. The slight deviation at low volume fraction could be due to several factors; influence of the droplet size, of the interfaces or some experimental artifact. However, more experiments will be necessary to check whether this deviation is significant or not.

The diffusion coefficient of mandelic acid in water is $8.2\,e^{10}m^2s^{-1}$. This value is close to any of the values of D_1 obtained from the fit. So far we have not measured the diffusion coefficient of these molecules either in decane or in a microemulsion of the composition present in the emulsion. This data would be necessary to discern between an effect of the interfaces and that of the continuous phase. The major role that the diffusion coefficient and solubility of the added molecule in the continuous phase plays in the global diffusion coefficient is clearly shown by the comparison of the concentration profiles obtained from method 1 for mandelic acid and NaCl shown in Fig. 4. The diffusion coefficient for the NaCl can be estimated to have a maximum value of 10^{-11}, that is, two orders of magnitude lower than that of mandelic acid in the same conditions. Application of Eq. (2) gives an estimate for the

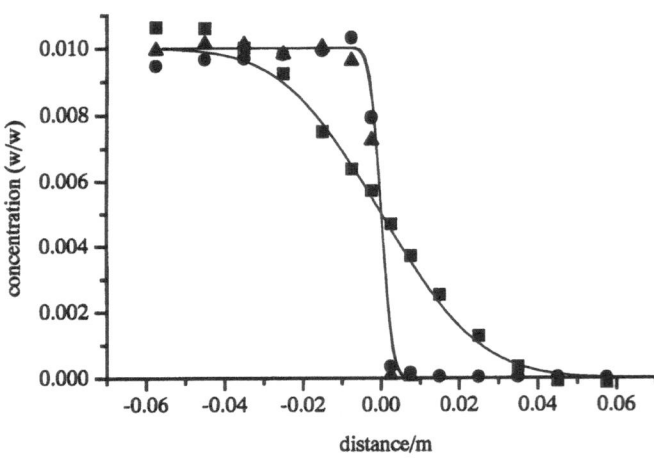

Fig. 4 Concentration as a function of time comparing the results for three molecules. ■ Mandelic acid ($\phi = 0.90$), ● Citric acid ($\phi = 0.90$) and ▲ Fructose ($\phi = 0.95$). The curves are best fits of Eq. (1)

diffusion coefficient of NaCl in the continuous phase of 10^{-13}, 3×10^{-13} for citric acid and 1.5×10^{-13} m²s⁻¹ for fructose (these values are upper limit estimates, no diffusion was detected during the experimental time). These differences in diffusion coefficient should be explained by differences in diffusion coefficient in the continuous phase because their diffusion coefficients in water are of the same order of magnitude than that of mandelic acid.

Conclusions

We have shown that diffusion coefficients can be measured in highly concentrated emulsions by the two methods presented. Method 1 gives good accuracy in the determination of the diffusion coefficients and could be applied to W/O or to O/W emulsions. In normal conditions reproducibility is within 1%. Method 2 gives somewhat less accuracy and cannot be applied to test the release from an O/W emulsion to water. The influence of system variables on the diffusion coefficient of added molecules can be monitored and interpreted taking into account the structure of the emulsion studied.

Acknowledgments The authors would like to acknowledge support from DGICYT (grant PB 92-0102).

References

1. Kunieda H, Solans C, Shida N, Parra JL (1987) Colloids and Surfaces 24:225
2. Solans C, Azemar N, Parra JL (1988) Progr Colloid Polym Sci 76:224
3. Ravey JC, Stebe MJ (1989) Physica B 394:156
4. Bampfield A, Cooper J (1988) In: Becher (ed) Encyclopedia of Emulsion Technology. Marcel Dekker, New York, Vol 3: 281
5. Pons R, Carrera I, Erra P, Kunieda H, Solans C (1994) Colloids and Surfaces A: Physicochemical and Engineering Aspects 91:259
6. Pons R, Ravey JC, Sauvage S, Stebe MJ, Erra P, Solans C (1993) Colloids and Surfaces: Physicochemical and Engineering Aspects 76:171
7. Pons R, Erra P, Solans C, Ravey JC, Stebe MJ (1993) J Phys Chem 97:12320
8. Koizumi T, Higuchi WI (1968) Journal of Pharmaceutical Sciences 57:87
9. Söderman O, Lönnqvist I, Balinov B (1992) in Sjöblom (Ed) Emulsions – A Fundamental and Practical Approach, Kluwer Academic Publishers, Netherlands, p 239
10. Cussler EL (1984) In Diffusion, Mass transfer in fluid systems. Cambridge University Press, Cambridge, Chapter 3

Progr Colloid Polym Sci (1996) 100:137–142
© Steinkopff Verlag 1996

J.-L. Salager

Quantifying the concept of physico–chemical formulation in surfactant–oil–water systems – State of the art

Prof. J.-L. Salager (✉)
Lab. FIRP
Ingeniería Química
Universidad de Los Andes
Mérida 5101, Venezuela

Abstract The properties of surfactant–oil–water systems either at equilibrium or in dispersed state depend upon a large number of formulation variables that include not only the nature of the three components, but also the influence of electrolytes, alcohols and other additives (type and concentration), as well as temperature and pressure.

These variables contribute to an overall affinity balance at interface, a fact that was recognized in Winsor's pioneering work as an attempt to interpret the experimental results rendered by the use of Banckoft's rule or Griffin's HLB. Then, Shinoda introduced the Phase Inversion Temperature (PIT), a characterization parameter that relies on an easily attainable experimental situation, whereas Beerbower and Hill proposed the Cohesive-Energy-Ratio (CER) approach as an attempt to develop a theoretical concept that could be easily linked with experimental results. In the 1970s, the Enhanced Oil Recovery research drive resulted in an extensive amount of experimental work dedicated to the development of multivariate empirical correlations for the attainment of three-phase behavior, a very well defined physico-chemical situation. The correlations contain a numerical contribution of the effect of each formulation variable that was later identified as an energy contribution to the surfactant affinity difference (SAD), a generalized formulation variable. The straightforward use of the correlations motivated experimentalists to extend its validity to very different systems, far beyond the original oil recovery scope. On the other hand, the surfactant partition coefficient between oil and water, an early measurement of the physico–chemical match, has been recently reinstated to a prominent place thanks to enhanced analytical methods; this approach is shown to lead to the same kind of quantitative description as the SAD, with a partial contribution of each group to the overall balance.

Key words Surfactant – HLB – PIT – formulation – microemulsion

Introduction

Surfactant–oil–water systems occur in hundreds of domestic applications such as personal care products, pharmaceuticals, foods and beverages, laundry and washing compounds, paints application, and so forth, as well as industrial processes such as waste water treatment, enhanced oil recovery, detergency and cleaning, paper making, ore separation, etc.

These systems are often formulated to provide many different properties such as tensioactivity, wettability, dispersion stabilization, solubilization in microemulsion, fancy rheological behavior, visual enhancement with liquid crystal or latex dispersion, etc. As a consequence, the number of components in a commercial product is often astonishingly high, for instance in the 40–50 range. No wonder that the formulation has been considered an art, in which a successful "magic recipe" is worth a lot of development time and money.

This paper is aimed at showing that the Art of formulation has evolved into a Science, and that much know-how is now available to reduce the uncertainties of the formulator alchemy.

What is the physico-chemical formulation?

The formulation deals with the nature of the components, while the proportions or respective amounts are considered as composition variables. In the simplest ternary system case, there will be three components, i.e., the surfactant (S), the oil (O) and the water (W), each with its chemical potential yardstick that quantitatively defines its physico-chemical state at a given temperature and pressure.

The physico–chemical formulation variables are thus at least five in a true ternary: three standard chemical potentials, temperature and pressure; in an actual SOW system there are many more formulation variables, since all components are often very complex mixtures, either in order to produce a synergy or to adjust some property. The aqueous phase always contains electrolytes that can be different in nature and concentration, while the oil phase can be as simple as a pure alkane or as complex as a crude oil. On the other hand, so-called additives, e.g., co-surfactants, co-solvents, hydrotropes, or protective colloids are introduced into the system for some reason.

HLB and PIT

The first approach was to implicitly reduce the number of variables to be taken into account down to only one or two. Obviously, this short cut has a chance to work only if the selected variables are the most important ones, and/or if the other formulation variables are either constant or have a negligible effect on the particular problem. Although this simplification can lead to serious inaccuracies, it is the simplest way to reduce the dimension of the formulation problem.

This was the premise taken in the two first empirical yardsticks: 1) Griffin's Hydrophile–Lipophile Balance or HLB introduced in 1949, and 2) Shinoda's Phase Inversion Temperature or PIT proposed in 1964 [3].

The HLB number depends essentially upon the surfactant, although the original experimental determination based on an emulsion stability maximum was taking into account the oil phase nature. However, these experiments are very tedious and inaccurate and are no longer into use. Today, the HLB is a surfactant characteristic parameter that is calculated directly from the relative weight of the hydrophile and lipophile part of the surfactant. For instance, the HLB number of polyethoxylated nonionic surfactants is estimated by:

$$HLB = \frac{100}{5}$$

$$\times \frac{\text{molecular weight of polyethylene oxide chain}}{\text{total molecular weight}} . \quad (1)$$

As a consequence, the HLB number cannot render the effect of the water phase salinity, or the influence of the temperature, or other variables which are known to affect the physico–chemical situation in SOW systems. This means that the inaccuracy can be as serious as several HLB units in some cases. Another drawback is that surfactants with the same HLB number can exhibit quite different behavior, in particular if they contain mixed products that exhibit a fractionation phenomenon.

Nevertheless, the HLB scale is still very widely used, probably because of its extreme simplicity and because it gives a "guesstimate" that many would consider good enough. However, it is believed that many experimental discrepancies and probably a considerable lost of research and development time can be credited to HLB inaccuracy.

The PIT was originally the temperature at which a non-ionic surfactant switched its dominant affinity from the aqueous phase to the oil phase, in a so-called phase transition process. Later on, it was related to the emulsion inversion, to be finally renamed HLB-temperature a few years ago to mean that it was the temperature at which the hydrophilic and lipophilic tendencies of the surfactant were balanced. This concept is linked with the dehydration of the polyethoxylated chain with increasing temperature that reaches a point where the surfactant is no longer soluble in water and a surfactant phase separates from the original aqueous solution (in absence of oil phase), and results in a turbidity occurrence at the so-called cloud-point. It has been known for quite a while that the presence of a small amount of dissolved hydrocarbon can change the cloud point, but this effect is not actually manageable enough for measurement purposes. On the contrary the PIT measurement results in much more reliable data, which take into account two formulation variables, i.e., the surfactant type and the oil phase nature. Of course variables such as the water salinity or the alcohol addition are not taken into account and can alter the PIT estimate.

Although the PIT is limited to ethoxylated nonionic surfactant systems and to the usual liquid water range of temperature which can be handled safely, it has a very important precursor position when related to the recent advances. In effect, the PIT experimental procedure was indeed the first unidimensional formulation scan in which the temperature is changed continuously with all other formulation variables held constant, until an experimentally detectable situation arises.

In the light of current knowledge, it can be said that this was a farsighted view and a precursor technique. Today, the PIT can be viewed as the optimum temperature in the multivariant SAD (Surfactant Affinity Difference) concept to be discussed later on. *A posteriori*, this explains why the PIT of surfactant mixtures follows a linear mixing rule with respect to the PITs of the base surfactants, a fact that could have surprised some thermodynamic specialists of the time. It is now known that this may be a quite approximated rule because of the fractionation phenomena that are likely to occur near the PIT.

Winsor R ratio

The second kind of approach presented by Winsor [4] in 1954 was theoretical in essence. The relationship between the different components was not measured through some experimental occurrence but it was estimated according to a balance of interaction energies between the surfactant molecules (located at interface) and the oil and water bulk phase molecules, per unit surface area.

The original Winsor interaction energies ratio was written

$$R = \frac{A_{CO}}{A_{CW}} \qquad (2)$$

later on, a more complete definition was preferred:

$$R = \frac{A_{CO} - A_{OO} - A_{LL}}{A_{CW} - A_{WW} - A_{HH}}, \qquad (3)$$

where the interaction subscripts C, O, W, L and H, respectively refer to the surfactant as a whole, the oil, the water, the surfactant lipophilic and hydrophilic group.

A change in the R ratio from $R < 1$ to $R > 1$ or vice versa was associated to essentially all phase behavior transitions and corresponding phenomena occurrence, a bulk of know-how that has been reviewed extensively [5].

Winsor's clever rationalization was some 20 or 30 years ahead of his time. It introduced a very important feature, i.e., the concept of a unique, global, and overall formulation parameter, e.g., the R ratio, to render the effect of all the distinct formulation variables actually manipulated by the formulator. In other terms, he stated that a physico–chemical situation in a SOW system could be handled with a single parameter instead of the score of known formulation variables. This was extremely important for the advancement of subsequent research, since it made it clear that the observed phenomena should not be related to the specific numerical values of the formulation variables (a premise which was unmanageable because of the large number of variables), but instead to some overall physico–chemical condition to be satisfied by and depending upon these variables.

Winsor failed to find a quantitative numerical expression for this R parameter as a function of the individual formulation variables, probably because the liquid state theory was not advanced enough to estimate accurately the molecular interactions; in any case, it would have lacked the high speed computer required to carry out such calculations, which by the way, are probably not yet sufficient. As a matter of fact, the level of approximation used in the current model is still relatively crude [6].

On the other hand, it can be said that Winsor missed a point that could have made its concept easier to link with experimental measurements. Winsor was dealing with energies, and as it is known from the first law of thermodynamics, the energies do sum up. It can be conjectured that had Winsor taken the algebraic sum of the interaction energies rather than their ratio, the translation of the generalized formulation concept into a numerical equation would have evolved earlier. Yet, Winsor's R ratio allowed him to make his point and to relate the generalized formulation, i.e., the R value, to the phase behavior of SOW systems [4–5].

Winsor's research work indicated that in a true SOW ternary there were only so-called I, II, and III, three types of diagrams, and that the type of diagram, and thus of multiphase region, depended only upon the interaction energies ratio R.

After that, it was clear that the formulation effect could be described, at least in the ternary approximation, through a single, although complex, parameter that can be referred to as the generalized formulation whatever its actual form, R, or as in other concepts to be discussed later on. This was quite a breakthrough since it reduced considerably the number of independent variables to be handled, and it made it possible to compare or compensate effects of the different formulation variables, and to find alternate physico–chemical situations without carrying out millions of experiments.

If it is understandable that Winsor's work was over-looked in his time and over the next 20 years, because it had no direct application beyond the pedagogical message, it is unfortunate that this concept had not permeated yet in industrial practice after the enhanced oil recovery research effort of the 1970s showed its importance in the resolution of actual problems [5, 7].

Cohesive energy ratio

In an attempt to attain a formulation concept with both the theoretical content of Winsor's R and the down the bench numerical data feature of the HLB, Beerbower and collaborators [8] introduced the Cohesive Energy Ratio (CER) approach in 1971. From the conceptual point of view it was very similar to Winsor's R, but this time it was the ratio between the adhesion energy of the surfactant "layer" with the oil phase, and the adhesion energy of the surfactant "layer" to the water phase. To express this ratio, first recall that the cohesion energy between molecules of a pure component system is calculated as:

$$\delta^2 = \frac{\Delta H_{\text{vap}}}{v_L}, \tag{4}$$

where ΔH_{vap} is the enthalpy of vaporization, and v_L the molar volume in the liquid state, both measurable quantities. δ is the so-called solubility parameter, that is a direct measurement of the intermolecular cohesion forces. When a mixed system is dealt with, the adhesion forces between the two kinds of molecules are calculated according to London's geometric mean relationship as follows.

If

$$\delta_{\text{AA}}^2 = \frac{\Delta H_{\text{A·vap}}}{v_{\text{A·L}}}$$

and

$$\delta_{\text{BB}}^2 = \frac{\Delta H_{\text{B·vap}}}{v_{\text{B·L}}} \quad \text{then } \delta_{\text{AB}}^2 = \delta_{\text{AA}} \delta_{\text{BB}} . \tag{5}$$

Solubility parameters have been measured and tabulated for hundreds of substances, and they are often used in relation with the regular solution model to estimate the activity coefficient in a mixture and the eventual separation into two phases. The adhesion between the surfactant and the oil phase was estimated by calculating a δ_{AB}^2 term where A represented the oil phase and B stood for the lipophilic part of the surfactant that was assumed to be similar to that of a hydrocarbon with the same chain length. Unfortunately, the corresponding adhesion energy on the water side of the interface was not easily calculated, in particular because of the lack of experimental data for the hydrophilic group of some surfactants.

As a consequence of this drawback, as well for the lack of accuracy of the geometric mean assumption in this case, the final numerical value of the cohesive energy ratio was as inaccurate as the HLB number and even worse in some cases, and the CER was no real help in practice.

Numerical correlations for optimum formulation

In 1973, the oil embargo triggered an intense research effort to develop enhanced recovery processes aimed at bringing to the surface the oil that remained trapped in the reservoir porous medium, after the conventional secondary recovery was carried out. Among the proposed methods were the so-called low-tension, micellar or surfactant flooding processes, in which a surfactant solution of appropriate formulation was injected in the reservoir to produce an ultra low interfacial tension between the crude and the water, so that the capillary forces could be easily overwhelmed by a water flooding [7]. The huge amount of research carried out an industrial and academic centers showed that an ultra low tension minimum could be attained when the formulation matched the Winsor III case, at which $R = 1$. The occurrence of a so-called optimum formulation (as it was labeled because it coincided with the ultralow tension occurrence) was thus linked with a precise physico–chemical situation in which the affinity of the surfactant for the oil phase exactly equilibrated its affinity for the water phase. Since these affinities could possibly change with any of the formulation variables, it was necessary to know how this could happen from the practical point of view, i.e., with numerical data related to bench variables such as the water salinity, the temperature or the oil phase nature.

Exhaustive experimental studies carried out for both anionic and nonionic surfactant systems showed that the optimum formulation was attained whenever a certain condition between the formulation variables, so-called correlation for optimum formulation, was satisfied.

Salager et al. found that for anionic surfactants the correlation can be expressed as [9]:

$$\ln S - K\,\text{ACN} - f(A) + \sigma - a_T\,\Delta T = 0 . \tag{6}$$

For nonionic surfactants Bourrel et al. [10] found a similar correlation:

$$\alpha - \text{EON} + b\,S - k\,\text{ACN} - \phi(A) + c_T\,\Delta T = 0 , \tag{7}$$

where S is the salinity in wt.% of NaCl, ACN or Alkane Carbon Number is a characteristic parameter of the oil phase, $f(A)$ and $\phi(A)$ are function of the alcohol type and concentration, σ, and α are parameters characteristic of the surfactant structure, and EON is the average number of ethylene oxide group per molecule of nonionic surfactant.

ΔT is the temperature deviation measured from a certain reference (25 °C), k, K, a_T and c_T are empirical constants that depends upon the type of system.

Surfactant affinity difference (SAD)

Let us call SAD the "Surfactant Affinity Difference", that is, the difference between the negative of the standard chemical potential of the surfactant in the oil phase and the corresponding term for the water phase.

$$SAD = -\mu_0^* - (-\mu_w^*) = \mu_w^* - \mu_0^* . \qquad (8)$$

The equivalence of the right-hand term of the correlation with SAD/RT was proposed several years after the finding of the empirical correlations, as a physico–chemical interpretation of their meaning [11]. Now, it is clear that SAD is a form of the generalized formulation parameter, hopefully expressed in terms of measurable and manipulable variables. It is worth noting that the correlations do not contain any cross-term, but on the contrary, are linear forms of independent terms, as in thermodynamics' first law. Each term can be viewed as an energetic contribution to the overall interaction balance.

The original purpose of these correlations was to relate numerically the effect of the different formulation variables in order to attain or to maintain an optimum formulation at SAD = 0 by changing two or more variables in a compensated way. In addition to the cited original works, several other studies contributed to extend the range of application of these correlations to other oils, other electrolytes and other surfactants, and some reviews are available [5, 11].

The by-product of these studies was that an optimum formulation can be taken as a reference state, that can be accurately pinpointed by experiment. As a consequence, an off-optimum formulation can be defined as a deviation from optimum formulation. Thus, it is possible to compare systems in a same physico-chemical state (same deviation from optimum formulation) although they have no single formulation variable with a common value. This deviation concept was successfully applied to describe the relationship between the formulation and the emulsion properties.

It should be mentioned that the same kind of reasoning was carried out by other investigators, with quite similar results from the conceptual point of view, even though they did not attain the numerical quantification level of the SAD expression.

Krugliakov [12] recently presented to western audience his concept of Hydrophile–Lipophile Ratio (HLR), which is the ratio of the energy of adsorption of the surfactant molecule from the water phase to its energy of adsorption from the oil phase; this was first published in the Russian literature 20 years ago. This is essentially the same overall presentation as Winsor's, but with an energy term whose variation can be traced back to the formulation variables, at least in some approximate way. In any case, it seems that the HLR suffers from the same drawback as Winsor's R ratio, i.e., it is a ratio of energy terms instead of an algebraic difference.

Free energy of transfer and partition coefficient between excess phases of a Winsor III system

In his 1957 analysis of the physical–chemistry of emulsifying agents, Davies [13] presented two different suggestions. First, he proposed to split the HLB into group contributions, probably with an implicit energy summation underneath this idea; and on the other hand, he related the distribution of the surfactant between oil and water to the HLB.

The free energy of transfer of a molecule of surfactant from water to oil is:

$$\Delta G_{(w->0)} = RT \ln(C_w/C_0), \qquad (9)$$

where the C_w and C_0 are the surfactant concentrations in water and oil, respectively. Davies related the partition coefficient to HLB and to the rate of coalescence of the two types of emulsions, according to his theory that the emulsion type was determined by the ratio of these rates. In his analysis, he took care to point out that the partitioning should be considered in absence of any micellar structure. This was quite a problem as far as the analytical technique was concerned. In effect, the CMC of nonionic surfactant is so low that the concentrations to be dealt with were beyond the range of separation-detection equipment availables in the 1960s.

The measurement of surfactant distribution in oil–water systems with extremely low surfactant concentration is not the only experimental path. There is another elegant way that works at high surfactant concentration, i.e., the analysis of surfactant in excess water and excess oil in three-phase systems which are at optimum formulation. This technique was used by Graciaa and colaborators [14] to interpret and predict the partitioning of ethoxylated nonionic oligomer mixtures and to compute the actual interfacial composition with a pseudophase model. Since the partitioning was measured at optimum formulation, there was as many partitioning data as optimum formulation. In other words, the partitioning depended not only upon the surfactant HLB, but also upon all formulation variables.

In a recent investigation aimed at screening different cases of optimum formulation SOW systems, Marquez [15] showed that the partition coefficient of a surfactant species is an excellent indicator of the concept of generalized formulation. In the assumption that the activity coefficient is unity, which is legitimized by the low surfactant concentration found in the excess phases of Winsor type III systems, then the equilibrium between the water and oil phases can be written in terms of the chemical potentials of the surfactant:

$$\mu_w = \mu_0 = \mu_w^* + RT \ln C_w/C_{w \cdot ref}$$
$$= \mu_0^* + RT \ln C_0/C_{0 \cdot ref} , \qquad (10)$$

where the standard chemical potentials are indicated with an asterisk, while the concentration references bear the subscript ref. If the partition coefficient is defied as $K = C_w/C_0$, then:

$$RT \ln K + \text{constant} = \mu_0^* - \mu_w^* = \Delta G_{(w \, - \, > \, 0)}$$
$$= - \text{SAD} . \qquad (11)$$

The constant is the partition coefficient between the two reference states. As a matter of fact, the value of this constant does not matter, since it does not change with the formulation variables that influence the μ^*'s. By measuring the partition coefficient with different systems exhibiting a variety of oils, surfactant characteristics, alcohol content, brine salinity, and temperature, Marquez was able to correlate the term $RT \ln K$ with the formulation variable in a linear relationship very similar to SAD expression by the correlations, according to a splitting-contribution technique suggested by Cratin [16]. For instance, these studies allow the experimental determination of the molar free energy of transfer $\Delta \mu^*(w - > 0)$ of a ethylene oxide group or to a methylene group by showing that:

$$RT \ln K(\text{EON, SACN}) = \text{EON} \times \Delta \mu^*(\text{EO group})$$
$$+ \text{SACN} \times \Delta \mu^*(\text{CH}_2 \text{ group}) + \Delta \mu^*(\text{remaining}) , \qquad (12)$$

where EON is the number of ethylene oxide groups in the oligomer molecule and SACN is the number of methylene groups in the surfactant hydrophobe. Marquez et al. [15, 17] were also able to determine the contribution of temperature, water salinity and alcohol content. The determination of these physico–chemical parameters is fairly straightforward if the formulator is able to reach an optimum formulation with the desired system, and provided that he may use one of the modern analytical technique to measure the concentration of the different surfactant species in the excess phases.

Conclusion

Winsor's R concept has now quantifyable equivalents, and the formulator should invest some time in knowing these new approaches that give more reliable results than the early yardsticks such as the HLB number.

References

1. Becher P (ed) (1985–88) Encyclopedia of Emulsion Technology, 3 volumes. Marcel Dekker, New York
2. Griffin WC (1949) J Soc Cosm Chem 1:311
3. Shinoda K, Kunieda H (1985) In: Becher P (ed) Encyclopedia of Emulsion Technology, Vol 1, Chap 5, Marcel Dekker, New York
4. Winsor P (1954) Solvent Properties of Amphiphilic Compounds, Butterworth, London
5. Bourrel M, Schechter RS (1988) Microemulsions and Related Systems, Marcel Dekker, New York
6. Bourrel M, Biais J, Bothorel P, Clin B, Lalanne P (1991) J Dispersion Sci Technology 12:531
7. Shah DO, Schechter RS (eds) (1977) Improved Oil Recovery by Surfactant and Polymer Flooding, Academic Press, New York
8. Beerbower A, Hill M (1971) In: Mac Cutcheon's Detergents and Emulsifiers, Allured Publishing Co.
9. Salager JL, Morgan J, Schechter R, Wade W, Vasquez E (1979) Soc Petrol Eng J 19:107
10. Bourrel M, Salager JL, Schechter RS, Wade WH (1980) J Colloid Interface Sci 75:451
11. Salager JL (1988) in Encyclopedia of Emulsion Technology, Becher P, Ed., Vol 3, Chap 3, Marcel Dekker
12. Kruglyakov P (1993) Worl Congress on Emulsion, Paris. References in Russian (1971) Dokladi Akademia Nauk CCCP, T. 197 (5) 1106, and Isveztia Sib Otd Akademia Nauk CCCP, N° 9, V 4, 11
13. Davies JT (1957) Proceedings 2nd International Congress Surface Activity, Butterworths, London, Vol I, 426
14. Graciaa A, Lachaise J, Sayous JG, Grenier P, Yiv S, Schechter RS, Wade WH (1983) J Colloid Interface Sci 93:474
15. Márquez N (1994) Dr. Dissertation, Univ de Pau PA, Pau, France
16. Cratin P (1971) In: Ross S (ed) Chemistry and Physics at Interfaces – II. American Chemical Society Pub., Washington
17. Márquez N, Antón R, Graciaa A, Lachaise J, Salager JL (1995) Colloids and Surfaces A 100:225

Progr Colloid Polym Sci (1996) 100:143–147
© Steinkopff Verlag 1996

G. Marion
K. Benabdeljalil
J. Lachaise

Interbubble gas transfer in persistent foams resulting from surfactant mixtures

G. Marion · K. Benabdeljalil
Dr. J. Lachaise (✉)
L.T.E.M.P.M.
Centre Universitaire de Recherche
Scientifique
Avenue de l'Université
64000 Pau, France

Abstract We study interbubble gas transfer in persistent foams resulting from relatively concentrated surfactant mixture solutions. The components of the surfactant mixtures have been carefully chosen to delay the breaking of the liquid films. The analysis of the phenomenon is performed as long as the liquid films are preserved from rupture.

Foams are obtained by independently fixing foaming solution and gas flows through a coarse porous structure. Precise measurements of these flows give the initial gas volume fractions of the foams while real time diffractometry gives their bubble size initial distribution. The drainage of the foam films is also measured to know the mean variation of the film thickness versus time.

The relative decrease of the global bubble area induced by the gas transfer is determined from reflectometry measurements. We calculate this transfer on the basis of the Lemlich theory, taking into account the liquid drainage in the films. We show that a detailed comparison between theory and experiment allows to determine the surface resistances and the volume resistances of the liquid films to the gas transfer. For the studied foams it is found that the volume resistance is much higher than the surface resistance. We attribute the high values of this volume resistance to a long-range structure created in the films by the excess of ionic surfactant molecules. A prefiguration of this structure is detected in the foaming solutions by elastic light scattering.

Key words Foam – drainage – gas transfer – reflectometry – light scattering

Introduction

In a wet foam, interbubble gas transfer can only be studied under good conditions if evaporation and breaking of the liquid films are minimized. Evaporation of the liquid can be stopped by putting the foam in a closed vessel. The breaking of the films can be delayed by using judicious surfactants.

Interbubble gas transfer produces the decrease of the relative interfacial area of the bubbles during the foam degradation. It has already been shown that this decrease can be measured by reflectometry [1]. It has also been shown that reproducible foams can be obtained by independently fixing foaming solution and gas flows through a coarse porous structure, and that the characterization of their initial structure can be obtained by real-time diffractometry [2]. The bubble size initial distributions of thus generated foams were found to be lognormal and the Lemlich's theory of interbubble gas transfer has been extended to this distribution type [3].

Until now, the influence of the films drainage was neglected, because reflectometry considers only the upper layers of the foam where drainage acts significantly during the first instants of the degradation. In this paper drainage is taken into account and we discuss the relative contributions of the surface and the volume of the films to the gas transfer resistance. The association of the surfactant molecules in the most performing foaming solutions is also discussed.

Interbubble gas transfer

Theory

First developed by Clark and Blackman [4] and De Vries [5], the interbubble gas transfer has been improved by New [6] and Lemlich et al. [7, 8]. Instead of viewing the gas as diffusing directly from bubble to bubble, Lemlich [7] suggested that it was more realistic to view the gas as first diffusing into the liquid region midway between the bubbles. The concentration of gas in this liquid can be considered as being equivalent (though Henry's law) to a gas pressure in the liquid. Then, by virtue of the classical law of Laplace and Young, this equivalent gas pressure can be viewed as that which would exist within a fictitious spherical bubble of radius r_{21} taken equal to the ratio between the second and first moments of the bubble radius distribution. Thus, the pressure difference between a bubble of radius r and the liquid is:

$$\Delta P = 2\gamma \left(\frac{1}{r_{21}} - \frac{1}{r} \right),$$ (1)

where γ is the interfacial tension.

The mass transfer rate:

$$\frac{dm}{dt} = J A \Delta P$$ (2)

is proportional to the interfacial area A through which the transfer takes place, to the pressure difference ΔP, and to the effective permeability J of the interbubble medium to gas transfer.

Assuming conservation of gaseous moles throughout the foam and considering the gas as perfect, it can be shown [7] that the variation with time of the size of a bubble radius is:

$$\Delta r = \frac{2J\gamma \mathscr{R} T}{M P_a} \left(\frac{1}{r_{21}} - \frac{1}{r} \right) \Delta t,$$ (3)

where $\mathscr{R}$ is the perfect gas constant, T the absolute temperature, M the gas molar mass, P_a the atmospheric pressure.
With the variable charges $r = r_0 R$, $r_{21} = r_0 R_{21}$ and

$$t = \frac{r_0^2 M P_a}{2 J_0 \gamma \mathscr{R} T} \theta,$$ (4)

where r_0 is the initial mean radius of the bubble distribution, J_0 the initial effective permeability, and θ an undimensional parameter, the relation (3) becomes:

$$\Delta R = \frac{J}{J_0} \left(\frac{1}{R_{21}} - \frac{1}{R} \right) \Delta \theta.$$ (5)

For an unit surface area, the effective permeability can be viewed as the reciprocal of the sum of two resistances, namely, the surface resistance $1/h$ of the membranes of the bubbles, and the volume resistance He/D. In this last expression, H is Henry's constant, D is the diffusion constant of the gas in the liquid, and e is the mean average liquid thickness between the bubbles. Thus, the instantaneous expression of the global resistance to gas transfer is,

$$\frac{1}{J} = \frac{1}{h} + \frac{He}{D}$$ (6)

while its initial value is,

$$\frac{1}{J_0} = \frac{1}{h} + \frac{He_0}{D}$$ (7)

with approximately,

$$e_0 = \frac{1}{3} r_0 \frac{1 - \varphi}{\varphi} \exp \left(\frac{5}{4} \sigma_0 \right)$$ (8)

for a foam the gas volume fraction of which is φ and the bubble radii of which are lognormally distributed with the standard deviation σ_0 around the mean value r_0 [3]. Of course the relations (6) and (7) assume that h and H/D are constant during the foam degradation, which is probably a first approximation.

To compare surface resistance and volume resistance with global resistance, it is useful to introduce the volume resistance fraction:

$$\Phi_{vr} = \frac{He_0/D}{1/J_0} = \frac{H}{D} e_0 J_0$$ (9)

or the surface resistance fraction:

$$\Phi_{sr} = \frac{1/h}{1/J_0} = \frac{J_0}{h} = 1 - \Phi_{vr}.$$ (10)

Interbubble gas transfer and drainage alter the liquid film thickness during the foam degradation. This alteration can be approximated by the relation,

$$\frac{e}{e_0} = \left(1 - \frac{V_{ld}}{V_{10}} \right) \frac{A_0}{A},$$ (11)

where V_{ld} is the drained liquid volume, V_{10} is the initial liquid volume in the foam, A_0 the initial interfacial area in the foam.

By using the relations (6), (7), (9), (10), (11), the expression (5) becomes the so-called relative radius evolution equation, which accounts simultaneously for interbubble gas transfer and drainage,

$$\Delta R = \left(\dfrac{1}{1 + \left[\left(1 - \dfrac{V_{ld}}{V_{10}} \right) \dfrac{A_0}{A} - 1 \right] \Phi_{vr}} \right) \left(\dfrac{1}{R_{21}} - \dfrac{1}{R} \right) \Delta\theta \ . \quad (12)$$

The fixed parameters are M, $\mathscr{R}$, T, P_a. The measurable foam parameters are γ (tensiometry measurements), φ (volume measurements), r_0 and σ_0 (granulometry measurements), V_{ld}/V_{10} (drainage measurements), A/A_0 (reflectometry measurements). e_0 is calculated with the φ, r_0 and σ_0 values from Eq. (8). The estimated parameter is H/D or h. The unknown parameters are Φ_{vr} and J_0.

The initial radius distribution is truncated and discretized. At each $\Delta\theta$ step its change is computed using the relative radius evolution Eq. (12), and the new value of A is calculated. Then the area is eliminated between calculated $A(\theta)/A_0$ and measured $A(t)/A_0$ to obtain a line in a t versus θ plot, in accordance with Eq. (4). This linearization is obtained using an automatic iterative method, which gives the value of Φ_{vs} with a relative precision higher than 5%. Finally, the slope of the line allows to obtain the value of J_0. It is then easy to calculate surface and volume resistances to gas transfer.

Experiments and discussion

We have obtained sufficient persistent foams with 5 g/l aqueous solutions of Montelane KRO (which will be noted KRO), of HPC or of their mixtures. KRO is an anionic surfactant: the sodium dodecylsulfate three times ethoxylated. HPC is a cationic surfactant: the hexadecyl pyridinium chloride. Foams have been generated and their initial structure has been measured using the methods described in [2]. Their mean characteristics are the following:

gas volume fraction: $\varphi \approx 0.85$
nature of the distribution: lognormal
mean radius $r_0 \approx 25\ \mu m$
standard deviation $\sigma_0 \approx 0.45$
initial liquid film thickness $e_0 \approx 4\ \mu m$

The decrease of the relative interfacial area of the bubbles has been evaluated by reflectometry according to the method presented in [1]. The drainage of the liquid films has been followed by carefully measuring the volume of the drained liquid. In Fig. 1 are given several examples of drainage. In foams generated with a single surfactant it is

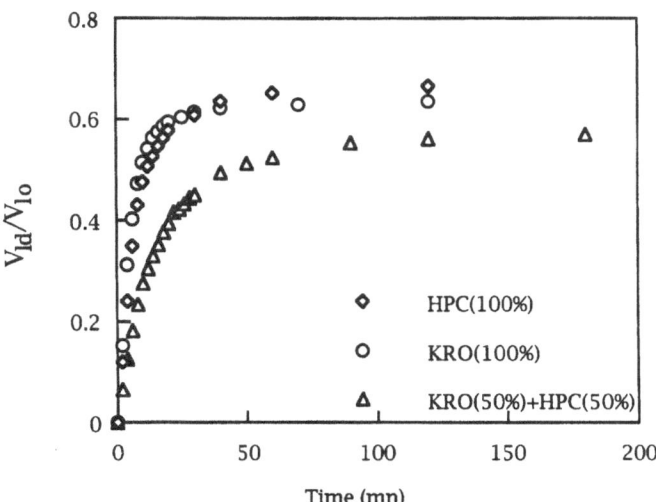

Fig. 1 Examples of variations versus time of the ratio of the drained liquid volume by the initial water volume in the foam

observed that drainage is slightly faster than in foams generated with surfactant mixtures.

When a single surfactant is used, we have supposed that the volume resistance of the liquid film may be assimilated to the water resistance. So we have taken:

$$H = 4.8 \times 10^{+6}\ kg^{-1}\ m^3\ Pa \quad \text{and} \quad D = 2 \times 10^{-9}\ m^2\ s^{-1}$$

which are Henry's constant for nitrogen in water and the nitrogen diffusion constant in water [9]. Then the comparison of the experimental results with Lemlich's model gives the initial effective permeability J_0 and the surface resistance of the liquid films (by means of the resistance fraction).

When mixtures of the two previous surfactants are used, the same calculation strategy leads for the film surface resistances to a synergism incompatible with the expected change of the surfactant molecule arranging induced by the composition change at the film surface. So for the surface resistances of the films obtained with these mixtures, we have adopted interpolated values (on the basis of molar fractions) between the surface resistances determined separately for the films formed with the two surfactants. Then the comparison of the experimental results with Lemlich's model gives J_0 and the volume resistance of the liquid films (always by means of the resistance fraction).

J_0 variations versus HPC molar fractions are reported in Fig. 2. For comparison these variations have been successively calculated without drainage correction and with drainage correction. Negligence of drainage leads to an overestimation of J_0 which can reach 30% for foams generated with a single surfactant. But the discrepancy becomes insignificant for surfactant mixtures close to equimolarity because the greatest part of interbubble gas

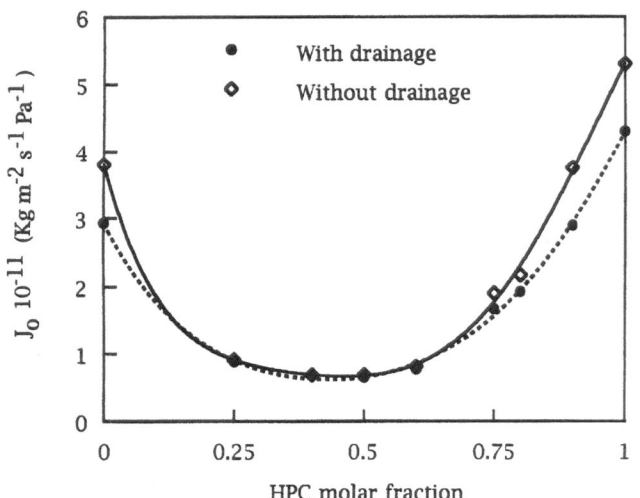

Fig. 2 Variation of the initial effective permeability versus the HPC molar fraction in the surfactant mixture

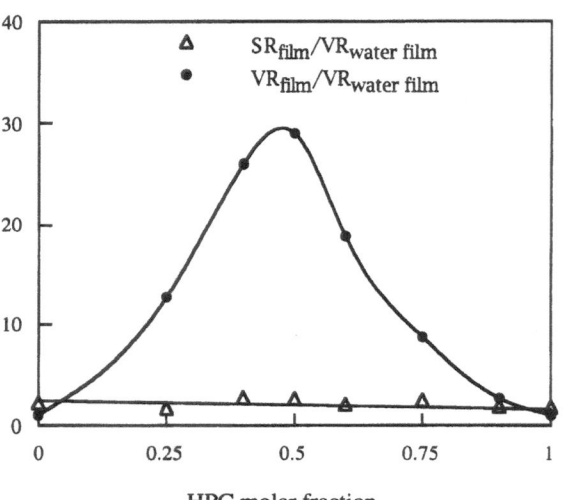

Fig. 3 Variations of the surface resistance (SR_{film}) and the volume resistance (VR_{film}) versus the HPC molar fraction in the surfactant mixture. (These resistances are related to the volume resistances of water films of same thicknesses.)

transfer takes place while drainage is practically achieved. In this concentration range, J_0 is four times lower than for a foam generated with HPC and three times lower than for a foam formed with KRO; this fact reveals an important synergism.

In Fig. 3 are reported the variations of the volume resistances and of the surface resistances of the liquid films versus mixture composition of the foaming solution. These resistances are related to the volume resistances of water films of same thicknesses. Film surface resistances in foams generated with KRO or HPC are approximately two times higher than the volume resistances of the same water films.

This result is in agreement with the fact that introduction of surfactant molecules at the air/liquid interface favors the foam stability by reducing the gas transfer. But the more striking result concerns volume resistances. In effect, in the equimolar concentration range, these resistances can be 30 times higher than the volume resistance of the water films of same thicknesses. This drastic increase which is responsible for the relative stability of the foams must probably be attributed to the particular association of the surfactant molecules in the films. In the next section the elements of this arranging are researched in the foaming solutions.

Surfactant molecule association in the foaming solutions

The studied surfactant solutions which give the more persistent foams are mixtures of an anionic surfactant and of a cationic surfactant. It is known that these mixtures have very low critical micellar concentration. As the studied mixtures are relatively concentrated (5 g/l), surfactant molecules must be organized in relatively large objects. The presence of these objects is revealed by a weak opalescence.

To obtain information on these objects the considered foaming solutions have been submitted to a static light scattering diagnostic. Measurements have been performed at the wavelength of 6328 Å. The sampling time was 1 μs and the experiment duration was 500 s. Scattered intensity was measured versus the scattering vector q. To avoid volume corrections, the scattered intensity was related to the intensity scattered by benzene at the same scattering angle.

In Fig. 4 is reported the curve relative to the equimolar mixture. It shows a scattering peak for $q = 10^{-3} \text{ Å}^{-1}$, the relative value of which is as high as 4000. Similar scattering peaks have already been observed in microemulsions [10, 11] and more recently in aqueous solutions of surfactant mixtures [12]. They were attributed to relatively large and concentrated objects (giant micelles, vesicles, etc.) which scatter the light following the relation:

$$I(q) = \frac{1}{a - bq^2 + cq^4} \tag{13}$$

where a, b and c are positive constants. These constants allow to determine the correlation length ξ and a distance d which, in first approximation, can be considered as the mean distance between the objects:

$$\xi = \left(\frac{c}{a}\right)^{\frac{1}{4}} \tag{14}$$

$$d = \frac{2\pi\sqrt{2}}{\sqrt{\frac{1}{\xi^2} + \frac{b}{2c}}}. \tag{15}$$

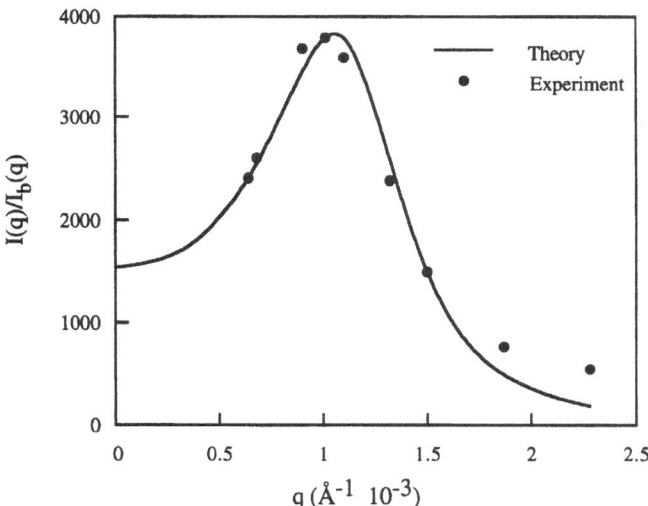

Fig. 4 Variation versus the scattering vector of the intensity of the light scattered by the foaming solution the surfactant mixture of which is equimolar. (This intensity is related to the intensity of the light scattered by benzene in the same experimental conditions)

For the equimolar mixture, a good agreement between theory and experiment is obtained with $\xi = 830$ Å and $d = 5560$ Å. (Fig. 4). A quantitative evaluation of the distribution of the surfactant molecules between the objects reveals that these molecules could be ordered in bi-layers. It is probably the confining of these particular objects in the foam films which increases their volume resistance by expanding a long-range structure in which the charges of the ionic surfactant molecules must play a leading part. This result is in agreement with the works of authors like Friberg who, a long time ago, drew attention to the favorable influence of liquid cristals on foam stability [13], or Wasan who, more recently, put in evidence micellar layers inside very thin films [14].

Conclusions

Interbubble gas transfer in persistent foams has been modeled on the basis of Lemlich's theory, taking into account the liquid drainage in the films. An automatic comparison between theory and experiment allows to determine the effective permeability, the surface resistance and the volume resistance of the films to gas transfer.

The introduction of the drainage leads to a decrease of the value of the effective permeability compared with the one evaluated by neglecting it. This decrease is particularly perceptible for foams generated with one single surfactant. It is negligible for foams generated with surfactant mixtures, the concentrations of which are close to equimolarity, because in these systems, the greatest part of interbubble gas transfer takes place while drainage is practically achieved.

Film surface resistances are approximately two times higher than the volume resistances of water films of same thicknesses. Film volume resistances of foams generated by surfactant mixtures are always higher than the volume resistances of water films of same thicknesses. In the equimolar concentration range they can be 30 times higher. This drastic increase is responsible for the relative stability of the studied persistent foams.

The most outstanding foaming solutions appear to be constituted by large and concentrated objects in which the surfactant molecules could be arranged in bi-layers. It is probable that confining these particular objects in the foam films increases their volume resistance by expanding a long-range structure in which the electric charges of the ionic surfactant molecules must play a leading part.

Acknowledgement The authors are indebted to the Société Seppic for the gift of Melanol KRO.

References

1. Lachaise J, Graciaa A, Marion G, Salager JL (1990) J Dispersion Sci Tech 11:409–432
2. Lachaise J, Sahnoun S, Dicharry C, Mendiboure B, Salager JL (1991) Prog Colloid Polym Sci 84:253–256
3. Marion G, Sahnoun S, Mendiboure B, Dicharry C, Lachaise J (1992) Prog Colloid Polym Sci 89:145–148
4. Clark NO, Blackman M (1948) Trans Farad Soc 44:1–7
5. De Vries AJ (1957) Foam Stability, Rubber-Stichling, Delft
6. New GE (1967) Proc Int Congress Surface Active Substances 1964, 2:1167
7. Lemlich R (1978) Ind & Eng Chemistry Fundam 17:89–93
8. Ranadive A, Lemlich R (1979) J Colloid Interf Sci 70:392
9. Abraham H, Sacerdote P (1913) Recueil de Constantes Physiques, Gauthier-Villars, Paris
10. Teubner M, Strey R (1987) J Chem Phys 87:3195–3199
11. Chen SH, Chang SL, Strey R (1990) Prog Colloid Polym Sci 81:30–35
12. Dufau P (1994) Thesis, University of Pau, France
13. Friberg S, Saito H (1976) in Foams, 33–36, edited by Akers RJ, Academic Press, London
14. Wasan DT (1992) Chem Eng Ed, 104–112

Progr Colloid Polym Sci (1996) 100:148–150
© Steinkopff Verlag 1996

Yu.V. Shulepov
S.Yu. Shulepov

Equilibrium states and structure factor of concentrated colloidal dispersions in optimized random phase approximation

Dr. Yu.V. Shulepov (✉)
Institute of Colloid and Water Chemistry
Ukrainian National Academy of Sciences
42 Vernadsky Ave.
Kiev 252142, Ukraine

Abstract In this work, the expression for correlation function of the concentrated dispersion of colloidal particles is calculated in optimized random phase approximation. We assume the time-averaged structure and thermodynamic states determined by the screened Coulomb pair potential and hard-sphere interaction between identical spherical colloidal particles immersed into electrolyte solution. The pair potential beyond the shortest separation of hard spheres obeys three models: a) equilibrium regime of charge regulation of particle, b) fixed potential on the surface of particle c) fixed charge of particle. Behavior of structure factor and corresponding correlation function shows increasing dispersion structure, eventually exhibiting solid-like state as we transfer from a) to the c) potential model.

Key words Concentrated dispersions – structure factor – x-ray, light or neutron scattering

In recent years, substantial progress has been made in the description of colloidal dispersions based on the theory of simple liquids (for example see [1–3]).

In the statistical theory of concentrated colloids the microscopic nature of the array of particles and the thermodynamic functions of the system are determined by calculation of the pair correlation function.

Here we propose the simplest model for calculation of the correlation function that is designated as the optimized random phase approximation (ORPA) in the physics of simple liquids [4–6]. According to this method the general correlation function of the system has the form

$$g(r) = g^{HS}(r) + \zeta^{ORPA}(r), \tag{1}$$

where the $g^{HS}(r)$ and $\zeta^{ORPA}(r)$ are the components of correlation function of hard spheres and in ORPA, respectively. The first we take is the Percus–Yevick approximation (PYA) [7]. The last term (1) can be derived by summation over all connected chain diagrams in cluster expansion of the correlation function [4]:

$$\zeta^{ORPA}(r) = \rho^{-1} \int \zeta^{ORPA}(k) \exp(i\vec{k}\vec{r}) d\vec{K}, \tag{2}$$

$$\zeta^{ORPA}(k) = \frac{[S^{HS}(k)]^2 \phi^*(k)}{1 - S^{HS}(k)\phi^*(k)}, \tag{3}$$

$$\phi^*(k) = \rho \int \phi^*(r) \exp(i\vec{k}\vec{r}) d\vec{r}$$

$$= \frac{24\eta}{y} \int_{-1}^{\infty} (1 + x) \sin[y(1 + x)] [-\beta_B U(r) + \phi(r)] dx$$

$$= -\beta_B U(k) + \phi(k), \tag{4}$$

$$\phi(r) = \begin{cases} \sum_{m=0}^{s} \alpha_m x^m, & -1 \le x \le 0; \\ 0, & 0 < x; \end{cases} \tag{5}$$

$$\phi^*(r) = -\beta_B U(r) + \phi(r), \tag{6}$$

$$x = \frac{r - d}{d}, \qquad y = kd.$$

Integration in expressions (2), (4) is carried out in the space of wave vectors $\vec{k}$ and particle coordinates $\vec{r}$, respectively.

Function $\zeta^{\text{ORPA}}(r)$ is the long-range component of general correlation function (1), $\zeta^{\text{ORPA}}(k)$, $\phi^*(k)$, $\beta_{\text{B}} U(k)$, $\phi(k)$ are the Fourier transformations of $\zeta^{\text{ORPA}}(r)$, $\phi^*(r)$, $\beta_{\text{B}} U(r)$, $\phi(r)$ functions, respectively. Function $U(r)$ is the long-range pair potential of interaction of particles immersed into electrolyte and interacting by their double layers. The quantities $\eta = \pi \rho d^3/6$ and ρ are the volume fraction and volume density of particles, respectively, $\beta_{\text{B}} = 1/k_{\text{B}}T$, k_{B} is the Boltzmann's constant, T is the temperature, d is the diameter of the particles. The structure factor of the hard sphere model, $S^{\text{HS}}(k)$, is determined by the relation

$$S^{\text{HS}}(k) = 1 + \rho \int \exp(i\vec{k}\vec{r})[g^{\text{HS}}(r) - 1]d\vec{r}. \qquad (7)$$

The optimized potential $\phi^*(r)$ coincides with $-\beta_{\text{B}} U(r)$ everywhere except inside the core where $\phi^*(r)$ is such that $\zeta^{\text{ORPA}}(r) = 0$ for all radii $r < d$ belonging to this core. In such a manner the polynomial coefficients α_m in Eq. (5) are actually dependent on density, temperature and other parameters of the system. The method of least squares has been used for calculations of α_m coefficients [8]. We took the initial polynomial of low degrees in (5) and then increased it up to negligible dependence of results of calculations on it. This is the so-called optimization procedure of determination of $\phi^*(r)$ potential.

The following expression holds for the PYA of the structure factor [9]:

$$S^{\text{HS}}(k) = \frac{1}{1 - \rho C^{\text{HS}}(k)}, \qquad (8)$$

$$-\rho C^{\text{HS}}(k) = \frac{24\eta}{y^6} \left\{ \lambda_2^2 y^3 (\sin y - y\cos y) \right.$$

$$- 6\eta(\lambda_1 + \lambda_2)^2 y^2 [2y\sin y - (y^2 - 2)\cos y - 2]$$

$$+ \frac{\eta\lambda_2^2}{2} [-(y^4 - 12y^2 + 24)\cos y + 4y^3\sin y$$

$$\left. - 24y\sin y + 24] \right\},$$

$$\lambda_1 = -\frac{3}{2}\frac{\eta}{(1-\eta)^2}, \qquad \lambda_2 = \frac{1 + 2\eta}{(1-\eta)^2}.$$

Finally, we determine the total interaction potential between the double layers of pair of identical colloidal particles $U(r)$.

In the case of charge-stabilized dispersion the repulsive screened Coulomb potential is generally larger than thermal energies at small separations, and three types of interaction potential can be considered in the framework of

DLVO theory [10–12]:

$$U^{\text{W}}(r) = -\frac{4q_s^2}{\text{æ}^2 d^3 \varepsilon} \ln[1 - \beta\exp(-\text{æ}dx)], \qquad (9)$$

$$U^{\varphi}(r) = \frac{\varepsilon d\varphi_s^2}{4} \ln[1 + \exp(-\text{æ}dx)], \qquad (10)$$

$$U^{\text{q}}(r) = -\frac{4q_s^2}{\text{æ}^2 d^3 \varepsilon} \ln[1 - \exp(-\text{æ}dx)], \qquad (11)$$

$$\beta = (1 - \delta)/(1 + \delta), \qquad \delta = \frac{4q_s ez\beta_{\text{B}}}{\text{æ}d^2\varepsilon},$$

$$\text{æ}^2 = 8\pi z^2 e^2 c\beta_{\text{B}}/\varepsilon, \qquad x \geq 0,$$

where the expressions (9)–(11) are applied in the cases of equilibrium regime of charge regulation [12], constant electric potential and constant charge on the surface of particles [10, 11], respectively. The quantities $q_s = \pi d^2\sigma_s$, φ_s are the surface charge and potential of the particle, respectively, σ_s is the surface charge density of the particle. The Debye–Huckel screening parameter æ is defined for $1-1$ electrolyte of concentration c and ion valence z, e is elementary charge, ε is the dielectric constant of dispersing fluid.

All expressions (9)–(11) are valid in the so-called case of small potentials which can be determined by condition $\delta < 1$. (The quantity δ is equal to the order of magnitude the dimensionless surface potential of the particle [12]). The further condition $\text{æ}d \gg 1$ is required in this case for application of the Derjaguin approximation [11].

Using the expression for general correlation function (1), we can evaluate the structure factor of the system

$$S(k) = 1 + \rho \int \exp(i\vec{k}\vec{r})(g(r) - 1)d\vec{r}$$

$$= S^{\text{HS}}(k) + \zeta^{\text{ORPA}}(k). \qquad (12)$$

In such a manner, we carry out the optimization procedure for the $\phi^*(r)$ potential when the temperature, density and other parameters of the system are fixed. In response we can calculate the structure factor of the colloidal dispersion with formula (12).

In order to relate the surface charge density of the particles, σ_s, to the surface potential, φ_s, we have employed the Stern and Grahame [13] model of the electric double layer, which provides the following equation in the case of small potentials

$$\sigma_s = \frac{\varepsilon\text{æ}\varphi_s}{4\pi}. \qquad (13)$$

The corresponding radial correlation function is derived from (12) through the equation

$$g(r) = 1 + \frac{1}{12\pi\eta\left(\frac{r}{d}\right)} \int_0^\infty [S(k) - 1]\sin\left(y \cdot \frac{r}{d}\right)y\,dy. \qquad (14)$$

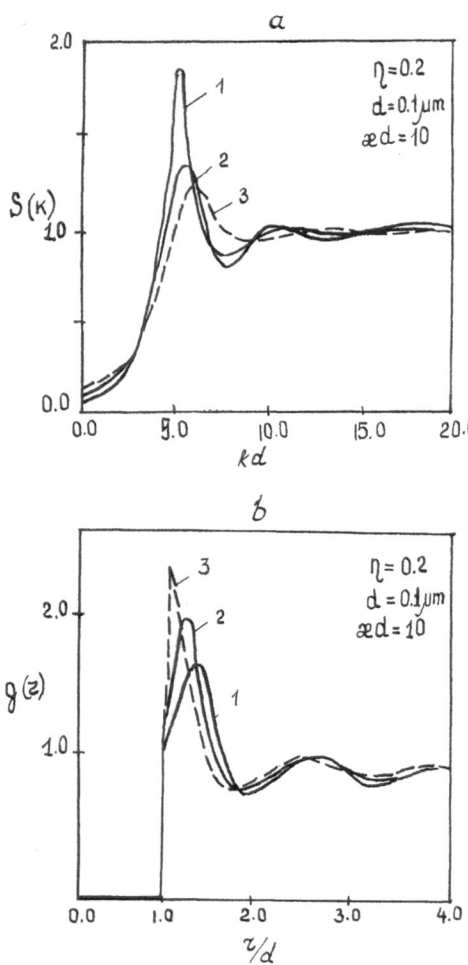

Fig. 1A $S(k)$ vs kd for parameters determined in the text. **B** The corresponding radial correlation functions. 1) the constant charge of the particle, 2) the constant electrical potential of the particle, 3) hard spheres' behavior which coincides graphically with the case of equilibrium regime of charge regulation (dashed curve)

Behavior of the structure factors (k) against kd and corresponding radial correlation function $g(r)$ vs r/d for various models of potential interactions (Eqs. ((9)–(11)) are pre-

sented in Fig. 1a, b ($T = 298\,\mathrm{K}$, $\eta = 0.2$, $\varepsilon = 78.3$, $d = 0.1\ \mu\mathrm{m}$, $q_s = 10^2 e$, $z = 1$, $c = 1\ \mathrm{mM/1}$).

Behavior of structure factors and corresponding radial correlation functions for various types of potential interactions (Fig. 1) reveals increasing structure, eventually exhibiting solid-like state as we transfer from equilibrium regime of charge regulation to regime with fixed charge on the particle surface. Decrease of isothermal compressibility, as reflected in $S(k = 0)$ (Fig. 1), provides support for this view.

As follows from Fig. 1, the curves for the cases of hard spheres and equilibrium regime of charge regulation are graphically indistinguishable (for the small electrical potentials of particles). Hence it follows that many of experimental data on x-ray, neutron or light scattering in dispersions can be described on the basis of hard sphere model.

The type of interaction potential can be established after comparison of theoretical $S(k)$ with experimental data.

Now we obtain the thermodynamic functions of the system on the correlation function. We define the internal energy per one particle, E, in the following way:

$$\beta_B E = \frac{\rho}{2} \int \beta_B U(r) g(r) d\vec{r} = \frac{\beta_B U(k = 0)}{2}$$

$$+ \frac{1}{24\pi\eta} \int_0^\infty [S(k) - 1]\, \beta_B U(k) y^2\, dy. \tag{15}$$

The excess osmotic pressure, p, can be calculated by using the thermodynamic relation

$$\beta_B p = \rho^2 \left(\frac{\partial \beta_B F}{\partial \rho}\right)_T = \rho^2 \frac{\partial}{\partial \rho}\left(\int_0^{\beta_B} E\, d\beta_B\right)_T, \tag{16}$$

where F is the Helmgoltz free energy per one particle. The derivatives in (16) are taken at constant T.

The excess osmotic pressure obtained directly by compression measurements can then be compared with these calculated by formula (16).

Acknowledgment One of the authors (Yu.V.) wishes to thank the International Association for Collaboration between the Scientists of the Independent States of the Former Soviet Union for financial support of this study (INTAS project 93-3372).

References

1. Pusey PN (1980) In: Hansen JP, Levesque D, Zinn-Justin J (eds) Liquids, Freesing and Glass Transition. North-Holland, Amsterdam, pp 768–942
2. Russel WR, Saville PA, Schowalter WR (1989) Colloidal Dispersions. Cambridge Univer Press, Cambridge
3. Hayter JR, Penfold J (1981) Mol Phys 42:109–118
4. Lebowitz JT, Stell G, Baer S (1965) J Math Phys 6:1282–1298
5. Carazza B, Parola A, Reatto T, Tau M (1982) Physica 116A:207–226
6. Yuchnovsky IR, Shulepov YuV (1985) J Stat Phys 38:541–572
7. Percus JK, Yevick GJ (1958) Phys Rev 110:1–13
8. Korn GA, Korn TM (1961) Mathematical Handbook for Scientists and Engineers: Mcgraw-Hill Book Co, NY
9. Aschoroft NW, Lekner J (1966) Phys Rev B 145:83–90
10. Verwey EJW, Overbeek JThG (1948) Theory of the Stability of Lyophobic Colloids. Elsevier, Amsterdam
11. Derjaguin BV (1989) Theory of Stability of Colloids and Thin Films. Consultants Burea, NY
12. Shulepov SYu, Dukhin SS, Lyklema J (1995) J Coll Int Sci 171:340–350
13. Reerink H, Overbeek JThH (1954) Disc Faraday Soc 18:74–82

Progr Colloid Polym Sci (1996) 100:151–155
© Steinkopff Verlag 1996

EMULSIONS AND CONCENTRATED SYSTEMS

A. Kasper
S. Kirsch
F. Renth
E. Bartsch
H. Sillescu

Development of core-shell colloids to study self-diffusion in highly concentrated dispersions

A. Kasper · S. Kirsch · F. Renth
E. Bartsch (✉) · H. Sillescu
Institut für Physikalische Chemie
Universität Mainz
Jakob-Welder-Weg 15
55099 Mainz, FRG

Abstract To study single particle motion in highly concentrated colloidal dispersions, a host-tracer colloid system was developed, consisting of crosslinked polymer micronetwork spheres placed in a good solvent. The host colloid is made invisible to the experimental probe by matching its refractive index to that of the solvent. For the tracer particles a core-shell structure was chosen to ensure the interaction potential to be identical to that of the host particles. Therefore the shell was made of the same polymer as the host. The core differs in refractive index from the solvent and is therefore visible due to scattered light.

The self-diffusive motion of a few strongly scattering tracer particles is then studied by dynamic light scattering (DLS) and optical dark field microscopy. DLS results give a proof for the "hard sphere" character of the

particles studied and also show that true single particle motion is detected. Mean squared displacements, determined via optical dark field microscopy, are compared with results calculated from DLS measurements. In combination with a third method, forced Rayleigh scattering, where dye-labeled tracers are used, self-motion was studied for nine decades in time. The results obtained from the three methods are fully consistent. In concentrated dispersions ($\phi = 0.54$) the transition from short-time to long-time diffusion is visible. This paper also presents first results obtained by optical microscopy for a sample in the glassy state ($\phi = 0.63$).

Key words Micronetwork spheres – core-shell particles – light microscopy – dynamic light scattering – mean squared displacements

Introduction

Colloidal dispersions have recently become important model systems for the study of the glass transition in quasiatomic systems [1, 2]. Very detailed theoretical predictions for the dynamical behavior of hard spheres at the glass transition have become available [3, 4], which have been quantitatively verified for hard sphere PMMA colloids with respect to *collective* dynamics [1]. It is therefore of special interest to investigate *single* particle motion of

colloidal particles in highly concentrated dispersions as well, and to check as to how far the predicted behavior with respect to the self intermediate scattering function $F_S(Q, t)$, the mean squared displacement $\langle \Delta r^2(t) \rangle$ and the long-time self diffusion coefficient D_S^L can be found in experiment.

In order to investigate single particle dynamics, a host-tracer system has to be developed in which the host particles are made invisible to the experimental probe by matching their refractive index to that of the surrounding solvent. The motion of suitable tracers can then be

152

A. Kasper et al.
Core-shell colloids: self diffusion at high concentrations

monitored by optical techniques like dynamic light scattering (DLS) [5], forced Rayleigh scattering (FRS) [6] and optical microscopy [7].

It is the aim of this contribution to report on the development of core-shell colloids that allow the extension of existing tracer diffusion results [5, 7] well into the glass transition regime. The host system consists of crosslinked polymer micronetwork spheres similar to those used in previous studies of the glass transition [2, 8]. However, t-butylacrylate (t-BuA) instead of styrene is chosen as monomer to allow perfect refractive index match. A crosslink density (inverse number of monomer units between crosslinks) of 1 : 10 was employed to guarantee hard sphere behavior. The tracer colloid (with the same degree of crosslinking as the host particles) consists of a polystyrene core and a poly-t-butylacrylate ($p(t$-BuA)) shell in order to achieve a large refractive index difference with respect to the solvent while ensuring the interaction potential to be identical to that of the host.

The self-diffusion of a few strongly scattering tracer particles is then studied as a function of overall volume fraction ϕ, using the scattered light by two different methods: i) The incoherent intermediate scattering function $F_s(Q, \tau)$ is measured by dynamic light scattering (DLS). ii) The scattered light is used to determine the trajectories of the tracer particles in real space via optical dark field microscopy. Mean squared displacements are then calculated and combined with data obtained via FRS [10] to cover 9 decades in time.

Experimental details

Synthesis and characterization

Both types of particles (host and tracer) were prepared by surfactant-free emulsion polymerization [9], using potassium peroxodisulfate as initiator. For all copolymerizations the mass ratio of comonomers was calculated in order to yield an average crosslink density of 1 : 10 (i.e., 10 monomer units between crosslinks).

For the host particles t-butylacrylate was copolymerized with ethanedioldiacrylate as crosslinker. The core-shell particles were synthesized in a two-step process. In the first step, the cores were prepared by radical copolymerization of styrene and diisopropenylbenzene (crosslinker). In the second step a mixture of t-butylacrylate and ethanedioldiacrylate was added and polymerized in the same way as described above, using the core particles as seeds. Because of the presence of the seed particles, no second nucleation occurred and the newly prepared copolymer formed a shell around the crosslinked polystyrene particles. 4-Fluorotoluene was used as isorefractive

solvent. Due to its good solvent properties, it causes the particles to swell. However, by virtue of the crosslinks the particles remain spherical. The resulting size (hydrodynamic radius R_H) of the particles is related to their initial size (unswollen, i.e., in emulsion) by the swelling ratio S, defined as

$$S = \frac{R_H^3(\text{swollen})}{R_H^3(\text{unswollen})} \, . \tag{1}$$

Due to different experimental requirements, the particles prepared for the DLS studies differed in size from the particles synthesized for observation by light microscopy. Table 1 gives a summary of the sizes of the particles used in this study.

Figure 1 shows a TEM picture of a sample of tracer particles. The shell is too thin to be clearly visible in this picture. It can nevertheless be surmised from the distance of particles in contact being slightly larger than twice the radius of the cores. However, the spherical shape and the low polydispersity of the core-shell particles are obvious.

Dynamic light scattering

The experimental setup is described in detail elsewhere [11]. Here, we only recall briefly the relevant theoretical relations. DLS measures the normalized, time averaged intensity auto-correlation function $g_T^{(2)}(Q, t) = \langle I(Q, t) I(Q, 0) \rangle_T / \langle I(Q) \rangle_T^2$ ($\langle \cdots \rangle_T$ denotes the time average) which for homodyne conditions yields the normalized intermediate scattering function $f^M(Q, t)$ [12] via the Siegert relation [13]

$$f^M(Q, t) = \frac{F^M(Q, t)}{F^M(Q, 0)} \, . \tag{2}$$

Here $Q = (4\pi n / \lambda) \sin(\theta/2)$ is the scattering vector, with n being the refractive index of the medium, λ the wavelength of the laser light and θ the scattering angle. The quantities $F^M(Q, t)$ and $F^M(Q, 0) = S^M(Q)$ denote the measured dynamic and the measured static structure factor for a polydisperse colloidal dispersion [12], respectively. The initial decay of $f^M(Q, t)$ is related to an apparent diffusion coefficient $D_{app}(Q)$ via [12]

$$\left[\frac{\partial \ln f^M(Q, t)}{\partial t} \right]_{t \to 0} = \frac{1}{F^M(Q, 0)} \left[\frac{\partial F^M(Q, t)}{\partial t} \right]_{t \to 0}$$
$$= - D_{app}(Q) Q^2 \, , \tag{3}$$

with

$$D_{app}(Q) = \frac{H^M(Q)}{S^M(Q)} \bar{D}_0 \, , \tag{4}$$

Table 1 Characterization of the colloid systems investigated by three different optical methods: dynamic light scattering (DLS), optical video microscopy (VM) and forced Rayleigh scattering (FRS). R_H is the hydrodynamic radius as measured by DLS

Method	Host	Tracer	Solvent
DLS	poly-t-butylacrylate $R_H = 290$ nm	shell: poly-t-butylacrylate core: polystyrene $R_H = 255$ nm	4-fluorotoluene
VM	poly-t-butylacrylate $R_H = 480$ nm	shell: poly-t-butylacrylate core: polystyrene $R_H = 455$ nm	4-fluorotoluene
FRS [10]	polystyrene $R_H = 81$ nm	polystyrene, dye-labeled $R_H = 81$ nm	2-ethyl-naphthalene

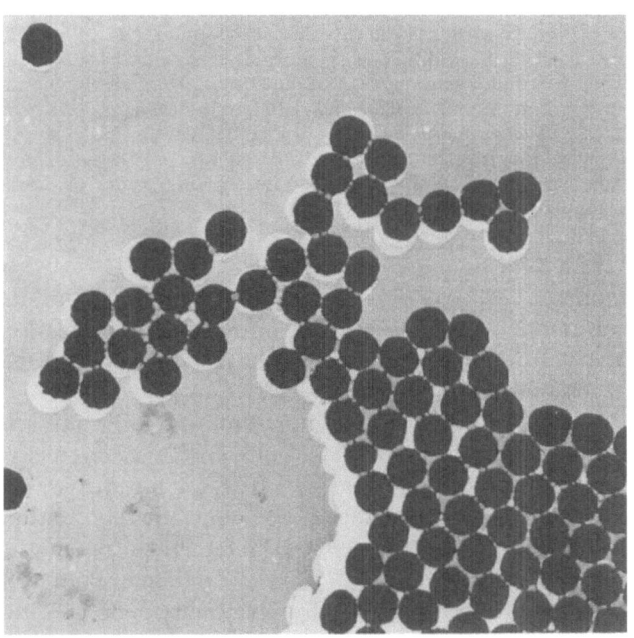

Fig. 1 TEM picture of the core-shell tracer particles used for optical microscopy. The picture shows an area of 9 μm · 9 μm in size. Note that the shell material is not fully stable under TEM conditions and is partially degraded by the electron beam

where $H^M(Q)$ is the hydrodynamic factor and $\bar{D}_0$ the average free particle (Stokes–Einstein) diffusion coefficient. For a system consisting of slightly polydisperse spherical particles the measured dynamic structure factor can be separated into two terms [12]:

$$F^M(Q,t) = \left(\frac{\overline{b(Q)^2}}{\overline{b^2(Q)}}\right) F_{coll}(Q,t) + \left(1 - \frac{\overline{b(Q)^2}}{\overline{b^2(Q)}}\right) F_s(Q,t) . \quad (5)$$

Here $b(Q) = V(n_{colloid} - n_{solvent}) f(Q)$ denotes the scattering amplitudes (V is the particle volume and $f(Q)$ is the normalized scattering amplitude with $f(0) = 1$). $F_{coll}(Q,t)$ represents the collective part of the dynamic structure

factor and $F_s(Q,t)$ the self part given by

$$F_s(Q,t) = \left\langle \exp\{i\vec{Q}\ [\vec{r}_i(t) - \vec{r}_i(0)]\} \right\rangle . \quad (6)$$

For an optically bidisperse system, consisting of a large number of host colloids which are isorefractive with the solvent and of small amounts of strongly scattering tracer particles, Eq. (5) reduces to $F^M(Q,t) = F_s(Q,t)$, and DLS essentially measures the self-diffusion in this case. From Eq. (6) the mean squared displacement can be calculated as

$$\langle \Delta r^2(t) \rangle = -\frac{6 \ln F_s(Q,t)}{Q^2} = 6D_{Tracer} t . \quad (7)$$

Furthermore, the long (D_S^L) and short-time (D_S^S) self diffusion coefficients can be obtained from Eq. (7) in the limits $t \to \infty$ and $t \to 0$, respectively [12] (cf. Eq. (8): $4D_S^L$ replaced by $6D_S^L$).

Optical microscopy and digital image processing

To track single colloidal particles in a concentrated dispersion, an optical microscope (Olympus BX-60, Objective 100x/Oil immersion) with dark field illumination was used. The tracer particles scatter light and appear as bright spots on a dark background. Due to refractive index match the host particles are invisible.

Time-resolved digital image processing (performed on an Apple Macintosh IIvx with a Data Translation Quick Capture DT2255-60 frame grabber board [7], time resolution 0.25 s) of the pictures obtained by a CCD-camera (Proxitronic) connected to the microscope enables the determination of two dimensions of the particle trajectories. From these data, mean square displacements can be calculated which in the limit $t \to \infty$ yield the long-time self diffusion coefficient D_S^L:

$$D_S^L = \lim_{t \to \infty} \frac{\langle \Delta r^2(t) \rangle}{4t} = \lim_{t \to \infty} \frac{\langle [\vec{r}(\tau + t) - \vec{r}(\tau)]^2 \rangle_\tau}{4t} . \quad (8)$$

154
A. Kasper et al.
Core-shell colloids: self diffusion at high concentrations

Forced Rayleigh scattering

With this technique a holographic grating is created via the interference pattern of two coherent laser beams that are crossed in a sample containing polystyrene micronetwork spheres ($R_H = 81$ nm, crosslink density 1:10) as host and the same spheres, labeled with a photochromic dye, as tracers. The bleaching of the dye molecules at the positions of constructive interference results in a grating of the absorption coefficient or the refractive index and yields a concentration grating with respect to the labeled particles which is then detected by an attenuated laser beam at the Bragg angle. Since this concentration grating decays due to self-diffusion of the tracer particles, the self-intermediate scattering function is related to the intensity decay of the scattered light, according to $I(Q,t) = [C_1 + F_s(Q,t)]^2 + C_2$ with C_1 and C_2 being the coherent and the incoherent background, respectively. More details and a brief description of the experimental setup can be found in an accompanying publication [10]. From $F_s(Q,t)$ the mean squared displacement can be obtained as stated above. Thus, FRS can yield the same information on single particle dynamics as DLS. DLS has the disadvantage that a small refractive index mismatch between host particles and solvent may cause strong coherent scattering. However, the two techniques, working in different time scales, are complementary. A detailed comparison of data obtained by DLS and FRS, demonstrating that both methods yield equivalent information, will be discussed in a forthcoming publication [14].

Results and discussion

In order to check if the $p(t\text{-BuA})$ micronetwork particles with a crosslink density of 1:10 behave as hard spheres, the apparent diffusion coefficients extracted via Eq. (3) from DLS measurements on the host system have been used to calculate an experimental $S^M(Q)$ according to Eq. (4). In these calculations the hydrodynamic factor $H^{hs}(Q)$ for a monodisperse hard sphere system at volume fraction $\phi = 0.34$, interpolated from literature data [12], was used for $H^M(Q)$. The experimental $S^M(Q)$ determined in this way is consistent with theoretical values obtained from the Percus–Yevick approximation for a hard sphere system [15] for size polydispersities $\sigma = [\langle R^2 \rangle - \langle R \rangle^2]^{1/2}/\langle R \rangle$ between 0.04 and 0.09, as shown in Fig. 2. Deviations from the theory at high and low wave-vectors are a consequence of multiple scattering effects (most pronounced around the minimum of the particle form factor) and of incoherent scattering caused by size polydispersity, respectively. The σ-values derived from the Percus–Yevick calculations compare favorably with results derived from

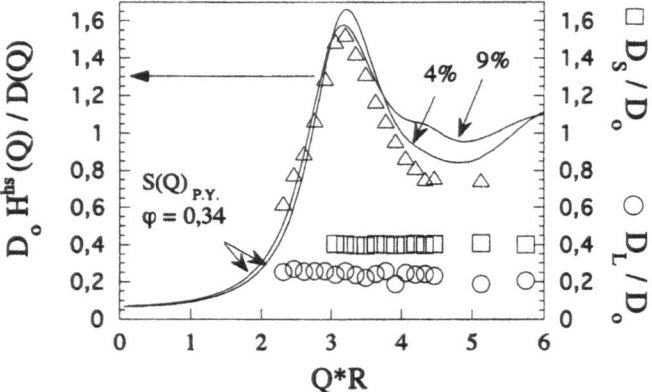

Fig. 2 Q-independence of the short-time (D_S^S, squares) and the long-time (D_S^L, circles) self-diffusion coefficients, normalized by the free particle diffusion coefficient D_0, are illustrated. In contrast, the inverse apparent diffusion coefficients ($D(Q)$, triangles), monitoring the collective motion, rescaled with D_0 and $H^{hs}(Q)$, are in good agreement with the theoretical $S(Q)$ calculated for polydisperse hard spheres according to ref. [15] with size polydispersities of 4 and 9 percent (solid lines)

an analysis of the Q-dependence of $D_{app}(Q)$ in the dilute solution following to ref. [16]. A more detailed discussion of the hard sphere character of the host system will be presented elsewhere [17].

Included in Fig. 2 are the short-time and long-time self diffusion coefficients obtained from DLS measurements of the host-tracer system at $\varphi = 0.34$ for various wave-vectors. The Q-independence of the quantities shows that self diffusion is, indeed, probed. Note that even in the region of the maximum of $S(Q)$, where coherent contaminations due to non-perfect refractive index match would clearly show up, no significant Q-dependence is observed. Furthermore, the volume fraction dependence of D_S^S/D_0 [17] and of D_S^L/D_0 [10] are in good agreement with theoretical predictions for hard spheres and data obtained from other hard sphere colloids [12]. Thus, on the basis of the present data, our host-tracer system is a good candidate for studying single particle dynamics of hard spheres at the colloid glass transition.

In Fig. 3, we present first results for the mean squared displacements $\langle \Delta r^2(t) \rangle$ as determined by DLS, optical microscopy and FRS. To compare data obtained for particles of different sizes (adapted to the requirements of the respective experimental technique), mean squared displacements are given in units of the particle radius squared and the time axis is rescaled to $(D_0/R_H^2) \cdot t$.

The figure shows that the mean squared displacements obtained from the three techniques for comparable volume fractions are fully consistent (note that no fudge factor is involved), which again proves that all three methods properly monitor self-diffusion. Furthermore, a combination

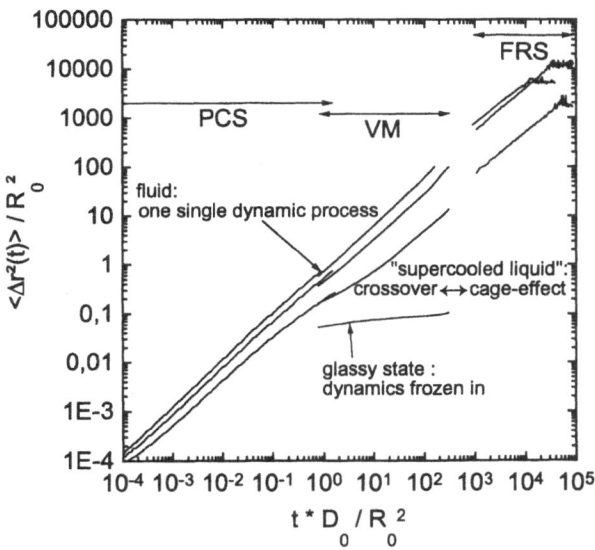

Fig. 3 Rescaled mean squared displacements versus rescaled time: a comparison of results obtained from DLS, optical video microscopy (VM) and FRS. Volume fractions are (from top line to bottom line): 0.34, 0.42, 0.54, 0.63. For volume fraction $\phi = 0.63$, only VM data are available so far. Slight mismatches between data sets obtained from different methods are due to slight differences in the volume fractions of the respective samples

of the different techniques allows to determine $\langle \Delta r^2(t) \rangle / R_H^2$ for nine decades in time, which is important for studying the dynamics of the colloidal glass transition. At $\phi = 0.54$, the transition from short-time diffusion to long-time dynamics (the onset of the cage effect) can be seen in the log-log plot as a transition range connecting two straight lines of slope one. At $\phi = 0.63$, the dynamics seems to be almost frozen, as would be expected for a sample in the glassy state. Here, corresponding data from DLS and FRS are not yet available.

In summary, we have shown that a host-tracer system of micronetwork collioids, employing core-shell or dye-labeled particles as tracers, allows the study of self diffusion of hard sphere-like colloids for nine decades in time, using a combination of three optical techniques: DLS, FRS and optical microscopy. Future work will focus on the dynamics in the glass transition region to obtain a more detailed (physical) picture of the processes involved.

Acknowledgments The authors thank A. Doerk for preparation of the colloid particles, S. Stoelken, G. Weber and Dr. G. Lieser for their help in performing transmission electron microscopy measurements.

Financial support by the Stipendienfonds der Chemischen Industrie and the Sonderforschungsbereich 262 of the Deutsche Forschungsgemeinschaft is gratefully acknowledged.

References

1. van Megen W (1995) In: Yip S (ed) Relaxation Kinetics in Supercooled Liquids – Mode Coupling Theory and its Experimental Tests, Transp Theory Stat Phys 24:1017–1051
2. Bartsch E (1995) In: Yip S (ed) Relaxation Kinetics in Supercooled Liquids – Mode Coupling Theory and its Experimental Tests, Transp Theory Stat Phys 24:1125–1145
3. Götze W (1991) In: Levesque D, Hansen JP, Zinn-Justin in Liquids, Freezing and the Glass Transition, Les Houches Session L1, Elsevier, Amsterdam, pp 287–504
4. Fuchs M (1995) In: Yip S (ed) Relaxation Kinetics in Supercooled Liquids – Mode Coupling Theory and its Experimental Tests, Transp Theory Stat Phys 24: 855–880

5. van Megen W, Underwood SM (1989) J Chem Phys 91:552–559
6. Bartsch E, Frenz V, Möller S, Sillescu H (1993) Physica A 201:363–371
7. Schaertl W, Sillescu H (1993) J Coll Interface Sci 155:313–318
8. Bartsch E, Frenz V, Sillescu H (1993) J Non-Cryst Solids 172–174:88–97
9. Kirsch S, Doerk A, Bartsch E, Sillescu H, Landfester K, Boeffel C, Spiess HW, Mächtle W, in preparation
10. Renth F, Bartsch E, Kasper A, Kirsch S, Stoelken S, Sillescu H, Koehler W, these proceedings
11. Bartsch E, Frenz V, Baschnagel J, Schaertl W, Sillescu H, in preparation

12. Pusey PN (1991) In: Levesque D, Hansen JP, Zinn-Justin in Liquids, Freezing and the Glass Transition, Les Houches Session L1, Elsevier, Amsterdam, pp 763–942
13. Berne BJ, Pecora R (1976) Dynamic Light Scattering, Wiley, New York
14. Kirsch S, Renth F, Bartsch E, Sillescu H, work in progress
15. van Beurten P, Vrij A (1981) J Chem Phys 74:2744–2748
16. Pusey PN, van Megen W (1984) J Chem Phys 80:3513–3520
17. Kirsch S, Bartsch E, Sillescu H, work in progress

Progr Colloid Polym Sci (1996) 100:156–161
© Steinkopff Verlag 1996

M. Olteanu
S. Pertz
V. Raicu
O. Cinteza
V.D. Branda

Concentrated graphite suspensions in aqueous polymer solutions

Dr. M. Olteanu (✉) · O. Cinteza
V.D. Branda
University of Bucharest
Department of Physics Chemistry
13 Republicii Blvd.
70346 Bucharest, Romania

Abstract The present paper reports the interactions of several poly-acrylamides having high molecular weight and variable content of ionizable groups with concentrated colloidal suspensions of graphite spherical particles. The method of electric conductivity dispersion with frequency has shown that within the range of small KCl concentrations, the adsorbed polymer layer expands, and it then contracts, with the increase of ionic strength. The KCl concentration, at which the contrac-tion of polymer begins, decreases with molecular weight at the same hydrolysis degree and with the increase of hydrolysis degree for the same molecular weight. The expansion of the adsorbed layer increases with number of hydrolyza-ble groups on the chain and with the amount of adsorbed polymer.

Key words Graphite suspensions – polyacrylamide adsorption – disper-sion of electric conductivity with frequency

Introduction

The interactions between polymers in solutions and col-loidal suspensions, made up of solid particles, have not been sufficiently investigated yet, due to both the system complexity and to the great number of interfering para-meters. Finally, the polymer–solid particle interactions may lead to the stabilization of the initial particles, to the formation of some redispersible aggregates or to the total destabilization of the colloidal system.

The present paper reports interactions of several poly-acrylamides having high molecular weight and variable content of ionizable groups with dilute and concentrated colloidal suspensions of graphite containing spherical par-ticles [1]. Using the dispersion of the electric conductivity, the influence of electrolytes was followed on the concen-trated suspensions with particles having, on their surface, layers of adsorbed polymer, of thickness comparable with the size of the particles. The behaviour of the suspensions in electric field, within the range of low frequencies, gives interesting information on the solid–solution interface [2, 3]. The study of the polyacrylamides adsorption, large-ly used polymers, onto solid surfaces of different nature is a topic still under investigation [4–7].

Materials and methods

The polymers used were poly(acrylamides) (PAM), having high molecular weights and varied degrees of hydrolysis obtained from Sanpoly–Sankyo Chem. Ind. Ltd. Tokyo (Table 1).

Viscosity-average molecular weights M_v were deter-mined with an Ubbelohde viscometer in aqueous sodium nitrate 1 mol/l at 300 K and calculated by means of relation:

$$[\eta] = 3.73 \; 10^{-4} \, M_v^{0.66} \, ,$$

were $[\eta]$ is the intrinsic viscosity. Degrees of hydrolysis (DH) were determined by conductometric and pH-metric

Progr Colloid Polym Sci (1996) 100:156–161
© Steinkopff Verlag 1996

Table 1 Average diameter of graphite particles calculated from turbidimetric data

No	Additive	M_v $\times 10^{-6}$	DH %	K $10^{-17}\,cm^{-4}$	D nm	Maximum amount of PAM adsorbed $\times 10^4$ g/g
1	PAM 500–20	5.0	20	137.9	55.4	5.78
2	PAM 600–0	6.2	0	177.8	59.7	2.90
3	PAM 600–20	6.1	20	194.4	70.0	5.20
4	PAM 600–40	6.1	40	142.9	64.3	8.72
5	PAM 1000–25	10.3	25	138.9	52.6	10.40
6	water	–	–	129.5	50.6	–

PAM 0.004%.

methods [8]. In water and electrolyte solutions, the hydrolyzed polyacrylamides behave like weak polyelectrolytes.

Graphite suspensions with spherical-shaped particles (Fig. 1a) were obtained by a previously described method [1]. Figure 1a shows the geometrical shape by "visualizing" the biggest graphite particles.

Before the contact with the PAM solutions, the graphite particles were centrifuged and washed with bidistilled water for removing the grinding additives.

Adsorption of the polymer onto the solid–liquid region was measured using the solution depletion method: 1.73 g graphite was dispersed in 100 cm^3 of polymer aqueous solution of different concentration with or without required salinity (KCl). The suspensions were gently shaken from time to time during 48 h. The amount of polymer in solution was determined interferometrically with an ITR-2 device (Russia), at 25 °C, from a calibration curve, after ultracentrifugation of suspensions.

Sedimentation analysis in gravitational field was made with a Sartorius automatic balance, when the maximum PAM adsorption was reached, in the presence and absence of KCl. Comparatively, graphite dispersions with lamellar-shaped particles, obtained through a different method, were studied under same conditions [9]. Microscopy under reflected light of the resin-included particles revealed anisotropic, lamellar shapes having micrometric sizes (agglomerated lamellae) (Fig. 1b). Analysis by x-ray diffractions confirmed the hexagonal structure of the graphite. Both the concentrated graphite suspensions with spherical particles and those having lamellar particles contained an appreciable amount of particles of submicronic colloidal sizes which did not sediment in the gravitational field.

Turbidimetric method (Specol spectrophotometer) was applied to the dilute spherical graphite suspensions. The dilution was made in aqueous solutions at constant PAM concentration (0.004%).

The average diameter of graphite particle D, was calculated from the equation:

$$\lim_{c \to 0} \mathfrak{I}/c = KD^3 \tag{1}$$

where: $\mathfrak{I}$ is turbidity, c is graphite suspension concentration (g cm^{-3}). K was calculated according to the equation:

$$K = \frac{16}{9} \frac{\pi^4 d}{\lambda^4 d_P} \left(\frac{n_P}{n} - 1\right)^2 , \tag{2}$$

where: λ is the wavelength of the incident light (450 nm), d and d_P are the densities of the suspension medium and

Fig. 1 Micrograph of graphite particles (a) spherical (SEM-Leitz REM 1600 T); (b) lamellar (microscopy under reflected light)

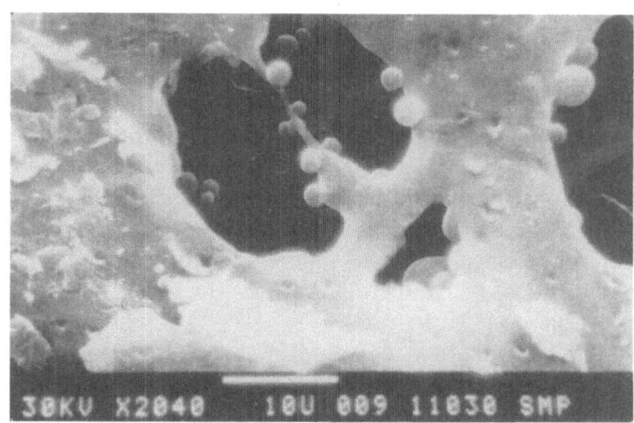

(a)

(b)

dispersed phase, respectively, n and n_P are the refractive indices of the suspension medium and dispersed phase, respectively. The values of K and D are listed in Table 1.

Conductivity dispersion of concentrated polymer additives graphite suspensions (8.7% graphite) was carried out within the 10^4–10^8 Hz frequency range at $25\,°C$. The measuring method has been treated in detail elsewhere [10]. Basically, it consists of an open-ended coaxial probe connected to an HP 4194 A Impedance Analyzer. The cell constant was determined by immersing the probe in a liquid of well-known conductivity [11]. Having determined the cell constant ($k = 0.002 \pm 2 \times 10^{-5}$ m), the conductivity of the sample σ is simply related to the measured conductance G by equation:

$$\sigma = G/k \, . \tag{3}$$

The determination errors of sample conductivity were less than 2% over the whole frequency range.

The experimental data points fit an equation introduced by Kijlstra et al. [12]:

$$\sigma(\omega) = \sigma(0) + \Delta\sigma \, \frac{\omega\tau(\omega\tau)^{1/2}}{(1 + \omega\tau)[1 + (\omega\tau)^{1/2}]} \, , \tag{4}$$

where: $\sigma(\omega)$ is the sample conductivity as a function of frequency of the applied field, ω the angular frequency, τ the relaxation time and $\Delta\sigma = \sigma(\infty) - \sigma(0)$ the conductivity increment. Equation (4) is a good approximation of Fixman's theory [13] and predicts broader conductivity dispersions than does the Debye equation.

The choice of KCl as the indifferent electrolyte has been made since the small difference between the friction coefficients of K^+ and Cl^- simplifies the theoretical interpretation.

The conductivity was determined immediately (10 s), following a slight stirring of the samples.

Results and discussion

Turbidimetric study of the dilute graphite suspensions showed that in the presence of polyacrylamides at constant concentration, high enough to ensure a maximum adsorption, the values of the graphite particles diameters (Table 1) are rather close to the initial ones. Submicronic dimensions are not modified in the KCl presence up to concentrations of 0.5 g/l when aggregation processes occur.

Since the turbidimetric method does not reveal the layer of the adsorbed polymer, the values in Table 1 correspond to the diameter of the graphite particles and are not essentially different as a function of the type of the polymers present in the solution. It might be considered

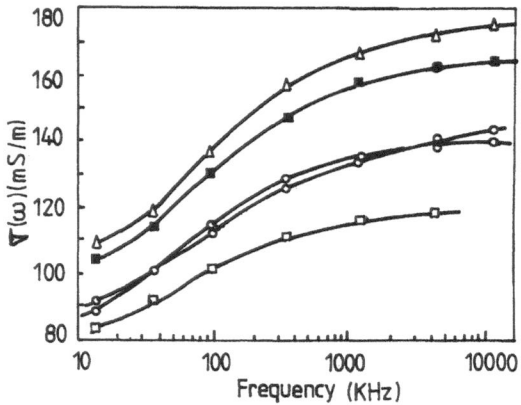

Fig. 2 Dependence of electric conductivity on the frequency of the electric field applied for the graphite suspensions with spherical particles in aqueous PAM solution (0.004%) and varied KCl concentration: 0 g/l (○); 0.07 g/l (●); 0.15 g/l (□); 0.30 g/l (■); 1.2 g/l (△); concentration of graphite 8.7%

that in dilute system, in the presence of high molecular weight polyacrylamides, irrespective of the DH, no important association process takes place among particles.

The maximum PAM adsorbed amount, expressed in g PAM/g graphite increases with the hydrolysis degree for the same M and with the molecular weight for the same DH (Table 1). The maximum value remain unchanged, within the range of the experimental errors, and in presence of KCl up to concentrations of 3 g/l. Concentration of 0.004% (in weight) for PAM and of maximum 3 g/l for KCl in the dispersion medium were used for the study of the graphite concentrated suspensions.

The dispersion of the electric conductivity with frequency might give information on the size of the graphite particles and on the adsorbed polymer layer. It might be also applied under relatively higher concentration, when other methods cannot be used. KCl concentration (0–3 g/l) was thus selected to correspond to the range within which the highest contraction of the random coil conformation occurs in the volume of the polyacrylamide solution [8, 14]. Figure 2 gives the curves of variation of the electric conductivity with frequency for PAM 1000-25. Similar curves are obtained for all polyacrylamides used in the present experiment.

The conductivity increment in Eq. (4) depends on the equilibrium double-layer structure, types of ions in the system, the particle radius (r), and the Debye length. Because of the higher complexity of such a function, we believe $\Delta\sigma$ is not a relevant parameter in our analysis. Unlike, the relaxation time, τ seems to be more useful in interpreting the results.

Progr Colloid Polym Sci (1996) 100:156–161
© Steinkopff Verlag 1996

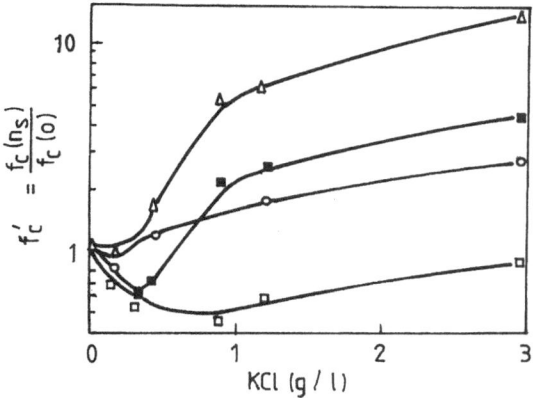

Fig. 3 Variation of the relative characteristic frequency, f_C', with KCl concentration for graphite suspensions with spherical particles in aqueous PAM solutions (0.004%): 500–20 (■); 600–20 (△); 600–0 (□); 1000–25 (○); $f_C' = f_C(n_S)/f_C(0)$

Table 2 Characteristics of graphite suspensions in KCl aqueous solutions calculated from conductivity data

No	Additive	KCl concentration g/l		z^a
		at f_{Cm}'	at f_{Ci}'	
1	PAM 500-20	0.35	1.05	22.727
2	PAM 600-0	0.90	–	–
3	PAM 600-20	0.30	1.05	27.272
4	PAM 600-40	0.16	0.40	54.545
5	PAM 1000-25	0.15	0.55	56.818

[a] Number of possible charges per polymer chain.

The parameter is given by the following equation [13]:

$$\tau = \frac{1}{2\pi f_C} = \frac{\beta r^2}{2k_B T}, \tag{5}$$

where: f_C is the characteristic frequency, k_B Boltzman's constant, T absolute temperature, and β ion friction coefficient in the background electrolyte solution. For a univalent electrolyte, β is given by:

$$\beta = \frac{e^2}{\varepsilon k_B T} n_S, \tag{6}$$

where: e is the unity charge, ε permittivity of medium and n_S the number of ions in the volume unity.

Plotting of the relative characteristic frequency, f_C', defined as the ratio between $f_C(n_S)$ in the presence, and $f_C(0)$ in the absence of electrolyte, as a function of KCl concentration, evidences specific behaviors (Fig. 3). Equations (5) and (6) give:

$$f_C' = \frac{f_C(n_S)}{f_C(0)} = \frac{r_0^2 n_{S0}}{r^2 n_S}. \tag{7}$$

The curves in Fig. 3 (based on processing of experimental data given in Fig. 2) present two distinct ranges: the former, where f_C' decreases and reaches a minimum value f_{Cm}', and the latter where f_C' increases with KCl concentration evidencing a slope change of f_{Ci}'. After that, the increase is very slow. The shape of curves shows characteristics for each type of the polyacrylamide studied (Table 2).

Figure 3 and Table 2 show that the values of KCl concentration at f_{Cm}' depend on the molecular weight, but mainly on the hydrolysis degree of PAM in the dispersion medium.

The f_C' decrease with KCl concentration may be explained, based on Eq. (7), as an expansion of the polymer adsorbed layer on the surface of graphite particles. For hydrolyzed PAM, in the presence of of small amounts of K^+ and Cl^- ions, the relaxation time increases. In the absence of KCl ($n_S = 0$ in Eqs. (5) and (6)), f_C should be infinite. In fact, the values obtained are finite and this explains the superficial electric charge of the graphite particles due to the adsorption of the hydrolyzed polymers. In the case of nonhydrolyzed PAM (600–0), f_C' variation is less obvious. KCl retention in the polymer coil might lead to the increase of solvation, to the extension of the adsorbed layer, or the polymer contains very few ionizable groups, nondetected through our methods. The literature mentions that a content less than 0.3% of acid groups might give a polyelectrolyte character to the high molecular weight polyacrylamides if the polymer is adsorbed [14]. In the adsorbed state, the average density of the adsorbed layer is much higher than the coil in solution, and interaction between ionic groups are greatly favored. KCl concentration, at which the minimum relatively critical frequency, f_{Cm}', is obtained for the polyacrylamides studied by us, decreases linearly with the increase of the number of possibly ionizable groups, z, per polymeric chain (Table 2) and with the amount of the adsorbed polymer (Table 1).

Increase of f_C' after reaching the minimum (Fig. 3) corresponds to the shielding effect produced by microelectrolyte, which, in the cases studied, acts at lower concentrations than those where the random coil contraction occurs in bulk solution. The contraction of the adsorbed PAM layer is complete around the value of the inflexion points, f_{Ci}' where the KCl concentration ranges within 0.4 and 1.05 g/l versus the 37.25–52.15 g/l concentrations required the random coil contraction in bulk solution [2].

Attempts to measure the dispersion of electric conductivity on graphite suspensions with lamellar particles (Fig. 1b) emphasized only the existence of the contraction range of the adsorbed PAM layer. The experimental data are processed by Cole–Cole equation [15]. These

160
M. Olteanu et al.
Concentrated graphite suspensions in aqueous polymer solutions

Table 3 Characterization of flocculation process of the graphite suspensions calculated from sedimentometry: PAM 0.004%; Graphite 0.173%; KCl 0.596 g/l

No	Additives	Amount of graphite deposited % at different time (s)		r^a (μm)
		300	1200[b]	
1	water	34.92	42.46	7.00
2	PAM 500-20	22.61	27.77	5.50
3	PAM 500-20-KCl	24.60	38.47	9.15
4	PAM 600-0	21.82	16.78	6.40
5	PAM 600-0-KCl	23.81	38.09	7.50
6	PAM 600-20	14.68	18.65	3.60
7	PAM 600-20-KCl	21.82	36.51	7.00
8	PAM 600-40	15.87	19.84	3.75
9	PAM 600-40-KCl	23.81	39.28	7.10
10	PAM 1000-25	12.30	15.87	9.25
11	PAM 1000-25-KCl	25.00	28.57	7.00
12	water	6.75	6.75	
13	PAM 1000-25	11.65	12.03	
14	PAM 1000-25-KCl	23.30	24.06	

[a]Particle radius from the peak of differential distribution curves.
[b]Maximum deposited amounts.
Samples 12–14 were obtained for the lamellar graphite (Fig. 1b).

preliminary studies show the influence of the geometry of the particles on the dielectric relaxation. No experimental data are given.

As f_C depends on the $r^2 n_S$ product (Eq. 7), the presence of some agglomeration of particles might lead to significant change of their sizes, a fact that would shift f_C' to much lower value than those found by us. Decrease of f_C' within the range of low KCl concentration cannot be explained by the formation of aggregates, but it can be justified by increase of n_S and/or by the expansion of the adsorbed polymer layer, whose size in the case of high molecular weight PAM may equal and even overcome the diameter of the graphite particles [6]. Layer expansion may lead to an increase of $r^2 n_S$ in the presence of KCl versus $r_0^2 n_{S0}$ in the electrolyte absence.

Sedimentation experiments in gravitational field of the graphite suspensions did not evidence changes of the hydrodynamic radius in the KCl concentration range corresponding to the f_C' decrease. Only the formation of the adsorbed PAM layer considerably decreasing the sedimentation rate in comparison with that of the graphite particles in water was recorded (Table 3). Viscosity of the 0.004% PAM solutions used ranges within 1.1 and 1.4 CP

for all the polymers studied and it does not influence significantly the increase of the Stokes friction coefficients. Decrease of aggregation, of the sedimentation rate, depends mainly on M; efficiency in the suspension stabilization decreases in the order: PAM1000-25 > PAM600-20 = PAM600 − 40 > PAM600 − 0 > PAM500-20.

The range within which f_C' increases with KCl concentration to f_{Ci}' should correspond to the condition $f_C(n_S) > f_C(0)$, i.e., $r^2 n_S < r_0^2 n_{S0}$. As in that range $n_S \gg n_{S0}$, it results that $r < r_0$, therefore the layer of the adsorbed polymer is contracted. The consequence of the decrease of the PAM layer sizes is an aggregation process of the particles produced in time and includes a small percent of the whole of the suspension particles. Gravitational sedimentation studies proved that the formation of the aggregates from primary particles is favored by the KCl concentration corresponding to the area within which f_C' increases (Table 3). Processing of the sedimentation curves and analysis of the differential distribution curves gave micrometric values for the average radius of the aggregates, higher in KCl PAM solution that in the polymer solution.

Conclusions

The present paper has shown the efficiency of the dispersion method of the electric conductivity with frequency for the study of interaction of high molecular weight polymers (of the PAM type) with solid dispersions concentrated in aqueous medium of variable salinity.

Polyacrylamides with M of the 10^6 order and various hydrolysis degrees (Table 1) are adsorbed onto the surface of the spherical graphite particles [1]. Within the range of small KCl concentrations, the adsorbed layer expands, and it then contracts, with the increase of ionic strength (Fig. 3, Table 2). The expansion of the layer increases with the number of hydrolyzable groups on the chain and with the amount of adsorbed PAM.

PAM stabilizes the spheric graphite suspensions and destabilizes the suspensions of lamellar graphite, the higher the molecular weight is. The amount of aggregate particles practically remains unchanged in time and increases with KCl concentration of medium (Table 3).

The study will be extended by interpretation of the data of electric conductivity dispersion in concentrated graphite suspensions with lamellar particles.

References

1. Olteanu M, Peretz S, Popescu G (1993) Colloids Surfaces A 80:127–130
2. DeLacey EHB, White LR (1981) J Chem Soc Faraday Trans 2, 77:2007–2039
3. Carrique F, Delgato AV (1995) Progr Colloid Polym Sci 98:140–144

3. Carrique F, Delgato AV (1995) Progr Colloid Polym Sci 98:140–144
4. Lecourtien J, Lee LT, Chauveteau G (1990) Colloids Surfaces 47:219–231
5. Meadows J, Williams PA, Garvey MJ, Harrop R (1990) J Colloid Int Sci 139:260–267
6. Spalla O, Cabane B (1993) Colloid Polym Sci 271:357–371
7. Lee LT, Somasundaran P (1989) Langmuir 5:854–860
8. Vilcu R, Leca M, Olteanu M (1978) Rev Roumaine Chim 23:1171–1178
9. Popescu Pietris V (1988) Utilaje din Materiale Carbografitice, Editura Tehnica, Bucuresti, pp 103
10. Raicu V (1995) Meas Sci Technol 6:410–414
11. Stogryn A (1971) IEEE Trans Microwave Theory Tech 19:733–736
12. Kijlstra J, van Leeuwen HP, Lyklema J (1993) Langmuir 9:1625–1633
13. Fixman M (1980) J Chem Phys 72:5177–5186
14. Gramain PH, Myard PH (1981) J Colloid Interface Sci 84:114–126
15. Cole KS, Cole RH (1941) J Chem Phys 9:341

Progr Colloid Polym Sci (1996) 100:162–169
© Steinkopff Verlag 1996

Experimental investigation of the structure of nonionic microemulsions and their relation to the bending elasticity of the amphiphilic film

M. Gradzielski
D. Langevin
B. Farago

M. Gradzielski (✉)
Lehrstuhl für Physikalische Chemie 1
Universität Bayreuth
95440 Bayreuth, FRG

D. Langevin
Centre de Recherche Paul Pascal
Avenue Schweitzer
33600 Pessac, France

B. Farago
Institute Laue-Langevin
B.P. 156X
38042 Grenoble Cedex 09, France

Abstract In this investigation we studied ternary microemulsions of the system nonionic surfactant (C_iE_j)/hydrocarbon/water. In this system droplet microemulsions were investigated by means of SANS experiments and measurements of the interfacial tension. The obtained structural quantities of the droplet microemulsion such as radius, polydispersity index, and macroscopic interfacial tension can be directly related to the bending constants (bending modulus κ and Gaussian modulus $\bar{\kappa}$) of the system.

The sum of the elastic constants $2\kappa + \bar{\kappa}$ can be deduced independently from interfacial tension and polydispersity index (from SANS data in the shell contrast) and both are in good agreement for all systems. This sum increases with increasing thickness of the amphiphilic film, i.e. with increasing length of the surfactant chain. In contrast, variation of the chain length of the hydrocarbon does not influence the elastic properties of the surfactant film. A dilution series also allows for a determination of $2\kappa + \bar{\kappa}$ from the concentration dependence of the deduced radii and polydispersity indices. Again, good agreement is observed with the values obtained before. Experiments both on oil-in-water (O/W) and water-in-oil (W/O) droplets indicate that $2\kappa + \bar{\kappa}$ decreases with increasing temperature (W/O at higher temperature).

In general, one can state that the properties of these microemulsions are well described by the elastic theory for the amphiphilic film. Structural parameters like radius and polydispersity are interrelated with the interfacial tension and if two of these quantities are known the third can be predicted reliably.

Key words Microemulsions – bending energy – SANS – interfacial tension – polydispersity

Introduction

Microemulsion are systems in which a mixture of oil water is dispersed in a thermodynamically stable way by the presence of an amphiphile [1, 2]. The size of their structural units is typically in the range of 30–300 Å which explains why they are transparent. In principle, three structural types have been shown to exist in such systems [2–4]:

- oil-in-water (O/W) droplets
- water-in-oil (W/O) droplets
- bicontinuous (sponge-like) structure.

Although such systems have been known now for more than 50 years [5], they are still an active field of research since a detailed understanding of their properties is still not fully achieved. Most theories that try to explain them are based on the bending energy of the amphiphilic film that separates the hydrophilic and hydrophobic regions of the micro-emulsion [6–10]. Each surfactant film has a tendency to realize a preferred spontaneous curvature c_0 in its structures. The elastic properties of this film are described by two bending moduli: the bending modulus κ and saddle-splay modulus $\bar{\kappa}$. Unfortunately, the experimental determination of these elastic moduli is not simple and a variety of different experimental methods has been applied in order to measure them [11–14]. However, these different methods yield values that often show large discrepancies even for identical systems. Therefore a reliable determination of the elastic moduli still remains an important task in order to reach a more profound understanding of the properties of microemulsions.

The simplest microemulsions are those that contain droplet structures. Here the elastic theory gives clear predictions regarding size and polydispersity, as well as for the macroscopic interfacial systems of the corresponding systems. Hence, we decided to investigate these structural properties in connection with the interfacial tension in the simplest system one can imagine for a microemulsion, i.e. one that is made up from water, oil, and a surfactant. Such systems can be made with nonionic surfactants of the C_iE_j type (alkyloligoglycolethers) and their phase behavior has been studied in considerable detail [15–17].

All our investigations were carried out on droplet systems with typical sizes of 40–100 Å which are very well suited for small-angle neutron scattering (SANS) experiments where structural features such as size and polydispersity can be well resolved. In the following we will analyze scattering curves for various O/W microemulsions as well as interfacial tension data and interpret the results in terms of the elastic theory of microemulsions. In our investigations we also changed the molecular composition of the systems in a systematic way in order to see how the choice of surfactant and/or oil influences the properties of the corresponding amphiphilic film.

Experimental part

The SANS experiments were performed on the PAXE instrument at the Laboratoire Léon Brillouin, Saclay (Laboratoire Commune CEA-CRNS), France and covered a q-range of 0.008–0.023 1/Å. All samples were prepared in such a way as to contain a small amount of excess phase in order to make sure that all samples were in two-phase equilibrium, i.e., the microemulsion was saturated with the solubilizate.

The interfacial tensions were measured by means of the surface laser light-scattering technique as has been described in detail before [18].

Theoretical background

Elastic theory of microemulsions

Most current theories of microemulsions concentrate on the bending elasticity of the amphiphilic film as the key factor to explain their properties [6–10]. The bending energy is in general determined by the shape of the amphiphilic film and two elastic moduli, κ and $\bar{\kappa}$ [6]. Droplet type structures are the simplest case possible and have been treated in much detail [19–22]. If one just takes into account the bending energy and the entropic contribution that arises from the entropy of mixing, one can write the free energy F of the system as:

$$F = \int dA \cdot ((\kappa/2) \cdot (c_1 + c_2 - 2 \cdot c_0)^2 + \bar{\kappa} \cdot c_1 \cdot c_2)$$
$$+ N \cdot k \cdot T \cdot f(\Phi), \tag{1}$$

if one only takes into account the bending energy and the entropic contribution that arises from the entropy of mixing, and where c_1 and c_2 are the two principal curvatures and c_0 the spontaneous curvature of the surfactant film. It should be noted here that in the following we will always employ for the entropy of mixing term $f(\Phi)$ a random mixing approximation that yields [20]:

$$f(\Phi) = (1/\Phi) \cdot \{\Phi \cdot \ln \Phi + (1 - \Phi) \cdot \ln (1 - \Phi)\}. \tag{2}$$

Other expressions for the entropy have also been tested by us, but we found that this approximation yielded the most self-consistent results [23].

For the case of spherical droplets of radius R Eq. (1) reduces straightforwardly to:

$$F/A = 2 \cdot \kappa \cdot (1/R - 1/R_0)^2 + \bar{\kappa}/R^2$$
$$+ (k \cdot T/(4\pi \cdot R^2)) \cdot f(\Phi), \tag{3}$$

where A is the total surface of the droplets and R_0 the spontaneous radius of curvature.

The situation becomes particularly simple if the microemulsion is in equilibrium with the excess phase of the solubilizate. Here the droplets have attained their maximum radius R_m and this situation where the surfactant system has reached its solubilization capacity has also been named the emulsification failure [8]. For that case the minimization of the F directly relates R_m to the spontaneous radius R_0, and the elastic moduli κ and $\bar{\kappa}$:

$$\frac{R_m}{R_0} = \frac{2 \cdot \kappa + \bar{\kappa}}{2 \cdot \kappa} + \frac{k \cdot T}{8\pi \cdot \kappa} f(\Phi). \tag{4}$$

Analogously, one may deduce an expression for the macroscopically observed interfacial tension γ at the planar interface between microemulsion and excess phase that depends only on the maximum droplet size R_m and the elastic moduli [24]:

$$\gamma = \frac{2 \cdot \kappa + \bar{\kappa}}{R_m^2} + \frac{k \cdot T}{4\pi \cdot R_m^2} f(\Phi) . \tag{5}$$

Finally, one can also relate the polydispersity of the droplets to the bending constants of the systems. An analysis of the thermal fluctuations (i.e., deformations) of the microemulsion droplets in terms of spherical harmonics leads to an expression where the polydispersity index p can be expressed as a function of the elastic moduli (Eq. (6)) [20, 23]. In addition, we have shown recently that in a scattering experiment the scattering curve should mainly be determined by this $l = 0$ term [23].

$$p^2 = \frac{k \cdot T}{8\pi \cdot (2 \cdot \kappa + \bar{\kappa}) + 2 \cdot k \cdot T \cdot f(\Phi)} . \tag{6}$$

In addition it is straightforward to see that one can eliminate the volume fraction dependent term that is due to the entropic contribution by combining Eqs. (5) and (6). This then yields a relation between all experimentally observable quantities that should give a constant value W:

$$W = 8\pi \cdot p^2 \cdot \gamma \cdot R_m^2 / k \cdot T = 1 . \tag{7}$$

In summary, this means that the elastic model of the amphiphilic film yields precise predictions for structural properties of microemulsion droplets such as the maximum particle radius R_m, the polydispersity index p, and also for the macroscopic interfacial tension γ. Moreover, it also predicts how these properties should change as a function of concentration. However, all these quantities are experimentally accessible and therefore these relations (Eqs. (4–6)) can be used to obtain quantitative information of the elastic moduli of the respective systems.

Scattering theory for shell structured particles

In order to determine size and polydispersity from the SANS experiments the shell contrast for the microemulsion droplets is especially suited since in particular the polydispersity can be obtained much more accurately from the shell spectra than in comparison with samples in bulk contrast [25]. In our system shell contrast means that both hydrocarbon and water were used in deuterated form and the surfactant was the only hydrogenated species.

For that case the microemulsion droplets may be described by a model that contains three distinct regions for the scattering length density (Fig. 1): the core that contains the solubilizate, the shell that contains the surfactant, and

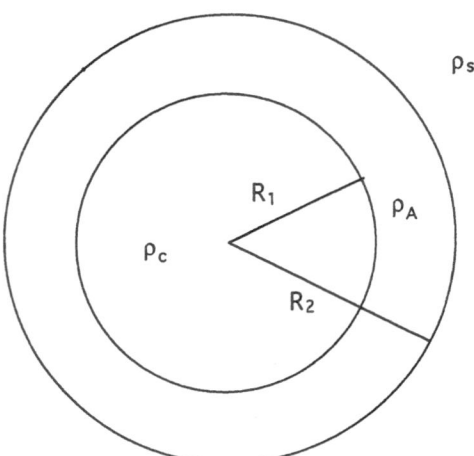

Fig. 1 Schematic drawing of a shell structured microemulsion droplet (symbols described in the text)

the solvent. The particle form factor $P(q)$ for such a shell structured aggregate can be written as:

$$P(q) = 16 \cdot \pi^2 \cdot (\rho_A - \rho_S)^2 \cdot \{R_2^3 \cdot f_0(q \cdot R_2) - A \cdot R_1^3 \cdot f_0(q \cdot R_1)\}^2 \tag{8}$$

with: $f_0(x) = (\sin x - x \cdot \cos x)/x^3$; $q = (4\pi/\lambda) \cdot \sin(\Theta/2)$; $A = (\rho_A - \rho_C)/(\rho_A - \rho_S)$; R_1, R_2: inner and outer radius of the droplet shell, respectively; scattering length densities: ρ_C: core; ρ_A: shell; ρ_S: solvent.

For all fits described later in the text we always identified ρ_C, ρ_A and ρ_S with the scattering length densities of hydrocarbon, surfactant, and D_2O, respectively (for the case of O/W microemulsions).

Equation (8) is valid for monodisperse particles but can easily be generalized for the case of a polydisperse system (since any real microemulsion will possess some degree of polydispersity). A simple way to account for this is to employ a Schulz distribution $f(r)$ (Eq. (9)) for the radii (with $t + 1 = 1/p^2$) which is centered around the mean radius R_m [26].

$$f(r) = \left(\frac{t+1}{R}\right)^{t+1} \frac{r^t}{\Gamma(t+1)} \exp\left(-\frac{t+1}{R_m} r\right) . \tag{9}$$

Now, the scattering intensity of the polydisperse system can be evaluated by:

$$I(q) = {}^1N \cdot \int_0^\infty dr \cdot f(r) \cdot P(q, r) \cdot S(q) \tag{10}$$

with: $r = (R_1 + R_2)/2$; ${}^1N$: number density of the aggregates.

In this model the scattering intensity is described by three parameters that can be obtained from fitting the theoretical expression to the experimental data:

- mean shell radius R_m ($= (R_1 + R_2)/2$)
- shell thickness d ($= R_2 - R_1$, which was kept constant since it should only depend on the chain length of the surfactant and vary much less than the particle size itself)
- polydispersity index p ($p^2 = \langle R^2 \rangle / \langle R \rangle^2 - 1$).

In Eq. (10), we also included the interparticle structure factor $S(q)$ that accounts for changes in the scattering intensity that are due to correlations between the different aggregates. However, our droplets are uncharged and therefore their interaction potential and hence also their $S(q)$ should be well described by a hard sphere model for which an analytical expression for $S(q)$ exists in the Percus–Yevick approximation which has been shown to be reliable [27]. This $S(q)$ depends on two additional parameters, the effective hard sphere diameter σ and the volume fraction Φ (or correspondingly, the number density of the system, where both are fixed by the sample composition). For less concentrated systems, as in our investigations, $S(q)$ will mainly influence the scattering intensity in the low q-range. Therefore we did not use the hard-sphere diameter σ as an individual fit parameter but fixed it always to the particle radius as used for the form factor $P(q)$ and we also checked that the choice of σ did not significantly influence the outcome of our data analysis.

It should be mentioned here that the instrumental resolution function (that is mainly given by the wavelength spread of the incident neutrons) has a similar effect on the scattering curves as does the polydispersity. In all our analyses, we approximated this resolution function by a Gaussian function of 10% standard deviation.

Results and discussion

Influence of the surfactant chain length

Of course the most important component in a microemulsion is the surfactant itself and it appears to be interesting how its choice influences the elastic properties of the amphiphilic film. Therefore we studied O/W microemulsions in the system C_iE_j/D_{22}–decane/D_2O where the surfactant chain length was varied. Here C_8E_3, $C_{10}E_4$, and $C_{12}E_5$ were employed in order to keep the balance between hydrophilic EO head group and hydrophobic alkyl chain approximately constant. Correspondingly, mainly the total thickness of the amphiphilic film will change in this experiment. All samples had a volume fraction of ~ 0.08 and the SANS spectra were recorded at $10\,°C$ and are given in Fig. 2 together with the fitted curves obtained with the shell model.

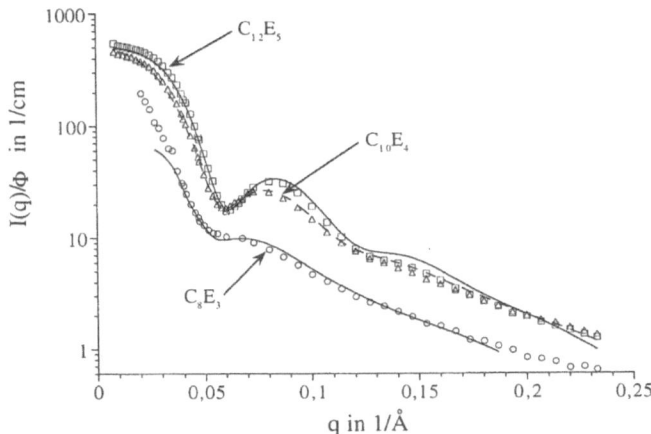

Fig. 2 SANS intensity curves for O/W microemulsions of C_8E_3 (○), $C_{10}E_4$ (△), and $C_{12}E_5$ (□) with D_{22}-decane as oil in D_2O (taken at $T = 10\,°C$). The solid lines are fits according to Eqs. (8–10)

Since the minimum in the scattering curves is located at about the same q-value, one can already conclude that the particles should possess similar radii, as for shell structured particles, this minimum is related to their size ($R \approx \pi/q$). Furthermore, one finds that this minimum that is related to the polydispersity becomes more pronounced with increasing chain length, i.e., the polydispersity decreases. Finally, it should be mentioned that the shell thickness d is mainly determined by the absolute intensity since d does not significantly change the shape of the scattering curve but, of course, depends very strongly on the total scattering intensity.

The experimental data were fitted with the model described earlier, and the obtained fitted parameters are summarized in Table 1 together with the macroscopic interfacial tension γ as determined for the identical systems (where only hydrogenated oil instead of the deuterated has been used). It should be mentioned here that for the $C_{12}E_5$

Table 1 Mean shell radius R_m, shell thickness d, polydispersity index p for the oil saturated microemulsions in the system C_iE_j/D_{22}-decane/D_2O. In addition the value for the interfacial tension γ at $10\,°C$ is given (for these measurements D_2O was used, i.e., normal decane instead of the deuterated decane) together with the values for $2\kappa + \bar{\kappa}$ derived via p and γ, and the factor W (Eq. (7))

Surfactant	C_8E_3	$C_{10}E_4$	$C_{12}E_5$
Φ	0.0841	0.0784	0.0765
R_m in Å	55.6	50.4	49.1
d in Å	11.0	13.0	15.2
p	0.207	0.141	0.113
γ in mN/m	0.0904	0.317	0.520
$(2\kappa + \bar{\kappa})(p)/k \cdot T$	1.20	2.28	3.42
$(2\kappa + \bar{\kappa})(\gamma)/k \cdot T$	0.90	2.34	3.50
W	0.772	1.028	1.022

system we also determined the polydispersity index p by means of a contrast variation method that is independent of the instrumental resolution function, and that yielded a very similar value [23]. This confirms that in our analysis the experimental instrumental resolution function has been properly accounted for.

These parameter allow us now to compute the sum of the elastic moduli $2\kappa + \bar{\kappa}$. There are two independent ways of accessing this sum, first directly from the polydispersity index p (Eq. (6)), and secondly from the interfacial tension γ together with the particle radius R_m (Eq. (5)). The calculated values are also given in Table 1 and although here one starts out from completely independent experimental quantities, one finds values for $2\kappa + \bar{\kappa}$ that are in very good agreement.

We observe that the value for $(2\kappa + \bar{\kappa})/k \cdot T$ increases with increasing surfactant chain length from 1.0 for C_8E_3, to 2.3 for $C_{10}E_4$, to 3.5 for $C_{12}E_5$. However, theory just predicts such an increase of the elastic constants with increasing chain length of the surfactant [28, 29]. Accordingly, it should scale with the surfactant chain length to the power 2.5–3, a scaling law that is well obeyed by our experimentally deduced values, i.e., the thicker the amphiphilic film the more rigid it becomes.

Finally, we also used Eq. (7) which interrelates γ, R_m, and p in order to calculate the dimensionless factor W that should be unity. For all samples W is in the range of 0.78–1.03, a very good agreement with theory, in particular if one considers that it is calculated from three independent experimental quantities. This means that by knowing two of these quantities one is able to predict the third one with good precision. Judging from this, we may state that our results already prove that the elastic theory of microemul-

sions is a powerful tool to explain the experimentally observed properties and can also be used to deduce the corresponding elastic moduli.

Variation of the hydrocarbon chain length

Of course, not only the surfactant may influence the microemulsion properties, but it also appears to be interesting to see how the employed hydrocarbon influences the bending properties of the amphiphilic film. In order to study this effect we always used $C_{10}E_4$ as surfactant together with hexane, heptane, octane, and decane as hydrocarbons, where the concentration of $C_{10}E_4$ was kept constant around 3.3 wt%. The SANS spectra are given in Fig. 3 together with the fitted curves as solid lines. As before, all experimental data are well described by our model of polydisperse shell-structured aggregates.

Already from the scattering curves it is evident that the minimum in the curves is similarly pronounced for all samples and therefore they should have similar polydispersities. The obtained fit parameters are given in Table 2, and one finds that with decreasing chain length of the hydrocarbon increasing amounts become solubilized and correspondingly larger droplets are formed. The mean shell radius R_m increases from 50 Å for decane to 90 Å for hexane which means that for hexane about twice the amount (by volume) can become solubilized in comparison to decane. As already expected the polydispersity index p remains unchanged at 0.14 independent of the oil employed. This implies that $2\kappa + \bar{\kappa}$ as determined from p will have identical values for all samples. However, this also applies to the $2\kappa + \bar{\kappa}$ value determined from γ. This is

Fig. 3 SANS intensity curves for O/W microemulsions composed from $C_{10}E_4$ in D_2O and various hydrocarbons: D_{14}-hexane, D_{16}-heptane, D_{18}-octane, and D_{22}-decane (taken at $T = 10\,°C$). The solid lines are fits according to Eqs. (8–10)

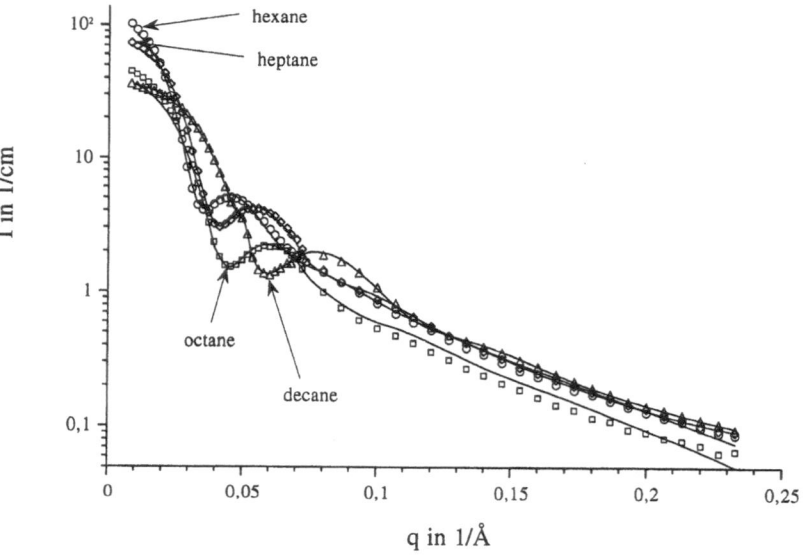

Table 2 Mean shell radius R_m, shell thickness d, polydispersity index p for the oil saturated microemulsions in the system $C_{10}E_4$/D-hydrocarbon/D_2O for various deuterated hydrocarbons. In addition, the value for the interfacial tension γ at $10\,°C$ is given (for these measurements D_2O was used i.e., normal decane instead of deuterated decane) together with the values for $2\kappa + \bar{\kappa}$ derived via p and γ, and the factor W (Eq. (7))

Oil	Decane	Octane	Heptane	Hexane
Φ	0.0784	0.0802	0.0948	0.1045
R in Å	50.4	67.2	74.8	88.7
d in Å	13.0	11.2	13.1	12.6
p	0.141	0.140	0.140	0.140
γ in mN/m	0.317	0.214	0.139	0.0852
$(2\kappa + \bar{\kappa})(p)/k \cdot T$	2.28	2.32	2.30	2.27
$(2\kappa + \bar{\kappa})(\gamma)/k \cdot T$	2.34	2.75	2.26	1.97
W	1.028	1.211	0.980	0.850

somewhat less obvious since γ varies by about a factor 4 and still in combination with the R_m values leads to values that are roughly constant, i.e., in the range of 2–$2.75\,k \cdot T$.

Therefore, we find that the sum $2\kappa + \bar{\kappa}$ of the bending moduli is not affected to any larger degree by the variation of the chain length of the hydrocarbon. This appears to be reasonable since the bending elasticity of the interface should mainly be determined by the surfactant film that forms this interface and only depends to a smaller degree on its solvent environment. It might be added here that fluorescence experiments in the system $C_{12}E_5$/decane or heptane also showed very similar polydispersities for both oils [30]. However, it should be mentioned that there exist situations in which the oil chain length may significantly change the elastic constants of the film, as for instance has been found by ellipsometric measurements in the AOT/brine system where large changes of κ have been observed with increasing chain length of the hydrocarbon [31].

The concentration dependence for O/W droplets in the system $C_{10}E_4$/D_{22}-decane

The entropic term in the expression for the free energy (Eq. (1)) introduces a concentration dependence. Therefore, theory will also predict how the parameters R_m, γ, and p should vary as a function of the volume fraction Φ of the dispersed phase of the microemulsion. This means that the measurement of these properties in a dilution series is still another independent method to determine the elastic moduli of the corresponding system. To study this effect, we chose the O/W system of $C_{10}E_4$/D_{22}–decane and the samples were simply prepared by diluting the most concentrated sample with D_2O. Again, all samples were

saturated with oil, i.e., in equilibrium with a small amount of excess phase, and the measurements were done at $10\,°C$.

As before, the scattering curves were analyzed with the shell model and the obtained values for R_m and p are plotted in Figs. 4a and 4b as a function of the volume dependent function $f(\Phi)$, where Φ is the volume fraction of surfactant plus oil. One finds that the radii decrease with decreasing concentration and at the same time the polydispersity index p increases from 0.133 to 0.209. Both effects are in good agreement with the theoretical predictions and in Figs. 4a and b the corresponding expressions (Eqs. (4) and (6)) have been fitted to the experimental data. From these fits we could deduce values of $2.02\,k \cdot T$ (from R_m) and $2.75\,k \cdot T$ (from p) for $2\kappa + \bar{\kappa}$, i.e., this completely independent method yields values for the sum of the elastic moduli that agree well with the values obtained before via different routes. All this shows that the interpretation of

Fig. 4A The droplet radii R_m as a function of $f(\Phi)$ for the system $C_{10}E_4$/D_{22}-decane/D_2O at $10\,°C$ (solid line: fit according to Eq. (4)); **B** The polydispersity index p as a function of $f(\Phi)$ for the system $C_{10}E_4$/D_{22}-decane/D_2O at $10\,°C$ (solid line: fit according to Eq. (6))

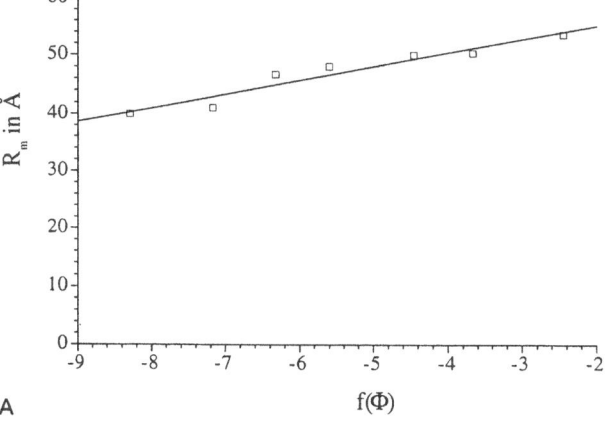

A

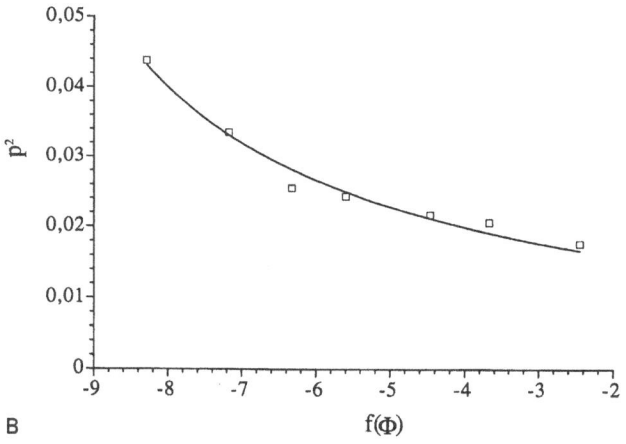

B

168

M. Gradzielski et al.
Bending energy and structure of nonionic microemulsions

the microemulsion properties by means of the elastic theory leads to a self-consistent picture that explains well the experimentally observed quantities.

Temperature dependence-W/O droplets

At higher temperatures the same system forms W/O microemulsions and we did an analogous experiment as described before in order to see how temperature influences the bending properties. For that reason a dilution series in the system $C_{10}E_4/D_{18}$-octane/D_2O was studied. However, with the same analysis as before, we obtained values of 0.9–1.2 k·T for $2\kappa + \bar{\kappa}$, i.e., they are lower than in the corresponding O/W system. This means that evidently the elastic constants appear to become smaller with rising temperature, an effect that will be looked into in more detail in future investigations.

Conclusions

We have investigated droplet-type microemulsions composed of nonionic surfactants of type C_iE_j/hydrocarbon/water. It has been shown that structural properties such as mean radius R_m and polydispersity index p, and the macroscopic interfacial tension γ can very well be described in terms of the bending properties of the amphiphilic film. SANS experiments in the shell contrast could be fitted accurately with a model of polydisperse spherical shells, and this method allows for precise determination of R_m and p.

These quantities as well as the interfacial tension γ at the planar interface can be related to the bending modulus κ and the saddle-splay modulus $\bar{\kappa}$ of the amphiphilic film, or to be more precise, to their sum $2\kappa + \bar{\kappa}$. In principle, this sum can be determined in completely independent ways, first directly from the polydispersity index p, secondly from the interfacial tension γ together with the radius R_m. Finally, a third independent determination can be done by applying the predicted concentration dependence for R_m and p to the experimentally determined quantities. All three different methods for the determination of $2\kappa + \bar{\kappa}$

yield values that are in good agreement, which shows that this approach is a reliable and self-consistent way to evaluate the elastic properties of the amphiphilic film.

In order to investigate the influence of the molecular structure of the constituent molecules on the bending properties, studies were done on systems where both the chain length of surfactant and hydrocarbon were varied in a systematic way. These experiments showed that for a variation of the chain length of the surfactant $2\kappa + \bar{\kappa}$ increases with increasing surfactant chain length and scales approximately with the chain length to the power of 3, a result that has been predicted from theory for increasing thickness of the amphiphilic film. Contrary to the experiments that employed various hydrocarbons for a given surfactant system showed increasing radii with decreasing chain length, whereas the polydispersity remained constant. The evaluation of $2\kappa + \bar{\kappa}$ yielded constant values for all cases, which means that the choice of the hydrocarbon has no influence on the elastic constants in our systems.

In addition, one can show R_m, p, and γ to be interrelated via the elastic theory and this relation has also been shown to be well fulfilled for our systems studied. This means that for these microemulsions the elastic theory allows for the calculation of one of these three quantities from the knowledge of the other two quantities.

Taking all this together, it is evident that the elastic theory of the amphiphilic film is a powerful tool for describing and understanding microemulsions and their properties. In particular, small-angle scattering is a very useful method for the precise determination of the structural parameters which in turn allow for a determination of the elastic properties of the amphiphilic film. These experiments, in conjunction with measurements of the interfacial tension, lead to a comprehensive understanding of the properties of droplet-type microemulsions.

Acknowledgments This project has been supported by a grant of the CEC (ERB4050PL920671). In addition, M.G. is grateful for a fellowship in the program "Human Capital and Mobility" (ERB4001 GT931413), also granted by the European Community. We are grateful to J. Texeira, A. Brulet, and L. Noirez for help with the SANS experiments. In addition, we would also like to thank R. Strey, S. Safran, and P.D. Fletcher for valuable discussions.

References

1. Prince LM (1977) Microemulsions: Theory and Practice. Academic Press, New York
2. Langevin D (1992) Ann Rev Phys Chem 43:341
3 Chevalier Y, Zemb T (1990) Rep Progr Phys 53:279
4. Langevin D (1992) Ann Rev Phys Chem 43:341
5. Hoar TP, Schulman JH (1943) Nature 152:102
6. Helfrich W, Naturforsch Z (1973) 28, 693
7. de Gennes PG, Taupin C (1982) J Phys Chem 86:2294
8. Safran SA, Turkevich LE (1983) Phys Rev Lett 50:1930
9. Andelman D, Cates ME, Roux D, Safran SA (1987) J Chem Phys 8:7229
10. Langevin D (1991) Adv Colloid Interface Sci 34:583

169

11. van der Linden E, Geiger S, Bedeaux D (1989) Physica A 156:130
12. Binks BP, Meunier J, Abillon O, Langevin D (1989) Langmuir 5:415
13. Borkovec M, Eicke HF (1989) Chem Phys Lett 157:457
14. Kummrow M, Helfrich W (1991) Phys Rev A 44:8356
15. Shinoda K, Friberg S (1975) Adv Colloid Interface Sci 4:281
16. Kahlweit M, Strey R (1985) Angew Chem Int Ed 24:654
17. Kahlweit M, Strey R, Firman P (1986) J Phys Chem 90:671
18. 'Laser Scattering by Liquid Surfaces and Complimentary Techniques', ed. Langevin D. Surfactant Science Series Vol. 41, Marcel Dekker, Inc., New York, 1992
19. Safran SA (1983) J Chem Phys 78:2073
20. Milner ST, Safran SA (1987) Phys Rev A 36:4371
21. Borkovec M (1989) J Chem Phys 91:6268
22. Safran SA (1991) Phys Rev A 43:2903
23. Gradzielski M, Langevin D, Farago B, Phys Rev E to be published
24. Meunier J, Lee LT (1991) Langmuir 7:1855
25. Sicoli F, Langevin D, Lee LT (1993) J Chem Phys 99:4759
26. Schulz GV (1939) Z Phys Chem B 43:25
27. Ashcroft NW, Lekner J (1966) Phys Rev 145:83
28. Szleifer I, Kramer D, Ben-Shaul A, Roux D, Gelbart W (1988) Phys Rev Lett 60:1966
29. Safinya CR, Sirota EB, Roux D, Smith GS (1989) Phys Rev Lett 62:1134
30. Fletcher PD, Johannsson R (1994) J Chem Soc Faraday Trans 90:3567
31. Binks BP, Kellay H (1991) J Meunier Europhys Lett 16:53

Progr Colloid Polym Sci (1996) 100:170–176
© Steinkopff Verlag 1996

F. Bordi
C. Cametti
P. Codastefano
F. Sciortino
P. Tartaglia
J. Rouch

Effect of salinity on the electrical conductivity of a water-in-oil microemulsion

F. Bordi · Dr. C. Cametti (✉)
P. Codastefano · F. Sciortino · P. Tartaglia
Dipartimento di Fisica
Università di Roma "La Sapienza"
Rome, Italy

J. Rouch
Centre de Physique Moleculaire
et Hertzienne
University Bordeaux I
Talence, France

Abstract The low-frequency electrical conductivity of water-in-oil microemulsions in presence of NaCl electrolyte solution has been measured in the temperature range from -15 to $60\,°C$, for different volume fractions of the surfactant + water phase, in the interval from $\Phi = 0.1$ to $\Phi = 0.8$. The effects induced by the presence of the added salt, concerning the phase diagram in the $(\Phi - T)$ plane and the conductivity regimes far and close to the percolation, are discussed on the basis of the Eicke–Hall theory and a new approach to the conductometric behavior of microemulsions recently developed, which gives a more detailed and accurate description of the percolation characteristics.

Key words Microemulsion – conductivity – percolation threshold – Eicke–Hall Theory

Introduction

Microemulsions are homogeneous, transparent, isotropic and thermodynamically stable solutions composed of large volume fractions of water and oil separated by a monomolecular layer of amphiphilic molecules. This definition encompasses a large variety of different structures depending on the composition of the system which can be built up of an inverted micellar phase consisting of water-in-oil droplets, a bicontinuous phase due to water channels in an oil phase, a lamellar structure, up to an inverted bicontinuous phase due to oil channels in an aqueous phase. Microemulsions have attracted attention, not only because of their practical interest in application, but also because of their aggregation patterns which span a wide variety of morphologies and thus make them a very interesting model system to study complex fluids.

Water-in-oil microemulsions have been the subject of numerous studies concerning their unusual physico–chemical, rheological and structural properties. Over the past few years, these systems have been widely investigated by means of different experimental techniques such as static [1] and dynamic [2] light scattering, small-angle neutron scattering [3], x-ray scattering [4] and time-resolved luminescence methods [5], besides frequency-domain or time-domain dielectric spectroscopic [6] and viscosimetric measurements [7] and a great deal of theoretical work [8] has been devoted to understanding their very complex behavior.

Recently [9], the phase behavior of water–decane, ionic amphiphile (sodium bis(2-ethylhexyl) sulfosuccinate [AOT]) ternary system has been extensively studied by means of low-frequency electrical conductivity measurements, since we have shown that a strong correlation exists between a transport property such as electrical conductivity and the microemulsion structure. That this type of correlation indeed exists is convincingly exemplified by the observed behavior illustrated in Fig. 1, where the low-frequency electrical conductivity is shown as a function of temperature for the particular system H_2O–decane–AOT at the fractional volume $\Phi = 0.400$ and water to AOT molar ratio of $X = 40.8$. As can be seen, for each structure

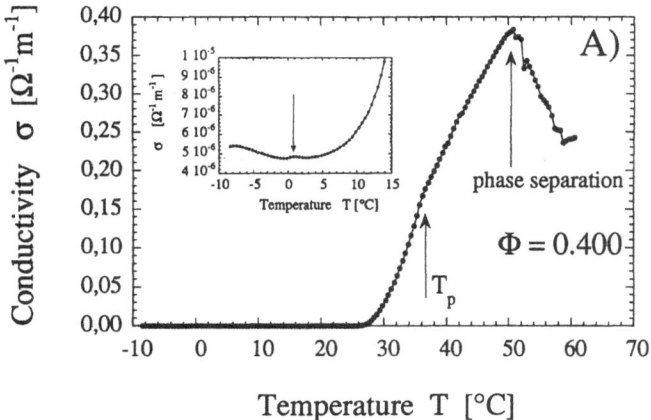

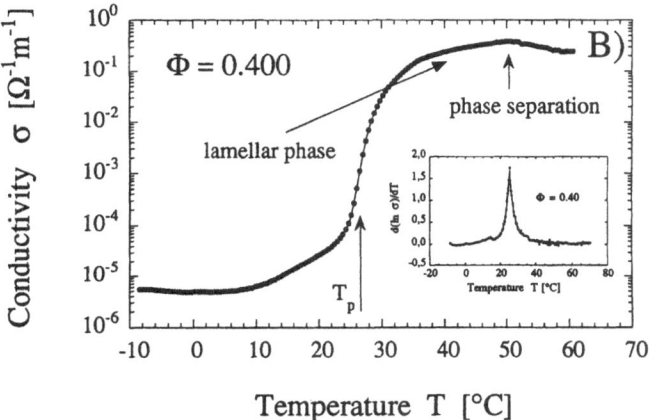

Fig. 1 Typical electrical conductivity spectra of a H₂O–AOT–decane microemulsion at a volume fraction $\Phi = 0.400$, as a function of temperature. A): the conductivity in a linear plot; the arrows mark the percolation temperature T_p and the temperature at which the phase separation occurs. The inset shows the behavior in the low temperature range. B): the conductivity in a logarithmic plot; the arrows mark the percolation temperature T_p, temperature interval where the lamellar phase occurs and the temperature of the phase separation. The inset shows the maximum in the $d(\log \sigma)/dT$ vs. T, indicating the percolation temperature

percolation threshold [10], considering in each region an appropriate theoretical model, which separately describes the transport process occurring in the system.

When the composition of the microemulsion in further varied by adding to the water phase a simple electrolyte solution, this complex phenomenology becomes more complicated, the overall system behaves differently and many approaches have been used in estimating the main parameters which influence the nature and the extension of the resulting structures. Among these, the approaches based on electrical conductivity measurements are extremely sensitive to evaluate a number of yet incompletely understood behaviors related to the different structures occurring in the different regions of the phase diagram.

In this note, we report on accurate electrical conductivity measurements of a water–decane–AOT microemulsion system at water-to-surfactant molar ratio of $X = 40.8$ over a wide interval of fractional volume, from $\Phi = 0.1$ to $\Phi = 0.8$ and in the temperature range from -15 to $70\,°C$. The aqueous phase of the system has been modified by adding a small amount of a uni-univalent salt, NaCl, at a concentration $C_s = 10^{-2}$ mol/l, so that each water droplet contains, beside the Na^+ ions produced by the ionization of the surfactant molecules, a further amount of Na^+ and Cl^- ions produced by the salt ionization, without any alteration of the whole electroneutrality of the system.

change of the system, there is a corresponding change in the electrical conductivity. In the present case, starting from low- to high-temperature, the system, initially composed of water-in-oil micelles, undergoes a percolation transition, followed by the formation of a bicontinuous structure which evolves, passing through a lamellar phase, from a water-in-oil to an oil-in-water aggregate, up to the phase separation, at higher temperatures.

As far as the low-frequency electrical conductivity is concerned, it is possible to explain this behavior quantitatively taking into account at least two different conductometric regimes, below the percolation and above the

Experimental section

Microemulsions composed of H₂O–AOT–decane were prepared at a water + surfactant volume fraction of $\Phi = 0.80$ by mixing appropriate fractions of low-conductivity, deionized water ($\sigma < 10^{-6} \Omega^{-1} cm^{-1}$ at 20 °C) and reagent-grade decane, using surfactant AOT purchased from Sigma Chem. Co. The surfactant was used as received, without any further purification. Microemulsions of lower concentrations, from $\Phi = 0.80$ to $\Phi = 0.10$, were obtained by dilution. The microemulsions were characterized by the molar ratio of water to surfactant $X = [H_2O]/[AOT] = 40.8$ in the one-phase region, where the water core has a size distribution of the Schultz type with an average size of about 50–60 Å and a polidispersity of about 30%, as determined by small-angle neutron scattering experiments [2]. The hydrodynamic radius of the surfactant coated water droplets, as determined by dynamic light scattering experiments [2] is of the order of 80 Å, largely independent of the volume fraction Φ. The electrical conductivity of the microemulsions was measured at a frequency of 10 kHz by means of a Hewlett–Packard mod. 4192A Impedance Analyzer, in connection with

172

F. Bordi et al.
Conductivity of a W/O microemulsion

a temperature-controlled parallel-plane capacitance cell. The cell constants were determined using standard solutions of known conductivity and permittivity. The temperature was varied from -15 to $70\,°C$, within $0.1\,°C$.

Results and discussion

In what follows, we will give a schematic description of the changes induced by the addition of simple salt (10^{-2} M NaCl) to the microemulsion system, mainly in the phase diagram and in the two electrical conductivity regimes, well below the percolation, where the charged water droplets contribute to conduction by means of Brownian movement, and close to percolation, where charge carriers jump between different clusters or transient merging of connected droplet occurs.

The phase diagram of the H₂O–AOT–decane microemulsion

The phase diagram in the T–Φ plane of the microemulsion investigated in presence of 10^{-2} M NaCl electrolyte solution is shown in Fig. 2, where each boundary reflects a change in the slope of the electrical conductivity. A comparison with the corresponding phase diagram of the ternary system, i.e., in absence of the added salt [9, 10], shows that noticeable changes occurred. These changes are very surprising taking into account the small value of the salt concentration. Considering that a full ionization of the AOT molecule might occur, a molar fraction $X = 40.8$ corresponds to about $100\,Na^+$ ions within each water droplet which yields a molar concentration of Na^+ in the water core of about 1.3 mol/l. We have only slightly alter-

ed the ion concentration, without changing the local electroneutrality; nevertheless, this change results in a deep alteration of the overall electrical conductivity behavior of the system. The main features evidenced by the inspection of the phase diagram can be summarized as follows. The region of the lamellar structure is strongly reduced compared to the salt-free case and, although the percolation line does not seem to be deeply altered, in the low fractional volume range a new behavior appears. From $\Phi = 0.30$ to the lowest fractional volume investigated, it becomes difficult to determine unambiguously the percolation temperature T_p. T_p was generally identified as that corresponding to the inflection point of the curve $\ln \sigma$ vs. T, i.e., the choice of T_p is made choosing the value that maximizes $d/dt\,(\ln \sigma)$ vs T. This criterion fails for fractional volumes lower than $\Phi = 0.30$. In this volume fraction interval, indeed, the curves $d\,\ln\sigma/dt$ vs. T display two separate maxima indicating that the conductivity curve as a function of temperature experiences two distinct inflexion points. Moreover, the amplitude of the maxima are different and have a different behavior as a function of T: the first maximum has an approximately constant amplitude, whereas that of the second maximum increases with decreasing Φ. This situation is clearly depicted in Fig. 3, at three different volume fractions from $\Phi = 0.175$ to $\Phi = 0.300$. Moreover, in the low-temperature range, in the sub-zero interval, the conductivity shows an anomalous behavior. As the temperature is decreased from high to low values, the conductivity curve is characterized by a progressive decrease until a minimum is reached, followed by a further slow increase and then by a sudden jump of about one or two orders of magnitude. The origin of this peculiar behavior is unclear without the hypothesis of a structural rearrangement of the system, since the microemulsion, built up in this region of composition of water

Fig. 2 The (Φ, T) phase diagram of water–AOT–decane microemulsion at the water to surfactant molar ratio $X = 40.8$, deduced from low-frequency electrical conductivity measurements. open symbols: microemulsion without added salt, full symbols: microemulsion with added salt (10^{-2} M NaCl). The lines define different regions of different microemulsion structures

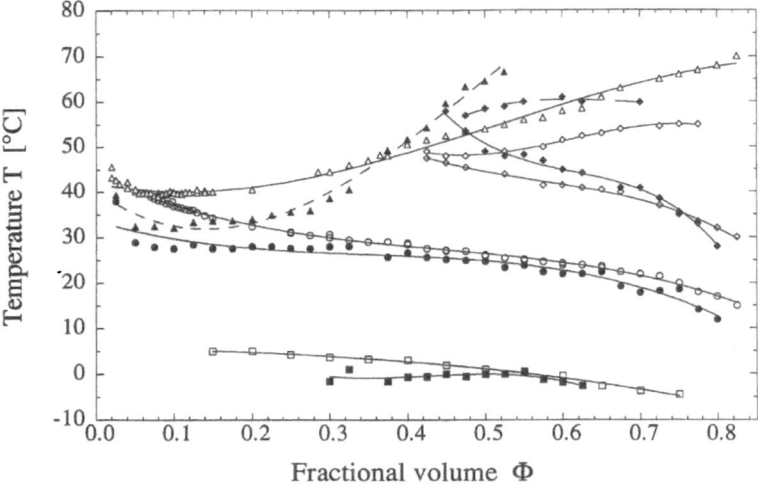

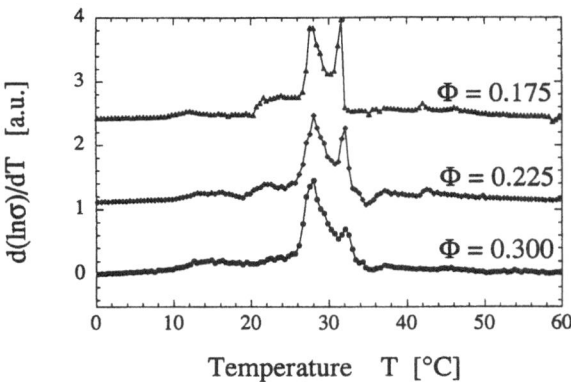

Fig. 3 The anomalous behavior of the electrical conductivity σ close to percolation at low- to moderate fractional volume Φ, revealed by the presence of two maxima in the behavior of $d(\log \sigma)/dT$ vs. T

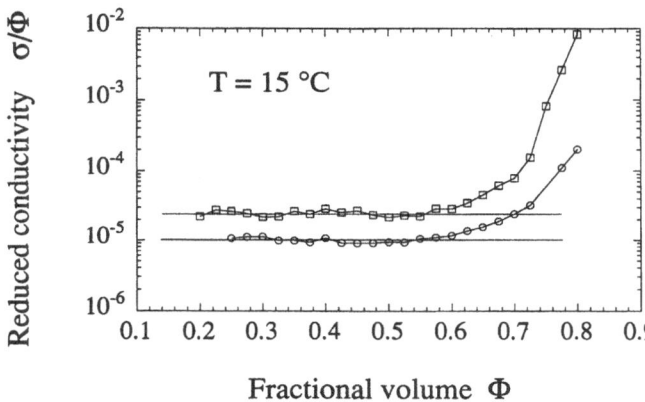

Fig. 4 The reduced low-frequency electrical conductivity σ/Φ as a function of fractional volume Φ at the temperature of 15 °C. (□) in presence of added salt (10^{-2} M NaCl); (○) in absence of added salt. The full lines are the calculated values according to the Eicke–Hall theory, with an hydrodynamic radius R of 80 Å (in absence of added salt) and of 60 Å (in presence of added salt). The deviation from a constant value as the fractional volume is increased indicates the onset of percolation

droplets dispersed in an oil matrix, should experience a progressive reduction of the overall conductivity due to a decrease of the droplet mobility, accompanied by a decrease of the equivalent conductance of Na^+ ions, inside the water core. Finally, the intermediate region between the two bicontinuous phases characterized by water-in-oil and oil-in-water structures occurs in a region of the $T-\Phi$ phase diagram somewhat shifted towards higher temperatures and higher volume fraction values.

The conductivity behavior well below the percolation threshold

As stated above, below the percolation threshold, the transport process occurs by means of movement of charged water droplets, whose ionization is due to the fluctuating exchange of the surfactant molecules. Within this theory [11, 12], the electrical conductivity is given by

$$\sigma = \Phi \frac{e^2 \sum_{-\infty}^{+\infty} z^2 \exp(-z^2 e^2/2\varepsilon R K_B T)}{8\pi^2 R^4 \eta \sum_{-\infty}^{+\infty} \exp(-z^2 e^2/2\varepsilon R K_B T)}, \quad (1)$$

where ε and η are the dielectric constant and viscosity of the oil phase (decane in our case) respectively, ze the electric charge of each water droplet, $K_B T$ the thermal energy and R the hydrodynamic radius of the dispersed particles. We have recently shown [9, 10] that for this microemulsion, Eq. (1) predicts the reduced conductivity σ/Φ in excellent agreement with the measured values well below the percolation threshold, confining the values of z to $z = \pm 1$. This means that, an average, each droplet bears an electronic charge. In other words, although the transport of an ionized surfactant from one droplet to another should not be energetically favorable, only small

deviations from the overall particle neutrality are expected in order to justify the measured behavior.

The effect of the salt addition on the conductivity behavior of the microemulsion below the percolation threshold is clearly shown in Fig. 4, where the conductivities of two water–AOT–decane microemulsions at the same composition with and without salt addition are compared as a function of the fractional volume Φ. As can be seen, the addition of salt produces a marked increase of the electrical conductivity that can be justified in the light of the Eicke–Hall theory (Eq. (1)), considering that, in the case of salt addition, a smaller hydrodynamic radius R of the water droplets is required. The full lines in Fig. 4 refer to $z = \pm 1$ both in the case of absence of salt and in the presence of 10^{-2} M NaCl, whereas we have assumed $R = 80$ Å and $R = 60$ Å respectively in the two cases. It must be noted that a very good agreement is obtained, for values of Φ well below the percolation, where the Eicke–Hall theory holds. The smaller hydrodynamic radius in presence of added salt can be due to a reduction of the head-group interactions in the surfactant layer as a consequence of the increased electrostatic screening. When a clusterization process takes place, enhanced by the increase of the droplet concentration, deviation from a linear dependence of the conductivity occurs and a different transport mechanism prevails.

The conductivity behavior near the percolation threshold

Electrical conductivity probes the connectivity of dispersed droplets over a macroscopic length scale and when this connectivity is achieved by charge hopping or transient merging of different droplets or by the formation of stable water channels, a very pronounced increase of the conductivity is observed. This process has been analyzed as a percolation process and is due to the transition from a conductivity mechanism due to Brownian motion of the charged water droplets to a conductivity regime dominated by the motion of charged carriers within a connected clusters of water droplets, extended on large scale length. On approaching the percolation threshold, as a function of temperature T, the electrical conductivity σ obeys a power-law divergence above and below the percolation temperature T_p

$$\sigma \approx (T - T_p)^{\mu} \quad T \geq T_p$$

$$\sigma \approx (T_p - T)^{-s} \quad T \leq T_p . \qquad (2)$$

The values of critical exponents for AOT–water–decane microemulsions reported in literature and deduced by locating the percolation threshold as a function of temperature, have been generally found in the range $1.1 \leq s \leq 1.6$ and $1.6 < \mu < 2.2$. The spread of these values is essentially due to the difficulty in locating exactly the percolation temperature T_p and in defining the extension of the power law region not too far above and below T_p. As an example, a typical plot of $\ln \sigma$ vs. $\ln |T - T_p|$ in the vicinity of the percolation threshold for the microemulsion at $\Phi = 0.525$ (10^{-2} M NaCl) is shown in Fig. 5, where the power-law behaviors of the static and dynamic percolation with exponents $\mu = 1.84$ and $s = 1.16$ respectively are also shown.

In the present case, the determination of the percolation temperature T_p is made as that value that maximizes the function $d\ln\sigma/dT$ vs. T. Because of the more or less flattening of the maximum, some uncertainty may occur,

resulting in an ambiguous choice of the threshold value T_p that might produce, in turn, quite different values of critical exponents. To overcome this difficulty and to establish the percolation temperature within a well-defined procedure, a detailed analysis was carried out, consisting in the fitting of Eqs. (2) to the experimental values, giving different arbitrary values of T_p near the probable threshold. For each value, the correlation coefficients r of the two straight lines in the two branches of the conductivity were calculated. The percolation temperature T_p and consequently the exponents μ and s are assumed to be those to which correspond the maximum of the correlation coefficient.

This analysis concerning the two branches of the conductivity σ as a function of $\ln |T - T_p|$ for both the microemulsion systems investigated (water/AOT/decane in presence of 10^{-2} M NaCl and in absence of added salt, at the same water to surfactant molar ratio $X = 40.8$) has been carried out over Φ ranging from $\Phi = 0.1$ to $\Phi = 0.80$. In presence of added salt, the exponent of the dynamic percolation s assumes a value of $s = 1.16$ approximately constant from $\Phi = 0.20$ to $\Phi = 0.55$. Above this value, marked deviations occur and a value of $s = 1.80$ is reached. In absence of added salt, s is approximately constant over the whole fractional volume range to the value $s = 1.64$. On the contrary, the static percolation μ does not experience any apparent change and the exponent assumes a value of about $\mu = 1.77$, constant over the whole Φ interval, both in presence and in absence of added salt. The indices s and μ are shown in Fig. 6 as a function of the volume fraction Φ.

A different approach to this problem yields a different and more consistent picture of the static and dynamic percolation indices. We have recently developed a new approach to the dielectric and conductometric behavior of

Fig. 5 A typical plot of log σ vs. log$|T - T_p|$ for the microemulsion at $\Phi = 0.525$ in presence of added salt (10^{-2} M NaCl). The two straight lines represent the power laws above and below percolation with exponents $\mu = 1.80$ (static exponent) and $s = 1.16$ (dynamic exponent). Deviations from a power law behavior very close to percolation are also evident

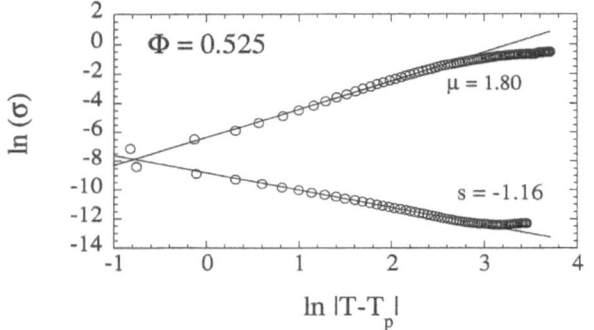

Fig. 6 The dynamic μ and static s exponents of the power laws close to percolation as a function of the fractional volume Φ. (●) in presence of added salt (10^{-2} M NaCl); (○) in absence of added salt

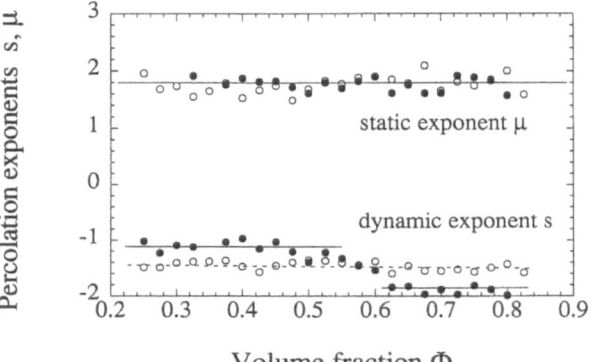

microemulsions close to percolation [13], which gives a more detailed description of the characteristic of this phenomenon. In particular, we are able to write a closed form expression for the complex conductivity $\sigma^*(\omega) = \sigma + \imath\omega\varepsilon$ of the system which, in the low-frequency limit, yields the static conductivity σ, including both the power law behavior on approaching the threshold and its saturation very close to it. In this respect, we overcome the difficulties, as above stated, inherent to a simple analysis of the conductivity curves in terms of a pure power law.

In order to derive the expression for $\sigma^*(\omega)$, we use the following simple model: the microemulsion droplets, characterized by a complex conductivity $\sigma_B(\omega) = \sigma_B + \imath\omega\varepsilon_B$, dispersed in a continuous oil phase of conductivity $\sigma_A(\omega) = \sigma_A + \imath\omega\varepsilon_A$, have a weak short-ranged attractive potential which causes the formation of large inhomogeneous clusters, characterized by a fractal dimension $D = 2.53$. According to the dynamic percolation model of conductivity in microemulsions below the percolation threshold, the conduction mechanism is due to charge migration in the bulk of cluster since the charge carriers perform an anomalous diffusion motion. The steps we use are the evaluation of the complex conductivity of a single cluster by means of an effective medium approximation, supplemented by the fractal and dynamic properties mentioned above, and the appropriated distribution of cluster sizes, typical of percolation phenomena. The final result concerning the static conductivity, which depends only on the dielectric properties of the components (σ_A, ε_A, σ_B, ε_B) and the relaxation properties of the single monomer (its dielectric increment Δ_1 and the relaxation time τ_1) is

$$\sigma(\omega = 0) = N\sigma_A\left(\frac{3}{D} - 1\right)\Delta_1\tau_1(\omega_A - \omega_B)k_c^{1-1/D}$$

$$\times \int_{k_c^{-1}}^{\infty} dz \frac{e^z z^{-1/D}}{\left(1 + \dfrac{1 - \omega_B\tau_1}{\omega_A\tau_1 - 1}\dfrac{\omega_A}{\omega_B}k_c^{2/d}\right)z^{2/d}}. \quad (3)$$

Here, D is the fractal dimension of the cluster, d the anomalous diffusion index, the frequencies ω_A and ω_B are defined as $\omega_A = \sigma_A/\varepsilon_A$, $\omega_B = \sigma_B/\varepsilon_B$ and finally N is the total number of monomers initially present in the solution. The quantity k_c has the meaning of cut-off cluster size and it is the parameter that drives the system to the percolation threshold, where it diverges. It is thermodynamically related to the dimensionless distance from the threshold $|(T - T_p)/T_p|$ by the relation

$$k_c = \left(\frac{\xi_0}{R_1}\right)^D \left|\frac{T - T_p}{T_p}\right|^{-\nu D},$$

where R_1 is the monomer radius, ξ_0 is the percolation correlation length amplitude, and ν is the index of its

divergence

$$\xi = \xi_0 \left|\frac{T - T_p}{T_p}\right|^{-\nu}.$$

Exactly at the threshold, we get

$$\sigma_p = N\left(\frac{3}{D} - 1\right)\Delta_1\tau_1\sigma_B\frac{\varepsilon_A}{\varepsilon_B}\frac{(1 - \omega_A\tau_1)(\omega_B - \omega_A)}{(\omega_B\tau_1 - 1)^2}$$

$$\times {}_2F_1\left(1, u; 1 + u; -\frac{\omega_B(\tau_1\omega_A - 1)}{\omega_A(1 - \tau_1\omega_B)}\right),$$

where $u = \mu/(s + \mu)$ and ${}_2F_1$ is the Gauss hypergeometric function. It is easy to show from the general expression

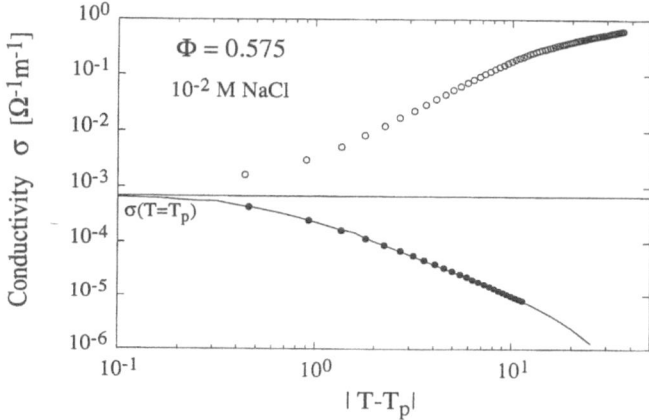

Fig. 7 The two branches of the electrical conductivity above and below the percolation for the microemulsion at $\Phi = 0.575$ in presence of added salt (10^{-2} M NaCl). The full line, below the percolation, represents the calculated values according to Eq. (3), giving exactly the measured value up to the percolation threshold

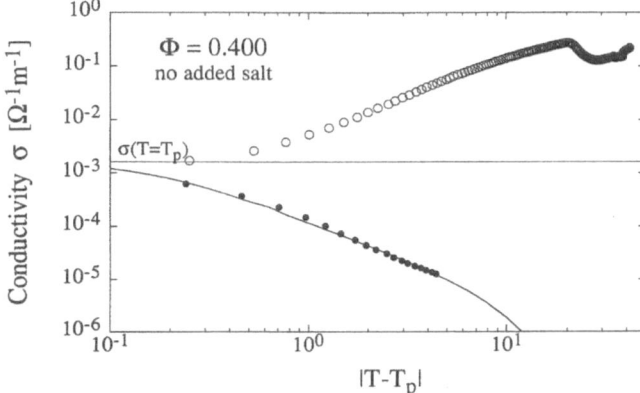

Fig. 8 The two branches of the electrical conductivity above and below the percolation for the microemulsion at $\Phi = 0.400$ in absence of added salt. The full line, below the percolation, represents the calculated values according to Eq. (3), giving exactly the measured value up to the percolation threshold

that, in a region not too close to the threshold, the conductivity σ behave as

$$\sigma \approx |T - T_{\mathrm{p}}|^{-s},$$

where $s = v(D - 1)$ is the index of the divergence of the static conductivity at percolation. The conductivity below the percolation threshold and very close to it, for both the microemulsion systems investigated, has been calculated, according to Eq. (3), assuming for the percolation indices the following values: $D = 2.53$, $v = 0.88$, $d = 1.36$ independent of the fractional volume Φ, the dynamic percolation index s assuming the value $s = 1.35$. The result of the above analysis, shown in Figs. 6 and 7, gives the full description of the conductivity behavior of the system below and up to the percolation, with very good agreement with the measured values over the whole measurement range we have investigated, with a single value of the index s. It must be noted, moreover, that the four parameters entering in Eq. (3), obtained from the fitting procedure, i.e., the two frequencies ω_{A} and ω_{B}, the relaxation time τ_1 of the single monomer and the cut-off cluster size k_{c}, assume values very close to the expected ones for the microemulsion investigated.

References

1. Candau SJ (1987) In: Zana R (ed) Surfactant Solutions Dekker
2. Rouch J, Tartaglia P (1991) In: Chen SH, Huang JS, Tartaglia P (eds) Structure and Dynamics of Strongly Interacting Colloids and Supramolecular Aggregates in Solution NATO ASI Series, Vol 369
3. Kotlarchyk M, Chen SH, Huang JS, Kim MW (1984) Phys Rev A29:2034
4. Cabos C, Marignan J (1985) J Phys Lett 46:267
5. Ganz AM, Boeger BE (1986) J Colloid Interface Sci 109:504
6. Peyrelasse J, Boned C (1988) Phys Rev A38:904
7. Bedeaux D, Koper GJM, Smects J (1993) Physica A 194:105
8. Chen SH, Lim TL, Huang JS (1987) In: Safran S, Clark NA (eds) Physics of Complex and Supramolecular Fluids, Wiley, NY
9. Di Biasio A, Cametti C, Codastefano P, Tartaglia P, Rouch J, Chen SH (1993) Phys Rev E47:4258
10. Cametti C, Codastefano P, Tartaglia P, Chen SH, Rouch J (1992) Phys Rev A 45:R5358
11. Eicke HF, Borkovec M, Das-Gupta B (1989) J Phys Chem 93:314
12. Hall DA (1990) J Phys Chem 94:429
13. Cametti C, Sciortino F, Tartaglia P, Rouch J, Chen SH (1995) Phys Rev Lett 75:569

Progr Colloid Polym Sci (1996) 100:177–181
© Steinkopff Verlag 1996

MICROEMULSIONS

Dielectric relaxation of microemulsions

M. Camardo
M. D'Angelo
D. Fioretto
G. Onori
L. Palmieri
A. Santucci

M. Camardo · M. D'Angelo
D. Fioretto · G. Onori (✉)
L. Palmieri · A. Santucci
Dipartimento di Fisica
Universitá di Perugia
Via A. Pascoli
06100 Perugia, Italy

Abstract The complex permittivity of $AOT/H_2O/CCl_4$ and $AOT/H_2O/n$-heptane microemulsions has been measured by a frequency domain coaxial technique in the range 0.02/20 GHz as a function of the molar ratio $[H_2O]/[AOT]$ (W). A relaxation phenomenon was observed whose behavior vs. W has been interpreted, in the dilute limit, in terms of two coexisting mechanisms: the reorientation of the whole micellar aggregate and the "free" rotational diffusion of the completely hydrated AOT ion pairs.

Key words AOT – dielectric relaxation – microemulsions – IR spectra

Introduction

It is well-known that sodium bis (2-ethylhexyl) sulfosuccinate (AOT) enables a high degree of solubility of water in oil without the addition of any other component. There is considerable evidence that over wide ranges of composition and temperature AOT forms microemulsions consisting of small surfactant-coated water droplets in Brownian motion [1, 2]. It is also known that, in some systems, the particles show attractive interactions [3, 4].

In previous papers [5, 6], infrared spectroscopy in the O–H stretching region has been used to study the structure of water in AOT reverse micelles. The behavior of IR spectra as a function of the molar ratio $[H_2O]/[AOT]$ (W) was explained in terms of an initial hydration of the surfactant headgroups followed by the formation of a water-pool where a continuous equilibrium between "bound" and "bulk" water exists.

More recently, a detailed dielectric investigation of the influence of both surfactant hydration and interparticle interactions on the dynamical properties of AOT reverse micelles has been reported [7, 8]. Even though information on structure and dynamics of both AOT shell and of water confined into the reverse micelles could be obtained from dielectric relaxation measurements, systematic studies of dielectric properties of AOT microemulsions are very scarce and the theoretical interpretation of the results rather unsatisfactory.

Our dielectric investigation was focused on the diluted region of the system, where the model of dispersed droplets in a continuous medium is still valid, and to the region of small amounts of water where the properties of the system change strongly with water content. A relaxation phenomenon was observed in the microwave region, whose behavior as a function of W, in absence of interparticle interactions, was successfully interpreted in terms of two coexisting mechanisms: the reorientation of the whole micellar aggregate and the "free" rotational diffusion of the completely hydrated AOT ion pairs. On the whole, our data reveal that the dynamics of these systems is richer than that usually reported in the literature and suggest a fruitful complementarity of IR and dielectric spectroscopy for understanding the role of hydration and interactions in the dynamics of reverse micelles.

In the present paper we report the most significant results of our previous investigation presenting in some detail the model proposed to interpret the dielectric data in the dilute limit.

Results

The complex dielectric function of $AOT/H_2O/CCl_4$ and $AOT/H_2O/n$-heptane systems was measured by a frequency domain coaxial technique in the range 0.02/20 GHz at low molar ratios, W. A relaxation phenomenon, located in the 0.02/1 GHz region was observed whose properties are markedly affected by interparticle interactions. These interactions depend on different parameters and in particular on the nature of dispersing medium [3, 9]. It was shown [8], however, that when the volume fraction ϕ of dispersed phase (water and AOT) tends to zero, the relaxation process exhibits a similar behavior in the two solvent oils, thus suggesting a negligible role of the interactions. Figure 1a, b show the values of the relaxation time τ_1, as a function of W, relative to samples at $\phi = 0.05$ in CCl_4 and n-heptane, respectively. In both systems, τ_1 rapidly decreases with increasing W up to W close to 6 keeping nearly constant at higher molar ratios. In spite of the strong dependence of τ_1 on the hydration degree of the micelles, the dielectric increment $\Delta\varepsilon_1$ measured in the

Fig. 1 Dielectric relaxation time τ_1 vs. W. (a): $AOT/H_2O/CCl_4$, (b): $AOT/H_2O/n$-heptane system, (———): calculated according to Eq. (3). See text

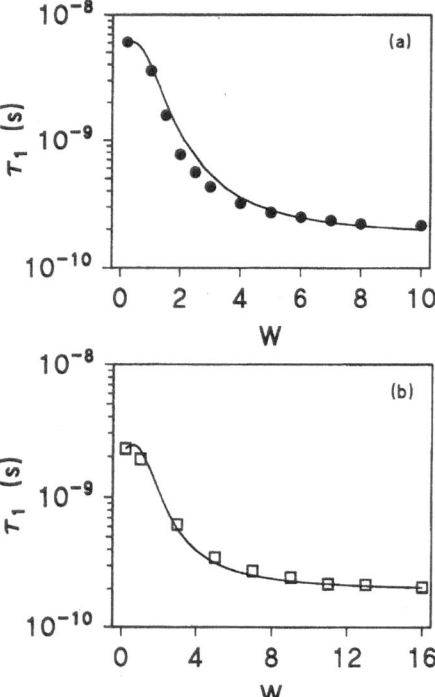

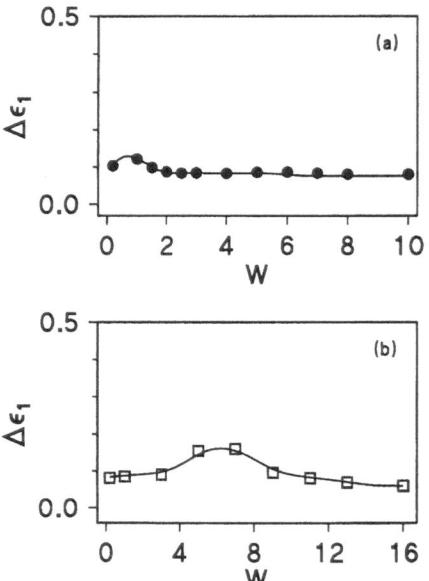

Fig. 2 Dielectric relaxation strength $\Delta\varepsilon_1$ vs. W. (a): $AOT/H_2O/$ CCl_4, (b): $AOT/H_2O/n$-heptane system. Lines are a guide for the eyes

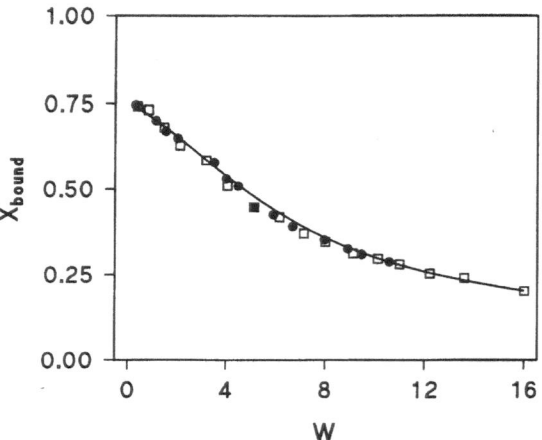

Fig. 3 Fraction of bound water molecules X_{bound} as a function of W for $AOT/H_2O/CCl_4$ ($\bullet$) and $AOT/H_2O/n$-heptane ($\square$) microemulsions. (———): calculated assuming the existence of three independent binding sites per AOT molecule [5]

two solvent oils does not undergo appreciable changes as water is added to the samples (see Fig. 2a, b).

The results of previous IR studies [5, 6] on the hydration process of AOT reverse micelles in CCl_4 and n-heptane are reported in Fig. 3 where the fraction of "bound water" X_{bound} is plotted versus W. In both systems the values of X_{bound} show a decrease in the whole range of W, which is particularly steep at the lowest molar ratios. The behavior is qualitatively similar to that of τ_1 versus W suggesting a connection between the two observed phenomena.

Model

The relaxation process observed at the lowest values of W ($W = 0.2$), where the almost dehydrated reverse micelles exhibit a nearly rigid structure, was attributed to the rotational diffusion of the individual micelles [7, 8]. Supposing a spherical shape for the microaggregate, we calculated the micellar radius R from the Debye–Stokes formula:

$$\tau = \frac{4\pi\eta R^3}{K_B T},\qquad (1)$$

where τ is the experimental relaxation time, η the viscosity and $K_B T$ the thermal energy.

Using for η the value of the viscosity of the solution, the values $R = (12.0 \pm 0.8)$ Å and $R = (11.8 \pm 0.7)$ Å were obtained for the sample in CCl_4 and in n-heptane, respectively. These results are mutually consistent within the experimental errors and consistent with the results from the literature [10–12].

To further support the proposed description, the dipole moment $\mu_{\text{mic.}}$ of a dehydrated microaggregate was estimated by using the Debye extension of the Clausius–Mossotti equation and compared with the data in the literature. The Debye equation is written in terms of the molar fractions of micelles f_1 and of solvent molecules f_2 as:

$$\frac{\varepsilon - 1}{\varepsilon + 2}\left[\frac{f_1}{N_{01}} + \frac{f_2}{N_{02}}\right] = f_1 \cdot \left[\frac{\varepsilon_\infty - 1}{\varepsilon_\infty + 1} \cdot \frac{1}{N_{01}} + \frac{\mu_{\text{mic.}}^2}{9\varepsilon_0 K_B T}\right]$$
$$+ f_2 \cdot \left[\frac{\varepsilon_{\text{solv.}} - 1}{\varepsilon_{\text{solv.}} + 2} \cdot \frac{1}{N_{02}}\right],\qquad (2)$$

where ε is the low frequency dielectric permittivity of the solution, N_{01} and N_{02} the numerical density of micelles and solvent molecules, respectively, ε_0 the absolute dielectric constant of free space, ε_∞ the high-frequency dielectric constant of micelles and $\varepsilon_{\text{solv.}}$ the dielectric constant of the solvent oil. Using the value $N_a = 17$ [13, 14] for the mean aggregation number of monomeric surfactant molecules in the two dispersing media, the apparent dipole moment $\mu_{\text{app.}}$ of a surfactant molecule in the micelles, calculated dividing $\mu_{\text{mic.}}$ by N_a, is 0.77 Debye and 0.78 Debye for the sample in CCl_4 and n-heptane, respectively. The results, practically independent of the solvent oil, are in very good agreement with that determined by Eicke et al. [15].

It is known, however, that increasing the water content inside the micelles, the micellar radius R increases almost linearly with W [11] so that τ_1 is expected to increase approximately with W^3, according to the Debye–Stokes equation (1). Different from this, the values of τ_1 decrease in the initial range of W and keep nearly constant at higher

W values (see Fig. 1a, b). The striking resemblance between the behavior of τ_1 and X_{bound} as a function of W suggests that the departure of the experimental values of τ_1 from those predicted by Eq. (1) can be connected with the hydration process of the micellar aggregates. We thus supposed that, as the water content increases inside the micelle, an increasing number of AOT ion pairs could achieve a sufficient mobility to reorientate independently from the whole microaggregate. On these bases, the relaxation phenomenon observed at the highest values of W was attributed to the rotational diffusion of the "free" AOT ion pairs at the water–surfactant interface. As a further support to this hypothesis, we notice that, at the highest W's, the investigated dielectric dispersion is located close to a relaxation process observed in concentrated electrolyte solutions and attributed to the rotational diffusion of dipolar solute species (ion pairs) [16, 17].

With this in mind, the existence of two fractions of AOT ion pairs was supposed for each value of W: $(1 - X(W))$ referring to the fraction of surfactant ion pairs reorienting with the whole micelle with the Debye diffusion time $\tau_D(W)$ and $X(W)$ referring to the fraction of the "free" ion pairs whose reorientation is independent from that of the whole micelle and is characterized by a lower relaxation time, τ_0. The value of τ_0 was chosen as the limiting value of τ_1 at the highest molar ratios. Since the exchange of water molecules among hydration layers causes fluctuations in AOT ion pair aqueous environment characterized by a rate which is about two orders of magnitude faster than $1/\tau_1$ [18] (fast exchange condition [19]), the relaxation time τ_1 was calculated by the equation:

$$\frac{1}{\tau_1} = \frac{X(W)}{\tau_0} + \frac{1 - X(W)}{\tau_D(W)}.\qquad (3)$$

In the previously mentioned IR study [5, 6], the experimental data were treated on the bases of a model assuming the existence of three independent binding sites per AOT molecule. This model gives the opportunity of evaluating the fractions of AOT molecules corresponding to the three possible hydration degrees of AOT ion pairs. It was shown [7, 8] that by replacing the fraction of completely hydrated AOT ion pairs (three water molecules per AOT ion pair) in $X(W)$ of Eq. (3) the resulting trend of τ_1 as a function of W (solid lines in Fig. 1a, b) satisfactorily reproduces the experimental data. On the other hand, supposing that the partially hydrated AOT ion pairs (i.e., with at least one or two hydration water molecules) were able to freely reorientate, too, the values of τ_1 calculated from Eq. (3) in terms of these two species significantly depart from the experimental points. From these results it appears that only the completely hydrated surfactant ion pairs acquire a sufficient mobility to provide a separate contribution to the relaxation process.

According to the previous discussion, the relaxation phenomenon observed at the highest values of W was expected to rise from the fluctuating electric dipole moment imparted to the microaggregate by the rotational diffusion of the "free" AOT ion pairs. As a further support to the proposed description for the dynamics of these systems, we interpreted the experimental data of relaxation strength at the highest values of W in terms of the rotational diffusion of "free" AOT ion pairs and compared the AOT ion pair dipole moment μ, calculated from the experimental data of relaxation strength, with the value reported in the literature. The dispersed particles were sketched as spherical surfactant-coated water droplets with AOT ion pairs located at the surfactant–water interface. The dipole moment μ of an AOT ion pair was related to the orientational contribution $\Delta\varepsilon_{ip}$ to the equivalent permittivity ε_p of a spherical particle suspended in an oil which models the microaggregate. The surface polarizability due to the AOT ion pairs rotational diffusion at the water core surface was transformed into a volume polarizability of a reverse micelle substitute homogeneous sphere and $\Delta\varepsilon_{ip}$ was written according to the equation:

$$\Delta\varepsilon_{ip} = \frac{N_a \mu^2 g}{4\pi R^3 \varepsilon_0 K_B T} , \qquad (4)$$

where N_a is the aggregation number of surfactant molecules, R is the external micellar radius, and g the "ion-pairs" orientation correlation and/or local field correction factor. $\Delta\varepsilon_{ip}$ was related to the equivalent permittivity ε_p by means of the equation:

$$\Delta\varepsilon_{ip} = \varepsilon_p - \varepsilon_{ph} , \qquad (5)$$

where ε_{ph} is the high frequency permittivity of the heterogeneous sphere calculated by means of the Maxwell relation [20]:

$$\varepsilon_{ph} = \varepsilon_s \frac{2\varepsilon_s + \varepsilon_w - 2\varphi(\varepsilon_s - \varepsilon_w)}{2\varepsilon_s + \varepsilon_w + \varphi(\varepsilon_s - \varepsilon_w)} , \qquad (6)$$

where φ is the volume fraction of the aqueous core, ε_w the water static permittivity and ε_s the surfactant layer

substitute permittivity due to hydrocarbon electronic polarizability.

The equivalent permittivity ε_p was derived [11] from the polarizability per unit volume of an homogeneous sphere suspended in an oil, α_p, through the equation:

$$\alpha_p = \frac{\varepsilon_p - \varepsilon_m}{\varepsilon_p + 2\varepsilon_m} \qquad (7)$$

in which ε_m is the permittivity of the dispersing medium. In the dilute limit, when the dispersed particles can be considered independent, α_p can be obtained [11] from the microemulsion polarizability α by means of the Clausius–Mossotti relation:

$$\alpha = \frac{\varepsilon - \varepsilon_m}{\varepsilon + 2\varepsilon_m} = \alpha_p \cdot \Phi , \qquad (8)$$

where ε is the measurable low-frequency permittivity of the mixture. Assuming for ε the relaxed value of the real part of the complex permittivity, corresponding to the value measured at 10^7 Hz, we obtained from Eq. (8) the values $\alpha_p = 0.63$ for the sample at $W = 10$ in CCl_4 and $\alpha_p = 0.72$ for that at $W = 16$ in n-heptane. If the values of the aggregation number N_a and of the micellar radius R are known, it is possible to estimate the AOT ion pair dipole moment μ from (4), (5), (6), (7). Assuming $N_a = 59$ and $R = 27$ Å for the system in CCl_4, $N_a = 135$ and $R = 35$ Å for that at $W = 16$ in n-heptane [11], $g = 1$, $e_w = 80$ [21] and $e_s = 3.2$ [20], a dipole moment μ of about 10 Debye was obtained in both the considered cases. The resulting value of μ is in good agreement within the limits of the used approximations, with the value reported in ref. [15].

In previous papers [20, 22, 23] the occurrence of the relaxation phenomenon here investigated was tentatively explained in terms of different polarization mechanisms such as the triphasic model of interfacial polarization and the counterion diffusion polarization. The obtained results, however, differ by four for orders of magnitude from the expected ones. In contrast, the model of AOT ion pairs rotational diffusion proposed here provides a consistent interpretation for both relaxation strength and relaxation time at high values of W. Moreover, these results contribute to clarify the role that the hydration process plays in the dynamics of the micellar aggregates.

References

1. Eicke HF (1980) Top Curr Chem 87: 85–120
2. Chevalier Y, Zemb T (1990) Rep Prog Phys 53:279–371
3. Huang JS (1985) J Chem Phys 82: 480–484
4. Lemaire B, Bothorel P, Roux D (1983) J Phys Chem 87:1023–1028
5. Onori G, Santucci A (1993) J Phys Chem 97:5430–5434
6. D'Angelo M, Onori G, Santucci A (1994) Il Nuovo Cimento 16D:1601–1611
7. D'Angelo M, Fioretto D, Onori G, Palmieri L, Santucci A (1995) Phys Rew E 52:R4620–R4623
8. D' Angelo M, Fioretto D, Onori G, Palmieri L, Santucci A (1995) Phys Rew E: in press

9. Bedwell B, Gulari E (1984) J Coll Int Sci 102:88–100
10. Yoshioka HJ (1992) Coll Int Sci 150: 195–200
11. Van Dijk MA, Joosten JGH, Levine JK, Bedeaux D (1989) J Phys Chem 93: 2506–2512
12. Robinson BH, Steyler DC, Tack RB (1977) J Chem Soc Faraday Trans 45: 481–496
13. Ueno M, Kishimoto H (1977) Bull Chem Soc Jpn 50:1631–1637
14. Kon-no K, Kitahara A (1971) J Coll Int Sci 35:636–641

15. Eicke HF, Christen H (1974) J Coll Int Sci 48:281–290
16. Barthel J, Hetznauer H, Buchner R (1992) Ber Bunsenges Phys Chem 96: 988–997
17. Barthel J, Hetznauer H, Buchner R (1992) Ber Bunsenges Phys Chem 96: 1424–1432
18. Giese K (1972) Ber Bunsenges Phys Chem 76:495–500
19. Anderson JE (1967) J Chem Phys 47: 4879–4883

20. Peyrelasse J, Boned C (1985) J Phys Chem 89:370–379
21. Gabler R (1978) In: Gabler R, Electrical interaction in molecular biophysics. Academic Press New York-S. Francisco-London
22. Cametti C, Codastefano P, Tartaglia P (1990) Ber Bunsenges Phys Chem 94:1499–1503
23. D'Angelo M, Fioretto D, Onori G, Palmieri L, Santucci A (1995) Colloid Polym Sci 273

Progr Colloid Polym Sci (1996) 100:182–185
© Steinkopff Verlag 1996

MICROEMULSIONS

M.G. Giri
M. Carlà
C.M.C. Gambi
D. Senatra
A. Chittofrati
A. Sanguineti

Percolation in fluorinated microemulsions: A dielectric study

M.G. Giri · M. Carlà
C.M.C. Gambi (✉) · D. Senatra
Department of Physics
University of Florence
L.E. Fermi 2
50125 Firenze, Italy

A. Chittofrati · A. Sanguineti
Ausimont
R & D Centre
Colloid Laboratory
Via San Pietro 50
20021 Bollate, Milan, Italy

Abstract This work deals with the dielectric investigation of fluorinated water-in-oil microemulsions with per-fluoropolyether (PFPE) compounds. The microemulsion ohmic conductivity and dielectric constant at different temperatures and frequencies have been studied at water to surfactant molar ratio W/S = 11 for monophasic samples having volume fractions of the dispersed phase, ϕ, in the range 0.2 to 0.6. The study as a function of temperature in the interval $-10°$ to $40°C$ showed a percolation phenomenon (Phys Rev E (1994) 50:1313). The analysis in terms of scaling laws and the calculation of the scaling exponents indicated that the percolation process is mainly dynamic. Preliminary results are here presented of the study as a function of frequency.

Key words Microemulsion – fluorinated compound – percolation – dielectric study – frequency dependence

Introduction

Since the pioneering work of ref. [1] the percolation phenomenon has been demonstrated to take place in water-in-oil microemulsions with hydrogenated oil for both water-based [1–9] and waterless microemulsions [10–12], and for both ionic [1–13] and non-ionic surfactants [13] (in this last case an electrolyte is added to the system). Recently, the percolation phenomenon has also been found in fluorinated microemulsions [14] composed of per-fluoropolyether compounds, of which both the surfactant tail and the oil are perfluorinated.

When microemulsions are composed of aggregates of more or less spherical shape in Brownian motion in a continuous medium and sufficiently close to each other so that an efficient transfer of charge carriers between aggregates can take place, the percolation transition corresponds to the appearance of a cluster of infinite size. In practice, for a sample of finite size the increase of the number density of dispersed aggregates in the continuous medium forms one cluster which connects the system from one side to the opposite side (for example, the two electrodes of a dielectric cell containing the sample) at the percolation threshold. We recall that for hard-spheres dispersed randomly in an insulating continuous medium, the percolation threshold corresponds to the close-packing limit which is 0.65 expressed as volume fraction of the dispersed conducting phase. Microemulsions composed of dispersed aggregates in a continuous medium are systems of interacting objects. Therefore the interaction between the aggregates does play an important role [15, 16] in the correlation of positions leading to percolation thresholds at volume fractions lower than 0.65. Moreover, the aggregates give rise to a cluster of infinite size, or percolate, either for an increase of the number density of the aggregates themselves [2–6, 10, 11], or for a temperature increase [3, 7, 8, 14], or a pressure increase [12], etc. The actual description of the percolation phenomenon corresponds to what is called dynamic percolation [15, 16]; it differs from static percolation. The latter implies a coalescence of droplets, as the percolation threshold is approached, leading to the formation of one continuous channel of the dispersed medium (bicontinuous structure [15, 16, 9, 17]).

In ref. [14] we reported about the behavior of the static dielectric constant and the conductivity versus temperature of a fluorinated water-in-oil microemulsion at constant water to surfactant molar ratio W/S = 11 for which previous light scattering investigations at $T = 25 °C$ [18] demonstrated that water droplets of hydrodynamic radius 30 Å interact via an attractive potential (second virial coefficient $\alpha = -8$). Analysis in terms of scaling laws and the calculation of the scaling exponents indicate a dynamic percolation process with an exponent $s \simeq 1.2$ below the thermal threshold for all the samples studied and above the thermal threshold for intermediate ϕ values $0.33 < \phi < 0.48$, where ϕ is the volume fraction of the dispersed phase (water + surfactant)/(water + surfactant + oil). The exponents of static percolation $s \simeq 0.7$ were found above the threshold at $\phi < 0.33$ and $\phi > 0.48$. For all the samples μ exponents $\simeq 2$ were found as expected for both static and dynamic percolation.

In this paper the percolation phenomenon exhibited by the system at W/S = 11 is analyzed by measuring the dielectric constant as a function of frequency.

Materials

Water was taken from a Millipore Milli-Q system. The oil and the surfactant are PFPE compounds of general formula reported in ref. [14]. The surfactant molecular weight is 710, with a molecular weight distribution of 95% by gaschromatographic analysis. The surfactant polar head is a $-COO^-NH_4^+$ group. The oil molecular weight is 900, the density 1.8 g/cm³ and the viscosity 6.2 cp. The ternary system water, fluorinated oil and fluorinated surfactant systematically studied in the range 10° to 80 °C shows a monophasic domain throughout this range: the extension of the monophasic region changes as a function of temperature. The phase diagram behavior in the range 20° to 50 °C as well as the binary phase diagrams related to the system under investigation are reported in [19]. The experimental procedure used to study the dielectric behavior of the PFPE microemulsions is described in ref. [20].

Results and discussion

From the point of view of the dielectric properties, one microemulsion sample can be considered a conductor-insulator mixture, where the conductive part is the aque-

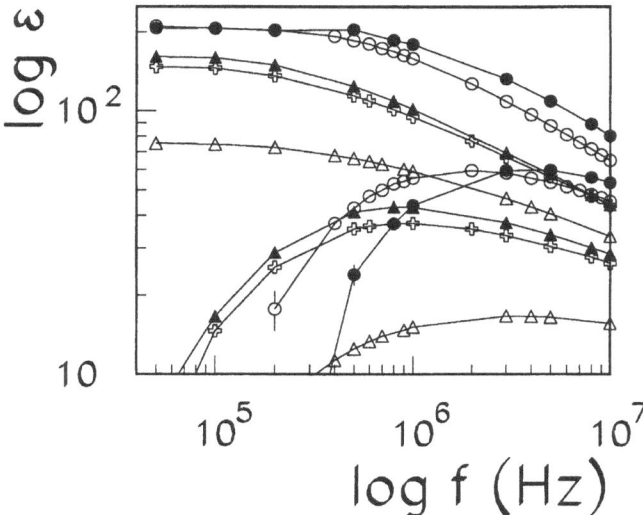

Fig. 1 The ε' and ε'' frequency dependence at constant temperature $T = 20 °C$ for samples with $\phi = 0.501$ (full circles), $\phi = 0.462$ (open circles), $\phi = 0.395$ (full triangles), $\phi = 0.361$ (open plus) and $\phi = 0.327$ (open triangles). The logarithmic scales are decimal for all the figures

ous region and the insulating part the oily one. Thus the microemulsion complex permittivity ε^* at constant temperature and pressure, is a function of the complex permittivities of the dispersed phase (water) ε_1^*, of the continuous phase (oil) ε_2^*, of the ϕ value, of the geometry of the system and of the interactions taking place within the system. The models based on mean-field theories that yield good results when interactions are weak, i.e., for low ϕ values, and when the microemulsion can be considered as macroscopically homogeneous, no longer apply when the dispersed particles cannot be considered isolated, i.e., when clusters of varying sizes form. In this case the concept of percolation can successfully be used.

In this paper we present the experimental frequency dependence of the complex permittivity for samples close to and at the percolation threshold. In Fig. 1 the real (ε') and the imaginary (ε'') parts of the complex permittivity ($\varepsilon^* = \varepsilon' - j\varepsilon''$) are reported as a function of frequency at $T = 20 °C$ for samples with different ϕ values. In Fig. 2 the same parameters are reported as a function of frequency for one sample with $\phi = 0.327$ at different temperatures in the range 2° to 26 °C. We point out that the study as a function of ϕ performed at $T = 20 °C$ gives a percolation threshold $\phi_p = 0.401$ [21].

The frequency dependence of the complex permittivity for a conductor-insulating mixture in the frame of the percolation theory takes the form [3, 22]:

$$\varepsilon^*(\omega) = E \cdot \exp[j\pi(1-u)/2] \cdot \omega^{u-1} , \qquad (1)$$

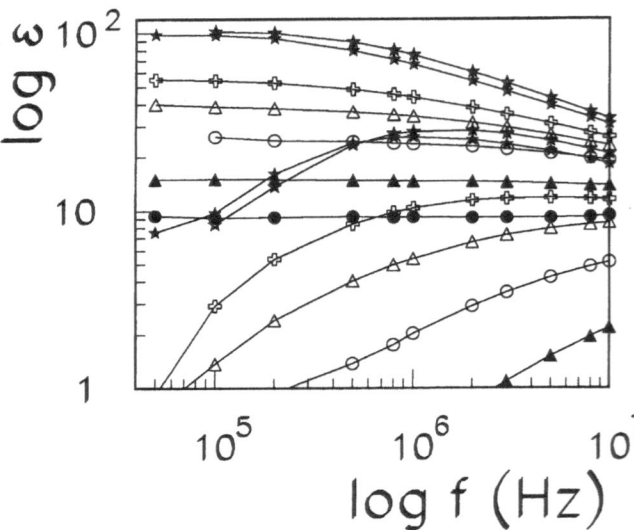

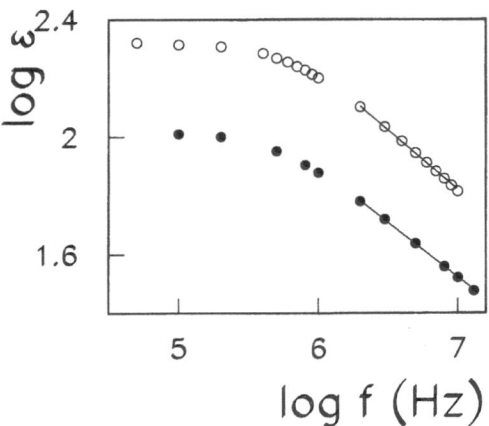

Fig. 3 Experimental frequency dependence of ε' and straight line for the calculation of the critical exponent u. The error bars are standard deviations. Sample with $\phi = 0.462$ at $T = 20\,°C$ (open circles). Sample with $\phi = 0.327$ at $T = 24\,°C$ (full circles)

Fig. 2 The ε' and ε'' frequency dependence for the sample with $\phi = 0.327$ at different temperatures: 2 °C (full circles), 8 °C (full triangles), 12 °C (open circles), 16 °C (open triangles), 18 °C (open plus), 22 °C (full stars), 24 °C (open stars). The $\varepsilon''(\omega)$ curve at $T = 2\,°C$ is lower than the scale of the picture

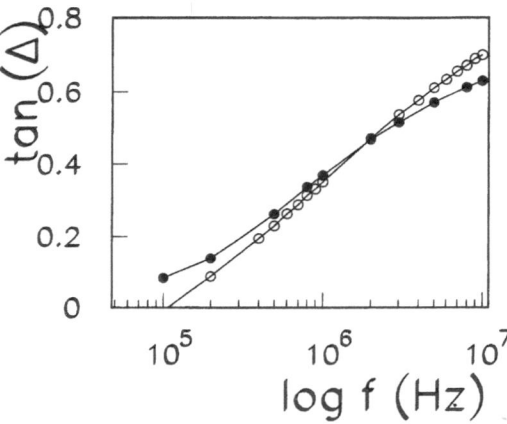

Fig. 4 The tangent of the loss angle as a function of frequency for the samples of Fig. 3 with identical symbol meaning

where E is a real value and u is a critical exponent. Equation (1) implies that both $\varepsilon'(\omega)$ and $\varepsilon''(\omega)$ exhibit a ω^{u-1} frequency dependence and that the loss angle Δ defined by $\tan\Delta(\omega) = \varepsilon''(\omega)/\varepsilon'(\omega)$ is independent of frequency and equal to $\pi(1 - u)/2$. We recall that, for such a conductor-insulator mixture, the conductivity as well as the static dielectric constant scaling laws as a function of $|\phi - \phi_p|$ are characterized by two critical exponents μ and s whose value is $\mu \simeq 2$ for both static and dynamic percolation, while is $s \simeq 0.7$ for static and $\simeq 1.2$ for dynamic percolation. The exponent u is related to μ and s by the equation $u = \mu/(\mu + s)$. This implies a value of about 0.625 for dynamic percolation and of about 0.741 for static percolation. These two u values give loss angle tangent of 0.668 and 0.431 for dynamic and static percolation, respectively.

The ε' and ε'' curves of Fig. 1 have a similar trend for the different ϕ values. In the frequency range investigated, the analysis in terms of scaling laws can be performed only for the $\varepsilon'(\omega)$ functions. Increasing the ϕ value, a linear dependence in logarithmic scale is found at $\phi = 0.462$ according to Eq. (1); from the slope of the straight line, reported in Fig. 3, the u exponent is deduced to be 0.59, a value indicating a dynamic percolation regime. The ε' and ε'' curves of Fig. 2 have a similar trend at the different temperatures and, as for the curves of Fig. 1, in the frequency range investigated only the $\varepsilon'(\omega)$ curves can be analyzed in terms of scaling laws. Increasing the temperature, a linear frequency dependence in logarithmic scale is found at $T = 24\,°C$; from the slope of the straight line (see

Fig. 3) a u exponent equal to 0.628 is deduced, a value which indicates a dynamic percolation regime. In Fig. 4 the tangent of the loss angles increases in both cases as a function of frequency up to values higher than 0.6. The u exponents and the $\tan\Delta$ values support the interpretation that a dynamic percolation process occurs in the microemulsions studied.

By comparing the latter results with those of ref. [14] the following conclusion can be drawn. For the sample with $\phi = 0.462$, we found a thermal threshold of 16.5 °C (from permittivity) and of 12.5 °C (from conductivity) with exponents typical of dynamic percolation below and above the thermal thresholds. As the curve of Fig. 3 is at 20 °C the sample is above threshold; furthermore as $\phi_p = 0.401$ the sample is above threshold also from the point of view of concentration. For the sample with $\phi = 0.327$ for which the thermal threshold is 22.8 °C (from permittivity) and 19.3 °C (from conductivity), we found a dynamic percolation process below and a static percolation process above the thermal threshold. However, the results of this paper at $T = 24$ °C (see Figs. 3 and 4), i.e., above the thermal threshold, indicate a dynamic percolation process. Further work is required.

In conclusion, the study reported here confirms that the percolation phenomenon taking place in fluorinated microemulsion with perfluoropolyethers is of dynamic type.

Acknowledgments The authors acknowledge useful discussions on percolation theory with R. Livi and thank Dr. P. Gavezzotti for the preparation of surfactants. This work was supported by Ausimont S.P.A. and by MURST 40% and 60% funds. MG Giri thanks Ausimont for support during this research work.

References

1. Lagues M, Ober R, Taupin C (1978) J Phys Lett (Paris) 39:487–491
2. Van Dijk MA (1985) Phys Rev Lett 55:1003–1005
3. Van Dijk MA, Casteleijn G, Joosten JGH, Levine YK (1986) J Chem Phys 85:626–631
4. Peyrelasse J, Moha-Ouchane M, Boned C (1988) Phys Rev A 38:904–917
5. Peyrelasse J, Boned C (1990) Phys Rev A 41:938–953
6. Cametti C, Codastefano P, Di Biasio A, Tartaglia P (1989) Phys Rev A 40:1962–1966
7. Cametti C, Codastefano P, Tartaglia P, Rouch J, Chen S-H (1990) Phys Rev Lett 64:1461–1464
8. Cametti C, Codastefano P, Tartaglia P, Chen S-H, Rouch J (1992) Phys Rev A 45:R5358–R5361
9. Knackstedt MA, Ninham BW (1994) Phys Rev E 50:2839–2843
10. Mathew C, Saidi Z, Peyrelasse J, Boned C (1991) Phys Rev A 43:873–882
11. Boned C, Peyrelasse J, Saidi Z (1993) Phys Rev E 47:468–478
12. Boned C, Saidi Z, Xans P, Peyrelasse J (1994) 49:5295–5302
13. Eicke H-F, Meier W, Hammerich H (1994) Langmuir 10:2223–2227
14. Giri MG, Carlà M, Gambi CMC, Senatra D, Chittofrati A, Sanguineti A (1994) Phys Rev E 50:1313–1316
15. Safran SA, Webman I, Grest GS (1985) Phys Rev A 32:506–511
16. Grest GS, Webman I, Safran SA, Bug ALR (1985) Phys Rev A 33:2842–2845
17. Di Biasio A, Cametti C, Codastefano P, Tartaglia P, Rouch J, Chen S-H (1993) Phys Rev E 47:4258–4264
18. Sanguineti A, Chittofrati A, Lenti D, Visca M (1993) J Colloid Int Sci 155:402–408
19. Chittofrati A, Lenti D, Sanguineti A, Visca M, Gambi CMC, Senatra D, Zhou Z (1989) Progr Colloid Polymer Sci 79:218–225
20. Giri MG, Carlà M, Gambi CMC, Senatra D, Chittofrati A, Sanguineti A (1993) Meas Sci Technol 5:627–631
21. Giri MG, Carlà M, Gambi CMC, Senatra D, Chittofrati A, Sanguineti A (work in progress)
22. Luck JM (1985) J Phys A 18:2061–2078

Progr Colloid Polym Sci (1996) 100:186–190
© Steinkopff Verlag 1996

MICROEMULSIONS

S. Amokrane
P. Bobola
C. Regnaut

Adhesive spheres mixture model of water-in-oil microemulsions

Dr. S. Amokrane (✉) · C. Regnaut
Laboratoire de Physique des milieux
désordonnés
Faculté des Sciences et de Technologie
Université de Paris 12 Val de Marne
Av. du Gal de Gaulle
94010 Créteil, France

P. Bobola
Laboratoire de Chimie Physique
URA 176 CNRS
Université Pierre et Marie Curie
11 rue Pierre et Marie Curie
75231 Paris Cedex 05, France

Abstract The multicomponent adhesive spheres mixture model is used for investigating the role of the direct intermicellar interaction in the effective adhesion of reverse micelles with different surface composition. A good correlation is found between the self-stickiness parameters deduced from a fit of the zero wave vector limit of the structure factor and those determined from a model interaction potential proportional to the overlap volume. A simple basis for the energetic proportionality constant is suggested.

Key words Microemulsions – overlap potential – adhesive spheres

Introduction

The analysis of structural data from neutrons, x-rays or light scattering measurements (see for e.g. refs. [1–3] and refs. therein) indicates that the interaction between colloidal particles is – among others factors – strongly influenced by the structure and composition of their surface. This is particularly evident in the case of reverse microemulsions [3]. But since the data are usually analyzed by using the effective one-component fluid model, the interaction so deduced reflects also the effect of the suspending medium. For instance, model potentials taking into account some aspects of the structure and composition of the solute surface and some characteristics of the suspending medium have been proposed for coated silica particles [1] or water in oil microemulsions [4]. An extreme idealization of this interaction is Baxter's sticky potential [5] in which the attractions are characterized by a single parameter – the effective stickiness τ^{-1}-incorporating all the physico–chemical specificities of the particles forming the dispersion (see for e.g. [6] for applications of Baxter's model).

Though these one component fluid models are certainly useful as a first step in the analysis, they do not allow a simple understanding of the various factors which determine the effective solute–solute interaction. In the case of microemulsions for example, the variation of the adhesiveness with the surfactant or co-surfactant chain length is explained by invoking a complex mechanism in which the precise value of the volume per CH_2 group at the interface plays a crucial role [4]. One may then think that a simpler picture might arise when the suspension is modeled as a truly multicomponent mixture. By doing so, one may indeed examine the role of solvent specificities such as the size, solvent–solvent and solvent–solute attractions, besides that of the direct solute–solute interaction. This approach has been used for instance in refs. [7–9] in combination with the Percus Yevick approximation (PYA) for computing the distribution functions of the mixture. In our previous works [8–9] the solute–solute effective adhesiveness was investigated in the limit of vanishing solvent to solute size ratio. Some general trends were pointed out and the crucial role of hetero-adhesions emphasized. Calculations at non zero size ratio [10] outlined the subtle balance between solvent–solvent and solute–solvent adhesions which might explain the temperature dependence of the structure factor for silica particles dispersed in benzene [1]. However, we did not make the connection between the parameters of the model and the physico–chemical parameters which might be responsible for the observed behaviors. In the present work, some steps in this direction are made. The water/dodecane microemulsions investigated in ref. [3] are considered as an illustration. An attempt will be made to correlate the variation of the micelles self-stickiness parameter with the composition of their surface (surface layer made of sodium dodecyl sulfate (SDS) and alcohol). The effective adhesion being investigated from the structure factor deduced from light scattering measurements, a brief summary of the formulae needed for computing the theoretical structure factors is thus made in the second section. The relation between self-stickiness and surface composition is discussed in the final section.

Adhesive spheres mixture in the Percus Yevick approximation

The equations for computing in the PYA the structure factors for an N-component mixture of adhesive spheres are well known in the literature [11]. The main steps are briefly recalled here. Let D_k, ρ_k, and $\eta_k = \frac{\pi}{6}\rho_k D_k^3$ be respectively the diameter, number density and packing fraction of the kth species. The Baxter sticky potentials $u_{mn}(r)$ between species m and n are defined by: $e^{-u_{mn}(r)/kT} = (D_{mn}/12\tau_{mn})\delta(r - D_{mn-}) + \Theta(r - D_{mn-})$, where $\delta(r - D_-)$ is the asymmetrical Dirac distribution, $\Theta(r - D_-)$ the unit step function, τ_{mn}^{-1} the inverse stickiness parameter and $D_{mn} = (D_m + D_n)/2$. In the PYA, the partial direct correlation functions can be written as: $c_{mn}(r) = \Theta(D_{mn-} - r)c_{mn}^R(r) + (\lambda_{mn}D_m D_n/12D_{mn})\delta(r - D_{mn-})$, where $c_{mn}^R(r)$ is a regular function. This defines the adhesiveness factors λ_{mn}. The partial structure factors S_{mn} are computed from the inverse partial structure factors matrix $\mathbf{S}^{-1} = \mathbf{Q}^\dagger\mathbf{Q}$ where $\mathbf{Q}^\dagger$ is the transposed and complex conjugated of the matrix $\mathbf{Q}$ whose elements $q_{mn}(q)$ are the Fourier transforms of the Baxter's functions. Their expressions in the PYA are given for instance in [9, 12]. For computing $q_{mn}(q)$, the factors λ_{mn} are obtained from D_{mn}, τ_{mn} and η_m by solving the system of $N(N + 1)/2$ coupled quadratic equations [11]:

$$\lambda_{mn}\frac{\tau_{mn}}{D_{mn}} = x\frac{D_{mn}}{D_m D_n} + \sum_{k=1}^{N}\left(\frac{\eta_k}{2D_k}\right)$$
$$\times\left[3x^2 - x(\lambda_{mk} + \lambda_{kn}) + \frac{\lambda_{mk}\lambda_{kn}}{6}\right], \quad (1)$$

where $x = (1 - \sum_{k=1}^{N}\eta_k)^{-1}$. For a solute–solvent mixture ($N = 2$; $1 \equiv$ solvent, $2 \equiv$ solute), λ_{11}, λ_{12} and λ_{22} are obtained from the input parameters τ_{11}, τ_{12}, τ_{22}, and the diameter ratio $y = D_1/D_2$ by solving Eq. (1). For the forthcoming discussion, we need the zero wave vector limit of the solutes structure factor, $S_{22}(0)$, computed from the following expressions:

$$S_{22}(0) = \frac{|q_{11}(0)|^2 + |q_{21}(0)|^2}{|\det \mathbf{Q}(0)|^2}$$

$$q_{11}(0) = 1 + \eta_1[x + (1 + x\eta_1)\Lambda_{11} + xy\eta_2\Lambda_{12}]$$

$$q_{21}(0) = \sqrt{y\eta_1\eta_2}[xy(1 + \eta_2\Lambda_{22}) + (1 + x\eta_1)\Lambda_{12}]$$

$$\det \mathbf{Q}(0) = x\{1 + \eta_1\Lambda_{11} + \eta_2\Lambda_{22} + \eta_1\eta_2 \det\Lambda\}, \quad (2)$$

where $\Lambda_{ij} = 3x - \lambda_{ij}$ and det means determinant. Particular cases of Eqs. (2) were also considered in ref. [7]. We note here that for a given set of τ_{ij}, the packing fractions η_1 and η_2 for which the suspension can be homogeneous must obey the condition $\det \mathbf{Q}(0) > 0$.

Change in self stickiness and surface composition of the reverse micelles

The following discussion concerns some of the microemulsions investigated by Brunetti et al. [3]. The selected systems are reverse micelles of water in dodecane with a surface layer made of either SDS or methyl-SDS and an alcohol of variable chain length. Their parameters are given in Table 1. These micelles were choosen because only their surface composition changes (the size ratio in particular is nearly constant).

In order to establish a comparison between theory and experiment, we must specify the solvent and solute hard sphere diameters, the stickiness parameters τ_{ij} and take into account that experiments are performed at constant pressure and temperature. Following ref. [4], the effective hard sphere diameter of the micelles was taken as $D_2^{HS} = D_2^{ap} - 21\Delta n_c$, where Δn_c is the difference in the number of

Table 1 Micellar Parameters*, R^{ap}: apparent radius in nm. B'_2: twice the osmotic second virial coefficient in units of the apparent micellar volume. Δn_c: difference in number of CH groups between surfactant and alcohol. ρ_b: solvent average bulk number density in nm^{-3}. D_2^{HS}/D_1^{HS}: solute/solvent hard sphere diameter ratio. Surfactant: sodium dodecylsulfate (SDS) (C_6^2, C_7^a) and methyl-SDS (C_7^b) Alcohols: hexanol (C_6^2) and heptanol (C_7^a, C_7^b) *R^{ap} and B'_2: C_6^2, C_7^a: from ref. [3]; C_7^b: from Roux D (1984) Thesis, Université de Bordeaux I

	C_6^2	C_7^a	C_7^b
R^{ap}	6.3 ± 0.5	6.1 ± 0.2	6.0
B'_2	−3.8 ± 1	+3 ± 2	−4
Δn_c	6	5	6
ρ_b	2.919	2.864	2.793
D_2^{HS}/D_1^{HS}	16.67	16.34	15.54

188
S. Amokrane et al.
A model of water in oil microemulsions

Table 2 Stickiness parameters of the micelles C_7^a and C_7^b vs stickiness parameter of C_6^2. τ_{22}^{eq}: stickiness determined from the overlap potential; τ_{22}^f: stickiness determined by fitting $S_{22}(0)$. τ_{11}: solvent stickiness parameter. Values in parentheses correspond to the second definition of D_2^{HS}

$\tau_{22}(C_6^2)$	C_7^a			C_7^b		
	τ_{22}^{eq}	$\tau_{22}^f\ (\tau_{11}=1.)$	$\tau_{22}^f\ (\tau_{11}=0.09)$	τ_{22}^{eq}	$\tau_{22}^f\ (\tau_{11}=1.)$	$\tau_{22}^f\ (\tau_{11}=0.09)$
0.9	1.83	2.3	10.	0.91	0.85	0.81
0.5	1.11	1.18	2.8	0.51	0.48	0.44
0.4	0.92 (1.08)	0.97 (0.84)	2.25	0.41	0.38	0.36
0.25	0.63 (0.68)	0.57 (0.52)	1.	0.26	0.25	0.22

CH groups (of length $l = 0.126$ nm) between surfactant and alcohol and $D_2^{ap} = 2R^{ap}$ the apparent diameter (see however the discussion of Table 2). Experimental volume fractions ϕ_2 are then converted into hard sphere packing fractions as $\eta_2 = \phi_2 (D_2^{HS}/D_2^{ap})^3$. The solvent molecules were represented as effective spheres of diameter $D_1^{HS} = (6\eta_1^b/\pi\rho_1^b)^{1/3}$. A value $\eta_1^b \approx 0.45$ appropriate for liquids at normal pressure and temperature conditions was taken for the bulk solvent packing fraction. Since the bulk contains traces of alcohol, the mean bulk density ρ_1^b was taken as the average of the respective densities for the pure substances. In the mixture, the actual solvent packing fraction η_1 differs from η_1^b. For instance, in constant pressure experiments, η_1 and η_2 are not independent variables. From the theoretical point of view, imposing a strictly constant pressure is numerically difficult. It can however be shown [13] that when the diameter ratio is vanishingly small, the pressure computed in the PYA from the compressibility equation is dominated by the fluid of small particles, as for binary mixtures of hard spheres [14]. Keeping thus the packing fraction $\eta_1^* = \eta_1(1 - \eta_2)^{-1}$ of the small particles relative to the free volume left by the large ones is roughly equivalent to keeping the pressure constant. For diameter ratios $y \leq 0.1$ and for significant solvent stickiness (τ_{11} less than about 0.1) we have checked numerically that the pressure computed with η_1^* fixed does indeed not vary too much in the experimental volume fraction range [13]. We thus took $\eta_1^* = \eta_1^b \approx 0.45$, that is $\eta_1 \approx 0.45(1 - \eta_2)$.

Among the stickiness parameters τ_{ij}, the solvent stickiness parameter τ_{11} can in principle be estimated as usual by fitting the experimental second virial coefficient or – for instance – the compressibility of an equivalent square well fluid [13]. But since the range estimated in this way is fairly large and τ_{12} and τ_{22} are still unknown, we preferred adopting a different strategy in this work. By selecting micelles with different surface composition in the same solvent (i.e., same τ_{11} and possibly same τ_{12}) one would expect the *variation* of the effective interaction to reflect mostly the variation of the direct adhesiveness characterized in the model by τ_{22}. If this were true, a simple correlation between the variation of τ_{22} and the surface composition could be found.

The procedure is as follows. For a series of values of τ_{11} and τ_{22}, the experimental structure factors for the micelle C_6^2 (see Table 1) are fitted with the aid of τ_{12}. By keeping τ_{11} and τ_{12} fixed, we next determine the new set of values of τ_{22} which fit the data for the micelles C_7^a and C_7^b (actually, since there is a slight change in the diameter ratio, the parameter kept constant is $t_{12} = \tau_{12}D_1/D_{12}$, which is a more intrinsic property of the hetero-adhesion than τ_{12} [9]). A typical fit for the micelles C_6^2 and C_7^a is shown in Fig. 1 (the "experimental" values of $S_{22}(0)$ shown were computed from Vrij's formula which reproduces very

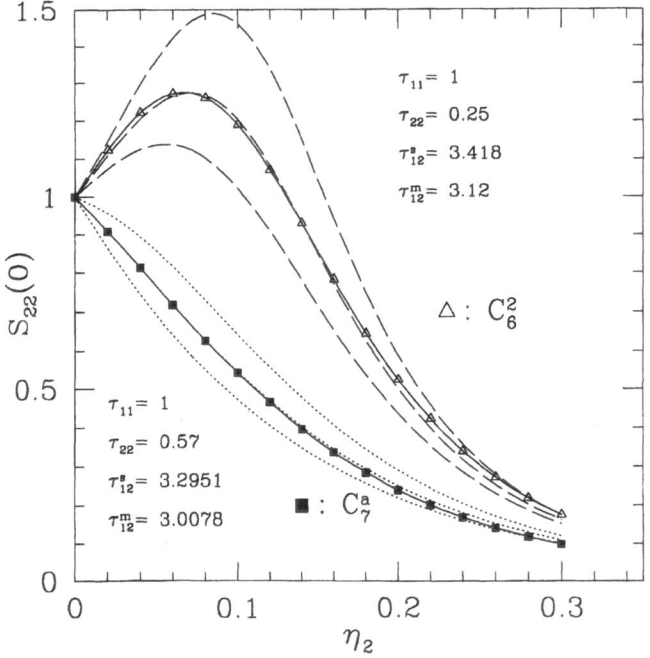

Fig. 1 Theoretical and experimental solute structure factor at zero wave number $S_{22}(0)$ versus hard sphere packing fraction η_2 for the micelles C_6^2 and C_7^a. Theoretical $S_{22}(0)$ computed from Eqs (1) and (2): full curves. Experimental $S_{22}(0)$ from Vrij's formula, with account of experimental error in the osmotic virial coefficient (Table 1). Dashed curves: C_6^2. Dotted curve: C_7^a

$\tau_{11} = 1$
$\tau_{22} = 0.25$
$\tau_{12}^* = 3.418$
$\tau_{12}^m = 3.12$

$\triangle : C_6^2$

$\tau_{11} = 1$
$\tau_{22} = 0.57$
$\tau_{12}^* = 3.2951$
$\tau_{12}^m = 3.0078$

$\blacksquare : C_7^a$

Progr Colloid Polym Sci (1996) 100:186–190
© Steinkopff Verlag 1996

well the original data [3]). The feature to be noticed is that good fits of $S_{22}(0)$, especially at high η_2 could not be obtained unless t_{12} is allowed to vary with η_2. The curves shown were obtained by assuming a linear variation between the values $\tau_{12}^s (\eta_2 = 10^{-5})$ and $\tau_{12}^m (\eta_2 = 0.3)$ used as the fitting parameters. For the moment it is not possible to determine whether this is an artifact of the model or this corresponds to some real physical effect. Given the adopted strategy, this is however of no consequence since t_{12} is unchanged from one micelle to the other. It is also stressed that fits of the same quality can be obtained for a series of values of τ_{11} and τ_{22}, so that these parameters cannot be determined from the fits. A series of values of τ_{11} and τ_{22} leading to a good fit are shown in Table 2. Two values of τ_{11} were selected for simulating a low and high solvent stickiness. For the reference micelle C_6^2, values of τ_{22} between 0.25 and 0.9 are shown (these values should not be taken literally). Below the lower limit, the quality of the fits deteriorates significantly and above the upper limit, τ_{22} may also reflect the defects of the PYA (see below). For the micelles C_7^a and C_7^b, τ_{22}^f is the value which fits $S_{22}(0)$ and τ_{22}^{eq} is the stickiness of a sticky potential giving the same second virial coefficient B_2 (not to be confused with the osmotic virial coefficient B_2') than a model intermicellar interaction potential $U_{22}(r)$. This potential consists of a hard sphere repulsion for a center to center micellar separation $r \leq D_2^{HS}$ and an attractive part assumed to be proportional to the overlap volume, $U_{22}^{at}(r) = \alpha V^0(r)$, where the overlap volume is $V^0(r) = \frac{\pi}{6}(D_2^{ap} - r)^2 (D_2^{ap} + \frac{r}{2})$ for $D_2^{HS} \leq r \leq D^{ap}$, and 0 otherwise. In this picture, $U_{22}^{at}(r)$ is assumed to arise from the Van der Waals interaction between CH groups belonging to two micelles when their surface layers interpenetrate and α is a proportionality constant related to the average of this interaction in the overlap volume and in the presence of the solvent. If only the chain length of surfactant and/or co-surfactant is changed, α should be constant when the same solvent is considered. α was thus determined by imposing the stickiness τ_{22} for the micelle C_6^2 to give the same B_2 than $U_{22}(r)$, the overlap volume being computed with the parameters D_2^{HS} and D^{ap} for C_6^2. This value of α was then used to compute τ_{22}^{eq} from the overlap volumes for C_7^a and C_7^b. Table 2 shows then a good correlation between τ_{22}^f and τ_{22}^{eq}, especially at low τ_{22}^f and high τ_{11}, that is when the effective micellar adhesion is determined by the direct adhesion and not by that induced by the solvent. One

important thing to notice is that the marked difference in self-stickiness between C_6^2 or C_7^b and C_7^a results from a change of one unit in Δn_c. Though the associated change in the interaction range is moderate, it affects significantly the magnitude of $U_{22}(r)$. In particular, its contact value being roughly proportional to the square of the range $21\Delta n_c$, it is multiplied by about 1.5 when going from C_7^a to C_6^2. This illustrates well the importance of the interpenetration length and more generally of the overlap volume. It should now be stressed that τ_{22} may also partly reflect the artificial adhesion which is necessary for forcing the PYA to reproduce the correct effective adhesion associated with the purely steric effect, since it otherwise underestimates it [9]. The equivalence between τ_{22} and an overlap potential is thus correct only if τ_{22} reflects mostly the real adhesion, since there is no obvious reason for a proportionality between the artificial adhesion and the overlap volume. Though both effects are difficult to separate, a simple calculation shows that the lower τ_{22}, the more it reflects the real adhesion [13], as it should be the case when a good correlation between τ_{22}^f and τ_{22}^{eq} exists. Finally, this correlation worsens when the hard sphere diameter is redefined as $D_2^{HS} = D_2^{ap} - l\Delta n_c$ (some values are given in parentheses in Table 2). This definition which corresponds to the end of a surfactant of one micelle and the end of an alcohol of the other one being in contact is less favorable to the direct adhesion since it corresponds to a smaller interaction range and hence smaller overlap volumes. However, the general trends are the same as with the former definition of the hard sphere diameter.

In summary, this study of the relation between the effective adhesion and the composition of the micelles surface confirms the role of the overlap volume. But it also suggests an important contribution of the direct micellar adhesion, that is the one related to the interpenetratation of the surface layers. Of course the interpretation proposed by Brunetti et al. [3] in terms of effective interaction potential is an alternative one. This potential is to a large extent determined by the exact value of the volume per CH_2 group in the aliphatic layer, $V_{CH_2}^L$. A precise understanding of its change from the bulk to the interface is however not so immediate. Although the parameter α introduced above plays a role – in the *direct adhesion* – in some ways similar to that of $V_{CH_2}^L$ in the *effective* one, our approach suggests a simpler physical basis for its interpretation.

References

1. De Kruif CG, Rouw PW, Briels WJ, Duits MHG, Vrij A, May RP (1989) Langmuir 5:422–428

2. Robertus C, Joosten JGH, Levine K (1990) Phys Rev A 42:4820–4831 and (1990) J Chem Phys 93:7293–7300

3. Brunetti S, Roux D, Bellocq AM, Fourche G, Bothorel P (1983) J Phys Chem 87:1028–1034

4. Lemaire B, Bothorel P, Roux D (1983) J Phys Chem 87:1023–1028
5. Baxter RJ (1968) J Chem Phys 49:2770–2774
6. Penders MHGM, Vrij A (1992) Phys Rev A 45:1282
7. Penders MHGM, Vrij A (1991) Physica A 173:532–547
8. Heno Y, Regnaut C (1992) Compte Rendus Académie des Sciences, 315:163–168
9. Regnaut C, Amokrane S, Heno Y (1995) J Chem Phys 102:6230–6240
10. Regnaut C, Amokrane S, Bobola P (1995) Progr Colloid Polym Sci 98:151–154
11. Perram JW, Smith ER (1975) Chem Phys Lett 35:138–140
12. Robertus C, Philipse WH, Joosten JGH, Levine K (1989) J Chem Phys 90:4482–4490
13. Amokrane S, Regnaut C (to be published)
14. Biben T, Hansen JP (1991) Phys Rev Lett 66:2215–2218

Progr Colloid Polym Sci (1996) 100:191–194
© Steinkopff Verlag 1996

MICROEMULSIONS

C. Vázquez-Vázquez
J. Mahía
M.A. López-Quintela
J. Mira
J. Rivas

Preparation of Gd$_2$CuO$_4$ via sol–gel in microemulsions

C. Vázquez-Vázquez · J. Mahía
Prof. M.A. López-Quintela (✉)
Department of Physical Chemistry
University of Santiago de Compostela
15706 Santiago de Compostela, Spain

J. Mira · J. Rivas
Department of Applied Physics
University of Santiago de Compostela
15706 Santiago de Compostela, Spain

Abstract We have synthesized microparticles of Gd$_2$CuO$_4$ via sol–gel technology in microemulsions, carrying out the sol–gel process inside the microdroplets of the microemulsion. The microemulsion was a mixture of cyclohexane/brij 96/ aqueous phase (70:20:10% weight). Urea was used as gelificant agent. The method is based on removing solvents from the mixture applying vacuum in order to obtain a gel when cooling down. By calcinating the gel at low temperatures, we have attained the crystallization of Gd$_2$CuO$_4$ micro-particles. Samples were characterized by x-ray diffraction (XRD), photon correlation spectroscopy (PCS), and dc-magnetization (vibrating sample magnetometer (VSM)).

Key words Sol–gel – microemulsions– microparticles – Gd$_2$CuO$_4$ – magnetic properties

Introduction

The discovery of the new family of electron-doped superconductors R$_{1.85}$M$_{0.15}$CuO$_{4-y}$ (with R = rare earth, and M = Ce, Th) [1, 2] has pointed out the great importance of the undoped R$_2$CuO$_4$ [3–5] because of the possible relation between magnetism and superconductivity. Gd$_2$CuO$_4$, a perovskite-like compound with Nd$_2$CuO$_4$-type structure (T'-phase) [6], does not become superconductor when doped with Ce or Th [7], contrary to the lighter rare earths (R = Pr, Nd, Sm, Eu), and shows a complex magnetic behavior different from the other members of the family.

Until now, all the studies of these compounds have been performed on single crystals and powders obtained by solid state reaction [8]. This technique, which uses metal oxides as starting materials and needs several annealings at high temperatures during long time with frequent intermediary grindings, has however several problems, like, for example, poor homogeneity and high porosity of the samples, no control on the particle size, etc. To avoid these problems, which are common to the

synthesis of other type of high temperature superconductors, several sol–gel techniques have been developed, showing different advantages compared with the conventional ceramic fabrication techniques. For example, with sol–gel techniques [9–12] high purity and good homogeneous materials are achieved, they require lower processing temperatures and shorter annealing times than conventional techniques, they present high reproducibility, and they offer good control of the stoichiometry, size and shape of the particles obtained.

The study of fine and ultrafine particles has prompted increasing interest due to the new properties the materials present when grain size is reduced and, as a consequence, the new applications which can appear. The microemulsion method is a suitable way to synthesize monodisperse and nanosize particles.

A microemulsion is a thermodynamically stable system composed of (at least) three components: two immiscible and a surfactant. Taking advantage of the different solubility of salts between an aqueous and an organic phase, we can use water-in-oil (W/O) microemulsions as reaction

192

C. Vázquez-Vázquez et al.
Preparation of Gd$_2$CuO$_4$ in microemulsion

medium. W/O microemulsions are made up by nanodroplets of water dispersed in a continuous oily phase with the surfactant at the interface [13]. Surfactant molecules control nanodroplets size, and these can be used as microreactors [14–18].

In the present work we describe the synthesis of Gd$_2$CuO$_4$ microparticles via sol–gel technology in microemulsions, which allow to obtain polycrystalline powders at lower temperatures than those used by solid state reaction.

Experimental procedure

In all synthesis procedures carried out in this work the chemicals employed were Aldrich h.p. In order to obtain a polycrystalline reference pattern, Gd$_2$CuO$_4$ was synthesized by solid state reaction. For this purpose, Gd$_2$O$_3$ and CuO were used as starting materials. Gd$_2$O$_3$ was previously dried 3 h at 650 °C. A stoichiometric mixture of these reagents was ground in a ball mill during several hours and then annealed in several steps with intermediary grindings: 12 h at 950 °C, 20 h at 1000 °C and, finally, 24 h at 1080 °C with a slow cooling rate of 2 °C/min.

For the synthesis of the microparticles we started with a microemulsion which contained 70% in weight of cyclohexane (organic phase), 20% of Brij 96, or decaethyleneglycol oleyl ether (surfactant), and 10% of an aqueous solution containing the metallic nitrates and urea, used as gelificant agent ([urea]/[salts] = 10). The sol–gel method with urea is explained elsewhere [12]. This system is not a single phase microemulsion at room temperature, but it becomes a single phase above 40 °C. Solvents were removed applying vacuum at 45 °C yielding to the formation of gels inside the droplets when cooling down. Vacuum periods range from 6 to 48 h. When the microemulsion was cooled down the mixture became a gel below 20 °C. Gels were calcinated at several temperatures (600°, 700° and 800 °C) during various periods of time (3, 15 and 39 h) in order to remove organic compounds and to get the crystallization of Gd$_2$CuO$_4$.

The structural characterization of the polycrystalline powders was carried out by x-ray powder diffraction, using a diffractometer Philips PW-1710 with Cu anode (radiation CuK$_{\alpha 1}$ of $\lambda = 1.54060$ Å). The measurements were performed in air at room temperature. In order to determine the size distribution, measurements of photon correlation spectroscopy (PCS) were carried out. These measurements were performed with an Ar laser Liconix series 5000 of 5 W operating at $\lambda = 488$ nm and a goniometer ALV-SP80 controlled automatically by means of an ALV-LSE unit. For these measurements, particles were dispersed in water. Magnetization of the samples was measured with a vibrating sample magnetometer (VSM).

Results and discussion

By x-ray diffraction the ceramic sample was identified as pure and well crystallized Gd$_2$CuO$_4$. For the sol–gel in microemulsion method it was observed that at low temperatures cubic Gd$_2$O$_3$ and CuO were the predominant phases but longer calcination periods, or higher temperatures, resulted in more proportion of Gd$_2$CuO$_4$. In Table 1, we summarize the peak area ratio between the mean peak of Gd$_2$O$_3$ ($d \approx 3.122$ Å) and the mean peak of Gd$_2$CuO$_4$ ($d \approx 2.770$ Å). The best results at 600 °C were obtained for 6 h under vacuum (area ratio approx 1 : 1). It was observed that, in general, the proportion of Gd$_2$CuO$_4$ increases as the annealing temperature increases and also better crystallizations are obtained annealing for longer periods of time. In Fig. 1 is shown the evolution of the samples with the annealing temperature for 6 h under vacuum and 3 h of thermal treatment. An average crystallite size, D_{hkl}, was determined by applying the Debye–Scherrer's formula using the half maximum width, β, of x-ray diffraction peak at $d \approx 2.770$ Å:

$$D_{hkl}(\overset{\circ}{A}) = \frac{0.9 \cdot \lambda}{\beta \cdot \sin \Theta}$$

where λ is the x-ray wavelength and Θ the Bragg angle. Results are shown in Table 2. We observe, as expected, a tendency to a great increase of crystallite size as the annealing temperature and annealing time increase.

By photon correlation spectroscopy (PCS) we obtained the size of the aggregates. The results show that the average hydrodynamic radius of the aggregates ranges between 250 and 450 nm regardless of annealing temperature and annealing time.

Initial magnetization of the samples was measured at different temperatures. The measurements of magnetization as a function of the applied magnetic field show a linear dependence but this linearity presents some deviations near zero magnetic fields. Extrapolating the linear part to zero magnetization we obtain a magnetic field which has been reported before in monocrystals [3–5] and ceramic samples [7] as an internal magnetic field intrinsic of the sample and related to the canting of the antiferromagnetic order of the copper ions. In Fig. 2 the internal field of the samples of 48 h under vacuum and the ceramic sample are represented as a function of the temperature. The internal fields for the ceramic sample agree well with those reported in the literature: maximum $H_{int} = 815$ Oe. However, for the microemulsion samples this H_{int} strongly diminishes as the annealing temperature decreases, becoming nearly zero for samples annealed at 600 °C. The same phenomenon had been reported before by Chevalier and Mathieu for α-Fe$_2$O$_3$ [19] and was related with a decrease

Progr Colloid Polym Sci (1996) 100:191–194
© Steinkopff Verlag 1996

Table 1 Peak area ratio between the mean peak of cubic Gd_2O_3 ($d \approx 3.122$ Å) and the mean peak of tetragonal Gd_2CuO_4 ($d \approx 2.770$ Å) as a function of evaporation time (in %)

	45 °C 6 h	45 °C 12 h	45 °C 24 h	45 °C 48 h
600 °C 3 h	110	130	201	135
600 °C 15 h	82	100	153	150
600 °C 39 h	98	108	153	164
700 °C 3 h	83	89	56	111
700 °C 15 h	72	82	36.7	100
700 °C 39 h	48	59	21.4	79
800 °C 3 h	34.0	28.6	23.4	50
800 °C 15 h	23.4	25.6	16.4	33.5
800 °C 39 h	17.6	22.6	07.9	16.6

Table 2 Crystallite sizes of Gd_2CuO_4 particles obtained by the Debye–Scherrer's formula using the mean peak of Gd_2CuO_4 (in nm)

	45 °C 6 h	45 °C 12 h	45 °C 24 h	45 °C 48 h
600 °C 3 h	19.2	14.0	18.0	14.4
600 °C 15 h	20.4	16.9	23.2	19.0
600 °C 39 h	25.8	21.4	27.7	21.4
700 °C 3 h	83	69	42.4	78
700 °C 15 h	118	122	42.8	106
700 °C 39 h	142	109	40.5	125
800 °C 3 h	145	125	93	64
800 °C 15 h	295	276	153	120
800 °C 39 h	1380	413	122	176

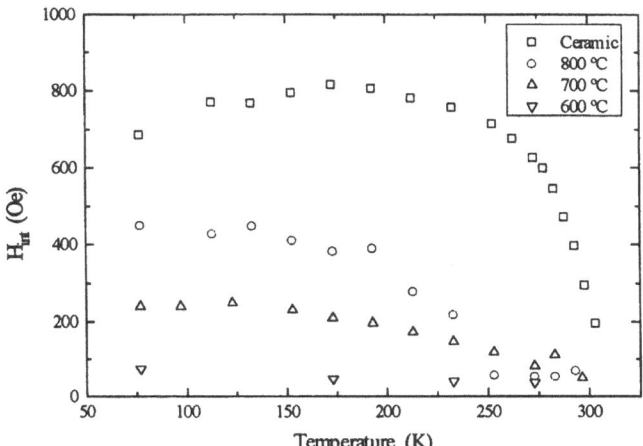

Fig. 2 Internal fields vs. temperature for the samples at 48 h under vacuum at different annealing temperatures

Fig. 1 Phase evolution as a function of the annealing temperature for the samples at 6 h under vacuum and 3 h of thermal treatment

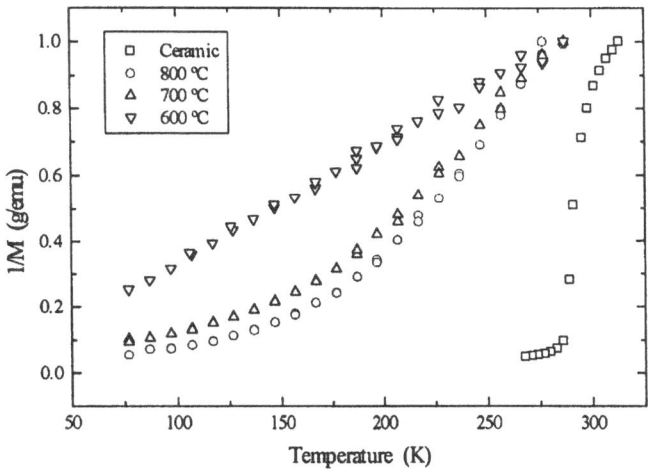

Fig. 3 Normalized $1/M$ vs. temperature for the samples at 48 h under vacuum at different annealing temperatures

of particle size. So, as the annealing temperature decreases, particle size also decreases and this reduction of size modifies the magnetic properties of Gd_2CuO_4.

Magnetization of the samples was measured as a function of the temperature. A plot of $1/M$ vs temperature is shown in Fig. 3. $1/M$ was normalized for all calcination temperatures for best comparison. By extrapolating the linear part of $1/M$ at high temperatures, we have observed a shift of the extrapolated temperature to lower temperatures as the calcination temperature decreases, becoming

194

C. Vázquez-Vázquez et al.
Preparation of Gd$_2$CuO$_4$ in microemulsion

almost zero for 600 °C, 140 K for 700 °C, 170 K for 800 °C and 287 K for the ceramic sample. These results can again be explained by the size reduction of the particles because the size constraints lower the long range interaction between the copper magnetic moments, modifying the magnetic properties of the material.

Conclusions

We have synthesized Gd$_2$CuO$_4$ microparticles by a new method which consists of a sol–gel process carried out inside the nanodroplets of the microemulsion. This method improves the conventional solid state reaction techniques, lowering considerably the annealing temperature required to obtain well-crystallized materials. Smaller size particles have been obtained than those synthesized by ceramic techniques. Great differences have been observed on the magnetic properties of particles prepared by those techniques.

Acknowledgments The authors wish to acknowledge financial support from DGICYT, PB92–1086; Fundación Ramón Areces; NSF-DMR-91172122 and NATO, CRG920255. C.V. also thanks Fundación Segundo Gil-Davila for its financial support.

References

1. Tokura Y, Takagi H, Uchida S (1989) Nature 337:345
2. Markert JT, Maple MB (1989) Solid State Commun 70:145
3. Thompson JD, Cheong SW, Brown SE, Fisk Z, Oseroff SB, Tovar M, Vier DC, Schultz S (1989) Phys Rev B 39(10): 6660–6666
4. Oseroff SB, Rao D, Wright F, Tovar M, Vier DC, Schultz S (1989) Solid State Commun 70(12):1159–1163
5. Oseroff SB, Rao D, Wright F, Vier DC, Schultz S, Thompson JD, Fisk Z, Cheong SW, Hundley MF, Tovar M (1990) Phys Rev B 41(4):1934–1948
6. Kubat-Martin KA, Fisk Z, Ryan RR (1988) Acta Cryst C 44:1518–1520
7. Butera A, Caneiro A, Causa MT, Steren LB, Zysler R, Tovar M, Oseroff SB (1989) Physica C160:341–346

8. Yan MF, Ling HC, O'Bryan HM, Gallagher PK, Rhodes WW (1988) IEEE Transactions on Components, Hybrids, and Manufacturing Technology 11(4): 401
9. Kordas G (1992) J Non-Cryst Solids 121:436–442
10. James PF (1991) en The 2nd European Conference on Advanced Materials and Processes. EUROMAT'91, 350–351. University of Cambridge, UK
11. Sakka S (1989) Mat Res Soc 221–232
12. Mahía J, Vázquez-Vázquez C, Basadre-Pampín MI, Mira J, Rivas J, López-Quintela MA (1996) J Amer Ceramic Soc 79:407–411
13. Degiorgio V, Corti M (Eds) (1985) in Physics of Amphiphiles: Micelles, Vesicles and Microemulsions, North Holland, Amsterdam

14. Kurihara K, Kizling J, Stenius P, Fendler JH (1983) J Amer Chem Soc 105: 2574
15. Kandori K, Kon-no K, Kitahara A (1984) Bull Chem Soc Jpn 57:3419
16. Fendler JH (1987) Chem Rev 87:877
17. Lianos P, Thomas JK (1986) Chem Phys Lett 125:299
18. López-Quintela MA, Rivas J (1993) J Colloid Interface Sci 158:446
19. Chevalier RM, Mathieu S (1943) Ann Phys (Paris) 18:258

Progr Colloid Polym Sci (1996) 100:195–200
© Steinkopff Verlag 1996

MICROEMULSIONS

S.M. Andrade
S.M.B. Costa

Fluorescence studies of the drug *Piroxicam* in reverse micelles of AOT and microemulsions of Triton X-100

S.M. Andrade · Dr. S.M.B. Costa (✉)
Centro de Química Estrutural
Complexo I
Instituto Superior Técnico
1096 Lisboa Codex, Portugal

Abstract The microenvironment of the polar core of reverse micelles of the anionic surfactant AOT in isooctane and the microemulsions of the non ionic surfactant Triton X-100 in cyclohexane/hexanol were studied by electronic absorption and steady-state fluorescence spectroscopy incorporating an antiinflamatory nonsteroidal drug *Piroxicam*. Two acid-base equilibria of this probe in water ($pK_a = 3.1$ and $pK_a = 4.8$) are strongly and differently affected by the nature of the two interfaces and inner pools of each microheterogeneous system.

In AOT, fluorescence quantum yields vary up to $w_0 = 10$ reflecting changes in the microviscosity sensed by the probe and intramicellar pH gradients, reaching a nearly constant value up to $w_0 = 40$. In Triton X-100 the interface is polar and protic and interacts with the probe as a Lewis base. Three different spectroscopic species were detected: one attached to the interface at $w_0 = 0$, the second one with structured water up to $w_0 = 8$ and a third one in free water at large w_0 values ($w_0 = 16$), with increasing microviscosity values.

Fluorescence anisotropy studies with other probes in the same micellar systems are in good agreement with the patterns of microviscosity found in both systems.

Key words Reverse micelles – microemulsions – fluorescence – *Piroxicam*

Introduction

Microheterogeneous systems such as micelles and microemulsions have been considered as model systems which provide a suitable microvicinity for inclusion of macromolecules such as proteins, polynucleotides and drugs. Unique interfacial properties confer a great potential interest to micellar aggregates in various fields, for example, paints, pharmaceutical drugs and enhanced oil recovery [1, 2].

Reverse micelles and microemulsions formed in organic solvents incorporate a pool of water which has properties clearly distinct from those of bulk water [3, 4]. Various techniques have been applied in the study of reverse micelles, e.g., light-scattering and N.M.R. which make use of intrinsic properties of the system. On the other hand, there are methods which are based on the fluorescence of a probe molecule incorporated in the system. The latter enable the correlation of photophysical data with some parameters associated with the structure of the aggregate and therefore have been used extensively [5].

Several drugs are interesting in this context. Piroxicam, *Prx*, (4-hydroxy-2-methyl-N-pyridil-2H-1,2 benzothiazine-3-carboxamide-1,1 dioxide) (Fig. 1) is a non steroidal antiinflamatory drug which induces photosensitive action. Photophysical studies of this molecule exhibit excited state intramolecular proton transfer (ESIPT) in aprotic solvents [6]. This effect is perturbed in protic solvents through the formation of an anion and/or an open conformer in the ground state.

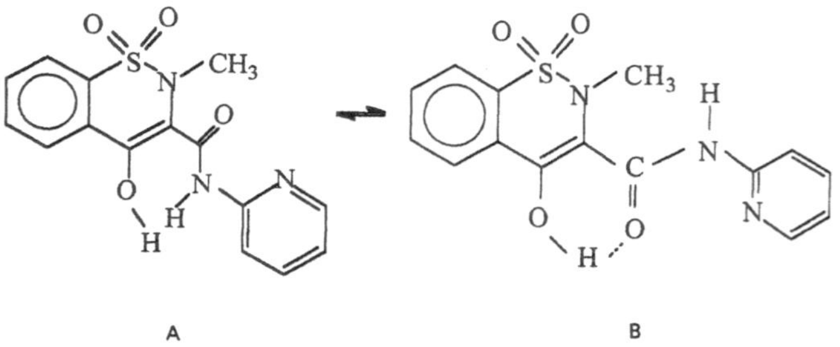

Fig. 1 Equilibrium of two isomers of Piroxicam (*Prx*): **A** open conformer; **B** closed conformer

In this paper, the absorption and emission properties of *Prx* incorporated in an anionic reverse micelle, sodium 1,4-bis(2-ethylhexyl)sulfosuccinate, (AOT) and a nonionic microemulsion of (iso-octylphenoxy-poly(oxyethylene) glycol, Triton X-100, are compared and applied to extract information on the micropolarity and microviscosity of the respective interfaces and inner pools.

Experimental

Reagents

Prx was a generous gift from "Laboratórios Medinfar" and Acridine Orange was obtained from Sigma. Triton X-100 (Trx) was purchased from Riedel-de-Haan and used as supplied, AOT was from Sigma and used without purification as well. All solvents used were spectroscopic grade, except for 1-hexanol. Buffer solutions were made up using recommended procedures [7].

Apparatus

Absorption spectra were recorded with a Jasco V-560 UV/VIS spectrophotometer. Steady-state emission spectra and fluorescence anisotropy measurements were recorded with a Perkin–Elmer LS 50 B spectrofluorimeter. All data were stored in a computer. The instrumental response at each wavelength was corrected by means of a curve obtained using the appropriate fluorescence standards (until 400 nm) together with the one provided with the apparatus. The fluorescence quantum yields of *Prx* were measured using quinine sulphate (QS) in $1\,N\,H_2SO_4$ solution ($\phi_f \approx 0.546$) as standard. The integration of the corrected spectra was made over the emission wavelength range and corrections for changes in the respective refractive index were carried out [8].

Sample preparation

Microemulsions of Trx were prepared by adding 1-hexanol in cyclohexane, in order to have a solution of 0.2 M in Trx, 3:2 (w/v) ratio of Trx/Hexanol and $w_0 = [H_2O]/[Surf] = 0$. Buffered water was then added to vary w_0. The incorporation of *Prx* was achieved by adding a small amount of *Prx* in acetone, bubbling a stream of N_2 to remove the acetone, the final concentration of *Prx* being $5 \times 10^{-6}\,M$. AOT micelles were prepared by dissolving it in isooctane and then adding the buffer with continuous shaking. The buffer concentrations used were 50 mM and 25 mM respectively in Trx and AOT.

Results and discussion

Prx in aqueous solutions

The absorption of *Prx* in aqueous solutions is strongly pH dependent, reflecting the equilibrium of two forms, neutral ($\lambda_{abs}^{max} = 335$ nm) and anionic ($\lambda_{abs}^{max} = 360$ nm) with an isobestic point at $\lambda = 348$ nm. The emission of the neutral form is much stronger than the anionic one. On the basis of the variation of the absorption spectra with the pH, a value of $pK_a = 3.1$, for the formation of the monoanion was obtained in good agreement with $pK_a = 3.3$ previously reported [6]. At higher pHs another species is formed, possibly a di-anion which is also in equilibrium with the neutral form, the $pK_a \approx 4.8$.

The photophysical parameters obtained for the neutral form are nearly identical to those found for the closed conformer in aprotic solvents [6]. However, an open conformer would be more likely to exist in aqueous media due to hydrogen bonding solute-solvent.

Prx in AOT reverse micelles

The absorption and emission of *Prx* in reverse micelles of AOT in isooctane were studied as a function of the water content (w_0) and at different pHs (Fig. 2). In the absence of water, $w_0 = 0$, the absorption spectra is similar to that obtained in the apolar solvent, showing that no association exists between this probe and the surfactant. The increase of w_0 leads to a batochromic shift at a given pH = 3.8 with the same isobestic point at $\lambda = 348$ nm showing the same equilibrium neutral form $\rightleftarrows$ monoanion with the neutral form being dominant at smaller w_0 values. The same equilibrium persists at pH = 6.8, whereas for higher pHs (pH = 10.2) even at $w_0 = 2.5$ only the di-anion form exists with the maximum shifted to

Fig. 2A Absorption spectra and **B** emission spectra of *Prx* in reverse micelles of AOT, pH = 3.8 at several w_0 values. $1 - w_0 = 0$, $2 - w_0 = 2.5$, $3 - w_0 = 5$, $4 - w_0 = 10$, $5 - w_0 = 20$, $6 - w_0 = 40$

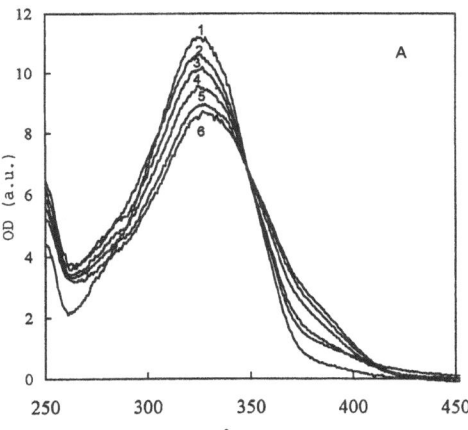
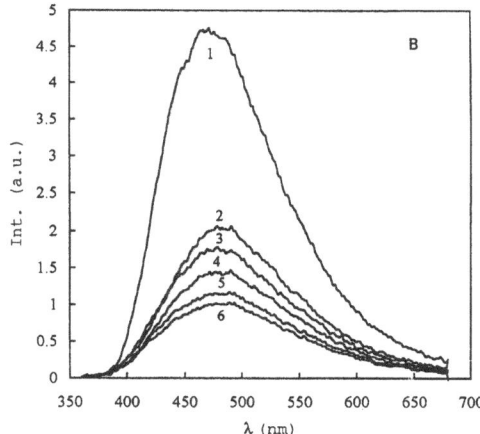

$\lambda = 356$ nm (identical to that observed in water at the same pH).

The explanation for this behavior may lie in the fact that the intramicellar pH is rather different from the one measured in the starting aqueous solution. The negative head groups in AOT induce a large intramicellar gradient of pH where the interface would be more acidic and the aqueous pool more alkaline.

The existence of a well defined isobestic point at acidic and neutral pHs supports the presence of two spectroscopically distinct species corresponding to the equilibrium defined by

$$Prx_i + bH_2O \rightleftarrows Prx_w + bH_2O_w \,, \tag{1}$$

where Prx_i and Prx_w represent the drug associated to the interface or free in the aqueous pool respectively and b is the number of water molecules needed to displace the *Prx* from the interface as proposed earlier [9]. The equilibrium constant will be given by

$$K_p = \frac{[Prx]_w}{[Prx]_i [H_2O]^b} \,. \tag{2}$$

However, *Prx* is also soluble in isooctane and therefore it is necessary to consider the partition between the organic solvent and the interface defined by $K = [Prx]_i/[Prx]_s$. Hence the equation which will account for the quantum yield is given by [10]

$$\phi_f = \frac{\phi_{f_s} + \phi_{f_i} \dfrac{a_i}{V_i} K + \phi_{f_w} \dfrac{V_w}{V_i} K_p [H_2O]^b}{1 + \dfrac{a_i}{V_i} K + \dfrac{V_w}{V_i} K_p [H_2O]^b} \,, \tag{3}$$

where a_i, $V_i e V_w$ represent the area and volumes of the interface and the aqueous pool respectively. The values

obtained by fitting are $K = 6.7 \times 10^{-8}$, $K_p = 75$ M^{-1}, $b = 2$, $\phi_{f_i} = 1.6 \times 10^{-3}$, $\phi_{f_w} = 7.3 \times 10^{-4}$, showing that in fact K is negligible and *Prx* solubilizes preferentially in water. The quantum yield at the interface is higher than in the water pool and similar to that obtained in isooctane suggesting that the probe localization may be near the hydrophobic moieties filled with the organic solvent.

Prx in microemulsions of Trx-100

The absorption and emission spectra of *Prx* in microemulsions of Trx-100 at pH = 3.8 are given in Fig. 3. The existence of the same isobestic point at $\lambda = 348$ nm supports the equilibrium between the form with a maximum at $\lambda = 365$ nm ($w_0 = 0$) which shifts to the blue $\lambda = 335$ nm at $w_0 = 16$ with the increase of the water content. This effect is exactly the opposite of that found in AOT reverse micelles and is very likely triggered by the Trx-100 acting as a Lewis base [11] inducing the formation of the anionic form, Prx^-

$$Trx-OH + PrxH \rightleftarrows Trx-OH_2^+ + Prx^-$$

The interface effect is very important leading at $w_0 = 0$ to the same spectroscopic species as in pure Trx-100. At higher w_0, there is a competition between the *PrxH* and H_2O to hydrate the head groups of polyoxyethylene in Trx-100 and therefore the anionic form Prx^- is converted into *PrxH*.

The absorption spectra presented in Fig. 3 can be reproduced by fitting parameters given in Eq. (4)

$$OD_t = [Prx]_t \frac{\varepsilon_i + \varepsilon_e K + \varepsilon_w K_p [H_2O]^b}{1 + K + K_p [H_2O]^b} \,, \tag{4}$$

Fig. 3A Absorption spectra and **B** emission spectra of *Prx* in microemulsions of Trx-100, pH = 3.8 at several w_0 values. $1 - w_0 = 0$, $2 - w_0 = 4$, $3 - w_0 = 8$, $4 - w_0 = 10$, $5 - w_0 = 12$, $6 - w_0 = 14$

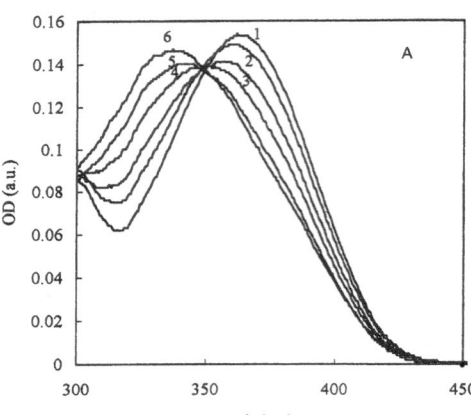 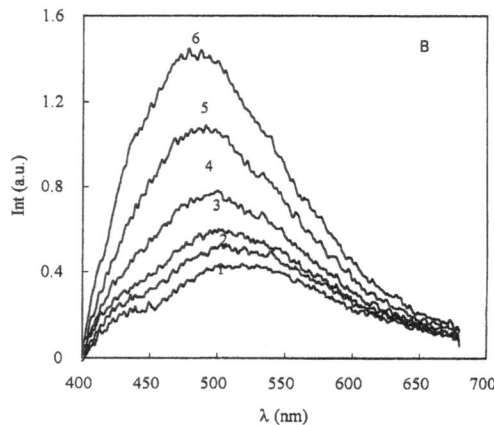

where ε_i, ε_e, $e\ \varepsilon_w$ represent the molar absortivities in the interface, structured water and free water respectively and K, K_p and b have the same meaning as stated in the preceding section.

Using the values obtained it was possible to carry out a decomposition of the absorption spectra in three species: one attached to the interface (given by the spectra at $w_0 = 0$), the second one with structured water (shifted to the blue) and a third one in free water (similar to the one obtained in aqueous solution at low pH).

The structure of this aggregate changes above $w_0 = 8$, as is shown from the spectra. At pH = 3.8 the nature of the intramicellar water is likely to be quite different in the non ionic system inducing a rather viscous and low polarity microvicinity which may favor the closed conformer.

Microviscosity in AOT and Trx-100

The photophysical study of *Prx* in various solvents leads to a correlation of the quantum yield with the viscosity, following the concept of free volume [12], reflected in relation

$$\phi_f = c\eta^{2/3} . \qquad (5)$$

Indeed, a linear correlation

$$\mathrm{Ln}\ \phi_f = (-8.6 \pm 0.3) + (0.65 + 0.05)\ \mathrm{Ln}\ \eta \qquad (6)$$

is found for a wide variety of (11) protic and nonprotic solvents, showing that the molecular relaxation of the first excited singlet is dependent on the free volume.

From the linear correlation obtained it is then possible to derive several "microviscosity" values sensed by the probe molecule at each w_0 for both systems studied. In the

reverse micelle of AOT the microviscosity decreases as the water content is increased, since at the interface the water is strongly bounded and rather structured. Therefore, one may say that the microviscosity trend follows the structural alterations of the system until $w_0 = 10$ and reaches a limiting value at $w_0 = 40$ higher than the viscosity in water.

By contrast, in the microemulsions of Trx-100, the "microviscosity" changes as the probe progresses from the interface to the pool, increasing rapidly above $w_0 = 8$. This result is in good agreement with literature reports [13] of direct viscosity measurements in this microemulsion. The authors propose that up to $w_0 = 15$ there is an isotropic region of Newtonian fluid which becomes anisotropic and non-Newtonian for large water quantities.

A confirmation of these patterns of "microviscosities" found in these systems was sought by studying the fluorescence anisotropy of another probe Acridine Orange (AO) in AOT and Trx-100, since its photophysics in these systems was previously studied [14].

In Fig. 4a) are plotted the microviscosities found for AO using the following equation [15]

$$\frac{1}{\bar{r}} = \frac{1}{r_0}\left(1 + \frac{RT}{V_h \eta_s}\tau_f\right)\left(1 + \frac{3\tau_f}{\rho_i}\right), \qquad (7)$$

where $\bar{r}$ is the average ratio of fluorescence anisotropy, r_0 is the limiting anisotropy observable in the absence of any molecular rotation, τ_f is the fluorescence lifetime, V_h is the micellar hydrodynamic volume, η_s is the solvent viscosity and $\rho_i = 3V_p\eta_w/RT$ is the relaxation time which relates with the probe hydrodynamic volume and the microviscosity of the pool.

In Fig. 4A are also plotted the data obtained for other fluorescence probes [16], Rhodamine 6G (Rod 6G), Rhodamine B (Rod B) and Fluorescein (Fluorc) obtained in the

Fig. 4 Variation of pool microviscosities with w_0.
A Reverse micelles of AOT
◆ Rod 6G □ Rod B △ Fluorc (from ref. [16]), -AO (pH = 7)
● *Prx* (pH = 7);
B Microemulsions of Trx-100, *Prx* ◆ pH = 1.5, △ pH = 3.8, ● pH = 10.1; AO-pH = 7

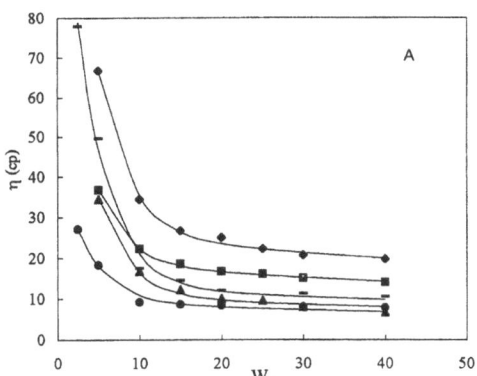
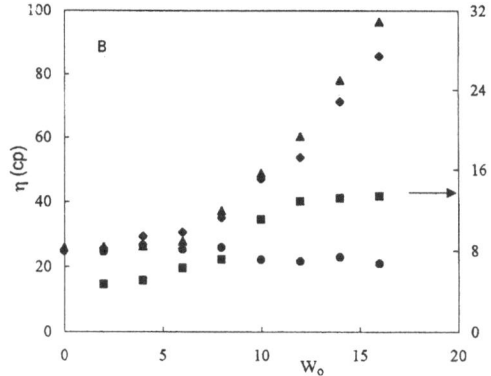

same reverse micelles of AOT, as well as the microviscosities values obtained for *Prx* as described earlier. All curves can be fitted with the same empirical equation:

$$\eta_w = a \cdot \exp(-b \cdot w_0) + c \cdot \exp(-d \cdot w_0) \tag{8}$$

with $b = 0.27$, $d = 0.01$ and the magnitude of the pre-exponential factors a and c reflect the radial probe localization. Cationic dyes such as AO and Rod 6G will locate nearer the negative interface, sensing higher microviscosities, whereas *Prx* with a negative charge will locate closer to the aqueous pool sensing lower viscosities.

Figure 4B shows the opposite trend observed for the same probe (AO) in Trx-100 compared with the "microviscosities" sensed by *Prx* at three different pHs.

Conclusions

The microenvironment of the polar inner core of reverse micelles of AOT and microemulsions of Trx-100 was studied by steady-state fluorescence spectroscopy applied to a drug, Piroxicam, *Prx*. A parallel study in aqueous media shows the existence of two acid-base equilibria with $pK_{a1} = 3.2$ and $pK_{a2} \approx 4.8$.

In reverse micelles of AOT the probe partitions favorably towards the pool sensing the pH of the aqueous inner pool. At acidic and neutral pHs and lower w_0 values

$(w_0 < 10)$, the probe interacts with the interface. At higher w_0 it is located essentially in the bulk water. The microviscosity decreases rapidly until $w_0 = 10$ reaching a limiting value at larger w_0 values. This trend of variation was independently confirmed by fluorescence anisotropy using another fluorescent probe, AO.

In microemulsions of Trx-100 the interface is polar and protic. In the absence of water the probe *Prx* exists essentially as an anion (open form), whereas at larger content of water the interface effect decreases due to the competition of water molecules for the surfactant head groups, leading to the neutral form of Prx, as a closed conformer. Structural alterations in the system occur at $w_0 = 8$ as sensed by drastic alterations in the microviscosity. The same variation was confirmed by fluorescence anisotropy studies using AO.

Using one antiinflamatory fluorescent drug this study shows the possibility of its microencapsulation in the inner aqueous pool of microemulsion systems. Furthermore, the anionic and nonionic type show contrasting characteristics, namely reflecting differences in the micropolarity and microviscosity of these aggregates.

Acknowledgments This work was supported by CQE/4, IST. Andrade S. acknowledges J.N.I.C.T. for a grant Praxis XXI/BD/3649/94. The authors also thank "Laboratórios Medinfar" for the gift of *Prx*.

References

1. Lindman B, Wennerström H (1980) Top Curr Chem 87:1–83
2. Tanford C (1980) The Hydrophobic Effect, John Wiley & Sons, New York
3. Sunamoto J, Hamada T, Seto T, Yamamoto S (1980) Bull Chem Jpn 53:583–589
4. Wong M, Thomas JK, Grätzel M (1976) J Am Chem Soc 98:2391–2397
5. Gehlen MM, De Schryver FC (1993) Chem Rev 93:199–221
6. Vogel AI (1978) Textbook of Quantitative Inorganic Analysis, 4th Edn, Longman, London
7. Parker CA (1968) In: Photoluminescence in Solutions, Elsevier, Amsterdam, 252
8. Kim YE, Cho DW, Kang SG, Yoon M, Kim D (1994) J Lumin 59:209–217
9. Costa SMB, Velázquez MM, Tamai N, Yamazaki I (1991) J Lumin 48:341–351
10. Velazquez MM, Gonzalez-Blanco C, Costa SMB, Laia CT, Medeiros GMM, to be published
11. Andrade SM, Mendonça CM, Costa SMB (1993) Book of Abstracts "1° Encontro de Química-Física da Sociedade Portuguesa de Química", Lisboa pp P19

12. Foster T, Hoffmann GZ (1971) Phys Chemie NF 75:63–76
13. Kumar C, Balasubramanian D (1979) J Colloid Interface Sci 69(2):271–279
14. Andrade SM, Costa SMB (1994) Book of Abstracts "XIV Encontro Nacional da Sociedade Portuguesa de Química", Aveiro, pp A32
15. Weber G (1971) J Chem Phys 55: 2399–2407
16. Hasegawa M, Sugimura T, Suzaki Y, Shindo Y (1994) J Phys Chem 98: 2120–2124

Progr Colloid Polym Sci (1996) 100:201–205
© Steinkopff Verlag 1996

V. Degiorgio
R. Piazza
G. Di Pietro

Depletion interaction and phase separation in mixtures of colloidal particles and nonionic micelles

Prof. Dr. V. Degiorgio (✉) · R. Piazza
G. Di Pietro
Unità INFM di Pavia
Dipartimento di Elettronica
Università di Pavia
via Ferrata 1
27100 Pavia, Italy

Abstract Reversible flocculation is observed in aqueous suspensions of polymer colloids when a sufficient amount of nonionic amphiphile is added. The separation process is driven by depletion forces originated by the presence of the nonionic micelles. The minimum amphiphile concentration needed to induce flocculation, Φ_{Smin}, decreases upon increasing the size of the colloidal particles. We also find that Φ_{Smin} decreases as the temperature of the suspension approaches the cloud point temperature of the water – amphiphile system.

Key words Colloids – nonionic surfactants – phase separation

Introduction

Depletion forces induced among colloidal particles by high-molecular-weight solutes are important in the industrial preparation of latices and are likely to play some role in biological processes involving vesicles and other colloidal-size aggregates within the cell. In recent years the attention has been mainly focused on phase-separation phenomena induced by adding polymers to a colloidal suspension [1]. The segregation process originates from the effective interparticle attraction produced by the exclusion of the polymer coils from the region between closely spaced colloidal particles [2], and yields a concentrated colloidal phase in equilibrium with a very dilute suspension. An equivalent approach relates the phase-separation process to the minimization of the configurational entropy of a binary mixture of large spheres (the colloidal particles) and small spheres (the polymer coils), and allows to sketch a simplified phase diagram of the system [3]. Effects similar to those observed in colloid–polymer mixtures take place also in colloid–amphiphile systems. As suggested by Ma [4], in such systems the role of the "small spheres" is taken by the micellar aggregates formed by the am-

phiphilic molecules. We have recently presented [5] experimental results concerning a phase separation process in a colloidal suspension induced by the addition of a nonionic amphiphile, and we have compared our results with those reported for the colloid–polymer mixture and with theoretical predictions. Particular attention was paid in ref. [5] to the structure of the concentrated phase. It was found that the very large size ratio between the colloids and the micelles is responsible for a dynamic freezing of the segregation process which yields a gel-like separated phase having an internal fractal structure and a particle concentration considerably smaller than that of the colloidal crystal.

Most aqueous solutions of nonionic amphiphiles present in the phase diagram a demixing region, bounded by a lower consolute curve (called cloud curve in the surfactant literature), where the system separates into two phases having different micellar concentration [6]. This process is analogous to the phase separation of a simple binary liquid mixture, with the difference that the large size ratio between micelle and solvent molecule pulls the critical point at quite low amphiphile concentrations and yields a rather flat coexistence curve. Effective attractive interactions between the micelles can then be triggered by raising the temperature and approaching the consolute curve. Indeed, the light-scattering study of solutions of nonionic micelles (without colloidal particles) shows that the osmotic virial coefficient monotonically increases as the temperature is raised, presenting, very far from the critical region, a negative value comparable to that expected for hard-spheres, and becoming positive some tens of degrees below the critical temperature [6]. Our previous investigations [5] concerned a specific colloidal particle, and was limited to room temperature, where the virial coefficient of the used micellar solution is almost vanishing. Such a condition is equivalent to the θ-point for a polymer solution, and corresponds to very weak intermicellar interactions, thus permitting a better comparison of our results with the existing models. In this paper we study the temperature dependence of the separation process for various colloidal particles having different size or chemical composition. We find that the minimum surfactant concentration required to induce the reversible flocculation does not depend on the nature of the particles, but is considerably dependent on both particle size and operating temperature.

Experimental

We have used two kinds of model colloidal particles, fluorinated latex spheres made of MFA, a polytetrafluoroethylene copolymer, and standard polystyrene (PS) latex spheres. The MFA particles, having a radius of 110 nm and low polydispersity [7], were kindly donated by Ausimont, Milano. The same particles were also used in our previous work [5] together with the commercial nonionic amphiphile Triton X-100 (Rohm & Haas). Polystyrene latex particles with radius varying in the range 19–120 nm were used to investigate the dependence of the segregation process on the colloidal particle radius. Most of the measurements were performed with Triton X-100. A set of data was obtained with the nonionic amphiphile octaoxyethylene glycol dodecyl ether ($C_{12}E_8$), a monodisperse preparation obtained from Fluka. The properties of $C_{12}E_8$ solutions have been thoroughly investigated [6], so that they represent a well-suited system for a detailed study of the temperature dependence of the phase-separation effects. Triton X-100 comes from an industrial preparation and is a mixture of amphiphile molecules with different numbers of oxyethylene groups. Although not so well characterized from a chemical point of view, Triton X-100 presents the advantage of being about three orders of magnitude cheaper than pure C_iE_j amphiphiles, and is therefore particularly helpful for preliminary experiments requiring large amounts of material. All the colloidal

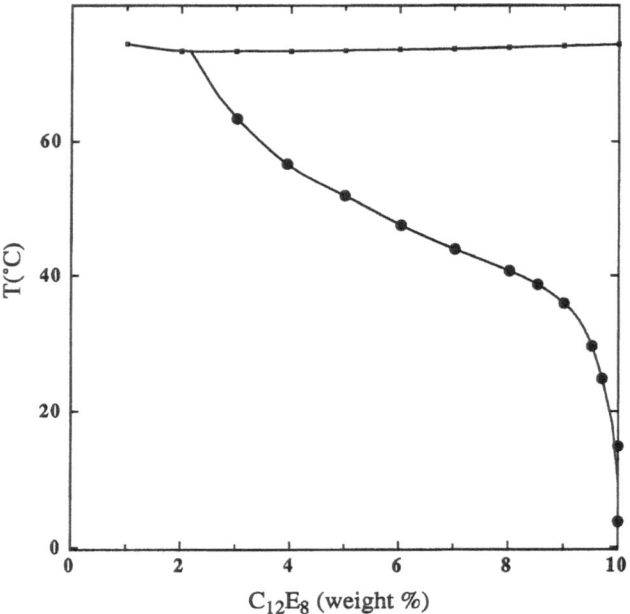

Fig. 1 Reversible flocculation line for a suspension of MFA particles at a volume fraction of 1% with added nonionic amphiphile ($C_{12}E_8$) and 100 mM NaCl as a function of the temperature. The dots represent the observed instability points, the curve connecting the dots is only drawn to guide the eye. The upper curve is the measured cloud point curve

suspensions were prepared in presence of 100 mM NaCl, in order to screen adequately the electrostatic interparticle interactions which arise from the presence of electric charges on the particle surface [7]. We have shown in a previous light-scattering investigation that the nonionic surfactant is adsorbed on the particle surface forming a monolayer which, in the case of Triton X-100, has a thickness of 2 nm [7]. The volume fraction of nonionic surfactant which is needed to fully cover the colloidal particles is given by: $\Phi_{SA} \approx 0.05 \, \Phi_P$, where Φ_P is the volume fraction of colloidal particles. Therefore the volume fraction occupied by the nonionic micelles is given by: $\Phi_{Sm} = \Phi_S - \Phi_{SA} - \Phi_0$, where Φ_S is the total volume fraction of surfactant, and Φ_0 is the volume fraction of surfactant which remains dispersed in monomeric form.

We observed that a homogeneous MFA suspension becomes unstable when a sufficient amount of surfactant is added, and that a dense colloidal phase separates out by gravitational settling in a time scale of a few hours. The dense phase can be completely redispersed by simply shaking the cell, and no sign of irreversible aggregation is observed. We denote by Φ_{Smin} the minimum surfactant volume fraction which induces phase separation. We show in Fig. 1 a set of data obtained by using $C_{12}E_8$ and $\Phi_P = 1\%$. As can be seen, Φ_{Smin} remains approximately

Progr Colloid Polym Sci (1996) 100:201–205
© Steinkopff Verlag 1996

constant at low temperatures, and begins to decrease appreciably above 25 °C. The upper curve in Fig. 1 is the experimentally determined consolute curve for the binary system $C_{12}E_8$–H_2O. The minimum (critical point) is at $T_c = 73.5$ °C and $\Phi_S \approx 3\%$. The limiting stability line for the colloidal suspension approaches the coexistence curve not far from the critical point.

Although we did not perform a detailed study of the composition of the separated particle-rich phase as a function of temperature, we have made some qualitative observations which can be summarized as follows. We find that the segregated phase has a particle volume fraction of about 0.5 only when the phase separation process develops at a temperature which is just beyond the instability value. Phase-separating at temperatures exceeding this minimum temperature value by some degrees yields instead expanded sediments with a particle volume fraction which can be much smaller than 0.5 and is monotonically decreasing as the temperature increases. It is interesting to note that the structure of the compact sediment which forms just beyond the instability temperature strongly depends on the value of the surfactant volume fraction. While for Φ_S larger than about 0.07 the concentrated particle phase appears as completely amorphous, sediments obtained just beyond the instability line for $\Phi_S \leq 0.07$ show the typical iridescence of an ordered colloidal crystal phase.

To get a more complete view of the temperature dependence of the separation process, we have studied, besides the MFA suspension, a series of suspensions of polystyrene latex spheres of variable radius by using Triton X-100 as the nonionic surfactant. The critical point of aqueous solutions of Triton X-100 at 100 mM NaCl ionic strength is at $T_c \approx 63.4$ °C, $\Phi_C \approx 0.02$. Figure 2 presents the limiting stability lines obtained for four distinct polystyrene particle radii a between 19 and 120 nm, and for the MFA particles with $a = 110$ nm. A first qualitative observation is that, while far from the consolution curve the amount of surfactant needed to induce particle segregation strongly depends on the particle size, all lines get closer in the critical region. Moreover, the comparison between the instability lines for MFA and 120 nm polystyrene particles clearly indicates that Φ_{Smin} does not appreciably depend on the chemical nature of the latex spheres.

We call Φ_{S0} the value of Φ_{Smin} measured at 25 °C, that is, where critical effects are still very weak. The inset in Fig. 3 shows that the inverse of Φ_{S0} is approximately proportional to the particle radius. If we choose to scale Φ_{Smin} to the value Φ_{S0} relative to each particle radius, all instability lines tend to collapse on a single curve, as shown in Fig. 3. This suggests the existence of some scaling property in the mechanism leading to the instability of the colloidal dispersion.

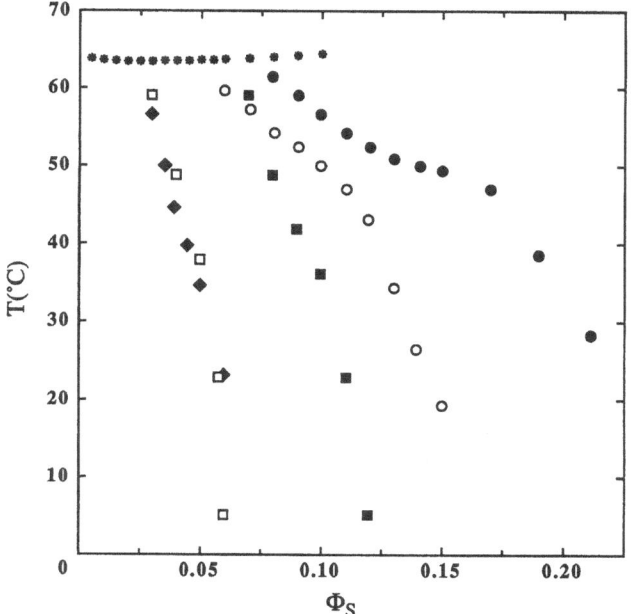

Fig. 2 Reversible flocculation as a function of the temperature for various colloidal suspensions at a volume fraction of 1% with added nonionic amphiphile (Triton X-100) and 100 mM NaCl (diamonds) MFA, $a = 110$ nm; (full dots) PS,19 nm; (open dot) PS, 35 nm; (full square) PS, 60 nm; (open squares) PS, 120 nm. (stars) are measured cloud points

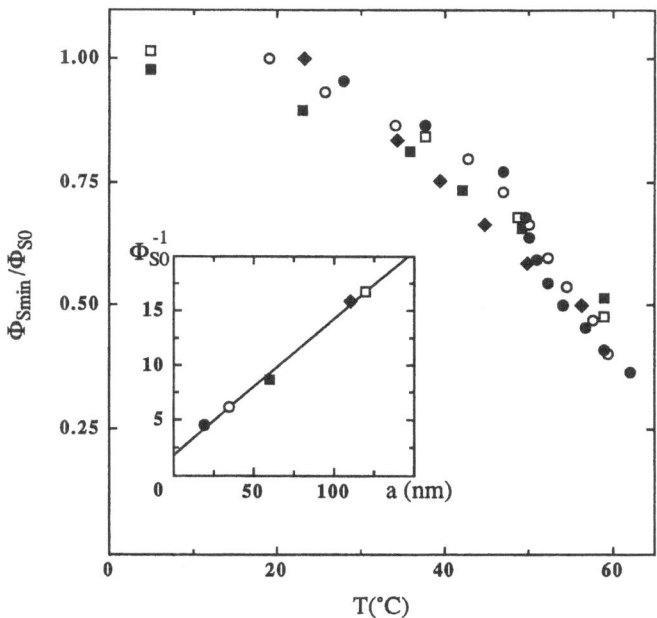

Fig. 3 The ratio Φ_{Smin}/Φ_{S0} reported as a function of T for various colloidal suspensions by using the data of Fig. 2. The inset shows the dependence of Φ_{S0}^{-1} on the particle size

Discussion

We discuss first the room temperature results. It was already noted in our previous paper [5] that the measured value of Φ_{S0} is much smaller than the value predicted by the model of ref. [3]. Such a discrepancy could be due in some part to the inadequacy of the simple free-volume approach, which, compared to the integral-equation calculations, tends to overestimate the concentration of small spheres needed to reach the spinodal instability [8]. Moreover, in our case the depletion mechanism arises not only from volume restriction, but also from osmotic effects due to the presence of the surfactant layer on the particle surface, as discussed by Vrij [2b]. The van der Waals attractive interactions among the colloidal particles should play a negligible role in the reversible flocculation process, as confirmed by the observation that Φ_{S0} is independent from the chemical composition of the particles.

The observed dependence of Φ_{S0} on the size of the colloidal particles is in qualitative agreement with that obtained by Bibette et al. [9] studying the phase separation of oil-in-water emulsion droplets induced by ionic micelles. Reference 9 presents a simple model which seems to account for the observed increase of Φ_{S0} as the size ratio δ between small and large particles is increased. A recent investigation performed on a mixture of monodisperse emulsion droplets of two different sizes [10] also finds that Φ_{S0} is strongly increasing as δ increases.

The dependence of Φ_{Smin} on the temperature T can be due to several effects: i) the intermicellar potential becomes attractive as T increases, as shown by the fact that the system presents an upper critical point [6]; ii) the micelle size and shape might change somewhat with T [6]; iii) the interaction potential among the colloidal particles coated by the nonionic surfactant becomes also attractive when T is increased [11].

Concerning effect iii), it should be stressed that, according to our results, the phase separation cannot simply be induced by fully covering the colloidal particles with an amphiphile layer, but a sufficiently concentrated micellar phase is always needed for the phase separation to take place. The basic mechanism that leads to the temperature effects is probably to be found in changes of the micellar phase, although effect iii) might contribute to slightly modify the phase boundary.

Point ii) was raised in ref. [6] and is still considerably debated in the literature, but, at least in the case of $C_{12}E_8$, a rather detailed small-angle-neutron-scattering study [12] excludes the existence of large changes in the micellar structure.

Effect i) is then probably the most relevant one for explaining the temperature dependence of Φ_{Smin}. Attractive interactions among the micelles give rise to a negative contribution of the second virial term to the osmotic pressure. Apparently, one would then expect that the amount of surfactant needed to induce segregation should rather increase with temperature, since the depletion potential scales as the osmotic pressure [13]. However the simple approach proposed in ref. [2] and used in ref. [13] neglects any correlation between the micelles. Since attractive interactions imply a large value of the micellar pair correlation function at short distances, one could guess that a micelle excluded from the gap between two colloidal particles "carries around" with itself a region where the local surfactant concentration is larger than the average. This corresponds to an increase of the effective micellar size, yielding a larger depletion volume which could more than counteract the decrease of the micellar osmotic pressure. Such an argument is supported by a model developed for a suspension of hard-spheres in a bad solvent [14], where attractive interactions between the solvent particles are introduced by modeling them as small adhesive hard spheres. The model shows that depletion forces between the colloidal particles noticeably increase as a function of the stickiness between the small spheres.

The model of ref. [14] might also suggest some connection between our results and the segregation of colloidal particles in critical mixtures of simple liquids [15], which was observed in several experiments but is not clearly understood. In the case of critical mixtures the role of the depleting agent could be taken by the component of the mixture which is less preferred by the particles. Although the molecular size of the component is not sufficient to give depletion effects in ordinary conditions, the correlation effects present near the critical point could perhaps be described in terms of an "effective" size of the component larger than the molecular size. However, further investigation is needed to assess the plausibility of this hypothesis.

Finally, it is interesting to recall that crystals of membrane proteins can be grown in aqueous solutions of nonionic amphiphiles which are near the coexistence curve on the amphiphile-rich side [16]. It is not unlikely that depletion effects similar to those described in this article might play an important role in such a process.

Acknowledgments We thank Ausimont (Milano, Italy) for the gift of MFA particles, and A. Vrij for useful discussions. This work was partially supported by the Italian Ministry for University and Research (MURST 40% funds).

References

1. See: Illett SM, Orrock A, Poon WCK, Pusey PN (1995) Phys Rev E 51:1344 and references quoted therein
2. a) Asakura S, Oosawa F (1954) J Chem Phys 22:1255; b) Vrij A (1976) Pure Appl Chem 48:471
3. Lekkerkerker HNV, Poon WCK, Pusey PN, Stroobants AN, Warren PB (1992) Europhys Lett 20:559
4. Ma C (1987) Colloid Surf 28:1
5. Piazza R, Di Pietro G (1994) Europhys Lett 28:445
6. Degiorgio V (1985) In: Degiorgio V, Corti M (eds) Physics of Amphiphiles: Micelles, Vesicles, and Microemulsions, North-Holland, Amsterdam, p 303
7. Degiorgio V, Piazza R, Bellini T, Visca M (1994) Adv Colloid Interface Sci 48:61
8. Lekkerkerker HNV, Stroobants AN (1993) Physica A 195:387
9. Bibette J, Roux D, Nallet F (1990) Phys Rev Lett 65:2470
10. Steiner U, Meller A, Stavans J (1995) Phys Rev Lett 74:4750
11. Long JA, Osmond DWJ, Vincent B (1973) J Colloid Interface Sci 42:545 Thompson L, Pryde DN (1981) J Chem Soc Faraday Trans I 77:2405
12. Zulauf M, Weckström K, Hayter JB, Degiorgio V, Corti M (1985) J Phys Chem 89:3411
13. Poon WKC, Selfe JS, Robertson MB, Ilett SM, Pirie AD, Pusey PN (1993) J Phys II France 3:1075
14. Penders MHGM, Vrij A (1991) Physica A 173:532
15. Beysens D, Estève D (1985) Phys Rev Lett 54:2123; Narayanan T, Kumar A, Gopal ESR, Beysens D, Guenoun P, Zalczer G (1993) Phys Rev E 48:1989; Gallagher PD, Kurnaz ML, Maher JV (1992) Phys Rev A 46:7750
16. Zulauf M (1985) In: Degiorgio V, Corti M (eds) Physics of Amphiphiles: Micelles, Vesicles, and Microemulsions, North-Holland, Amsterdam, p 663

Progr Colloid Polym Sci (1996) 100:206–211
© Steinkopff Verlag 1996

The effect of monovalent and divalent cations on sterically stabilized phospholipid vesicles (liposomes)

K. Kostarelos
Th.F. Tadros
P.F. Luckham

K. Kostarelos · Th.F. Tadros
Zeneca Agrochemicals
Formulations Section
Jealott's Hill Research Station
Bracknell
Berkshire RG12 6EY, United Kingdom

Dr. K. Kostarelos (✉) · Th.F. Tadros
P.F. Luckham
Imperial College of Science
Technology of Medicine
Chemical Engineering and Chemical
Technology Department
University of London
London SW7 2BY, United Kingdom

Abstract When phospholipids are dispersed in water and sonicated for approx. 240 min, vesicles (liposomes) of 40–45 nm in diameter are formed. Their steric stabilization is achieved by the addition of (tri)-block copolymers of the A-B-A type (A is PEO and B is PPO). Importance is placed on the way the copolymer was added to the vesicles. Stabilization is thought to be optimized when adding the copolymer initially, therefore allowing participation between the latter and phospholipid molecules towards vesicle formation, leading to the physical anchoring of the hydrophobic homopolymer (PPO) into the bilayer. In contrast, when the copolymer molecules were added after vesiculation was complete, stabilization was not optimum. The effect of a monovalent (NaCl) and a divalent ($MnCl_2$) cation on the vesicle systems was systematically studied by dynamic light scattering (PCS) and turbidity measurements. The effect of electrolytes was studied by: a) increasing the cation concentration only in the outer vesicle aqueous phase, monitoring the occurring flocculation of the vesicles, and b) forming the vesicles in high electrolyte concentration and diluting, therefore creating an osmotic gradient between the two aqueous phases. Vesicle flocculation caused by NaCl or $MnCl_2$ occurred in all systems, but the ones coated with copolymers required higher electrolyte concentrations. When an osmotic gradient developed between the vesicle inner and outer aqueous phases, resistance against swelling was observed only in the case of vesicle systems formed in the presence of copolymers. In both electrolytes (NaCl and $MnCl_2$) coating the vesicle surface with polymer molecules resulted in increased protection against flocculation. Moreover, the initial addition of block copolymers produced vesicles that could sustain osmotic swelling.

Key words Liposomes – vesicles – steric stabilization – cations

Introduction

Model membranes consisting of, or containing acidic phospholipids carry negative charges on their surface due to the dissociation of the $-POO_4^-$ and $-COO^-$ groups. The effect of cations onto phospholipid vesicles of varying lipid composition, carrying negative charges on their surface, has been the scope of considerable research almost since the description of liposomes as simple model membrane systems [1]. It has been shown that Ca^{2+} induces a fusion process with a time-course of several hours. [2]. The flocculation/aggregation of vesicles as a possible primary step in the fusion process has also been suggested.

Progr Colloid Polym Sci (1996) 100:206–211
© Steinkopff Verlag 1996

Since flocculation of liposomes can be seen as a prerequisite of bilayer fusion, numerous studies have been concerned with this effect. Monovalent cations have been reported to cause aggregation of acidic phospholipid vesicles without being able to induce any fusion [3,4], due to the low binding affinities of these ions for the phospholipid surface, compared with those of divalent cations. This was attributed to the compression of the double layer [5,6].

It is widely accepted that the aggregation behavior of vesicles induced by divalent cations involves a much more complicated process, not well explained to date. A particularly important effect occurring when divalent cations are added to the solution leading to reduction of the electrostatic repulsive forces is surface ion binding, where the cations can bind to the ionizable groups and decrease the apparent negative charge density on the membrane surface [7]. The binding of multivalent cations may well lead to the reversal of sign of the zeta-potential values. There have been interesting reports of further short-range repulsive hydration forces between membranes, due to the interaction between the hydration layers of the polar phospholipid headgroups [8]. Generally, the stability of vesicles against divalent cations can be described by the DLVO theory, with the addition of repulsive hydration (short-range) and/or steric (longer-range) forces, according to the surface characteristics (e.g. nature of phospholipid headgroup, presence of macromolecules) of the particular vesicle system.

In the present study, use of the flocculation of phospholipid vesicles in the presence of monovalent and divalent cations was thought to provide valuable information on the stabilization and permeability of vesicle systems coated with A-B-A (where A is polyethylene oxide (PEO) and B polypropylene oxide (PPO)) block copolymer molecules. The coating of the liposome surface was carried out in two different ways, resulting in two different vesicle systems, as described elsewhere [9]. The intermixing of phospholipid and polymer molecules followed by hydration and SUV formation, was thought to lead to incorporation of the hydrophobic PPO blocks of the copolymer inside the lipid bilayer. While the addition of the copolymer molecules after vesiculation resulted in a poor Langmuirian-type adsorption onto the liposome surface [9]. The effect of monovalent and polyvalent cations on the vesicle systems was monitored using dynamic light scattering and turbidimetry.

Experimental

Materials

The vesicle systems were formed by using a mixture a soybean lecithin lipids (approx. 50% D-α-dimyristroyl-

phosphatidylcholine (DPPC)), purchased by Sigma. The aqueous dispersion medium was always double distilled, deionized water. The A-B-A type (tri)-block copolymer used was of the Synperonic PE family, namely the PF127, where A is polyethylene oxide (PEO) and B is polypropylene oxide (PPO), supplied by ICI Surfactants, Belgium; This polymer has the following structure:

$$(EO)_{99} - (PO)_{65} - (EO)_{99} .$$

The salts used in this study NaCl and $MnCl_2 \cdot 4H_2O$ were purchased by Sigma. Both the monovalent and divalent cations were of standard laboratory grade and were used without further purification Mn^{2+} was used in order to offer alternative information on the effect of divalent cations other than Ca^{2+} (by far the most extensively investigated cation).

Preparation of vesicles

Liposomes were prepared following the hydration-sonication method [10], using a Kerry ultrasonic bath (50 Hz). Two methods have been used to include the A-B-A copolymers into the vesicle systems. According to the first method the vesicles were prepared (hydration and sonication steps) in the presence of copolymer molecules at the desired concentration. Following the second technique, the liposomes were formed, and after sonication were diluted with copolymer aqueous solutions in order to reach the required final concentrations. These systems were left to stand for at least 24 h until may measurement was carried out.

Dynamic light scattering

In a dynamic light-scattering experiment, the z-average diameter is determined from the diffusion coefficient, D, of the vesicles as they randomly move due to the Brownian motion. The principles of operation are described in more detail elsewhere [11]. The way in which photon correlation spectroscopy (PCS) was used to determine the mean diameter of the vesicles studied in the present work is also mentioned elsewhere [9].

The instrument used was the Malvern (UK) 4700 PCS apparatus, with an Argon laser beam (50 mW) emitting at a wavelength of 488 nm. The scattered intensity fluctuations were recorded at a 90° angle and 25 °C. The estimated error of the dynamic light scattering results (from the standard deviation) presented in this study was between 1–2% of the z-average diameter value directly obtained from the instrument.

Turbidity measurements

Relative turbidities were determined against the bare liposome system at physiological buffer conditions. Variations in the amount of scattered light at 450 nm were in this way used to monitor the flocculation and swelling of the vesicle systems. All turbidity measurements were carried out using the Pye-Unicam SP1700 Ultraviolet Spectrophotometer.

Results and discussion

From previous investigations [9] on the structural and surface characteristics of the vesicles, two important conclusions were drawn: a) two structurally different vesicle systems seem to occur depending on the way in which the block copolymer molecules are added; b) the vesicle surface in both cases is coated with copolymer molecules which most probably will attain different chain conformations. These two conclusions served as a guideline for studying the stability of the vesicle systems towards flocculation in the presence of salts.

Effect of monovalent cations (Na^+)

The systems on which most importance was to be placed were the bare liposome system, the vesicles with copolymer molecules (1% wt/wt) added initially, and the vesicles with copolymer molecules (1% wt/wt) added after liposome formation. These three vesicle systems were comparatively studied in order to determine possible differences in their behavior towards cations, exposed to the cations under identical experimental conditions.

The mean vesicle size was monitored while increasing the NaCl concentration in the dispersion medium (outer aqueous phase). As can be seen from Fig. 1, the bare liposome system started flocculating between 0.1 and 0.5 M NaCl, while the two block copolymer-containing systems demonstrated higher resistance to the cations, with increase in their mean size between 2–3 M NaCl. The flocculation of the bare liposome system by Na^+ was reversible. By gradual dilution of the sample, deflocculation of the liposomes occurred. This finding agrees with previously reported effects of monovalent cations onto acid surface liposomes [6]. The weak nature of the flocculation of liposomes caused by the monovalent cations indicated that its most probable cause was the depression of the electrostatic double layer at the vesicle surface, the "screening effect" of the Na^+ cations, without any considerable binding of the cations onto the phospholipid

Mean Vesicle Diameter (nm)

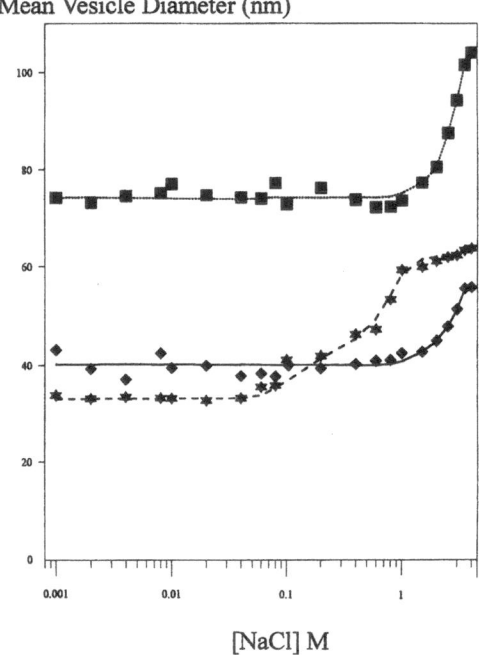

[NaCl] M

Fig. 1 Mean vesicle diameter determination by dynamic light scattering for increasing concentrations of NaCl at the external aqueous phase. The final lipid concentration was 0.02% wt/wt at all times. (☆) bare liposomes; (◆) vesicles with copolymer added after (A) formation; (■) vesicles formed in the presence of copolymer. This notation is kept for all figures

surface. The copolymer-coated liposome surfaces seem to resist aggregation due to this thinning of their double layers, probably due to increased steric repulsions between the PEO chains of approaching vesicles.

Next, the vesicle systems were prepared in high electrolyte aqueous buffer, followed by dilution and therefore reduction of the NaCl concentration only in the external aqueous phase. Figure 2 shows how the mean vesicle diameter is changing with a decrease of the NaCl concentration from 1 M to 0.001 M. There seems to be some deflocculation occurring as mean vesicle size decreases with dilution for all three systems. However, the most important feature of this study is the final vesicle sizes obtained after dilution of the external aqueous phase to 0.001 M NaCl, where the bare liposomes and the vesicles with added copolymer after formation (A) present remarkably increased sizes (70 and 75.4 nm, respectively) compared with the ones in the normal aqueous buffer (36.4 and 42.5 nm). In contrast, for the system where the polymer may be incorporated inside the bilayer, almost no mean diameter increase was obtained (82 nm compared with 78 nm in pure water). This result can be interpreted as an indication that the PPO block is indeed included inside

Mean Vesicle Diameter (nm)

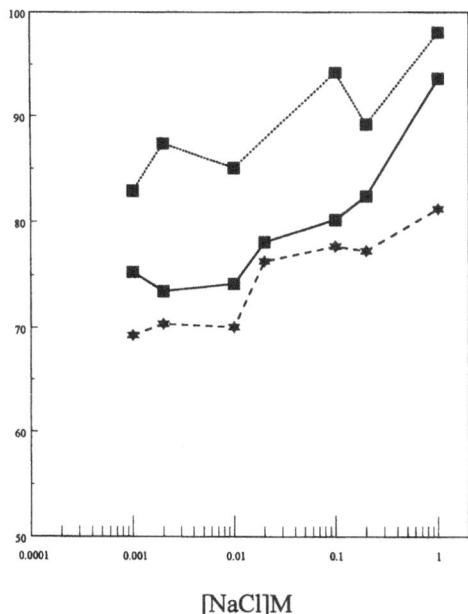

[NaCl]M

Fig. 2 Mean vesicle diameter of vesicles when diluting the outer aqueous phase

Mean Vesicle Diameter (nm)

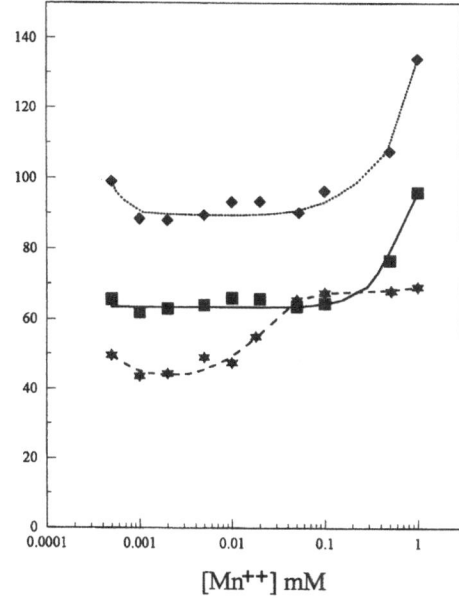

[Mn⁺⁺] mM

Fig. 4 Mean vesicle size for increasing concentrations of $MnCl_2$ at the external aqueous phase

Relative Turbidity

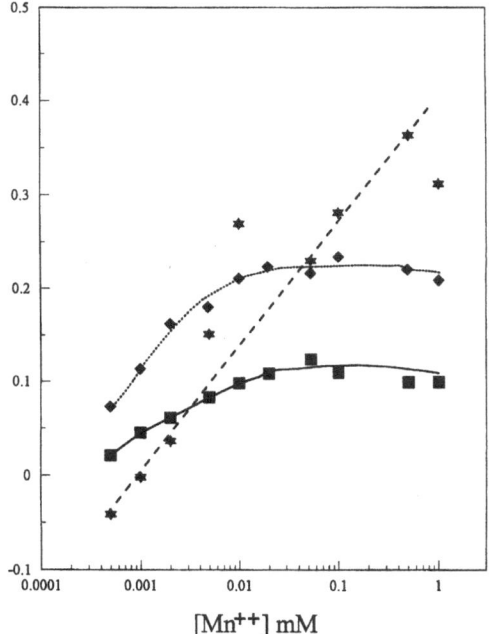

[Mn⁺⁺] mM

Fig. 3 Turbidity relative to bare liposome system in pure water, for increasing concentrations of $MnCl_2$ at the external aqueous phase

the alkyl lipid phase resulting in a more "rigid" bilayer, being less prone to swelling/shrinking effects.

Effect of divalent cations (Mn^{2+})

Studies using $MnCl_2$ to investigate the interaction with vesicles were undertaken. The concentration of electrolyte was increased in the outer aqueous phase, and the behavior of the different systems was examined by turbidity and PCS. It has to be noted that much lower concentrations were used in the case of divalent cations than in the similar study using NaCl, because of the much higher affinity that these ions exhibit for the liposome surface.

When increasing the electrolyte concentration on the vesicle outer aqueous phase, the turbidity of the bare liposome system increases sharply, indicating the severe effect of increasing concentrations of added Mn^{2+} ions (Fig. 3). The copolymer containing vesicle systems, though, exhibit different behavior from the bare liposome system, with turbidity values reaching a plateau above 0.01 mM Mn^{2+}. The mean vesicle diameter determination for the three vesicle systems offered a picture very similar to the turbidity measurements (Fig. 4). The underlying effect is the analogy in the observed behavior of the copolymer containing vesicle systems, where particle size increases around 0.5 mM $MnCl_2$, while the bare liposomes seem

Relative Turbidity

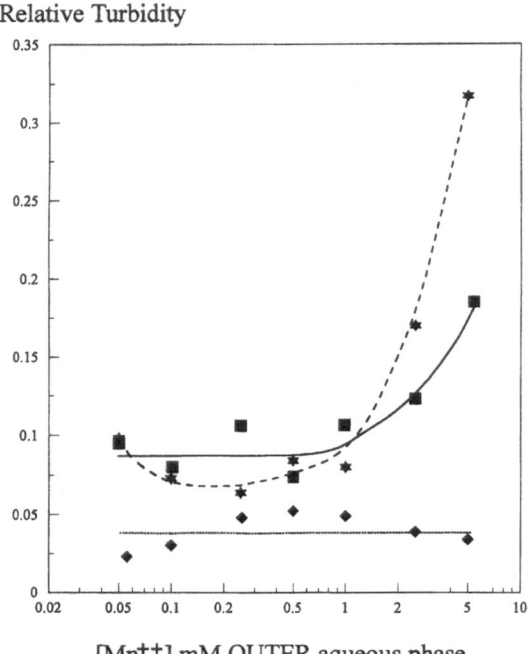

[Mn++] mM OUTER aqueous phase

Fig. 5 Relative vesicle turbidity for diluting the outer aqueous phase. Vesicles prepared in 5 mM MnCl₂

Mean Vesicle Diameter (nm)

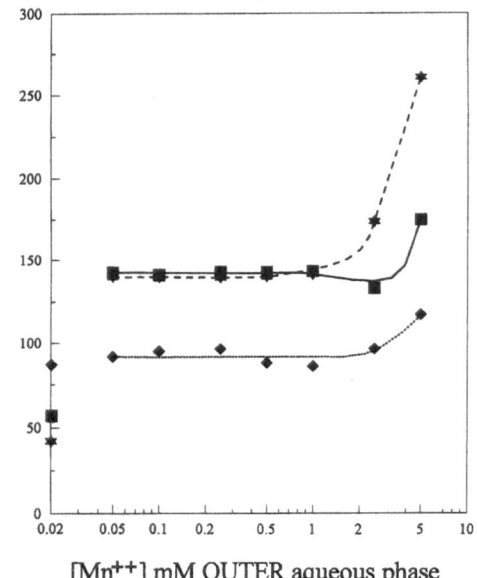

[Mn++] mM OUTER aqueous phase

Fig. 6 Mean vesicle size when diluting the outer aqueous phase. Note that the points on the y-axis indicate the mean vesicle diameter for each vesicle system in pure distilled water

much more susceptible to flocculation with increasing electrolyte concentrations.

The preparation of the vesicles in high $MnCl_2$ concentration provided further information on their flocculation and response to osmotic effects. The electrolyte concentration in which the vesicle systems were formed was 5 mM $MnCl_2$ and the final dilution step resulted in 0.05 mM $MnCl_2$ at the external aqueous phase. From both the turbidity and the sizing data obtained (Figs. 5 and 6), the vesicle system with the block copolymer added initially (I) exhibited distinctly different behavior from the two other systems. This vesicle system seemed to be almost completely resistant to the effects of Mn^{2+}. Turbidity values for this system remained unchanged throughout the 100-fold decrease in the electrolyte concentration of the vesicle outer aqueous phase, while its mean vesicle diameter increases minimally (approx. 90 nm) compared to the mean vesicle diameter values obtained when forming the vesicle in pure water (77.7 nm). On the other hand, the liposome system and the vesicles with added copolymer after (A) formation of liposomes, demonstrated almost identical behavior of much higher mean diameter values (approx. 140 nm compared with 42.5 and 61.3 in pure water) and significantly lower stability against flocculation, particularly at high concentrations of Mn^{2+} in the outer water phase. These results served as indications that the formation of vesicles in high electrolyte ($MnCl_2$) was improved in the presence of block copolymer molecules. Moreover, the response of the three vesicle systems to the osmotic gradient created after dilution of the outer aqueous phase, agrees with the observations from the study of monovalent cations. Only the (I) system did not respond to almost any change, displaying increased stability.

Conclusions

From the obtained results, different vesicle structures of varying stability were formed, depending on the way in which the block copolymer molecules were added to the vesicle system. The stability of the vesicle systems against flocculation and osmotic swelling was determined by monitoring the mean vesicle diameter and the dispersion turbidity. Improved resistance against flocculation caused by the presence of monovalent (NaCl) and divalent ($MnCl_2$) cations at the external vesicle aqueous phase was observed when coating the liposome surfaces with block copolymers. Bilayer rigidity was improved only for the vesicle systems prepared in the presence of polymer.

Progr Colloid Polym Sci (1996) 100:206–211
© Steinkopff Verlag 1996

References

1. Bangham AD (1968) Progr Biophys Mol Biol 18:29–95
2. Papahadjopoulos D, Poste G, Schaeffer BE, Vail WJ (1974) Biochim Biophys Acta 352:10–28
3. Ohki S (1982) Biochim Biophys Acta 689:1–11
4. Niz S, Bentz J, Wilschut J, Düzgünes N (1983) Prog Surf Sci 84:266–269
5. Day EP, Kwok AYW, Ho JI, Vail WJ, Bentz J, Niz S (1980) Proc Natl Acad Sci USA 77:4026
6. Ohki S, Roy S, Ohshiama H, Heayards K (1984) Biochemistry 23:6126–6132
7. Matsumura H, Furusawa K (1989) Adv Coll Int Sci 30:71–109
8. Ohsawa K, Oshima H, Ohki S (1981) Biochim Biophys Acta 648:206–214
9. Kostarelos K, Tadros TH F, Luckham PF, in preparation
10. Huang C (1969) Biochemistry 8(1):344
11. Ostrowsky N (1993) Chem Phys Lipids 64:45–56

Progr Colloid Polym Sci (1996) 100:212–216
© Steinkopff Verlag 1996

N. Fauconnier
A. Bee
J. Roger
J.N. Pons

Adsorption of gluconic and citric acids on maghemite particles in aqueous medium

Dr. N. Fauconnier (✉) · A. Bee
J. Roger · J.N. Pons
Laboratoire de Physicochimie Inorganique
Université Pierre et Marie Curie
4, place Jussieu
75252 Paris Cedex 05, France

Abstract Adsorption of gluconic and citric acids, separately or simultaneously, on maghemite (γ-Fe$_2$O$_3$) nanoparticles in aqueous medium is studied at 25 °C. The concentrations of adsorbed acids are determined by HPLC. The stability range is determined by spectrophotometry. It is strongly dependent on pH and on concentration of adsorbed acids. The concentration of adsorbed ligands can be monitored to obtain stable colloidal solutions in the range of pH 6–8, with particles coated simultaneously by the two ligands. In particular, these magnetic liquids are interesting in view of medical applications.

Key words Magnetic fluids – adsorption – gluconic acid – citric acid – stability range – HPLC

Introduction

Ferrofluids are magnetic liquids constituted of nanoparticles of iron oxide [1]. To develop their applications in the biomedical area, biological molecules such as antibodies, drugs or enzymes, must be grafted on to the magnetic particles. Usually, to strengthen the binding (biological molecule-particle) a crosslinker is used. The latter must fulfill the following conditions: to be strongly adsorbed on the particles, to form a stable binding with the biological molecule and to possess free ionizable functions to stabilize the sols in physiological medium by electrostatic repulsions between the particles. Many ligands of iron can be used; for example adsorption of dimercaptosuccinic acid has been studied in our laboratory [2], to graft antibodies on maghemite particles. Sometimes another chelating agent must be adsorbed on the particles to stabilize the colloidal solution by providing additional surface charges or to be used as a second crosslinker if two different biological molecules must be simultaneously grafted. With this aim in view, we have recently developed the synthesis of ferrofluids made up of particles coated by two chelating agents. In this paper, adsorption of the mixture (gluconic-citric acids) on maghemite (γ-Fe$_2$O$_3$) particles has been studied by observing changes in colloidal stability in aqueous media of varying pH. Citric acid is added to stabilize the sol, its free functions wearing the charge necessary for electrostatic repulsions between particles. The adsorbed gluconic acid allows the grafting of the biological molecule by way of its free alcoholic functions. The adsorption of citric acid and gluconic acid can be understood in terms of surface coordination reactions at the oxide water interface. Their functional groups, such as carboxylate functions, substitute for the amphoteric surface hydroxyl groups to form complexes with the structural metal ions of the oxide surface. The stability range of the magnetic solutions is obtained by spectrophotometry. It appears to be strongly dependent on the pH and on concentration of added organic acids. The stability behavior of the system is correlated with adsorption isotherms, the quantity of adsorbed acids being determined by HPLC measurements. Our objective in this paper is to obtain a ferrofluid stable in physiological medium, made up of particles carrying the appropriate functions in order to graft an anticancer drug. To elucidate the mechanism of surface complexes, an IR study is now in process and the results will be published later.

Materials and methods

The ionic aqueous ferrofluids used in this present work are cationic sols of γ-Fe_2O_3 macro-ions synthesized according to a method described elsewhere [3], by alkalizing an aqueous mixture of iron (II) chloride and iron (III) chloride. The magnetite so obtained is then acidified and oxidized in maghemite. The polydisperse system is constituted of roughly spherical particles of which the mean diameter, obtained by x-ray diffraction, is 8.3 nm. The electric charge of an oxide arises from the amphoteric behavior of the surface hydroxyl groups. Their number consequently depends on the pH of the solution. In acidic medium (pH = 2.7), the molar ratio of superficial protonated sites to total iron is $2.44 \cdot 10^{-2}$ [4].

The maghemite complexes are prepared by stirring during 10 min a constant quantity of maghemite and solution of the organic acids, the final concentration of iron being $4.42 \cdot 10^{-2}$ mol$\cdot$L^{-1} in all the samples. The molar percentage of organic acids added to total iron of the ferrofluid is noted R. The values of pK of carboxylic functions are: 2.79, 4.30, 5.65 for citric acid, and 3.56 for gluconic acid [5].

The quantity of anions adsorbed on the surface is determined by HPLC. This method and the preparation of the samples are described in detail elsewhere [6]. The separation is carried out on a 8% cross-linked sulphonated divinyl benzene styrene copolymer in the hydrogen form 300×7.8 mm (OA 2216 Benson Polym. INC NV). The mobile phase is a sulfuric acid solution whose pH is 2.6, used at a flow rate of 0.5 mL/min.

The flocculation of the ferrofluid is detected by visible spectrophotometry with an Hitachi U2000 spectrophotometer. The experimental device is described elsewhere [7]. After stirring, samples are circulated into quartz cells (1 cm), their absorbance is measured at 700 nm as a function of the pH and they are reinjected into the reactor. In absence of flocculation, sol absorbance follows Beer Lambert's law, the flocculation is determined by an important increase of absorbance which reflects variation of the medium turbidity.

All the experiments are carried out at 25 °C with an ionic strength equal to $5.5 \cdot 10^{-3}$ mol$\cdot$L^{-1}.

Results and discussion

Adsorption isotherms

The effect of pH on the adsorption of organic acids was measured for a quantity of acid added corresponding to a value of R equal to 4%. The maximum adsorption

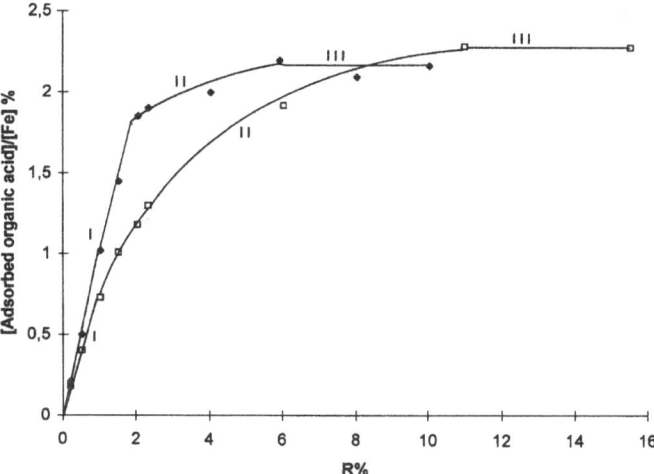

Fig. 1 Quantities of adsorbed citric (♦) (pH = 2.8) and gluconic (□) (pH = 3.6) acids versus the quantity of acids introduced (R%). The three regions are noted I, II, III

occurs at a pH equal to the pK of the first carboxylate function. Hingston et al. have shown that this maximum in the pH-adsorptivity profiles of weak acids is a general phenomenon [8]. For the following adsorption isotherms, the experiments are then carried out at a pH near the pK.

Citric acid

Figure 1 shows three regions for the adsorption of citric acid on the maghemite particles. This curve looks like isotherms obtained by Parfitt et al. with oxalate ions on goethite [9]. Initially, about 1.85% of citric acid is strongly adsorbed (I), in the second region (R > 2%) an additional quantity corresponding to 0.35% of citric acid is adsorbed more weakly. In the third region (R > 6%), the adsorption maximum is reached and citric acid added is no longer complexed on the surface. The adsorption maximum (2.2%) appears to be equal to the amount of protonated surface hydroxyl groups.

Gluconic acid

In the case of gluconic acid the adsorption curve is similar (Fig. 1) and shows also three regions, but the quantity of organic acid adsorbed in the first region is smaller (1%) and a more important quantity of gluconic acid must be added to reach the maximum (R > 10%). This result indicates that the stability of the surface complexes is weaker for gluconic acid than for citric acid. This trend is in

214

N. Fauconnier et al.
Adsorption on maghemite particles

Fig. 2 Quantities of citric (♦) and gluconic (□) acids adsorbed for samples with varying concentrations of citric acid $R\%$ and 6% of gluconic acid. Quantities of adsorbed citric acid (△) when only citric acid is added to the ferrofluid

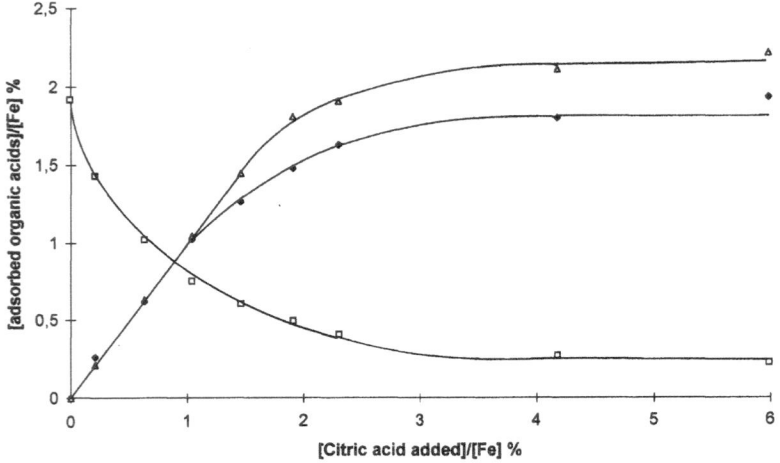

Citric and gluconic acids mixture

To study the interaction of the mixture (citric-gluconic acids) with colloidal iron oxide increasing quantities of citric acid are added to the ferrofluid (R varying up to 6%) and for all the samples the same quantity of gluconic acid is then introduced (6%) (Fig. 2). In this case, there is a competition for the surface sites between the two acids. By HPLC measurements, the adsorbed quantity of each ligand is precisely known. Without citric acid, the quantity of gluconic acid adsorbed is 1.92%, this quantity decreases with increasing amounts of citric acid. For R below 1%, all the citric acid introduced is adsorbed, gluconic acid complexing the vacant sites. But when the amount of citric acid added is more important, the quantity adsorbed on maghemite is smaller than previously for citric acid added alone; gluconic acid displaces a part of citric acid [15]. The total quantity of acids adsorbed (gluconic + citric acids) is equal to the number of coordination sites of the maghemite particles. In our experimental conditions, the two ligands are always present on the particles, even for a high quantity of citric acid added (6%). In this case, 9% of the ionizable sites are complexed by gluconic acid. Measurements of citric and gluconic acids adsorbed for up to 30 days showed that the quantity of each adsorbed organic acid remained the same. It is interesting to see that the proportion of each ligand may be adjusted and then it will be possible to fix the appropriate amount of biological molecule on the adsorbed gluconic acids.

agreement with the stability constants of the corresponding complexes in solution [5].

Stability domain

The study of the curves $A = f(pH)$ obtained for different amounts of ligands added, allows a precise determination of the pH value of the beginning and the end of the flocculation. These experimental results allow to establish a stability range of the ferrofluid as a function of pH and quantity of organic acid added. To understand the mixed systems with the two different anions fixed simultaneously on the particles, it is necessary to elucidate at first the stability of the maghemite sol in the presence of the two ligands added separately. In all the experiments, the mixture (acidic ferrofluid + organic acids) is first alkalinized until pH 11 with tetramethylammonium hydroxide (TAMOH), and nitric acid is then added until pH 2. The absorbance and pH measurements are studied for different quantities of organic acids introduced as a function of the concentration of HNO_3 added.

Citric acid

The flocculation range obtained for ferrofluid with increasing quantities of citric acid introduced, is progressively displaced to the acidic medium (Fig. 3) [11]. For a value of R above 1.5% the stability range does not evoluate any more. This result agrees with HPLC measurements which show that for this value of R a high proportion of surface coordination sites are complexed by citric acid (60%). The pH value of the flocculate redissolution is about 3.5. For this pH, the free carboxylate functions are deprotonated and ensure the stability of the sol.

Progr Colloid Polym Sci (1996) 100:212–216
© Steinkopff Verlag 1996

Fig. 3 Stability range of the system (maghemite + citric acid $R\%$). Samples are first alkalinized by TAMOH and secondly acidified by HNO_3. The stability range corresponds to the decreasing in pH

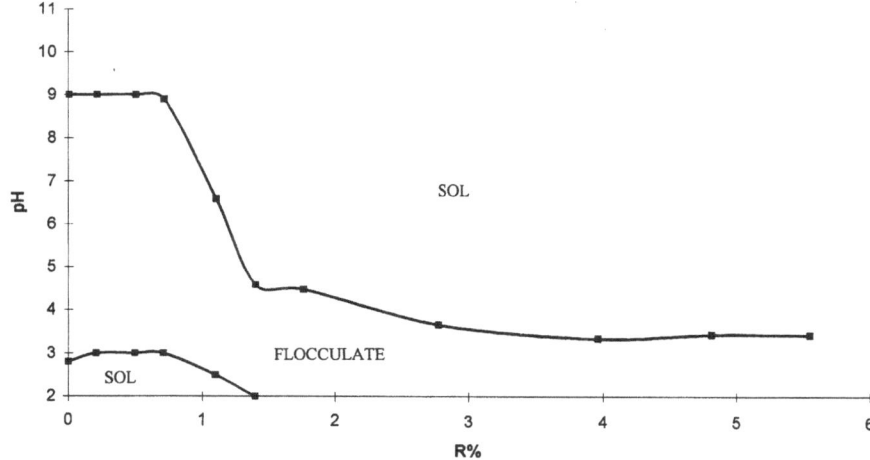

Fig. 4 Stability range of the system (maghemite + gluconic acid $R\%$) by decreasing the pH by HNO_3

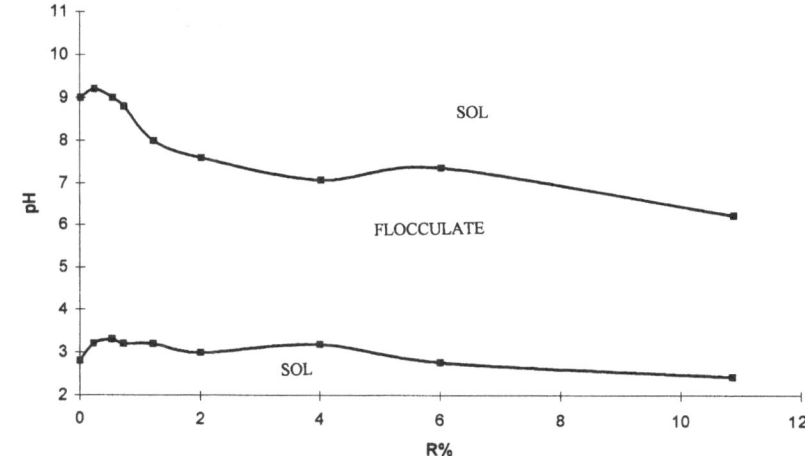

Gluconic acid

Adsorption of gluconic acid seems to have little effect on the range of the stability of the ferrofluid. By increasing quantities of gluconic acid, the stability range of the ferrofluid is weakly displaced to the acidic medium (Fig. 4). But the acidity constant of the carboxylate function having a pK of 3.56, the adsorption of gluconic acid is very weak in acidic medium. In that case, the stability of the ferrofluid is due to the charges of surface hydroxyl groups remaining on the particles. Flocculation occurs then when more gluconic acid is adsorbed. When the maximum adsorption is reached ($R > 10\%$) the stability range evoluates no further. An infrared study is in process to elucidate the mode of adsorption of gluconic acid.

Gluconic and citric acids mixture

Adsorption of only gluconic acid on maghemite particles does not allow the formation of a stable ferrofluid in physiological medium. Consequently, another chelating agent carrying ionizable functions at pH 7 must be introduced to obtain a biocompatible ferrofluid. That is why the stability of maghemite sol in presence of the mixture (gluconic-citric acids) has been studied in media of varying pH. The experiments are realized in the same conditions as previously for the adsorption measurements: increasing amounts of citric acid are first introduced and a constant quantity of gluconic acid (6%) is then added. The curves $A = f(pH)$ allow to establish the stability range of the two ligands ferrofluids (Fig. 5). Without citric acid ($R = 0\%$) only 6% of gluconic acid are added and the stability range

Fig. 5 Stability range of the system (maghemite + citric acid $R\%$ + gluconic acid 6%) by decreasing the pH by HNO_3

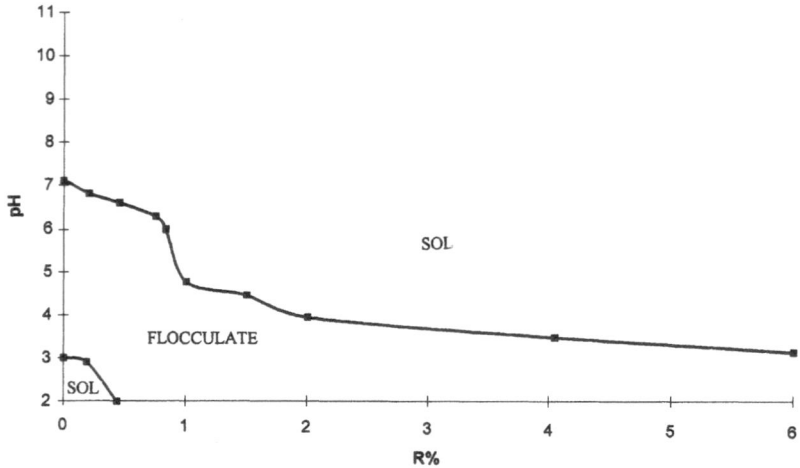

is the same as gluconic one. When R rises, the stability range is shifted to the acidic medium, and for $R > 1\%$, it is the same as for citric acid alone. The ferrofluid made of particles coated by the two ligands is stable from pH 3.5 to 11. This result agrees with HPLC measurements, for $R > 1\%$ the quantity of citric acid adsorbed becomes more important than the gluconic one, and the sol is then stabilized by the charge of free functions of citric acid.

Conclusion

Adsorption of the mixture citric-gluconic acids on maghemite particles allows to obtain stable ferrofluids in

a range of pH 6–8. Particles are coated simultaneously by the two ligands which complex surface iron of the oxide. It has been seen that the quantity of adsorbed gluconic acid is strongly dependent on the concentration of the added citric acid. Adsorption isotherms and spectrophotometric measurements show that it is possible to monitor the quantity of each ligand adsorbed. The amount of gluconic acid on the particles varies from 0.8% to 0.2% according to the quantity of citric acid added. In this case, the sols are stable from pH 3.5 to 11. The second part of this work will be the grafting of the active molecule on the particles by using free functions of gluconic acid. It will be then guided on the target cells by an electromagnet [12–17].

References

1. Bacri JC, Perzynski R, Salin D, Cabuil V, Massart R (1990) J Magn Mater 85:27–32
2. Massart R, Roger J, Cabuil V (1995) Brazilian J Physics 25:1–7
3. Massart R, Brevet Français 7918842, 1979. US Patent 4329241 (1982)
4. Bouchami T, Thesis Université Pierre et Marie Curie, Paris (1990)
5. Martell AE (1964) In: Stability constants of metal-ion complexeses
6. Fauconnier N, Bee A, Massart R, Dardoize FJ, Liq Chromatogr, accepted, in press

7. Roger J, Pons JN, Massart R (1989) Eur J Solid state Inorg Chem 26:475–488
8. Hingston FJ, Atkinson RJ, Posner AM, Quirk JP (1967) Nature (London) 215:1459
9. Parfitt RL, Farmer VC, Russel JD (1977) J Soil Sci 28:29–39
10. Waite TD, Morel FMM (1984) J Colloid and Int Sci 102:121–137
11. Zhang Y, Matijevic E (1984) Colloid Polym Sci 262:723–726
12. Widder KJ, Senyei AE, Scarpelli DG (1978) Proc Soc Exp Biol Med 58:141–146

13. Widder KJ, Senyei AE, Ranney DF (1980) Cancer Res 40:3512–3517
14. Widder KJ, Morris RM, Poore G, Howard DP, Senyei AE (1981) Med Sci 78:579–581
15. Ibrahim A, Couvreur P, Roland M, Speiser P (1982) J Pharm Pharmacol 35:59–61
16. Papisov MI, Savelyev VY, Sergienko VB, Torchilin VP (1987) Int J Pharm 40:201–206
17. Papisov MI, Torchilin VP (1987) Int J Pharm 40:207–214

Progr Colloid Polym Sci (1996) 100:217–220
© Steinkopff Verlag 1996

MIXED COLLOIDAL SYSTEMS

D. Bastos-González
R. Hidalgo-Alvarez
F.J. de las Nieves

Influence of heat treatment on the surface properties of functionalized polymer colloids

D. Bastos-González · R. Hidalgo-Alvarez
Biocolloid and Fluid Physics Group
Department of Applied Physics
Faculty of Sciences
University of Granada
18071 Granada, Spain

Dr. F.J. de las Nieves (✉)
Department of Applied Physics
Faculty of Experimental Science
University of Almeria
04120 La Canada, Almeria, Spain

Abstract In this work the electrophoretic mobility of two polystyrene latexes was measured versus the concentration of NaCl. A conventional sulfate latex obtained by emulsion polymerization of Styrene and an aldehyde latex prepared by emulsion copolymeri-zation of Styrene and Acrolein were used throughout this work. The latexes were subjected to a heat treatment (at 115 °C for 17 h) and after this were characterized from the surface and electrokinetic point of view. The electrophoretic mobility results of the heat-treated latexes were transformed into zeta-potential data by the O'Brien and White theory and the results compared with the zeta-potential values previously found for the original latexes. The heat treated latexes showed a lower zeta-potential values although with similar tendency than that showed by the original ones.

Key words Heat treatment – functionalized polymer colloids – zeta-potential

Introduction

The aim of studying electrophoretic mobility from a non-equilibrium double layer is principally to find the electrokinetic potential (ζ-potential) in non-equilibrium electrosurface phenomena. Monodisperse spherical polystyrene particles have proved to be very useful model systems for testing the most recent theoretical approaches [1–3]. As most of the theories deal with spherically shaped particles of identical size, the introduction of monodisperse latexes appeared to offer excellent chances for experimental verification of these theories. However, it is well known that electrophoretic mobility curves pass through a maximum as a function of increasing electrolyte concentration (or ionic strength) [4]. The current electrophoretic theories used in the conversion of mobility into ζ-potential give rise to a maximum in ζ-potential as well. This behavior contradicts the current layer models which predict a continuous decrease in potential. Various explanations for this maximum have been proposed [5–8], and some authors [9] have even pointed out that a maximum mobi-

lity value does not necessarily imply a maximum in ζ-potential, indicating that the conversion of mobility in ζ-potential of polystyrene microspheres/electrolyte solution interface should be done by means of a theoretical approach which takes into account all possible mechanisms of double-layer polarization. The term polarization implies that the electric double layer (e.d.l.) around the particles is regarded as being distored from its equilibrium shape by the motion of the particle. Overbeek [10] and Booth [11] were the first to incorporate polarization of the e.d.l. into the theory. They consider that the transfer and charge redistribution processes involved only the mobile part of the e.d.l. Also, O'Brien and White [1], starting with the same set of equations of Wiersema [12], have more recently published a theoretical approach of electrophoresis, which takes into account any combination of ions solution and the possibility of very high ζ-potential (up to 250 mV), far enough from the values to be expected in most experimental conditions.

On the other hand, there are different theories that try to explain, qualitatively, this anomalous behavior on the

218
D. Bastos-González et al.
Electrokinetic behavior of heat treated latexes

basis of the surface characteristics. One of the more accepted mechanisms is the called "hairy layer" model [4–6]. This model postulates that on the surface of the particles there is a layer of flexible polymer chains that have terminal ionic groups. These chains are extended into solution for a distance which varies depending on the electrolyte concentrations. At low concentrations the extended chains reduce the electrophoretic mobility by displacing the shear plane farther than if all the charge was on the surface. This allows an increase of the surface conduction inside the shear plane and therefore a decrease in the ζ-potential. As the electrolyte concentration increases the chains collapse over the surface, reducing the distance between the shear plane and the surface; the reduction in that distance could also yield the diminution of the surface conductance, so then the mobility increases. At higher concentrations, the classical behavior is observed, since the double-layer compression provokes the complete collapse of the chains over the surface. In a previous paper [13], we presented a study with polystyrene particles with different surface groups, in which we subject a part of the samples to high temperatures (heat-treatment) with the purpose to remove the presence of a possible "hairy layer": we tried to demonstrate, comparing results from surface and electrokinetic characterization, if the responsible mechanism of the anomalous behavior was due to the presence of a hairy layer. From the results obtained, we concluded that only the latexes with surface groups susceptible to changes (reflected in the variation of surface charge density), showed a different behavior in the electrokinetic answer with 1:1 electrolyte, but we could not attribute this difference to the presence of a hairy layer.

The main purpose of this work is convert the electrophoretical data for 1:1 electrolyte into ζ-potential using the O'Brien–White theory for latexes that suffered any change after the heat treatment and to compare the results with the original ones.

Materials and methods

Two monodisperse surfactant-free polystyrene latexes were used: a sulfate conventional latex, called DBG-0 [14] and an aldehyde latex, called AD-1, copolymer of styrene/acrolein [15], in both cases potassium persulfate was employed as initiator. The latexes were cleaned by repeated cycles of centrifugation/decantation/redispersion in a centrifuge (Kontron Instruments), followed by serum replacement for several days until the conductivity of the supernatant was constant and similar to the constant of the double-distilled water. The heat treatment was carried out following the procedure of Chow and Takamura [6]. The samples were introduced into Pyrex tubes, sealed with special caps and were then placed in an oven at 115 °C for 17 h. Due to a slight increase in the conductivity of the sample after heat treatment, these had to be cleaned by use of serum replacement during 1 day. The latexes with heat treatment have the same name as the original ones with double T added.

The average of the particles was estimated by means of Photo Correlation Spectroscopy (PCS) with a commercial device (Malvern II 4700).

Surface charge densities (σ_0) of the latexes were determined by conductimetric and potentiometric titrations. Since the aldehyde groups have no charge, we have followed the method of the hydroxilamine hydroclorate reported by Yan et al. [15]. The titrations for AD-1 latex, are shown and described in a previous paper [16].

Table 1 summarizes the particle size, p.d.i. and surface charge densities of the latexes with and without heat treatment. It can be seen that the latexes were monodisperse and the mean particle diameter and uniformity were not significantly altered by heat treatment. The latex DBG-0 presented weak and strong acid on its surface [17, 18], after the heat treatment there was an increase in the surface charge density and a total conversion of the strong in weak acid (DBG-0TT). Similar results were obtained for the aldehyde latex. The original AD-1 presented on its surface strong acid and aldehyde groups, while the AD-1TT latex increased significantly its σ_0, but transformed in weak acid, and it had a diminution in the amount of aldehyde groups. The reasons for these changes have been explained in a previous paper [13]. Table 1 also includes the functional groups presented on the particle surface.

The electrophoretic mobility was carried out with a commercial device (Zeta-Sizer 4, Malvern Instruments, England). For each electrolyte concentration, first the

Table 1 Characteristics of original and heat-treated latexes: type of surface groups, particle size PCS, polydispersity index (p.d.i.), surface charge density (σ_0) and amount of aldehyde groups

Latex	Surface Groups	PCS d (nm)	PCS p.d.i.	σ_0 ($\mu C/cm^2$)	Aldehyde Groups ($10^{-4}\ \mu eq/cm^2$)
DBG-0	$-COO^- + -SO_4^-$	399 ± 10	0.058	10.4 ± 0.3	–
DBG-0TT	$-COO^-$	385 ± 8	0.076	18.4 ± 0.8	–
AD-1	$-SO_4^- + -CHO$	349 ± 7	0.055	2.9 ± 0.2	1.17 ± 0.05
AD-1TT	$-COO^- + -CHO$	361 ± 6	0.069	13.4 ± 0.6	0.67 ± 0.03

original latex was measured and immediately after the heat-treated latex in the same conditions.

Results and discussion

The influence of electrolyte concentration on the particle electrophoretic mobility is summarized in Table 2, along with the corresponding electrokinetic radius for each concentration, κa, where κ is the reciprocal Debye–Hückel length and a is the particle radius. It can be observed that the maxima in μ_e for DBG-0 and AD-1 latexes are in a concentration of $5\,10^{-3}$ or 10^{-2} and $5\,10^{-3}$ M, respectively. After the heat treatment, the values of mobility at very low concentrations remain similar to the originals and decrease strongly from 10^{-3} M in both heat-treated latexes. Therefore, as a result of the heat treatment a displacement of the maximum to lower concentrations and low values of mobility at higher concentrations have been obtained.

Figures 1 and 2 show the ζ-potentials as a function of electrokinetic radius obtained from the O'Brien–White theory. The DBG-0 (Fig. 1) and AD-1 (Fig. 2) latexes display a maximum in ζ-potential when κ is around 40 in both cases. For DBG-0TT and AD-1TT latexes the maximum fully disappears, indicating that with the changes suffered on the surfaces of the heat-treated particles the considerations made in the O'Brien–White theory to transform μ_e in ζ-potential values are capable to eliminate this maximum. However, the most outstanding and controversial result is that on the basis of hairy layer model, the effect of the heat treatment is to restructure the surface,

creating smooth particles. Surface smoothing should move the shear surface inward and increase the ζ-potential at low and intermediate electrolyte concentrations and to converge with not heat-treated latexes at high electrolyte concentrations. Our results show that the ζ-potential values in heat-treated latexes are lower than their originals, the AD-1TT maintain this tendency in all range of concentrations, while the DBG-0TT present slightly high values of ζ-potential at low concentrations, but it is clearly observed that at higher concentrations the values do not tend to converge. The interpretation of the data is even more difficult to understand by the fact that the heat-treated latexes have higher surface charge densities that their originals. Similar results, were obtained by Rosen and Saville working with poly (methil methacrylate)/

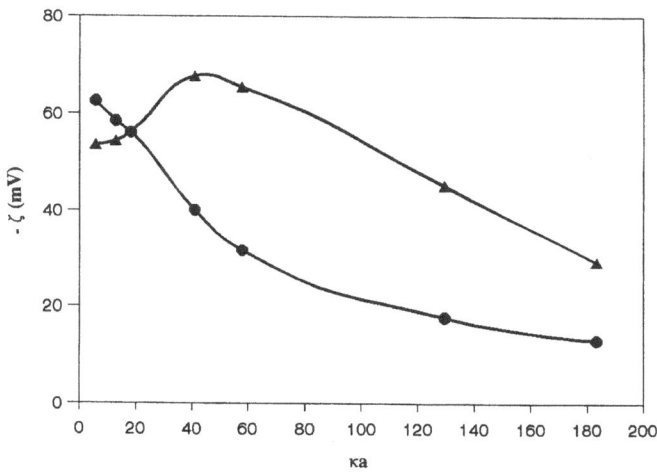

Fig. 1 ζ-potentials from O'Brien–White theory versus κa for DBG-0 and DBG-0TT latexes. (▲, DBG-0; ●, DBG-0TT)

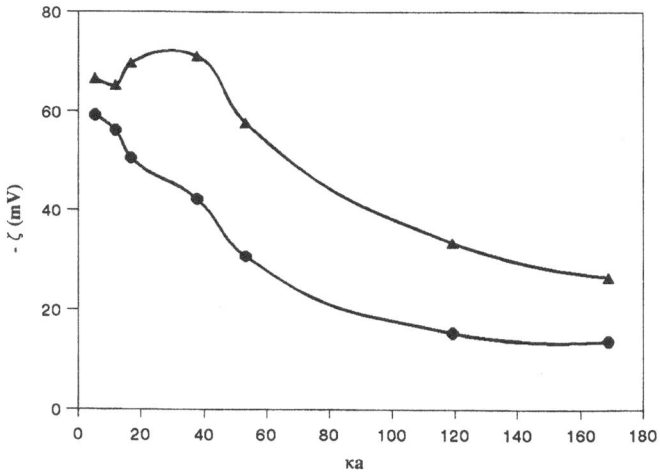

Fig. 2 ζ-potentials from O'Brien–White theory versus κa for AD-1 and AD-1TT latexes. (▲, AD-1; ●, AD-1TT)

Table 2 Electrophoretic mobilities (μ_e) and electrokinetic radius (κa) for original and heat-treated latexes in various NaCl solutions

[NaCl] (M)	κa	μ_e (DBG-0) (10^{-8} m²/Vs)	μ_e (DBG-0TT) (10^{-8} m²/Vs)
10^{-4}	5.8	−2.91	−3.25
$5\,10^{-4}$	12.9	−3.29	−3.49
10^{-3}	18.3	−3.56	−3.55
$5\,10^{-3}$	40.9	−4.58	−2.87
10^{-2}	57.9	−4.61	−2.32
$5\,10^{-2}$	129.4	−3.41	−1.35
10^{-1}	183.3	−2.23	−1.00

[NaCl] (M)	κa	μ_e (AD-1) (10^{-8} m²/Vs)	μ_e (AD-1TT) (10^{-8} m²/Vs)
10^{-4}	5.4	−3.35	−3.10
$5\,10^{-4}$	11.9	−3.72	−3.34
10^{-3}	16.9	−4.12	−3.63
$5\,10^{-3}$	37.7	−4.71	−3.00
10^{-2}	53.4	−4.10	−2.25
$5\,10^{-2}$	119.3	−2.53	−1.17
10^{-1}	168.7	−2.03	−1.06

acrolein particles and using potassium persulfate as initiator [19]. They hint that if the ion adsorption exists, a possible explanation could be found. Works of Elimelech and O'Melia [7] as well as Voegtli and Zukoski [20] suggest that the electrokinetic charge of surfactant free latexes is strongly influenced by ion adsorption. Elimelech and O'Melia measured the electrophoretic mobility of polystyrene particles in various types of inorganic electrolytes (1:1, 2:1 and 3:1) and concluded that competitive processes involving counter and coion adsorption at the interface determine the shape of the mobility curves. Voegtli and Zukoski devised a nonelectrokinetic experiment for determining the total amount of ion adsorption. Their results imply that both counterions and coions bind in significant numbers to the surface, and their nonelectrokinetic adsorption densities agree with predictions based on a dynamic Stern layer analysis [21] of complementary electrophoretic mobility and static conductivity data [20].

If the ion adsorption occurs, the decrease in ζ-potential after the heat treatment could be related with a change in the ion adsorption and this would be reflected in some property of the heat treated particles. In the previous paper mentioned before [13], we carried out the adsorption of a non-ionic surfactant and found that after heat treatment the latexes had suffered a decrease in their hydrophobicity. Nevertheless, it is difficult to ensure that these variations (a decrease around 12.5% for DBG-0TT and 27.4% for AD-1TT) could explain the great differences encountered in the ζ-potential. Moreover, as Rosen and Saville also point out if surface transport inside the shear plane also occurs to varying degrees, the interpretation of mobility data is further complicated.

Acknowledgments This work is financially supported by the Comisión Interministerial de Ciencia y Technología (CICYT), proyect MAT 93-0530-C02-01.

References

1. O'Brien KH, White LR (1987) J Chem Soc Faraday Trans 2 77:1607
2. Semenikhin NM, Dukhin SS (1975) Kollidn Zh 37:1127
3. Hunter RJ (1981) In: Zeta Potential in Colloid Science Ed Matijevic E, Vol 7, Wiley, New York
4. Hidalgo-Alvarez R (1991) Adv Colloid Interface Sci 34:217
5. Goff RJ, Luner P (1984) J Colloid Interface Sci 99:468
6. Chow RS, Takamura K (1988) J Colloid Interface Sci 125:1
7. Elimelech M, O'Melia ChR (1990) Colloids Surfaces 44:165
8. Verdegan BM (1992) Ph D Thesis, Wisconsin-Madison University
9. van der Linde AJ, Bijsterbosch BH (1990) Croatica Chim Acta 63:455
10. Overbeek JThG (1943) Kolloidchem Beih 54:287
11. Booth F (1950) Proc Roy Soc (London) Ser A 203:504
12. Wieserma PH, Loeb AL, Overbeek JThG (1966) J Colloid Interface Sci 22:78
13. Bastos-González D, Hidalgo-Alvarez R, de las Nieves FJ (1996) J Colloid Interface Sci 177:372
14. Goodwin JW, Hearn J, Ho CC, Otewill RO (1973) Brit Polym J 5:347
15. Yan ch, Zhang X, Sun Z, Kitano H, Ise N (1980) J Applied Polym Sci 40:89
16. Bastos D, Santos R, Forcada J, Hidalgo R, de las Nieves (1994) Colloids Surfaces A: Physicochem Eng Aspects 92:137
17. Kamel AA-M, Ph D Thesis Lehigh University 1981
18. Van del Hul HJ, Vanderhoff JW (1971) Brit Polym J 2:121
19. Rosen LA, Saville DA (1992) J Colloid Interface Sci 149:542
20. Voegtli LP, Zukoski CF (1991) J Colloid Interface Sci 141:92
21. Zukoski CF, Saville DA (1986) J Colloid Interface Sci 114:32

Progr Colloid Polym Sci (1996) 100:221–229
© Steinkopff Verlag 1996

N. Stubičar
K. Banić
M. Stubičar

Kinetics of crystal growth of α-PbF₂ and micellization of non-ionic surfactant Triton X-100 at steady-state condition

Dr. N. Stubičar (✉) · K. Banić
Laboratory of Physical Chemistry
Chemistry Department
Faculty of Science
University of Zagreb
Marlićev trg 19/II
P.O. Box 163
10001 Zagreb, Croatia

M. Stubičar
Physics Department
Faculty of Science
University of Zagreb
Bijenička C. 32
P.O. Box 162
10001 Zagreb, Croatia

Abstract A competition between the crystal growth rate of orthorhombic lead fluoride at 25 °C and the micellization of non-ionic surfactant Triton X-100, due to its adsorption/desorption and interaction with the lattice ions, was investigated at steady-state conditions, using pF-stat method. The relatively low super-saturated, equimolar solutions of constant composition at isoelectric pH 5.2–5.3 were used for growth of well-defined seeds, in order to modify the kinetics of growth without affecting the anisotropic optical properties of these crystals.

The promotion of rate up to five times, in two steps, took place if Triton X-100 micellar solution is in the range between CMC and c_1 (10 × CMC), which is characterized with the ratio between the activity of free fluoride ions and Triton X-100 concentration greater than 1. Above the c_1 of Triton X-100 the inhibition of crystallization rate, two to three times, is evidenced, caused by the depletion effect, and the mentioned ratio is smaller than 1. The main role in the proposed mechanism plays α-PbF₂ crystal surface with F⁻ active sites being attached by spherical (below c_1) and above it by elongated bilayer chains of primary micelle aggregates. This is confirmed by IR spectra of solids and was directly visualized by polarizing microscope examinations. Both of these effects are quantitatively expressed as a degree of promotion/inhibition, using the rates in the absence and presence of non-ionic surfactant. The pF-stat kinetic data were also analyzed in the sense of adsorption/desorption process through the adapted Langmuir adsorption isotherm. The mutual changes in Gibbs free energies of micellization and micellar bilayer formation, the adsorption/desorption (attachment) of Triton X-100 and crystal growth of α-PbF₂ are discussed.

Key words Kinetics of crystal growth – lead fluoride – micellization of Triton X-100 – non-ionic surfactant – pF-stat measurements

Introduction

In sufficiently supersaturated solution of sparingly soluble salts like lead fluoride phases at constant temperature, ionic strength, pH, constant degree of supersaturation and specific surface area of seeds, the presence of non-ionic surfactant micelles may greatly affect kinetics and mechanism of crystal growth. Namely, the effect may be either to reduce or also to enhance the growth rate. Generally, the same non-ionic surfactant do act as *inhibitor and/or promoter* depending on colloidal characteristics of both: ionic crystals and non-ionic surfactant and on their mutual mass ratio. The promotion/hindrance of either the growth rate or aggregation of colloidal dispersions is traditionally reviewed as arising from either electrostatic or steric interactions; we mention here only some older papers [1–4]. From the fundamental point of view, it is especially interesting that there is an intermediate space region in which both electrostatic and steric effects contribute to the overall growth rate like in the present case, where these two repulsions are not independent of each other. At the dynamic conditions as it is steady-state crystal growth of lead fluorides using the pF-stat method, known as the constant composition method [5], non-ionic surfactant Triton X-100 acts through accumulation at the solid/solution interface and through micellization in the bulk of a solution, which is in turn affected by the supersaturated solution as regards to the sparingly soluble salt present. The non-ionic surfactant's adsorption and micellization alters both: i) dynamic surface properties of lead fluoride microcrystals, and ii) composition and structure of the supersaturated solution used for growth. In fact, surface properties of lead fluoride having an amphoteric character play a dominant role in the kinetics of growth by changing the mechanism of growth (thermodynamic effect) [6–8]. Namely, it was shown previously that changing the ratio between lead nitrate to potassium fluoride concentration (activity), the attained equilibrium pH has changed from the acidic, due to a hydrolysis of lead nitrate in the case of its large excess, to neutral or mildly alkaline in the excess of potassium fluoride over it, and the result is a change, not only in size and size distribution, then also in morphology and even in the crystal structure of solids grown. Besides the known orthorhombic, α-PbF$_2$ phase and cubic β-PbF$_2$ phase determined by XRD analysis, the unknown structure of needle-like phase was detected.

In this paper, we present the influence of Triton X-100 on the kinetics of crystal growth and aggregation of orthorhombic α-PbF$_2$ phase at 25 °C, and some of the other facts such as microscopic examinations and IR spectroscopy necessary to explain the mechanism of these processes. The micellization of Triton-X-100, p-(1,1,3, 3-tetramethylbutyl)phenoxy-poly(oxyethylene)glycol, was studied in detail in many laboratories including ours, and has attracted great interest since 1954. It has been shown by us [9, 10] that it forms primary spherical micelles at relatively low solute concentration (3×10^{-4} mol dm^{-3} or 0.188 mg cm^{-3}), having a hydrophobic core of $r \sim 1.4$ nm, surrounded by an outside hydrophilic shell of $r \sim 5.6$ nm, i.e., radius of primary spherical Triton X-100 micelles in water at 20 °C is about 7 nm, calculated from the radius of gyration $R_g = 4.3 \pm 0.3$ nm, and the micellar molar mass is $(8.80 \pm 0.13) \times 10^4$ Da corresponding to $N_w \sim 140$, as was determined by SAXS measurements. With increasing either concentration, at about $c_1 = 10 \times$ CMC, or temperature, or the composition of solution, these soft spheres grow to become the so-called giant micelles and, finally, they transform into elongated bilayers, as we will show on photomicrographs.

As we have proved in the above-mentioned papers [6–8], the formation of lead hydrolytic species, hydroxyl complexes and ion pairs in solution alters the mechanism and consequently the kinetics of growth. Here it is necessary to consider the interactions of Pb(NO$_3$)$_2$ and KF electrolytic solutions with the NSAA micelles and monomers as the influents of the mechanism of electrosteric reduction/acceleration of crystal growth. Dynamic light scattering (DLS) and other relevant methods were employed for this purpose, but the details will be described elsewhere. In this paper, we present the relative role which have some of these various factors in solution on the kinetics and mechanism of growth of well characterized lead fluoride seeds at the dynamic steady-state conditions. That means stationary activity of fluoride ions and unchangeable parameters during the course of process such as: supersaturation, ionic strength, pH, non-ionic surfactant concentration, temperature and conditions such as the preparation procedure and the way and speed of mixing.

Experimental section

Materials

The chemicals used: KF (Merck, Darmstadt), Pb(NO$_3$)$_2$, KNO$_3$ (Kemika, Zagreb), Triton X-100 for scintillation measurements (Merck, Darmstadt, No 11869), viscous liquid, $\rho = 1.05$ kg/dm^3, and standard pH buffers: pH = 4.00, 5.00, 6.00 (Kemika, Zagreb), were of reagent grade. The solutions were prepared with twice-distilled water filtered through membrane filters (0.1 mm, Millipore, Bedford, MA). The stock KF solution was standardized potentiometrically using the standard Ca(NO$_3$)$_2$ solution in 75 vol.% EtOH. Concentrated KF solutions were stored in plastic bottles to avoid interactions with silicates. The stock lead (II) nitrate solution was standardized volumetrically with standard EDTA and xylenol-orange indicator.

Methods

The potentiometric pF-stat set-up consisted of Titrator Radiometer (Copenhagen, Denmark), which governs two autoburettes: ABU-12 Radiometer and Multi-dosimat 645 with printer Dosigraph 625, fluoride ion-selective electrode, Metrohm, Herisau, Switzerland, coupled with a saturated calomel electrode, Iskra, Kranj, Slovenia, in a double-walled reaction vessel (300 cm^3 volume) thermostat with circulating water (25.0 ± 0.1) °C, as was described earlier [6–8]. The details of the procedure are given below.

The characterization of the system was done by polarizing microscopy (with λ plate) for checking anisotropy, shape and rough estimation of crystal size, by pH measurements of suspension to check its constancy and by taking IR spectra of solid phase, as was described previously [8]. The Leitz–Wetzlar microscope (Germany) equipped with crossed polars and λ plate, connected to an automatic camera was used. IR spectra were scanned in the region 4000–200 cm^{-1} using Perkin–Elmer Model 783 spectrophotometer and KBr pellet technique. Always 10 mg of samples were used per 100 mg of KBr.

Procedure

The experimental procedure was as follows: the supersaturated solution was prepared by slow addition of titrant KF first, with checking the Nernstian response of F-electrode in each run, and then addition of titrant Pb(NO$_3$)$_2$ into the water or aqueous solution of Triton X-100 of the final appropriate concentration. The opposite order of addition was by performing potentiometric titration for the determination of the activity product of PbF$_2$ in T-X-100 micellar solution, which was computed using the known Gran plot analysis. In pF-stat measurements 1 ml of seed slurry was added with Justor micropipette (always about 50 mg of α-PbF$_2$) into such stable supersaturated solution (after checking the constancy of E_{MF} for 30 min). The seeds were characterized by the activity ratio 1/1.993, pH = 5.3, orthorhombic α-PbF$_2$ crystal structure (XRD analysis), specific surface area $A_s = 2.7$ m^2/g (BET method), and have anisotropic voluminous bipyramidal shape and all spectral colors under polarized light. An aliquot 10 ml of suspension was withdrawn, after each 10 ml of KF and 1 ml of Pb(NO$_3$)$_2$ addition, so that the mass increment of PbF$_2$ grown was $\Delta m = m/m_0 = 2, 3, \ldots, 7$. The change in T-X-100 concentration of the working solution due to the titrants addition during the course of crystallization was up to 4%. These samples were used for microscopic examination and pH measurements. At the end of the run the suspension was pressure-filtered (or left 1 day to settle completely), dried in a desiccator over silica-gel at room temperature, and then the IR spectra of solid phase were taken.

Analysis of pF-stat measurements

The rate of crystal growth, calculated from the perfect linear change of volume added in time, which is proportional to the rate, and normalized to the mass of seeds, is given by an empirical equation:

$$j/(\text{mol s}^{-1}\,\text{g}^{-1}) = k_j\,A_s(m/m_0)^q\,(K_{S_0}^\ominus)^{p/3}\sigma^p . \tag{1}$$

This implies that for a particular crystal phase (constant thermodynamic activity product $K_{S_0}^\ominus$), at constant relative supersaturation (σ), the rate is proportional to the increment of mass (m/m_0) on the exponent $q = (2/3)$, (see e.g. ref. [11]), times the specific surface area (A_s). If this product is constant, then it is possible to calculate the apparent order of the overall growth rate (p). It was found out in our experiments with lead fluorides without T-X-100 [8] that this was achieved after 100–200% of growth (i.e., $m/m_0 = (2–3)$). This is also very similar to the results on the other systems as calcium phosphates [12], where constant A_s was achieved after about 100% of growth using also the constant composition study. The relative supersaturation, σ, for 2:1 salt is defined by:

$$\sigma = [\text{AP}/K_{S_0}^\ominus]^{1/3} - 1 , \tag{2}$$

where $K_{S_0}^\ominus$ is the thermodynamic solubility product at $I \to 0$ and 25 °C. The value $(1.20 \pm 0.08)_{14} \times 10^{-8}$ was taken, as an average of our experimentally determined values for α-PbF$_2$ in water. The activity product is defined as:

$$\text{AP} = \text{a}(\text{Pb}^{2+})\,\text{a}^2(\text{F}^-) = [\text{Pb}^{2+}][\text{F}^-]^2 y_\pm^6 . \tag{2a}$$

Mean activity coefficient $y_\pm$ of univalent species may be computed using the simple Debye–Hückel equation, so that $y(\text{Pb}^{2+}) = y^4(\text{F}^-) = y_\pm^4$, if all relevant equilibria (as stated in ref. [7]) and their stability constants are known, which is not the case for the micellar T-X-100 solutions. The potentiometric titration of Pb(NO$_3$)$_2$ in 2×10^{-2} mol dm^{-3} T-X-100 solution (well above c_1) with KF, gave 3 to 5 times smaller activity product compared to that in water: AP (PbF$_2$ in micellar T-X-100, at 25 °C) $= 4.2 \times 10^{-9}$. Hence, the particular activity products (for σ calculation) were calculated from the experimentally determined activities of free F$^-$ ions in supersaturated solution; from the measured potential of F electrode, using Nernst equation with the slope (58.94 ± 0.21) mV and the intercept (243.3 ± 0.6) mV (correlation coefficient was 0.999). In the experiments described in this paper with the same concentration of constitutive ions in each run, the change in σ is caused exclusively because of lattice ions interaction with T-X-100.

The driving force for crystallization is the Gibbs free energy change per mole of formula unit: $-\Delta G_{\text{crys}}/\text{kJ mol}^{-1} = 3RT \ln(\sigma + 1)$.

The degree of promotion and inhibition, i, of crystal growth rate we defined as:

$$i = (j_0 - j_i)/j_0 , \tag{3}$$

where j_0 and j_i are the rates of crystallization in the absence and presence of non-ionic surfactant, respectively. So it holds: $0 < i < 1$ for inhibition, and $0 < i < -1$ for promotion of growth.

The pF-stat kinetic data may be analyzed in terms of the adsorption/desorption process too, and may be transformed to the Langmuir adsorption isotherm in the form:

$$j_0/(j_0 - j_i) = (1 - b)^{-1} + [K(1 - b)\, C]^{-1}, \qquad (4)$$

where $K = k_{ads}/k_{des}$ is also assigned as the measure of the surfactant "affinity" for the surface (see e.g. ref. [13]).

Results and discussion

pF-stat kinetics

Since we were aware that surface state of lead fluoride crystals plays a dominant role in the kinetics of growth [8], we have studied the kinetics in the presence of non-ionic surfactant T-X-100 at high ionic strength (excess of $Pb(NO_3)_2$, $I = 0.1$ mol dm^3, further referred to as exp. VIII) i.e., the situation where positive charge is compensated by the counterions (NO_3^-), and at relatively low ionic strength ($I = 0.018$ mol dm^{-3}, referred to as exp. IV), i.e., stoichiometric activity ratio, where negative charge is a result of certain small amount of F^- ions because of its comparatively greater absorbability. The rates of growth of these systems and normalized rates to the mass increment as a function of mass increment in respect to the initial mass of seeds at time zero, are presented in Figs. 1, and 2 (top and bottom). It is obvious that the growth rates are in all cases linear functions and the normalized rate curves are second order functions with the correlation coefficients always above 0.96 (lines in graphs present fitted experimental points). The maximum deviation up to 4% includes a slight change in T-X-100 concentration due to dilution during the growth process. The growth rate of systems performed at high I decreases with T-X-100 concentration, and normalized rate exponential curves coincide for all T-X-100 concentrations and normalized rate exponential curves coincide for all T-X-100 concentrations (a small deviation at larger masses is also a result of dilution). The mechanism taking place in this case is known in colloid chemistry as a steric stabilization effect at high counter ion concentrations [14]. That means that the same mechanism is involved below and above c_1 of T-X-100 in solution, only the stabilization effect increases with T-X-100. However, at low ionic strengths the rate of growth increases with T-X-100 concentration and reaches the limiting value at the concentration $c_1 = 3 \times 10^{-3}$ mol dm^{-3} (referred to exp. IV.N), and thereafter it decreases. Also the normalized rate curves do not

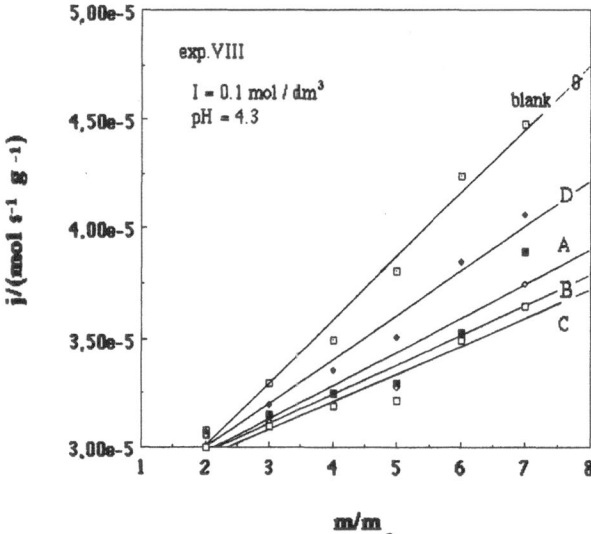

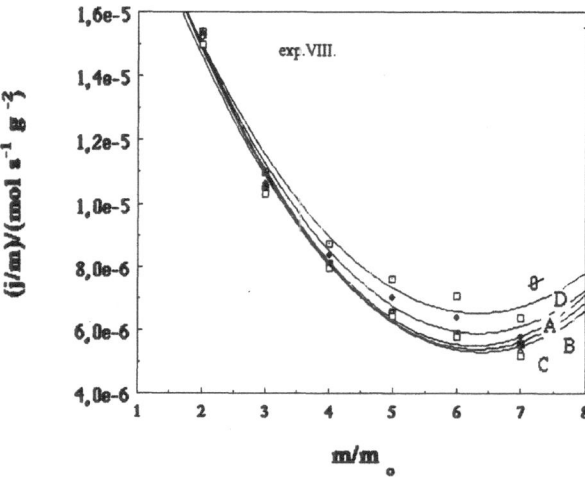

Fig. 1 pF-stat rate (top) and normalized rate (bottom) of lead fluoride crystal growth in supersaturated solution: $c(Pb(NO_3)_2 = 0.1$ and $c(KF) = 4 \times 10^{-3}$ mol/dm^3, pH = 4.3 without Triton X-100 (O), and with Triton X-100: (D) 3×10^{-4} mol/dm^3 and pH = 3.8, (A) 4×10^{-3} mol/dm^3 mol/dm^3 and pH = 3.8, (B) 0.01 mol/dm^3 and pH = 3.75, (C) 0.1 mol/dm^3 and pH = 4.05. Needle-like crystals. Experimental dots and fitted lines

coincide, indicating that different processes are involved depending on T-X-100 concentration. This is presented in Fig. 3, where two distinct maxima of the growth rates are seen: a side maximum at lower and a main maximum at higher T-X-100 concentrations. The promotion of the growth rate in the T-X-100 concentration region between CMC and c_1 is up to five times, which is exactly the same as an enhancement of lead fluoride growth caused by high electrolyte concentration as is demonstrated with 0.1 mol dm^{-3} KNO$_3$ published in ref. [8]. The reduction

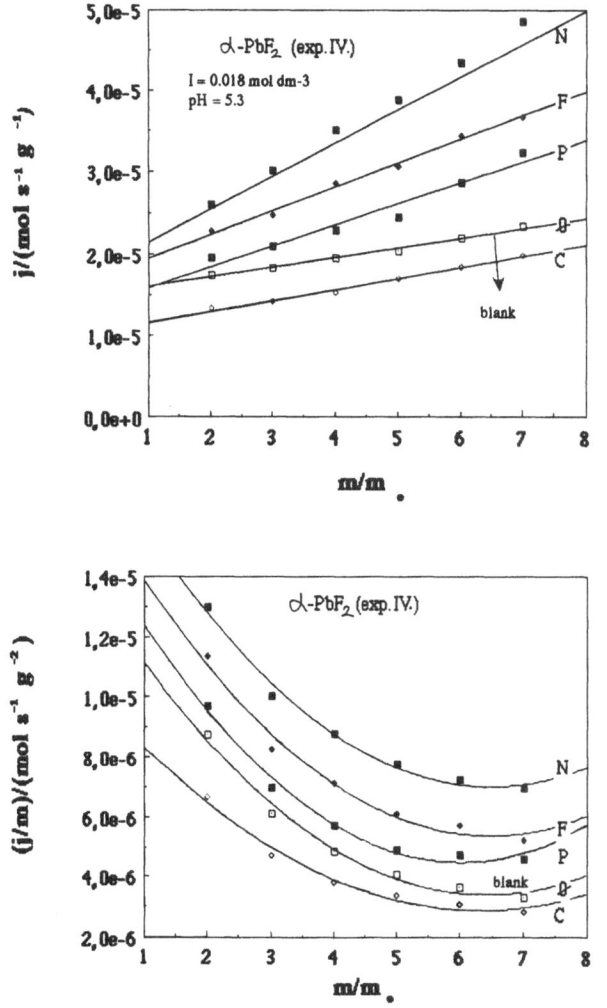

Fig. 2 pF-stat rate (top) and normalized rate (bottom) of α-PbF$_2$ in supersaturated solution: $c(Pb(NO_3)_2) = 3.5 \times 10^{-3}$ and $c(KF) = 5.5 \times 10^{-3}$ mol/dm^3, pH = 5.25 without Triton X-100 (O), and with Triton X-100: (F) = 6×10^{-4} mol/dm^3, (N) 3×10^{-3} mol/dm^3, (P) 8×10^{-3} mol/dm^3 and (C) 0.1 mol/dm^3, pH = 5.2 or 5.3 in all cases

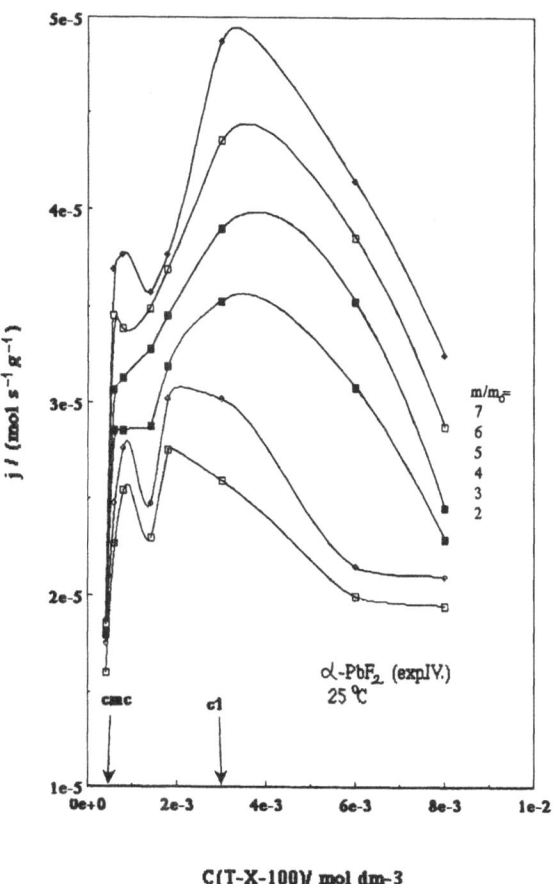

Fig. 3 pF-stat rate of growth of α-PbF$_2$ as a function of Triton X-100 concentration, showing two maxima referring to two different processes. (Numbers denote increment of mass of PbF$_2$ grown as regards to the initial mass of seeds)

of the growth rate by about two to three times is evident above c_1 up to 0.01 mol dm^{-3} T-X-100. The inhibitory effect is accompanied and presumably attributed to the decrease of the solubility product of grown phase in micellar solution for two to three times. The experimentally determined ratios between the activity of F$^-$ ions and T-X-100 concentration are given in Table 1 and are used as a guide for the elucidation of the different processes taking place.

Interestingly, we found a direct correlation between the effective F$^-$ ion activity and T-X-100 concentration (and state) in supersaturated solution, leading to the promotion of growth if this ratio is greater than 1, and inhibition of growth if it is lower than 1, i.e., experiments with T-X-100

concentration up to 3×10^{-3} mol dm^{-3} and above it. Exponential decay of the ratio between the activity of free F$^-$ ions and Triton X-100 concentration as a function of Triton X-100 concentration is proved. Surface properties as a consequence of the non-ionic surfactant state in solution are responsible for the phenomena observed. Experimentally determined relative supersaturations, σ, fluctuate with Triton X-100 in the same manner as the pF-stat rate normalized to the mass of PbF$_2$ does, (not presented), reaching maximum below c_1. At c_1 there is a minimum in σ value and afterwards it increases with Triton X-100 concentration. Accordingly, the driving force for crystallization, expressed as Gibbs free energy change per formula unit of PbF$_2$, increases above c_1, process becomes more spontaneous (last column in Table 1).

Quantitatively the degree of promotion/inhibition of α-PbF$_2$ crystal growth, i, is presented in Fig. 4, as a function of mass increment (top) and of Triton X-100 concentration (bottom). It has negative values between 0 and 1 for

Table 1 Composition of the supersaturated solution used for α-PbF$_2$ crystal growth at 25°C. c(KF) = 5.5×10^{-3} mol/dm^3, c(Pb(NO$_3$)$_2$) = 3.5×10^{-3}, pH$_{eq}$ = 5.2 or 5.3 in all exp., c(Triton X-100) var

Exp.	c(TX-100) mol/dm^3	σ	a(F)/c(TX-100)	ΔG_{crys} kJ/mol
IV	0	0.50	0	−3.01
IVE	4×10^{-4}	0.10	6.3	−0.71
IVF	6×10^{-4}	0.33	5.1	−2.12
IVG	8×10^{-4}	0.25	3.6	−1.66
IVH	1×10^{-3}	0.49	3.4	−2.96
IVL	1.4×10^{-3}	0.53	2.5	−3.16
IVM	1.8×10^{-3}	0.60	2.0	−3.49
IVI	2×10^{-3}	0.51	1.7	−3.06
IVN	3×10^{-3}	0.34	1.0	−2.17
IVO	6×10^{-3}	0.65	0.6	−3.72
IVP	8×10^{-3}	0.76	0.5	−4.20

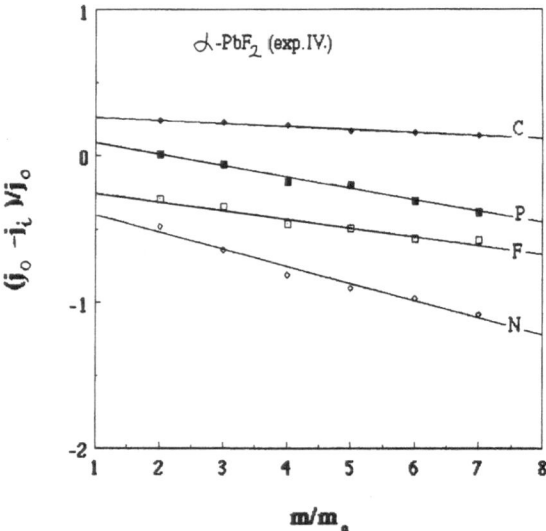

promotion and positive values for inhibition. The degree of promotion increases with T-X-100 concentration for all mass increments, and reasonably it is greater for larger amounts of PbF$_2$ and T-X-100. On the contrary, the inhibition effect is greater for the smaller amounts of PbF$_2$.

If we analyze the pF-stat kinetic promotion effect of T-X-100, according to Eq. (4), we get the linear relationships, especially for smaller mass increments and larger T-X-100 concentrations. We evaluate $K/\Delta m = 241 \pm 10$ and $\Delta G_{ads} = -13.6 \pm 0.2$ kJ/mol for systems denoted with IVF to IVN, and $\Delta G_{ads} = -15.0 \pm 0.2$ kJ/mol for systems IVL to IVN (on the left side of the main maximum in Fig. 3). For systems exhibiting stabilization effect (above c_1 we calculated $K/\Delta m = -35.1$, and $\Delta G_{ads, des} = 8.8 \pm 0.7$ kJ/mol, respectively, which means that desorption is here a spontaneous process.

If we compare these three competitive processes taking place at the dynamic steady-state, for the systems between CMC and c_1, the most energetically favored process is the formation of spherical micellar aggregates ($\Delta G_{mic} = -20.1$ kJ/mol) then adsorption, ($\Delta G_{ads} = -13.6$ and -15.0 kJ/mol, respectively) and finally crystallization ($\Delta G_{crys} \approx -2$ kJ/mol). Above c_1 the transformation to elongated micellar aggregates is comparably less favorable ($\Delta G_{mic1} = -14.4$ kJ/mol) adsorption is unfavorable ($\Delta G_{ads, des} \approx 8.8$ kJ/mol, from the kinetic pF-stat results), and $-\Delta G_{crys}$ starts to increase with Triton X-100.

Two conflicting effects are involved here: reduction of electrostatic repulsive interactions, reduction of crystal-solution interfacial tension (thermodynamic effect) tends to increase aggregation or crystal growth; on the contrary, second effect, the blocking of surface active sites, impedes an advance of new crystal surface nuclei incorporation, and thus reduces the rate of growth.

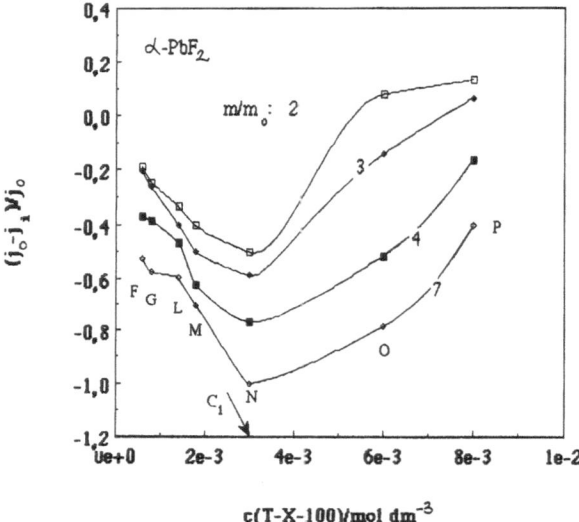

Fig. 4 Degree of promotion/inhibition (see Eq. 3) of α-PbF$_2$ crystal growth as a function of Triton X-100 concentration (bottom) and mass of PbF$_2$ (top)

Microscopic observations

Microscopic observations using polarized light and λ plate, shown on the photomicrographs Figs. 5b–h, are excellent direct visual evidence, which gives an explanation of the mechanism of rate promotion described in pF-stat kinetic study (maximum in Fig. 3). The spherical Triton X-100 giant micellar aggregates of primary micelles consisting of the anisotropic bilayer (or probably trilayer chains), with attached lead fluoride crystals are seen (b, c, d). Many examples of optical interference, Fig. 5b are seen on the pictures of these systems: black cross with two blue and two yellow fields vis-à-vis, if optic axis coincides with microscope axis (see e.g. ref. [15]). It can be directly observed (due to dehydration on microscopic glass) that curvature of the spherical Triton X-100 micellar aggre-

gates has become reduced, so that very sudden aggregation of crystals took place (e, f); and finally, very thin bilayers are formed which produce regular net-like structure with lead fluoride trapped either between two layers, along

Fig. 5 Photomicrographs of α-PbF$_2$ crystals and Triton X-100 micellar aggregates in 3×10^{-3} mol/dm^3 Triton X-100 solution (exp. IVN; for the other data see Table 1), which show the promotion of rate (micrographs 5b–h). Micrograph 5a shows crystals grown in 0.1 mol/dm^3 Triton X-100 (exp IVC), which display blue or yellow color under the polarizing light. These crystals are about three times smaller in size as compared to the no-surfactant system (taken at the same % of mass increment, which is 600% in all presented cases)

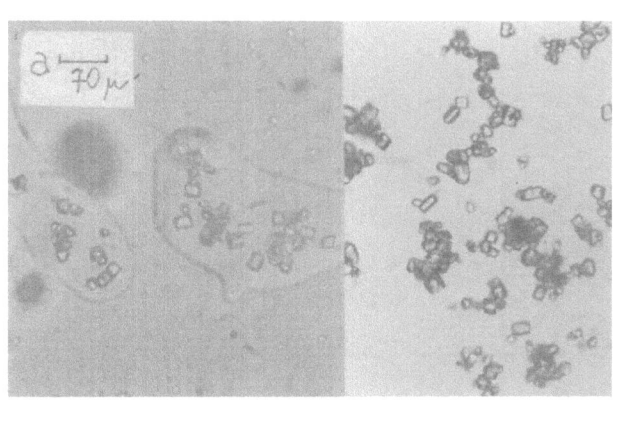
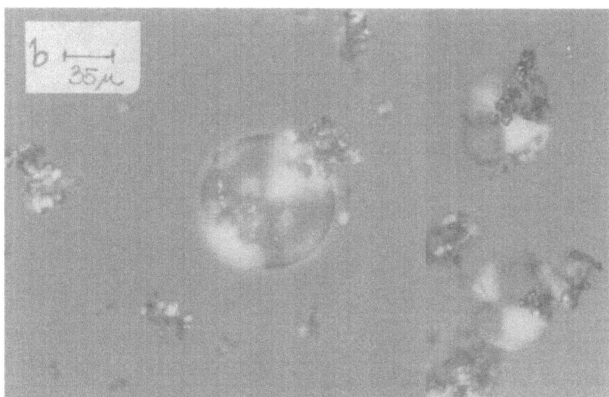
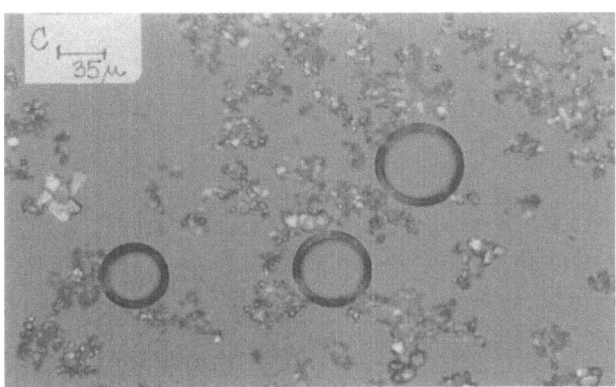
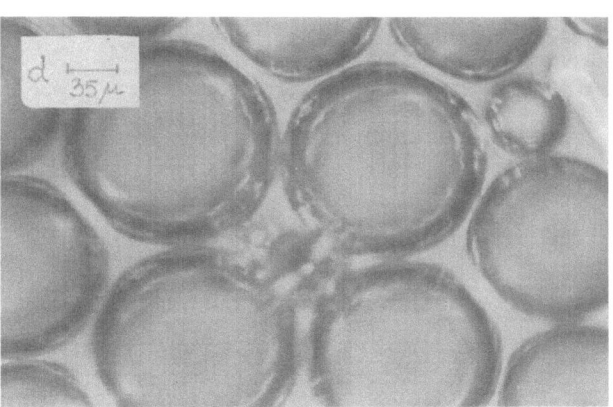
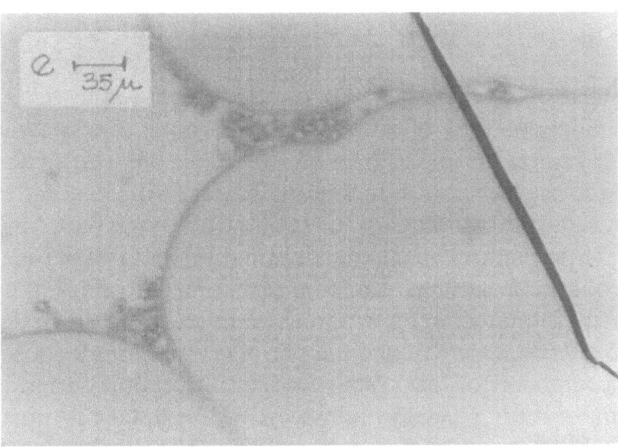
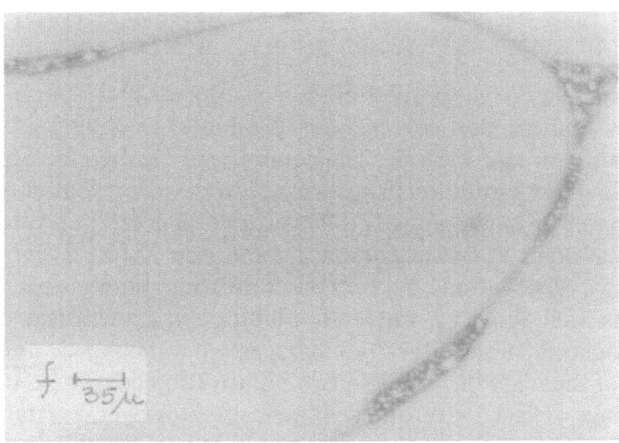

228

N. Stubičar et al.
Kinetics of lead fluoride crystal growth

Fig. 5 Continued

Fig. 6 IR spectra of α-PbF$_2$ grown in supersaturated solution: C(KF) = 5.5 × 10^{-3}, c(Pb(NO$_3$)$_2$) = 3.5 × 10^{-3} mol/dm^3, pH = 5.3 (spectrum 1), the same as in 1 plus 0.1 mol/dm^3 KNO$_3$ (spectrum 2), as in 1 plus 1.5 × 10^{-3} mol/dm^3 KOH pH = 6.3 (spectrum 3), as in 1 plus 2 × 10^{-3} and 2 × 10^{-1} mol/dm^3 Triton X-100, respectively (spectrum 4 and 5)

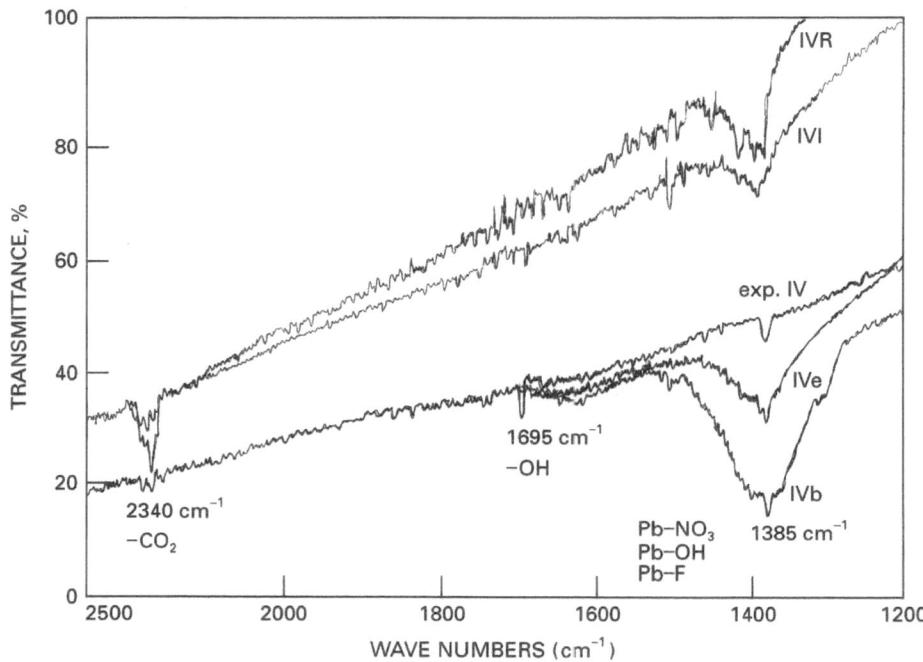

them (g), or at the corners of these net (h); the hydrophobic environment between bilayers promotes secondary nucleation of lead fluoride. The equilibria are shifted toward solid phase formation in accordance to the lower solubility in micellar media above c_1. This situation corresponds to the systems at concentration c_1 of Triton X-100. Microscopic observations, 5a, of crystals grown in micellar solutions well above c_1, where an inhibition of growth rate is evidenced, show rather monodisperse parallelpipeds and all of them are displayed in two polarization colors: either orange/yellow or turquoise/blue (a, b). However, crystals grown from the same equimolar solutions without Triton

X-100 show a whole spectrum of colors (from red to violet), depending on their thickness. The aggregation of crystalline units causes the increase of pF-stat rate, but size of the crystals is smaller with increasing Triton X-100 concentration in the range from CMC up to about 100 times the CMC.

These pictures are also a confirmation that this particular non-ionic surfactant is, to a certain extent, efficient as growth inhibitor; it, by two to three times slows down the rate of crystal growth and the size of crystals, but it causes no contamination of crystals through incorporation inside them, so it can be almost fully recycled; the uptake is rather reversible (the adsorption data will be presented elsewhere).

FT-IR examinations

Infrared spectroscopy was used to get information about the nature of Triton X-100 adsorption on the α-PbF$_2$ crystal surface. In Fig. 6, five spectra of lead fluoride grown from the equimolar supersaturated solution of lattice ions are presented: pure α-PbF$_2$ (spectrum 1), the same grown from the $0.1 \, mol \, dm^{-3} \, KNO_3$ solution (spectrum 2), grown from the $1.5 \times 10^{-3} \, mol \, dm^{-3} \, KOH$ i.e., pH = 6.3 (spectrum 3), and grown from the 2×10^{-3} and $2 \times 10^{-1} \, mol \, dm^{-3}$ Triton X-100 solution, respectively, i.e., about c_1 and well above it (spectra 4 and 5). The main peak at about $1385 \, cm^{-1}$, assigned to the Pb-NO$_3$, Pb-OH and Pb-F ion-pairs respectively, become more pronounced due to KNO$_3$ and KOH (spectra 2 and 3). It is accepted in the literature ([16] and references therein) that lead nitrate aqueous solution is structured as an octahedral complex consisting of four nitrate groups (in x, z plane) and two water molecules (along with y axis). The peak at about $1385 \, cm^{-1}$ show all metal fluorides. A great resemblance is obvious of the peaks, in spectra 2 and 3, with those in spectra 4 and 5, although they are much broader and shifted to somewhat greater wavenumbers, and of course they are stronger for higher Triton X-100 concentration. Also, a very weak peak at about $1695 \, cm^{-1}$ assigned to the $-OH$ groups is evident in spectra 3, 4, 5. We must stress that the difference between the spectra of solids without and with Triton X-100 is not detectable for smaller amounts of the samples with respect to KBr in pellets, as it was mentioned in the experimental part, especially not for smaller Triton X-100 concentrations. These spectra show that Triton X-100 bonding to the α-PbF$_2$ crystal surface is through $-OH$ groups, either deliberated in Triton X-100 micellar solutions, or through its $-OH$ endgroups. Namely, the steady linear increase of pH with Triton X-100 concentration above c_1 and its change to protonated form is measured in 0.03 and $0.003 \, mol \, dm^{-3}$ Pb(NO$_3$)$_2$ solution and also in $0.1 \, mol \, dm^{-3}$ KF solution.

Conclusion

The competition between three processes taking place in mixed system: ionic crystal – non-ionic surfactant, have been studied at dynamic condition. They are: crystal growth of orthorhombic lead fluoride from the supersaturated solution taking place on well defined seeds, formation of spherical giant micelles and afterwards elongated micellar bilayer aggregates and, finally, their mutual interactions through surface active sites, as well as with ionic species in bulk of solution (this last point will be presented elsewhere). In this paper the results of kinetic study at 25 °C are presented, with special attention to explain the mechanism, using pF-stat method, which was found very suitable for such investigations. The micellar solution of non-ionic surfactant makes the solution more hydrophobic, causing the reduction of the PbF$_2$ solubility by about five times, as was determined by potentiometric titration. The pF-stat rate was accelerated to the same extent, reaching the limiting value just at about c_1 ($c_1 \approx 10 \times CMC$), followed by an inhibition of rate for about two to three times (depends on mass of PbF$_2$), as regards the non-surfactant system. The rate of promotion is *de facto* aggregation of crystals of relatively equal size, which is facilitated by their attachment to the bilayer micellar aggregates of Triton X-100. This is directly visualized on photomicrographs. The IR spectra of solids show the nature of Triton X-100 attachment to the lead fluoride surface. Also the PbF$_2$ crystals incorporated inside the corners of three bilayers (net-like structure) are seen, due to the secondary nucleation which was promoted by hydrophobic environment.

Acknowledgment This work was supported by the Ministry of Science and Technology, Republic of Croatia (Project No: 1-07-223).

References

1. Matai KG, Ottewill RH (1966) Trans Faraday Soc 62:750, ibid 759
2. Ash SG, Clayfield EJ (1976) J Colloid Interface Sci 55:645
3. Glazman Yu, Blashchuk Z (1977) J Colloid Interface Sci 62:158
4. Puzigaća-Stubičar N, Težak B (1977) Croat Chem Acta 49:395
5. Tomson MB, Nancollas GH (1978) Science 200:1059
6. Stubičar N, Ćavar M, Škrtić D (1988) Progr Colloid Polym Sci 77:201
7. Stubičar N, Ščrbak M, Stubičar M (1990) J Crystal Growth 100:261
8. Stubičar N, Marković B, Tonejc A, Stubičar M (1993) J Crystal Growth 130:300
9. Stubičar N, Petres JJ (1981) Croat Chem Acta 54:255
10. Stubičar N, Matejaš J, Zipper P, Wilfing R (1989) In: Mittal KL (ed) Surfactants in Solution, Vol 7. Plenum, New York, p 181
11. Christoffersen J, Christoffersen MR (1990) J Crystal Growth 100:203
12. Koutsoukos P, Amjad Z, Tomson MB, Nancollas GH (1980) J Amer Chem Soc 102:1553
13. Amjad Z (1993) Langmuir 9:597
14. Hunter RJ (1989) Foundations of Colloid Science, Vol I and II, Oxford University Press, London
15. Muir ID (1977) In: Zussmann J (ed) Physical Methods in Determinative Mineralogy, Chapter 2, Academic Press, London
16. Frostemark F, Bengtsson LA, Holmberg B (1994) J Chem Soc Faraday Trans 90:2531

Progr Colloid Polym Sci (1996) 100:230–234
© Steinkopff Verlag 1996

O. Lopez
A. de la Maza
L. Coderch
J.L. Parra

Selective solubilization of the stratum corneum components using surfactants

O. Lopez · A. de la Maza (✉)
L. Coderch · J.L. Parra
Departamento de Tensioactivos
Centro de Investigacion y Desarrollo
(C.I.D.)
Consejo Superior de Investigaciones
Cientificas (C.S.I.C.) C/Jordi Girona 18–26
08034 Barcelona, Spain

Abstract The selective solubilization of pig stratum corneum (SC) by the nonionic surfactant octyl glucoside (OG) was investigated. The solubilized components (lipids and proteins) were freed of surfactant by dialysis. The lipid composition was obtained by thin-layer chromatography (TLC) coupled to an automated flame ionization detection (FID) system and the proteinic analysis was carried out using an amino acid autoanalyzer. The TEM technique was used to visualize the type of structures formed. The use of OG concentrations lower than its critical micelle concentration (CMC) led to a progressive solubilization of SC membranes, whereas when using OG concentrations slightly higher than the CMC a clear increase in its solubilizing power was obtained. In these conditions, some amino acids building the SC keratinocyte envelopes and bound to the lipids (predominantly ceramides) by covalent bounds were extracted. TEM revealed during and after surfactant incubation the formation of open multilayered structures and concentric conformations which may be correlated with the previous steps of mixed micelle formation between SC components and OG. The OG may be considered as a more suitable solubilizing agent with respect to the conventional solvents in order to obtain more specific and selective solubilization of SC components without denaturing effect on protein structures.

Key words Octyl glucoside – stratum corneum solubilization – stratum corneum dissociation

Introduction

Although a number of investigations have been devoted to the understanding of the principles governing the interaction of different surfactants with native stratum corneum (SC) or simplified SC membrane models, the mechanisms of this interaction are far from understood, since a detailed description of the process has yet to be given [1–4]. One of the most commonly used amphiphilic compounds in membrane solubilization and reconstitution experiments is the nonionic surfactant octyl glucoside, which is believed to be a "mild" surfactant with respect to its denaturing effect on proteins and shows a relatively high critical micelle concentration (CMC), this characteristic being suitable for bilayer reconstitution via dialysis [5–7].

In earlier papers, we studied some parameters involved in the subsolubilizing and solubilizing interactions of different surfactants with simplified membrane models as phospholipid or stratum corneum lipid bilayer in order to establish a criterion for the evaluation of surfactant activity in these structures [8–10]. In the present work we seek

Progr Colloid Polym Sci (1996) 100:230–234
© Steinkopff Verlag 1996

to extend our investigations by characterizing in detail the specific and selective interaction of the nonionic surfactant octyl glucoside with pig stratum corneum at surfactant concentrations lower and higher than its CMC. To this end, we present the lipid and the amino acid compositions for each solubilized fraction after the surfactant separation by dialysis and TEM pictures of SC during and after incubation with octyl glucoside at subsolubilizing level. This information may enhance our understanding of the complex mechanism involved in the process of solubilization of the stratum corneum components by this specific nonionic surfactant.

Materials and methods

The nonionic surfactant n-Octyl β-D-Glucopyranoside (OG) was purchased from Sigma Chemicals Co. (St. Louis, MO). Tris(hydroxymethyl)-aminomethane (TRIS buffer) obtained from Merck was prepared as 10 mM TRIS buffer adjusted to pH 7.40 with HCl, containing 100 mM of NaCl. The standards used for the lipid analysis; ceramide type III (Cer), cholesteryl palmitate, triglycerides and cholesterol were supplied by Sigma Chemical Co. (St. Louis, MO) and palmitic acid (reagent grade) was purchased from Merck. The cholesteryl sulphate was prepared by reaction of cholesterol with excess chlorosulphonic acid in pyridine and purified chromatographically. The lipids of the highest purity grade available were stored in chloroform/methanol 2:1 under nitrogen at $-20\,^{\circ}\text{C}$ until use.

Isolation of stratum corneum and solubilization by octyl glucoside

The SC was isolated following the method described by Wertz and Downing [11, 12]. To solubilize SC membranes, ten milliliters of 10.0 mM OG buffered solution were added to the SC sheets (150 mg) and the mixture was sonicated in a bath sonicator (514 ECT, Selecta) at $25\,^{\circ}\text{C}$ for 15 min. The preparations were then incubated at $25\,^{\circ}\text{C}$ for 24 h under nitrogen atmosphere [13, 14]. The mixture was then filtered to remove the supernatant (S_1) from the SC residues (R_1). These residues were again treated with the same OG concentration and in the same conditions, resulting in the supernatant S_2 and the residue R_2.

The initial treatment of SC sheets (150 mg) was repeated using a OG concentration of 20 mM in the same conditions and after filtration the supernatant S_3 and the residue R_3 were obtained. The SC sheet solubilization percentages were determined as a difference between the SC weight before and after the surfactant treatment in each case.

Surfactant removal

To remove the surfactant from the solubilized material, the supernatants were dialyzed at $25\,^{\circ}\text{C}$ using 500 ml of TRIS buffer. On the basis of previous dialysis experiments we estimated a removal of at least 99.9% of the OG, in accordance with the results reported by Jackson et al. [15]. Dialysis was carried out using a Dianorm Equilibrium Dialyzer (Dianorm, Geräte, Munich, Germany). Pieces of dialysis tubing (Visking type 18/32) were used as diaphragms to separate the chambers into two compartments [16].

Extraction of lipids

To determine the lipids unbounded to the corneocytes, the supernatants freed of OG by dialysis (S_1, S_2 and S_3) were concentrated to dryness and extracted for 2 h and 1 h respectively with mixtures of chloroform: methanol (2:1 and 1:1, v/v) [17, 18]. The different extracts were then combined, concentrated to dryness, weighed and redissolved in chloroform/methanol (2:1) prior to analysis.

To determine in the supernatant S_3 the lipids covalently bound to the proteic material of corneocyte envelope, the unbound lipids were previously extracted by organic extractions and the insoluble material (R_4) was then heated at $60\,^{\circ}\text{C}$ for 1 h with 1.0 M NaOH in methanol: water 9:1 [17, 19, 20]. The mixture was then acidified to pH 4.0 with 2.0 M HCl and shaken with chloroform. After filtration through a coarse sintered glass filter, the chloroform layer was separated and the aqueous phase was reextracted with chloroform. The chloroform extracts were recombined before evaporation to recover the lipids freed by alkaline hydrolysis. These lipids were then redissolved in a small volume of chloroform/methanol 2:1, prior to analysis.

Both lipid extractions also removed some amino acids building the SC corneocyte envelopes [17, 20]. As a consequence, proteins were directly analyzed from each extraction residue.

Analysis of lipids and proteins

The quantitative lipid analysis of each fraction was carried out using thin-layer chromatography (TLC) coupled to an automated flame ionization detection (FID) system (Iatroscan MK-5, Iatron lab. Inc. Tokyo, Japan) [21]. Lipid fractions were directly spotted onto silica gel-coated Chromarods (type S-III) in 0.5, 1 and 1.5 μl using a SES 3202/IS-02 semiautomatic sample spotter with a precision two microliter syringe. The rods (in sets of 10 mounted

232
O. Lopez et al.
Solubilization of stratum corneum by surfactants

semipermanently on stainless steel racks) were developed using the following mixtures: 1°) chloroform/methanol/water (57/12/0.6) for a distance of 2.5 cm (twice), 2°) hexane/diethyl ether/formic acid (50/20/0.3) to 8 cm, 3°) hexane/benzene (35/35) to 10 cm. A total scan was performed to quantify all the lipid components. The same procedure was applied to different standard solutions to obtain the calibration curves for the quantification of each compound.

The low molecular weight proteins removed in the extractions of bound and unbound lipids were previously hydrolyzed with HCl 6.0 N for 24 h at 110 °C to obtain their amino acid compositions. These determinations were carried out using an Automated amino acid Analyzer (Biotronik).

Electron microscopy

To negative staining of supernatants samples, carbon-coated copper/palladium grids G-400 mesh, 0.5 Taab with 0.5 % E 950 collodium films in *n*-amylacetate were employed. A drop of each solution was sucked off the grid after 1 min with filter paper down to a thin film. Negative staining of the samples with a drop of a 1% solution of uranyl acetate was performed. After 1 min this drop was again removed with filter paper and the resulting stained film was dried in a dust-free place. Samples were also examined in a Hitachi H-600 AB transmission electron microscope operating at 75 kV.

OG critical micelle concentration

The surface tensions of buffered solutions containing increasing concentrations of OG were measured by the ring method [22] using a Krüss tensiometer. The CMC of the OG was determined from the abrupt change in the slope of the surface tension values versus surfactant concentration showing a value of 19.0 mM.

Results and discussion

Treatment of SC membranes with OG concentration lower than its CMC

A systematic investigation of pig SC membrane solubilization due to the presence of OG at lower concentration than its CMC was carried out. Pieces of SC (150 mg) were treated with OG 10.0 mM in TRIS buffer. This treatment led to the solubilization of 14% of the initial material,

Table 1 Weight lipid percentages in the fractions L_1, L_2, L_3 and L_4 determined by TLC/FID and obtained from the calibration curves of the standard compounds

Lipids composition of SC treated with 10 mM OG (% weight)				
	L_1	L_2	L_3	L_4
Ceramides	18.40	22.80	30.51	74.47
Cholesterol	15.50	18.20	15.71	–
Free fatty acids	22.16	19.53	13.30	9.89
Cholesteryl sulphate	7.20	6.87	9.54	–
Cholesteryl esters	19.57	15.01	14.67	15.70
Triglycerides	17.17	17.59	16.27	–

which consisted predominantly in proteins and lipids, proteins being the predominant species.

It is noteworthy that, despite the fact that OG surfactant is one of the most commonly used compounds in membrane solubilization, its solubilizing effect in pig SC membranes has not been reported to date. In fact, previous studies involving the structural organization of SC based in the selective membrane solubilization were carried out using organic solvents and alkaline hydrolysis to free lipid and protein material linked to corneocyte envelopes [17–20]. After obtaining the calibration curves for each standard compound, lipid quantification was performed and the lipid content in the supernatant S_1 expressed in % in weight is indicated in Table 1 (L_1). The corresponding content in amino acids due to the proteins solubilized in the supernatant S_1 also expressed in % in weight, is indicated in Table 2 (AA_1).

The remaining SC tissue residues (R_1), were treated again with 10 mM OG and the resulting percentage of solubilized material was the 28% in weight with respect to the treated SC rests. The contents in lipids and amino acids solubilized in the supernatant S_2 are also indicated in Tables 1 and 2 (L_2 and AA_2).

From these results, we may assume that a selective solubilization of both lipid and protein components of SC occurred at OG concentrations lower than its CMC. The ability of OG to solubilize SC structures increased in the second treatment with respect to the first one.

The need for successive chloroform: methanol extractions to determine the lipid composition of supernatant S_1 and S_2 may be attributed to the fact that OG surfactant solubilized lipids linked to proteins by hydrophobic interactions, thus it is necessary to break these links prior TLC-FID analysis [21].

To elucidate the structural transitions involved in the solubilization of SC membranes by OG, TEM observations were carried out in the supernatants S_1 during (sample 1) and after the incubation (sample 2) with OG 10 mM. The corresponding microphotographs are shown in Fig. 1. TEM pictures revealed in both cases open

Table 2 Weight amino acid percentages in the fractions AA_1, AA_2, AA_3 and AA_4

| Amino acid composition of SC treated with 20 mM OG (% weight) | | | |
	AA_1	AA_2	AA_3	AA_4
ASP	9.67	7.60	9.82	–
GLU	18.23	16.24	17.52	9.01
SER	12.03	11.99	11.43	8.94
GLY	11.12	12.68	12.98	8.51
HIS	3.77	5.03	2.68	–
ARG	3.22	5.10	4.72	3.98
THR	3.81	3.61	4.04	4.28
ALA	3.77	3.78	3.41	4.94
PRO	2.96	3.13	3.96	18.84
TYR	5.61	3.28	2.96	2.99
VAL	4.94	4.71	4.76	4.35
MET	1.00	2.09	0.89	0.90
CYS	–	–	2.82	–
ILE	3.90	3.54	3.76	3.89
LEU	7.17	7.93	5.75	10.98
PHE	0.92	2.29	1.92	3.37
LYS	7.59	6.99	6.70	15.03

multilayered structures and concentric conformations, which are probably related to the previous steps of mixed micelle formation between SC components and OG. These structures were more clearly observed after 24 h of OG incubation. Hence, the self-assembly organizations between OG and SC components depended on the level of solubilization of the system and consequently, on the composition of the systems.

Treatment of SC membranes with OG at a concentration higher than its CMC

When the experiments described in the anterior section were carried out at surfactant concentration slightly higher than its CMC (20.0 mM), a clear increase in the extent of SC solubilization occurred. Thus, about 70% in weight of the SC material was solubilized. The solubilized material (S_3) consisted predominantly of proteins and a minor percentage in lipids (compositions given in Tables 1 and 2, L_3 and AA_3). In order to know if additional surfactant amounts could increase its solubilizing effect in SC structures, treatments with high OG concentrations were carried out. From these investigations we conclude that the use of OG concentrations between 21 and 30 mM do not significantly increase the percentage of material solubilized.

Separation of bound lipids

In order to elucidate if the lipids covalently bound to the corneocyte envelopes were solubilized by OG at higher

Sample A

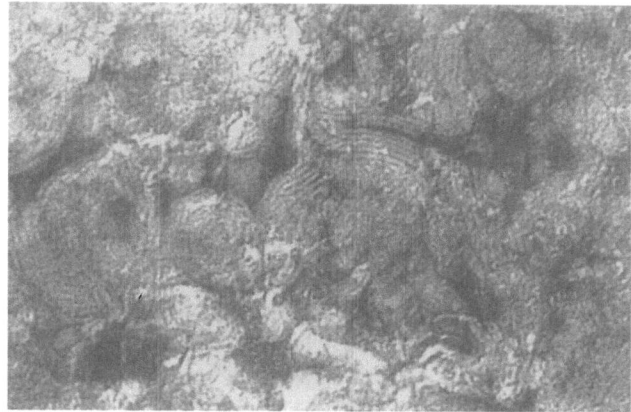

Magnification |———| 100nm

Sample B

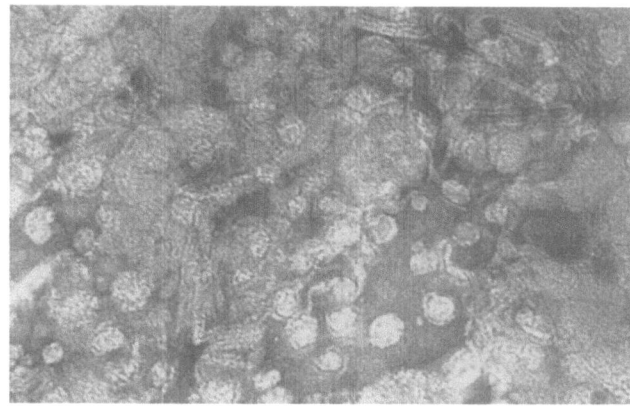

Magnification |———| 100nm

Fig. 1 TEM during (sample A) and after (sample B) the incubation of SC membranes with OG 10.0 mM. The magnifications are given in the microphotographs

concentration that its CMC (20.0 mM), the residue R_4 was submitted to a mild alkaline hydrolysis (1 h with 1.0 M NaOH in methanol:water 9:1) followed by chloroform extractions. The corresponding lipid and amino acid compositions are given in Tables 1 and 2 indicated as L_4 and AA_4. It can be seen that additional amounts of lipids covalently linked to the corneocyte structures were determined. These lipids mainly consisted in ceramides, in agreement with results reported by Wertz and Downing [20]. Table 2, also shows the presence of new amino acids in the hydrolyzed material, Pro and Lys being the predominant species. Although glutamic acid is the dominant species in the SC the presence of high concentrations of Pro and Lys in the fraction AA_4 may be considered as a new approach with respect to results reported by Wertz and Downing [18].

Conclusions

Despite the fact that OG surfactant is one of the most commonly used compounds in membrane solubilization, its solubilizing effect in pig SC membranes has not been reported to data. From our findings we may conclude that the presence of OG concentration lower than its CMC led already to the solubilization of the 14% of the SC membrane, whereas the use of OG concentrations slightly higher than its CMC solubilize approx. 70% of the SC membrane. In these conditions some amino acids building the corneocyte envelopes and bound to the lipids (predominantly ceramides) by covalent bonds were extracted.

These findings are in agreement with those reported by Wertz and Downing, who indicated that the extracted amino acid residues were linked to the lipids (predominantly ceramides) by means of their carboxylic group. However, the presence of higher concentrations of Pro and Lys in the fraction AA_4 may be considered as a new approach with respect to the results reported by Wertz and Downing. TEM during and after incubation of SC membranes with OG (10.0 mM) revealed the formation of open multilayered structures and concentric conformations. These self-assembly organizations may be correlated with the previous steps of mixed micelle formation between SC components and OG.

References

1. Moon KC, Maibach HI (1991) In: Menné T, Maibach HI (eds) Exogenous Dermatoses: Environmental Dermatitis. CRC Press, Boca Raton, FL, pp 217–226
2. Braun-Falco O, Korting HC, Maibach HI (1992) In: Braun-Falco O, Korting HC, Maibach HI (eds) Liposome Dermatitis, (Griesbach Conference) Springer-Verlag, Berlin, p 301
3. Wilhelm KP, Surber C, Maibach HI (1991) J Invest Dermatol 97:927–932
4. Downing DT, Abraham W, Wegner BK, Willman KW, Marshall JL (1993) Arch Dermatol Res 285:151–157
5. Almog S, Litman BJ, Wimley W, Cohen J, Wachtel EJ, Barenholz Y, Ben-Shaul A, Lichtenberg D (1990) Biochemistry 29:4582–4592
6. Cully DF, Paress PS (1991) Mol Pharmacol 40:326–323
7. Lummis SCR, Martin IL (1992) Mol Pharmacol 41:18–26
8. De la Maza A, Parra JL (1994) Biochem J 303:907–914
9. De la Maza A, Parra JL (1995) Langmuir 11:2435–2441
10. De la Maza A, Manich MA, Coderch L, Bosch P, Parra JL (1995) Colloids Surfaces 101:9–19
11. Wertz PW, Downing DT (1983) J Lipid Res 24:753–758
12. Swartzendruber DC, Kitko DJ, Wertz PW, Madison KC, Downing DT (1988) Arch Dermatol Res 280:424–429
13. Dencher NA, Heyn MP (1982) Methods in Enzymology 88:5–10
14. Paternostre MT, Roux M, Rigaud JL (1988) Biochemistry 27:2668–2677
15. Jackson ML, Schmidt CF, Lichtenberg D, Litman BJ, Albert AD (1982) Biochemistry 21:4576–4582
16. Kragh-Hansen U, le Marie M, Nöel JP, Gulik-Krzywicki T, Møller JV (1993) Biochemistry 32:1648–1656
17. Wertz PW, Swartzendruber DC, Kitko DJ, Madison KC, Downing DT (1989) J Invest Dermatol 93:169–172
18. Downing DT (1992) J Lipid Res 33:301–314
19. Wertz PW, Downing DT (1986) Biochem Biophys Res Commun 137:992–997
20. Wertz PW, Downing DT (1987) Biochim Biophys Acta 917:108–111
21. Ackman RG, Mc Leod CA, Banerjee AK (1990) J of Planar Chrom 3:450–490
22. Lunkenheimer K, Wantke D (1981) Colloid and Polymer Sci 259:354–366

Progr Colloid Polym Sci (1996) 100:235–240
© Steinkopff Verlag 1996

Influence of surfactant–gelatin interaction on microcapsule characteristics

V. Sovilj
P. Dokić
M. Sovilj
A. Erdeljan

Dr. V. Sovilj (✉) · P. Dokić · A. Erdeljan
Department of Applied Chemistry
Faculty of Technology
University of Novi Sad
Bul. Cara Lazara 1
21000 Novi Sad, Yugoslavia

M. Sovilj
Department of Chemical Engineering
Faculty of Technology
University of Novi Sad
Bul. Cara Lazara 1
21000 Novi Sad, Yugoslavia

Abstract Investigations of influence of surfactant-gelatin interaction on microcapsule characteristics were undertaken.

In the first step of microencapsulation 20% (v/v) paraffin oil emulsions were prepared in the solution of SDS of different concentrations with 3.0 % of gelatin. Concentration of SDS was varied in order to provide both types of gelatin-SDS interactions (ionic and hydrophobic ones). Microencapsulation was performed by addition of the crosslinking agent, formaldehyde, to the emulsion. The emulsions were spray dried afterwards and the microcapsules were obtained in a powder form.

Microcapsule characteristics such as redispersibility, mean size, wall thickness, wall density, kinetic of core release and wall permeability have been determined.

The results show that at prevailing hydrophobic mechanism of interaction, microcapsules have smaller wall thickness, wall density and greater permeability. Rate of paraffin oil release was measured by continuous flow method, and release kinetics was correlated with Hixon–Crowell equation. Release constant calculated by the equation depends on the interaction mechanism. An increase in the value of the constant was obtained if the hydrophobic mechanism of interaction is the prevailing one. This fact shows that conformational changes on gelatin molecules cause formation of permeable microcapsule wall and therefore faster release of core material.

Key words SDS-gelatin interaction – microcapsule characteristics – release kinetic – permeability

Introduction

Ionic surfactants can modify the properties of protein by forming surfactant-protein complexes. There are two possible mechanisms for the surfactant–protein molecular interaction: the ionic and the hydrophobic one [1–3]. Surface active properties of such complexes, as well as their adsorption at the phase interface, depend on the mechanisms of interaction [4–6].

The most used protein in the food and pharmaceutical industry is gelatin and therefore surfactant–gelatin interaction is very important and intensively investigated by various methods [7–11]. The conformational changes on gelatin molecules, caused by the formation of complexes, affect the structure of adsorption layers and ability of the film formation around the dispersed oil droplets [5]. That is not only of fundamental interest but of practical interest, as well, making surfactant–gelatin interactions an important factor for microencapsulation processes and microcapsule characteristics. Namely, the changes in the adsorption layer structure due to the interactions can be supposed as a factor influencing porosity and permeability of the wall of the microcapsules and regulating release of the microcapsule content.

In the present paper, an influence of anionic surfactant–sodium dodecyl sulfate and gelatin interaction on microcapsule characteristics has been investigated.

Materials and methods

Materials

In microcapsule preparation, as wall-forming material gelatin was used (type B (225 Bloom), product of Sigma-USA) with an isoelectric point of 5.12, determined by viscosity and turbidity measurements. Anionic surfactant, sodium dodecyl sulfate (SDS) was supplied from Merck (Germany), purity > 99% and critical micelle concentration of $8.8 \cdot 10^{-3}$ mol/l at 40 °C. Paraffin oil, pharmaceutical purity, density of 0.843 g/cm^3 at 25 °C, product of Jugolek (Yugoslavia), as core material, and formaldehyde (35%) product of Kemika (Croatia) as crosslinking agent were used. All other chemicals were of analytical grade.

Microcapsule preparation

Paraffin oil (20% volume) was dispersed at 40 °C in the solutions of SDS of different concentrations by means of homogenizer Ultraturrax K-45 (Janke & Kunkel, Germany) at 6000 rpm during 7.5 min. After that, gelatin (3.0% in continuous phase) was added to the emulsions and homogenization was continued until total time of 15 min. Concentrations of SDS were: 0; 2.7; 4.2; 9.0; 12.5 and 17.0 g/l. The emulsions were left for 1 h at 40 °C to stabilize the adsorption layer. After that a crosslinking agent, formaldehyde (3% on gelatin mass), was added to the emulsions. Microencapsulation was performed by spray drying of the emulsions in a Mini Spray Dryer (Büchi, Switzerland) afterwards. The microcapsules were obtained in a powder form.

Microcapsule characteristics

Microcapsule characteristics such as redispersibility in water, mean size, wall thickness, wall density, kinetic of core release and permeability were determined.

Redispersibility

After spray drying microcapsules were obtained in the form of white crisp powder. Under optical microscope smaller or larger aggregates could be seen. After addition of the water to microcapsule powder, redispersibility, i.e., ability to disaggregate of microcapsules was observed.

Mean size

Microcapsule size distribution was determined by electronic counter, Type B (Coulter Counter Electronics, England) with measuring tube orifice of 50 μm. From microcapsule size distributions data, mean diameters were calculated.

Microcapsule density

Density of dry microcapsules was determined by picnometer at 25 °C using the method for powder material. Displacement fluid was ethyl alcohol.

Wall thickness

In determination of wall thickness, core material, i.e., paraffin oil was extracted from a known mass of microcapsules with petrolether [12]. After total extraction, petrolether was evaporated and mass of paraffin oil determined.

Mass of microcapsules taken for extraction could be expressed as:

$$M = N v \rho \, (\text{g}) \qquad (1)$$

v – volume of single microcapsules, ρ – density of microcapsules, N – number of particles.

From Eq. (1), number of particles in M grams of microcapsules is:

$$N = \frac{6M}{d^3 \pi \rho}, \qquad (2)$$

d – mean size of microcapsules determined from particle size distribution.

If the whole mass of extracted paraffin oil from M gram of microcapsules is M_c, then single core mass m_c is:

$$m_c = \frac{M_c}{N} \, (\text{g}) . \qquad (3)$$

Mean size d_c, of the core is:

$$d_c = (6 m_c / \pi \rho_c)^{1/3} \, (\mu\text{m}) \qquad (4)$$

ρ_c – density of core (paraffin oil).

Wall thickness was calculated as:

$$h = \frac{d - d_c}{2} \; (\mu m) . \tag{5}$$

Wall density

After oil extraction, wall material was dried and measured. Wall density was calculated as:

$$\rho_w = \frac{M_w/N}{v - v_c} \; (g/cm^3) \tag{6}$$

M_w – mass of dried wall material, v_c – volume of single core.

Kinetic of core release

Rate of paraffin oil release from microcapsules was measured by continuous flow method [13] in which microcapsules were immersed in petrolether and with peristaltic pump petrolether was continuously added and withdrawn with the same rate. Release of paraffin oil with time was followed with an UV detector (Pharmacia Fine Chemicals, Sweden) at 280 nm and automatically recorded:

From the obtained data release kinetics was calculated and correlated with Hixon–Crowell cube-root law [14]:

$$W^{1/3} = W_0^{1/3} - Kt \; (g^{1/3}) , \tag{7}$$

where W is amount of paraffin oil remaining in the microcapsules at time t, W_0 is the initial amount and K is the release constant.

Wall permeability

After the extraction of paraffin oil had been performed, empty dry microcapsules were added to the water and their swelling, i.e. increasing of size due to water diffusion was observed under optical microscope. The degree of swelling was proportional to the wall permeability.

Results and discussion

The concentrations of SDS corresponding to certain mechanisms of SDS–gelatin interaction were previously determined by conductometric and potentiometric titration of gelatin solutions of various concentrations with SDS solution [15]. From the result of the titrations at gelatin concentration of 3.0%, SDS concentrations were calculated. The concentration of SDS when interaction with gelatin is finished was also calculated (Table 1).

Spray drying of the emulsions gave white crisp powder of gelatin microcapsules in the form of aggregates at all SDS concentrations (Fig. 1).

Microcapsule characteristics depending on SDS concentrations, i.e., on mechanisms of interaction were presented in Table 1.

Redispersibility of microcapsules in water increase with higher SDS concentration. At prevailing ionic mechanisms of SDS–gelatin interaction SDS molecules bonds ionically to the $-NH_3^+$ groups accessible on the gelatin molecules, forming unionizable surfactant–gelatin complex [1, 3]. Therefore aggregates can be redispersed in water by a moderately intensive manual mixing. As interaction take place microcapsules aggregates can easily redisperse in water, because of dissociation of head groups of hydrophobically bonded SDS and electrostatic repulsion between charged groups (Fig. 2).

Microcapsules mean size does not depend on the mechanisms of interaction. They have the same mean diameter (about 9 μm) as the droplets in the emulsion. The decrease of microcapsule size after completion of interaction has some explanation. It can be due to very high concentration of SDS. In that case adsorption layer is

Table 1 Characteristics of gelatin microcapsules

SDS (g/l)	2.7	4.7	9.0	12.5	17.0
Mechanism of Interaction	Prevail ionic		Prevail hydrophobic		Finished interaction
Microcapsule characteristics					
Redispersibility	Intensive mixing		Moderate mixing redispersed		Easily
Mean size d (μm)	8.96	9.62	9.96	8.60	6.13
Wall thickness h (μm)	0.530	0.645	0.745	0.637	0.493
Wall density ρ_w (g/cm^3)	0.697	0.497	0.480	0.423	0.376
Release const. $K \; 10^3$ ($g^{1/3}/s$)	1.29	1.46	1.71	1.84	1.92
r	0.993	0.996	0.992	0.986	0.991
Permeability	low		high		dissolve

Fig. 1 Photomicrograph of gelatin microcapsules after spray drying (1 mm = 0.8 μm)

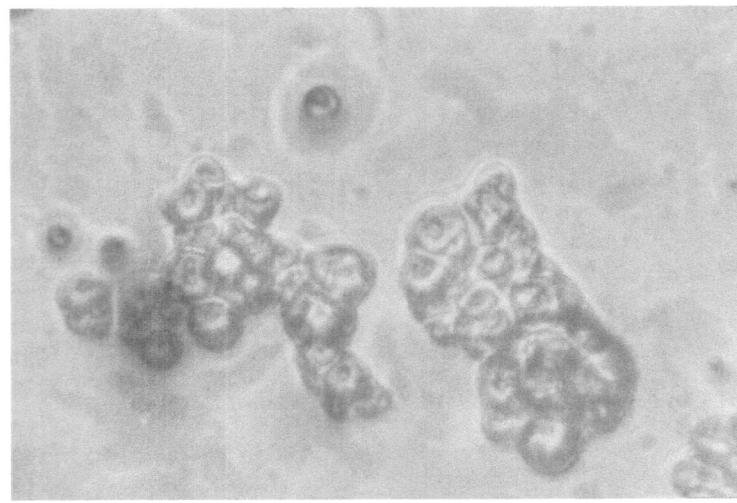

Fig. 2 Photomicrograph of gelatin microcapsules after dispersing in water (1 mm = 0.8 μm)

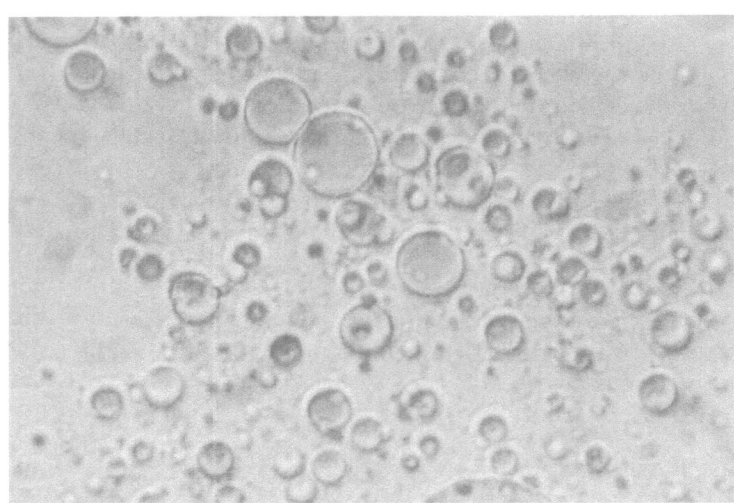

not sufficiently compact and large particles failed to encapsulate.

Wall thickness of microcapsules increases until SDS concentration of 9.0 g/l, at prevailing ionic and partly hydrophobic mechanism of interaction, and after that decreases. This is in good agreement with theoretical consideration that at high amount of hydrophobic bonded SDS molecules to gelatin, desorption of SDS–gelatin complexes occurs.

Wall density decreases with the increasing of SDS concentration. At ionic mechanism of interaction, gelatin concentration in adsorption layer is high [4], conformation of gelatin molecules is undisturbed, molecules are tight and densely packed and wall structure is compact.

When hydrophobic mechanism of interaction prevails, changes in the conformation of gelatin molecules take place, structure is not sufficiently compact [5]. Further increasing of SDS concentration produce significant changes in conformation, unfolding of gelatin molecules and low wall density. Therefore, when the interaction is finished (SDS concentration of 17.0 g/l) wall density becomes two times less then at ionic mechanism of interaction (SDS concentration of 2.7 g/l) although wall thickness is almost the same in both cases.

Rate of paraffin oil release from microcapsules in dependence on SDS concentration is presented in Fig. 3.

Release kinetics was correlated with Hixon–Crowell equation and release constant K and coefficient of

Fig. 3 Rate of paraffin oil release from gelatin microcapsules obtained at various SDS concentrations: 2.7 (□), 9.0 (▲) and 12.5 (*) g/l

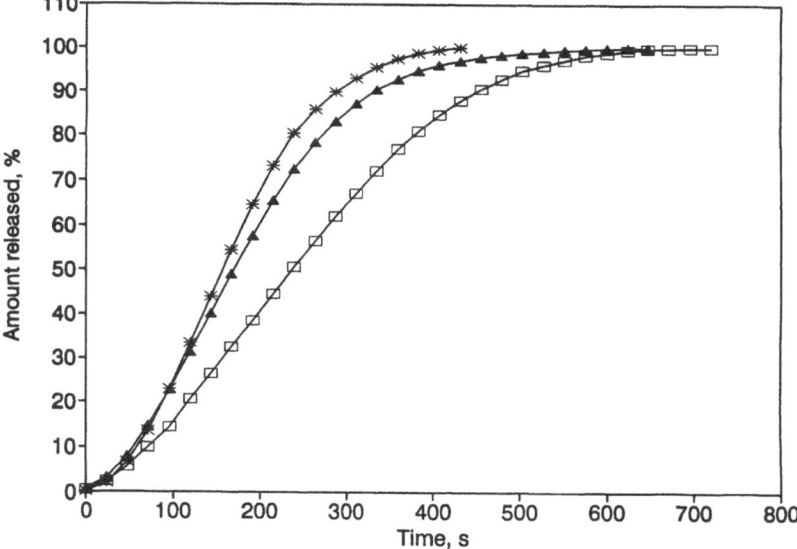

Fig. 4 Hixson-Crowell release kinetics of paraffin oil from microcapsules obtained at SDS concentration of 2.7 (□) and 12.5 (▲) g/l

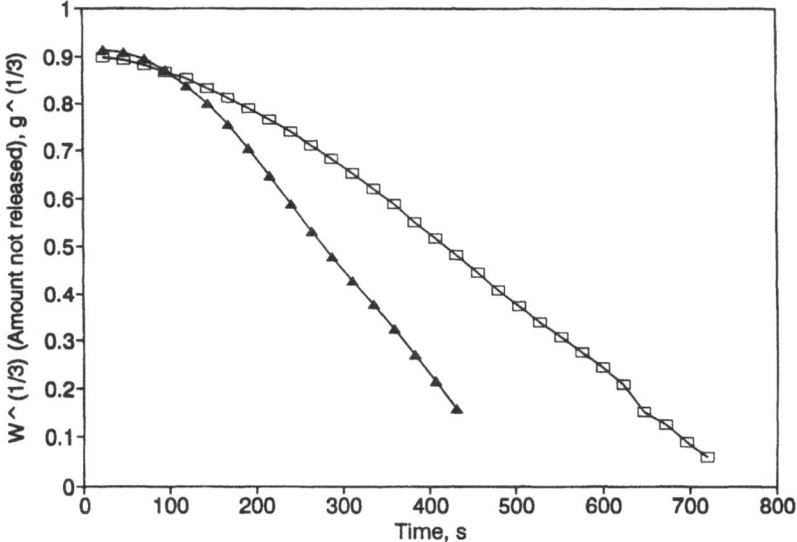

correlation r were calculated (Table 1). An increase in the value of the constant K was obtained if the mechanism was changed from ionic to hydrophobic one. This fact shows that changes in the conformation of gelatin molecules at hydrophobic mechanism of interaction cause the formation of permeable microcapsule wall and therefore faster release of the core material (Fig. 4).

Water was put onto the empty and dry microcapsule. The swelling, i.e., the increase in size due to water diffusion was observed under optical microscope. Degree of swelling which is proportional to the wall permeability, increases with the change of mechanism of interaction from ionic to hydrophobic one, Fig. 5.

At SDS concentration of 17.0 g/l, when interaction is finished, microcapsule walls were invisible under the microscope, which indicates very intensive swelling or perhaps dissolving.

This investigations show that surfactant–gelatin interaction has significant influence on microcapsules wall permeability, which is important for practical application in microencapsulation process.

Fig. 5 Suspension of empty
microcapsules in water at SDS
concentration of 2.7 (a) and 12.5
(b) g/l (1 mm = 0.8 μm)

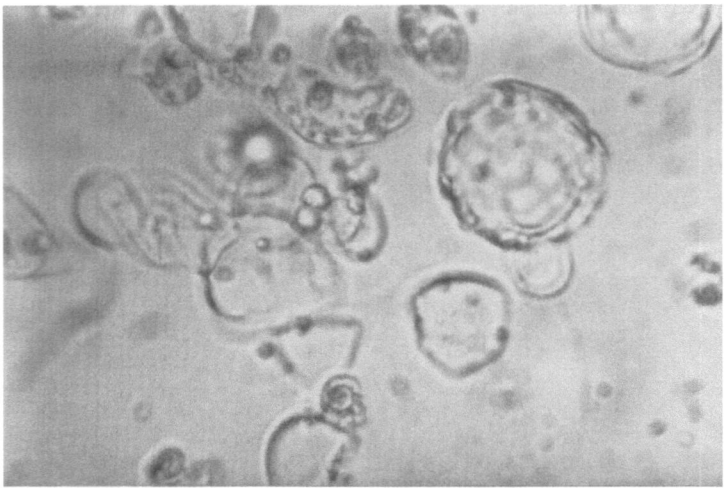

(a)

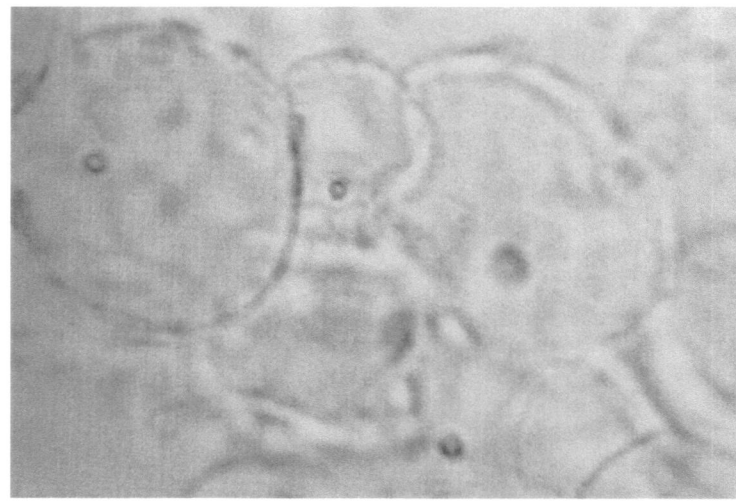

(b)

References

1. Knox WJ, Wright JF (1965) J Colloid Interface Sci 20:177–186
2. Arora JPS, Soam D, Singh SP, Kumar R (1984) Tenside Deterg 21:87–90
3. Wüstneck R, Zastrov L, Kretzschmar G (1985) Kolloidn Zh 47:462–471
4. Goddard ED (1986) Colloids Surfaces 19:301–329
5. Sovilj V, Djaković LJ, Dokić P (1993) J Colloid Interface Sci 158:483–487
6. Tadros TH F (1974) J Colloid Interface Sci 46:528–540
7. Arora JPS, Pal C, Dutt D (1991) Tenside Deterg 28:215–218
8. Henriquez M, Abuin E, Lussi E (1993) Colloid Polym Sci 271:960–966
9. Fruhner H, Kretzschmar G (1992) Colloid Polym Sci 270:177–182
10. Whitesides TH, Miller DD (1994) Langmuir 10:2899–2909
11. Wüstneck R, Wetzel R, Buder E, Hermel H (1988) Colloid Polym Sci 266:1061–1067
12. Madan PL, Luzzi LA, Price JC (1974) J of Pharmaceutical Sci 63:280–284
13. Washington C (1990) Int J of Pharmaceutics 58:1–12
14. Benita S, Donbrow M (1982) Int J of Pharmaceutics 12:251–264
15. Sovilj V, Djaković LJ, Ristanović D (1994) Hem ind 48:151–154

Progr Colloid Polym Sci (1996) 100:241–245
© Steinkopff Verlag 1996

S. von Hünerbein
M. Würth
T. Palberg

Microscopic mechanisms of non-linear rheology of crystalline colloidal dispersions

S. von Hünerbein (✉) · M. Würth
T. Palberg
University of Konstanz
Faculty of Physics
Postfach 55 60 M675
78434 Konstanz, FRG

Abstract One of the most intriguing properties of colloidal systems is their ability of thinning or thickening under shear. The underlying physical mechanisms of non-linear rheometry may be the shear-induced formation and destruction of long-range positional and orientational order. Since only in rare cases comprehensive structure and velocity information is accessible from inside a viscosimeter, generally, homogeneous samples are assumed. From recent experiments, however, there are indications of inhomogeneous phase and flow behaviour in colloidal model systems, in particular for denser systems of strongly interacting particles. We here investigate a well characterized suspension of spherical particles interacting via a screened electrostatic potential. Their equilibrium phase is a body-centered cubic crystal. Upon flow through an optical model capillary viscosimeter an overall shear thinning is observed in the apparent viscosity. Superimposed, however, dilatancy is observed at three different overall fluxes. We give a detailed study of the local structures and shear rates in the tube. The non-Newtonian behavior is explained in terms of a succession of non-equilibrium phase transitions with the most striking feature being the simultaneous existence of up to four concentrically arranged phases under conditions of stationary flow.

Key words Colloidal dispersions – shear flow – non-equilibrium phase transitions – nonlinear rheology

Introduction

For most simple fluids the shear stress σ within the fluid is proportional to the shear rate $\gamma = dv/dx$ times the viscosity η: $\sigma = \eta\gamma$. This is called Newtonian or linear rheology. Complex fluids like colloidal dispersions often show significant deviations from this simple law, and this property finds a number of interesting applications. In painting, thixotropy or shear-thinning is due to the parallel arrangement of the pigment platelets under the shearing forces of the brush. The particles again become orientationally disordered due to Brownian motion after cessation of shear.

Often non-Newtonian rheology is coupled to shear induced phase transitions including shear induced ordering of fluids and shear induced melting of crystalline phases or shear modified transitions from the isotropic to the nematic phase [1–6]. A discontinuous decrease of the apparent viscosity over many orders of magnitude was observed upon shear melting [7], while the extreme dilatancy observed in other cases was attributed to cluster formation [3].

Although shear induced phase transitions thus show pronounced effects on the rheology of colloidal suspensions, only very few papers exist giving detailed connections between the suspension structure and the colloidal flow properties [3, 8, 9]. Most studies on shear melting focussed on structural aspects alone or discuss the underlying microscopic mechanisms [5, 8–11]. Some of the experiments have been supported by computer simulations and non-equilibrium phase diagrams have been derived [10, 12, 13].

In most studies homogeneous stationary states had to be assumed. This assumption may not be true once processes of nucleation and growth are involved, or complicated flow geometries are used. Only very recently observations of stationary two phase coexistence for simple geometries have been reported on charge stabilized silica [9] and polystyrene spheres [14].

Results and discussion

We here examine a particularly simple colloidal system in a simple flow geometry, i.e., charged latex spheres of well defined shape, radius and charge in a capillary viscosimeter. Such monodisperse suspensions have already acquired the state of model systems within the heterogeneous class of colloidal dispersions [15]. Due to their long ranged interaction they show an equilibrium phase transition from the fluid to a crystalline ordered state at micromolar concentrations of screening electrolyte even at packing fractions Φ well below 1%. Since the typical interparticle spacing is then on the order of the wavelength of visible light the systems are optically accessible by light scattering or microscopy. The particles under investigation here and the equilibrium suspension properties have been carefully characterized in both the fluid and the solid state (Lot #2011 M9R, Seradyn, U.S.A., radius $a = 51$ nm, bare charge $Z = 620$; [16]). The shear modulus of the solid formed at $\Phi = 0.0035$ and under completely deionized conditions is on the order of $G = 0.4$ Nm^2. The system therefore may be shear molten simply by shaking the sample. Once shearing is terminated the system instantaneously relaxes to a metastable state of fluid order, from which it crystallizes back to the solid state on the time scale of some seconds to minutes.

All these investigations were performed at mechanical equilibrium. Here we subject the system to continuous flow through a horizontally mounted capillary viscosimeter of 40 mm inner diameter and a length of 50 cm. It flows under the hydrostatic pressure difference between two reservoirs of fixed height difference Δh. In both the suspension is kept under inert gas atmospheres connected by separate tubings. A peristaltic pump refills the upper reservoir through a closed teflon tubing system containing an ion exchange chamber and a conductivity measurement to maintain and control conditions of complete deionization at a packing fraction fixed at $\Phi = 0.0035$ [17].

In Fig. 1a we present the apparent viscosity η_{APP} as deduced from Δh and the overall flux V, as integrated from the spatial velocity distribution $v(r)$ measured by high resolution Doppler velocimetry. Four regions of apparent shear thinning are clearly discernible separated by seemingly discontinuous increases in η_{APP}. We note that in the vicinity of the viscosity steps it sometimes was impossible to obtain conditions of a stationary flux. For high fluxes (completely shear molten state) a limiting viscosity is approached. Comparison to measurements with pure water at 18 °C give a value of η_{APP}(melt)$/\eta_{WATER} = 1.17$ for the melt. Thus shear thinning of colloidal suspensions as observed in a capillary rheometer may be a complex multistep process and an interpretation in terms of a single first order phase transition will not suffice for its explanation.

The suspension enters the capillary in the shear molten state and relaxes into a partly crystalline flow pattern. A stationary state is usually reached well before the end of the capillary. Only for the highest fluxes the tube length is too short to allow for complete relaxation within the passage time. Depending on the overall flux up to four different phases are observed and the structure of the individual phases was confirmed by static light scattering [11, 18]. These are polycrystalline bcc. solid, hexagonally packed sliding layers of two different sliding mechanisms and an isotropic shear melt. While the relative volume of each phase is a strong function of the overall flux V, their radial sequence seems to be fixed.

We therefore carefully analysed the structural evolution of the suspension along and across the tube by light scattering and video microscopy. We further analysed the local shear rates. The local velocities are measured by Laser Doppler velocimetry using a fiber optic detection scheme to gain a sufficiently high spatial resolution of 10 μm. From each velocity profile the local shear rate is precisely determined by differentiating with respect to r. Simultaneously the distribution of individual phases is determined using high resolution Bragg microscopy. A laser beam crossing the capillary is observed side on by a video camera equipped with a macro lens. The camera is placed at the upper Bragg spot of the layer-phases. The image of the laser beam then appears with different brightness, since the static structure factor of the three (four) regions differs significantly. Layer-phase regions appear extremely bright, the fluid shows a much lower intensity and the core appears almost dark, since only rarely an individual crystallite fulfills the orientational Bragg condition.

An example of the observed very complex behavior is given in Fig. 2 together with the locally measured velocities. In this remarkable case of a four phase coexistence the local shear rate shows a corresponding step-like behavior. The core region remains unsheared and flows as a solid plug. The outer phases each show different shear rates. The boundaries seem to be diffuse which, however, is an effect of the long integration time for the velocity measurements used to enhance the signal/noise ratio. With short integration times, say of less than 10 s, we observe a sharp (< 50 μm) structural boundary slowly moving in and outward over a distance of $\Delta r \approx 250$ μm.

The layer-phases do not merge completely but stay separated into two regions of different flow behavior. Note that the nearly linear decrease of velocity with the radius gives a shear rate practically independent of the position (stretched Hagen–Posseuille flow). The inner layer region shows the lower gradient, which only very slightly increases with further increasing flux, while the layer-phase adjacent to the wall shows a strong increase in the velocity gradient. At the same time the scattered intensity of this outer region is decreased. This is attributed to a transition from a state of planes registered between single sliding events to a state of continuous gliding as suggested by several authors [5, 7, 11]. As compared to the case of layer-phases the parabolic velocity profile of the fluid phase is more pronounced. For radially small fluid regions, however, it is well approximated by a constant shear rate. In Fig. 2 it has a considerably lower value than the shear rate in the layer-phase-regions. The interesting conclusion is that at $V = 76$ mm^{3}s^{-1} the fluid phase is lubricated on both sides by a phase of higher order and lower viscosity.

The evolution towards the stationary states is an interesting subject in itself. We here only give some examples and annotate some important points. In Fig. 3 we show the evolution of stationary states for four different overall fluxes. Note first that for all fluxes a stationary state is reached well before the outlet of the tube. Further note the strong dependence of the observed scenario on the overall flux. In all cases, however, the core region evolves by homogeneous nucleation and subsequent growth, while the layer-phases need a solid substrate to heterogeneously nucleate on the wall or the core (best visible in Fig. 3d). Growth velocities of the layer phase are slower than for the

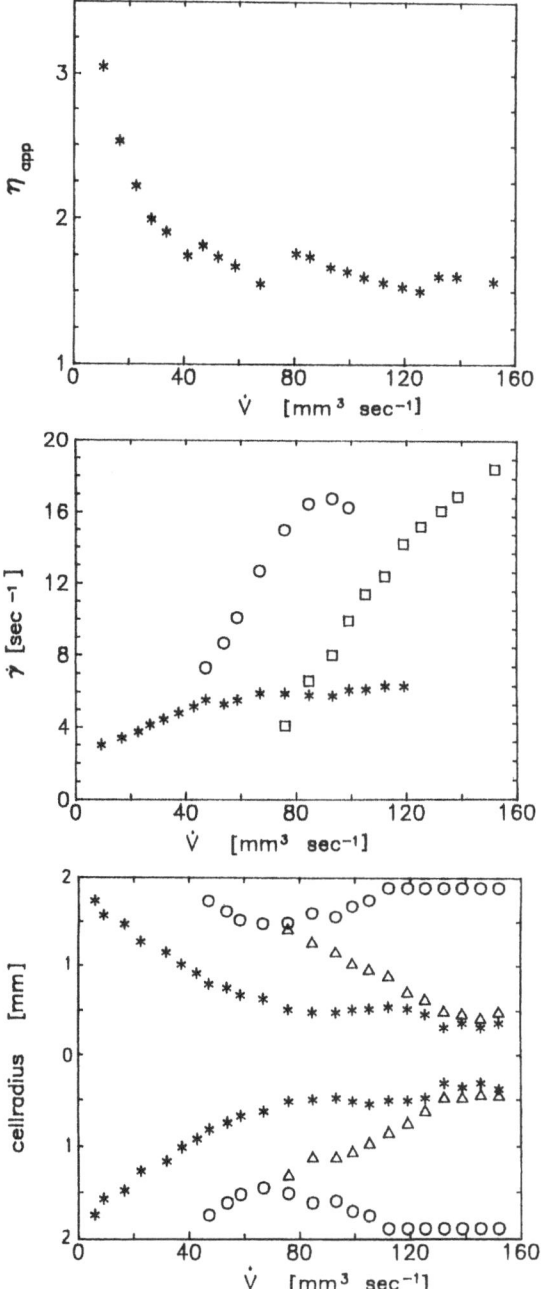

Fig. 1 Evolution of several quantities in dependence on the overall flux V, as measured by integrating the observed velocity profiles $v(r)$. For technical reasons the flux was preadjusted by increasing the speed of the refilling pump in equidistant steps; then the height differences of the reservoirs were fixed to specific values yielding stationary flow. The suspension and shear parameters are: $\Phi = 0.0035$, $c = 0$ μmol l^{-1}, $a = 51$ nm, $Z = 620$. The measured quantities are: a) Apparent viscosities η_{APP} as deduced following the procedures of Laun et al. [3]. Note the strong overall tendency for shear thinning, the saturation behavior at large fluxes and the three steplike increases. b) Shear rates in the stationary state as deduced from local velocity measurements. Stars: Layer structure with registered sliding; open circles: layer structure showing continuous sliding; open squares: fluid phase. c) Radial distribution of the respective phases. Stars denote the boundary of the polycrystalline core to other phases; circles denote the boundary of the continuous sliding phase to other phases and triangles the boundary between registered sliding layer-phase and fluid phase. Note that for each phase the shear rate increases linearly with increasing flux until a further phase is observed and the shear rate saturates

polycrystalline phase (Fig. 3c). In Fig. 3b one recognizes an equilibration of phase volumina at the cost of the core region. It seems possible to explain the balance between the volumina based on energy fluxes, accounting for transition enthalpies and energy dissipated in viscous flow. This however will be studied in detail in a forthcoming paper.

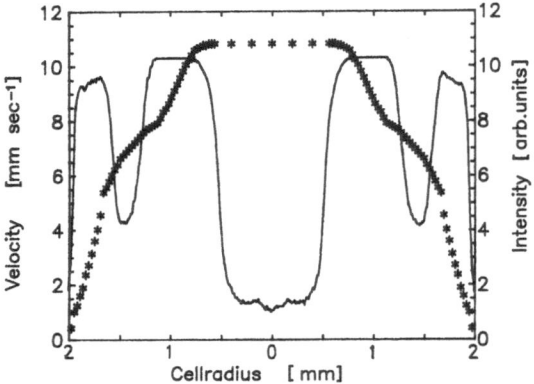

Fig. 2 Representative example of the shear-induced phenomena observed in the stationary state at $V = 76 \text{ mm}^3 \text{s}^{-1}$. Solid lines show the radial profiles of the scattered intensity as measured by Bragg microscopy. High intensities correspond to phases of hexagonal sliding layers, medium intensities to the fluid phase and low intensity to the polycrystalline core. Stars denote the local velocity as measured by laser Doppler velocimetry

To compare all the flow data collected in the stationary states and in dependence on the overall flux we plot the shear rates measured within the different phases in Fig. 1b. The solid core representing the equilibrium stable phase remained unsheared in all experiments and was lubricated by one or two layer-phases and/or a fluid phase. Note that for all fluxes at least two phases coexist. While the possibility of a coexistence between two phases has recently been confirmed experimentally [9, 14], this is the first demonstration of a multiphase coexistence.

Within all phases the shear rates increase nearly linearly until a new phase is observed. The saturation values are approximately 6 s^{-1} and 17 s^{-1} for the two layer-phases. As shown in Fig. 1b, for our capillary viscosimeter the typical sequence of phases appearing under increasing shear is polycrystalline, registered sliding layers, continuous sliding layers, and fluid. From intuitive considerations, one would expect that smallest shear rates are found in the cell center and increase towards the cell walls. The concentric arrangement therefore should also reflect the abovementioned succession of nonequilibrium phases with increasing perturbation of the mechanical equilibrium. Clearly this is not the case. One may speculate that the unexpected radial arrangement of phases is determined by the kinetics of heterogeneous nucleation and growth of the layer-phases on the wall and the polycrystalline core, respectively.

Figure 1c shows the corresponding distributions of phases. With increasing flux the registered sliding layer-phase grows at the expense of the solid core, then the

Fig. 3 Evolution of the phase distribution towards the stationary state for
a) $V = 9 \text{ mm}^3 \text{s}^{-1}$;
b) $V = 60 \text{ mm}^3 \text{s}^{-1}$;
c) $V = 99 \text{ mm}^3 \text{s}^{-1}$;
d) $V = 138 \text{ mm}^3 \text{s}^{-1}$. Circles denote the boundary between wall layer-phase and other phases, triangles the boundary between fluid phase and core layer-phase, and squares the core boundary

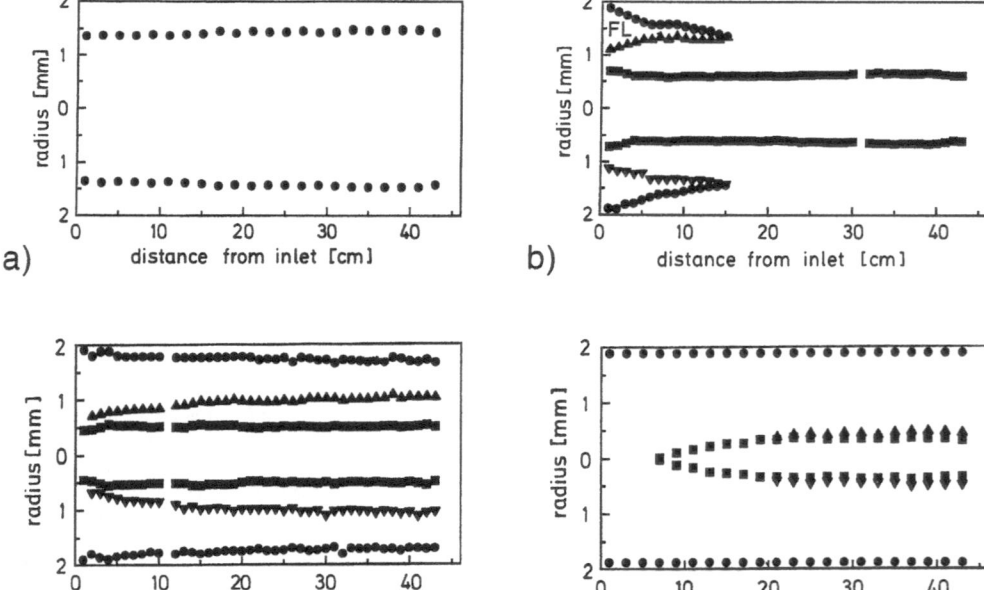

Progr Colloid Polym Sci (1996) 100:241–245
© Steinkopff Verlag 1996

continuous sliding layers grow at the expense of the solid core, while the volume of the registered sliding phase stays nearly constant. Between $V \approx 80$ mm^3 s^{-1} and $V \approx 120$ mm^3 s^{-1} the fluid phase increases at the expense of the layer-phases and above $V \approx 120$ mm^3 s^{-1} at the expense of the solid core. The observed stepwise melting process therefore seems to be a succession of first order transitions. A detailed explanation of the observed behavior, however, remains a challenge. In addition to thermodynamic and kinetic considerations it may have to include geometrical influences, too. To stress the latter point, we note that a 150 μm-thick slab of the layer-phase remains rather unchanged even at highest fluxes. This stability of the wall based layer is not understood. Layering of colloidal particles next to a wall has been observed but under mechanical equilibrium it does not exceed some ten ordered sheets [19]. We further note that a solid–fluid transition without the intermediate formation of layer-phases was recently observed for a colloidal crystal flowing through a parallel plate arrangement [13].

We finally compare Figs. 1b and c to the non-Newtonian behaviour depicted in Fig. 1a in order to explain both the overall shear thinning and the seemingly discontinuous increases observed at fluxes of approximately 40, 80 and 120 mm^3 s^{-1}. The registered sliding layer-phase shows a lower viscosity than the continuous sliding layer-phase. The viscosity of the fluid phase is somewhat larger but still much smaller than for the polycrystalline solid. Therefore the regions of monotonously decreasing η_{APP} may be interpreted as the increase in volume of low viscosity phases at the expense of the solid core, while each increase in η_{APP}

represents the stabilization of a further, less ordered phase of relatively higher viscosity. In the core layer the long-range positional order is still present for three dimensions, since the particles remain registered between sliding events. It is lost in radial direction upon the transition to the continuously sliding, wall based state. It is completely lost for the fluid phase.

We have presented an example of the possibilities to quantitatively characterize the very complex flow behavior of a suspension of well-defined colloidal spheres by means of a combination of light scattering and microscopic methods. In an optical model capillary viscosimeter the combined evolution of the sample's local structure and shear rate was monitored under carefully controlled shear conditions. For the first time it was possible to attribute the observed strongly non-linear and non-monotonous viscosity behavior to a succession of non-equilibrium phases, which for themselves seem to obey a Newtonian rheology. Up to four different phases were observed to coexist which strongly differ in their degree of long-range order.

The emerging qualitatively consistent picture demonstrates the value of opto-rheometric experiments and should stimulate enhanced experimental and theoretical efforts to quantify existing microscopic models of both phase transitions under shear, and their correspondence to the complex flow behavior observed in colloidal systems.

Acknowledgments The authors thank G. Porte, G. Nägele and P. Leiderer for critical discussions on the issue of non-equilibrium phase diagrams. Financial support by the Deutsche Forschungsgemeinschaft is gratefully acknowledged.

References

1. Hoffmann RL (1972) Trans Soc Rheol 16:155; J Coll Interface Sci 46:491 (1974); Adv Coll Interface Sci 17:161 (1982)
2. Berret JF, Roux DC, Porte G (1994) J Phys II (France) 4:1261
3. Laun HM, Bung R, Hess S, Loose W, Hess O, Hahn K, Hädicke E, Hingmann R, Schmidt F, Lindner P (1996) J Rheol 36:1057
4. Ackerson BJ (1983) Physica A 128:221
5. Ackerson BJ, Clark NA (1981) Phys Rev Lett 46:123
6. Ackerson BJ, Hayter JB, Clark NA, Cotter L (1986) J Chem Phys 84:2344
7. Chen LB, Zukoski CF (1990) Phys Rev Lett 65:44
8. Chen LB, Chow MK, Ackerson BJ, Zukoski CF (1994) Langmuir 10:2817
9. Imhoff A, van Blaaderen A, Dhont JKG (1994) Langmuir 10:3477
10. Stevens MJ, Robbins MO, Belak JF (1991) Phys Rev Lett 66:3004
11. Loose W, Ackerson BJ: Preprint
12. Stevens MJ, Robbins MO (1993) Phys Rev E 48:3778
13. Lahiri R, Ramaswamy S (1994) Phys Rev Lett 73:1043
14. Palberg T, Streicher K (1994) Nature 367:51
15. Pusey PN in Hansen JP, Levesque D, Zinn-Justin J (1989) "Liquids, freezing and glass transition", 51st summer school in theoretical physics, Les Houches (F) 1989, Elsevier, Amsterdam 1991, pp 763
16. Palberg T, Würth M, Simon R, Leiderer P (1994) Prog Colloid Polym Sci 96:62
17. Palberg T, Härtl W, Wittig U, Versmold H, Würth M, Simnacher E (1992) J Phys Chem 96:8180
18. Versmold H, Lindner P (1994) Langmuir 10:3034
19. Grier D, Murray C (1994) J Chem Phys 100:9088; Murray CA, Sprenger WO, Wenk RA (1990) J Phys: Condens Matter 2: SA 385

Progr Colloid Polym Sci (1996) 100:246–251
© Steinkopff Verlag 1996

Linear viscoelasticity of O/W sucrose-palmitate emulsions

A. Guerrero
P. Partal
M. Berjano
C. Gallegos

Dr. A. Guerrero (✉) · P. Partal
M. Berjano · C. Gallegos
Departamento de Ingeniería Química
(Universidad de Sevilla)
c/P. García González
s/n. 41012 Sevilla, Spain

Abstract Linear dynamic viscoelastic properties of concentrated O/W emulsions containing sunflower oil (60–80% O), water and a sucrose palmitate (1–5%) of HLB = 15 were obtained in this study. Droplet size distribution data was also analyzed.

All the emulsions presented a significant linear viscoelastic range that increased with oil and SE concentrations. The linear viscoelastic behaviour obtained is typical of a highly concentrated dispersion with strong elastic interactions that lead to an extensive flocculation and confer a high stability to the system. A plateau region, characteristic of an elastic network, is observed at intermediate frequency or time scale. After a critical shear stress the structural network broke down very rapidly.

The linear viscoelastic functions of these emulsions were reproduced by using the generalized Maxwell model having up to seven relaxation elements. The discrete relaxation spectrum obtained from this model turns out to be useful in relation to the microstructure of the system.

An increase in SE or oil concentration increased the linear viscoelastic functions of the emulsions studied. This increase cannot be explained only in terms of droplet size and polydispersity. An increase in droplet interactions should be also considered.

Key words Sucrose ester – viscoelasticity – emulsion – rheology – emulsifier

Introduction

There is an increasing interest in the research on the rheological behaviour of concentrated oil-in-water emulsions in view of its many industrial applications (food products, cosmetics, pharmaceutical, agrochemicals, etc.) Moreover, the information provided by rheological measurements may be useful to characterize the microstructure of the system, as well as in many practical fields such as quality control, storage stability, sensory evaluations, product development, design of unit operations, etc. However, this usefulness depends on two critical

stages: selection of the rheological technique to be used and interpretation of the results obtained.

Most of the rheological studies carried out on concentrated o/w emulsions refer to the viscous response of these materials under flow conditions. Much less research was devoted to their linear and non-linear viscoelastic properties. However, knowledge of the linear viscoelastic properties is of great importance since it allows to obtain information on the structure of the system in conditions close to the unperturbed state.

A powerful technique for studying the material response in such conditions is the oscillatory shear test. This kind of test must be performed at small strain or stress

Progr Colloid Polym Sci (1996) 100:246–251
© Steinkopff Verlag 1996

amplitude so that the material functions are independent of the shear strain or stress applied (linear viscoelastic region).

The present paper focused on the study of the linear viscoelastic properties of food quality o/w emulsions containing a sucrose ester of high HLB which has proven to form highly stable emulsions over a wide range of oil and emulsifier concentrations [1].

Experimental

Ryoto sucrose palmitate (P-1570, HLB = 15) from Mitsubishi–Kasei Food Corporation (Tokyo, Japan) was used as received. Sunflower oil was purchased in a local store. Oil-in-water emulsions were prepared using an Ultra-Turrax T-50 homogeneizer, at 5000 rpm and during 5 min. The sucrose ester concentration (SE) was varied from 1 to 5% (w/w) and the oil content (O) from 60 to 80% (w/w) (ϕ = 0.59–0.77). Emulsions were stored at 5 °C.

The oscillatory shear measurements were performed in a Haake RS100 RheoStress rheometer, using a cone and plate sensor system C35/4 (4°, 35 mm). All the samples had the same recent past thermal and rheological history. Three replicates of each test were done. All the rheological tests were performed at 25 °C.

Droplet size measurements were carried out in a Coulter Counter model Zb.

Results and discussion

Linear viscoelasticity range

The strain or stress range for linear viscoelasticity must be well defined in order to obtain meaningful oscillatory shear data. This linear range may be determined by a dynamic stress sweep at a constant frequency as shown in Fig. 1 for different values of the sucrose palmitate concentration. The behaviour illustrated in this figure is typical of many viscoelastic materials. The viscoelastic material functions (i.e., the complex modulus, G^*) remain constants. However, above a critical shear stress, τ_{cr}, a decrease in G^*, and other viscoelastic functions, is produced. An increase in τ_{cr} is obtained when both sucrose ester and oil concentrations are raised. These results reflect that the structure formed at rest becomes more resistant to shear forces. A similar effect was obtained by decreasing temperature [2]. The onset of the non-linear range is particularly dramatic for these emulsions which puts forward the fragility of the structure achieved at rest.

Influence of frequency

Figure 2 shows a typical frequency sweep behaviour for these emulsions. This kind of test is performed at a constant stress amplitude which is lower than τ_{cr}. The storage modulus, G', presents always higher values than the loss

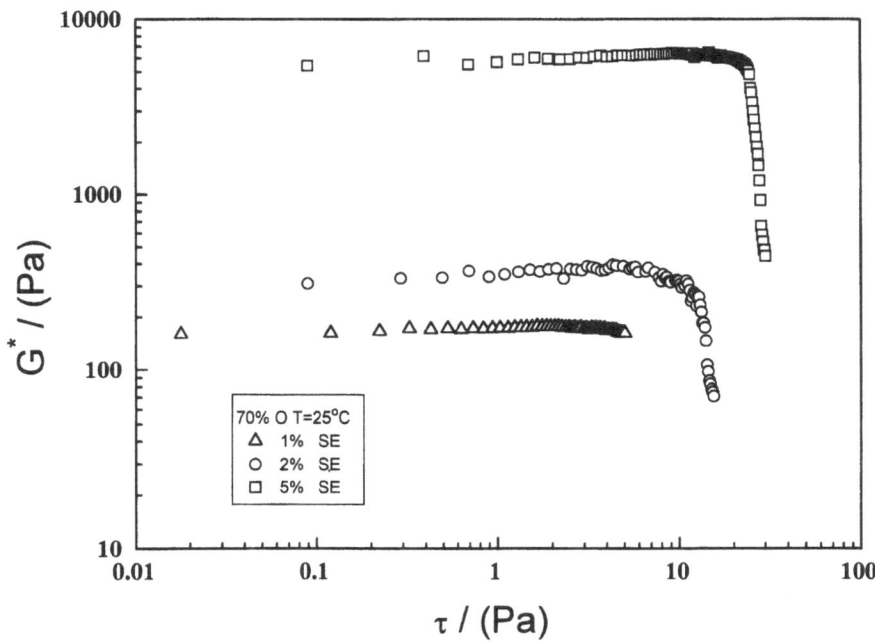

Fig. 1 Stress sweep test for emulsions containing 70% O and different SE content (T = 25 °C)

248
A. Guerrero et al.
Linear viscoelasticity of O/W sucrose-palmitate emulsions

modulus, G''. (Figs. 2, 3 and 4). Thus, the elastic behaviour dominates always over the viscous component so that more energy is stored than dissipated by the emulsion. This behaviour suggests the occurrence of a structural network due to extensive flocculation of the oil droplets produced by strong interdroplet interactions [3]. Similar results have been reported for concentrated emulsions with strong steric repulsions [4], especially when some degree of flocculation takes place [5–7].

A plateau region may be observed in G'' curves (Figs. 2, 3 and 4). Alternatively, the plateau region may be more clearly illustrated by a minimum in the loss tangent,

Fig. 2 Storage and loss moduli (G' and G'') and loss tangent ($tg\delta$) as a function of frequency for the emulsion 3% SE and 70% O ($T = 25\,°C$)

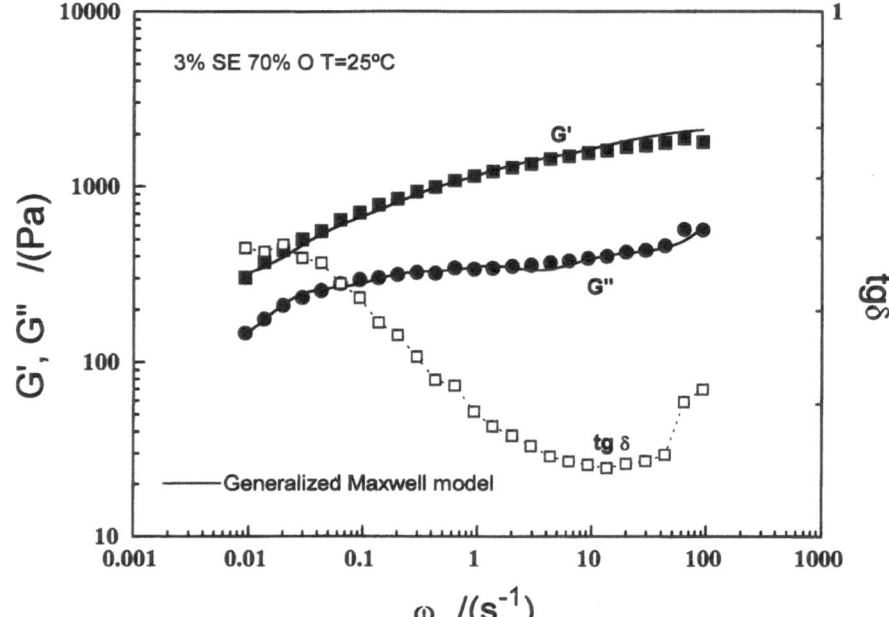

Fig. 3 Storage and loss moduli (G' and G'') as a function of frequency for emulsions containing 70% O: influence of SE content ($T = 25\,°C$)

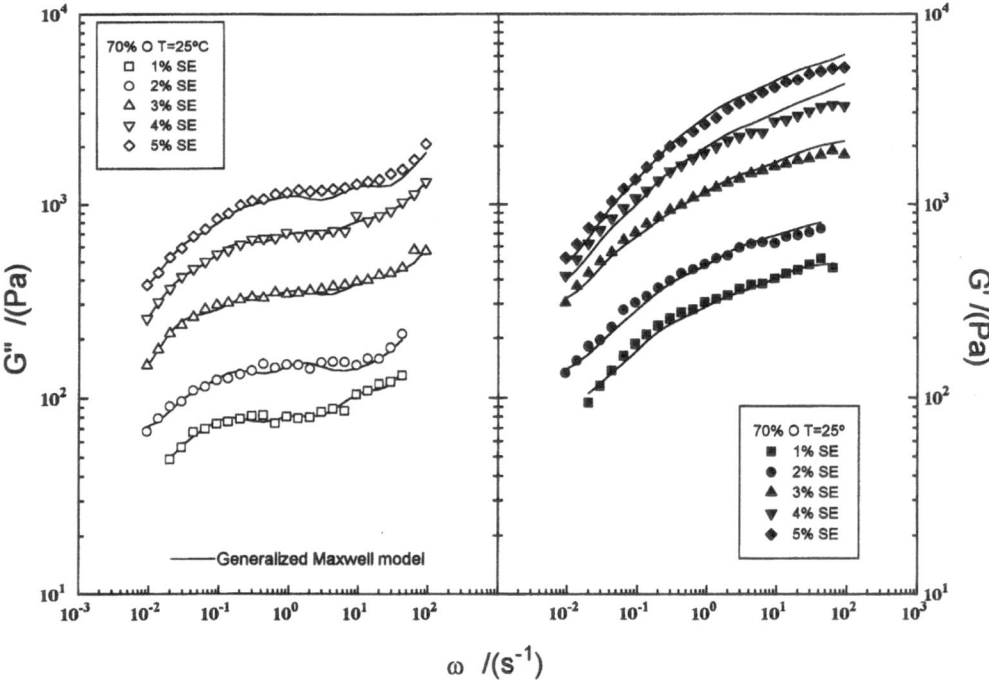

Progr Colloid Polym Sci (1996) 100:246–251
© Steinkopff Verlag 1996

Fig. 4 Storage and loss moduli (G' and G'') as a function of frequency for emulsions containing 3% SE: influence of oil content ($T = 25\,°C$)

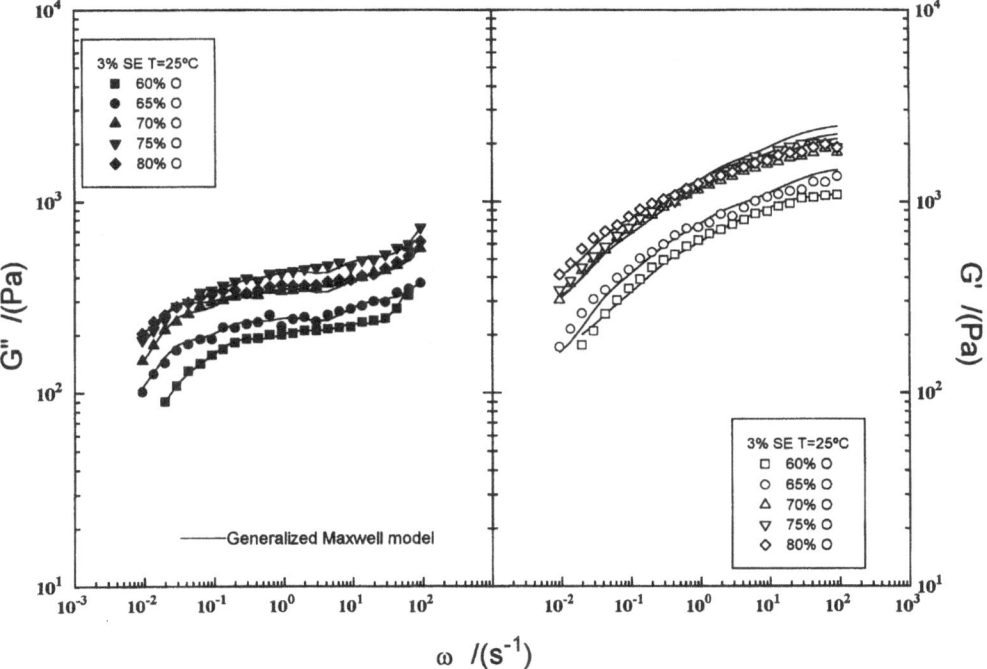

$tg\delta = G''/G'$ (Fig. 2). A plateau region has been largely described in polymer rheology, above a critical molecular weight, in terms of an entanglement between polymer molecules [8, 9] leading to an "entanglement network" in which the molecular motion is considerable restricted by neighbour molecules. This is a long-range phenomenon that has been also found for worm-like micellar solutions [10, 11]

In emulsion rheology, a structural elastic network analogous to the so-called entanglement network could be considered. Some authors investigated the linear viscoelastic properties of concentrated emulsions and used the term gel-like structure for this kind of behaviour [12, 13]. In any case this elastic network must be related to the above-mentioned extensive flocculation.

Linear viscoelastic model

A generalized Maxwell model may describe the linear viscoelastic behaviour of a material. This model considers a superposition of a series of n independent relaxation processes, each process having a relaxation time λ_k, and a relaxation strength g_k. According to this model, the shear stress $\tau(t)$ may be written as a function of the strain deformation $\gamma(t, t')$:

$$\tau(t) = \int_{-\infty}^{t} \sum_{k=1}^{n} \frac{g_k}{\lambda_k} \exp\left(-\frac{t-t'}{\lambda_k}\right) \gamma(t, t')dt' . \tag{1}$$

The linear material functions (i.e., G' and G'') may be obtained from this linear viscoelastic model [8, 14] as follows:

$$G'(\omega) = \sum_{k=1}^{n} g_k \frac{\omega^2 \lambda_k^2}{1 + \omega^2 \lambda_k^2} \tag{2}$$

$$G''(\omega) = \sum_{k=1}^{n} g_k \frac{\omega \lambda_k}{1 + \omega^2 \lambda_k^2} . \tag{3}$$

Equations (2) and (3) may be used to calculate the discrete relaxation spectrum by selecting a set of relaxation times λ_k (provided that $\lambda_{max}\omega_{min} > 1$ and $\lambda_{min}\omega_{max} < 1$) [14] and calculating the values of g_k that minimize the sum of deviation squares between predicted and experimental moduli. The consistency of the model may be deduced from the agreement with data as shown in Figs. 2, 3 and 4.

Figure 5 shows the discrete linear relaxation spectrum for emulsions at different values of SE concentration. The above-mentioned plateau region may be also observed in this figure at an intermediate time scale. A rapid decrease in the spectrum was obtained at low time scale which may be seen as the beginning of the transition region [8].

Influence of SE and oil concentration

An increase in SE and oil content produces an increase in both moduli G' and G'' (Figs. 3 and 4) as well as in the relaxation spectrum (Fig. 5). However, the influence of oil

250
A. Guerrero et al.
Linear viscoelasticity of O/W sucrose-palmitate emulsions

Fig. 5 Discrete relaxation spectra (g_k, λ_k) for emulsions with different SE content

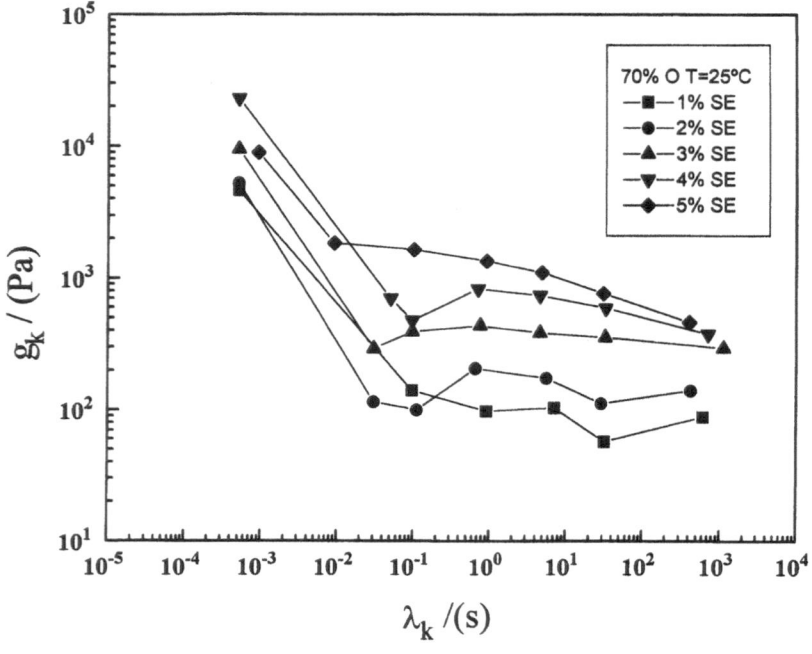

Fig. 6 Droplet size distribution curves: influence of SE content

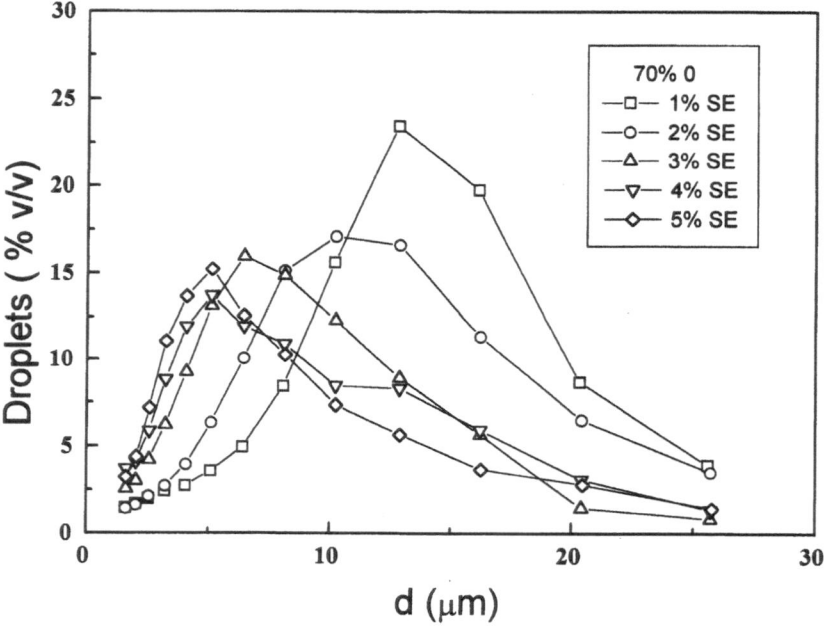

concentration is quantitatively lower under the experimental conditions. Thus, similar values of G' or G'' were obtained at high oil concentrations, which is somehow unexpected. A slightly more extended plateau region for the emulsion containing 75% and 80% oil may, however, explain the low values of G'' at intermediate frequencies. An increase in the plateau region with oil concentration

has been previously reported for salad dressing emulsions [15].

Oil and SE concentration also produces a decrease in the droplet mean diameter (i.e., the Sauter mean diameter) and in the polydispersity of these emulsions as may be seen in Fig. 6. This effect contributes to the increase in the viscoelastic functions, however an increase in droplet interaction has to be also taken into consideration.

Conclusions

The emulsions studied show a linear viscoelastic behaviour which is typical of a highly concentrated dispersion with strong elastic interactions. Such characteristics lead to an extensive flocculation and confer a high stability to the system. In spite of the high polydispersity of the emulsions, which allows a tight packing of oil droplets, a plateau region, characteristic of an elastic network, is observed at an intermediate frequency (G'' or $tg\delta$) or time (g_k) scale. After a critical shear stress the structural network breaks down very rapidly.

The linear viscoelastic functions of these emulsions may be adequately reproduced by using the generalized Maxwell model having up to seven relaxation elements. The discrete relaxation spectrum obtained from this model provides information that may be related to the microstructure of the system.

An increase in SE or oil concentration increases the linear viscoelastic functions of the emulsions studied as well as the resistance of the structural network. This increase cannot be explained only in terms of droplet size and polydispersity. An increase in interdroplet interactions should be also considered.

Acknowledgment This work is part of a research project sponsored by the CICYT, Spain (research project ALI90-0503). The authors gratefully acknowledge its financial support.

References

1. Partal P, Guerrero A, Berjano M, Muñoz J, Gallegos C (1994) J Texture Studies 25:331–348
2. Partal P (1995) Ph D Thesis. Universidad de Sevilla, Sevilla
3. Cao Y, Dickinson E, Wedlock DJ (1990) Food Hydrocolloids 4:185–195
4. Tadros Th F (1993) 1st World Congress on Emulsion, Paris, 4:237–266
5. Liang W, Tadros ThF, Luckham PF (1992) J Colloid Interface Sci 153:131
6. Gallegos C, Berjano M, Choplin L (1992) J Rheol 36:465–478
7. Franco JM, Guerrero A, Gallegos C (1995) Rheol Acta 34:513–524
8. Ferry JD (1980) Viscoelastic Properties of Polymers John Wiley & Sons New York
9. Wu S (1989) J Polym Sci 27:723–741
10. Hoffman H (1994) In: Structure and Flow in Surfactant Solutions. ACS Symposium Series 578 pp 2–31
11. Clausen TM, Vinson PK, Minter JR, Davis HT, Talmon Y, Miller WG (1992) J Phys Chem 96:474–484
12. Van Vliet T, Lyklema J, Van Den Tempel M (1978) J Colloid Interface Sci 65:505–508
13. Rivas HJ, Sherman P (1983) J Texture Studies, 14:251–265
14. Bird RB, Armstrong RC, Hassager O (1987) Dynamics of Polymeric Liquids John Wiley & Sons New York
15. Franco JM, Berjano M, Guerrero A, Muñoz J, Gallegos C (1995) Food Hydrocolloids 9:111–121

Progr Colloid Polym Sci (1996) 100:252–258
© Steinkopff Verlag 1996

RHEOLOGY

Structural models
to describe thixotropic behavior

J. Llorens
E. Rudé
C. Mans

Dr. J. Llorens Llacuna (✉)
E. Rudé · C. Mans
Department d'enginyeria Quimica
i Metal.lurgia
Facultat de Química
Universitat de Barcelona
C/. Martí i Franquès, 1
08028 Barcelona, Spain

Abstract Rheometric data of colloidal suspensions, which were prepared with a blend of typical raw materials for ceramic manufacture and deflocculants, were tested with different thixotropic models. Good agreement between theory and experiments has been found for a proposed thixotropic model.

Key words Rheology – thixotropy – ceramics – deflocculants

Introduction

Thixotropy phenomena are described as viscosity-time relationships that appear in some fluids after changes in shear rate. Thixotropic phenomena are very common in particulate colloidal systems (paints, inks, clay slurries, pharmaceuticals, cosmetics, agricultural chemicals, etc.). The viscosity in these systems depends on the floc size and shape. The most frequent structural changes that produce the phenomenon named thixotropy – brought about by shear rate changes – are that the structure breaks down under high shear rate but recovers under low shear rate or at rest. Therefore, thixotropy arises from changes on floc structural arrangement due to forces acting between suspended particles and breakdown due to the shear rate. Other less common structural changes are also possible: the structure builds up under shear, but disintegrates at rest. In these circumstances the associated rheological phenomena are named antithixotropy, which describes the opposite of thixotropy.

Interparticle forces provide the essential quality for thixotropic phenomenon. These forces are derived from the interatomic and intermolecular interactions on surface of particles. Electrostatic interactions come from the electrical double layer around the particle. Steric interactions are derived from adsorbed non-ionic macromolecules onto the particle surfaces. When two particles meet, the adsorbed macromolecular layers increase the osmotic pressure and the volume restrictions in the mixing zone and tend to push the particles apart. In many cases, small quantities of specific substances (deflocculants) are added into the system in order to increase these repulsive forces between the two particles. Typical deflocculants added in clay slurries are: sodium silicate, sodium polyphosphates and sodium polyacrylates. Polyacrylates are macromolecules with ionic end-groups (polyelectrolytes) that provide electrostatic and steric effects.

In engineering design it is appropriate to know the effect of thixotropy on the flow of materials during manufacture and in the final use. A typical engineering problem is to calculate the necessary torques to agitate a thixotropic colloidal suspension after a rest. The question can only be answered if we know the total thixotropic behavior. For this purpose, we require quantitative mathematical models of thixotropy. Much work has been conducted in this field [1–6], but very few attempts have been made to express the data in a consistent set of parameters, covering the total behavior.

The purpose of this paper is to compare the existing theories and to propose an appropriate theory that describes all the thixotropic behavior of particulate colloids systems.

Theory

Thixotropy comes under the field of generalized thermodynamic theory of relaxation phenomena [7]. This theory with continuum mechanics presents significant mathematical complexities. More functional theories are based in generalized structural models. These theoretical models of thixotropic constitutive equations can be derived from details of floc formation and break-up, and the effect of floc size and shape on viscosity. The viscosity, η, is evaluated in terms of a structural parameter, λ, that accounts for floc size and shape, and the time-dependent behavior is described by a rate equation for λ. The constitutive equations are:

$$\eta = \eta(\dot{\gamma}, \lambda) \qquad (1)$$

$$\frac{d\lambda}{dt} = g(\dot{\gamma}, \lambda) . \qquad (2)$$

When the fluid is at rest ($\dot{\gamma} \to 0$) for enough time, the viscosity is the zero shear viscosity, η_0, and the structure is fully built up, $\lambda = 1$. On the other hand, at high shear rate ($\dot{\gamma} \to \infty$) the viscosity is η_∞ and the structure is fully broken-down, $\lambda = 0$.

The Cheng model

In this model the viscosity is given by:

$$\eta = \frac{\tau_0}{\dot{\gamma}} + \frac{\tau_p}{\dot{\gamma}} \lambda + k\dot{\gamma}^{N-1} , \qquad (3)$$

where k, N and τ_p are constants and τ_0 is the yield stress.

To describe the structural changes, it is postulated that the breakdown rate depends on shear rate and also the current structure [8].

$$a(1 - \lambda) - b\lambda\dot{\gamma} = \frac{d\lambda}{dt} , \qquad (4)$$

where a and b are kinetic constants.

It can be expected that transient measurements provide useful information to measure the thixotropic behavior. Numerous transients have been published in the literature. However, three types of suitable transient experiments have been proposed: a step function in shear rate or shear stress [9, 10], a consecutive linear increase and decrease in shear rate [11], and a sinusoidal change in shear rate [12].

If step function in shear rate is produced, there is a decrease in viscosity with time under constant shear rate when shear rate is changed from a low to a high value (LS/S). On the other hand, the viscosity recovers with time when the shear rate is changed from a high to a low value (HS/LS). In all cases, the equilibrium flow curve is obtained when the shear stress attains the steady state value, Fig. 1. Under constant shear rate, it can be deduced from Eqs. (3) and (4) that the viscosity changes obey the

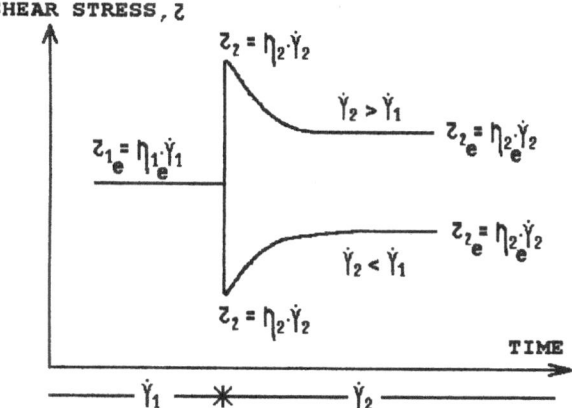

Fig. 1 Shear stress transients after step changes in shear rates

following equation:

$$\frac{d\eta}{dt} = \frac{a\tau_p}{\tau_0 + k\gamma^N - \eta_e\dot{\gamma}} (\eta - \eta_e) , \qquad (5)$$

where η_e is the equilibrium viscosity.

The viscosity transients are obtained from the solution of the differential equation (5):

$$\frac{\eta - \eta_e}{\eta_2 - \eta_e} = \exp(- t \cdot \theta) , \qquad (6)$$

where η_2 is the immediately registered viscosity after the shear rate has been suddenly changed from $\dot{\gamma}_1 \to \dot{\gamma}_2$ (Fig. 1), and $\theta = a + b \cdot \dot{\gamma}$.

When the net rate of change of structure is zero the equilibrium flow curve is obtained:

$$\eta_e = \frac{a \cdot \tau_p}{\dot{\gamma} \cdot (a + b \cdot \dot{\gamma})} + \frac{\tau_0}{\dot{\gamma}} + k \cdot \dot{\gamma}^{N-1} . \qquad (7)$$

Tiu–Boger model

In this model the constitutive equations are: a state equation analogous to Casson equilibrium flow curve

$$\sqrt{\eta} = \sqrt{\lambda} \cdot \frac{\sqrt{\tau_0} + \sqrt{\eta_\infty} \cdot \sqrt{\dot{\gamma}}}{\sqrt{\dot{\gamma}}} \qquad (8)$$

and a second order rate equation with a kinetic constant that depends on shear rate

$$A \cdot \dot{\gamma}^B \cdot (\lambda - \lambda_e)^2 = \frac{d\lambda}{dt} . \qquad (9)$$

In this case, from Eqs. (8) and (9), the viscosity changes under constant shear rate are given by

$$\frac{d\eta}{dt} = \frac{A \cdot \gamma^{B+1}}{(\sqrt{\tau_0} + \sqrt{\eta_\infty} \cdot \sqrt{\dot{\gamma}})^2} \cdot (\eta_e - \eta)^2 . \qquad (10)$$

If step function in shear rate is produced, the viscosity transients obey the following equation

$$\frac{\eta - \eta_e}{\eta_2 - \eta_e} = \frac{1}{\theta \cdot t + 1} \qquad (11)$$

$$\theta = A \cdot \dot{\gamma}^{B+1} \cdot \frac{\eta_e - \eta_2}{(\sqrt{\tau_0} + \sqrt{\eta_\infty} \cdot \sqrt{\dot{\gamma}})^2} . \qquad (12)$$

In this model, it is not possible to deduce an equilibrium flow curve, because the Eq. (9) does not explicitly specify the rate of formation and destruction of flocs.

Proposed model

It is known that the viscosity of suspension made with individual particles (completely deflocculated) depends on volume fraction of the dispersed phase, particle shape and particle size distribution. The literature contains theoretical and experimental results on viscosity of these suspensions that can be used as the basis for modeling the equilibrium flow curve or state equation. In the thixotropic modeling the flocs would take the place of the individual particles. The flocs contain occluded liquid which is immobilized and so adds to the effective volume fraction of the dispersed phase. A general conclusion from published literature [13, 14] is that the viscosity of concentrated suspensions completely deflocculated can often be satisfactorily described by equations like:

$$\eta = C_1 \cdot (1 - \Phi \cdot C_2)^{-\beta} , \qquad (13)$$

where Φ is the volume fraction of the dispersed phase and C_1, C_2 and β are parameters to fit experimental data.

Therefore, in this model Φ is taken as a structural parameter or an internal state variable. Because the floc shape depends on shear rate it seems reasonable to postulate that

$$\beta = A\dot{\gamma}^B , \qquad (14)$$

where A and B are constants.

The rate of formation and destruction of flocs is governed by Brownian and shear. To describe the structural change, it is postulated that the breakdown rate depends on shear rate and that the build-up rate depends on the amount of structure to be recovered. The net rate of change of structure is then given by the rate equation:

$$a(1 - \Phi) - b\Phi\dot{\gamma}^n = \frac{d\Phi}{dt} , \qquad (15)$$

where the breakdown rate is a power law function of the shear rate.

Under constant shear rate, it can be deduced from Eqs. (13) and (15) that the viscosity changes obey:

$$\frac{d\eta}{dt} = \frac{a}{\alpha} \cdot \eta^{(1-\alpha)} \cdot C_2 \cdot C_1^\alpha \frac{\eta_e^\alpha - \eta^\alpha}{\eta_\infty^\alpha - \eta_e^\alpha} , \qquad (16)$$

where $\alpha = -1/\beta$.

The solution of Eq. (16) gives the viscosity transients:

$$\frac{\eta^\alpha - \eta_e^\alpha}{\eta_2^\alpha - \eta_e^\alpha} = \exp(-t \cdot \theta) , \qquad (17)$$

where $\theta = a + b \cdot \dot{\gamma}n$.

This model gives transients of Cheng model when $\alpha = 1$.

When the net rate of change of structure is zero, a kind of generalized Cross equilibrium flow curve is deduced:

$$\frac{\eta_e^\alpha - \eta_0^\alpha}{\eta_\infty^\alpha - \eta_e^\alpha} = \frac{a}{b} \dot{\gamma}^n . \qquad (18)$$

When $\alpha = 1$ the Cross equilibrium is obtained.

When the shear rate is suddenly changed a significant inference can be obtained about the value of η_2 from Eqs. (13) and (14):

$$\eta_2 = C_1^{(1 - (\dot{\gamma}_2/\dot{\gamma}_1)^B)} \cdot \eta_1^{(\dot{\gamma}_2/\dot{\gamma}_1)^B} . \qquad (19)$$

Experimental

Some ceramic slips were prepared and their thixotropic behavior was tested with a step function in shear rate. The mineralogical characteristic of the ceramic material is reported in Fig. 2. It consists of a blend of typical raw

Fig. 2 Mineralogical characteristic of the ceramic material

Q-Quartz, A-Albite, M-Microcline, K-Kaolinite, T-Talc, Mu-Muscovite,

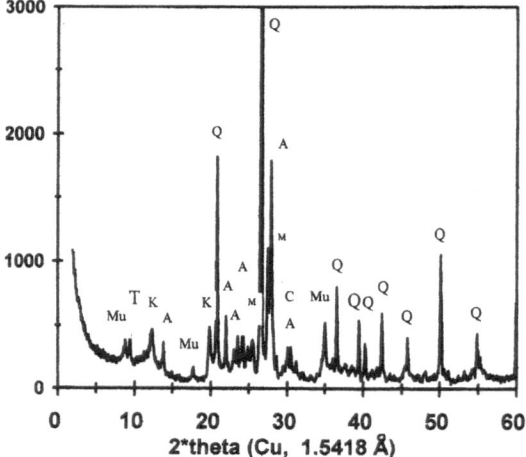

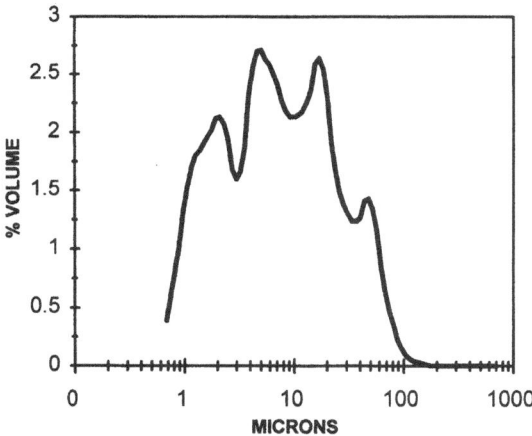

Fig. 3 Particle size distribution of the ceramic material

materials used in tile manufacture: clay, feldspar, kaolin, talc and quartz. The particle size distribution is shown in Fig. 3.

Four slips (32% by weight of water content) were prepared with different quantities of various deflocculants: sodium tripolyphosphate ($Na_5P_3O_{10}$), sodium silicate ($Na_2O \cdot SiO_2$), sodium polyacrylate Noramer 9210 (homopolymer of acrylic acid from NorsoHass with a 42% of solids, $M_w = 2000$) and sodium polyacrylate Norasol 410 N (homopolymer of acrylic acid from NorsoHass with a 40% of solids, $M_w = 10\,000$). The used nomenclature and the deflocculant content is reported in Table 1.

Table 1 Nomenclature and deflocculant content

Name	% by weight of deflocculant on dry-material basis			
	$Na_5P_3O_{10}$	$Na_2O \cdot SiO_2$	Noramer 9210	Norasol 410 N
5020-B	0.10	0.04	–	0.06
5020-A	0.10	0.04	0.06	–
5010-A	0.10	0.02	0.08	–
4020-A	0.08	0.04	0.08	–

Table 2 Step function in shear rate to analyze the rheological behavior

γ_1 [s^{-1}]	γ_2 [s^{-1}]	Time [min.]
2.7	27	2.5
27	270	2.5
270	460	2.5
460	270	2.5
270	27	2.5
27	135	2.5
135	460	2.5
460	135	2.5
135	27	2.5
27	2.7	2.5

To measure the thixotropic behavior the samples were sheared under constant shear rate ($\dot{\gamma} = 2.7$ s^{-1}) until no further changes on shear stress were observed. Then, they were tested with the step function in shear rate shown in Table 2. Equilibrium was reached because along the test changes are reversible: for each shear rate always the same shear stress was reached after enough elapsed time.

The rheological tests were carried out with a rotational viscometer HAAKE CV20. The measuring device was the ME35 bob/cup sensor system (inner cylinder diameter 28.93 mm, beaker diameter 30.0 mm, length 24.0 mm, and sample volume 1.8 cm^3). The test were carried out at 25 $\pm$ 0.1 °C. The shear stress transients were registered on a computer.

Results and discussion

The steady-state flow curve $\tau = \tau(\dot{\gamma}$, time $\rightarrow \infty)$ for the four slips is shown in Fig. 4. It can be deduced that the sodium polyacrylate with $M_w = 10\,000$ (Norasol 410 N) provides higher viscosity than the sodium polyacrylate with $M_w = 2000$ (Noramer 9210). There are no significant differences in the equilibrium viscosities among the slips that contain the same sodium polyacrylate (Noramer 9210).

The thixotropic models reported above were tested to fit the viscosity transients obtained experimentally with the rheological test. The Levenber–Marquardt iterative method [15] was used for the fittings and to obtain the parameters (θ, β). Only the Tiu–Boger and proposed models were able to fit the experimental data. Typical plots

Fig. 4 Steady-state flow curve for the slips

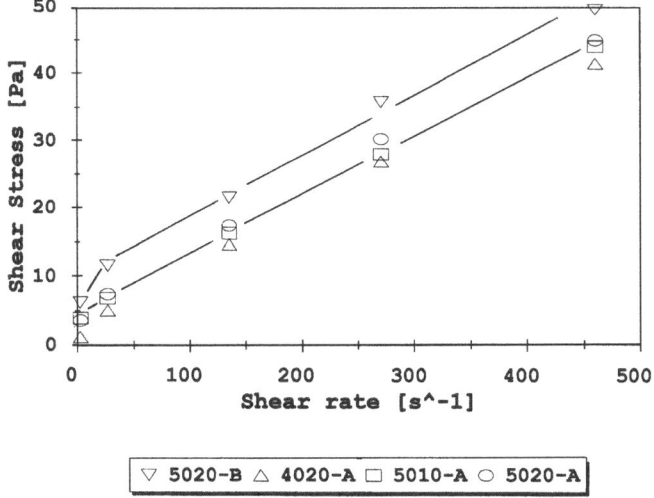

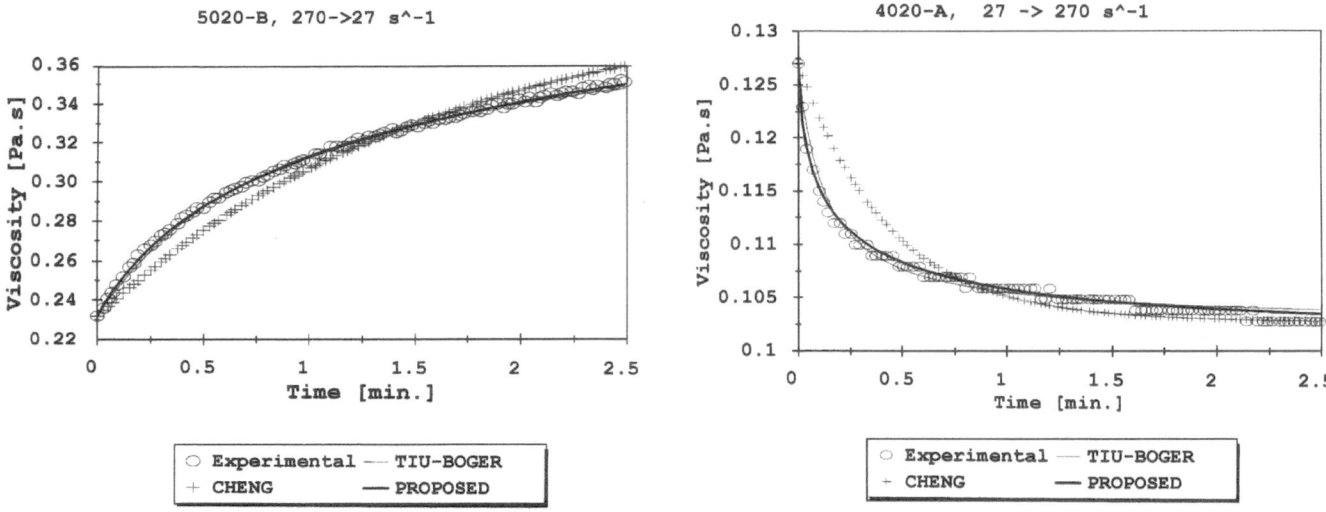

Fig. 5 Comparison between the experimental viscosity transients (slip 5020-B) and calculated curves for the models for a step change in shear rate from 270 to 27 s^{-1}

Fig. 6 Comparison between the experimental viscosity transients (slip 4020-A) and calculated curves for the models for a step change in shear rate from 27 to 270 s^{-1}

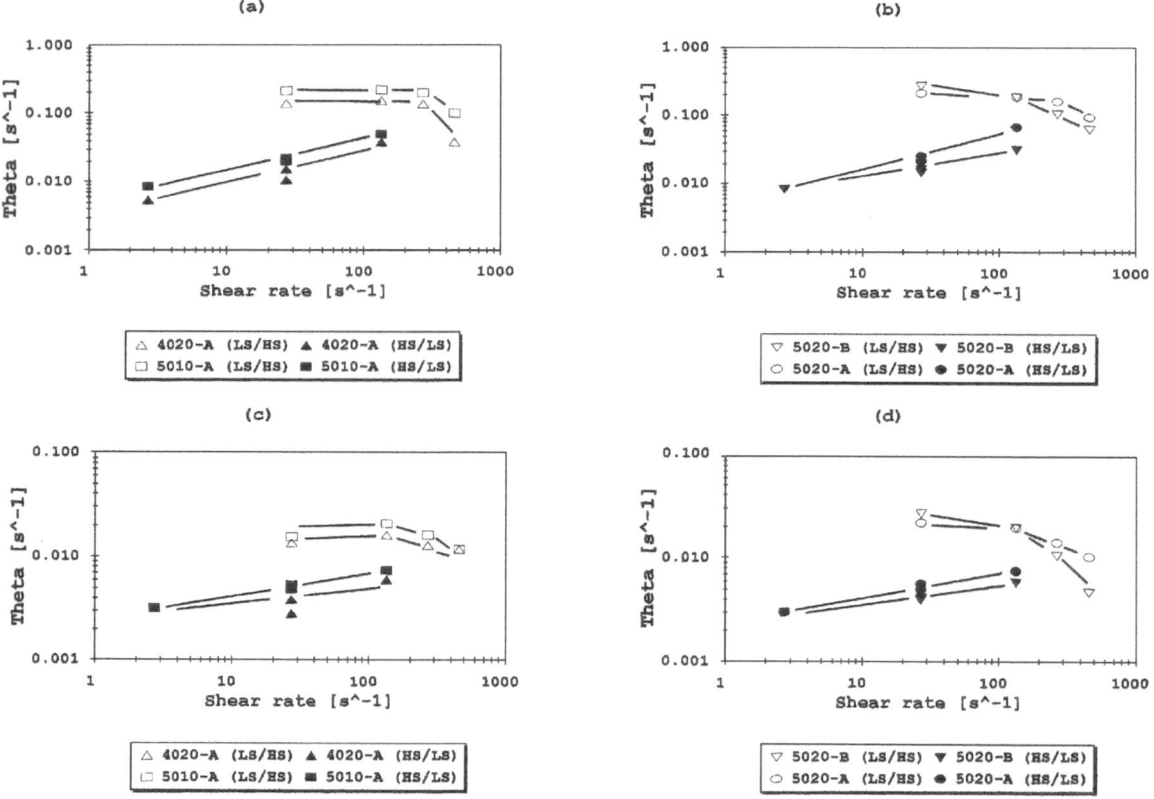

Fig. 7 Values obtained for θ. (**A**) and (**B**) for the Tiu–Boger model. (**C**) and (**D**) for the proposed model

with comparison between the experimental viscosity values and calculated curves for the models are shown in Figs. 5 and 6. The Tiu–Boger and proposed model provide similar agreement between theory and experimental data.

The values obtained for β in the proposed model were positive when the shear rates were changed from low to high values (LS/HS). However, the values of β were negative when the shear rates were changed from high to low values (HS/LS). In both conditions, the absolute values obtained for β were the same for the same values of the shear rates, $\dot{\gamma}_2$. That means that the values of C_2 in Eq. (13) are positive for the LS/HS steps and are negative for the HS/LS steps. Therefore, the relationship between the volume fraction of the dispersed phase, Φ, and the viscosity is changed when the structure breaks down or builds up.

The values of θ represent the intrinsic time scale of the thixotropic behavior. When the values of θ are low, the time needed to reach the equilibrium is high and therefore the slip is more thixotropic. The values obtained for θ in the Tiu–Boger and proposed models are function of the shear rates, $\dot{\gamma}_2$. In both models the same pattern was found: when the viscosity transients come from LS/HS steps and the values of $\dot{\gamma}_2$ increase, the values of θ decrease, but when the viscosity transients come from HS/LS steps and the values of $\dot{\gamma}_2$ increase, then the values of θ also increase. The values obtained for θ for the Tiu–Boger and proposed models are shown in Fig. 7. Significant differences in the values of θ were obtained for different deflocculant compositions. Some conclusions can be deduced from the measurements: the slip 4020-A is more thixotropic than the slip 5010-A, when the values of the shear rate are high the slip 5020-B is more thixotropic than the slip 5020-A, but are similar in thixotropy when they are low.

Equation (8) from the Tiu–Boger model was not able to fit the steps of viscosity (or shear stress) responses from step changes in shear. However, Eq. (19) from the proposed model was able to give a good estimate of these steps changes in viscosity. Only for the lowest shear rate ($2.7\,s^{-1}$) were the predictions unsatisfactory. The parameter B coming from Eq. (14) was close to 0.5; only small differences were found for the different slips, Fig. 8.

The proposed model can express the complete thixotropic behavior in a consistent set of parameters: $\theta = \theta(\dot{\gamma})$, $\beta = \beta(\dot{\gamma})$. Therefore, it is possible to reconstruct the shear stress response from a step function in shear rate. For example, it is shown in Fig. 9 the experimental shear stress vs. time from the rheological test for the slip 5020-A and the outlook from the proposed model. Similar responses were found for the other slips.

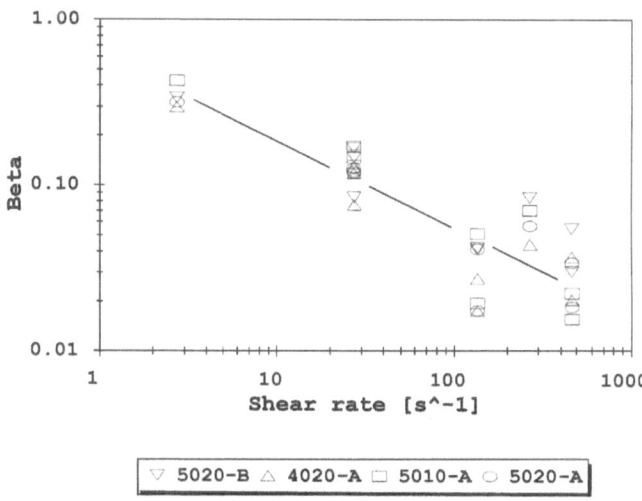

▽ 5020-B △ 4020-A □ 5010-A ○ 5020-A

Fig. 8 Values of β for the proposed model

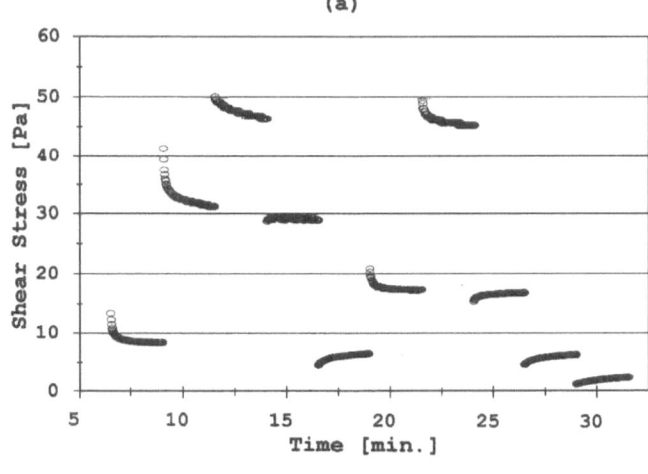

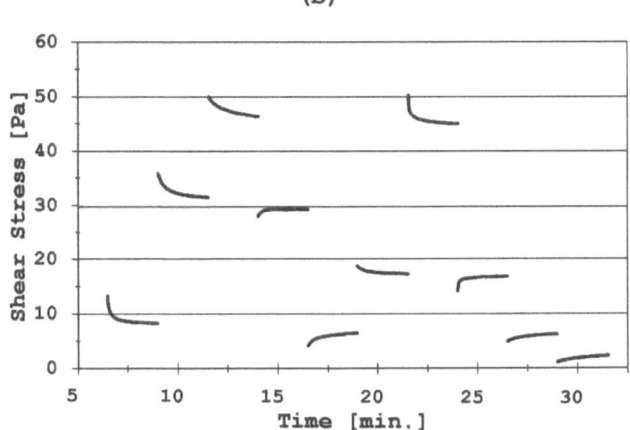

Fig. 9 (A) Experimental shear stress vs. time from the rheological test for the slip 5020-A. (B) Outlook from the proposed model for the same slip

Concluding remarks

The Tiu–Boger and the proposed model were able to fit the experimental data from the rheological test. The fittings made with the Cheng model were unsatisfactory.

The relationships between the intrinsic time scale of the thixotropic behavior, θ, and the shear rate were similar for the Tiu–Boger and the proposed model.

Only the proposed model was able to express the complete thixotropic behavior in a consistent set of parameters: $\theta = \theta(\dot{\gamma})$, $\beta = \beta(\dot{\gamma})$.

Synopsis

Rheometric data of colloidal suspensions, which were prepared with a blend of typical raw materials for ceramic manufacture and deflocculants, were tested with different thixotropic models. Good agreement between theory and experiments has been found for a proposed thixotropic model.

Acknowledgments The authors express their gratitude to Norso-Haas S.A. for providing the sodium polyacrilates.

References

1. Cheng DCH (1986) "Thixotropy" Symposium on Gums, Polymers, Thickeners and Resines 24–25 Nov. Harrogate, UK
2. Mewis J (1979) Journal of Non-Newtonian Fluids Mechanics 6:1–20
3. Cheng DCH, Evans F (1965) Brit J Appl Phys 16:1599–1617
4. Alessandrini A, Lapasin R, Sturzi F (1982) Chem Eng Commun 17:13–22
5. Sestak J, Houska M, Zitny R (1982) Journal of Rheology 26(5):459–475
6. Amemiya JI (1992) Journal of Food Engineering 16:17–24
7. Astarita G (1989) In: Thermodynamics Plenum Press, New York, pp 131–153
8. Carleton AJ, Cheng DCH, Whittaker W (1974) Technical paper No IP74-009, Institute of Petroleum, London
9. Pryce J (1941) J Sci Instrum 18:19–24
10. McMillen EL (1932) J Rheol 3:164–179
11. Green H, Weltmann RN (1943) Ind Eng Chem Anal De 15:424–429
12. Astbury NF, Moore F (1970) Rheol Acta 9:124–131
13. Krieger IM (1972) Adv Colloid Interface Sci 3:111–118
14. de Kruif CG, van Iersel EMF, Vrij A, Russel WBJ (1985) J Chem Phys 83:4717–4725
15. Guide to Standard Mathematica Packages (1993) Wolfram Research, Inc, pp 396–399

Progr Colloid Polym Sci (1996) 100:259–265
© Steinkopff Verlag 1996

RHEOLOGY

The effect of surface friction on the rheology of hard-sphere colloids

J. Castle
A. Farid
L.V. Woodcock

J. Castle · A. Farid
Prof. L.V. Woodcock (✉)
Department of Chemical Engineering
University of Bradford
Bradford BD7 1DP, United Kingdom

Abstract Intuition plus experimental evidence indicate that an essential, but missing, ingredient from all previous theoretical and simulation models of *experimental* idealised hard-sphere monodisperse suspensions is surface friction. The modelling of real, i.e., rough, spheres requires at least one additional system parameter, to account for contact rotational friction, in the dimensional analysis leading to corresponding states scaling laws.

Idealised latex particles with variable surface roughness have been produced by a dispersion polymerisation in which the molecular weight of the polymeric steric surface stabilisers is varied. A degree of surface roughness of the dry particles is characterised experimentally using an argon adsorption technique. This differential adsorption technique also measures inhomogeneities in surface roughness, on a particular distance scale, "holes and crevices", which arise from the uptake of the PVP steric stabiliser on the surface. This technique, however, does not discriminate interparticle roughness. Suspended particles show a solvation effect whereupon the higher the molecular weight of the steric stabiliser, the greater the surface friction.

Controlled stress rheological flow curves for three suspensions differing only in the extent of surface roughness are reported. The results show a substantial increase in yield stresses and apparent viscosities of dense suspensions with surface roughness.

Key words Surface friction – rheology – colloids

Introduction

In order to predict flow processes of a colloidal suspension in a given geometry, two ingredients are required: i) the constitutive transport properties of the material, and ii) solutions to the conservation equations of fluid mechanics [1]. Because it is not possible to maintain homogeneity of particle density and velocity gradient in conventional rheometic devices, such as cone-and-plate or coaxial cylinder rheometers, the requisite constitutive rheological relations, such as the stress vs. strain rate flow curves, are not directly accessible experimentally [2]. Computer simula- tions of colloidal dynamics, in which driving surfaces can be replaced by periodic boundary conditions which main- tain homogeneous shear [3], do offer a potential route to accurate rheological flow curves. Such simulations have hitherto been confined to idealised model systems in which the particle interactions and hydrodynamic equations of motions can be prescribed.

Even for hard-sphere models, little progress has been made in testing computer simulation predictions against experimental rheometric fluid mechanics because there are uncertainties in what are the essential ingredients in both the equations of motion of the particles, and the minimum requirements in the model. Indeed, almost all

260
J. Castle et al.
Rheology of hard-sphere colloids

RELATIVE SHEAR VELOCITY

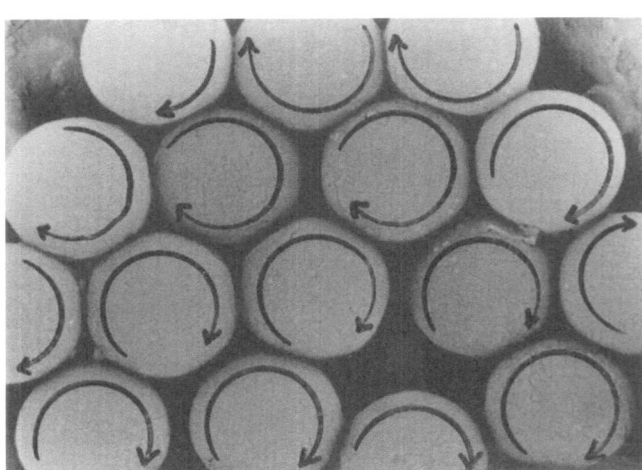

Fig. 1 Scanning electron micrograph of a sample of the PVP sterically-stabilised monodisperse polystyrene colloidal particles showing the surface roughness; the superimposed arrows demonstrate how, when particles rotate with the flow-field, the stress arising from particle–particle rotational friction forces becomes dominant

the theoretical research, going right back to Einstein, to date, and all the computational models relate to the ideal suspension of monodisperse, *frictionless* hard spheres in a Newtonian medium. Both intuition and experimental evidence suggest that this model is an inadequate representation of the current experimental range of "hard-sphere" colloids because of the neglect of surface friction.

During flow, additional to hydrodynamic resistance forces, frictional resistances arise from the repulsive or osmotic forces acting between the solid particles themselves. In the idealised hard-sphere model, interactions between the particles themselves are collisional, and, on collisions the particles exchange momentum thereby creating an osmotic pressure. This leads to shear-rate dependent osmotic pressure and granular "temperature" (particle kinetic energy) gradients. When two spheres which have surface friction collide, or, interact at contact, there is not only a transfer of linear momentum, but also a transfer of rotational or angular momentum. The former gives rise to an osmotic contribution to the normal and off-diagonal elements of the osmotic pressure tensor, causing dilatancy and shear thickening respectively, but the frictional interaction also has a contribution to the osmotic shear stress. The objective of this research is to obtain an estimate of the effect of this contribution in experimental well-characterised suspensions.

The present experimental system is a monodisperse polystyrene latex; an electron micrograph of a typical sample is shown in Fig. 1. In frictionless (smooth) hard-sphere models, the particles are deemed to rotate freely with the flow field as shown. It is clear, however, that neighbouring particles rotate against each other, and that the transfer of rotational momentum on contact cannot be neglected in modelling the rheology for rough spheres. In the following section evidence will be presented to suggest that rotational interparticle surface friction contributes substantially to the rheology of experimental micron-range colloidal suspensions, and that it may ultimately dominate the rheological behaviour at higher packing fractions.

Dimensional analysis

Since the earliest treatment of suspension viscosity by Einstein [4], most previous theoretical and simulation models of colloidal suspensions assume that the suspended particles are monodisperse, frictionless hard-spheres [5–7]. The lack of surface friction implies that the particles will rotate freely with the flow field but will not exchange any rotational momentum on colloisional contact.

Monodisperse frictionless spheres in a suspension, which collide elastically, are characterised by the particle variables diameter σ, mass m, medium viscosity η_0 and medium density ρ_0. For a suspension undergoing homogeneous laminar Couette shear flow, three system-state variables are required to specify the state of the system, the particle concentration ϕ, the thermal energy per particle $k_B T$, and the rate of shear strain γ. This is the simplest theoretical model suspension that can approximate reality. The objective is to predict the requisite functions of the model system and system-state variables such as the suspension viscosity, or the particle diffusivities.

In the case of the viscosity, the suspension viscosity (or stress) may be written as a function

$$\eta = f(\sigma, m, \eta_0, \rho_0, \phi, k_B T, \gamma) . \tag{1}$$

Properties of the system are expressible in dimensions of mass, length and time. Krieger [5] first employed the method of dimensional analysis whereupon the variables in Eq. (1) can be expressed as dimensionless groups so that according to Buckingham's Pi Theorem, viscosity, for instance, can be expressed in terms of $8-3 (= 5)$ dimensionless groups. In the case of a sheared suspension, the five convenient system and system-state variables are: suspension viscosity ($\eta^* = \eta/\eta_0$), solids fraction (ϕ^*), medium density ($\rho_0^* = \rho_0 \sigma^3/m$), and two alternative reduced shear rate scales, a Peclet number and a Reynolds number, i.e., $\gamma_{Pe}^* = \gamma \eta_0 \sigma^3/k_B T$ and $\gamma_{Re}^* = \gamma \eta_0 \sigma^2/\eta_0$ respectively.

For the model of frictionless spheres therefore, an exact result of dimensional analysis is that the reduced suspension viscosity is a function only of the solids fraction and

Progr Colloid Polym Sci (1996) 100:259–265
© Steinkopff Verlag 1996

Fig. 2 Flow curves of monodisperse polystyrene latex suspended in isopropyl alcohol at
different temperatures; the upper curve shows the reduced
suspension effective viscosity
plotted against the shear stress
according to the Peclet scaling
of Krieger [4]; the lower set of
curves shows the same
rheometric data for the reduced
viscosity plotted against the
reduced shear-rate or Reynolds
number [2]

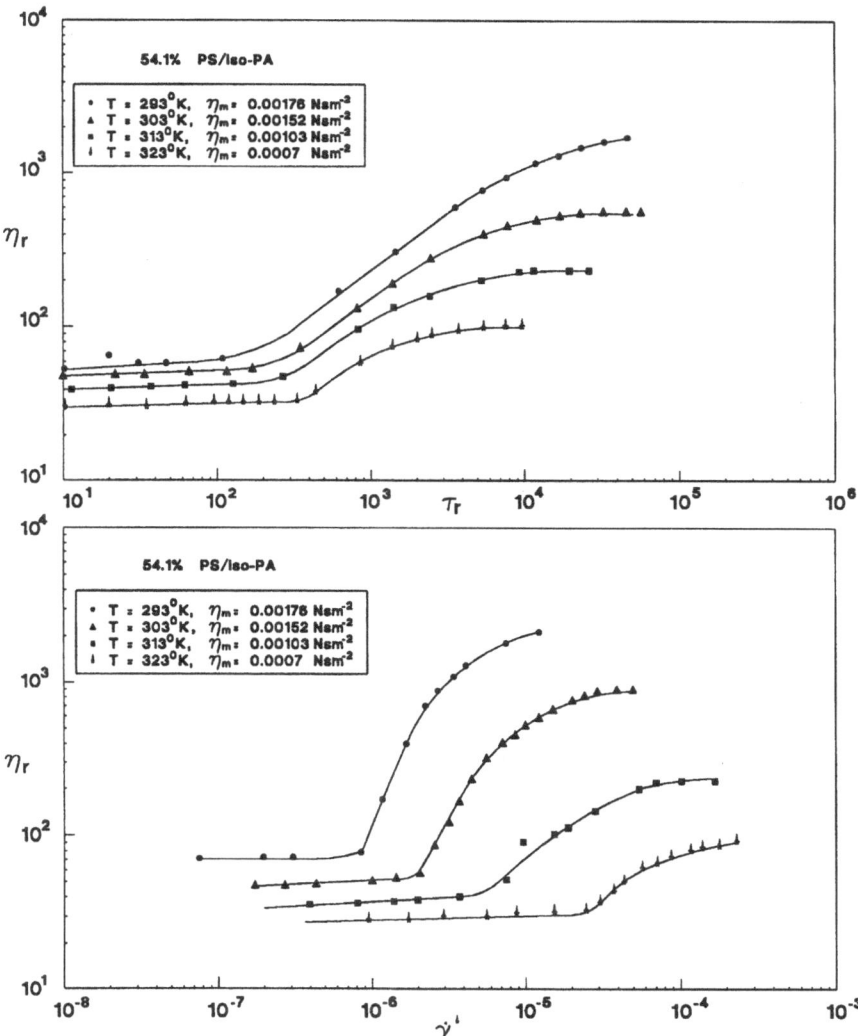

two dimensional shear rates, i.e.,

$$\eta^* = f(\phi, \gamma^*_{\text{Pe}}, \gamma^*_{\text{Re}}) \ . \tag{2}$$

When inertial effects can be neglected Eq. (2) gives Peclet
(Krieger) scaling

$$\eta^* = f(\phi, \gamma^*_{\text{Pe}}) \tag{3}$$

this is the model appropriate to most previous simulations
in which inertial terms in the equations of motion are
neglected [6]. When the particles are sufficiently large,
or the medium sufficiently viscous, however, so that
Brownian motion is recessive and thermal energy plays
no direct role in the physics, then the exact scaling law
for a suspension of monodisperse frictionless spheres is
Reynolds scaling

$$\eta^* = f(\phi, \gamma^*_{\text{Re}}) \ . \tag{4}$$

Starting with these formally exact results, Eqs. (3) and (4),
two pieces of experimental evidence can be adduced to
suggest that real suspensions will not obey these scaling
laws, whichever domain of scaling is appropriate, because
of the neglect of surface friction.

The first evidence comes from experimental measurements of the apparent viscosity- shear rate flow curves of
a monodisperse polystyrene (hard-sphere) suspension in
media in which only the temperature is varied. In the
idealised model, without surface roughness, the only effect
of temperature would be to change the medium viscosity
and $k_B T$, the thermal energy, and, to a lesser extent, the
densities. When the flow curves are plotted in reduced
units as a function of the Peclet and Reynolds numbers
(Fig. 2), the reduction fails to comply with either Eq. (3) or
Eq. (4). Compliance with either of these scaling laws would
require all the reduced flow curves to collapse on to

Table 1 Synthesis of latex samples with variable surface roughness

Sample	Stabiliser	Alcohol	Initiator (in 500 g styrene)	Median diameter μm	Median volume μm^3
A	K15	methyl	8.0 g	2.164	5.277
B	K25	methyl	8.0 g	1.536	2.572
C	K25	methyl	12.0 g	1.706	4.712
D	K30	amyl	3.5 g	2.069	5.885
E	K60	isopropyl	5.0 g	2.235	5.299
F	K90	t-butyl	8.0 g	3.082	4.189

a single line. The temperature dependence of the reduced flow curves is direct experimental evidence that an additional system parameter is required; intuition suggests that it is a temperature-dependent molecular property of the sphere surfaces; the reduced viscosity decreasing with increasing T suggesting a surface roughness.

The second evidence arises from comparisons with experiment of predictions from corresponding states scaling laws which uniquely relate the suspension viscosity when Brownian motion is recessive, to the isokinetic flow curve of the classical hard-sphere fluid [7]. The predictions show a discrepancy of 2 to 3 orders of magnitude suggesting a substantial defficiency in the model. The neglect of surface friction *inter alia* has been suggested [2, 7]; alternative explanations are the neglect of surface/interfacial solvation effects, or the assumption of elastic collisions for the hard-sphere particle–particle interactions.

Particle synthesis and surface roughness

Monodispersed polystyrene particles in the 1–10 micron range, with a standard deviation of $< 2\%$, have been produced by dispersion polymerisation in organic media. These sterically stabilised particles can be readily dispersed in water so that the rheometric flow curves of concentrated suspensions of these particles, up to the maximum packing fraction, may be easily measured. The method of Almog et al. [9] has proved to be the most effective for the production of monodispersed latexes. The detailed method of synthesis has been described elsewhere [10].

The particle size and distribution were determined by a Coulter Counter using an electrical impedance technique. The dry powder density was measured using a Helium Multivolume Pycnometer 1305. The average density calculated by this method was 1.07 g/cc. The literature value [11] for the density of pure polystyrene is 1.047 g/cc. at 20 C. As the particles used are not purely polystyrene, some deviation was expected. Particle morphology was also looked at under the scanning electron microscope; a Joel JSM 6400 scanning electron microscope was used. This method can be used to size the particles, using an

enlarged photograph of a field, however (see Fig. 1) it has its limitations because of the small sample size, magnification scale error, and the labour intensity.

Recent improvements in the resolution of the available scanning electron microscopy techniques allow for better particle visualisation. Particles that had previously appeared to have perfectly smooth surfaces now showed surface roughness. Published work by Paine et al. [12] agrees with these findings and shows excellent visualisation of the surface of sterically stabilised polystyrene particles produced by a similar reaction to the one used here. There was distinct evidence of surface roughness. The PVP steric stabiliser is clearly seen to be grafted to the surface providing a stabilising layer of 10 to 20 nm in thickness.

The effect of altering the molecular weight of the stabiliser was investigated using PVP at 10 000, 24 000, 40 000, 160 000 and 360 000 Daltons. PVP of 40 000 Daltons produced the most monodispersed colloid, but 24 000 and 160 000 were deemed acceptable, being in the 5% limit. The molecular weight of the polymeric stabiliser alters the rate of monomer adsorption so an increase in molecular weight should result in a decrease in particle size. The highest molecular weight PVP (360 000 Daltons) however, could be adsorbed onto more than one colloid, thus promoting coagulation. When the stabiliser molecular weight was too low (10 000 Daltons) a broader size distribution and some coagulum were observed as the stabilising power of the smaller chains was lower.

As the moelcular weight of the stabiliser was increased the particles appeared rougher but they were also smaller. This is due to the bulky stabiliser reducing the swelling of the particle nuclei during production. It was necessary to also change the reaction conditions to produce similar sized particles with different stabiliser layers but with the same degree of monodispersity. The objective of this series of syntheses (Table 1) was to produce three different samples with the same particle size, the same density, and differing only in the degree of surface roughness. The following particle size analysis results were obtained for the range of syntheses.

Samples A, D and E, were selected suitable particles to fulfil the similar size and monodispersity specification.

Table 2 Surface properties of three different latex samples of nearly same size

Sample	Stabiliser mw Dtns.	Density g/cm^3	Diameter (μm)	BET area (m^2/g)	Calc. area (m^2/g)
A	24 000	1.074	2.164	4.269	2.5816
D	40 000	1.070	2.069	3.079	2.7130
E	160 000	1.070	2.235	2.709	2.5122

This can be shown by comparing the volumes of the particles given in the above table.

The surface areas of the three similarly sized particles with different stabiliser layers were measured by determining the BET nitrogen adsorption plots (Table 2). These turn out to be close to the calculated values from size and density measurements. The BET method and calculation obtains the adsorbant surface area value but determining the monolayer volume of adsorbed gas at lower (near ambient) pressures. It is only valid if the relative pressure range (p/p_0) is between 0.05 and 0.35. If the plot is linear the surface area can easily be calculated from the slope and the intercept. This method does not force the gas into the pores. This dry powder analysis gives no indication of the extent of surface roughness of the solvated, suspended particles. The densities of the three selected samples were also measured.

Experimental rheometry

The three selected samples were then suspended and dispersed at several different weight percentages in distilled water. The samples were sonicated several times over a 1-month period to ensure complete wetting and 100% dis-

persion. The suspensions were prepared at 40, 45, 50, 55, and 60% by weight. Since the densities of water and polystyrene are almost the same this is effectively the solids volume packing fraction. Rotational viscometry was conducted using a Carri Med CSL100 controlled stress rheometer fitted with the Couette geometry. After careful loading of the sample into the rheometer, an immiscible light oil layer was placed on the surface of the sample to prevent water loss during the experiment.

All the samples were then subjected to the same programed increase in applied shear stress and the rates of shear flow were monitored. All the rheometric measurements were done at 25 °C. The rheological flow curves are similar for the different sets of suspensions but occur at different positions depending on the molecular weight of the stabiliser.

A comparison between the flow curves obtained from the three selected samples at 50% packing is shown in Fig. 3. In this range of shear stress, at 50%, the three curves show a linear stress – strain rate relationship; the measured Newtonian viscosities are sample A 0.01429 Pa s, sample D 0.02086 Pa s, and sample E 0.08375 Pa s. This result shows that as the molecular weight of the steric stabiliser layer on the particles increases, there is a corresponding increase in the viscosity of

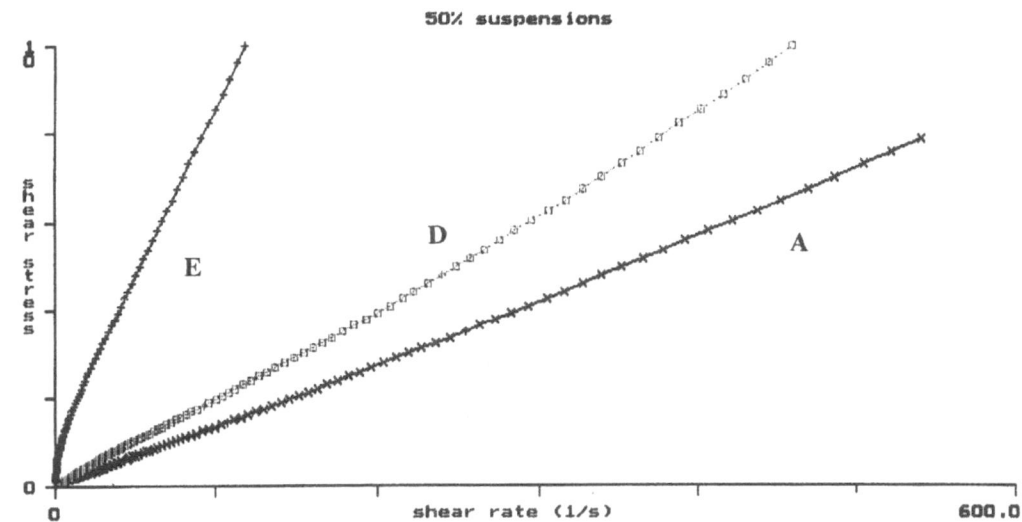

Fig. 3 Rheometer printouts for the shear stress as a function of the rate of shear strain for controlled (constant) stress experiments using coaxial cylinder geometry in a Cari-Med rheometer. A comparison of the flow curves of the three different suspensions at 50% solids in the Newtonian region; the calculated viscosities at low shear of the three samples are respectively, and are consequently designated rough, rougher and roughest

264
J. Castle et al.
Rheology of hard-sphere colloids

Fig. 4 Concentration
dependence of the effect of
surface friction; only the
molecular weight of the steric
stabiliser varies from one
sequence to another. The
highest particle density studied
is 60% and the yield stress of
the two rougher samples at this
concentration (bottom) are so
high that the stresses are off the
scale of the instrument. The
results shown correspond to the
concentrations of 60%, 55%,
50%, 45% and 40% packing
fractions

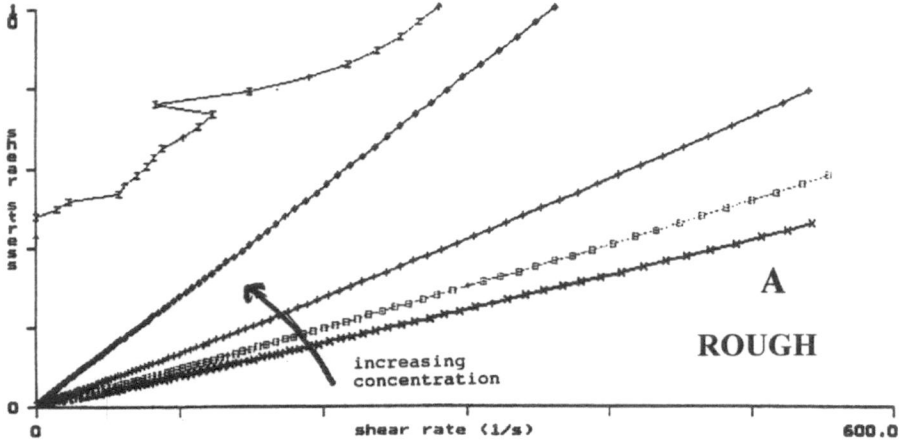

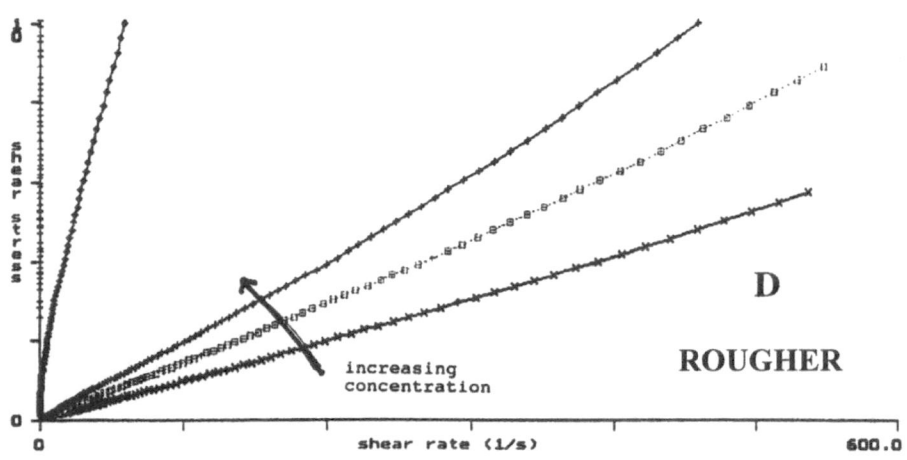

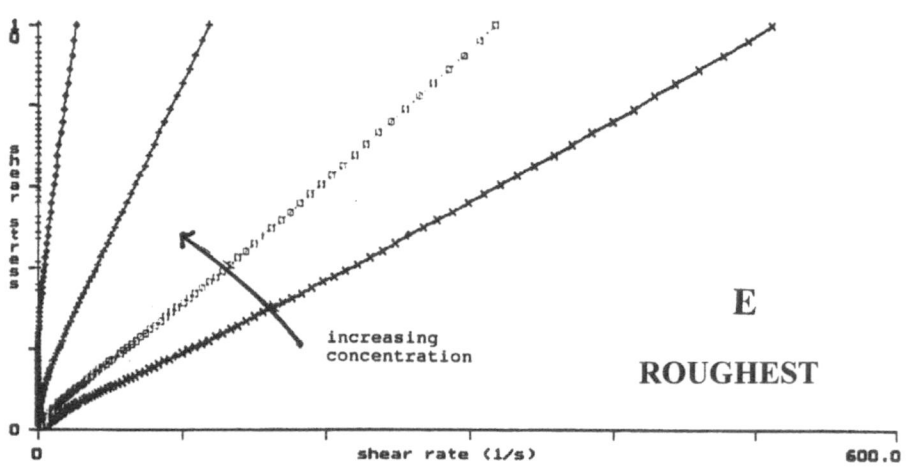

Progr Colloid Polym Sci (1996) 100:259–265
© Steinkopff Verlag 1996

a fixed concentration of suspension, all other parameters being the same. In suspension, the particles are solvated and the longer chain polymer gives rise to enhanced surface friction. On the basis of these comparisons, the three samples have been designated "rough", "rougher" and "roughest", respectively. The full set of rheometric flow curves for each individual sample over the concentration range are shown in Fig. 4.

The results at the higher concentrations all show the same trend. The effect of the higher molecular weight PVP stabiliser, or surface roughness, becomes even more pronounced as the concentration increases. Except for the 60% sample, however, the viscosity over this stress range appears to remain Newtonian. At 60% all three samples show a yield stress but the yield stress for the rougher and roughest samples are so high that they are off the scale of the instrument. The curve at 60% for sample A is rather erratic, suggesting non-steady time-dependent effects.

Conclusions

i) Particle–particle rotational frictional forces must be included in theoretical treatments of hard-sphere suspension rheology, or in computer simulation models, irrespective of the equations of motion, if connection is to be made with experimental colloidal suspension rheology. It seems that an additional parameter quantifying surface friction is required in the dimensional analysis and to improve the corresponding states scaling predictions based upon simple hard-sphere scaling laws.

ii) Model hard-sphere latex suspensions with similar particle size and densities, but with tailored variable roughness, can be produced by varying the molecular weight of the steric stabiliser.

iii) The roughness needs to be characterised for a solvated system, not dry. The present rheometric results do not correspond with the BET surface area measurements on the dry samples. It appears that surface roughness may be associated with solvation effects and the property really needs to be characterised and quantified for wet particles which are fully solvated.

iv) Rheometry, nevertheless, shows a large increase in the effective viscosity of hard-sphere suspensions with surface roughness; the roughness effect becomes more pronounced with increased solids fraction.

Acknowledgments We wish to thank the UK Engineering Physical Science Research Council for the award at a research grant and to acknowledge the advice and support of Dr. Andrew Cluely and the UK Polymer Engineering Directorate.

References

1. see e.g. Schowalter W (1978) Mechanics of Non-Newtonian Fluids. Pergamon, Oxford
2. Jomha AI, Merrington A, Woodcock LV, Barnes HA, Lips A (1991) Powder Technology 65:343–370
3. Lees A, Edwards SF (1972) J Phys C 5: 1921–1929
4. Einstein A (1956) Theory of Brownian Motion Dover Publications (New York)
5. Krieger IM (1972) Adv Coll Int Sci 3: 111–135
6. Bossis G Brady JL (1984) J Chem Phys 80:5141–5155
7. Hopkins A, Woodcock LV (1990) J Chem Soc Faraday Trans 86: 2121–2132
8. Almog Y, Reich S, Moshe L (1982) British Polymer Journal 14:131–136
9. Castle J (1995) Proceedings of Conference on Multi-phase Materials made by Emulsion Polymerisation (Lancaster: UK)
10. Paine AJ, Deslandes Y, Gerroir P Henrissat B (1990) J Coll Int Sci 138:170–189
11. Weast RC (Ed.) Handbook of Chemistry and Physics 68th Ed. CRC Press (Florida, USA 1987)

Progr Colloid Polym Sci (1996) 100:266–270
© Steinkopff Verlag 1996

RHEOLOGY

A. Cerpa
M.T. García-González
P. Tartaj
J. Requena
L.R. Garcell
C.J. Serna

Rheological properties of concentrated lateritic suspensions

A. Cerpa · P. Tartaj · J. Requena
Dr. C.J. Serna (✉)
Instituto de Ciencia de Materiales
de Madrid
CSIC, Cantoblanco
28049 Madrid, Spain

M.T. Garcia-González
Instituto de Ciencias Medioambientales
de Madrid
CSIC, Serrano 115 Dpdo
28006 Madrid, Spain

L.R. Garcell
Facultad de Ingenieria Química
Sede Julio A. Mella
Universidad de Oriente
Ave. Las Américas s/n
Santiago de Cuba, Cuba

Abstract The rheological and surface properties of concentrated suspensions of lateritic minerals from Cuba have been studied. The physico–chemical analyses indicated that the samples were mainly composed by different proportions of serpentine–goethite mixtures. The rheology of each lateritic suspension was determined as function of solid concentration and pH. Flow curves for samples containing 25 to 36% by weight of solid were interpreted following the Bulkley Hershel or Casson models. For each sample, the maximum in the yield stress and viscosity occurs at pH values close to the isoelectric point (IEP), which takes place between 4 and 9 depending on the serpentine–goethite ratio. This wide range in stability with pH explains the variability observed in the suspensions depending on their mineralogical composition. At pH values outside the IEP a low viscosity is obtained in each sample and the suspensions are well dispersed.

Key words Concentrated suspensions – rheology – isoelectric point – serpentine – goethite

Introduction

Colloidal processing is an important research area which has found a wide range of industrial applications [1]. However, the complex behavior of suspensions is not well understood since predictive theories are available only for dilute suspensions of single phase components.

Thus, the behavior of aqueous concentrated suspensions which are in many cases those of industrial interest continue to be an important problem from both experimental and theoretical point of view.

Our aim is to have a better knowledge about the different behavior of lateritic soils in sedimentation processes, when they are treated in leaching plants for nickel and cobalt extraction. For this purpose, a detailed chemical and mineralogical characterization of different lateritic samples was conducted by: x-ray diffraction, infrared spectroscopy, x-ray fluorescence, TEM.

Previous works [2–10] have pointed out the presence of several factors that influence the observed changes in viscosity of the suspensions. Thus, factors such as particle size and shape, suspension pH, chemical and mineralogical composition of the solid phase are very important in determining the flow properties of the suspensions [6].

This work presents the relationship between rheological and surface properties of concentrated lateritic suspensions with their mineralogical composition.

Experimental procedure

Origin and sample preparations

Three lateritic samples from Cuba, corresponding to different deposits: Yamanigüey and Atlantic "Perfil Nipe" were chosen. Samples SG and G come from Yamanigüey, while GS was extracted from Atlantic deposit. In each case, the fraction selected for the study was less than 2 microns.

In order to determine the mineralogical composition of the lateritic suspensions, a pure natural serpentine (SP) and a pure synthetic goethite (GP) prepared according to reference [11] were also included in this study as reference samples.

X-ray diffractograms of pure serpentine show its most intense reflections at 7.3 and 3.65 Å which correspond to the 001 and 002 basal planes. For pure goethite the two more intense reflections occur at 4.18 and 2.45 Å.

The IR spectrum of pure serpentine shows Si–O stretching bands at 1080 and 966 cm^{-1} and an intense band at lower frequency (445–460 cm^{-1}) for Si–O–Mg [12, 13]. In addition, there is a band around 620 cm^{-1} that corresponds to an OH deformation mode. The IR spectrum of pure goethite shows two OH deformation bands at 892 and 800 cm^{-1} and a Fe–O band at 400 cm^{-1}.

Sample characterization

Mineralogical identification was assessed by powder x-ray diffraction (Philips PW1130) and infrared spectroscopy (Nicolet 20SXC). X-ray diffraction patterns were registered between 10° and 60° (2θ) employing CuKα radiation. To obtain the infrared spectra the samples were diluted in KBr matrices.

Quantitative determinations of elements were performed by x-ray fluorescence spectrometry on glass disks with international rock standards being used for calibration. The Kα lines were measured with a Siemens SRS300 sequential spectrometer (Rh end-window tube).

The morphology of powders was examined by transmission electron microscopy (TEM Philips EM300) and their size was determined by counting about 100 particles.

Colloidal and rheological properties

The isoelectric point (IEP) of the solids was determined by measuring electrophoretic mobilities of aqueous dispersions as a function of pH, in a Delsa Coulter 440 apparatus. For this, 15 mg of samples were dispersed in 100 cm^3 of a 0.01 mol·dm^{-3} NaCl solution to keep the ionic strength constant and the pH varied by adding HCl or NaOH, as needed.

The flow properties were measured at different concentrations and suspensions pH, using the Haake Rotovisco RV20 concentric cylinder viscosimeter.

Results and discussion

Structural and morphological characterization

X-ray diffractograms of the lateritic samples SG, GS and G are shown in Fig. 1 along with the pure synthetic goethite (GP) and pure serpentine (SP). It can be seen that the ratio of serpentine/goethite decreases in the order SG, GS and G. X-ray diffractogram of the samples also display the presence of small amount of gibbsite.

The infrared spectra of the same samples (Fig. 2) are in accordance with the results obtained by x-ray diffraction, showing only the presence of bands characteristic of serpentine and goethite. Absorption bands due to gibbsite were not detected in the infrared spectra.

Chemical analyses of the samples (Table 1) reveal the presence in sample SP of a high iron content (13.36%), probably forming a solid solution in the serpentine lattice which in principle can occupy both octahedral and tetrahedral positions [12]. However, the ratio between silicon and magnesium (Mg/Si = 0.81) lower than in the ideal serpentine (Mg/Si = 1.5) suggests that the iron is mainly found in octahedral positions, since quartz was not detected in the samples. The amount of aluminum content in

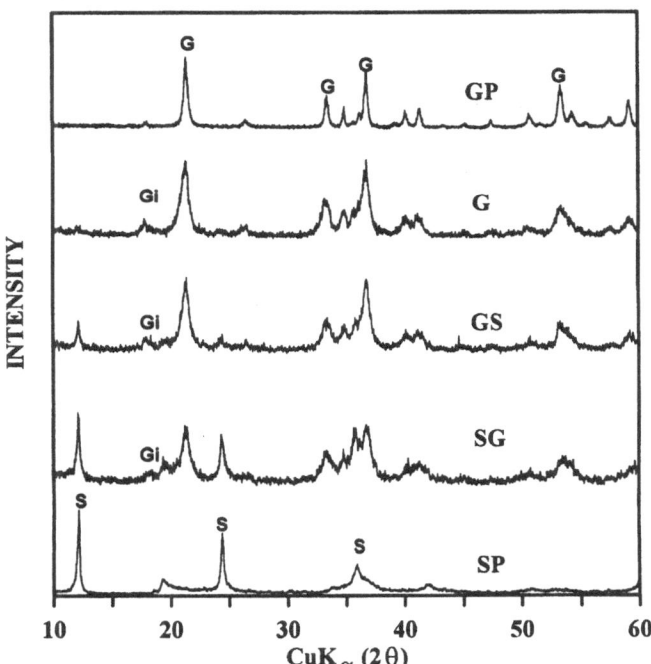

Fig. 1 Random powder x-ray diffraction patterns for lateritic samples SG, GS and G compared with those of pure goethite (GP) and serpentine (SP). The symbols inside the plot represent: S = Serpentine, Gi = Gibbsite, G = Goethite

268
A. Cerpa et al.
Rheology of lateritic suspensions

the lateritic samples remains approximately constant (between 7–9%). This amount is clearly greater than that estimated by the relative intensity of the diffraction peak corresponding to gibbsite, that appears at 4.85 Å,

suggesting that a slight amount of aluminum forms a solid solution with the serpentine structure. The presence of aluminum in the serpentine phase was confirmed by EDX analyses.

The TEM shows that the sample SG contains mostly platy-shaped particles of serpentine with a slight amount of fibrous particles, suggesting the existence of various serpentine phases (chrysotile, lizardite, and antigorite) [13]. In the GS sample the platy particles of serpentine predominate. The G sample contains elongated lath-like particles and also star-shaped twin crystals which correspond to goethite according to EDX analyses (data not shown).

The size of platy-shaped serpentine and lath-like goethite particles are in the range of 0.1–0.5 μm.

Rheological properties

Effect of solid concentration on flow properties of suspensions

In Fig. 4, we illustrate for sample SG the effect of solid concentration on the flow properties of lateritic suspensions. The natural pH of the suspension was about 7.3. A similar behavior was manifested by samples GS and G (data not shown).

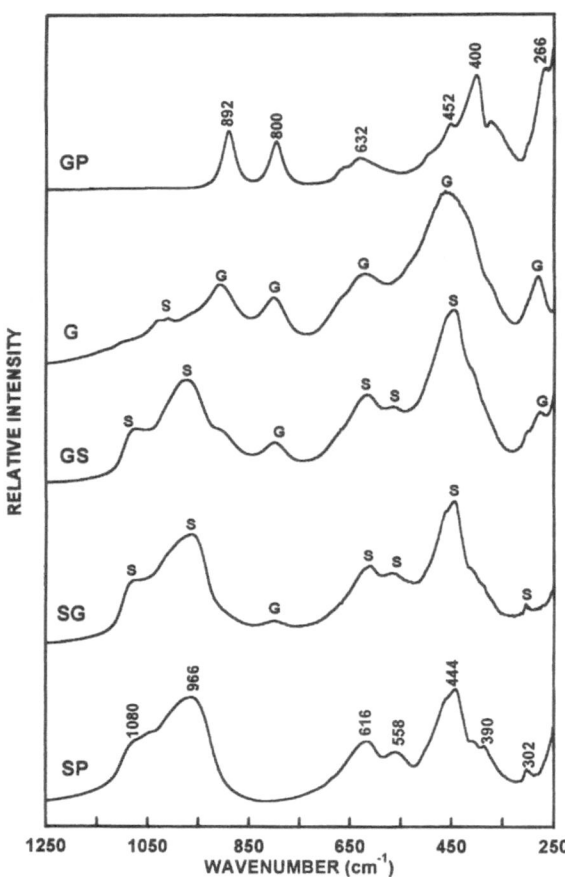

Fig. 2 Infrared spectra for lateritic samples SG, GS and G compared with those of pure goethite (GP) and serpentine (SP). Bands attributed to serpentine and goethite are labeled S and G, respectively

Table 1 Composition (% by weight) of majority elements in samples SP, SG, GS y G

Sample	SiO$_2$	Al$_2$O$_3$	Fe$_2$O$_3$	MgO
SP	37.55	2.05	13.36	30.31
SG	22.48	7.24	38.37	16.34
GS	13.37	6.36	55.28	8.55
G	3.74	9.20	68.84	1.34

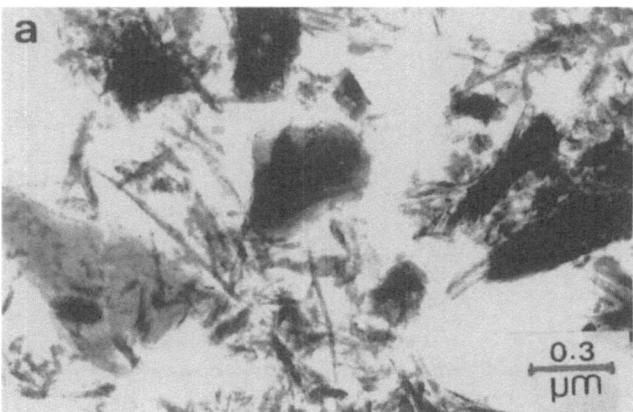

Fig. 3 TEM micrographs of sample GS (a) and sample G (b)

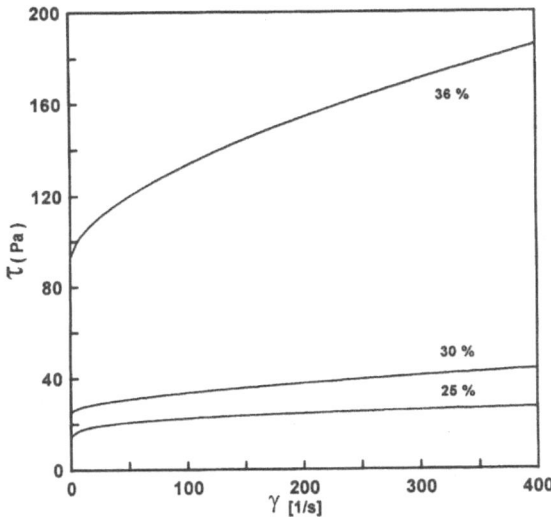

Fig. 4 Flow curves for sample SG containing 25, 30 and 36% by weight of solids concentration. The pH of the suspension was about 7.3

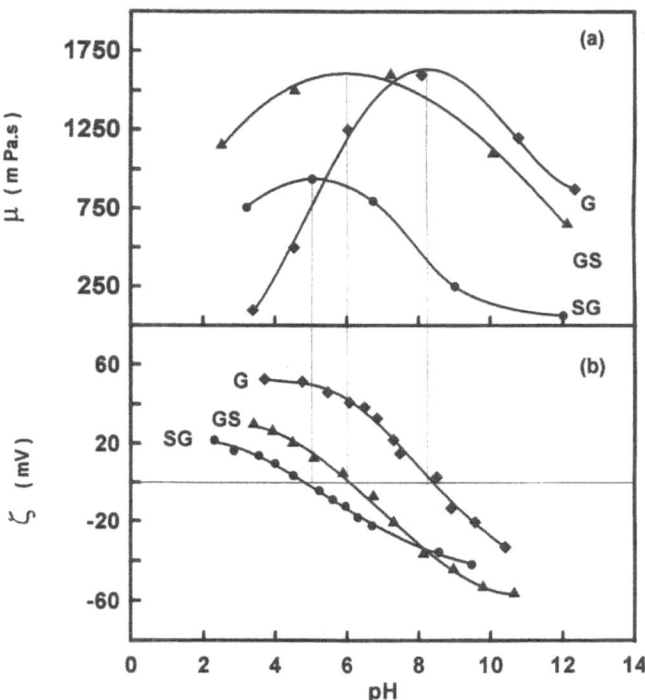

Fig. 5 (a) Apparent viscosity ($\gamma = 200$ s^{-1}) for suspensions containing a 36% by weight of solids concentration and (b) zeta potential, for samples SG, GS and G as a function of pH suspensions

The flow curves were interpreted following the Bulkley Hershel or Casson models [14]. These models are characteristics of plastic flow behavior in which the viscosity varies with the shear rate. In addition, it is observed that an increase in concentration suspension results in a greater shear stress due to the increase in particle to particle interactions as they became closer [8, 10, 15].

Effect of pH on the flow and surface properties of mineral suspensions

Figure 5(a) is a plot of apparent viscosity as a function of pH for samples SG, GS and G containing a 36% (by weight) in solids concentration. The important features in the plot are the pH of complete deflocculation and the pH of apparent viscosity maximum. It can be seen that the apparent viscosity maximum is shifted toward basic pH as the amount of goethite increases. In addition, the value of the apparent viscosity increases with the amount of goethite in the samples, suggesting that goethite has a stronger interaction than serpentine as a result of its smaller particle size or different structure. The maximum of shear stress and apparent viscosity should correspond to a state of maximum flocculation and it should appear near to the isoelectric point of each sample [16]. On the contrary, complete deflocculation is defined as the state where the suspension has zero yield stress [16]. It is noted that complete deflocculation occurs near pH = 12 for the sample SG, suggesting that the presence of serpentine

favours the dispersion or deflocculation of the lateritic samples.

Figure 5(b) shows the effect of pH on the zeta potential for samples SG, GS and G. For sample SG, the maximum value of viscosity appears at pH values close to the isoelectric point = 4.8, where maximum flocculation takes place. For sample GS this maximum appears at pH (IEP) = 6.1, while in sample G it appears at pH (IEP) = 8.4. It can be seen that the lateritic samples present a wide range of IEP depending on their mineralogical composition (serpentine/goethite ratio).

The particles in Fig. 5(b) are positively or negatively charged at pH values above or below the IEP and therefore the suspensions should experience progressive deflocculation due to the increase in interparticle repulsion forces. It is also noted that samples SG and GS present lower viscosity values at pH > 10, where the maximum value of zeta potential occurs. Nevertheless, sample G has its lower viscosity in the acidic region (pH = 3–5). These results are in agreement with the data obtained in Fig. 5(a).

Acknowledgments One of us (A. Cerpa) acknowledges a fellowship of the Agencia Española de Cooperación Iberoamericana (A.E.C.I).

References

1. Kostas SA, Raffi MT (1991) J Colloid Interface Sci 143:54–68
2. Atsushi T, Kunio Y (1994) Powder Technology 78:165–172
3. Avotins PA (1979) International Lateritic Symposium. Chapter 31:610–635
4. Valdés GF (1984) Revista Technológica XIV:44–50
5. Cheng DC (1980) Chem and Ind: 403–406
6. Hunter RJ (1987) Foundations of Colloidal Science. Clarendon Press, Oxford, pp 546–551
7. Cheng DC (1990) J Mat Sci 25:353–373
8. Garcell LR, Cerpa A (1992) Rev Tecnología Química 1:63–68
9. Nikumbh AK (1990) J Mat Sci 25:15–21
10. Leong YK, Boger DV (1990) J Colloid and Interface Sci 136:249–255
11. Schwertmann U, Cornell RM (1991) Iron Oxides in the Laboratory. Preparation and Characterization. Weinheim, Basal, New York, pp 64
12. Brindley GW, Brown G (1980) Crystal structures of clays minerals and their X-ray identification. Mineralogical Society, London, pp 2–24
13. Wilson MJ (1987) A Handbook of determinative methods in clays mineralogy. Blackie, New York, pp 234–235
14. Whorlow RW (1992) Rheological Techniques. Second Edition. Ellis Horwood. New York, pp 16–18
15. Coussot P, Piau JM (1994) Rheol Acta 33:175–184
16. Leong YK, Boger DV, Parris D (1991) Trans I Chem E 69:381–385

Progr Colloid Polym Sci (1996) 100:271–275
© Steinkopff Verlag 1996

BIOCOLLOIDS

N.L. Burns
K. Holmberg

Surface charge characterization and protein adsorption at biomaterials surfaces

N.L. Burns (✉) · K. Holmberg
Institute for Surface Chemistry
Box 56 07
114 86 Stockholm, Sweden

Abstract In an effort to more fully understand the nature of protein adsorption, and the role adsorbed proteins have in mediating interfacial processes in biomaterial application, effect of surface charge is being explored. For this purpose, a novel electrokinetic technique whereby electroosmosis is measured at flat plates is being used to characterize surfaces with respect to effective surface charge and origin of charge. Protein adsorption is monitored by in situ ellipsometry or by an enzyme linked immunosorbent assay (ELISA). Correlations can then be made between the observed surface properties and biological response to the surface, including bacterial adherence and complement activation of blood proteins. Surfaces studied include a wide variety of radio frequency plasma polymers, and polysiloxane modified surfaces; all of interest as biomaterials. The plasma polymer surfaces are being used as model surfaces to study protein mediated bacterial adherence. Siloxane polymer coatings are being used to interfere with salivary pellicle and plaque formation on teeth, with different patterns in the adherence of oral bacteria observed with different polysiloxanes. In the above systems, patterns in protein adsorption related to surface charge and functionality, as well as other surface properties, can be identified. It appears that surface charge and charging properties are important factors in formation of biofilms.

Key words Electroosmosis – surface charge – protein adsorption – bacterial adherence

Introduction

An understanding of the adsorption behavior of proteins at solid surfaces is of tremendous import in elucidating many biological processes and in the development of a large number of biotechnical and pharmaceutical applications [1]. For example, absorbed proteins are known to mediate bacterial adhesion [2], as in the formation of dental plaque, and the biological response to foreign materials [3], as in blood contact devices. In biotechnology, the controlled adsorption of biologically active substances is exploited to "sense" a response [4], for example, one associated with a specific disease as in the case of solid phase diagnostics, or to retain a substance for bioavailability [5], as in drug administration. In all of these processes, control over the primary adsorptive layer is essential, being the stimulus for any following effect.

Adsorption of proteins to solid surfaces is known to depend on general physico-chemical properties such as wettability and surface charge density, with a large number of papers discussing the relation between surface hydrophilicity and adsorbed amount of various proteins. Studies using ellipsometry and surface fluorescence to monitor adsorption have shown that many proteins, particularly blood proteins, adsorb in higher amounts on hydrophobic than on hydrophilic surfaces, although there are exceptions, e.g. lysozyme [6, 7]. Conversely, studies have shown that extremely hydrophobic low energy materials, e.g. perfluorinated or polydimethylsiloxane surfaces, exhibit a low tendency to adsorb proteins [8, 9]. It seems that the relation of wetting to protein adsorption is not a clear one.

The role of surface charge has also been addressed in the literature, however, not to a great extent. This is perhaps due to limitations in quantification of surface charge. Traditionally, the measurement of electrokinetic phenomena has been used to determine surface charge. Electrophoresis, streaming potential and electroosmosis have all been employed to this effect [10]. Unfortunately, electrophoresis is limited to colloidal particles. Streaming potential has been useful outside the colloidal range for materials that can form porous plugs, e.g. fibers. Electroosmosis has also been subject to limitations in the past with transparent surfaces being required in order to visualize fluid flow [11].

In this paper substrate surfaces are characterized with respect to surface charge using an electrokinetic technique whereby electroosmosis is measured at flat plates. The technique, developed by the authors, is applicable to nontransparent, nonconducting surfaces; in principle any that can be machined into small plates [11]. In combination with this electroosmosis technique, measurements of protein adsorption and bacterial adherence can be performed to help elucidate the role of surface charge in these processes. The electroosmosis technique used in combination with other surface characterization techniques, e.g. ESCA and contact angle, can provide a more complete account of the competing factors influencing protein adsorption. Following are a few examples of using the electroosmosis technique in such a way.

Experimental

Surface characterization

Surfaces used in the protein adsorption and bacterial adherence studies were characterized electrokinetically by measuring the pH dependence of electroosmotic fluid flow at 1×2 cm sample plates in 1 mM NaCl at 25 °C, adding 1 mM HCl or NaOH to obtain the desired pH. A detailed discussion of the method is given elsewhere [11].

Protein adsorption measurements

Ellipsometrically determined protein adsorption experiments were carried out on plasma polymer surfaces on polished silicon slides (Okmetic, Finland). 1,2-Diaminocyclohexane (DACH) (99% Aldrich, Germany), acrylic acid (AA) (99% Aldrich, Germany) and hexamethyldisiloxane (HMDSO) (98 + % Aldrich, Germany) were deposited and polymerized on the surface by radio frequency plasma deposition. Proteins in the ellipsometry study include human serum albumin (HSA), human immunoglobulin (IgG) and human fibrinogen (Fgn), obtained from Sigma Chemicals, USA. Details of the ellipsometric technique and plasma surface preparation are given in the literature [12].

Enzyme linked immunosorbent assay (ELISA) experiments were carried out on DACH modified polystyrene plates. HSA and human lysozyme (Lys) from Sigma were used in the study. The ELISA technique for the systems studied are elaborated in a separate paper [13].

Oral bacterial adherence studies were performed in vitro on five different polysiloxane surfaces: Wacker Silicone BS 1306, a polydimethylsiloxane containing aminoalkyl substituents (Wacker Chemie, Germany); Wacker Silicone 1311, a polydimethylsiloxane containing partially neutralized aminoalkyl substituents; Wacker Stone Strengthener OH, a low molecular weight ethyl silicate; Wacker Silicone BS 15, a low-molecular weight potassium methyl siliconate; and Water Glass, a sodium silicate (Eka Nobel, Sweden). Electroosmotic characterization of the polysiloxanes were carried out on 1×2 cm polysiloxane coated glass slides cut from microscope slides (Kebo Lab, Stockholm, Sweden). The bacterial adherence experiments were performed using polysiloxane coated hydroxyapatite beads and a ^{35}S-labelled *Streptococcus sanguis*. Details of the bacterial adherence study are given elsewhere [14].

Results and discussion

Electroosmosis is an electrically induced fluid flow at an interface due to electrical field forces acting upon the diffuse part of the electrical double layer. The effective surface charge (σ_0) required to produce an observed electroosmotic fluid mobility (U_{eo}) can be calculated by combining the Smoluchowski equation and solutions to the Poisson–Boltzmann equation for a 1:1 electrolyte

$$\sigma_0 = -(8c\varepsilon kT)^{1/2} \sinh\left(\frac{e\eta U_{eo}}{2\varepsilon KT}\right) \qquad (1)$$

given bulk electrolyte concentration (c), permittivity of the medium (ε), temperature (T), the Boltzmann constant (k), Coulombic charge (e), and viscosity of the medium (η). pH at the surface (pH$_s$), differing from the bulk solution pH as a result of electrical double layer effects can be taken

combining the Smoluchowski equation with the Boltz-mann equation to yield

$$pH_s = pH_{bulk} - 0.434\left(\frac{e\eta U_{eo}}{\varepsilon KT}\right).$$ (2)

It follows from Eqs. (1) and (2) that surface pH is highly dependent on electrolyte concentration, having a greater effect at low ionic strength. Since electroosmosis must be measured at low ionic strength to produce a significant effect (in this work 0.001 M) and protein adsorption usually at a much higher ionic strength (1 M or more), where the electrical double layer effect is screened, it is reasonable to assume that surface pH is the appropriate value to use in discussion of protein adsorption measurements. Likewise, surface charge calculated from electroosmosis measurement is appropriate to the protein adsorption conditions in the absence of specific electrolyte adsorption.

Figure 1 shows the pH dependence of surface charge calculated from electroosmosis measurements for three different plasma polymer surfaces. This type of measurement is essentially a surface titration and the figure shows that the freshly prepared polymerized DACH surface has a surface charge that originates from a single basic group ($pKb \approx 10$). This is consistent with amino functional groups present on the surface. For the polymerized acrylic acid surface charge appears to originate from a single acid group ($pKa \approx 3$), consistent with a carboxyl group. As for the polymerized HMDSO surface, charge appears to originate from two distinct acid groups, one in larger numbers having weak acid characteristics ($pKa \approx 7.5$) and one a stronger acid group of much lower density ($pKa \approx 4.5$). This is consistent with a high concentration of silanols at the surface, and a smaller amount of carboxyl groups, perhaps from oxidation products in the organic components of the plasma [15].

Ellipsometrically determined in situ protein adsorption measurements were performed on the plasma polymer surfaces above with HSA, IgG and Fgn at pH 7.2 [12]. All three proteins adsorbed to the greatest extent at the polymerized DACH surface, slightly less on the polymerized HMDSO surface. With the polymerized AA surface much less protein adsorbed and the adsorption kinetics were slower. These observations can be explained in terms of the respective charges at the surface and the protein. At pH 7.2 all of these proteins carry a net negative charge. The positively charged polymerized DACH surface would attract the protein. For the polymerized AA surface the negative charge would repel the protein and slow adsorption kinetics. These two surfaces have similar wetting properties, both hydrophilic, suggesting that surface charge and/or specificity of the functional group are

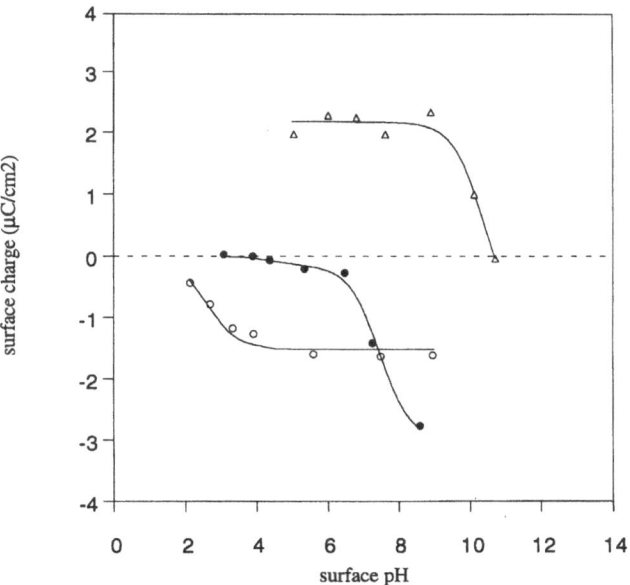

Fig. 1 The pH dependence of surface charge for a freshly prepared PP-DACH surface (($\triangle$)), a PP-AA surface (($\bigcirc$)) and a PP-HMDSO surface ($\bullet$). Drawn lines correspond to theoretical fits using a single-acid, single-base site dissociation model

governing properties with respect to the differences in protein adsorption. As for polymerized HMDSO, the surface has a very low charge at pH 7.2 and is quite hydrophobic; however, there are a high number of "ionizable" groups, the silanol. This suggests that, in addition to hydrophobicity, specificity of functional group at the surface governs protein adsorption.

It has been observed that the polymerized DACH changes its surface composition upon standing under ambient conditions. Previously, this aging effect was presumed to be due to oxidation initiated by free radical process [16]. However, recent evidence suggests that the change in surface composition is due to the formation of alkyl ammonium carbamates, a phenomenon common to amino-functional surfaces [13]. As mentioned above the freshly prepared polymerized DACH surface behaves as possessing a single basic charge determining species (Fig. 1). In Fig. 2 the effect on charging properties upon storing the polymerized DACH surface under ambient conditions is shown. Upon aging there is a reduction in basic group density, with a corresponding appearance of an acid group ($pKa \approx 3-5$) to a point where the acid to base ratio is one to one. This behavior is consistent with the formation of alkyl ammonium carbamates at the surface.

Since the polymerized DACH surface spontaneously changes its acid to base ratio and charging characteristics with time, it serves as an interesting model substrate for

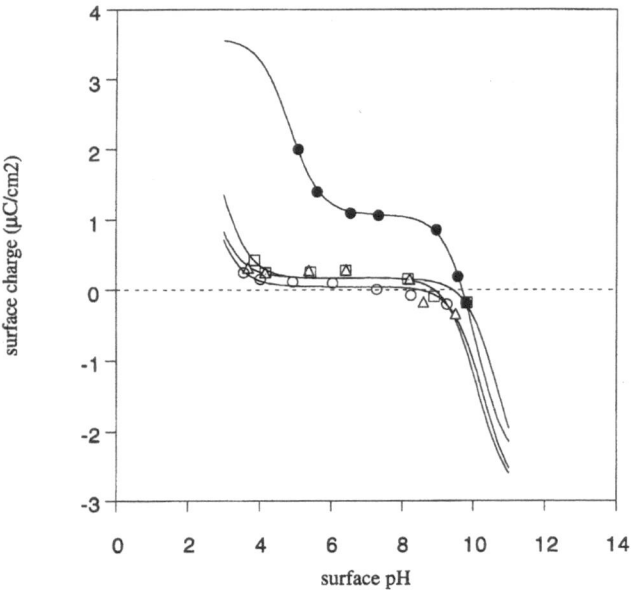

Fig. 2 The pH dependence of surface charge for the DACH surface after exposure to atmosphere for 1 day (●), 8 days (△), 15 days (□), and 64 days (○) as calculated from electroosmotic data. Drawn lines correspond to theoretical fits using a single-acid, single-base site dissociation model

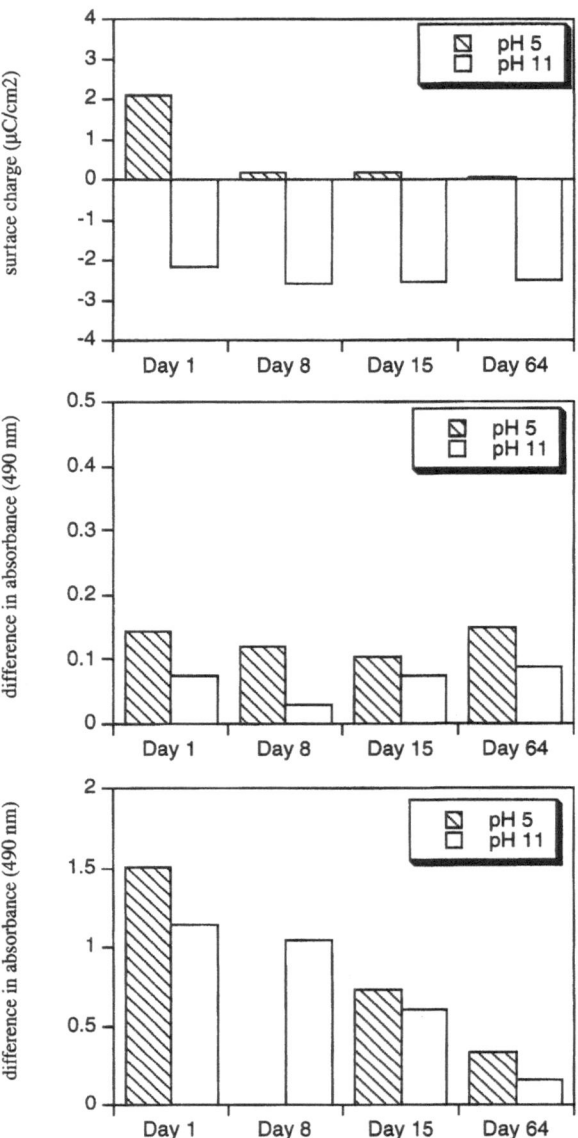

Fig. 3 Changes in surface charge (a) at pH 5 and 11 with time for PP-DACH surfaces and the corresponding adsorbed amounts, as recorded by ELISA, for HSA (b) and Lys (c) compared to untreated polystyrene

protein adsorption. As such, protein adsorption experiments were performed with HSA and Lys using an ELISA technique [13]. Figure 3 shows the results of the protein adsorption experiments. For both proteins adsorbed amounts were higher at pH 5 than 11. This effect is not surprising for HSA having an i.e.p of pH 5. However, Lys has its i.e.p. at 11 and is thus positively charged at pH 5. There is enhanced adsorption for a positively charged protein to a positively charged surface as compared to a neutral protein at a negatively charged surface. The reasons for this are not clear; however, the same effect was observed by Bos et al. for Lys at a positively charged tin oxide substrate [17]. In this case, factors other than charge considerations must be considered, e.g. conformational stability of the protein with regard to pH.

Another interesting aspect of Fig. 3 is the change in adsorption with time, i.e., change in charge and surface composition. HSA adsorption was not dependent on the change. However, adsorption of Lys decreased at both pH's with aging, even at the high pH where change in surface is not pronounced. Again, this anomaly cannot be attributed to surface charge effects. Another consideration is the hydrophobicity of the surface. With aging the PP-DACH surface becomes slightly less hydrophilic [18]. It appears in this context that Lys adsorption is quite sensitive to small changes in wetting properties and/or surface composition.

In a final example of using surface charge measurement to elucidate protein adsorption mechanisms, the results of a study on saliva mediated oral bacterial adherence at various polysiloxane surfaces is presented [14]. Figure 4 shows the charging properties of the five polymers studied, along with the glass control, demonstrating a wide range of properties among the compounds. Interestingly, all five compounds significantly reduced saliva mediated bacterial adherence. Most prominent was the reduction of adherence with the Wacker Silicone BS 15, both with and without saliva. From the characterization this compound

Progr Colloid Polym Sci (1996) 100:271–275
© Steinkopff Verlag 1996

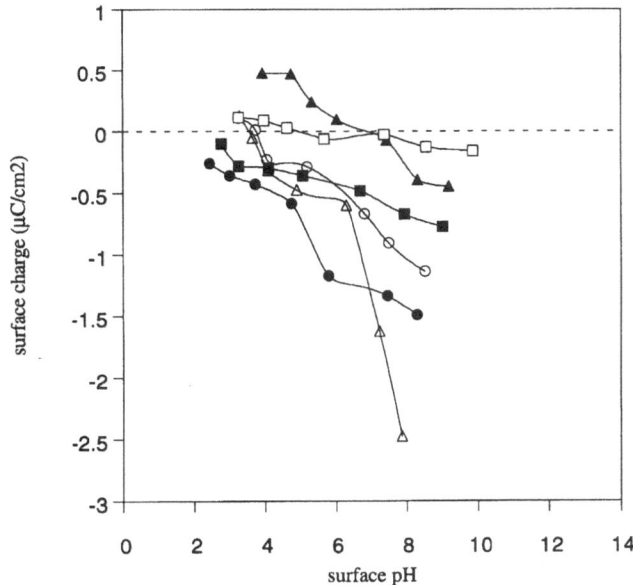

Fig. 4 The pH dependence of surface charge for glass surface treated with different polysiloxanes: Wacker Silicone BS 1306 (○), Wacker Silicone 1311 (▲), Wacker Stone Strengthener OH (△), Wacker Silicone BS 15 (□), Water Glass (■). Clean glass (●) was used as control

was unique in that it showed the least amount of charge across the entire pH range, and was the most hydrophobic (advancing contact angle 90°). This suggests that the lack of ionizable (polar) surface groups is an important factor. Conversely, the least effective surface at reducing bacterial

adherence was the Wacker Silicone 1311. This compound was unique in that it contained a significant number of amino-functional groups, evident from the positive charge at low pH. This surface was also quite hydrophobic (contact angle 83°) and uncharged at pH 7. Again specific surface functionality appears to play an important role, more so than net charge or wetting property.

Conclusions

From the adsorption studies presented it is clear that surface charge plays a role in protein adsorption processes, though not predictably, and only as one of many competing factors. These studies have shown that from the pH dependence of electroosmosis it is not only possible to determine surface charge but also semi-quantitatively identify ionizable surface groups, this also being a major factor in specific protein interactions with a surface. Though not all inclusive, electrokinetic characterization of macroscopic surfaces is a useful complement to other characterization techniques in correlating surface properties with the behavior of proteins at interfaces.

Acknowledgments The authors wish to thank and acknowledge Dr. Bo Lassen for the ellipsometry measurements, Dr. Carina Brink for the ELISA measurements, and Dr. Jan Olsson and Annette Carlén for the bacterial adherence measurements. Dr. Martin Malmsten is thanked for his helpful comments on the manuscript.

References

1. Gristina AG (1987) Science 237:1588
2. Lassen B, Holmberg K, Brink C, Carlén A, Olsson J (1994) Colloid Polym Sci 272:1143
3. Andrade JD (ed) (1985) Surface and Interfacial Aspects of Biomedical Polymers. Vol. 2 Plenum Press, New York
4. Malmsten M, Lassen B, Holmberg K, Thomas V, Quash G (1996) J Colloid Interface Sci 177:70
5. Malmsten M (1994) J Colloid Interface Sci 168:247
6. Elwing H, Ivarsson B, Lundström I (1986) Eur J Biochem 156:359
7. Wahlgren MC, Arnebrant T, Askendal, A, Welin-Klinström S (1993) Colloids Surf A 70:151
8. Lyman DL, Muir WM, Lee IJ (1965) Trans Am Soc Artif Intern Org 11:301
9. Bruil A, Brenneisen LM, Terlingen JGA, Beugling T, van Aken WG, Feijen J (1994) J Colloid Interface Sci 165:72
10. Hunter RJ (1981) Zeta Potential in Colloid Science. Academic Press, London
11. Burns NL (submitted) J Colloid Interface Sci
12. Lassen B, Malmsten M (submitted) J Colloid Interface Sci
13. Burns NL, Holmberg K, Brink C (in press) J Colloid Interface Sci
14. Olsson J, Carlén A, Burns NL, Holmberg K (1995) Colloids Surf B 5:161
15. Yasuda H (985) Plasma Polymerization. Academic Press, London
16. Gölander C-G, Rutland MW, Cho DL, Johansson A, Ringblom H, Jönsson S, Yasuda HK (1993) J Appl Polym Sci 49:39
17. Bos MA, Shervani Z, Anusiem ACI, Giesbers M, Norde W, Kleijn JM (1994) Colloids Surf B 3:91
18. Bo Lassen (private communication)

Progr Colloid Polym Sci (1996) 100:276–280
© Steinkopff Verlag 1996

R. Bru
J.M. López-Nicolás
A. Sánchez-Ferrer
F. García-Carmona

Cyclodextrins as molecular tools to investigate the surface properties of potato 5-lipoxygenase

Dr. R. Bru (✉) · J.-M. López-Nicolás
A. Sánchez-Ferrer
F. García-Carmona
Departamento de Bioquímica y Biología
Molecular "A"
Facultad de Biología
Universidad de Murcia
30080 Espinardo, Murcia, Spain

Abstract The behavior of potato 5-lipoxygenase acting on both monomeric and aggregated linoleic acid has been studied. While the substrate preparation was transparent, lipoxygenase activity was determined by means of a spectrophotometric method, which was not useful to determine the activity in turbid samples. In the latter case a polarographic method was used. Cyclodextrins were used in order to increase the range of monomeric, and thus transparent, linoleic acid. This clearly revealed that potato 5-lipoxygenase can be saturated by linoleic acid monomers but when aggregates are formed at higher linoleic acid concentration, activity raises and stabilizes in a new saturation level. This was interpreted as an activation of lipoxygenase induced by the aggregation of its substrate, linoleic acid. Kinetic parameters were determined in each region – monomeric and aggregate – and are consistent with the surface activation hypothesis.

Key word 5-lipoxygenase – cyclodextrin – inclusion complex – linoleic acid – activation

Introduction

Lipoxygenases (LOXs) (EC 1.13.11.12) catalyze the dioxygenation of polyunsaturated fatty acids (PUFA) that contain one or more (1Z, 4Z)-pentadiene systems. Their products are chiral (E, Z) conjugated hydroperoxy fatty acids [1–3] which are the precursors of a number of physiological effectors in animal tissue [4] and possibly also in plants [5].

PUFA in aqueous medium form different polymorphic structures depending on their concentration, the protonation state and temperature. For instance, at concentrations below critical micelle concentration (CMC) only monomers of FA exist, but in higher concentration structures such as oily doplets, lamellae and micelles may be formed depending on the pH of the medium [6]. In LOX kinetic studies these structural arrangements have been avoided in the reaction media for both kinetic and technical reasons [7], so conditions have been set so that PUFA stay in its monomeric state, namely pH 9–10 with concentrations up to 100 μM. Unfortunately, very few LOXs, that of soybean being one, show activity in these conditions. Furthermore, most LOXs have an optimum pH between 6 and 7 [8] which are quite favorable conditions for the aggregation of PUFA.

The aim of the present study was to investigate whether the aggregation of the substrate without detergents has any effect on the enzyme and discuss the possible physiological implications. For this, the enzyme from potato tubers, named 5-LOX for its positional specificity, with its optimum pH 6.3 [9] was selected. At this pH the monomer-to-aggregate transition for linoleic acid occurs at 10 μM [10]. Below this concentration, spectroscopic methods are adequate for monitoring the enzyme activity, but at higher concentrations they fail due to the turbidity of the linoleic acid (LA) preparation. On the other hand, polarographic methods can be used with turbid samples but are very insensitive to low concentration of LA. We have shown that cyclodextrins (CDs) increase the concentration for the monomer-to-aggregate transition of PUFA [10, 11], seeming to overcome the above technical problems. CDs are natural water-soluble cyclic oligosaccharides, composed of 6, 7 or 8 glucose units and named α-, β- and γ-CD, respectively [12]. Their ability to interact with lipids and other hydrophobic molecules is based on the formation of so-called inclusion compounds, where the lipid is included in the hydrophobic torus, thus shifting the hydration water molecules of the internal groups in a thermodynamically favorable manner.

In this study we utilize CDs as a molecular tool to investigate the properties of potato 5-LOX in the presence of monomers and aggregates of LA.

Materials and methods

Materials

Linoleic acid was purchased from Cayman Chemical Co. (Paris, France). β-Cyclodextrin was obtained from Sigma (Madrid, Spain). Lipoxygenase was purified from potato tubes (var. Desireé) according to Mulliez et al. [13] (27 μmole/min/mg protein). Diphenylhexatriene (DPHT) was a product from Fluka (Madrid) and tetrahydrofurane (THF) was from Merck (Darmstadt, Germany). All the other chemicals used were of the highest purity.

Fluorimetric determination of CMC

The CMC value of LA was determined by means of a fluorescence spectroscopy method as described elsewhere [10, 14]. 2 ml samples contained 0.1 M phosphate buffer pH 6.3, 0.88 μM DPHT (supplied in 2 μl THF), 1% v/v ethanol and the indicated concentration of β-CD and LA. The samples were flushed with N_2 and incubated for 30 min in the dark at the desired temperature for equilibration and in order to reverse the photoisomerization of the fluorescent probe. Fluorescence intensity measurements were made at 25 °C at 430 nm (358 nm excitation wavelength) in a Kontron SFM-25 spectrofluorimeter equipped with thermostated cells. CMC was determined graphically from a plot of fluorescence relative values versus LA concentration (see Fig. 1A) as the intersection between the lines defining the fluorescence tendency in the pre- and post-micellar regions.

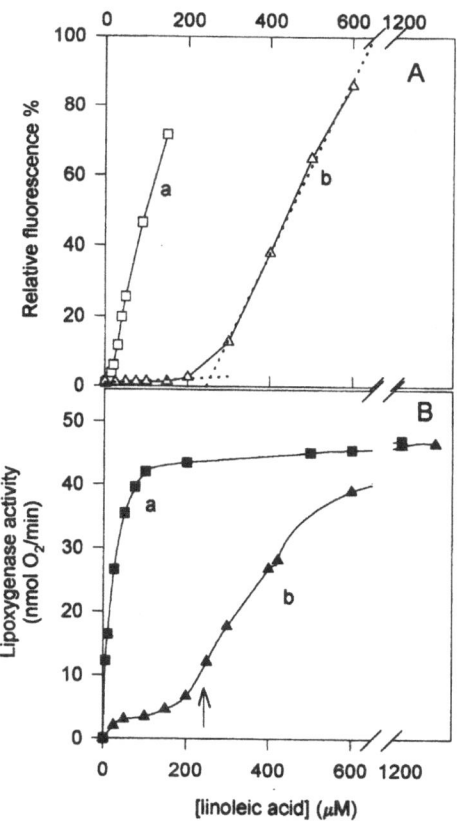

Fig. 1 A) Dependence of relative fluorescence intensity of diphenylhexatriene at 430 nm (excitation wavelength 358 nm) on linoleic acid concentration. B) Dependence of oxygen uptake rate by potato 5-lipoxygenase on linoleic acid concentration. Curves "a"; no β-cyclodextrin. Curves "b" 1 mM β-cyclodextrin. The arrow indicates the CMC in the presence of β-cyclodextrin. Enzymatic reactions were carried out using 1.8 μg/ml protein. Determinations were made at pH 6.3 and 25 °C

Spectrophotometric determination of LOX

LOX activity was assayed by monitoring the increase in absorbance at 234 nm ($\varepsilon_{234} = 25\,000$ M^{-1}cm^{-1}) of the forming hydroperoxides in an Konton Uvikon 940 spectrophotometer at 25 °C equipped with thermostated cells. The reaction was started by adding 5 μl enzyme to 1 ml of complex LA-β-CD preparation.

Polarographic determination of LOX

LOX activity was determined by monitoring O_2 consumption with a Clark-type electrode. Samples of 1 ml were prepared as for fluorescence spectroscopy but omitting the fluorescent probe and the solvent. They were then shaken vigorously before use in order to become air-saturated and finally transferred to the stirred, thermostated oxygraph chamber (Hansatech Ltd, Norfolk, UK). The reaction was started by injection of 10 μl of potato 5-LOX. The millivolts recording scale was transformed into oxygen concentration by using an oxygen calibration method [15]. Activity was expressed as the maximal amount of O_2 consumed per minute (maximal slope of the reaction progress curves).

Results and discussion

FA forms aggregates above a certain critical concentration. As shown in Fig. 1A, the appearance of these aggregates at increased concentrations of LA can be detected by measuring the fluorescence of a probe such as DPHT, whose quantum yield increases when surrounded by an apolar environment such as that of the aggregate core [14]. In the presence of CDs, the aggregation behavior of the FA changes to higher CMC values. Figure 1A shows such an effect of β-CD in different conditions.

The increase in CMC is due to the formation of inclusion complexes between LA and β-CD, which do not take part in the aggregation phenomenon. It has been shown by speed-of-sound measurements [16] that the apparent CMC of an amphiphile such decyltrimethylammonium bromide in the presence of β-CD is actually the sum of the CMC of the pure amphiphile (CMC_O) plus the concentration of inclusion complexes.

Thus, the pre-aggregate region consists of free LA monomers and inclusion complexes if CD is present, while the post-aggregate region also contains LA aggregates.

With such an enlargement of the premicellar region, it is possible to study the enzyme activity by both spectroscopic and polarographic methods. The spectroscopic method is especially useful for studying the pre-aggregate region since the samples are completely transparent. As can be seen in Fig. 2B, the effect of LA concentration on potato 5-LOX activity appears to be hyperbolic as long as LA concentration does not approach the monomer-to-aggregate transition region (indicated by an arrow). This result is clearly seen only in the presence of β-CD since in its absence a hyperbolic kinetics is difficult to discern. Close to the transition region there seems to be a burst of activity, but the turbidity of the samples does not permit the adequate determination of enzyme activity at increased concentrations of LA.

The polarographic method permits accurate enzyme activity determinations at high LA concentration in the post-aggregate region and also in the pre-aggregate region when β-CD is present. When no β-CD is present an apparently hyperbolic kinetics is observed but the activity level reached is much higher than that determined by the

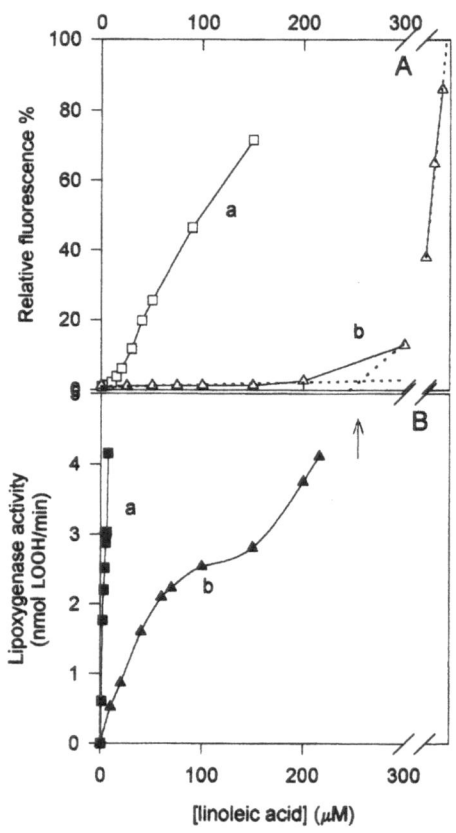

Fig. 2 A) Dependence of relative fluorescence intensity of diphenylhexatriene at 430 nm (excitation wavelength 358 nm) on linoleic acid concentration. B) Dependence of linoleic acid hydroperoxide (LOOH) formation rate by potato 5-lipoxygenase on linoleic acid concentration. Curves "a": no β-cyclodextrin. Curves "b" 1 mM β-cyclodextrin. The arrow indicates the CMC in the presence of β-cyclodextrin. Enzymatic reactions were carried out using 1.8 μg/ml protein. Determinations were made at pH 6.3 and 25 °C

spectroscopic method (Fig. 1B, curve a). In the presence of β-CD (Fig. 1B, curve b) a complex kinetic behavior is observed which can be described as follows: at low LA concentration, below the CMC, activity first increases and then reaches a plateau, displaying a typical hyperbolic behavior as when studied by the spectrophotometric method. At increased LA concentration, the activity further increases and finally reaches a new plateau. The activity level of the first plateau is approximately the same as that reached in the spectrophotometric method, while the activity level of the second plateau is the same as that reached in the polarographic method in the absence of CD.

As Fig. 1B, curve b and Fig. 2B demonstrate, potato 5-LOX activity is saturable by the monomeric form of LA, and follows a hyperbolic kinetics. However, the appearance of LA aggregates, as shown by the increase in

Table 1 Kinetic data for potato 5-lipoxygenase both in the absence and in the presence of β-cyclodextrin. The spectrophotometric method is based on the detection at 234 nm of the forming hydroperoxides, while the polarographic method is based on the oxygen uptake during the lipoxygenase reaction. Data are given with the standard deviations provided by the fitting method

	β-cyclodextrin	
	−	+
Spectrophotometer[a]		
V_{max} (nmole/min)	4.6 ± 0.2	4.5 ± 0.2
Km (μM)	3.2 ± 0.4	60 ± 5
Clarck electrode[b]		
V_{max} (nmole/min)	46 ± 1	$48 \pm 2 (5.9 \pm 0.8)$
Km (μM)	20 ± 1	$368 \pm 15 (49 \pm 3)$
C.M.C. (μM)	10 ± 2	248 ± 8

[a] Data obtained from Lineweaver–Burk plots fitted by linear regression.
[b] Data obtained by non-linear regression to the Hill equation $v = (V_{max} \cdot [S]^h)/(K^h + [S]^h)$ except those in brackets, which were obtained from Lineweaver–Burk plots of reaction rate data in the premicellar region.

fluorescence, is accompanied by a new burst of enzyme activity, which was also shown to be saturable at high LA concentration. Therefore, in the sole presence of monomers potato 5-LOX displays a kinetic behavior different from that displayed in the presence of LA aggregates: since the activity level at saturation using aggregates is much higher than when monomers are used (ca. 10-fold, see Table 1), it can be proposed that potato 5-LOX undergoes an aggregate-induced activation process. This phenomenon of surface activation has been well described for other enzymes that also act on surface active substrates. Phospholipase A_2 undergoes a time-dependent surface activation due to enzyme dimerization after binding to the aggregated substrate [17]. In this case the activity seems to be absolutely surface-dependent and so the activation phenomenon is clearly manifested. The fact that potato 5-LOX exhibits certain activity towards monomers obscures to a great extent the activation in the absence of β-CD. Conversely, β-CD highlights this process as it keeps LA monomeric at concentrations which can be detected by the polaro-graphic method. As regards the mechanism of surface activation it is merely possible to make some speculations that need further research to be confirmed. Work in this direction is currently being carried out in our

laboratory. If potato 5-LOX surface activation has any physiological role, it must be associated to situations in which its substrate forms aggregates. For instance, in the processes of senescence and aging of plant tissues, in which it has been proposed that LOX plays a central role [18], large amounts of PUFA are released from membranes by phospholipase action. Likewise, aggregated PUFA can occur in injured plant tissues, where the role of LOX in the biosynthesis of the so-called wound hormone is well established [19].

Table 1 shows the kinetic parameters as determined by both the spectrophotometric and polarographic method, in both the presence and absence of 1 mM β-CD.

β-CD causes the Km to increase and has almost no effect on V_{max}. Although this is the effect of a classical non-competitive inhibitor, it is more likely that β-CD would act by sequestering the LA substrate forming the inclusion complex, rather than binding to the LOX active center. This means as well that the LA-β-CD complex cannot be utilized as substrate. Using LA dispersions in Tween 20 as substrate for soybean LOX, an inhibition of the reaction by increasing amounts of β-CD has been observed, and this inhibition was explained according to formation of inclusion complexes [20].

The spectrophotometric method permits determination of the kinetic parameters of potato 5-LOX acting on monomers of LA (see Table 1). On the other hand, in the absence of β-CD, the polarographic method permits the determination of the kinetic parameters of potato 5-LOX acting on LA aggregates, and in the presence of β-CD, it permits the determination of the parameters acting on monomers of LA as well (see Table 1, values in brackets). As a result, we obtained good agreement as regards to V_{max} values for monomers when determined by both methods and, in addition, as regards Km for monomers determined by both methods in the presence of β-CD. As a conclusion, the kinetic data of potato 5-LOX acting on either monomers or aggregates of LA are consistent when determined by two different methods, and are also consistent with the hypothesis of the surface activation of potato 5-LOX.

Acknowledgments This work has been supported in part by research grants from CICYT (Proyecto BIO94-0541). J.M.L.N. is holder of a grant of FPI from MEC. R.B. holds a contract for Doctores Reincorporados. Support from J. García Carrión S.A. is gratefully acknowledged.

References

1. Veldkin GA, Vliegenthart JFG (1984) In: Eichhorn GL, Marzili LG (eds) Advances in Inorganic Biochemistry VI: p 139–161. Elsevier, Amsterdam

2. Schewe T, Rapoport SM, Kühn H (1986) Adv Enzymol Relat Areas Mol Biol 58:191–272

3. Kühn H, Schewe T, Rapoport SM (1986) Adv Enzymol Relat Areas Mol Biol 58: 273–311

4. Parker CW (1987) Annu Rev Immunol 5:65–84
5. Gardner HW (1991) Biochim Biophys Acta 1084:221–239
6. Cistola DP, Hamilton JA, Jackson D, Small DM (1988) Biochemistry 27:1881–1888
7. Schilstra MJ, Veldink GA, Verhagen J, Vlieghenthart JFG (1992) Biochemistry 31:7692–7699
8. Vick BA, Zimmerman DC (1987) in: The Biochemistry of Plants. A Comprehesive Treatise (Stumpf PK, Conn EE, eds) Vol 9
9. Sekiya J, Aoshima H, Kajiwara T, Togo T, Hatanaka A (1977) Agric Biol Chem 41:827–832
10. López-Nicolás JM, Bru R, Sánchez-Ferrer A, Garcıa-Carmona F (1995) Biochem J 308:151–154
11. Bru R, López-Nicolás JM, García-Carmona F (1995) Colloid Surf 97:262–269
12. Saenger W (1980) Angew Chem Int Ed Engl 19:344–362
13. Mulliez E, Leblanc JP, Girerd JJ, Rigaud M, Chottard JC (1987) Biochim Biophys Acta 916:13–23
14. Chattopadhay A, London E (1984) Anal Bichem 139:408–412
15. Rodriguez-López JN, Ros-Martínez JR, Varón R, García-Cánovas F (1982) Anal Biochem 202:356–360
16. Junquera E, Aicart E, Tardajos G (1992) J Phys Chem 96:4533–4537
17. Menashe M, Romero G, Biltonen RL, Lichtenberg D (1986) J Biol Chem 261:5328–5333
18. Kumar GNM, Knowles NR (1993) Plant Physiol 102:115–124
19. Siedow JN (1991) Rev Plant Physiol Plant Mol Biol 42:145–188
20. Laakso S (1984) Biochim Biophys Acta 792:11–15

Progr Colloid Polym Sci (1996) 100:281–285
© Steinkopff Verlag 1996

BIOCOLLOIDS

K. Holmberg
S.-G. Oh
J. Kizling

Microemulsions as reaction medium for a substitution reaction

Dr. K. Holmberg (✉) · S.-G. Oh · J. Kizling
Institute for Surface Chemistry
P.O. Box 56 07
114 86 Stockholm, Sweden

Abstract Synthesis of sodium decyl sulfonate from 1-bromodecane and sodium sulfite was performed in microemulsions based on nonionic surfactant, in liquid crystals and in two-phase systems with or without a phase transfer agent added. The reactions were fast in both bicontinuous and W/O microemulsion, slower in liquid crystal and very sluggish in two-phase systems also in the presence of a Q salt or a crown ether. Addition of a small amount of anionic surfactant to the microemulsion systems decreased reaction rate. Addition of cationic surfactant either increased or decreased the reaction rate depending on the choice of counterion.

Key words Microemulsion – surfactant – phase transfer agent – Q salt – crown ether – liquid crystal – organic synthesis – substitution reaction

Introduction

Incompatibility between inorganic salts and hydrophobic organic compounds is a commonly encountered challenge in preparative organic chemistry. Insufficient phase contact of reagents usually leads to low reaction rates and may give rise to unwanted side reactions. Hydrolysis of esters with caustic soda, oxidative cleavage of olefins with permanganate and periodate, addition of hydrogen sulfite to aldehydes and to terminal olefins, and synthesis of alkyl sulfonates by treatment of alkyl chloride with sodium sulfite are examples of organic reactions where compatibility problems between reactants are common [1].

There are several ways to overcome the problem of poor phase contact. The traditional approach has been to perform the reaction in an aprotic, polar solvent, such as dimethylformamide (DMF) or dimethylsulfoxide (DMSO). Such solvents are often capable of dissolving both the non-polar organic compound and the inorganic salt. However, for toxicity reasons many of these solvents are unsuitable for large scale work. In addition, they often create a work-up problem since their removal by low vacuum evaporation may be tedious.

Another approach is to use a mixture of water and a water-immiscible organic solvent as reaction medium. The interfacial area of the system is increased by agitation and some kind of phase transfer agent is usually added to improve phase contact. There are two main types of phase transfer agents, 'onium salts and crown ethers. The 'onium salt is most often a quaternary ammonium salt (Q salt) but phosphonium salts are also used. The lipophilic cation associates with the nucleophilic reagent – the inorganic anion. The ion pair formed, Q^+Nu^-, is transferred into the organic phase where reaction with the lipophilic compound, R–X, occurs. Chlorinated hydrocarbons are usually employed as the organic phase since solvation of Q^+ is particularly strong in such solvents. Strong cation solvation in the organic phase renders the anion unassociated which increases its nucleophilicity [2].

$$Q^+Nu^- + R-X \longrightarrow R-Nu + Q^+X^-$$

Crown ethers, which are cyclic polyethers, function in a similar way. They form complexes with inorganic cations that fit into their cavity and the lipophilic complex, together with the counterion, is transferred into the organic phase. Like with the 'onium salts, the anion will be unassociated and poorly solvated, and thus, of high nucleophilicity [3].

Use of microemulsions as reaction medium can be seen as yet another approach to overcome reactant solubility problems [4, 5]. Being microheterogeneous mixtures of oil, water and surfactant, they are excellent solvents for nonpolar organic compounds as well as inorganic salts. Representative examples of reactions performed in microemulsions include alkylation of 2-alkylindane-1,3-diones with benzyl bromide [4], oxidation of $Fe(CN)_6^{4-}$ by $S_2O_8^{2-}$ [6] and detoxification of "half-mustard," $CH_3CH_2SCH_2CH_2Cl$ [7]. Microemulsions are also being used as reaction medium for bioorganic synthesis, where advantage is taken of their ability to dissolve both a hydrophobic substrate and a water-soluble enzyme [8, 9].

In this work different types of microemulsions are evaluated as reaction medium for synthesis of sodium decyl sulfonate from 1-bromodecane (oil-soluble) and sodium sulfite (water-soluble). Reaction profiles are determined and compared with those obtained in two-phase systems with a Q salt or a crown ether as phase transfer agent.

Materials and methods

The surfactants used were from the following sources: penta(ethylene glycol) monododecyl ether ($C_{12}E_5$) from Nikko Chemicals, Japan, sodium dodecyl sulfate (SDS) from Sigma, USA and tetradecyltrimethylammonium bromide ($C_{14}TAB$) from Aldrich, USA. Tetradecyltrimethylammonium acetate ($C_{14}TAAc$) was prepared as previously described [10]. All other chemicals were reagent grade. Water was purified by decalcination, prefiltration and reverse osmosis followed by passing through a modified Milli Q unit.

Phase diagrams were constructed as described in a previous publication [11]. Self-diffusion coefficients of individual components of microemulsions were measured by the pulsed field-gradient spin-echo method [12] on a Jeol FX-100 apparatus operating at $v_o = 100$ MHz for protons. The length of the gradient pulse was 1–18 ms and the time separation of the gradient pulse was 140 ms.

Reactions in microemulsions were carried out as described in previous communications [10, 11].

Reaction in the presence of Q salt was performed as follows: The Q salt, $(C_4H_9)_4N^+HSO_4^-$, (2 mmol, Aldrich) and Na_2SO_3 (2 mmol) were dissolved in water (50 ml) and

added to a solution of decyl bromide in dichloromethane (2 mmol in 50 ml). Four samples were run at the same time, all under stirring at 25 °C and at the same rpm. Stirring was discontinued after 1, 2, 4 and 6 h, respectively. A sample (10 ml) of the organic phase was evaporated to dryness, the residue dissolved in decane (10 ml) and the solution washed four times with water (20 ml) to remove salts. Decyl bromide concentration in the decane solution was monitored by UV, as described previously (11). Degree of conversion was calculated from the decrease in decyl bromide concentration. Reaction in the presence of the crown ether 18-crown-6 (Aldrich) was performed as with the Q salt but using K_2SO_3 instead of Na_2SO_3.

Results and discussion

Phase diagram and structure determination

The pseudoternary phase diagram of the system $C_{12}E_5$/ dodecane/aqueous Na_2SO_3 solution is shown in Fig. 1a. Figure 1b shows a partial phase diagram of the same system with addition of a small amount of cationic surfactant (2 wt% $C_{14}TAB$ on total amount of surfactant). As can be seen, the main difference between the two phase diagrams is that in the system containing cationic surfactant a narrow two-phase region is present along the surfactant-oil axis. Very similar phase behavior was obtained when $C_{14}TAB$ was replaced by either $C_{14}TAAc$ or SDS. In the system with nonionic surfactant only, the L2 region extends all the way to the surfactant-oil axis.

Self-diffusion NMR was used to determine the microstructure at compositions A of Fig. 1a and B and C of Fig. 1b. Table 1 shows the values of the self-diffusion coefficients, D, for the main components at these compositions. From the $D/D°$ values of water and dodecane at points A and B it can be concluded that the long channel that extends from the L1 region towards the hydrocarbon corner is a bicontinuous microemulsion. The $D/D°$ value at point C is indicative of a W/O microemulsion.

As is seen from the table, the water D-value of microemulsion B is higher than that of A in spite of the fact that the water content is lower and the hydrocarbon content higher in B than in A. This is probably due to the higher surfactant concentration at composition A than B. The NMR method measures an average of "free" and surfactant-bound water. The surfactant-bound water has a low D-value; hence, the more surfactant at constant water-to-oil ratio, the lower the measured water $D/D°$ (13). Thus, the method used underestimates the D-values for free water and it does so more for A than for B. (The division in free and bound water should not be regarded as a static phenomenon; the two types of water coexist and exchange

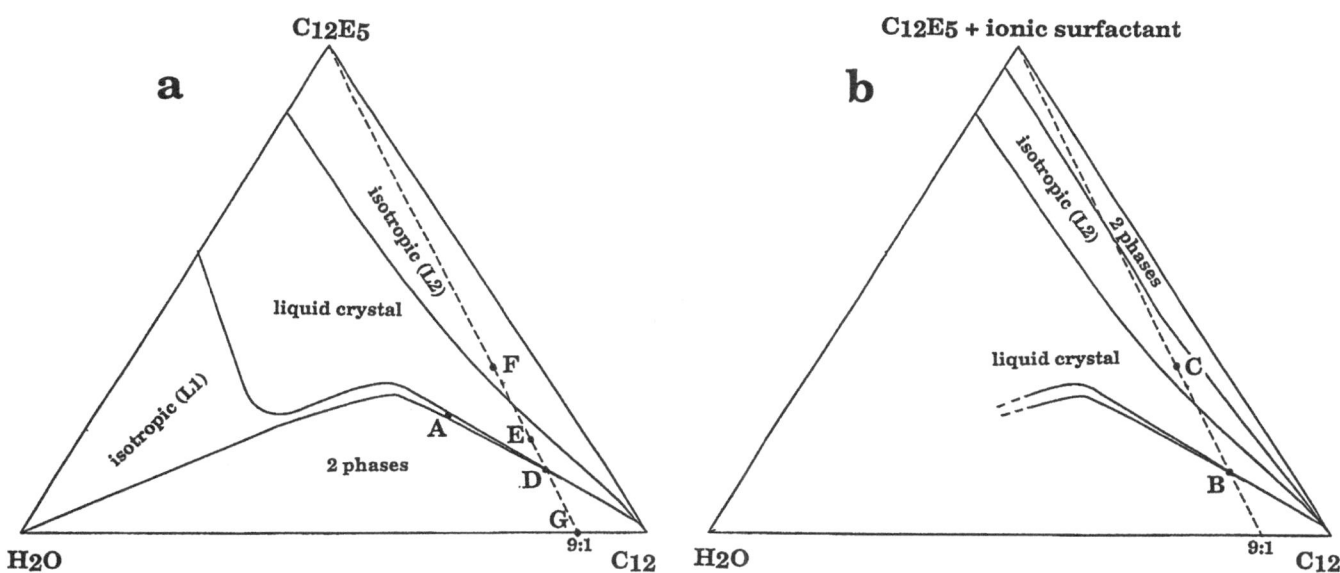

Fig. 1 Phase diagrams for the system $C_{12}E_5$/dodecane/aqueous Na_2SO_3 solution (a) and for the same system with addition of $C_{14}TAB$ (2 wt % on total amount of surfactant) (b) both at 25 °C

Table 1 Self-diffusion coefficients, D, (in m^2s^{-1}) for the main components of microemulsion A of Fig. 1a and microemulsions B and C of Fig. 1b at 25 °C. Self-diffusion of neat water, D_0 (HDO), was 1920×10^{-12} m^2s^{-1} and that of neat dodecane, D_0 ($C_{12}H_{26}$), was 596×10^{-12} m^2s^{-1}

	Microemulsion A			Microemulsion B			Microemulsion C		
	HDO	$C_{12}E_5$	$C_{12}H_{26}$	HDO	$C_{12}E_5$	$C_{12}H_{26}$	HDO	$C_{12}E_5$	$C_{12}H_{26}$
$D \times 10^{12}$	368	26.6	271	538	58.6	453	110	43.5	533
D/D_0	0.19		0.45	0.28		0.76	0.057		0.89

quickly.) The D-values recorded for dodecane at compositions A, B and C follow the expected trend, a fact that lends support to the above discussion about underestimation of the D-value of free water.

Reactivity in different media

The reaction studied is a typical second-order nucleophilic substitution reaction (S_N2), with the reaction rate in a homogeneous system being proportional to the concentration of both components. In a microheterogeneous system, kinetics will also depend on the transfer of reactants across the interface. A kinetic model has been worked out in which the three pseudophases; oil, water and interface, are taken into account [11].

$$C_{10}H_{21}Br + Na_2SO_3 \longrightarrow C_{10}H_{21}SO_3Na + NaBr$$

The reaction was carried out at points D, E and F of Fig. 1a, representing a bicontinuous microemulsion, a liquid crystalline phase and a W/O microemulsion, respectively, all having a 9:1 dodecane-to-water ratio. Note that these formulations contain only one surfactant, the nonionic $C_{12}E_5$.

Reactions were also conducted in three surfactant-free two-phase systems kept under stirring: dodecane-water 9:1 (point G of Fig. 1a), dichloromethane-water with Q salt added and dichloromethane-water with added crown ether. The reaction profiles obtained are shown in Fig. 2a. As can be seen, the reaction is relatively fast in the two microemulsions, slower in the liquid crystalline phase and extremely sluggish in all three two-phase systems. Neither the Q salt, nor the crown ether caused any pronounced increase of the reaction rate. We believe that the reason why the phase transfer agents are not effective in this reaction is that the product formed, decyl sulfonate, is

284
K. Holmberg et al.
Microemulsions as reaction medium

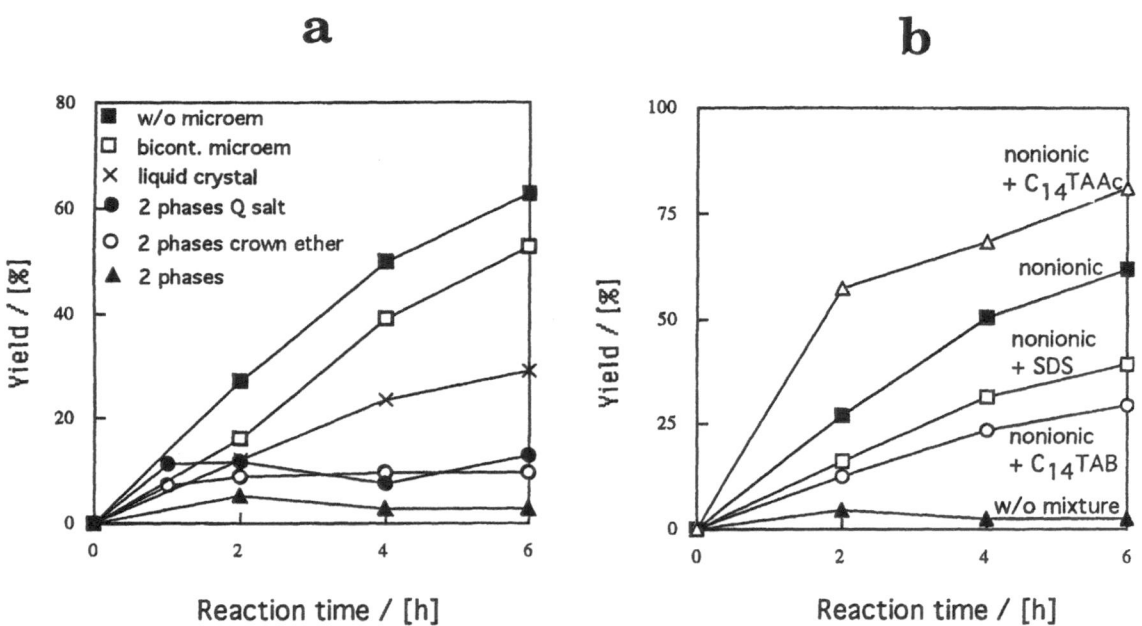

Fig. 2 Effect of type of system (a) and of addition of ionic surfactant to microemulsions based on $C_{12}E_5$ (b) on reaction profile

a much more lipophilic anion than the inorganic ions; hence, formation of the desired ion pair between the phase transfer agent and the nucleophilic reagent becomes less probable as the reaction proceeds.

Reactions in the two types of microemulsions and in the liquid crystalline phase were carried out also at dodecane-to-water ratios of 8:2, 6.5:3.5 and 5:5. The same general trend of higher reactivity in the microemulsions than in the liquid crystalline phase was obtained at all ratios [11]. The lower reaction rate in the liquid crystalline phase is probably related to the high rigidity of this structure as compared with microemulsions. Diffusion of reactants across the interface is likely to be favored by highly dynamic and flexible microstructures. Various types of liquid crystalline phases have been investigated as media for organic [14] and bioorganic [15, 16] reactions before, but no previous comparison between reaction rates in microemulsions and liquid crystalline phases based on the same components appears to have been carried out.

Reactions were also performed with addition of a small amount (2 wt % on total surfactant) of ionic surfactant added to the system. The reaction profiles obtained at composition C of Fig. 1b (W/O microemulsion) are shown in Fig. 2b. SDS, which produces negative charge at the oil–water interface causes a decrease in the reaction rate. This is the expected result since electrostatic double-layer forces will render approach of the sulfite ion into the interfacial region more difficult. Contrary to expectations, the cationic surfactant $C_{14}TAB$ also causes a decrease of reaction rate. The other cationic surfactant, $C_{14}TAAc$, on the other hand, gives a considerable increase in reactivity. Evidently, the choice of counterion is decisive for the reaction rate. The same trend was obtained for reactions at composition B of Fig. 1b (bicontinuous microemulsion) [10].

The difference in effect of the two cationic surfactants is likely to be related to the difference in polarizability of the counterions. Introduction of a cationic surfactant into the surfactant monolayer that separates oil and water domains in the microemulsion gives a catalytic effect, provided the counterion does not interact so strongly with the interface that approach of the reactant – the sulfite ion – is made difficult. Bromide ions bind strongly and prevent diffusion of sulfite into the reaction zone; acetate ions bind weakly and allow sulfite to diffuse into the interfacial region. The ion order of $Br^- > SO_3^{2-} > CH_3COO^-$ is in agreement with values of ion-exchange constants [17]. A very similar effect of bromide ions on rate of an S_N2 reaction involving a negatively charged nucleophile has been encountered before in a micellar system [18].

Acknowledgments The authors thank Ms. Sara Gutfelt for excellent technical assistance, Mr. Norman Burns for linguistic help and the Wennergren Foundation for financial support for SGO.

Progr Colloid Polym Sci (1996) 100:281–285
© Steinkopff Verlag 1996

References

1. Schomäcker R (1992) Nachr Chem Tech Lab 40:1344
2. Starks CM, Liotta CL (1978) Phase Transfer Catalysis, Academic Press, New York
3. Pedersen CJ, Frensdorff HK (1972) Angew Chem, Int Ed 11:16
4. Schomäcker R, Stickdorn K, Knoche W (1991) J Chem Soc, Faraday Trans 87:847
5. Holmberg K (1994) Adv Colloid Interface Sci 51:137
6. Lopez P, Rodriguez A, Gomez-Herrera C, Sanchez F, Moya MA (1992) J Chem Soc, Faraday Trans 88:2701
7. Menger FM, Elrington AR (1991) J Am Chem Soc 113:9621
8. Khmelnitski YL, Kabanov AV, Klyachko NL, Levashov AV, Martinek K (1989) In: Pileni MP (ed) Structure and Reactivity in Reversed Micelles, Elsevier, Amsterdam
9. Holmberg K (1995) In: Solans C, Kunieda H (eds) Industrial Applications of Microemulsions, Marcel Dekker, New York, in press
10. Oh SG, Kizling J, Holmberg K (1995) Colloids Surfaces A 104:217
11. Oh SG, Kizling J, Holmberg K (1995) Colloids Surfaces A 97:169
12. Stilbs P (1987) Prog Nucl Magn Resonance Spectrosc 19:1
13. Jonströmer M, Jönsson B, Lindman B (1991) J Phys Chem 95:3293
14. Ramesh V, Labes MM (1988) J Am Chem Soc 110:738
15. Klyachko NL, Levashov AV, Pshezhetsky AV, Bogdanova NG, Berezin IV, Martinek K (1986) Eur J Biochem 161:149
16. Miethe P, Gruber R, Voss H (1989) Biotechnol Lett 11:449
17. Bartet D, Gamboa C, Sepulveda L (1980) J Phys Chem 84:272
18. Thompson RA, Allenmark S (1992) J Colloid Interface Sci 148:241

Progr Colloid Polym Sci (1996) 100:286–289
© Steinkopff Verlag 1996

Lecithin W/O microemulsions as a host for trypsin. Enzyme activity and luminescence decay studies

S. Avramiotis
A. Xenakis
P. Lianos

S. Avramiotis · Dr. A. Xenakis (✉)
Institute of Biological Research
and Biotechnology
The National Hellenic Research Foundation
48, Vs. Constantinou Ave.
11635 Athens, Greece

S. Avramiotis · P. Lianos
Physics Section
School of Engineering
University of Patras
26500 Patras, Greece

Abstract The possibility of using a microemulsion system based on the natural surfactant lecithin for enzymic studies was examined. Trypsin was solubilized in water-in-oil microemulsions of lecithin in isooctane and tested for activity. The enzyme activity was found higher in the microemulsion than in an aqueous solution. This "superactive" behavior was attributed to high local concentrations of the substrate, induced by the presence of the surfactant. Luminescence quenching measurements using $Ru(biPy)_3^{2+}$ as a lumophore and $Fe(CN)_6^{3-}$ as a quencher showed that the system is compartmentalized and that the microemulsion is percolating. The presence of trypsin influences the dispersed microenvironment by further restricting it, whereas by increasing the water content the interphase composition is altered.

Key words Lecithin – microemulsions – reverse micelles – trypsin – luminescence decay

Introduction

Water in oil (w/o) microemulsions are a particularly attractive system for biotechnological applications [1–3]. It is well established that many enzymes can retain their catalytic ability when hosted in the aqueous core of microemulsions [4, 5]. So far, most of the studies involving enzymic systems have been performed by using well characterized artificial surfactants such as the anionic AOT or the cationic CTAB. Nevertheless, these molecules are generally toxic and cannot be used in potential pharmaceutical applications of enzyme containing microemulsions [6]. An attractive alternative could be the use of natural emulsifiers such as lecithin as surfactants. Furthermore, these reverse micelles formed with phospholipid surfactants are used as model systems that simulate biological membranes.

In this work we have studied the possibility of solubilizing trypsin in microemulsion systems formulated with soya bean lecithin in isooctane. The enzyme hosted in the above medium was tested for proteolytic activity. The effect of the presence of the enzyme on the structure of the reverse micelles was also investigated, both by static fluorescence quenching and by time-resolved luminescence spectroscopy. In all cases, $Ru(biPy)_3^{2+}$ was used as lumophore and $Fe(CN)_6^{3-}$ as quencher, reagents that are well known as probes of water-in-oil microemulsions [7]. The luminescence decay profiles were analyzed by using a "percolation" model [8], which is particularly adapted to water-in-oil microemulsions.

Experimental

Trypsin from bovine pancreas type III was purchased from Merck, Darmstadt. Stock trypsin solutions were prepared and stored in a freezer. Each set of experiments was carried out with the same enzyme stock solution within 4 h. The substrate L-lysine-p-nitroanilide (LNA) was from Sigma.

Isooctane and chloroform were purchased from Merck, while propanol-1 was from Ferak, Berlin.

Tris (2,2'-bipyridine) ruthenium dichloride hexahydrate, $Ru(biPy)_3^{2+}$, was from GFS Chemicals and potassium hexacyanoferrate from Merck. All chemicals were of the highest available degree of purity and doubly distilled water was used throughout this study.

Crude lecithin from soybean was from Serva, Heidelberg, containing 18–26% phosphatidylcholine. It was purified by column chromatography using a 35 cm × 3.2 cm column filled with basic alumina (type 5016A from Fluka) and a chloroform/methanol (9/1 v/v) solvent system as eluent [9]. The fractions were followed by TLC on silica plates using for the elution a chloroform/methanol/water (65/25/1 v/v) solvent system. The purified phosphatidylcholine was identified by NMR using a 300 MHz Bruker spectrometer. An average molecular mass of 800 daltons was determined.

The preparation of microemulsions was generally performed as follows: Stock solutions of 3–10% w/w of lecithin in isooctane were prepared and stocked under nitrogen at 4 °C. The solubilization was accomplished by stirring and occasionally by briefly heating at 40 °C. The content in water was periodically checked by Karl Fischer titrations. The amount of water (in general less than 1%) was taken into consideration in the calculation of the global water content. Reverse micelles were formed with the addition of the appropriate amounts of propanol and buffered stock enzyme solution. Solubilization was achieved by gentle shaking within less than 1 min. The total amount of water was adjusted to give the desired value of the ratio $w_o = [H_2O]/[lecithin]$.

In a typical experiment for an enzyme activity measurement, two microemulsions are prepared: 40 μl of 0.04 M LNA in methanol are deposited in a tube and the methanol is evaporated. 0.45 ml of lecithin/isooctane 5% w/w, 50 μl propanol-1 and 7 μl of 0.1 M Tris/HCl pH 8.5 buffer solution were then added and stirred for some minutes to solubilize the solid substrate. In another tube 0.45 ml of the same lecithin/isooctane solution are mixed with 50 μl propanol-1, 2 μl of the above-mentioned buffer, and 5 μl of 0.32 mM trypsin forming a microemulsion by gentle shaking for a few seconds. These microemulsions were thus identical regarding all components except the enzyme and the substrate respectively. The reaction was initiated by mixing the above fractions and the catalysis was followed continuously in the transparent microemulsion in a thermostatted cell. The reaction was monitored at 410 nm where the molar extinction coefficient of the product p-nitroaniline was found $\varepsilon = 3700$ M^{-1} cm^{-1}. An identical microemulsion without substrate was used as reference. The concentration range of the substrates was 0.68–9.60×10^{-3} M for the LNA. All substrate and enzyme concentrations are expressed with respect to the volume of the reaction medium, known as "overall" concentrations [4]. All measurements were carried out in duplicate, except the kinetic determinations that were the mean values of three independent experiments.

Nanosecond luminescence decay profiles were recorded with the photon counting technique using a specially constructed hydrogen flash and ORTEC electronics. A Melles–Griot interference filter was used for excitation (450 nm) and a cutoff filter (600 nm) for emission. All samples were deoxygenated by the freeze-pump-thaw method. The decay profiles were recorded in 500 channels at 5 ns per channel and were analyzed by least-square fits using the distribution of the residuals and the autocorrelation function of the residuals as fitting criterion [10]. The lumophore was $Ru(biPy)_3^{2+}$. Its concentration was maintained at 10^{-5} M. All measurements were performed in thermostatted cells at 30 °C. The decay time of free $Ru(biPy)_3^{2+}$ ranged then around 530 ns. The quencher was $Fe(CN)_6^{3-}$ at a constant concentration equal to 2×10^{-4} M.

Results and discussion

Enzyme activity studies

The effect of water content on the activity of trypsin was examined in the lecithin based microemulsions. Figure 1 shows the variation of the enzyme activity as a function of the water content expressed in terms of the molar ratio w_o. Increasing w_o the enzyme activity follows a bell-shaped pattern with a maximum at $w_o = 20$. Similar behavior has

Fig. 1 Effect of the water content, w_o, on the initial velocity of the hydrolysis of LNA by trypsin in 5% lecithin/isooctane/propanol-1/water microemulsions. pH = 8.5, $T = 30$ °C, [Trypsin] $= 10^{-6}$ M

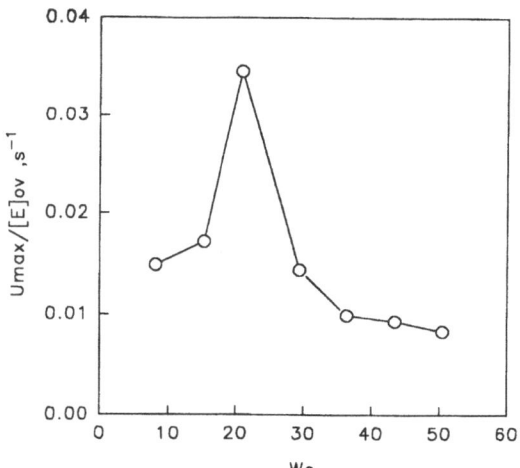

288

S. Avramiotis et al.
Lecithin W/O microemulsions as a host for trypsin

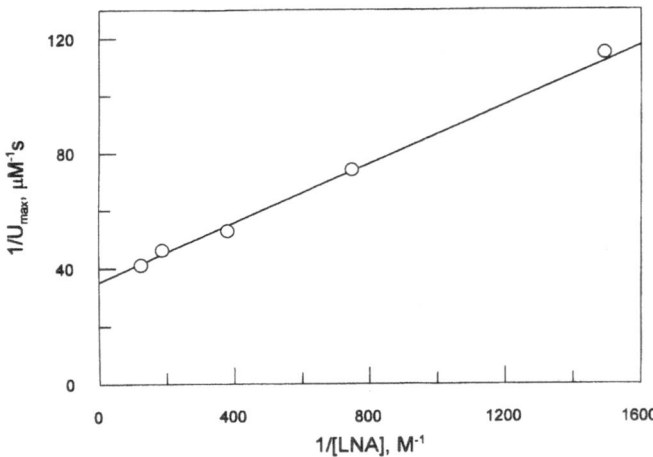

Fig. 2 Dependence of the rate of hydrolysis of LNA on substrate concentration in 5% lecithin/isooctane/propanol-1/water microemulsions. Lineweaver-Burk plot. pH = 8.5, $T = 30\,°C$, [Trypsin] = 10^{-6} M

Table 1 Kinetic constants of hydrolysis of LNA by trypsin in: Lecithin/Isooctane/Propanol-1/Water microemulsion system. Lecithin: Isooctane 5% w/w, Lecithin/Isooctane: Propanol-1 9:1 v/v. $T = 30\,°C$, pH 8.5

W_o	$K_{m,(ov)}$ mM	K_{cat} s^{-1}	$K_{cat}/K_{m,(ov)}$ M^{-1}s^{-1}
aq.sol.	0.63 ± 0.10	0.0057 ± 0.0006	9.05 ± 1.7
15.3	1.54 ± 0.01	0.0316 ± 0.0002	20.52 ± 0.19
20.9	2.80 ± 0.20	0.0905 ± 0.0060	32.32 ± 3.15
29.4	6.33 ± 0.19	0.0930 ± 0.0027	14.69 ± 0.61
37.5	3.04 ± 0.09	0.0320 ± 0.0009	10.53 ± 0.44
43.4	4.24 ± 0.10	0.0295 ± 0.0007	6.96 ± 0.23
50.4	8.85 ± 1.54	0.0530 ± 0.0080	5.99 ± 1.38

been encountered for trypsin activity in AOT microemulsion systems with an optimum w_o value of 10 [11, 12], and in synthetic model lecithin microemulsions with optimum w_o values of about 8 to 10, depending on the length of the fatty acid moiety of the model phosphocholines used [13]. The shift of the optimum w_o in our case could be related to the considerably higher mean length of the fatty acid moieties of the lecithin used in the present study. In addition, the value of w_o in our case can only be compared qualitatively with the model systems, since the presence of propanol affects the composition of the interface, which is the main feature of the molar ratio w_o.

The kinetics of the hydrolysis of LNA by trypsin in lecithin microemulsion systems followed the Michaelis–Menten model. A typical example is shown in Fig. 2. From such double reciprocal Lineweaver–Burk plots we have calculated the kinetic constants shown in Table 1. It can be noticed that the enzyme efficiency, expressed in terms of the ratio K_{cat}/K_m, increases as the water content increases up to the value corresponding to $w_o = 20$, while after this

value the efficiency decreases, confirming the findings of Fig. 1.

The ratio K_{cat}/K_m, was found higher than the one observed in aqueous solutions. This "unusual" superactive behavior of trypsin has already been encountered in other microemulsion systems such as CTAB/isooctane cationic systems [11] and has been attributed to a high local substrate concentration in the reverse micelles, induced by the electrostatic repulsions between LNA and CTAB. In the present study the amphoteric surfactant lecithin may also induce similar high local substrate concentrations.

Luminescence quenching data

In order to clarify the effect of the enzyme molecules on the reverse micelles, we have undertaken a luminescence quenching study. For this, lecithin based microemulsions with various water contents were studied in the presence of different trypsin concentrations. The analysis of the luminescence decay profiles was done with a "percolation" model of stretched exponentials given by the following equation that applies to infinitely short pulse excitation [14–16].

$$I(t) = I_0 \exp(-k_0 t) \exp(-C_1 t^f + C_2 t^{2f}) \qquad (1)$$

k_0 is the decay rate without quenching, C_1 and C_2 are constants, and f a non-integer exponent, $0 < f < 1$. The application of this model is made under the assumption that the dispersed phase constitutes a percolation cluster either below or above the percolation threshold. If we further assume that the percolation cluster is self-similar, i.e., fractal, then $f = d_s/2 = d_f/d_w$, where d_s is the spectral dimension, d_f the fractal dimension and d_w the fractal dimension of the random walk. The reaction between the minority species, i.e., the excited Ru(biPy)$_3^{2+}$ and the majority species, i.e., the quencher Fe(CN)$_6^{3-}$, occurs by diffusion within the percolation cluster. The analysis of the data with Eq. (1) allows the calculation of C_1, C_2 and f that are then used to obtain the first-order reaction rate through the following equation [14]:

$$K(t) = fC_1 t^{f-1} - 2fC_2 t^{2f-1} \; (s^{-1}) . \qquad (2)$$

The reaction rate $K(t)$ is time dependent, as expected for all reactions in restricted geometries.

Table 2 shows the results of the analysis of lecithin microemulsions with different water contents: 2 and 3% v/v and in the latter case for three different enzyme concentrations. In all cases C_2 was equal to zero, meaning that the microemulsions were percolating [14, 15]. Quenching is carried out not only between a lumophore and a quencher solubilized in the same droplet, but also by transfer of quenchers between different droplets, i.e.,

Progr Colloid Polym Sci (1996) 100:286–289
© Steinkopff Verlag 1996

Table 2 Data obtained by time-resolved analysis of luminescence quenching using Eqs. (1, 2) at fixed quencher concentration $[Q] = 2 \times 10^{-4}$ M. $[Ru(biPy)_3^{2+}] = 10^{-5}$ M. Lecithin: Isooctane: 5% w/w, Lecithin/Isooctane: Propanol-1:9:1 v/v. $T = 30\,°C$, pH 8.5

Water %	[Enzyme] $\times 10^6$	C_2	f	K_1 $\times 10^6\,s^{-1}$	K_L $\times 10^6\,s^{-1}$
2	0	0	0.25	60	0.18
3	0	0	0.88	1.22	0.71
3	0.9	0	0.91	1.37	0.72
3	3.0	0	0.90	1.33	0.78
3	6.0	0	0.77	2.41	0.71

droplet diffusion and collision. Probe exchange between droplets occurs during the excited state of the lumophore. This is reflected in the value of long-term $K(t)$, as it could be seen below. The values of the non-integer exponent f are a measure of the restrictions imposed on the reaction (quenching) by the reaction domain. f is smaller when the reaction is more restricted. In this sense we note that f largely increases on going from 2 to 3% v/v H_2O. A dramatic variation of the water/oil interface occurs upon such increases of the water content. At 3% v/v H_2O, f is around 0.9. The exchange of probes between droplets at that water concentration becomes almost as fast as in a continuous phase. Consequently, the reaction rate does not vary much with time. This, indeed, is reflected in the values of K_1 (at the beginning of the reaction) and K_L (at the end of the reaction). It is to be noted that there is an enormous difference between K_1 and K_L for 2% v/v H_2O, but K_1 is very close to K_L for 3% v/v H_2O. This means that in the first case most of the quenching is done at short times, while in the second case the reaction goes on with slightly decreasing probability. K_L is still different from zero even for 2% v/v H_2O. This means that even at long times reaction goes on, which justifies the above statement about percolating microemulsions.

Considering the effect of the presence of trypsin in the above studied microemulsions, we note that most of the corresponding data in the absence of enzyme are not much affected, except for the highest enzyme concentration. C_2 still remains equal to zero. f is still very high, except for $[Trypsin] = 6 \times 10^{-6}$ M, when it suffers substantial decrease. On the contrary, the reaction rate largely changes both at start and at long time. More specifically, K_1 increases and K_L decreases upon addition of enzyme. This is a repetition of an early phenomenon that enzymes as well as macromolecules, such as PEG, solubilized in w/o microemulsions result in increasing the short time quenching and impeaching diffusion [8, 16].

In conclusion, lecithin based microemulsions can be used for enzymic studies. The kinetics of the enzyme catalyzed reactions follow the classic Michaelis–Menten pattern, allowing the determination of the kinetic constants. Trypsin showed enhanced activity in the specific environment as compared to an aqueous solution. Luminescence quenching in these media showed that the microemulsions are percolating. The presence of trypsin affects the reaction domain, while increasing the water content dramatically affects the oil/water interphase.

Acknowledgment A.S. thanks the Greek Ministry of Education for a leave and D. Papoutsi for her technical assistance.

References

1. Luisi PL, Straub B (ed) (1986) Reverse Micelles. Plenum Press, London
2. Stamatis H, Xenakis A, Kolisis F (1993) Biotech Lett 15:471
3. Stamatis H, Xenakis A, Kolisis F (1995) An NY Acad Sci 750:237
4. Martinek K, Levashov AV, Klyachko NL, Khmelnitski YL, Berezin YV (1986) Eur J Biochem 155:453
5. Luisi PL, Magid L (1986) CRC Crit Rev Biochem 20:409
6. Attwood D, Florence A (Eds) (1983) Surfactant Systems. Their chemistry, pharmacy and biology. Chapman & Hall, London
7. Modes S, Lianos P, Xenakis A (1990) J Phys Chem 94:3363
8. Lianos P, Modes S, Staikos G, Brown W (1992) Langmuir 8:1054
9. Singleton WS, Gray MS, Brown ML, White JL (1965) J Am Oil Chem Soc 42:53
10. Grinvald A, Steinberg IZ (1974) Anal Biochem 59:583
11. Papadimitriou V, Xenakis A, Evangelopoulos A (1993) Coloids Surf B: Biointerfaces 1:295
12. Walde P, Peng Q, Fadnavis NW, Battistel E, Luisi PL (1988) Eur J Biochem, 173:401
13. Peng Q, Luisi PL (1990) Eur J Biochem 188:471
14. Lianos P, Modes S, Staikos G, Brown W (1992) Langmuir 8:1054
15. Papoutsi D, Lianos P, Brown W (1994) Langmuir 10:3402
16. Papadimitriou V, Xenakis A, Lianos P (1993) Langmuir 9:91

Progr Colloid Polym Sci (1996) 100:290–295
© Steinkopff Verlag 1996

BIOCOLLOIDS

L. Molina
A. Perani
M.-R. Infante
M.-A. Manresa
M. Maugras
M.-J. Stébé
C. Selve

Synthesis and properties of bioactive surfactants containing β-lactam ring

Dr. L. Molina (✉) · M.-J. Stébé · C. Selve
Université Henri Poincaré-Nancy I
Laboratoire d'Etudes des Systèmes
Organiques et Colloïdaux
INCM-CNRS FU 0008
LESOC URA CNRS 406
BP 239, 54506 Vandoeuvre-lès-Nancy
Cedex, France

A. Perani · M. Maugras
Groupe de Recherches sur les Interactions
Moléculaires aux Interfaces
Université Henri Poincaré
Nancy I, Batiment INSERM, CO 10
Plateau de Brabois
54511 Vandoeuvre-lés-Nancy Cedex, France

M.-R. Infante
Centro de Investgacio y Desarrollo (CSIC)
Departmento de Tensioactivos
c/Jorge Girona 18-26
08034 Barcelona, Spain

M.-A. Manresa
Laboratori de Microbiologia
Facultat de Farmacia
Universitat de Barcelona
08028 Barcelona, Spain

Abstract In this paper we report the synthesis of new molecules containing a β-lactam ring, a hydrophobic and a hydrophilic group. Physico–chemical studies demonstrated the surface activity of these new compounds and the formation of molecular organized systems for some of them. Moreover, aggressiveness of those molecules was evidenced by hemolysis tests and cell cultures. A classification was made to distinguish several groups of β-lactams in accordance with their biocompatibility. A tridimensional relation (structure/physico–chemical properties/biological properties) was therefore found out. The selective and significant antibiotic activity of those β-lactams clearly showed that they were not only surfactants but bio-surfactants.

Key words β-Lactam – surfactant – molecular, organized system – biocompatibility – apoptosis – antibiotic

Introduction

Surfactants have a great importance in numerous biological processes. For example [1,2], a correlation between biological properties, surface properties and molecular structure has been found. However, little is known about this tridimensional relation, particularly in the case of β-lactam compounds. For this purpose, new structures containing a β-lactam ring and two antagonistic groups, a hydrophilic and a hydrophobic one, were prepared. The study of physico–chemical and biological properties of these new compounds yielded to the definition of the above-mentioned link and the characterization of new biosurfactants.

Experimental and results

Synthesis

A new method was found to prepare these β-lactams containing both a hydrophilic and a hydrophobic group.

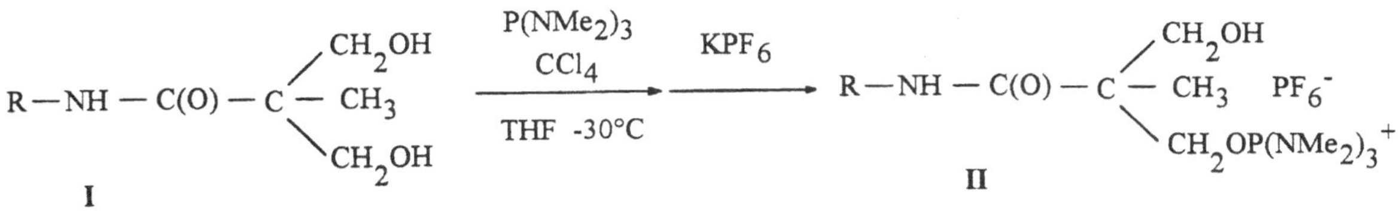

Scheme 1 Synthesis of alkyloxyphosphonium salts

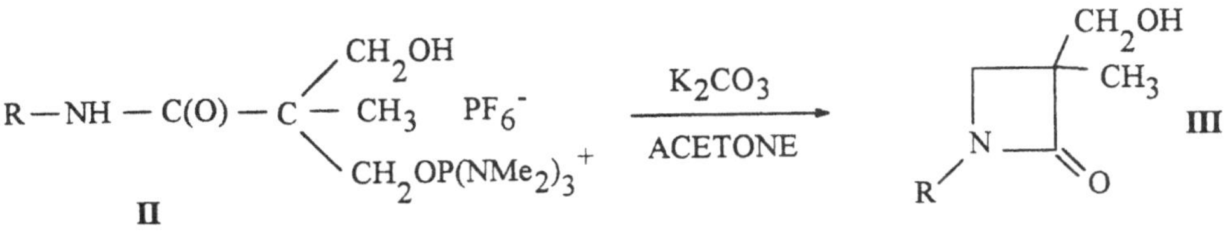

Scheme 2 Synthesis of β-lactams

Table 1 Synthesis results

		R	Cyclization yield (%)	Final yield[a] %	Number of steps[a]	Physical characteristics
III.a	R^H	C_8H_{17}	48	37	3	n_D^{20}: 1.473
III.b	R^H	$C_{10}H_{21}$	52	37	3	n_D^{20}: 1.478
III.c	R^H	$C_{12}H_{25}$	52	39	3	n_D^{20}: 1.477
III.d	R^H	$C_{14}H_{29}$	49	36	3	n_D^{20}: 1.481
III.e	R^F	$C_6F_{13}C_2H_4$	50	27	3	Pf(°C): 47
III.f	R^F	$C_8F_{17}C_2H_4$	45	24	3	Pf(°C): 51
III.g	R^*	$C_8H_{17}OCOCH(Bz)$	51	31	4	n_D^{20}: 1.502
III.h	R^*	$C_{10}H_{21}OCOCH(Bz)$	55	33	4	n_D^{20}: 1.490
III.i	R^*	$C_{12}H_{25}OCOCH(Bz)$	60	37	4	n_D^{20}: 1.498
III.j	R^*	$BzOCOCH(Bz)$	40	16	4	n_D^{20}: 1.553
III.k	R^*	$C_8H_{17}OCOCH(Ph)$	52	43	4	n_D^{20}: 1.504
III.l	R^*	$C_{10}H_{21}OCOCH(Ph)$	54	37	4	n_D^{20}: 1.500
III.m	R^*	$C_8H_{17}NHCOCH(Bz)$	50	22	5	n_D^{20}: 1.505

[a]: from fat chain.

In the first part of this process, compounds I and II were prepared as previously described [3] (Scheme 1) with excellent yields. The oxyphosphonium salts II were synthons for the preparation of β-lactams III through a 1–4 cyclization corresponding to an intramoleculary SN2 [4]. This reaction (Scheme 2) may be carried out in an anhydrous acetone media with a large excess (3 eq. mol.) of K_2CO_3. It was interesting that formation of azetidinone through a rapid and efficient 1–4 cyclization was allowed without involving strong bases or nitrogen activation by substitution (RO- for example). In this work, the nitrogen substitution groups R of salts II were kept in β-lactam III where they represent the hydrophobic moduli of the new structures. Hydrophily was brought in by both amide function of the ring and ring adjacent hydroxymethyl group.

In order to determine the tridimensional link related to these new molecules, several structures corresponding to different kinds of R groups were prepared. These groups were of three types: linear and hydrogenated (R^H), linear and perfluorinated (R^F) or fatty chiral aminoacid derivatives (R^*). Results of the synthesis of these compounds are summarized in Table 1.

Physico–chemical studies

Physico–chemical studies of β-lactams were performed in two sequences. First of all, surface properties were controlled by the determination of interfacial tension of several β-lactam aqueous solutions at different concentrations.

Table 2 Physico–chemical data of hydrosoluble compounds

	CMC(mol/l)	γ(mN/m)	Γ(mol/m²)	σ(Å²)
III.a	$0.4 \, 10^{-4}$	29	$3 \, 10^{-6}$	56
III.e	10^{-4}	16	$3.9 \, 10^{-6}$	43

Table 3 Physico–chemical data of few hydrosoluble compounds

	III.g	III.i	III.j	III.k	III.m
Solubility (mol/l)	10^{-6}	10^{-6}	10^{-6}	10^{-6}	10^{-6}
γ_{sat}(mN/m)	35	35	35	35	47

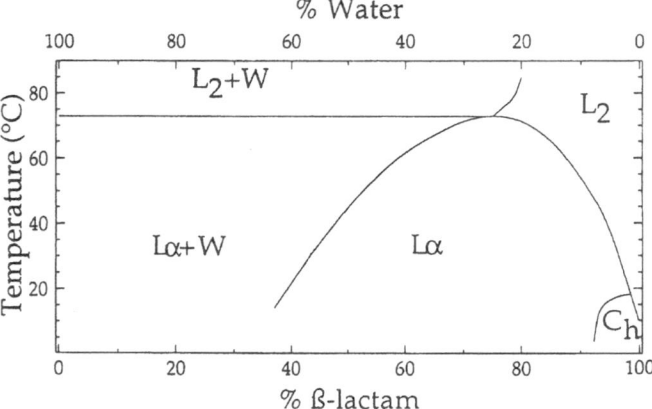

Fig. 2 Binary diagram of compound III.e in water

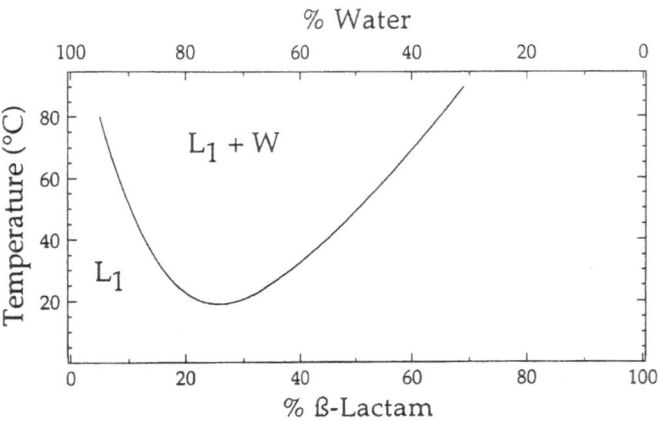

Fig. 1 Binary diagram of compound III.a in water

The diagram γ (interfacial tension in mN/m) versus $\log C$ (C in mol/l) of the more soluble surfactants was obtained. These afforded graphs with the same characteristics of those belonging to surfactants that aggregate in aqueous media [5]. Thus, we were able to measure the critical concentration and its related interfacial tension for they were the typical values of surfactants. Surface excess (Γ) and polar head surface (σ) were also found out from these graphs. All these values, corresponding to compounds III.a. and III.e., are collected in Table 2.

Most of the other compounds seemed to be not hydrosoluble enough to get the graph γ vs $\log C$. Therefore, we were not able to reach their critical concentrations. Actually, only the interfacial tension of saturated solution (γ_{sat}) was measured. However, it allowed us to confirm that these molecules also had surfactive properties since they lowered water interfacial tension from 72 down to ca. 35 mN/m (Table 3).

To complete our study, research on physico-chemical properties of compounds III.a. and III.e. was achieved.

For this purpose, their binary diagrams in water were investigated (Figs. 1 and 2). Although their structures were similar, they had fundamentally different behaviors. Hydrogenated compound III.a. had the same surface activity as hydrophilic hydrogenated surfactants [6]. Moreover, the formation of a cloud point is clear evidence for the existence of aggregates which were supposed to be spherical micelles. When binary diagram was performed with compound III.e., its shape was similar to that obtained with hydrophobic surfactants [7, 8]. Aggregates were observed through the formation of a lamellar phase that became vesicles at the lower concentrations when sonicated.

Biological studies

Two kinds of studies were performed to define the biological properties of those new β-lactams [9]. First, for a part of the experiments, only the biocompatibility of β-lactam surfactants was considered. Then, in a second part of the experiments, antibiotic activity was examined.

β-lactam's biocompatibility was determined by the measurement of their aggressiveness against living cells with hemocytes hemolysis tests and cell cultures. Hemocytes hemolysis in presence of β-lactam solutions led to a general interpretation about the cells states that were life or death by chemical lysis. Cell cultures (HF2 × 635 human/mouse hybridoma), after contact with aqueous β-lactam solutions, afforded supplementary information about the cell death mechanism. As a mater of fact, surfactants may induce either spontaneous death (necrosis) or programmed cell death (apoptosis) [10]. Four kinds of cells could then be distinguished: viable non apoptotic (VNA), viable apoptotic (VA), non viable apoptotic (NVA) and necrotic (N) cells. The results of these experiments led

Fig. 3 Cell culture with non aggressive surfactants

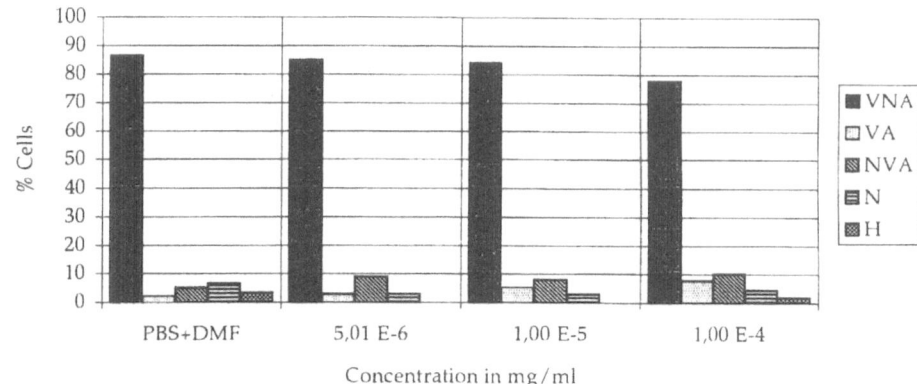

Fig. 4 Cell culture with few aggressive surfactants

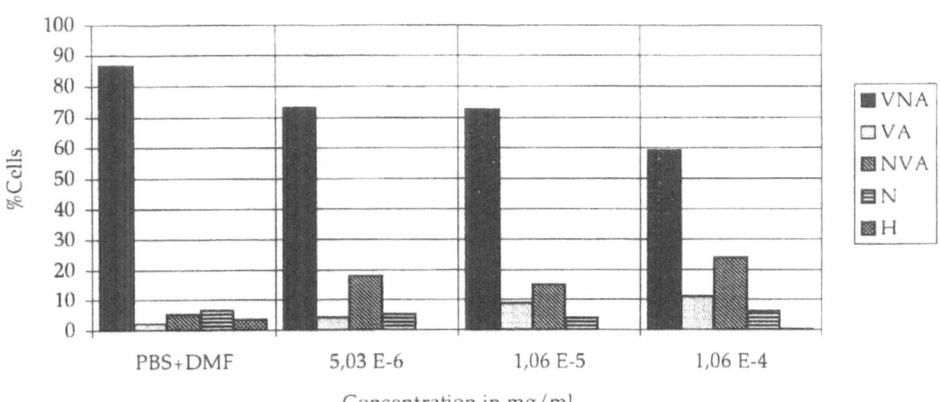

Fig. 5 Cell culture with very aggressive surfactants

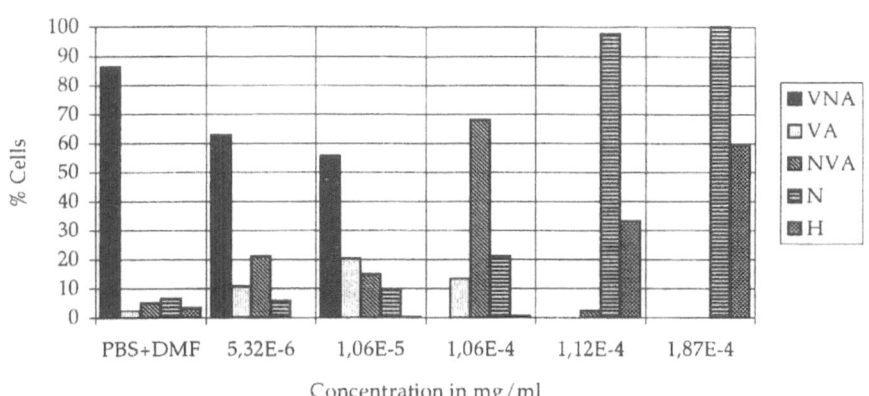

to the classification of β-lactams into three categories of growing aggressiveness molecules. The behavior of a non aggressive surfactant is represented in Fig. 3. In this case, most of the cells were viable non apoptotic. This first class included the compound III.i. A second group may be distinguished as representative of few aggressive surfac-

tants and its typical diagram is displayed in Fig. 4. Viable non apoptotic cells rate was then lower than previously in favor of apoptotic cells rate (viable or not). Compounds III.a, III.e. and III.j. were found therein. Finally, very aggressive molecules were separated in a third group (Fig. 5). An important rate of apoptosis was observed even

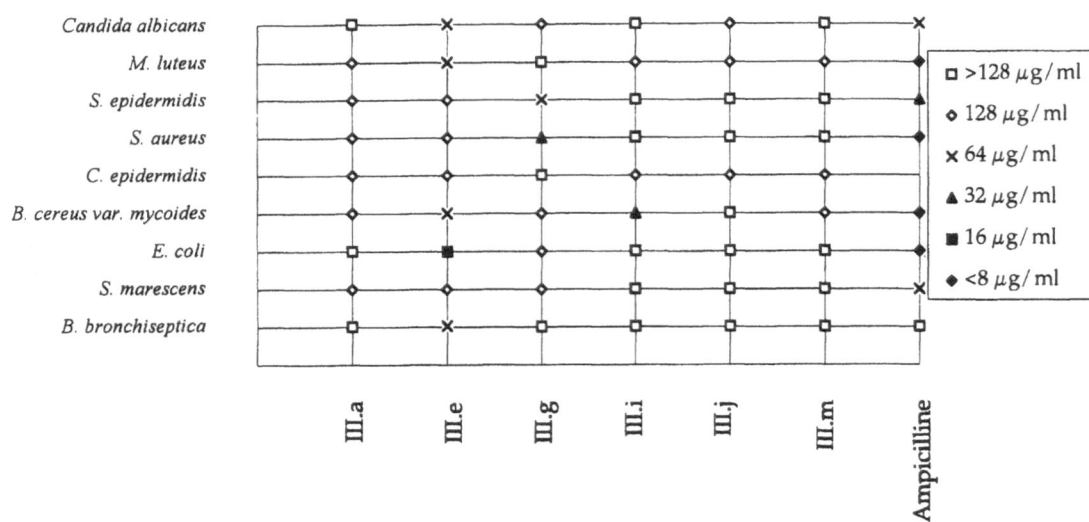

Fig. 6 Antibacterial activity (MIC in μg/ml)

at low concentrations of surfactants. When these concentrations exceed 10^{-4} mg/ml, 100% of necrosis was evidenced. Compounds III.g. and III.l. were considered as very aggressive surfactants.

Antibiotic properties of these new β-lactams were investigated in experiments where a various range of bacterias was used. Results are summarized in Fig. 6 together with ampicillin as reference. All the synthetic β-lactams had a significant and selective antibiotic activity. Three compounds should be particularly noticed:

1) III.i on *Bacillus cereus var. mycoides*,
2) III.e on *Echerichia coli* and a similar activity to that of ampicillin on *Candida albicans*
3) III.g on *Staphylococcus aureus*.

Discussion

All the results obtained in physical chemistry and in biology involved links between the different parts of the study. First of all, physico-chemical data evidenced that molecular structure was related to surface activity. When hydrophobic group (R) of β-lactams was linear and short (less than 10 carbons) molecules were hydrosolubles and afforded molecular organized systems (micelles, lamellae or vesicles). When R was linear but with higher carbonaceous chains or with aminoacid derivatives, therefore aqueous solubility was extremely low although a surface activity could be observed (lowering of water interfacial tension). Whatever the nature of the molecules may be, modifications of living cells' biological processes occurred while surfactant properties were shown. Concerning those new

β-lactams, a relation between structure, physico-chemical properties and HF2 × 635 hybridoma death process was demonstrated. Particularly, when hydrophobic group was sufficiently long (even containing an aminoacid), no disruption of the cells development was observed. This phenomenon has already been discussed and explained [2, 11] taking into account that a C12 chain was long enough to integrate into phospholipids of cells membranes without inducing weakness. Moreover, when R was linear and short, hydrogenated or perfluorated, cells growth was not basically modified and only apoptosis appeared at the higher surfactant concentrations. Low aggressiveness of those surfactants was correlated to their high solubility in water and their behavior in aqueous media (molecular organized systems formation).

On the other hand, when R was from an aminoacid with a C8 carbon chain, aggressiveness of the resulting surfactants against hybridoma was very important (apoptosis for low concentrations and necrosis for the higher ones). Low solubility in water of these surfactants could be responsible for their behavior for it was difficult to get homogeneous aqueous solutions of them. Moreover, it has been previously noticed that chain length [2, 11] (C8) induced disruptions in the cells functions when chains were incorporated into membrane.

Conclusions

Taking into account experimental results, it was clearly demonstrated that surfactive properties of the new synthetic β-lactams were related to their molecular structure. Cell death process was influenced by those properties and

Progr Colloid Polym Sci (1996) 100:290–295
© Steinkopff Verlag 1996

it was shown that β-lactam surfactants' structures had an important action on hybridoma cultures evolution. Therefore, the tridimensional relation was clearly evidenced. Up to now, we have not demonstrated the relation between antibiotic activity and molecular structure of the new synthetic β-lactams. But, through this characteristic, they can be defined as biosurfactants.

Acknowledgments This work was generously supported by CNRS and CSIC "Cooperation project 1940".

References

1. Ji L, Zhang G, Uematsu S, Hirabayashi Y (1995) FEBS Lett 358:211
2. Leblanc M, Riess JG, Poggi D, Follana R (1985) Pharm res 5:246
3. Molina L, Papadopoulos D, Selve C (1995) New J Chem 19:813
4. Molina L, Perani A, Infante MR, Manresa MA, Maugras M, Achilefu S, Stebe MJ, Selve C (1995) J Chem Soc Chem Commun, p 1279
5. Lindman B, Wennertröm H, Eicke HP (1980) "Micelles", Springer-Verlag, Heidelberg
6. Ravey JC, Stebe MJ (1994) Colloids and Surfaces A 84:11
7. Larsson K (1967) Z Phys Chem Neue Folge 56:173
8. Krog N, Larsson K (1968) Chem Phys Lipids 2:129
9. Perani A, Molina L, Stacey G, Infante MR, Selve C, Maugras M, FEBS Lett submitted for publication
10. Corcoran GB, Fix L, Jones DP, Molsen MT, Nicotera P, Oberhammer FA, Buttyan R (1994) Toxicol Appl Pharmacol 128:169
11. Maugras M, Stoltz JF, Selve C, Moumni M, Delpuech JJ (1989) Innov Tech Biol Med 10:145

Progr Colloid Polym Sci (1996) 100:296–300
© Steinkopff Verlag 1996

BIOCOLLOIDS

C. Otero
L. Robledo
M.I. del Val

Two alternatives: Lipase and/or microcapsule engineering to improve the activity and stability of *Pseudomonas* sp. and *Candida rugosa* lipases in anionic micelles

This work has been financed by Spanish CICYT (No PB92–0495) and a grant of the PFPI of the Spanish M.E.C. for L. Robledo

Dr. C. Otero (✉) · L. Robledo
M.I. del-Val
Instituto de Catálisis
C.S.I.C. Campus de la Universidad
Antónoma Cantoblanco
28049 Madrid, Spain

Abstract Different strategies were assayed to mitigate the inhibition effect of the activity of lipases by anionic reverse micelles, and to improve their stabilization in these media.

Isoenzyme B from *Candida rugosa* and *Pseudomonas* sp. lipases were studied in water, in simple and in mixed micelles. Considering that the lipase inhibition in anionic micelles had been previously related with the lipase hydrophobicity, and with the penetration degree of the protein into the interface, we varied the hydrophobic character of the protein surface, and decreased the charge density of the micellar interface.

The covalent modification of the *pseudomonas* lipase with the amphiphilic chains of polyethylene glycol increased its activity in water and in micellar systems, but decreased its stability in these media. The spectroscopic behavior of the protein showed conformational changes of this lipase after its modification, both in water and in micelles. The simultaneous use of the protein modification and of the AOT/Tween mixed micelles improved both its activity and stability. In optimal conditions, the modified lipase was 10 times more stable in mixed nanodroplets than in water.

The covalent immobilization of *Candida rugosa* lipase on a crosslinked polymer was also tested. The polymer size was greater than the miceller sizes. But, the protein immobilization did not decrease the inhibition effect of AOT micelles, because of the penetration of the active center into the interface was not increased in the presence of the microgel. The aggregates containing the enzyme-polymer system were of higher size than those that contained the native lipase. The immobilization decreased the stability of lipase B in micelles. It was attributed to the change of shape of the micellar aqueous pool from a sphere to a spherical layer, which minimized the micellar effect on the enzyme rigidity.

This work demonstrated that the increase of the lipase hydrophobicity and the reduction of the charge density of the interface are useful ways to mitigate the inhibition effect of anionic micelles. The results confirmed the negative effect of the anionic surfactant heads on the lipase activity, and the relevance of the interface nature on the lipase activation.

Abbreviation list – LB, lipase B from *Candida rugosa*; PSL, lipase from *Pseudomonas* sp.; PEG, polyethylene glycol (MW 5000); AOT, Sodium bis-(2-ethylhexyl) sulphosuccinate; $W_o = [H_2O]/$ [surfactant]; $t_{50\%}$, incubation time at which the biocatalyst lose 50% of its initial activity. IF, inhibition factor of the micellar system: [kcat/Km (water)]/[kcat/Km (micelles)].

Introduction

Lipases are a group of hydrolases that become activated at the water/oil interface [1]. Most of their x-ray structures showed that they have their catalytic centers covered by an amphiphilic lid, which is opened after the interfacial activation [2–4].

Although the AOT reverse micelles have high interfacial areas, the lipase catalyzed reactions in AOT nanocapsules are always slower than in water. We previously showed the influence of the hydrophobocities of the *C. rugosa* [5] and *Pseudomonas sp.* [6] lipases on their activities in AOT micelles. The inhibition factor of these lipases in AOT micelles (IF) was smaller when the lipase hydrophobicity increased, and this property seems to determine the degree of the lipase penetration into the micellar interface [5, 6]. The interfacial activation of lipases may also be unfavored in the presence of ionic surfactant heads at the micellar interface. However, the AOT nanocapsules may improve the enzyme stability [7], and/or solubilize hydrophilic (lipases) and hydrophobic (lipidic substrates) molecules. In consequence, their limitations should be mitigated.

In this work we will show how to mitigate the inhibition effect of the anionic AOT micellar system on the lipase hydrolytic activity. Different ways to enhance the enzyme activity/stability were assayed: i) the covalent linkage of the enzyme to a crosslinked polymer (microgel), which is solubilized in the micellar medium and its size is much higher than that of the AOT micelle; ii) the enzyme modification by covalent linkage of amphiphilic chains; iii) the modification of the AOT nanocapsules, reducing the density of charge of the micellar surface. The kinetic and spectroscopic studies of these enzyme derivatives were comparatively discussed.

Materials and methods

Pseudomonas sp. lipase was from AMANO. Its covalent modification with polyethylene glycol (MW 5000) was obtained following the method published in [8]. Seventeen percent of lipase amino groups were modified. The *isoenzyme* B was obtained from the crude *Candida rugosa* lipase (SIGMA) as we reported before [9], and its immobilization on an acrylic crosslinked polymer (microgel) was done as we reported in [10]. Microgels with sizes as high as 300 nm were obtained by the method previously described [10]. The methods followed for the spectroscopic and kinetic assays were as described in [6].

The water activities (Aw) of the micellar systems were determined with a Novasina apparatus at $30 \pm 0.1\,°C$.

A sample of 10 mL-containing or not the enzyme or the acrylic support was used, corresponding to the sample volume used in the activity and stability studies.

Results

Enzymatic activity in water and in micellar systems

The activity of PSL, PSL-PEG, LB and LB-microgel in the p-nitrophenyl buthyrate hydrolysis was compared in water, and in simple and mixed micelles of AOT/ TWEEN 85. The values of the kinetic parameters are in Table 1.

The specificity constant (kcat/Km) of the modified PSL was higher than that of the native lipase in aqueous and micellar systems. The chemical modification increased the kcat and the Km values in water, and *vice versa* in simple micelles of AOT. This effect was consistent with different conformations for PSL and PSL-PEG, both in water and in micelles, as consequence of the different interactions of the amphiphilic modifier (PEG) with the lipase and with each medium (aqueous or micellar).

Also, mixed micelles gave rise to an opposite variation in the values of the kinetic parameters of the native and modified lipases. Thus, the addition of Tween 85 (2% w/w) decreased the kcat and Km values of PSL, but increased those of PSL-PEG. The reduction of the concentration of charged surfactant heads at the interface by addition of the nonionic tensioactive, decreased the water activity (Table 1) and increased the droplets volume (from $4.3\,10^{-26}$ to $9.0\,10^{-26}$ L at Wo = 5, [10]). These variations – of the interface quality, of the micellar size and of Aw – favored the enzymatic activity of the modified lipase (lower inhibition factor (IF) in Table 1, where IF = [kcat/Km (water)]: [kcat/Km (micelles)]), although these variations were unfavorable for the native PSL.

The specificity constant, k_2^0 = kcat/Km, of isolipase B in water decreased by 3.4 times after it immobilization, due to the negative effect of the microgel linkage in the lipase affinity for the substrate (Km) and in the catalytic constant. This type of enzyme-support conjugation gave rise to a modest enhancement of k_2^0 in the AOT system. The microgel decreased the Km value by 25%, and the kcat value by 23%.

Spectroscopic behavior of the lipases

The emission spectra of PSL and PSL-PEG in buffer and in simple micelles are shown in Fig. 1. While in water the PSL-PEG spectrum (λmax = 333 nm) was blue shifted with respect to the PSL spectrum (λmax = 344 nm), in simple small micelles (Wo = 5) the chemical modification

Table 1 The kinetic parameters of the native, modified and immobilized lipases for the p-nitrophenyl butyrate hydrolysis in water and in simple and mixed AOT reverse micelles. Lipases were from Pseudomonas sp. (PSL) and from Candida rugosa (LB). Conditions for PSL and PSL-PEG: A) in aqueous systems: bicarbonate buffer 0.5M, pH 8.5; B) in micelles: [AOT] = 0.1M, Mes buffer 2 mM, pH = 5.5. Conditions for the native and immobilized LB: A) in aqueous systems: Mes buffer 0.1 M, pH 6.5 or Hepes 0.1 M, pH 7.5; B) in micelles: [AOT] = 0.1 M; Acetate buffer 2.2 mM, pH 5. Other conditions: 30 ± 0.1 °C

Enzyme	System	V_d^a $(10^{-26}\,L)$	Aw	K_m, ap (mM)	$kcat^b$	$kcat/Km^b, ap$	$I.F^c$
PSL	Water	∞	1.00	0.35	$6.0\ 10^{-3}$	17	–
	AOT (Wo = 5)	4.3	0.85	20	$1.4\ 10^{-3}$	$7.2 10^{-2}$	238
	AOT + TWEEN (2% w/w)	9.0	0.82	16	$0.8\ 10^{-3}$	$5.1\ 10^{-2}$	333
PSL-PEG	Water	∞	1.00	0.70	$1.6\ 10^{-2}$	23	–
	AOT (Wo = 5)	4.3	0.85	12	$9.3\ 10^{-4}$	$7.8\ 10^{-2}$	217
	AOT + TWEEN (2% w/w)	9.0	0.82	20	$5.2\ 10^{-3}$	$2.6\ 10^{-1}$	65
LB	Water	∞	1.00	0.25	1380	$5.59\ 10^{+6}$	–
	AOT (Wo = 5)	4.3	0.88	0.80	6.0	$7.51\ 10^{+3}$	744
LB-Microgel	Water	∞	1.00	0.75	1250	$1.66\ 10^{+6}$	–
	AOT (Wo = 5)	4.3	0.88	0.60	4.6	$7.60\ 10^{+3}$	735

[a] Volume of the empty aqueous droplets of simple and mixed micelles, determined as reported in [12].
[b] In the case of PSL: kcat in $Ms^{-1}/g\,prot\,L^{-1}$; kcat/Km in $L\,g^{-1}\,prot\,s^{-1}$. In the case of LB: Kcat in s^{-1} and kcat/Km in $M^{-1}\,s^{-1}$.
[c] Inhibition factor of the micellar system: [kcat/Km (native lipase in water)]/[kcat/Km (biocatalyst in micelles)]

Fig. 1 Fluorescence emission of the native and modified with polyethylene glycol lipases, in water and in micelles. Conditions: $\lambda_{ex} = 280$nm and $\lambda_{em} = 290$–450 nm; Emission and excitation slit widths were 5 mm; 30 ± 0.1 °C. In water: Sodium bicarbonate buffer 1 mM, pH 8.5. In micelles: [AOT] = 0.1 M; [buffer] = 2 mM, pH = 5.5 [PSL] = 1.2 10^{-4} g/L; [PSL-PEG] = 0.92 10^{-4} g/L

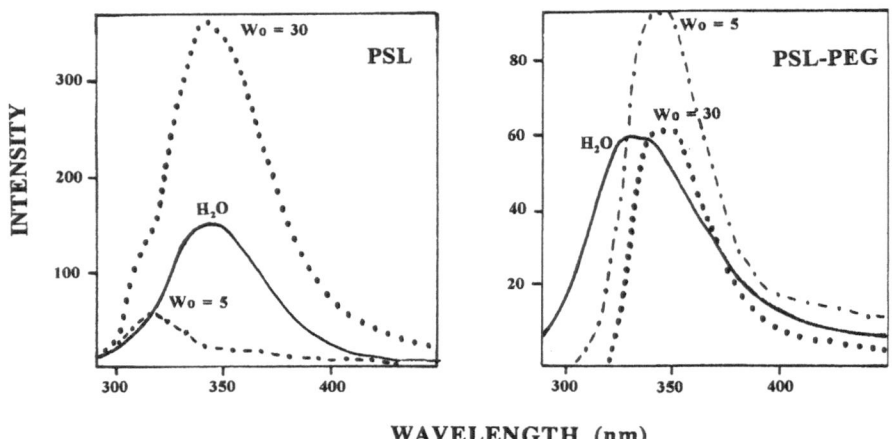

WAVELENGTH (nm)

gave rise to a red shift in tryptophan emission (λmax (PSL) = 316 nm and λmax(PSL-PEG) = 347 nm). Consequently, the existence of several chains of the amphiphilic PEG covalently linked to the protein surface, induced some conformational reorganization of the enzyme, and it was different in water and micellar systems.

Also the spectroscopic behavior of the native lipase in the micellar system tends to be similar to its spectrum in buffer when the hydration degree of the micellar system increase. That showed that PSL had a tendency to be placed in the aqueous core of large micelles (high Wo values). However, the enzyme-micelle interactions were very different when the enzyme was modified with PEG, and the value of λmax was independent of the micellar size (Wo). Considering that the PEG modification had been described as a method for the proteins solubilization in anhydrous organic media [11], we deduced that PSL-PEG was more in contact with the apolar region of the micelles (higher penetration degree into the interface) at any Wo value.

Table 2 Stabilities of the native, immobilized and the modified *Pseudomonas sp.* (PSL) and *Candida rugosa* B (LB) lipases, in simple and mixed micelles of AOT. Conditions for PSL: [PSL] = 7.4 10^{-4} g/L protein, [PSL-PEG] = 6.2 10^{-4} g/L protein, A) in aqueous systems: bicarbonate buffer 0.5 M, pH 8.5; B) in micelles: [buffer] = 2 mM. Conditions for the native and immobilized LB: A) in aqueous systems: [Protein] = 0.16 g/L; Mes buffer 0.1 M, pH 6.5 or Hepes 0.1 M, pH 7.5; B) in micelles: [protein] = 0.041 g/L; [AOT] = 0.35 M; buffer 5.25 mM. Other conditions: 30 ± 0.1 °C

Enzyme	System	Wo (pH[a])	Aw	V_{wp}^b (L)	t_{50}^c (h)
PSL	Aqueous	buffer (pH = 8.5)	1	∞	40
	AOT	5 (pH = 5.5)	0.85	4.3 10^{-26}	0.5
		70 (pH = 8.5)	1.00	–	216
	AOT + TWEEN (2% w/w)	5 (pH = 5.5)	0.82	9.0 10^{-26}	48
		70 (pH = 8.5)	0.94	–	170
PSL-PEG	Aqueous	buffer (pH = 8.5)	1.00	∞	15
	AOT	5 (pH = 5.5)	0.85	4.3 10^{-26}	0.8
		70 (pH = 8.5)	1.00	–	18
	AOT + TWEEN (2% w/w)	5 (pH = 5.5)	0.82	9.0 10^{-26}	7.2
		70 (pH = 8.5)	0.94	–	144
LB	Aqueous	buffer (pH = 7.5)	1.00	∞	26
	AOT	5 (pH = 5.0)	0.88	4.3 10^{-26}	2.0
		30 (pH = 5.5)	1.00	–	0.4
LB-Microgel	Aqueous	buffer (pH = 6.5)	1.00	∞	7.0
	AOT	5 (pH = 5.0)	0.88	4.3 10^{-26}	0.1
		30 (pH = 5.5)	1.00	–	0.4

[a] The optimal pH value of the enzyme in the p-nitrophenylbutyrate hydrolysis.
[b] Volume of the aqueous droplet of simple and mixed micelles, determined as reported in [12].
[c] Time at which the enzyme has 50% of its initial activity.

Enzyme stability

The stability of the native, modified and immobilized lipases in buffer and in simple and mixed micellar are compared in Table 2.

In general, PSL-PEG was less stable than PSL, except in low hydrated systems (Wo = 5) where this lipase showed its maximum activity. The native and modified lipases were less stable in small micelles than in water, but their stabilities enhanced when the Wo value did. Thus, they were higher than in water in systems of Wo = 70. The mixed micelles with Tween increased noticeably the stabilities of the two biocatalysts, PSL and PSL-PEG.

In the buffer system, the linkage of lipase B to the microgel reduced its stability. Isolipase B mainly solubilized into the water core of AOT micelles, resulted more stable in small droplets. In a similar case to lipase B (a *Pseudomonas sp.* lipase), the increase of stability when the enzyme did not penetrate the interface, have been associated to the increase of the enzyme rigidity in the small nanocapsules [12]. But, the opposite behavior was found with the immobilized lipase. That is, aggregates of low Wo values did not restrict the conformational mobility of the lipase. The destabilization of its immobilized derivative in micellar media was always high. This could be explained considering the formation of larger aggregates to include the microgel, and the change of shape of the micellar aqueous pool from a sphere to a spherical layer, which minimized the micellar effect on the enzyme rigidity.

Discussion

The enzyme and medium engineering were used here to improve the activities and stabilities of two lipases in reverse micelles. In the case of PSL and PSL-PEG, the kinetic and spectroscopic studies revealed different conformations and different kinetics parameters. The enzyme modification decreased the inhibition factor of AOT micelles from 238 to 217, due to the increase of the penetration degree of the lipase into the interface. The modified lipase was more active than the native form in both water and simple micelles (higher Kcat/Km). But it was much less stable than PSL in these two media. The micellar systems allowed stabilization of the two proteins at high Wo. Thus, values of $t_{50\%}$ of 216 h and 144 h for the native and modified lipases were obtained in optimal conditions, instead of the value of 44 h obtained with PSL in water. Moreover, the stabilization effect of micellar systems was improved using mixed aggregates, and thus the maximum value of $t_{50\%}$ could be obtained at lower Wo values (Wo = 20) than in the case of simple micelles (Wo = 70). This positive effect of mixed micelles on the stability was more important in the case of PSL-PEG.

This work demonstrated that the increase of the lipase hydrophobicty and the reduction of the charge density of the interface, are useful ways to mitigate the inhibition effect of anionic micelles. The results confirmed the negative effect of the anionic surfactant heads on the lipase activity, and the relevance of the interface nature on the lipase activation.

300

C. Otero et al.
Enhancement of lipases activities/stabilities in anionic nanoaggregates

References

1. Sarda L, Desnuelle P (1958) Biochim Biophys Acta 30:513–2
2. Grochulski P, Li Y, Schrag JD, Cygler M (1994) Protein Sci 3:82
3. Bradly L, Brozozowski AM, Dewerenda ZS, Dodson E, Dodson G, Tolley S, Turkenburg JP, Christiansen L, Huge-Hensen B, Norskov L, Thim L, Menge U (1990) Nature 343:767–770
4. Schrag JD, Li Y, Wu S, Cygler M (1991) Nature 351:761–764
5. Otero C, Rua ML, Robledo L (1995) FEBS Lett. 360:202–206
6. Otero C, Robledo L (1995) Prog Colloid Polym Sci 98:219–223
7. Gajjar L, Dubey RS, Srivastava RC (1994) Appl Biochem Biotech 49:101–112
8. Arcos JA, Otero C in J Am Chem Soc, (in press)
9. Rua ML, Diaz-Mauriño T, Fernández VM, Otero C, Ballesteros A (1993) Biochim Biophys Acta 1156:181–189
10. Otero C, Robledo L, Alcantara AR, in J Mol Catal B (in press)
11. Inada Y, Takahashi K, Yoshimoto T, Ajima A, Matsushima A, Saito Y (1986) Trends in Biotechnology 4:190–194
12. Otero C, Arcos JA, Robledo L Biocatalysis & Biotransformations. Accepted.

Progr Colloid Polym Sci (1996) 100:301–305
© Steinkopff Verlag 1996

BIOCOLLOIDS

T. Hianik
R. Krivánek
D.F. Sargent
L. Sokoliková
K. Vinceová

A study of the interaction of adrenocorticotropin-(1-24)-tetracosapeptide with BLM and liposomes

Dr. T. Hianik (✉) · R. Krivánek
L. Sokoliková · K. Vinceová
Department of Biophysics and Chemical
Physics MFF UK Mlynska dol. F1 84215
Bratislava, Slovak Republic

D.F. Sargent
Institut für Molekularbiologie und
Biophysik
Eidgenössische Technische Hochschule
8093 Zürich, Switzerland

Abstract Several biophysical methods have been used to study the interaction of the amphiphilic hormone derivative $ACTH_{1-24}$ with planar bilayer lipid membranes (BLM) and liposomes. Addition of the six-fold positively charged peptide (final concentration 10^{-4} M) to one side of the BLM of POPC leads to a more positive membrane intrinsic potential, an increase of BLM capacitance, a decrease of elasticity modulus $E_\perp$ perpendicular to the membrane plane, and faster dielectric relaxation time constants. The velocity number $[u]$, measured by ultrasonic velocimetry, decreased after addition of $ACTH_{1-24}$. This reveals an increase of volume compressibility of liposomes. The presence of $ACTH_{1-24}$ in liposomes of DMPC (molar ratio $ACTH_{1-24}/DMPC = 7.7 \times 10^{-4}$ mol/mol) resulted in a decrease of the $[u]$ value around the phase transition temperature of DMPC ($T_c \approx 24\,°C$) and an increase of absorption number $\Delta[\alpha\lambda]$ of ultrasound. This is additional evidence for interaction of $ACTH_{1-24}$ with the hydrophobic part of the membrane. The results obtained are consistent with the idea that the lipid bilayer may function as an antenna and mediator for the hormone-receptor interaction.

Key words ACTH – lipid bilayers – electrostriction – phase transitions – ultrasonic velocimetry

Introduction

The synthetic hormone derivative $ACTH_{1-24}$ is a linear, flexible, amphiphilic – tetracosapeptide. This derivative plays the main role in the physiological effect of adrenocorticotrophic hormone – ACTH [1]. Practical interest in $ACTH_{1-24}$ has been stimulated by its unique and valuable therapeutic properties. In addition, this amphiphilic peptide can be used as a model system for the study of protein-lipid interactions. The interaction of $ACTH_{1-24}$ with solvent-containing and solvent-free planar bilayer lipid membranes (BLM) was studied by use of the capacitance minimization method. A study with BLM [2] showed that the interaction of the peptide hormone with BLM leads to an increase of the intrinsic electric field (difference in surface potentials) across the membrane, and that the intrinsic potential becomes dependent on the ionic strength on the side of the membrane opposite to that to which $ACTH_{1-24}$ was added. It was suggested that adsorption of the hormone is followed by an insertion of part of the peptide, presumably the N-terminal region, across the bilayer. In contrast to this, analogous studies with solvent-free BLM [3] gave no indication that $ACTH_{1-24}$ spans the lipid bilayer. The question arises as to the mechanism of interaction of $ACTH_{1-24}$ with solvent-free membranes. Therefore, in addition to measurements of intrinsic potential we have also studied the dielectric relaxation and membrane compressibility perpendicular to the plane of the membrane

302

T. Hianik et al.
Interaction of ACTH with lipid bilayers

following adsorption of $ACTH_{1-24}$ to solvent-free BLM, as well as changes in ultrasound velocity number $[u]$ and absorption number $\Delta[\alpha\lambda]$ of liposomes in the presence of $ACTH_{1-24}$ using ultrasonic velocimetry.

Materials and methods

$ACTH_{1-24}$ (95% purity) was a gift from Ciba-Geigy (Basel) and was used without further purification. 1-Palmitoyl-2-oleoyl-sn-glycero-3-phosphocholine (POPC) was synthesized according to [4] and was a gift from Prof. A. Hermetter (Graz, Austria). A mixture of POPC + cholesterol (Merck) (4:1 w/w) was used for formation of BLM. BLM were prepared by the Montal-Mueller technique [5] in which two surface films of lipid plus n-hexane (Lachema) and n-hexadecane (Merck) (10 mg/ml, volume content of n-hexadecane was 20%) are apposed across a hole (~0.3 mm in diameter) in a thin copolymer septum (Slovnaft). After the films were spread, the hexane was allowed to evaporate for approximately 10–20 min before BLM were formed. The aperture of the copolymer septum on which the bilayers were formed was pretreated with the corresponding lipid solution. Large unilamellar liposomes (LUVET) with a diameter ≈ 100 nm were prepared from POPC, dimyristoylphosphatidylcholine (DMPC) (Sigma) and/or from a mixture of $ACTH_{1-24}$ + DMPC ($ACTH_{1-24}$/DMPC = $= 7.7 \times 10^{-4}$ mol/mol) by means of the extrusion method [6] using the LiposoFast-Basic (Avestin, Inc.). The concentration of lipids was 2 mg/ml. The aqueous phase for experiments with BLM and with LUVET prepared from POPC was a 9 mM KCl (Merck) solution buffered by 2 mM Hepes (Sigma) (pH 7.4). LUVET from DMPC were prepared in double distilled water. All chemicals were purity grade and used without further purification.

The interaction of $ACTH_{1-24}$ with the membranes was monitored using ac voltage electrostriction [7, 8], capacitance relaxation techniques [9] and ultrasonic velocimetry [10, 11] described in detail elsewhere.

Results

The time-course of the adsorption of $ACTH_{1-24}$ to a POPC + cholesterol membrane is shown in Fig. 1a (curve 1). At time A, upon addition of $ACTH_{1-24}$ (final concentration 0.11 mM) to one side ("cis") of the bilayer, a rapid initial rise of $\Delta\Phi_m$ is followed by a slower decline of the signal to a steady state value of 4.5 mV. The polarity of the signal corresponds to a binding of positive charges to the cis-side of the membrane ($ACTH_{1-24}$ carries six positive charges under normal physiological conditions). At

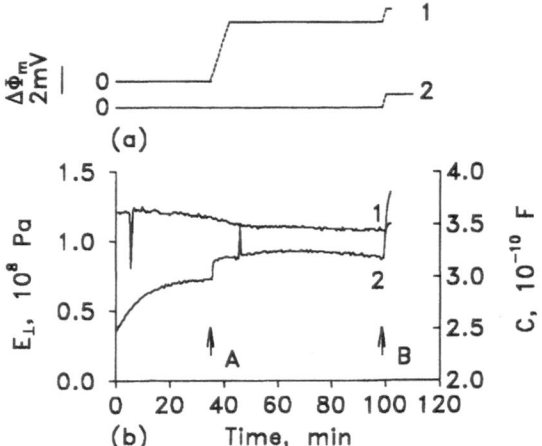

Fig. 1 Changes in intrinsic potential, $\Delta\Phi_m$, of solvent-free bilayers of POPC + cholesterol on addition of $ACTH_{1-24}$ to a final concentration of 0.11 mM (curve 1, time A), and a subsequent increase in the ionic strength on the opposite side from 10–110 mM by addition of KCl (time B). Curve 2 shows the change seen after increasing the ionic strength (time B) unmodified (pure POPC + cholesterol), and represents the baseline for curve 1. Changes in elasticity modulus $E_\perp$ (curve 1) and membrane capacitance (curve 2) for the same conditions as presented in Fig. 1a ($T = 20\,°C$, amplitude of ac voltage $E_o = 40$ mV, frequency $f = 1$ kHz)

time B the salt concentration on the opposite side ("trans") was increased from 10–110 mM KCl. This is done to test for the appearance of charges on the trans-side of the membrane (shielding effects), as would be expected from the incorporation reaction. An increase in $\Delta\Phi_m$ of 1 mV is observed. In curve 2 of Fig. 1a the effect on the pure POPC + cholesterol bilayers of increasing the ionic strength on the trans-side from 10–110 mM KCl is demonstrated. After equilibration $\Delta\Phi_m$ also reaches a value of + 1 mV (this change may be caused either by slight impurities in the lipid or be due to the pure lipid itself, e.g. a small head-group reorientation). Comparison of curves 1 and 2 shows that, within experimental error, the increase in ionic strength on the trans-side of the membrane results in the same change of the equilibrium value of $\Delta\Phi_m$ whether $ACTH_{1-24}$ has been added to the cis-side or not. These results are similar to those published for solvent-free BLM of DOPC at the same $ACTH_{1-24}$ concentration and pH [3]. Figure 1b illustrates the time-course of the elasticity modulus $E_\perp$ (curve 1) and membrane capacitance C (curve 2) following the addition of $ACTH_{1-24}$ to the cis-side at time A and KCl to trans-side of BLM of POPC + cholesterol at time B. Measurements of $E_\perp$ and C were performed simultaneously with those of $\Delta\Phi_m$. It is seen from Fig. 1b that the adsorption of $ACTH_{1-24}$ decreases the value of the modulus of membrane elasticity by about 10% (curve 1), and increase of membrane capacitance (curve 2) by about $\sim 8\%$. The increase of salt concentration on the opposite side ("trans") from 10–110 mM KCl caused only a small increase in $E_\perp$, while capacitance grows by a further 25%.

Capacitance relaxation results on pure, solvent-free POPC BLM before and after addition of 0.11 mM $ACTH_{1-24}$ to one side of the BLM are shown in Table 1. In this experiment current relaxation curves were averaged and the standard deviation taken as the experimental uncertainty. Only two relaxation components were detected: for pure POPC the slower component had a time constant of 2.33 ms and the faster component a value of 111 μs. These values are very similar to those obtained by Sargent [12] for phosphatidylcholines. Following addition of ACTH the faster component did not change, but the slower one decreased by about a factor of two to 1.1 ms.

Two parameters were measured by ultrasonic velocimetry: velocity number, $[u]$, and absorption number, $\Delta[\alpha\lambda]$ defined as: $[u] = (u - u_o)/u_o c$, $\Delta[\alpha\lambda] = (\alpha\lambda - \alpha_o\lambda_o)/c$, where c is the concentration of the solution expressed in g/ml and index "o" refers to buffer. Figure 2 shows the dependence of the changes velocity number $\delta[u] = [u] - [u]_o = \delta(\Delta u/u_o c)$ ($[u]_o$ is the velocity number in absence of $ACTH_{1-24}$, $[u]$ is

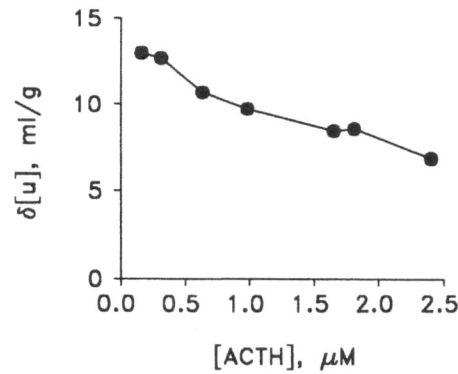

Fig. 2 The dependence of velocity number changes $\delta[u]$ on concentration of $ACTH_{1-24}$ in a suspension of POPC liposomes ($T = 25\,°C$)

the same in the presence of $ACTH_{1-24}$, Δu is the change of ultrasound velocity, u_o is ultrasound velocity in pure solvent and c is the concentration of lipid) on concentration of $ACTH_{1-24}$ in the liposome solution of POPC. We can see that $[u]$ decreases practically linearly with increasing peptide concentration.

The changes of velocity number $[u]$ (a) and absorption number $\Delta[\alpha\lambda]$ (b) of liposome suspensions during the temperature phase transition of pure DMPC (curves 1) and in the presence of small concentration of $ACTH_{1-24}$ ($ACTH_{1-24}/DMPC = 7.7 \times 10^{-4}$ mol/mol) (curves 2) are presented in Fig. 3. It is seen that around phase transition temperature $[u]$ decreases and $\Delta[\alpha\lambda]$ reaches a maximum for both pure and $ACTH_{1-24}$ modified lipsomes. This reflects the structural changes of liposomes during the

Table 1 Capacitance relaxation components of solvent-free POPC BLM following adsorption of $ACTH_{1-24}$ (final concentration in solution 0.11 mM)

Composition	τ (ms)	τ (μs)
POPC	2.33 ± 0.63	111 ± 25
POPC + $ACTH_{1-24}$	1.11 ± 0.55	113 ± 53

Fig. 3 The dependence of velocity number $[u]$ (a) and absorption number $\Delta[\alpha\lambda]$ (b) of ultrasound on temperature for a suspension of liposomes from pure DMPC (curves 1) and that modified by $ACTH_{1-24}$ (curve 2) (molar ratio $ACTH_{1-24}/DMPC = 7.7 \times 10^{-4}$ mol/mol). The values of mean $\pm$ S.D. calculated from three independent experiments are shown in the gel state ($T \approx 17\,°C$) and at the phase transition point ($T \approx 24\,°C$). The values of S.D. in the liquid crystalline state ($T \approx 38\,°C$) do not surpass the size of symbols. The differences between $[u]$ and/or $\Delta[\alpha\lambda]$ for pure and $ACTH_{1-24}$ modified DMPC liposomes were statistically significant according to the Student's t-test with $p < 0.05$ for $T \approx 24\,°C$ and $p < 0.01$ for $T \approx 38\,°C$. For the gel state these differences were statistically insignificant

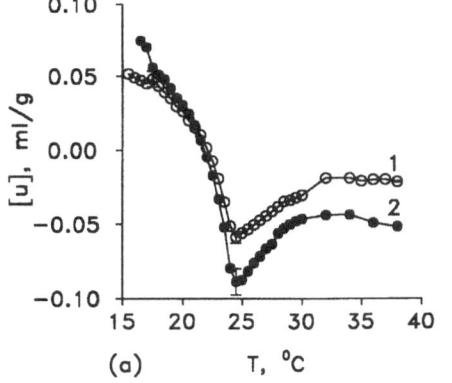

(a)

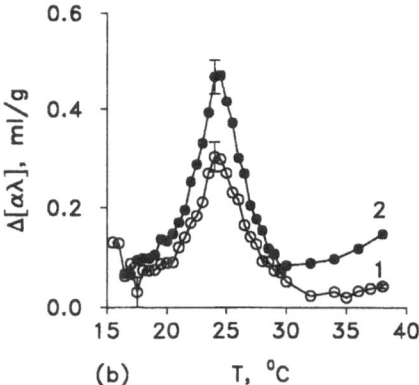

(b)

304
T. Hianik et al.
Interaction of ACTH with lipid bilayers

phase transition of lipids and corresponds to a increase of volume compressibility [13]. The presence of $ACTH_{1-24}$ in liposomes leads to decrease of velocity number by 33% and to an increase of absorption number by 32%.

Discussion

After addition of $ACTH_{1-24}$ the rise of $\Delta\Phi_m$ of BLM reflects a rapid adsorption of the peptide hormone. Our measurements of the change of intrinsic potential upon addition of $ACTH_{1-24}$ thus confirm the results obtained by Gremlich et al. [3]. Because the increase of the ionic strength on the trans-side causes a similar change in $\Delta\Phi_m$ to that seen with unmodified BLM, there is no evidence for a transfer of charges to the trans-side. If the ratio of incorporated to bound $ACTH_{1-24}$ is the same as that determined by Schoch et al. [2] for solvent-containing BLM, however, a quantitative analysis shows that the expected change in $\Delta\Phi_m$ upon increasing the ionic strength on the trans-side is only on the order of 1 mV. As this lies at the limits of the experimental accuracy, the negative result does not exclude the possibility of a translocation to the trans-side of solvent-free BLM.

The increase of membrane capacitance following addition of $ACTH_{1-24}$ is presumably due to a decrease of surface tension, γ, of the cis-monolayer. Differences in surface tension will lead to a buildup of a hydrostatic pressure gradient $\Delta p \approx 2\gamma/R$, ($R$ is the radius of the bulged membrane; see, e.g. [14]), and the resulting bulging will lead to an increase of membrane area and consequently of capacitance.

The value of elasticity modulus $E_\perp$ reflects the degree of compressibility of the hydrophobic region of the membrane [7]. The considerable decreases of $E_\perp$ following adsorption of $ACTH_{1-24}$ indicates a disordering of the membrane, predominantly in the hydrophobic core. This strongly suggests that part of the $ACTH_{1-24}$ molecule (presumably the N-terminus) inserts to some depth into the lipid bilayer.

The capacitance relaxation experiments support this model. As seen in Table 1, absorption of $ACTH_{1-24}$ leads to a decrease of one of the measured relaxation components, suggesting increased freedom of motion of molecular dipoles in the BLM, as could result from a disturbance of the lipid structure by penetration of $ACTH_{1-24}$ molecules. (That the time scale of this relaxation is in the millisecond range suggests that the observed motion stems from clusters of lipid molecules).

Let us discuss the reasons for the changes of ultrasound parameters of liposomes. The velocity number is connected with the apparent compressibility of a liposome suspension $\Phi_k = (\beta V - \beta_o V_o)/VC$, where β, β_o are the adiabatic compressibilities, V and V_o are sample volumes and C is the molar concentration of solution. For very dilute solutions

$$\varphi_k/\beta_o = \Phi_k/\beta_o M \approx -2[u] - 1/\rho_o + 2\varphi_V, \qquad (1)$$

where M is molecular mass of the dispersed particles, ρ_o is the density of the solvent, φ_k, φ_V are molar apparent compressibility and volume of solution, respectively. Several processes can contribute to the value of $[u]$, and consequently to φ_k. This can be changes of volume compressibility and volume of protein and/or liposomes as well as changes of the hydration of membranes and proteins [15, 16]. The changes of $\delta[u]$ induced by $ACTH_{1-24}$ in the concentration range 0–2.5 μM obtained from three independent experiments reaches value $\delta[u] = (3.7 \pm 1.4)$ ml/g. As reported elsewhere [16], we have shown that changes of $\delta[u]$ are connected mainly with changes of adiabatic compressibility of lipid bilayer $\delta\beta = -\beta_o\delta[u]/V_o = 1.42 \times 10^{-11}$ Pa^{-1}, where $V_o = 700$ ml/mol is the molar volume of POPC in the membrane and $\beta_o = 4.45 \times 10^{-10}$ Pa^{-1} is the adiabatic compressibility of pure solvent. The value of $\delta\beta$ represents an increase of adiabatic compressibility of the membrane by $\approx 2.4\%$. This result is in agreement with the increase of compressibility of BLM ($1/E_\perp$) perpendicular to the membrane plane. For the temperature dependence of $[u]$ and $\Delta[\alpha\lambda]$, we find an additional decrease of $[u]$ and increase of $\Delta[\alpha\lambda]$ of DMPC liposomes modified by $ACTH_{1-24}$. The decrease of $[u]$ at T = 15–25 °C indicates that $ACTH_{1-24}$ induces an increase of volume compressibility of liposomes. The additional increase of absorption coefficient at the phase transition temperature of DMPC shows that $ACTH_{1-24}$ induces additional perturbations of the colloid suspension. Given that proteins can induce a frequency shift of the maximum of the absorption number [17], then the changes of $\Delta[\alpha\lambda]$ induced by peptide and measured at the maximum of absorption number could be even higher. Such a comparison, however, requires the application of acoustics spectroscopy and is unavailable for single frequency method.

From the temperature dependence of $\Delta[\alpha\lambda]$ it is seen that, while in the gel state ($T < 24$ °C) there are practically no differences between pure and modified liposomes, in the liquid crystalline state ($T > 24$ °C) a considerable difference is found between absorption numbers of pure and $ACTH_{1-24}$ modified liposomes. According to molecular mechanisms of relaxation processes [18] the results imply that $ACTH_{1-24}$ influences the relaxation part of volume compressibility. In order to evaluate the changes of relaxation time of the system studied, one can use the equation for the effective relaxation time at single frequency ω: $\tau_{eff} = -[\alpha\lambda]_r/2\pi\omega[u]_r$ [19]. At 38 °C the changes of $\Delta\tau_{eff} = (\tau_{eff})_{DMPC+ACTH} - (\tau_{eff})_{DMPC}$ due to presence of

Progr Colloid Polym Sci (1996) 100:301–305
© Steinkopff Verlag 1996

$ACTH_{1-24}$ reach the value $\Delta\tau_{eff} = 3$ ns. This means that $ACTH_{1-24}$ induces an increase of effective value of relaxation time. We should note that an increase of relaxation time measured by molecular spectroscopic method in wide range of frequencies, was already observed for another short peptide – gramicidin A [17]. τ_{eff} represents, however, average characteristics of the whole spectrum of relaxation processes and can differ from the apparent relaxation time determined by acoustic spectroscopy [19]. On the other hand, the values of τ_{eff} are of the order of 10^{-8} s and are within the spectrum of relaxation times obtained by molecular spectroscopy method [20]. τ_{eff} of pure DMPC and $ACTH_{1-24}$ modified liposomes of DMPC increases as the temperature approaches the phase transition point. This agrees with results obtained for pure DPPC liposomes [19], and could be due to a temperature dependence of the size distribution of lipid clusters [21]. Changes of

τ_{eff} induced by $ACTH_{1-24}$ can thus be explained by an influence of peptide on the changes of the size of lipid clusters during phase transition of lipids.

Thus the measurements of $E_\perp$, relaxation time constant and ultrasound properties of liposome suspensions clearly demonstrate a considerable effect of $ACTH_{1-24}$ on the ordering of the lipid environment. These findings are indicative of an interaction of $ACTH_{1-24}$ with the hydrophobic core of the lipid bilayer.

Acknowledgment This work has been funded by Swiss National Foundation grant 7SLPJ041450, by the Commission of the European Communities in framework of the Copernicus program (Project No. CIPA-CT94-0213) and by Slovak Grant Agency grant. The authors would like to thank Dr. J. Dlugopolsky for expert technical assistance. We are grateful to Ciba-Geigy (Basel) for the generous gift of $ACTH_{1-24}$ and to Prof. Hermetter for the gift of POPC.

References

1. Schwyzer R (1977) Ann NY Acad Sci 297:3–26
2. Schoch P, Sargent DF, Schwyzer R (1979) Biochem Soc Trans 7:846–849
3. Gremlich HU, Sargent DF, Schwyzer R (1981) Biophys Str Mech 8:61–65
4. Hermetter A, Stütz H, Franzmair R, Paltauf F (1989) Chem Phys Lipids 50:57–62
5. Montal M, Mueller P (1972) Proc Nat Acad Sci USA 69:3561–3566
6. MacDonald RC, MacDonald RI, Menco BPM, Takeshita K, Subbarao NK, Hu L (1991) Biochim Biophys Acta 1061:297–303
7. Hianik T, Passechnik VI (1995) Bilayer Lipid Membranes. Structure and Mechanical Properties. Kluwer Academic Publishers, The Netherlands
8. Carius W (1976) J Coll Interface Sci 57:301–307
9. Sargent DF (1975) J Membrane Biol 23:227–247
10. Sarvazyan AP (1982) Ultrasonics 20:151–154
11. Hianik T, Piknova B, Buckin VA, Shestimirov VN, Shnyrov VL (1993) Progr Coll Polym Sci 93:150–152
12. Sargent DF (1975) In: Kaback HR, Neurath H, Radda GK, Schwyzer R, Wiley WR (eds) Molecular Aspects of Membrane Phenomena. Springer-Verlag, Berline, Heidelberg, New York, pp 104–120
13. Mitaku S, Ikegami A, Sakanishi A (1978) Biophys Chem 8:295–304
14. Kruglyakov PM, Rovin YG (1978) Physical Chemistry of Black Hydrocarbon Films. Nauka, Moscow (in Russian)
15. Sarvazyan AP (1991) Annu Rev Biophys Biophys Chem 20:321–342
16. Hianik T, Buckin VA, Piknova B (1994) Gen Physiol Biophys 13:493–501
17. Strom-Jensen PR, Magin R, Dunn F (1984) Biochim Biophys Acta 769:179–186
18. Stuehr J, Yeager E (1965) In: Mason WP (ed) Physical Acoustics. Academic Press, New York, London, Vol. 2, pp 351–462
19. Kharakoz DP, Colotto A, Lohner K, Laggner P (1993) J Phys Chem 97:9844–9851
20. Eggers F, Funck Th (1976) Naturwissenschaften 63:280–285
21. Mitaku S, Jippo T, Kataoka R (1983) Biophys J 42:137–144

Progr Colloid Polym Sci (1996) 100:306–310
© Steinkopff Verlag 1996

BIOCOLLOIDS

M. De Cuyper

Impact of the surface charge of magnetoproteoliposomes on the enzymatic oxidation of cytochrome c

Prof. De Cuyper (✉)
Interdisciplinary Research Centre
Katholieke Universiteit Leuven
Campus Kortrijk
8500 Kortrijk, Belgium

Abstract Upon immersing nanometer-sized Fe_3O_4 colloids in aqueous dispersions of phospholipid vesicles, a lipid bilayer is generated on the particle surface. The resulting "magnetoliposomes" can act as excellent hosts for membrane-bound enzymes, such as cytochrome c oxidase [De Cuyper and Joniau, Biotechnol. Appl. Biochem. 16, 201–210 (1992)].

In an attempt to tailor the catalytic properties of the immobilized enzyme, in the present study, we have explored the pivotal role played by the surface charge density of the magnetoliposome coat. In this respect, we have screened a series of bilayered phospholipid coatings consisting of anionic dimyristoylphosphatidylglycerol (DMPG), zwitterionic dimyristoylphosphatidylcholine (DMPC) or variable mixtures of the two. A cationic lipid coating, made of a heterogeneous mixture of DMPC and dioctadecyldimethyl-ammoniumbromide, was also tested. The profiles, representing the enzymatic activity which was measured spectrophotometrically at 550 nm and, if need be, corrected for scattered light due to clustering phenomena, showed that the highest degree of catalytic activity of lipid-embedded enzyme was found when moderately charged, anionic magnetoliposomes (5 to 10% DMPG) were used. The results are interpreted in terms of a different affinity of the substrate for the various membrane types.

Key words Cytochrome c – cytochrome c oxidase – dioctadecyldimethyl-ammoniumbromide (DODAB) – enzymes (membrane-bound) – magnetophoresis (high-gradient) – magneto(proteo)liposomes – membranes – phospholipids

Introduction

Cytochrome c is a water-soluble, highly cationic protein with an isoelectric point greater than 10 [1–3]. In vivo, it is found only in the mitochondrial intermembrane space, where it acts as an electron transporter between cytochrome b and cytochrome aa_3. The latter protein, commonly called "cytochrome c oxidase", is a multi-subunit protein that is vectorially embedded in the inner membrane of the mitochondrion. To determine those parameters crucial in the cytochrome c – cytochrome c oxidase interaction at the membrane interface, a wide variety of membrane models has been used and has made it possible to partially assess the relative importance of the type of phospholipid polar headgroup and the degree of membrane fluidity [4].

As a membrane model, we used in the present work so-called magnetoliposomes, which each consist of a nanometer-sized, magnetizable Fe_3O_4 core, surrounded by a phospholipid bilayer [5, 6]. The reason for using these structures, rather than classical vesicles or liposomes, is that their ability to be magnetized may have great potential in certain biotechnological areas. For instance, in magnetically-controlled bioreactors, magnetoproteoliposomes can be successfully used as immobilized biocatalysts [7–9].

As a continuation of our research to tailor the catalytic properties of cytochrome c oxidase-containing magnetoproteoliposomes, the aim of the present work was to obtain a comparative insight into the influence of membrane surface charge on the enzymatic oxidation of cytochrome c. Both anionic, zwitterionic and cationic lipids were used as membrane components.

Materials and methods

Materials

DMPC and DMPG were Avanti Polar Lipid products (Birmingham, Alabama, USA). 2-(Tris(hydroxymethyl) methylamino)-1-ethanesulfonic acid (TES), dioctadecyldimethylammoniumbromide (DODAB) and cytochrome c (Type III) were supplied by Sigma (Deisenhofen, Germany). All other chemicals were of the highest purity available. Cytochrome c oxidase was isolated from beef heart mitochondria and depleted of most of its annulus phospholipids, as previously reported [4]. Throughout this work, a 5 mM TES-buffer, pH 7.0, was used.

Preparation of magneto(proteo)liposomes

Magnetoliposomes were prepared essentially according to the procedure described in refs. [6, 8]. Briefly, the starting material was an aqueous-based magnetic fluid consisting of 15 nm sized, superparamagnetic Fe_3O_4 colloids, peptized with lauric acid. Upon incubating and dialyzing this magnetic fluid (0.37 mL containing 20 mg Fe_3O_4) in the presence of a five-fold weight excess of sonicated DMPC vesicles or mixed DMPC-DMPG vesicles, magnetoliposomes were formed, Separation of these biocolloids from non-adsorbed phospholipids was achieved by high-gradient magnetophoresis (Bruker electromagnet, Type BE-15). Since DODAB-containing vesicles induce rapid flocculation of the anionic Fe_3O_4-lauric acid colloids, a different protocol was used to generate cationic magnetolipo-

somes: DMPC magnetoliposomes (4 mL; phospholipid concentration: 1.43 mg/mL) were incubated for 2 days with an equal amount (with respect to lipid content) of DMPC-DODAB (8:2 w/w) vesicles. During this incubation period, equilibration of DODAB between vesicles and magnetoliposomes occurred via spontaneous transfer [10, 11]. Following the final high-gradient magnetophoresis step, the recovered magnetoliposomes had a coating consisting of DMPC and DODAB in the weight ratio of 9:1.

Incorporation of cytochrome c oxidase into the different coatings was achieved using a brief sonication step at a lipid-to-protein weight ratio of 10 [7, 8]. The resulting magnetoproteoliposomes suspensions were stored in ice.

Enzyme kinetics

The kinetic properties of the different types of magnetoliposomes were followed spectrophotometrically (see Results) for the first 15 min after their preparation. A 42 μM solution of cytochrome c, reduced for 75% by $Na_2S_2O_4$, was used as substrate. First order rate constants, k_1, were calculated from the initial data points of the kinetic plots [4].

Other assays

Phosphate and iron analyses were carried out as per the protocols previously described [8].

Results

Spectra of cytochrome c

The visible absorption spectra for cytochrome c (42 μM), showing various ratios of oxidized Fe^{3+}/reduced Fe^{2+}, are shown in Fig. 1. The profiles clearly demonstrate that oxidation of cytochrome c by cytochrome c oxidase can be easily monitored spectrophotometrically at 550 nm.

Activity of magnetoproteoliposomes with different surface charges

The enzyme was inserted in three different types of magnetoliposomes: i) positively-charged structures, constructed using intramembraneously mixed DMPC– DODAB

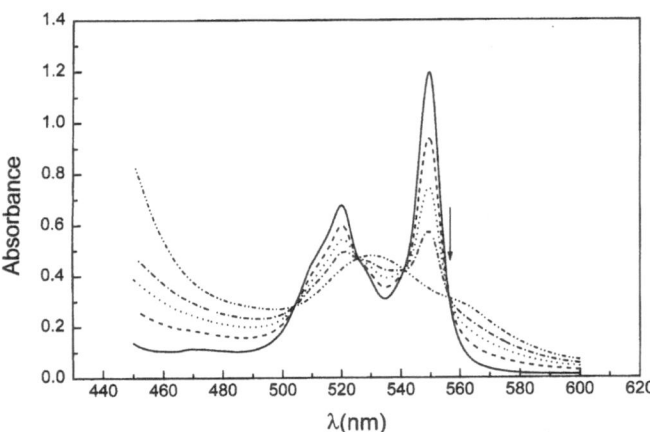

Fig. 1 Absorption spectra of cytochrome c (42 μM, 5 mM TES buffer, pH 7.0) in the fully oxidized (–··–··–) and fully reduced (———) states. Spectra at intermediate haem Fe^{2+}/Fe^{3+} contents are also shown (– – – – : 70% reduced; ····· : 47% reduced; and –·–·– : 27% reduced). Oxidation and reduction was performed using $K_3Fe(CN)_6$ and $Na_2S_2O_4$, respectively. The scan rate was 60 nm/min. The arrow indicates the isosbestic point at 556.3 nm

(9:1 weight ratio), ii) zero-charged magnetoliposomes, made with DMPC; and iii) negatively-charged structures, made using a mixture of DMPC-DMPG (9:1 w/w). (The chemical structures of the different membrane constituents are given in Fig. 2.) In all three cases, the amount of lipid adsorbed was in the range of 0.8–0.9 mmol/g Fe_3O_4, indicating that the particles were enwrapped with a lipid bilayer [8].

Upon incubation of the first type of magnetoproteoliposome (40 μl) with 2 mL of the stock solution of cytochrome c, no decrease in $A^{550\,nm}$ was observed. However, when the zwitterionic biocatalyst was used, the kinetic profile slowly decreased, following (pseudo) first order reaction rules. The halftime, calculated as $t_{1/2} = 0.69/k_1$, was 55.2 min. In contrast, in reaction mixtures containing anionic structures, the turbidity gradually increased due to aggregation, triggered by the cationic cytochrome c. To estimate the impact of scattered light on the overall absorption changes at the various time points, measurements were also performed at 556.3 nm, which is an isosbestic point in the absorption spectrum (i.e., the degree of absorption is independent of the oxidation state of cytochrome c) (Fig. 1). Experiments in which enzyme-free DMPC-DMPG (9:1) magnetoliposomes were incubated with cytochrome c under exactly the same conditions showed that the time-dependent $A^{550\,nm}$ changes parallel those seen at 556.3 nm) Consequently, absorption changes solely due to enzyme action could be estimated by subtracting the $A^{556.3\,nm}$ values from the measured $A^{550\,nm}$ values (Fig. 3A). Using these corrected values, linear first-order reaction plots were obtained ($r^2 = 0.999_3$) (Fig. 3B); in this case the calculated $t_{1/2}$ was 27.6 min.

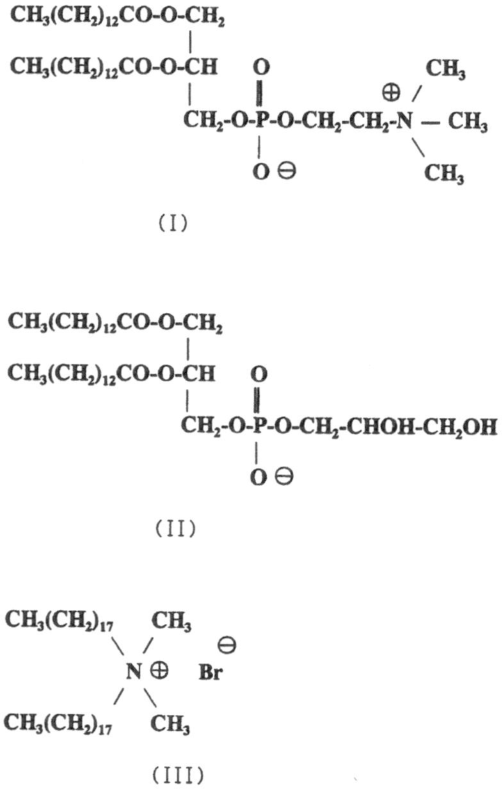

Fig. 2 Chemical structures of the lipid molecules used to construct the magnetoliposome envelope: (I) DMPC, (II) DMPG and (III) DODAB. Ionization is represented as it is thought to occur under the actual experimental conditions (pH 7.0)

Activity of mixed DMPC-DMPG magnetoproteoliposomes

We further investigated the catalytic performance of mixed DMPC-DMPG magnetoproteoliposomes at DMPG contents of 0, 5, 10, 50 and 100%. Within the timeframe of the kinetic experiments (3 min), the increase in ΔA/min due to turbidity was found to be proportiaonal to the DMPG content of the lipidic envelope. After correction for this disturbing phenomenon (see above), the calculated data showed an optimum activity for those magnetoliposomes whose coats fell in the lower charge range (Fig. 4).

Discussion

The rate of enzyme-catalyzed reactions depends, to a large extent, on the rate of diffusion of the substrate towards the catalytic site. In the case of membrane-embedded biocatalysts, such as cytochrome c oxidase, there is no doubt

Progr Colloid Polym Sci (1996) 100:306–310
© Steinkopff Verlag 1996

Fig. 3 Kinetics for cytochrome
c oxidase, immobilized in the
DMPC-DMPG (9:1) mag-
netoliposome coat. Part
A shows the kinetic curves,
measured at 550 nm (□) and at
556.3 nm (Δ). The ● data points
represent the $A^{550\ nm}-A^{556.3\ nm}$
values. The symbols on the far
right of the x axis indicate the
optical densities after complete
oxidation of the mixture with
$K_3Fe(CN)_6$. Part **B** shows the
first order reaction plots before
(□), and after (●) correction for
light scattering (see text for
details)

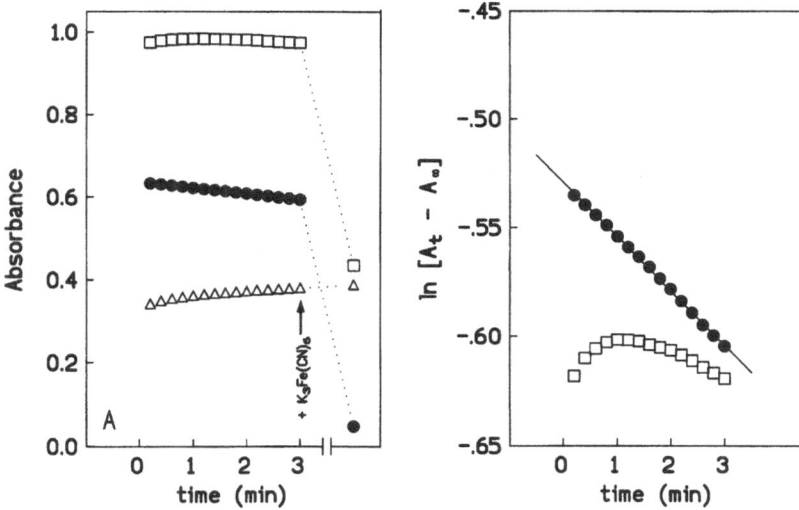

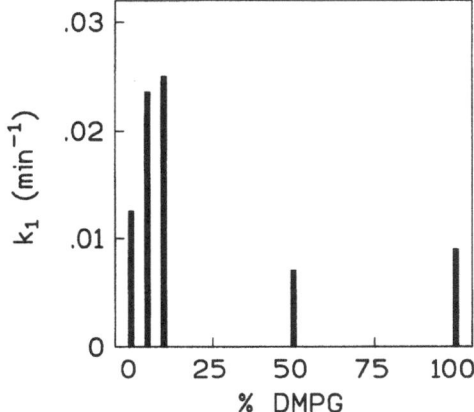

Fig. 4 First order rate constants calculated for cytochrome
c oxidase-magnetoliposome complexes varying in DMPC/DMPG
ratio. The kinetic runs were started within 15 min after preparing the
magnetoproteoliposomes

that the surrounding lipid molecules play a key role in
the overall transport process. In the present work, strong
proof for this is provided by the complete absence of
any activity if the enzyme is immobilized in cationic
DMPC-DODAB membranes. Electrostatic lipid-cyto-
chrome c repulsions, indeed, will prevent a close contact
between the cationic substrate and the oxidase-membrane
interface.

A clear-cut interpretation of the results obtained with
zwitterionic and anionic magnetoproteoliposomes (Fig. 4)
seems to be more complicated. Although the interaction of
cytochrome c with (partly) anionic or zero-charged membranes is well documented in the literature, a consensus on
the nature of this interaction has not yet been reached

[12–18] [1]. In addition, besides the fact that there have been
many studies describing static binding [12–18], data on
the kinetics of association, which are essential for the
understanding of dynamic enzyme processes, are scarce.
Our observations may, however, be explained by the
elegant kinetic binding studies of Kakinoki et al. [18],
who, by means of stopped-flow techniques, could clearly
divide the overall association phenomenon into two relaxation processes: an initial binding (occurring during
a simple collision), which is subsequently followed by
a partial penetration of cytochrome c into the membrane
bilayer. Since the initial binding rate constant was claimed
to be proportional to the anionic lipid content, the relatively high values for k_1, observed in the low DMPG range
(5 and 10%) magnetoproteoliposomes (Fig. 4), can be
readily understood in terms of a local increase in cytochrome c concentration at the membrane-enzyme interface.

In terms of magnetoproteoliposomes formed in the
higher PG zone (50 and 100%), it has been reported that
membrane saturation with cytochrome c is reached at
30 mol % of acidic lipid [20–21]. De Jongh et al. [22]
further suggested that, within this high PG range, an
anionic lipid-induced loosening and/or destabilization of
the native cytochrome c structure – especially its tertiary
structure (23, 24) – occurs. In this respect, it is interesting to

---

[1] In a recent paper, Rytömaa and Kinnunen [19] claimed that both
a suitable geometry of the phosphate group and the presence of two
acyl chains are required for tight binding of cytochrome c to acidic
phospholipids. During the binding process, the latter are assumed
to adopt an extended conformation so that one of the acyl chains is
able to become accommodated within the hydrophobic cavity of the
protein.

note that a change in coordination of the haem iron is also accompanied by opening of the porphyrine crevice as detected by resonance Raman spectroscopy [25–27]. Possibly, these putative conformational changes in cytochrome c may slow its oxidation rate.

The elucidation of the physiological significance of the present observations warrants further investigation. At physiological pH, about 25% of the inner mitochondrial membrane phospholipids bear a negative charge [28, 29].

On the basis of the results presented in Fig. 4, it is tempting to speculate that this charge content may play a critical regulatory role in the association of cytochrome c with the inner mitochondrial membrane "machinery" and thus in the control of the overall oxidative phosphorylation pathway.

Acknowledgments We would like to thank Mrs Linda Desender for assistance with the experimental work. The project was supported by a grant from the Belgian N.F.W.O.

References

1. Theorell H, Akesson A (1941) J Amer Chem Soc 63:1804–1827
2. Erecińska M, Vanderkooi JM (1975) Arch Biochem Biophys 166:495–500
3. Kim C, Keuppers F, Dimaria P, Faroqui P, Kim S, Paik WK (1980) Biochim Biophys Acta 622:144–150
4. De Cuyper M, Joniau M (1980) Eur J Biochem 104:397–405
5. De Cuyper M, Joniau M (1988) Eur Biophys J 15:311–319
6. De Cuyper M, Joniau M (1991) Langmuir 7:647–652
7. De Cuyper M (1996) In: Barenholz Y, Lasic DD, (eds) Handbook of Nonmedical Applications of Liposomes, Vol III: From Design to Microreactors, Chapter 18. CRC Press, Inc, Boca Raton, FL, USA, pp 323–340
8. De Cuyper M, Joniau M (1992) Biotechnol Appl Biochem 16:201–210
9. De Cuyper M, Joniau M (1995) In: (Cevc G, Paltauf F, Eds) Phospholipids: Characterization, Metabolism and Novel Applications, Chapter X, pp 101–110, AOCS Press, Champaign, IL, USA
10. De Cuyper M, Joniau M, Dangreau H (1983) Biochemistry 22:415–420
11. De Cuyper M, Joniau M, Engberts JBFN, Sudhölter EJR (1984) Colloids Surfaces 10:313–319
12. Papahadjopoulos D, Moscarello M, Eylar EH, Isac T (1975) Biochim Biophys Acta 401:317–335
13. Rietveld A, Sijens P, Verkleij AJ, de Kruijff B (1983) The EMBO Journal 2:907–913
14. Waltham MC, Cornell BA, Smith R (1986) Biochim Biophys Acta 862:451–456
15. Lee S, Kim H (1989) Arch Biochem Biophys 271:188–199
16. Snel MME, de Kruijff B, Marsh D (1994) Biochemistry 33:7146–7156
17. Choi S, Swanson JM (1995) Biophys Chem 54:271–278
18. Kakinoki K, Maeda Y, Hasegawa K, Kitano H (1995) J Colloid Interf Sci 170:18–24
19. Rytömaa M, Kinnunen PKJ (1995) J Biol Chem 270:3197–3202
20. Mustonen P, Virtanen JA, Somerharju PJ, Kinnunen PKJ (1987) Biochemistry 26:2991–2997
21. Rytömaa M, Kinnunen PKJ (1994) J Biol Chem 269:1770–1774
22. De Jongh HHJ, Ritsema T, Killian JA (1995) FEBS Lett 360:225–260
23. Muga A, Mantsch HH, Surewicz WK (1991) Biochemistry 30:7219–7224
24. Heimburg T, Marsh D (1993) Biophys J 65:2408–2417
25. Hildebrandt P, Stockenburger M (1989) Biochemistry 28:6710–6721
26. Hildebrandt P, Stockenburger M (1989) Biochemistry 28:6722–6728
27. Heimburg T, Hildebrandt P, Marsh D (1991) Biochemistry 30:9084–9089
28. Daum G (1985) Biochim Biophys Acta 822:1–42
29. Hovius R, Thijssen J, vd Linden P, Nicolay K, de Kruijff B (1993) FEBS Lett 330:71–76

Progr Colloid Polym Sci (1996) 100:311–315
© Steinkopff Verlag 1996

The direct measurement of the interfacial composition of surfactant/polymer mixed layers at the air-water interface using neutron reflection

J.R. Lu
J.A.K. Blondel
D.J. Cooke
R.K. Thomas
J. Penfold

Dr. J.R. Lu (✉) · J.A.K. Blondel
D.J. Cooke · R.K. Thomas
Physical Chemistry Laboratory
Oxford University
South Parks Road
Oxford OX1 3QZ, United Kingdom

J. Penfold
DRAL, ISIS
Chilton, Didcot OX11 0QZ
United Kingdom

Abstract The coverage and structure of caesium dodecyl sulphate (CsDS) and polyethylene oxide (PEO) in mixed monolayers formed at the air–water interface have been determined directly using neutron reflection in combination with deuterium labelling. At a fixed PEO concentration of 0.1 wt%, the variation of surface tension with concentration of CsDS showed two discontinuities, the first (x_1) corresponding to the onset of formation of surfactant/polymer aggregates and the second (x_2) corresponding to the onset of formation of pure surfactant aggregates. Neutron reflection has shown that there is a sharp decrease in the surface excess of PEO (Γ_p) with surfactant concentration, Γ_p reaching zero at x_1. The surface excess of CsDS was found to increase smoothly through the whole concentration range tending to a limit of 4.3×10^{-10} mol cm^{-2}, the limiting value for pure CsDS solutions. The results suggest that there is no significant interaction between surfactant and polymer at the interface and that the surface composition is determined by simple competitive adsorption.

Key words Surfactant – mixed layers – neutron reflection

Introduction

There has been considerable interest in the behaviour of surfactant/polymer mixtures in aqueous solutions because of their industrial significance, the effort focussing on the interactions between the two species. The interactions between a charged surfactant and a neutral polymer have been studied by a number of authors [1, 2] and the system that has been taken as a model has mainly been sodium dodecyl sulphate (NaDS) and polyethylene oxide (PEO). The techniques used to study this system have included surface tension, conductivity, activity measurements through ion specific electrodes, NMR, and small-angle neutron scattering (SANS). These techniques give information about different aspects of the aggregation in the bulk solution, but none of these methods offers information about the structure and composition of such mixtures at a flat interface. In this study, we show that the interfacial composition of a given species in the mixed layer can be directly determined using neutron reflection in combination with deuterium labelling.

Experimental

Deuterated caesium dodecyl sulphate (d-CsDS) was prepared by reacting deuterated dodecanol with chlorosulfonic acid in ether at a temperature less than 5 °C followed by neutralisation with aqueous CsOH. The mixture was dried by rotary evaporation and dissolved in dry ethanol to remove inorganic salt. The product was then extracted with hexane to remove unreacted dodecanol and recrystalised from ethanol and water several times until no minimum occurred around the critical micellar concentration (CMC). Protonated CsDS (h-CsDS) was made by ion exchanging NaDS (Polysciences) and was purified follow-ing the same procedure as for d-CsDS.

Both deuterated and protonated polyethylene oxides (d-PEO and h-PEO) were purchased from Polymer Laboratories. The reactions were initiated by methoxide ion and terminated by glycol. Thus, one end has an OH group and the other a methoxy group. The molecular weights, M_n, were 24 200 for h-PEO and 24 700 for d-PEO respectively with M_w/M_n of 1.02 and 1.01.

The reflection measurements were made on the reflectometer CRISP at the Rutherford–Appleton Laboratory (Didcot, UK) as described previously [3]. The measurements were made at 35 °C using an incidence angle of 1.5°.

Surface tension measurements were made on a Kruss K10 maximum pull tensiometer using a Pt/Ir ring at 35 °C. The ring was flamed in between each measurement.

Results and discussion

There are several stages to the interaction of a surfactant with a polymer, which are most clearly explained with reference to the surface tension–surfactant concentration plot, which is shown for the present system at a polymer concentration of 0.1% by weight in Fig. 1 together with the tension profile for pure CsDS for comparison. The discontinuity in the latter at x_0 marks the critical micelle concentration (CMC) of 6×10^{-3} M. The first discontinuity break point (x_1) in the plot for the mixture indicates the start of formation of surfactant aggregates on the polymer. The second discontinuity (x_2) indicates the saturation of the polymer and further addition of surfactant leads to the formation of pure surfactant aggregates. The picture obtained here is in general agreement with those reported in the literature, in particular the NaDS/PEO system studied by Cabane [4].

Below the CMC the surface excess of the pure surfactant can be obtained from the tension plot shown in Fig. 1 using the Gibbs equation since the activity coefficient can be obtained approximately from the known monomer concentration including, if necessary, the activity coefficient from the Debye–Huckel equation. It is, however, difficult to do this for the mixture because it is not clear either that the activity of the polymer is constant in a given region, say below x_1, or that the activity of the surfactant differs only from its concentration by the Debye–Huckel factor. A direct measurement of the activity coefficients of the surfactant in this type of system has not yet been done. It is thus not possible at present to derive the surface excess of surfactant in the presence of PEO from the tension plot. However, none of these factors presents any problem in neutron reflection when deuterium labelling is used.

Since the scattering length for protonated PEO is close to zero, the surface excesses of the surfactant can be determined by measuring the reflectivities of mixed monolayers containing h-PEO and d-CsDS. In this situation almost all the signal will be from the labelled surfactant. Figure 2 shows reflectivity profiles (plots of reflectivity against momentum transfer κ) at a fixed concentration of 0.1 wt% PEO and varying concentrations of d-CsDS. The isotopic composition of the water is adjusted to give null reflecting water (NRW) which contains about 10 wt% D_2O and makes no contribution at all to the specular reflectivity. The continuous lines are calculated using a model of a uniform layer in conjunction with the exact optical matrix formula [5]. The parameters obtained from a least squares fit to the data are the scattering length density ρ and the layer thickness τ. From these two parameters the area per molecule A can be derived using the following

Fig. 1 Surface tension of CsDS in the presence of 0.1 wt% PEO. Also shown is the tension plot for pure CsDS for comparison

Fig. 2 Reflectivities versus momentum transfer κ for deuterated CsDS in NRW in the presence of 0.1 wt% protonated PEO. The concentrations of CsDS are 1.32×10^{-4} ($\triangle$), 6.44×10^{-4} ($+$), 1.27×10^{-3} ($\times$), 2.51×10^{-3} ($\square$), 4.0×10^{-3} ($\triangledown$) and 4.0×10^{-2} M ($\bigcirc$)

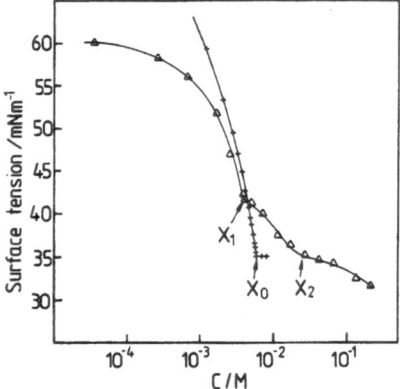

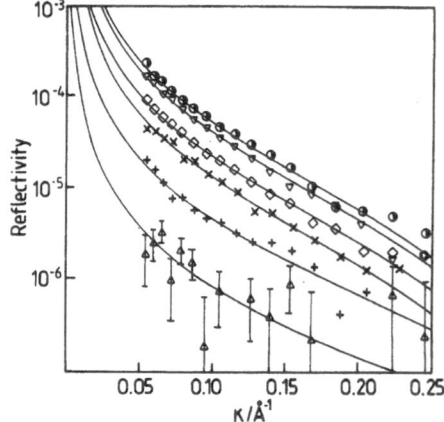

Progr Colloid Polym Sci (1996) 100:311–315
© Steinkopff Verlag 1996

equation

$$A = \frac{b_i}{\rho\tau} \qquad (1)$$

where b_i is the empirical scattering length of the molecule as a whole. The surface excess Γ can then be obtained from the following equation

$$\Gamma = \frac{1}{N_0 A} \qquad (2)$$

where N_0 is Avogadro's constant. All the calculated parameters are given in Table 1 and the scattering lengths for the surfactant and polymer given in Table 2. The steady increase of the reflectivity in Figure 2 with increasing surfactant concentration demonstrates the increasing surface concentration of surfactant. In calculating the surface excess of the surfactant the slight contribution to the signal from the protonated polymer can be accurately taken into account by using Eq. (1) in ref. [6]. The difference is negligible when the surface excesses for the deuterated sample are large but can be a few percent when the signal from the deuterated sample is weak. In comparison with experimental errors such a correction is small. Finally, it is worth commenting that, although we have used the model of a uniform layer to calculate the surface excess, the value obtained is almost independent of the structural model used [7].

Table 1 Structural parameters for CsDS adsorbed in the mixed monolayer

[CsDS]/M	A/Å^2	$\tau \pm 2$/Å	$\sigma \pm 2$/Å	$\Gamma \times 10^{10}$ mol cm^{-2}
1.32×10^{-4}	340 ± 50	10.0	8.0	0.5 ± 0.2
6.44×10^{-4}	160 ± 10	12.0	10.0	1.0
1.27×10^{-3}	90 ± 5	16.0	13.0	1.8
2.51×10^{-3}	65 ± 3	17.0	14.0	2.6
4.00×10^{-3}	49 ± 3	17.0	14.0	3.4
1.05×10^{-2}	45 ± 2	17.5	15.0	3.7
4.00×10^{-2}	40.5 ± 2	17.5	15.0	4.1 ± 0.2

Table 2 Physical Constants for Different Components

Unit	Scattering length $\times 10^5$/Å	Volume/Å^3
$C_{12}D_{25}$	246.6	350
$C_{12}H_{25}$	-13.7	350
SO_4Cs	31.5	45
C_2D_4O	45.8	65
C_2H_4O	4.1	65
D_2O	19.1	30
H_2O	-1.7	30

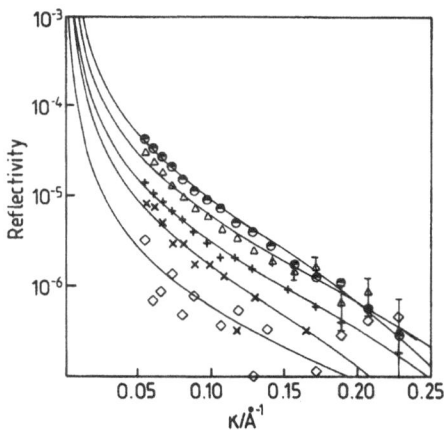

Fig. 3 Reflectivities versus momentum transfer κ for 0.1 wt% deuterated PEO in NRW at varying concentration of protonated CsDS. The concentrations of CsDS are 0 (○), 1.32×10^{-4} (△) 6.44×10^{-4} (+), 1.27×10^{-3} (×), and 2.51×10^{-3} (□)

The amount of PEO absorbed at the interface and its variation with surfactant concentration can be determined by using deuterated polymer and protonated surfactant because the signal is now predominantly from the labelled polymer. Figure 3 shows reflectivity profiles of a 0.1 wt% d-PEO solution with varying h-CsDS concentration. The decrease in the level of reflectivity shows that the polymer is gradually replaced by surfactant at the surface. The area per adsorbed segment can be calculated in the same way as for the surfactant and the values are given in Table 3.

The most direct way of obtaining structural information is to use the kinematic approximation [8]. For a layer of deuterated species adsorbed on the surface of null reflecting water, the reflectivity can be written approximately

$$R = \frac{16\pi^2}{\kappa^2} b_i^2 h_{ii} , \qquad (3)$$

where κ is the momentum transfer defined in terms of the grazing angle of incidence and the wavelength of the neutron ($\kappa = 4\pi \sin\theta/\lambda$), h_{ii} is the self partial structure factor of the surfactant or polymer, given by

$$h_{ii} = |n_i(\kappa)|^2 , \qquad (4)$$

Table 3 Structural parameters for PEO adsorbed in the mixed monolayer

[CsDS]/M	A/Å^2	$\tau \pm 2$/Å	$\sigma \pm 2$/Å	$\Gamma \times 10^{10}$ mol cm^{-2}
0	16 ± 1	20.0	17.5	10.4 ± 0.6
1.32×10^{-4}	21 ± 1	18.0	15.0	7.9 ± 0.3
6.44×10^{-4}	29 ± 2	17.5	15.0	6.2 ± 0.3
1.27×10^{-3}	45 ± 3	15.0	11.0	3.7 ± 0.2
2.51×10^{-3}	105 ± 15	14.0	10.0	1.6 ± 0.2

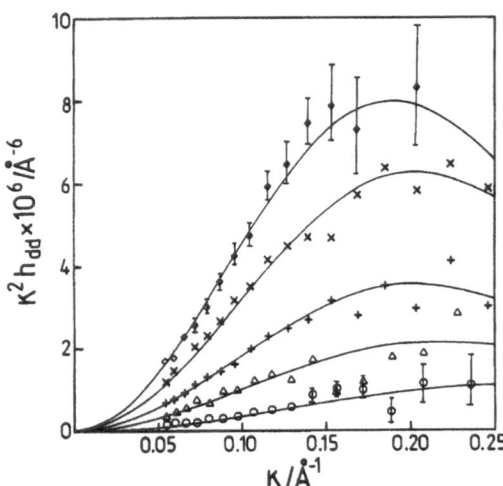

Fig. 4 Partial structure factors of surfactant versus κ at the concentration of CsDS of 6.44×10^{-4} ($\circ$), 1.27×10^{-3} ($\triangle$), 2.51×10^{-3} M ($+$), 4.0×10^{-3} ($\times$) and 4.0×10^{-2} ($\diamond$)

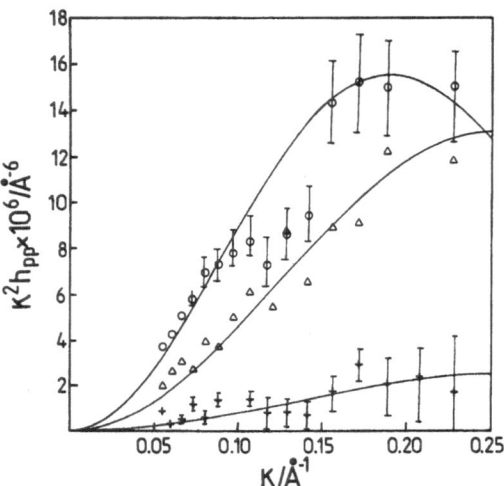

Fig. 5 Partial structure factors of PEO versus κ at the concentrations of protonated CsDS of 6.44×10^{-4} ($\circ$), 1.27×10^{-3} ($\triangle$) and 2.51×10^{-3} M ($+$)

where $n_i(\kappa)$ is the Fourier transform of the number density profile of the molecule in the direction normal to the surface. A Gaussian distribution has been shown to be a reasonable representation of loosely packed adsorbed layers, in which case, h_{ii} is [7]

$$h_{ii} = \frac{1}{A^2} \exp(-\kappa^2 \sigma_i^2) , \qquad (5)$$

where σ_i is the full width at $1/e$ of the maximum of the Gaussian distribution in the normal direction.

When the experiment is done using deuterated surfactant and protonated PEO in NRW the resulting h_{dd} is determined by the width and coverage for the surfactant molecule in the monolayer. Figure 4 shows the resulting plots of h_{dd} versus κ for CsDS at five different concentrations. The values for A and σ were calculated using Eq. (5) and are listed in Table 1. The values of A were found to be within error of those from the optical matrix calculation. The values for σ are 2–3 Å smaller than the thickness obtained from the uniform layer model. This results from the different shape of the distribution in the two cases. There is a marked variation in the thickness of the layer as the coverage changes, σ varying from 8 to 15 Å over the range studied.

A similar procedure can be used to obtain the width and coverage for the polymer in the mixed monolayer. Figure 5 shows the self-partial structure factor for the polymer, h_{pp}, at varying surfactant concentration. It should be noted that, although either a uniform layer or a Gaussian distribution fits the data well, the true segment distribution profile is not necessarily expected to follow either of them exactly. We have previously shown that the distribution of PEO on its own at the air–water interface

can be described as a combination of two half Gaussian distributions [9]. The more diffuse region extending into the subphase accounts for about 5% of the amount adsorbed and the effect of this fraction on the reflectivity occurs below 0.05 Å^{-1} in κ, which was not measured during the present experiment. This would then mean that the current measurement has underestimated the surface excess of the polymer by a percentage less than 5% and that the models adopted are only a close approximation to the actual distribution. In a future experiment we will measure the reflectivities below 0.05 Å^{-1} to check if there is a diffuse region of polymer layer in the presence of surfactant.

The other important information which can be obtained from the experiment is the separation of surfactant and polymer in the layer and this can be achieved through analysis of the partial structure factors. For the measurement using both deuterated surfactant and deuterated polymer, the reflectivity can be written as [6]

$$R(\kappa) = \frac{16\pi^2}{\kappa^2} \left(b_d^2 h_{dd} + b_p^2 h_{pp} + 2 b_d b_p h_{dp} \right) , \qquad (6)$$

where the self partial structure factors are as defined in (4), and h_{dp} denotes the cross-partial structure factor between the surfactant and the polymer and contains information about the relative distance of the two in the mixed layer. Accurate partial structure factors for the surfactant and polymer can be obtained by combining the reflectivities of the three isotopic species and solving the resulting three simultaneous equations derived from Eq. (6). To obtain an accurate h_{dp} it is best to normalise the surface coverage of the three reflectivities to the same value before applying the equations [6].

Progr Colloid Polym Sci (1996) 100:311–315
© Steinkopff Verlag 1996

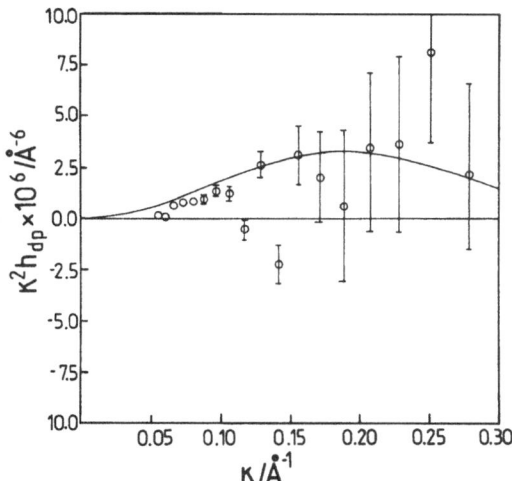

Fig. 6 Cross partial structure factor at 0.1 wt% PEO and 1.27×10^{-3} M CsDS with a calculated value of δ of 4 ± 1 Å

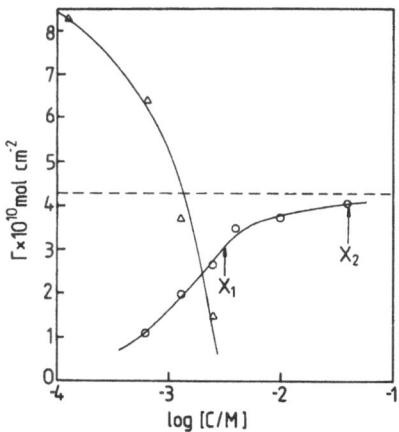

Fig. 7 Variation of surface excesses of PEO ($\triangle$) and CsDS ($\circ$) versus concentration of CsDS in solution. The solid lines are shown for guidance. The dashed line is the surface excess of pure CsDS for comparison

Figure 6 shows the plot of h_{dp} versus κ at a surfactant concentration of 1.27×10^{-3} M and polymer concentration of 0.1 wt%. The continuous line was calculated using

$$h_{dp} = \pm (h_{dd} h_{pp})^{1/2} \cos \kappa \delta_{dp} , \qquad (7)$$

where δ_{dp} is the separation between the centres of the chain and head distributions. The assumption underlying this equation is that the two distributions are symmetrical about their centres. This will be a fairly accurate approximation for the surfactant but less so for the polymer, as discussed above. However, a layer deviation from a symmetrical distribution would be needed to invalidate the approximation over the κ range being considered here [10]. Since all the terms except δ_{dp} are known δ_{dp} can be deduced directly from the data without any need for Fourier transformation or model fitting. The fit of Eq. (7) to $\kappa^2 h_{dp}$ shown in Fig. 6 gives a value of δ of 4 ± 1 Å. Since the full widths for the surfactant and polymer at this concentration are 13 and 11 Å respectively, the result indicates that there is a large overlap between surfactant and polymer layers. It is difficult to estimate the exact extent of

the overlap because the widths of the two distributions contain contributions from roughness.

The variation of the surface excess for each species at the air–water interface is summarised in Fig. 7. The coverage of pure CsDS at its CMC is shown for comparison. The polymer surface excess decreases dranmatically with surfactant concentration and drops to zero as x_1 is approached. The surface excess of the surfactant increases smoothly through x_1 and tends to approach the limiting value for pure CsDS as the concentration reaches x_2. The overall behaviour gives no sign of the two discontinuities which are seen in the tension plots and it must be concluded that these breaks are associated with discontinuities in the bulk behaviour. Since the individual surface excesses are dominated by their relative bulk concentrations the adsorption seems to be mainly a simple competition between the two species. It should be noted that for the NaDS/polyvinylpyrrolidone system we found an enhanced adsorption of surfactant below x_1 indicating a strong attractive interaction at the surface between polymer and surfactant [11].

References

1. Cabane B, Duplessix R (1982) J Physique 43:1529; also (1985) Colloid Surf 13:19
2. Fleer GJ, Cohen Stuart MA, Scheutjens JMHM, Cosgrove T, Vincent B (1993) Polymers at Interfaces, Chapman & Hall, London
3. Lee EM, Thomas RK, Penfold J, Ward RC (1989) J Phys Chem 93:381
4. Cabane B (1977) J Phys Chem 81:1639
5. Born M, Wolf E (1970) Principles of Optics, Pergamon, Oxford
6. Lu JR, Thomas RK, Binks BP, Fletcher PDI, Penfold J (1995) J Phys Chem 99:4113
7. Lu JR, Li ZX, Thomas RK, Staples EJ, Thompson L, Tucker I, Penfold J (1993) J Phys Chem 97:8012
8. Crowley TL (1984) DPhil Thesis, University of Oxford
9. Lu JR, Su TJ, Thomas RK, Penfold J (1995) Polymer, in press
10. Lu JR, Simister EA, Lee EM, Thomas RK, Rennie AR, Penfold J (1992) Langmuir 8:1837
11. Purcell IP, Thomas RK, Penfold J, Howe AM (1995) Colloid Surf A 94:125

Progr Colloid Polym Sci (1996) 100:316–320
© Steinkopff Verlag 1996

R. Miller
V.B. Fainerman
P. Joos

Dynamics of soluble adsorption layer studied by a maximum bubble pressure method in the μs and ms range of time

Dr. habil R. Miller (✉)
Max-Planck-Institut für Kolloid-und
Grenzflächenforschung
Rudower Chaussee 5
12489 Berlin, FRG

V.B. Fainerman
Institute of Technical Ecology
Blvd. Shevchenko 25
Donetsk 340017, Ukraine

P. Joos
Universitaire Instelling Antwerpen
Dept. Scheikunde
Universitetisplein 1
2600 Antwerpen, Belgium

Abstract The MPT1 allows to determine the bubble dead time and to calculate the effective surface age of a bubble. The developed theory for describing adsorption processes at the surface of a growing bubble and useful approximate solutions of this theory give access to a quantitative interpretation of experiments.

The standard version of the automated maximum bubble pressure tensiometer MPT1 has a time window from 2 ms up to 10 s. With a specially designed measuring cell of the MPT1 measurements down to 100 μs are possible.

The capacity of the bubble pressure tensiometer is demonstrated by measurements of the dynamic surface tensions of aqueous solutions of different Tritons in the entire time interval. An analysis is given for the applicability of the bubble pressure method to extremely short adsorption times.

Key words Maximum bubble pressure method – dynamic surface tension – surfactant solutions – temperature effects

Introduction

Studies of adsorption kinetics at liquid/gas interfaces are usually made by measuring the dynamic surface tension. For such studies many experimental techniques are available having different time windows from milliseconds to minutes and hours [1, 2]. The most advanced method to study adsorption kinetics at extremely short adsorption times is the maximum bubble pressure method which was already used by Rehbinder [3] and further developed by many authors [4–11]. Recent systematic improvements gave access to an accurate determination of the lifetime and the effective surface age of the bubble which allow now a quantitative interpretation of experimental data in term of adsorption mechanisms [12, 13].

In the present paper the applicability of the maximum bubble pressure tensiometer MPT1 (Lauda/Germany) for extremely short adsorption times is discussed with special emphasis of the effect of the initial conditions at the bubble surface after bubble detachment on the measured surface tension values. Experimental examples for different Triton surfactants and hexanol are given to demonstrate the capacity of the presented measuring technique.

Theory

To allow for a quantitative interpretation of maximum bubble pressure experiments two important prerequisites are needed: an accurate determination of the bubble dead time (time which a bubble needs to detach after the pressure has passed the maximum value) and the effective surface age (to the lifetime of the growing bubble equivalent time necessary to reach the same adsorption at a resting surface).

Progr Colloid Polym Sci (1996) 100:316–320
© Steinkopff Verlag 1996

The idea of the measuring procedure of the MPT1 is based on the existence of a critical point in the pressure–gas flow rate characteristic [14] which refers to the transition of the gas flow regime from a gas jet to a single bubble formation. The lifetime of a bubble τ from the moment of its formation to its hemispherical size can be calculated from the pressure P_c and the flow rate L_c in the critical point,

$$\tau = \tau_b - \tau_d = \tau_b \left(1 - \frac{LP_c}{L_c P} \right), \tag{1}$$

where τ_b is the interval between two bubbles, τ_d is the dead time, L is the flow rate and P is the pressure at $L < L_c$. Thus, from accurate measurement of the pressure-flow rate dependence and the determination of the critical point the calculation of the dead time and hence of the bubble lifetime is possible with good accuracy. We must note here that Eq. (1) is only an approximate relationship. Running theoretical work aims at an improved theory. First results show that especially for extremely high bubble frequencies small corrections are necessary [15].

The second important parameter to be known for a quantitative data analysis is the effective surface age τ_{ad} of a bubble in the moment of maximum bubble pressure. Fainerman et al. [12] presented an analysis of the adsorption process at the bubble surface. The basis of this analysis is the condition of constant pressure p at any moment of bubble life in the time interval $0 < t < \tau_b$. As a first approximation, one can assume that the bubble growth in the time $\tau < t < \tau_b$ is very fast, so that an almost bare surface can be assumed, $\gamma(t = 0) = \gamma_0$. A detailed analysis is shown in [2]. The final result is

$$\tau_{ad} = \frac{\tau}{2\xi + 1} \tag{2}$$

with

$$\xi = -\gamma^2(\tau) \int_{\sigma_0}^{\sigma(\tau)} \frac{d\gamma}{\gamma^3(1 + \sin\varphi)\sin\varphi} \approx \frac{\sin\varphi_0}{1 + \sin\varphi_0}, \tag{3}$$

and

$$\varphi_0 = \arccos\left(\frac{\gamma(\tau)}{\gamma_0}\right). \tag{4}$$

For solutions with a surface tensions γ close to that of the solvent γ_0, the effective surface age τ_{ad} coincides with the bubble life time τ. The function $1/(2\xi + 1)$ changes from 1 down to 0.5 at low surface tensions. As a rough estimate, we can say that the effective surface age τ_{ad} amounts to about 50% of the bubble lifetime.

Now we have to analyse the condition at which the assumption of a bare surface at the beginning of the bubble life is fulfilled. This condition depends on many para-

meters. We have to arrange the experimental conditions accordingly under which a bubble lifetime of $t \sim 0.1$ ms can be reached. With the MPT1 the flow rate L can be controlled with a precision of better than 1%. If the surface tension changes slowly, we can assume a constant pressure $P = $ const. For the fastest bubbles from Eq. (1), we obtain $t = 0.01 \tau_d$ and hence we need a dead time of $\tau_d \sim 0.01$ s to reach finally the 100 μs bubble lifetime interval.

The dead time τ_d depends on geometric parameters, such as the diameter d_{cap} and length l of the capillary and the distance between the capillary and the electrode d_{ce} which is mounted opposite the tip of the capillary and serves for the registration of the bubble frequency [12]. The situation at the tip of the capillary is shown schematically in Fig. 1. The distance between electrode and capillary d_{ce} can be changed by a screw so that the bubble size can be changed by small increments [12].

In the standard version of the measuring cell of the MPT1 the dead time is of the order of 30 ms. To decrease τ_d down to 10 ms it is necessary to decrease the length of the capillary l and the volume of detaching bubbles. As shown in [13] the use of a capillary of 7 mm length and 0.15 mm diameter and a bubble volume of 1 mm³ leads to a dead time of about 10 ms and hence allows dynamic surface tension measurements down to 100 μs bubble lifetime.

Another very important problem in maximum bubble pressure measurements at very short adsorption times is the initial amount of adsorbed material at the surface of a newly formed bubble. The larger this amount the higher is the initial surface pressure $\Pi = \Pi_0$. The time interval during which the bubble grows continuously from a hemispherical to the final size is τ_d. The relative expansion rate

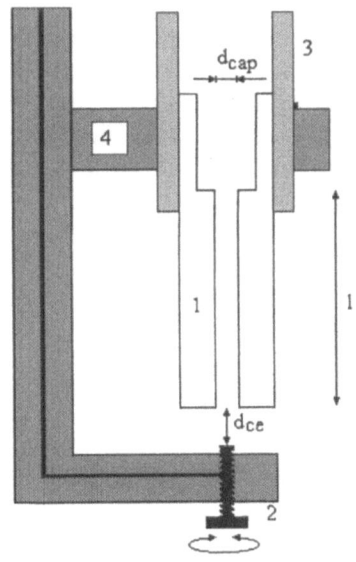

Fig. 1 Schematic description of the geometric parameters of the measuring cell of the MPT1: 1 – capillary, 2 – screw, 3 – tube, 4 – capillary holder, l – length of the capillary, d_{cap} – tip diameter, d_{ce} – distance between capillary and electrode (screw)

θ of the bubble surface area A is given by

$$\theta = \frac{d \ln A}{dt} = \xi / t . \tag{5}$$

The coefficient ξ, characteristic for the rate of expansion is $\xi = 1$. In the moment of bubble detachment this coefficient can even reach values $\xi > 1$. Hence, the effective lifetime of the bubble during this second stage of its growth (effective dead time) results in

$$\tau_{d,\,eff} = \frac{\tau_d}{2\xi + 1} \leqslant \frac{\tau_d}{3} . \tag{6}$$

To calculate the initial value of the surface pressure $\Pi = \Pi_0$ after a time $\tau_{d,\,eff}$ we use a diffusion controlled adsorption model and $\xi = 1$. In the range of short adsorption time the surface pressure is given by the approximate solution [2],

$$\Pi = 2RTc_0 \sqrt{\frac{Dt_{eff}}{\pi}} , \tag{7}$$

where D is the diffusion coefficient, c_0 is the surfactant bulk concentration, R is the gas law constant, and T is the absolute temperature.

The evaluation for a common surfactant ($D = 10^{-6}$ cm^2/s, $c_0 = 10^{-6}$ mol/cm^3) using Eq. (4) and a dead time of $\tau_d = 100$ ms yields $\Pi_0 = 5$ mN/m, while for $\tau_d = 10$ ms the initial surface pressure reduces to $\Pi_0 = 1.5$ mN/m. Thus, for surfactant concentration $c_0 < 10^{-6}$ mol/cm^3 we can assume that the initial load of the bubble surface is more or less negligible. For higher concentration we can expect a measurable shift of the measured short time values away from the surface tension of the pure solvent.

The analysis of the bubble formation and detachment process showed that there is a hydrodynamic influence at short bubble times [16, 17]. In regards to these effects, results were used given in [13] for surfactant solution having a viscosity similar to that of water. A detailed analysis of different effects (dynamics of the gas column in the capillary, movement of liquid meniscus after bubble detachment, flow conditions close to the capillary orifice) is under way [15].

Material and methods

The experiments were performed with the standard version of the maximum bubble pressure tensiometer MPT1 (Lauda, Germany), described in detail elsewhere [7]. The capillary provided with the MPT1 has an inner tip diameter of $d_{cap} = 1.5$ mm and a length of $1 = 10$ mm. In order to reach bubble lifetimes of 100 μs special capillaries

with the same inner tip diameter but lengths of 7 mm were used. Changing the position of the screw (2) (cf. Fig. 1), the distance between capillary and electrode d_{ce} could be controlled and thereby the volume of detaching bubbles.

The surfactants in this study, octylphenyl polyoxyethylene ether (C$_{14}$H$_{20}$O(C$_2$H$_4$O)$_n$H) with different numbers of oxyethylene groups, were purchased from Serva and Aldrich and used without further purification: Triton X-100 ($n = 10$), Triton X-114 ($n = 11.4$), Triton X-165 ($n = 16.5$), Triton X-305 ($n = 30.5$) and Triton X-405 ($n = 40.5$). The hexanol was purified by vacuum distillation. All surfactants solutions were prepared with doubly distilled water.

Results and discussion

The present studies are performed to demonstrate the applicability of the MPT1 in the time range from parts of

Fig. 2 Dynamic surface tension of two hexanol solutions as a function of $\sqrt{t}$ at comparatively high concentration; $c_0 = 5 \cdot 10^{-6}$ mol/cm^3 ($\square$), $c_0 = 10^{-5}$ mol/cm^3 ($\blacksquare$)

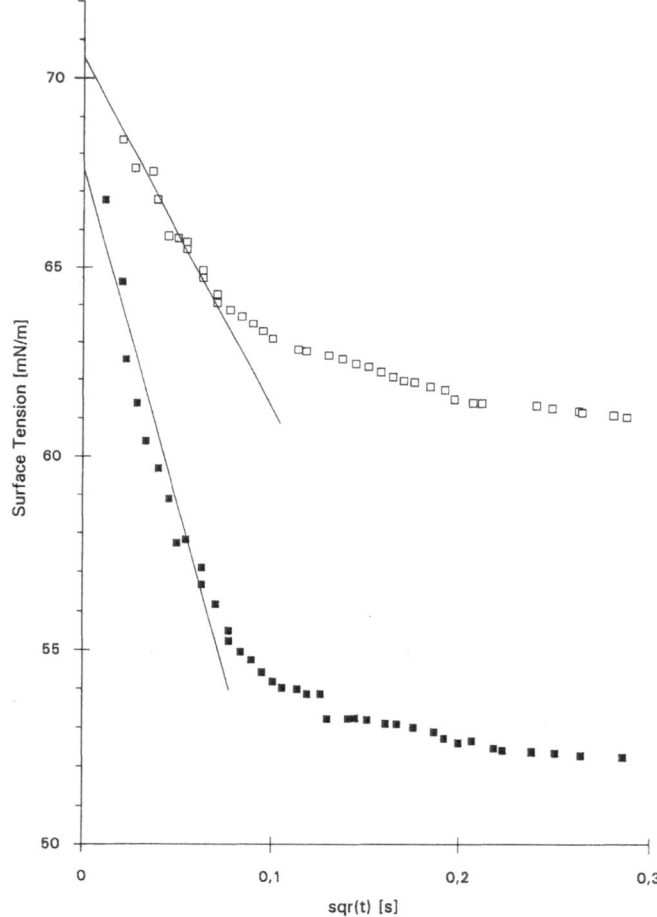

Progr Colloid Polym Sci (1996) 100:316–320
© Steinkopff Verlag 1996

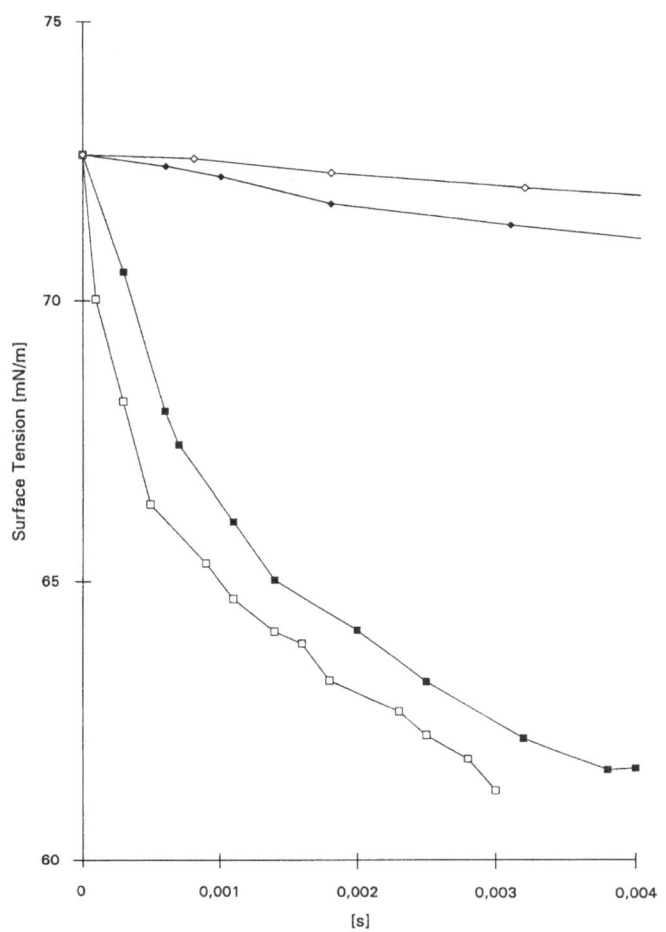

Fig. 3 Dynamic surface tension of different Triton X-n solutions as a function of τ_{ad} for the time interval between 0 and 5 ms at 20 °C; X-100 at $c_0 = 1.55 \cdot 10^{-7}$ mol/cm^3 ($\diamond$), X-114 at $c_0 = 4.23 \cdot 10^{-7}$ mol/cm^3 ($\blacklozenge$), X-165 at $c_0 = 1.07 \cdot 10^{-6}$ mol/cm^3 ($\square$), X-305 at $c_0 = 4.52 \cdot 10^{-7}$ mol/cm^3 ($\blacksquare$)

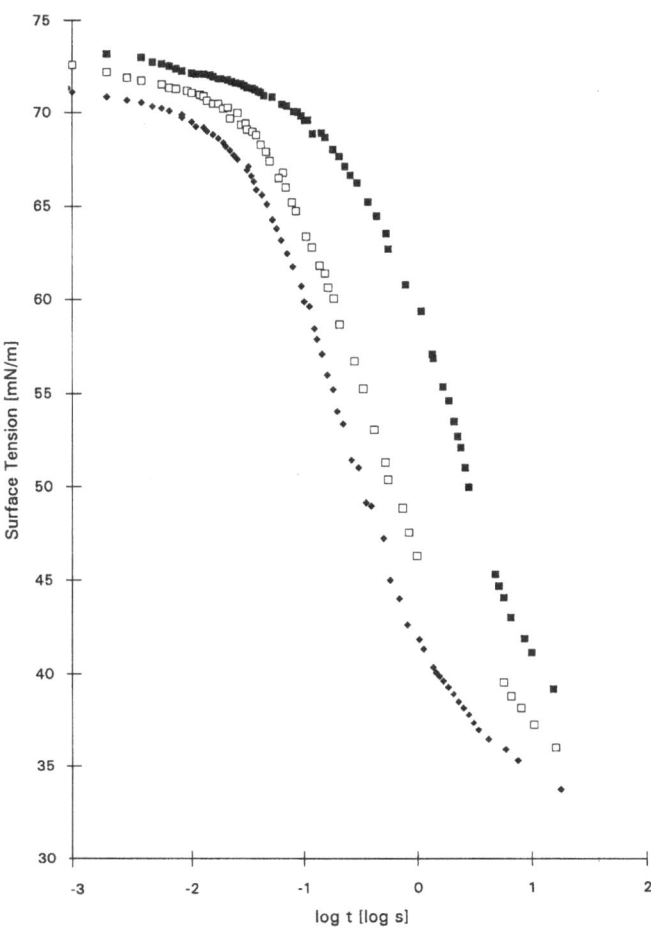

Fig. 4 Dynamic surface tension of different Triton X-114 solutions as a function of log t; $c_0 = 1.41$ ($\blacksquare$), 2.82 ($\square$), 4.23 ($\blacklozenge$) 10^{-7} mol/cm^3

a millisecond up to few seconds. The use of the MPT1 under stopped flow conditions, a modus under which long adsorption times up to few hundreds of seconds can be reached, was discussed elsewhere [19].

Measurements of aqueous hexanol solutions at concentrations above 10^{-6} mol/cm^3 confirm the estimate made by Eq. (7). The dynamic surface tensions for two hexanol concentrations are given in Fig. 2. For $c_0 = 5.10^{-6}$ mol/cm^3 the initial surface pressure value is about 1.5 mN/m and for $c_0 = 10^{-5}$ mol/cm^3 it is about 4 mN/m. Thus, the limit of $c_0 < 10^{-6}$ mol/cm^3 as the condition for zero adsorption at the newly formed bubble surface is realistic.

To demonstrate that for adsorption times 0.1 ms < τ_{ad} < 1 ms reliable experimental data can be measured, the dynamic surface tensions of four different Triton solutions are shown in Fig. 3. These results are very consistent with

the expected course of $\gamma(\tau_{ad})$. For the concentrations used in these experiments the extrapolated value of $\gamma(t \to 0)$ agrees very well with the surface tension of pure water γ_0.

Further experiments have been carried out with Triton X-114 (Fig. 4) and Triton X-100. The results demonstrate that very accurate surface tension data can be obtained in a time interval which is not accessible by other methods. Moreover, the apparatus even allows measurements in a broad temperature interval. The measurements of Fig. 5 were performed in an interval from 20 °C and 70 °C.

Conclusions

A comparatively small change in the design of the measuring cell of the MPT1 allows us to measure dynamic surface tensions of surfactant solutions down to 0.1 ms adsorption time. The data obtained for aqueous Triton and hexanol solutions agree well with the presented approximate theory.

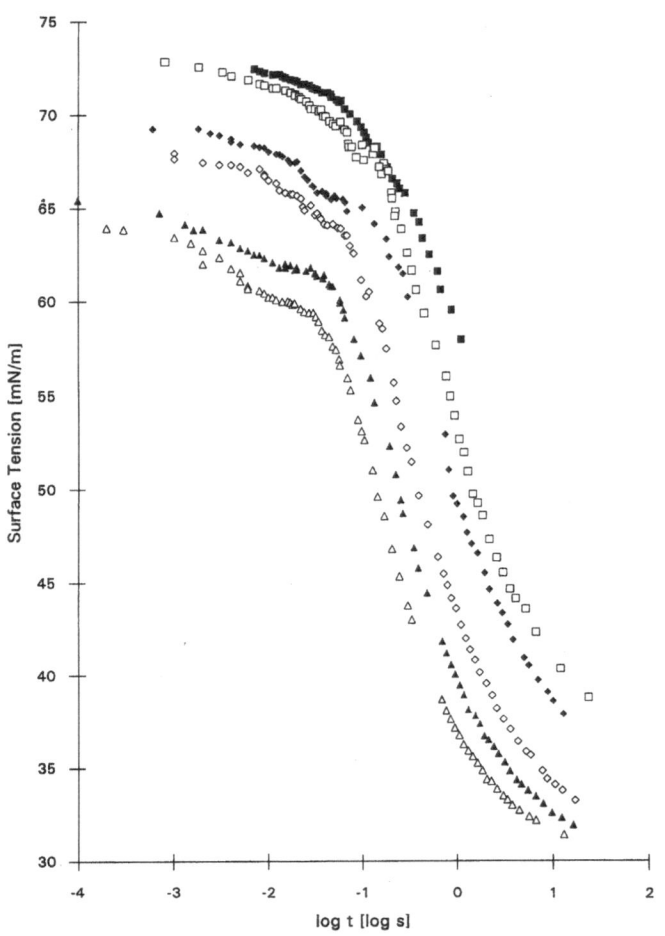

Fig. 5 Dynamic surface tension of a Triton X-100 solution as a function of $\ln t$ at different temperatures; $c_0 = 1.55 \cdot 10^{-7}$ mol/cm³; $T = 20\,°C$ (■), $30\,°C$ (□), $40\,°C$ (♦), $50\,°C$ (◇), $60\,°C$ (▲), $70\,°C$ (△)

For the initial surface pressure $t \to 0$ values close to zero are obtained for concentrations $c_0 < 10^{-6}$ mol/cm³.

The proposed changes of the cell give access also to very interesting studies, for example of micellar solutions of highly surface active substances. Measurements of dynamic surface tensions of solutions above the CMC give access to the micellar kinetics.

Acknowledgments The work was financially supported by projects of the European Community (HCMERBCHRXT930322 and INTAS 93-2463).

References

1. Miller R, Joos P and Fainerman VB (1994) Adv Colloid Interface Sci 49:249
2. Dukhin SS, Kretzschmar G, Miller R (1995) Dynamics of Adsorption at Liquid Interfaces. Theory, Experiment, Application. In: Möbius D, Miller R (eds) Studies of Interface Science Vol. 1, Elsevier, Amsterdam
3. Rehbinder PA (1924) Z Phys Chem 111:447
4. Austin M, Bright BB, Simpson EA (1967) J Colloid Interface Sci 23:108
5. Bendure RL (1971) J Colloid Interface Sci 35:238
6. Kloubek J (1972) J Colloid Interface Sci 41:1
7. Miller TE, Meyer WC (1984) American Laboratory 91
8. Woolfrey SG, Banzon GM, Groves MJ (1986) J Colloid Interface Sci 112:583
9. Hua XY, Rosen MJ (1988) J Colloid Interface Sci 124:652
10. Mysels KJ (1989) Langmuir 5:442
11. Markina ZN, Zadymova NM, Bovkun OP (1987) Colloids Surfaces 22:9
12. Fainerman VB, Miller R, Joos P (1994) Colloid Polymer Sci 272:731
13. Fainerman VB, Miller R (1995) J Colloid Interface Sci 174: in press
14. Fainerman VB (1992) Colloids Surfaces 62:333
15. Dukhin SS, Fainerman VB, Miller R, in preparation
16. Fainerman VB, Makievski AV, Miller R (1993) Colloids Surfaces 75:229
17. Kao RL, Edwards DA, Wasan DT, Chen E (1993) J Colloid Interface Sci 148:247
18. Fainerman VB, Makievski AV, Miller R (1995) Langmuir 11: in press
19. Fainerman VB, Zholob SA, Miller R, Loglio G, Cini R (1995) Tenside Surfactants Detergents in press

Progr Colloid Polym Sci (1996) 100:321–327
© Steinkopff Verlag 1996

Skin formation on liquid surfaces under non-equilibrium conditions

A. Prins
A.M.P. Jochems
H.K.A.I. van Kalsbeek
J.F.G. Boerboom
M.E. Wijnen
A. Williams

Dr. A. Prins (✉) · A.M.P. Jochems ·
H.K.A.I. van Kalsbeek
F.J.G. Boerboom · M.E. Wijnen
A. Williams
Department of Food Science
Wageningen Agricultural University
P.O. Box 8129
6700 EV Wageningen, The Netherlands

Abstract Skin formation on liquid surfaces is a well known phenomenon, especially when liquid surfaces of foodstuffs are subjected to mechanical disturbances. From the rheological point of view, skin formation indicates that a rigid surface layer, having a kind of network structure, is formed. This structure gives rise to pronounced dilational as well as pronounced shear properties. Proteins are found to promote such layers on aqueous solutions.

In spite of the fact that skins have been observed frequently, the physical properties of these layers are not well described in literature. In our investigation we have been able to make these skins under well defined physical conditions.

Skins of proteins such as BSA, β casein and egg white have been made under well defined physical conditions on the surface of aqueous solutions. Their surface dilational properties have been measured under different experimental conditions. The results demonstrate that depending on the method used pure elastic or viscoelastic behavior can be found.

This can be explained by assuming that the proteinaceous surface layer is an elastic gel filled with water. In a dynamic measurement viscous behavior is found due to the flow of water through the protein network. In a one-stroke compression experiment, when the flow of the water is finished, only the elastic properties of the protein network are measured.

Key words Skin formation – surface rheology – gel layer – proteins

Introduction

Skins on liquid surfaces are well known phenomena see [1, 2]. They can be observed in every day life at the surface of liquid food stuffs such as milk, beer and soup. Also, when the surface of natural waters in a river or a canal, is moving against an obstruction – a piece of wood for instance-it can be observed that the surface is motionless in front of the obstruction. This in spite of the fact that water underneath moves parallel to that surface. The conclusion is that these layers are able to compensate for a shear stress acting along that surface.

Liquid surfaces can be motionless as a result of a surface tension gradient acting along that surface which is large enough to compensate for the shear stress exerted by the moving liquid on that surface. Obviously, rheological surface properties play an important role in the appearance of rigid surface layers.

In the absence of relaxation processes, a surface can behave in a purely elastic way when, for instance, an insoluble monolayer is present in that surface. A phenomenon also well known from literature is that when such an immobile surface layer is compressed or the stress exerted by the streaming liquid is too large, the surface layer collapses. This means that the surface layer is broken into

322

A. Prins et al.
Skin formation under non-equilibrium conditions

bits and pieces which slide over each other, or the surface is wrinkling, indicating that it is pliable. Another possibility is that by compressing the monolayer the apparent surface tension can drop to such low values that the obstruction – the barrier-is not able to keep the layer in place: the surface dives under the barrier.

Another category of stagnant surface layers is the presence of a skin on the liquid surface. For the moment, a skin can be defined as a – compared to molecular dimensions – thick layer on the surface which has a certain amount of cohesion, rigidity. Components promoting skin formation are macromolecular surfactants such as proteins and polysaccharides.

A well known example of skin formation can be observed in beer (1). When a carbon dioxide bubble shrinks due to the dissolution of the gas in beer, at a certain moment further shrinkage is hampered due to the formation of a skin on the bubble surface. After some moments the skin is repelled from the bubble surface allowing the bubble to shrink further till a new skin is formed and the shrinkage stops again. This phenomenon can repeat itself several times. The repelled skins are not soluble any more in the beer and give rise to some turbidity.

Skins of this kind demonstrate a certain amount of cohesion: they can be picked up by a grid and taken out of the solution. From optical reflection it is deduced that these skins are relatively thick: orders of magnitude thicker than molecular dimensions.

From these observations the picture arises that a skin derives its coherence from the presence of a two-dimensional network, a kind of gel layer, in the surface. From this it is to be expected that mechanical properties of a skin can be found as well in a shearing as in a dilational deformation.

It is a matter of debate whether the concept of surface tension makes sense when the surface is covered with a skin. When the skin can be considered to be a thin bulk layer with rheological properties, then its two surfaces also come into play: the surface of the skin against air and the interface between the skin and the contacting liquid. Therefore it is to be expected that different results will be obtained when a skin is subjected to a dilational disturbance compared to the situation where a shearing deformation is applied. In dilation both bulk and surface tension effects will play a role, whereas in shear, bulk shear effects are supposed to be dominant.

The present paper describes the results of different rheological measurements carried out on surfaces of aqueous solutions for which it is to be expected that they can form a skin. To this end the following proteins have been used: β casein, bovine serum albumin, egg white and Teepol as an example of a non-skin forming surfactant.

The techniques used are the traditional surface dilational modulus measurement carried out with the newly developed ring trough using a sinusoidal small amplitude deformation [3]. The overflowing cylinder technique is used to measure the surface dilational viscosity [4–6]. Both techniques can be applied to air/water as well as to oil/water interfaces.

In addition, a so-called canal method has been used by which, analogous to the overflowing cylinder technique, dilational parameters far from equilibrium can be measured. The advantage of the canal method is that special attention can be paid to the compressed part of the liquid surface; this in contrast to the overflowing cylinder technique, which is especially suited for taking measurements at the expanding liquid surface. The canal method performed in a special way (more details will be given in the experimental part) can be used to measure the surface dilational elasticity of the compressed part of the liquid surface.

The purpose of the present paper is to demonstrate that depending on the technique used, different results will be obtained when dealing with liquid surfaces on which a skin is present.

Some of the results will be discussed in terms of a model in which the skin is considered to be a thin gel layer in which water can be transported when the layer is deformed. This transport gives rise to energy losses resulting in the presence of a viscous component in the surface dilational modulus.

Experimental

Materials

Bovine serum albumin (BSA) from Sigma, obtained by cold precipitation from alcohol, purity 96–90% used as such, pH = 5.6 acetate buffer

β casein from Eurial Poitouraine; analytical grade' 96% purity, used as such, pH = 6.7 imidazole buffer

Teepol from Centrale Aankoop FNZ, technical surfactant containing 15% active components

Egg white from raw egg; protein content about 1.2% w/w, sieved before use to get rid of inhomogeneities; dissolved in tap water.

Methods and results

The experimental techniques and the results will be described separately for each of the techniques used: the overflowing cylinder, the ring trough, and the canal method will be dealt with in this order.

Progr Colloid Polym Sci (1996) 100:321–327
© Steinkopff Verlag 1996

overflowing cylinder

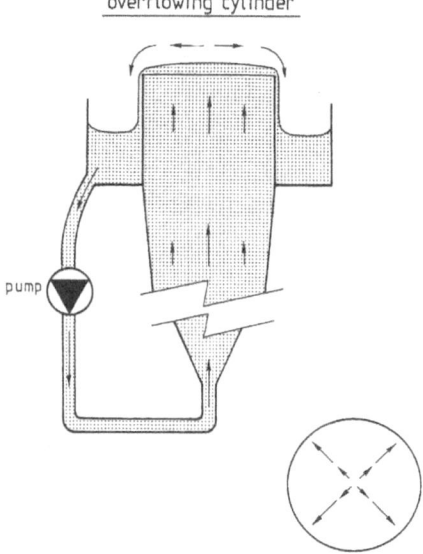

Fig. 1 Schematic representation of the overflowing cylinder technique including the flow pattern of the overflowing liquid

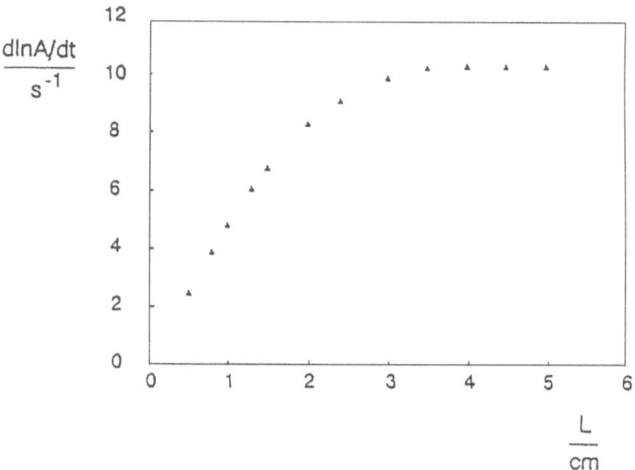

Fig. 2 Surface expansion rate expressed in $d\ln A/dt$ as a function of the falling film height l for a 0.3 vol% solution of Teepol in water

Overflowing cylinder

Because the overflowing cylinder technique is described in detail elsewhere (see [4–6]), for the sake of brevity only a few remarks will be made here.

Operating the overflowing cylinder with a dilute aqueous surfactant solution, the surface tension at the top of the free falling film at the outside wall of the cylinder is increased above the equilibrium surface tension, see Fig. 1. This increase in surface tension is caused by the expansion of the liquid surface at that very spot which in its turn is driven by the liquid falling down at the outside wall of the cylinder. This increased surface tension now pulls the horizontal liquid surface and some adhering liquid over the top rim of the cylinder.

The effectiveness of this mechanism follows from the observation that in the presence of a surfactant the rate of expansion, expressed in terms of $d\ln A/dt$, where A is the surface area of the liquid and t the time, of the horizontal surface can be increased by one order of magnitude until $10\ \mathrm{s}^{-1}$. For pure water, when no surface tension gradient is present, an expansion rate of about $1\ \mathrm{s}^{-1}$ is found under otherwise the same conditions.

Under steady state conditions which are usually obtained within a few seconds after starting the liquid flow, mechanical equilibrium at all the moving surfaces is ensured by equating the surface tension gradient

$$\frac{dy}{dr},$$

where y is the surface tension (Nm^{-1}) and r the distance from the center of the horizontal surface (m), to the shear stress exerted by the flow of the contacting liquid:

$$\frac{dy}{dr} = - \eta_b \left(\frac{dV_r}{dz} \right)_{z=0}, \tag{1}$$

where η_b is the bulk viscosity of the liquid, V_r the radial velocity of the liquid and z the vertical coordinate pointing downwards. The minus sign accounts for the fact that both stresses counteract each other.

Apart from being dependent on the pumping rate, in the presence of surfactants the expansion rate of the liquid surface is found to be strongly dependent on the height of the falling film at the outside wall of the cylinder – this in such a way that beyond a height of about 4 cm the effect is very small if any (see Fig. 2).

In short, the overflowing cylinder is a kind of Marangoni machine: at the top surface the surface tension gradient drives the liquid to flow over the rim, at the outside wall the down falling liquid drives the surface expansion and consequently the surface tension gradient.

The expansion rate of the overflowing surface together with the corresponding dynamic surface tension are an autonomous property of the surfactant solution. This means that these parameters are not only characteristic for the surfactant used, but also depend strongly upon the surfactant concentration, as shown in Fig. 3. In this graph the equilibrium and the dynamic surface tension, as measured by means of the overflowing cylinder, are plotted as a function of the surfactant (Teepol) concentration together with the measured surface expansion rates. At the very concentration where the slope of the dynamic surface

Fig. 3 Equilibrium surface
tension, dynamic surface tension
and corresponding surface
expansion rates as measured
with the overflowing cylinder
technique as a function of the
Teepol concentration in water

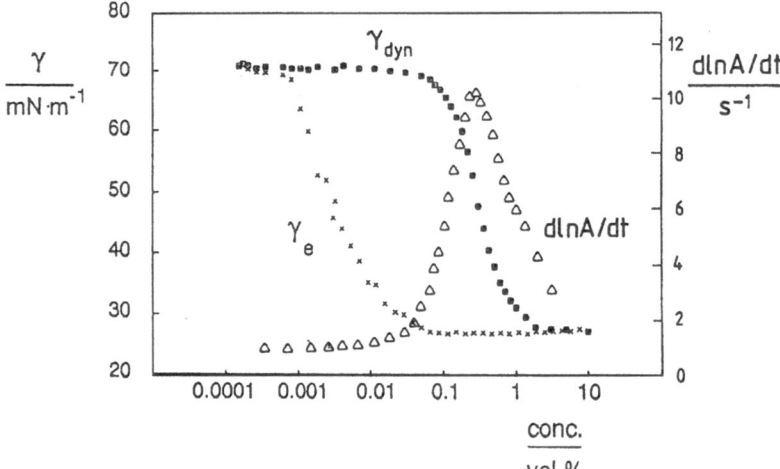

tension is at its maximum, the expansion rate reaches its maximum value.

Operating the overflowing cylinder with a low molecular surfactant solution such as Teepol, always expansion rates in excess of those of water (1 s^{-1}) are observed.

However, using skin forming surfactants such as proteins, the situation can be quite different: under certain conditions they operate like low molecular surfactants by increasing the expansion rate with respect to water. Under well chosen conditions adjusted by the falling film height and the protein concentration, surface expansion rates well below that of water can be observed, see Figs. 4 and 5. In the ultimate case completely motionless surfaces all over the overflowing surface can be found.

A typical experiment proceeds as follows: the overflowing cylinder is filled with a bovine serum albumin solution for 0.05 gram per liter and is allowed to overflow under condition that the height of the falling film is relatively large, say 2.5 cm. Under these conditions the surface expansion rate is about 2 s^{-1}, which is larger than the value for pure water: 1 s^{-1}; the solution behaves like a low molecular surfactant solution and the corresponding dynamic surface tension is measured to be about 67 mNm^{-1}.

Now, the height of the falling film is decreased to 1.5 cm. The compressed film at the bottom of the falling from then creeps slowly but steadily over the film surface to the top of the film, until it covers the whole surface including the horizontal surface.

The consequence for the mechanical equilibrium between the surface and the bulk of the liquid is that in the presence of such a "skin" the liquid drives the surface over the whole surface area of the overflowing cylinder. A kind of network structure in the adsorbed protein layer is supposed to be able to compensate for the viscous drag

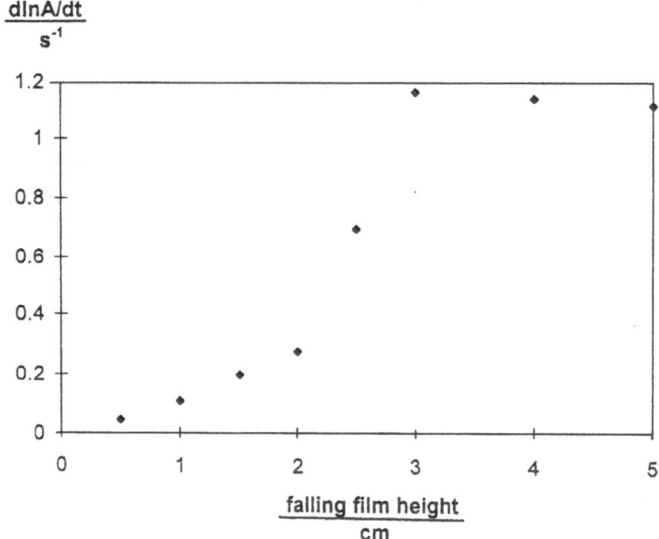

Fig. 4 Surface expansion rate expressed in $d\ln A/dt$ as a function of the falling film height for a 10^{-1} g/l solution of β casein in water demonstrating surface velocities lower than in the case of pure water. pH = 6.7 imidazole buffer

exerted by the liquid flowing below the surface. An example of this behavior is given for a 0.5 g/l BSA solution in Fig. 5.

The BSA surface layer is not motionless: it demonstrates a certain amount of relaxation behavior. From the obtained data the surface dilational viscosity can be calculated. The surface dilational viscosity is defined by

$$\eta_s^d = \frac{\Delta y}{d\ln A/dt},\qquad(2)$$

Progr Colloid Polym Sci (1996) 100:321–327
© Steinkopff Verlag 1996

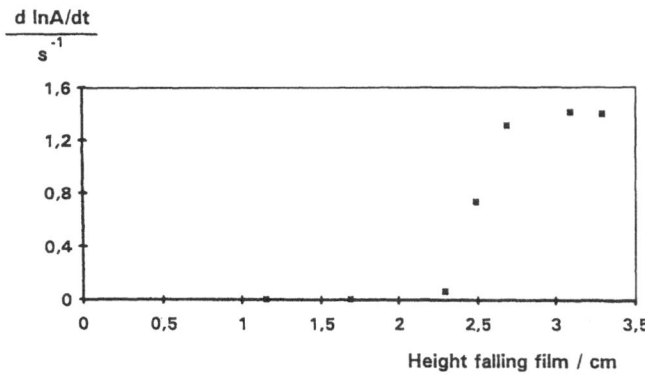

Fig. 5 Surface expansion rate expressed in $d\ln A/dt$ as a function of the falling film height for a 0.5 g/l solution of BSA in water. pH = 5.6 acetate buffer

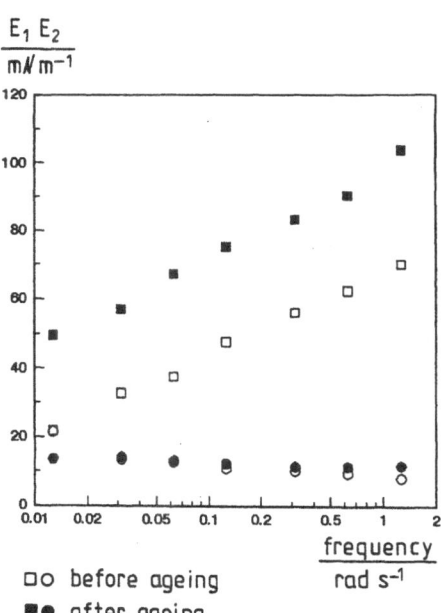

□○ before ageing
■● after ageing

Fig. 6 Elastic and viscous part of the surface dilational modulus, measured with the ring trough technique close to equilibrium for a 0.48% w/w aqueous solution for egg white. The effect of aging is indicated. The indicated egg white concentration is on basis of dry protein content

where Δy is the increase in surface tension when the surface is subjected to an expansion rate dlnA/dt. For BSA values of about 500 mNm^{-1} are obtained which are one order of magnitude larger than for Teepol: 30 mNm$^{-1}\cdot$s.

Experiments carried out with β-casein demonstrate the same behavior: at large film heights low molecular surface behavior and expansion rates larger than 1 s^{-1}, at smaller film heights skin formation and reduction for the expansion rate to values much smaller than 1 s^{-1} (see Fig. 4).

By using an aqueous egg white solution the skin formation results in a completely motionless surface, presumably due to surface denaturation and cross linking of the protein. The observation that the skin of egg white is not moving any more indicates that this skin is purely elastic. By increasing the film height again, the egg white skin ruptures at a certain moment, indicating that this skin has a yield strength which, in principle, is measurable.

Ring trough

A convenient method to measure the surface dilational modulus of a liquid surface is the ring trough method (for details see ref. [3]). In essence, a small amplitude sinusoidal area change is applied to the liquid surface and the resulting change in surface tension is measured including the phase shift between these two parameters when the surface is viscoelastic.

The surfaces dilational modulus E is defined as

$$E = \frac{dy}{d\ln A} \qquad (3)$$

For viscoelastic surfaces E depends on the frequency of the sinusoidal disturbance and in that case the modulus E can

be considered to consist of a storage E_1 and a loss modulus E_2 according to

$$E = (E_1^2 + E_2^2)^{1/2}. \qquad (4)$$

Because the surface of the egg white solution demonstrated pure elastic behavior in the overflowing cylinder method, the surface dilational modulus of this system was measured by means of the ring trough technique as a function of the frequency. The result of this measurement is given in Fig. 6, from which it appears that within this frequency range the surface behaves in a viscoelastic way, demonstrating that relaxation processes take place. It also demonstrates that by aging the modulus value increases somewhat. The modulus values obtained in this way have about the same value as published elsewhere.

The ring trough makes use of the Wilhelmy plate technique for the measurement of the change in surface tension. It may be that when a skin is present the Wilhelmy plate technique is not the right one: the forces are not transferred in the proper way to the plate for instance due to rupture of the skin close to the plate.

Therefore, it was decided to exert the stress on the liquid surface by means of a streaming liquid and measuring the resulting deformation by means of the displacement of marked spots on that surface. In this way no

326
A. Prins et al.
Skin formation under non-equilibrium conditions

Wilhelmy plate is required that can disturb the skin structure in the surface.

The canal technique

This method makes use of an open rectangular canal through which the liquid is allowed to flow. A part of the liquid surface is confined between two barriers in the surface, one upstream and one downstream, see Fig. 7. The liquid flow is controlled by means of a pump, tap and flow meter. The stress on the liquid surface which appears to be almost motionless, is calculated by means of the following relationship valid for a rectangular canal with square cross-section, see ref. [7]:

$$\sigma = 3.57 \eta Q \frac{b + h}{(b \cdot h)^2}, \tag{5}$$

here η is the bulk viscosity of the liquid, Q the liquid flux, and b and h are the width and the height of the wetted part of the canal. In this way stresses of the order of 10^{-3} Pascal can be exerted on the surface.

The stress exerted by the streaming liquid on the almost motionless surface is compensated by a surface tension gradient:

$$\sigma = \frac{dy}{dl}, \tag{6}$$

where l is the coordinate pointing upstream along the length of the canal. From this it follows that

$$y = l\sigma + y_0, \tag{7}$$

where y_0 is the surface tension measured at the point $l = 0$. From Eq. (7) it follows that

$$dy = l d\sigma + \sigma dl, \tag{8}$$

where it has to be noted that for the case considered the value of σdl is small compared to $l d\sigma$. Therefore, Eq. (8) can be written

$$dy = l d\sigma. \tag{9}$$

The area of the surface A confined between the two barriers is given by

$$A = b \cdot l, \tag{10}$$

where b is the width of the canal. Since b is constant, dA is given by

$$dA = b dl,$$

and

$$\frac{dA}{A} = d\ln A = \frac{b dl}{bl} = d\ln l. \tag{11}$$

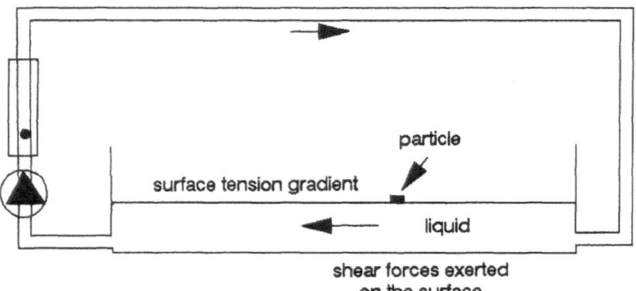

Fig. 7 Schematic representation of the canal method showing a cross-section through the canal, the position of a particle in the surface is determined by using a cathetometer. The dimensions of the canal are: length 50 cm, width 5 cm and height of the liquid level 5 cm

By introducing the expression for dy (Eq. (9)) and for $d\ln A$ (Eq. (11)) in the expression for E (Eq. (3)) this results in

$$E = \frac{l d\sigma}{d\ln l}. \tag{12}$$

A measurement of the surface dilational modulus proceeds now as follows: by changing the liquid flux Q through the canal by a known amount, the viscous drag on the surface changes by an amount $d\sigma$ according to Eq. (5). By means of a cathetometer the corresponding displacement dl of a particle on the liquid surface is measured. The value of l follows from the distance of the particle to the barrier. Finally the value of E follows from Eq. (12).

Applying this technique to an aqueous egg white solution, the following observations and measurements have been made: the first observation is that in spite of the fact that the solution is pumped through the canal the paper marker at the surface is completely motionless. This indicates already that no relaxation takes place: the surface behaves purely elastic under these conditions.

The second observation is that by changing the flow rate through the canal by a known amount the paper marker is displaced and is found to return to its original position when the original flow rate is reestablished. This also proves that the surface behaves purely elastic.

This observation can be used to measure the surface dilational elasticity according to Eq. (12). This has been done for various changes in the applied stress. The results of Fig. 8 demonstrate that within the accuracy of these measurements the values of E increase somewhat when the surface is allowed to age. The order of magnitude of E however is about the same as has been measured with the ring trough in a dynamic experiment, see Fig. 6.

The essential difference between the two measurements is that under a steady state compression of the surface in the canal the surface of the egg white solution behaves in

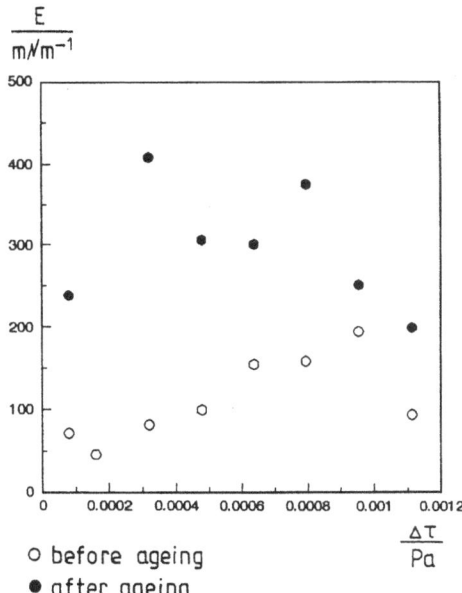

○ before ageing
● after ageing

Fig. 8 Surface dilational modulus as measured with the canal technique (equ. 12) as a function of the applied change in stress for a 0.48% w/w aqueous egg white solution. The effect of aging is indicated

an elastic way whereas in a dynamic experiment applying a sinusoidal disturbance the surface behaves viscoelastic.

Discussion

The resuts obtained with the egg white solution can be explained on the basis of the following speculative model: consider the skin formed on the surface of the egg white solution as a thin gel layer built up by the cross-linked protein molecules forming a network which is filled with water. Compressing and expanding such a layer results in the deformation of the protein network but also leads to the movement of water through that network. Transport of water through a network is accompanied by energy losses which means that viscous properties are coming into play. This can explain that by applying a periodical sinusoidal deformation to the surface, as takes place in the ring trough, finite values of the loss modulus are found. The fact that the value of the loss modulus does not depend on the frequency is in line with this model.

When in the canal method the surface layer is compressed and kept in this situation for some time to allow the water transport to come to an end, the protein network alone is responsible for the measured counteracting force and the layer behaves in an elastic way. This can explain the observation that by means of the canal method an elastic response to the applied deformation has been found.

The permanent network of the egg white skin can be the result of surface denaturation of the protein forming cross links of a chemical, s-s bonds, or a physical nature. The example of BSA demonstrates that in that network structure relaxation processes can take place which are held to be responsible for the relatively high value of the surface dilational viscosity η_s^d. In view of this it has to be noted that the pure elastic behavior of the egg white skin can be described by the surface dilational viscosity being infinitely large.

In the canal method the effect of shear on the surface behavior cannot be ruled out completely. It may be that the somewhat higher values of E obtained with the canal method compared to the ring trough can be explained by this phenomenon.

Conclusion

It is concluded that the surface dilational viscosity is a parameter that measures the cohesion in a protein skin formed at the surface of an aqueous solution.

In the ultimate case of a permanent network the elastic properties of a protein skin can be measured which seems not to depend on the experimental technique used.

The surface rheological behavior of an egg white skin can be explained by assuming that the skin consists of a protein gel filled with water.

References

1. Ronteltap AD (1989) "Beer foam physics". PhD thesis. Wageningen Agricultural University, Wageningen, the Netherlands
2. Fisher LR, Mitchell EE, Parker NS (1987) A critical role for interfacial compression and coagulation in the stabilization of emulsions by proteins. J Coll & Interf Science 119:592–594
3. Kokelaar JJ, Prins A, de Gee M (1991) A new method for measuring the surface dilational modulus of a liquid. J Coll & Interf Science 146:507–511
4. Bergink-Martens DJM, Bos HJ, Prins A, Schutte BC (1990) Surface dilation and fluid-dynamical behaviour of Newtonian liquid in an overflowing cylinder. 1. Pure liquids. J Coll & Interf Science 138: 1–9.
5. Bergink-Martens DJM, Bos HJ, Prins A, Schutte BC (1994) Surface dilation and fluid-dynamical behaviour of Newtonian liquid in an overflowing cylinder. 1. Pure liquids. J Coll & Interf Science 165: 221–228
6. Bergink-Martens DJM (1993) "Interface dilation. The overflowing cylinder technique" PhD thesis. Wageningen Agricultural University, Wageningen, the Netherlands.
7. Beek WJ, Mutzall KMK (1975) "Transport phenomena". John Wiley & Sons, London

Progr Colloid Polym Sci (1996) 100:328–329
© Steinkopff Verlag 1996

Dynamic surface properties of aqueous solutions

N.N. Kochurova
A.I. Rusanov

Dr. N.N. Kochurova (✉) · A.I. Rusanov
St. Petersburg University
Department of Chemistry
199034 St. Petersburg, Russia

Abstract The experimental data in a broad diapason of temperatures, concentrations and the surface age have been obtained. Equations of the thermodynamics of irreversible processes which describe the nonequilibrium properties of the surface are derived.

It is shown that the formation of a fresh surface may involve local breakdown of the electroneutrality other than the electric double layer.

Under equilibrium conditions the contribution to the surface tension from the orientation of the dipoles is proportional to the square of the jump of the electric potential created at the surface. When the ionic double layer is formed under nonequilibrium conditions, we find the term to be proportional to the first power of $\Delta\chi$.

A new discovery is that at small surface age (1–3 ms) the surface tension increases with the temperature. It may be that this effect is connected with decreasing of the surface entropy of fresh surface.

Key words Dynamic surface tension – nonequilibrium surface electric potential – aqueous solutions

Introduction

A great number of works are devoted to theoretical and experimental investigations of equilibrium bulk properties of liquids, whereas surface (especially nonequilibrium surface) properties have been studied to a lesser degree than bulk properties.

The results of experimental and theoretical studies of the dynamic surface tension γ and the nonequilibrium surface electric potential χ of aqueous solutions (nonorganic electrolytes and surfactants) are reviewed.

The experimental data in a broad diapason of temperatures (T), concentrations (C) and surface age (t) have been obtained [1–4]. The dynamic surface tension has been measured by the oscillating jet method and by the method of maximum pressure in the bubble. The measurements of the surface electric potential were carried out by the dynamic condenser method.

Correlation between $\Delta\gamma$ and $\Delta\chi$

Under equilibrium conditions the contributions to the surface tension from the orientation of the dipoles is proportional to the square of the jump of the electric potential created at the surface. When the ionic double layer is formed under nonequilibrium conditions we find a term proportional to the first power of $\Delta\chi$ [2, 4].

Local breakdown of the electroneutrality of the electric double layer

It is known [4], that the formation of a fresh surface may involve local breakdown of the electroneutrality of the electric double layer. The formation of surface charge is most likely at low concentration of electrolyte, when the electric double layer is relatively stretched and easily

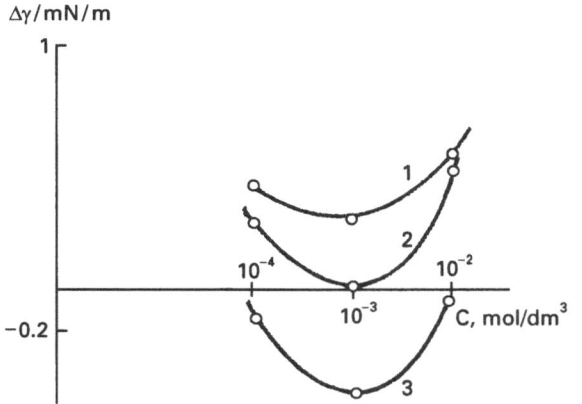

Fig. 1 Dependence $\Delta\gamma$ from concentration C and the surface age t for aqueous solutions of NaCl: (1) $t = 250 \div 500$ s (2) $t = 17$ s; (3) $t = 12$ s

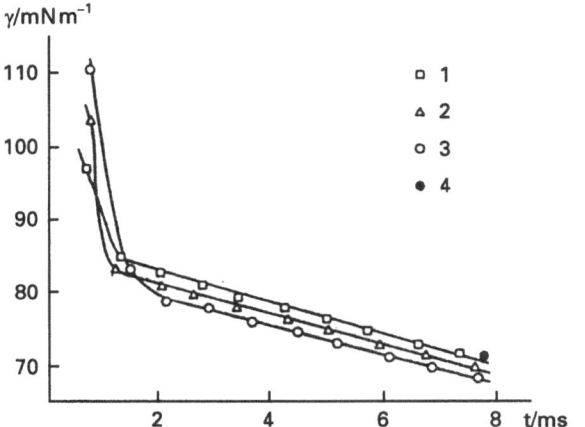

Fig. 2 Dynamic surface tension in water-cetyltrimethylammonium bromide system. $c = 5$, 6.10^{-4} mol dm^{-3}; $T/°C$: (1) 15, (2) 20, (3) 25, (4) from Ref. 4

deformed. Then the extremum of the surface tension arises (Fig. 1), shown by [3]:

$$\Gamma(t,c) = -\frac{c}{2RT} \cdot \frac{d\gamma(t,c)}{dc} - q\frac{d\chi(t,c)}{dc}, \qquad (1)$$

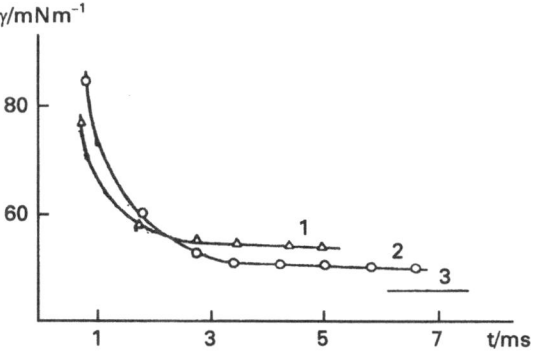

Fig. 3 Dynamic surface tension in water-sodium monobutylnaphthalenesulfonate system c/mol dm^{-3} = 2.8×10^{-4}: $T/°C$ (1) 15; (2) 25; (3) $\gamma = \gamma_\infty$ (equilibrium)

where Γ is the adsorption, and q is the surface density of charge.

Temperature inversion of the dynamic surface tension

The temperature inversion is observed at small age of surface; the surface tension increases with temperature, whereas on older surfaces it decreases with temperature [1, 4]. On the fresh surface the adsorption is still far from equilibrium, and the dynamic surface tension of the solution is close to the dynamic surface tension of water and largely determined by the character of the relaxation of the surface structure of water. Special experiments with a jet of water showed that heating the jet to 40–50°C restores the equilibrium value in 3 ms. In water a temperature 40°C is known as the Mendeleev temperature, at which the network structure of the water is loosened. This process strongly increases the rate at which the equilibrium structure of the water surface is established upon increasing the temperature (Figs. 2, 3).

References

1. Kochurova NN, Rusanov AI (1981) Kolloidn Zh 43:36–40
2. Kochurova NN, Rusanov AI (1984) Kolloidn Zh 46:9–13
3. Kochurova NN, Rusanov AI, Myrsakhmetova NO (1991) Dokl AN SSR 316:1425–1428
4. Kochurova NN, Rusanov AI (1994) Russian Chem Rev 62:1083–1095

Progr Colloid Polym Sci (1996) 100:330–337
© Steinkopff Verlag 1996

G. Bähr
P. Grigoriev
M. Mutz
E. John
M. Winterhalter

Electric potential differences across lipid mono- and bilayers

G. Bähr (✉) · M. Winterhalter
Biozentrum der Universität Basel
Abt. Biophysikalische Chemie
Klingelbergstraße 70
4056 Basel, Switzerland

P. Grigoriev
Institute of Cell Biophysics
of the Russian Academy of the Sciences
Puschino 142292, Russia

M. Mutz · E. John
Ciba Geigy AG
Physikalische Chemie
FD 3.12
4002 Basel, Switzerland

Abstract We investigate electric potential differences across lipid mono- and bilayers. Apart from externally applied electric fields, these differences are created by two sources: either by ions (dissociation of ionic groups or accumulation at the lipid layer) or by a net orientation of dipoles. The first leads to so-called surface potentials and is fairly described by the Gouy-Chapman theory. The second yields to the dipole potentials and a complete theoretical model is still lacking. Both sources change the electric potential in the lipid moiety. This internal electric potential provides a selective barrier for ion transport across membranes, and its control is the primary target of many drugs. Moreover, changes in the electric transmembrane gradient modify the elastic properties of a membrane and eventually even may induce shape transitions of liposomes or cells. We injected various concentrations of three different types of molecules either into the subphase below a lipid monolayer or into the phases adjacent to a planar lipid bilayer and recorded the successive change in electrical transmembrane potential difference. The first molecule was polyethylene glycol (PEG) which is supposed to modify the water structure at high concentrations. The second was Dibucaine, a local anesthetic, which has a high affinity, being accommodated between the lipid headgroups and the lipid tails. The third substance was Salmeterol, a bronchodilator, which has a higher affinity to the lipid phase than Dibucaine but the localization inside the lipid membrane is not known. The different action of the substances on mono- and bilayer potentials is discussed.

Key words Dipole potential – surface potential – lipid membrane – electric field – inner field compensation

Introduction

Electric fields play a major role in nature. External electric fields are powerful tools for cell characterization and manipulation [1, 2]. Important biotechnological applications are, e.g., electroinjection of macromolecules into living cells or laser trapping. More recently, pulsed electric fields have been used to treat skin cancer. Electric fields also can enhance the drug delivery pathway through the skin, a method known as iontophoresis [3]. Although the above-mentioned applications are in widespread use, little is known about the underlying processes. Besides externally applied fields, the electric field gradient across membranes is basically caused by two contributions:

One contribution is dissociated ionic groups [2–6]. It depends on the ionic strength of the surrounding electrolyte and is well described by the Gouy-Chapman theory.

This theory assumes a surface with a homogeneous smeared charge density in contact with a fully dissociated electrolyte. The counterions are taken as non interacting point particles. This theory describes astonishingly well the ion distribution opposed to the charged interface and delivers a relation between the surface charge density and the surface potential. The surface potential of a planar interface, $V_{surface}$, of less than 25 mV can be linearized (Debye-Hückel-approximation) and simplified for a 1:1 electrolyte to [7]:

$$V_{surface} = (\sigma/\varepsilon\varepsilon_0)\, \lambda_{Debye}$$

with $\lambda_{Debye} := (\varepsilon\varepsilon_0 kT/2n_0 e^2)^{1/2}$ as the Debye length, (1)

with σ: surface charge density, $\varepsilon \approx 80$: dielectric constant of the electrolyte, $\varepsilon_0 = 8.85 \cdot 10^{-12}$ As/Vm: dielectric permittivity of the vacuum, $e = 1.6 \cdot 10^{-19}$ C: elementary charge, n_0: ion number concentration, $k = 1.38 \cdot 10^{-23}$ J/K: Boltzmann constant and T: temperature. For larger surface potentials the non-linear Poisson-Boltzmann equation has to be solved. In this case the above relation writes [7, 8]:

$$V_{surface} = (2kT/e)\, ln(p + q)$$

with $q = (1 + p^2)^{1/2}$

and $p = (\sigma e/2\varepsilon\varepsilon_0 kT)\, \lambda_{Debye}.$ (2)

Interestingly, this theory still delivers satisfactory results even when the average distance between the individual surface charges is larger than the Debye-length [6, 9]. This was explained by the relatively rapid lateral movement of the lipid headgroups. In the other limit of high surface charge densities the Poisson-Boltzmann theory has to be modified to account for charge-charge interactions. The correspondence of the predicted values for surface charge densities and electric surface potentials by the Gouy-Chapman theory and measured values depends on the location of three planes. First on the lipid-water interface. It has a finite thickness, its location is not precisely defined and in reality shows some roughness on molecular dimensions. Second on the plane of the surface charges and third on the plane of measurement [6]. As the electric potential in the vicinity of the membrane changes rapidly with the distance, the actual plane of measurement is very crucial for the extrapolation on the surface potential.

The other contribution to electric field gradients across membranes is caused by a net orientation of dipolar groups [4, 10–14]. Unlike the origin of the first contribution, the source of the dipole potential is still a matter of discussion. The major part stems from the carbonyl groups in the lipids, another part is due to the terminal methyl end groups. Surprisingly the contribution from the headgroups is only minor and shows in some cases even the opposite sign of the total dipole potential [12–14]. Another candidate likely to show a strong influence on the dipole potential is the water layer next to the lipid membrane [15]. Several investigations in this direction were performed but a definite answer is still lacking [4, 15].

The internal electric potential contributes to the size of the energy barrier for charged particles in a membrane and thus controls the passive transport of ions across a membrane [4]. Typically lipophilic positive ions have a permeability of about four orders of magnitude less than their negative counterparts. The modification of this barrier is the primary target for many drugs and of great importance for cell signal trafficking. Furthermore, a modification of these internal potentials leading to a change of the electric transmembrane gradient can change the elastic properties of membranes and induce even shape transition of cells [16–18]. In recent years it was suggested that dipole potentials are responsible for the hydration force between membranes [11, 15]. A simplified theory suggested that the hydration pressure at the lipid water interface should be proportional to the square of the dipole potential [19–21]. Many experimental results support such a conclusion [4, 11, 20]. However, in a recent investigation based on a comparative study of lipids with and without carbonyl groups no difference in the hydration pressure was found although the dipole potentials were significantly different [15]. To date there is no explanation for these conflicting findings.

Electrical surface or dipole potentials of lipid moieties are usually measured on two model systems [10]. The first one is the monolayer at the air-water interface and the second one is the planar lipid bilayer [22]. Dense lipid monolayers at the so called bilayer equivalence surface pressure of 32 mN/m show potentials of about 250–600 mV [12, 13, 22] depending on their composition. But measurements on bilayers yield only 110–220 mV [22], for the same composition. Until now there is no reasonable model to explain these conflicting data. One reason for the higher potentials of lipid monolayers could be the higher asymmetry at the air-water interface than in the bilayer where the potential may decay on both sides into the water phase. However, this is not sufficient to explain the large discrepancy observed, which motivated us to investigate effects which possibly can lead to a better explanation. Therefore we added various membrane active macromolecules into the subphase of a monolayer and in the phases adjacent to a planar bilayer and studied their influence on the electric field gradient. In the following we report our preliminary results.

Materials and methods

Materials

Polyethylene glycol (PEG, Fig. 1A) with molecular weight 35 000 was purchased from Fluka (Buchs, Switzerland). A concentrated stock solution of 20% (w/v) was made by dissolving the PEG into ion-exchanged water (specific resistance $> 17\,\mathrm{M\Omega cm}$) and 150 mM NaCl (Fluka). Dibucaine (Fig. 1B) was purchased from Siegfried Handel (Zofingen, Switzerland). A 1M solution was prepared in a 10 mM phosphate buffer with 0.1 M NaCl at pH 7.4. The partition coefficient at the lipid-water interface was found to be $11\,000\,\mathrm{M^{-1}}$ (calculated from the data in [27]). In [23], a value of $660\,\mathrm{M^{-1}}$ (at pH 5.5, 0.1 M NaCl/50 mM buffer) was published, but this value takes only the partition due to the hydrophobic effect into account and not the interaction due to charges on lipids and Dibucaine. Salmeterol (Fig. 1C) was a generous gift from Ciba-Geigy

and used in a 1 mM solution containing 0.15 M NaCl and 10 mM phosphate buffer at pH 7.0. Dialysis measurements on this system yield a partition coefficient of about $12\,500\,\mathrm{M^{-1}}$. For the monolayer measurements a diluted lipid stock solution (0.1% (w/v)) was made by dissolving DPhPC (1,2-Diphytanoyl-sn-Glycero-3-Phosphocholine), POPC (1-Palmitoyl-2-Oleoyl-sn-Glycero-3-Phosphocholine) and POPC/DOPS (1,2-Dioleoyl-sn-Glycero-3-[Phospho-L-Serine]) (all from Avanti Polar Lipids, Alabaster, Al) in chloroform (Fluka). For bilayer measurements higher concentrated (1% (w/v)) stock solutions of the lipids were used.

Monolayer measurements

Lipid monolayers were spread with Hamilton Syringes (Hamilton, Bonaduz, Switzerland) at the air-water interface by adding about 10 µl of the stock solution to the Teflon trough (surface of about 30 cm^2, volume 32 cm^3). About 15 min were allowed for evaporation.

The surface potential was measured with the vibrating plate method as described previously [24]. First the surface potential of the pure air-water interface was recorded for control. In the following step the lipid film was spread. The voltage U_{min} now needed to prevent an electric current across the vibrating plate and which depends directly on the surface potential was used as reference. The surface active substance, either PEG, Dibucaine or Salmeterol was injected carefully from the side avoiding turbulence. The chamber was stirred for at least 5–15 min until a homogeneous distribution was assured. The subsequent change of U_{min} due to the action of the added substance we called ΔU_{min}. During all measurements the surface pressure was recorded with a Wilhelmy plate as described previously.

Bilayer measurements

We used a slightly modified version of the inner field compensation method first described by Sokolov and Kuzmin [25]. This method is based on the dependence of the lipid bilayer capacitance on the electric potential gradient across the membrane. A superposition of a sinusoidal AC voltage with frequency $\omega = 1010$ Hz and a DC voltage varied in the range of ± 100 mV is applied via Ag/AgCl electrodes on both sides of the bilayer. The capacity of the bilayer depends in a nonlinear way on the electric transmembrane gradient. This implies that the AC current through the membrane contains a component with the double frequency 2ω of the applied sinusoidal voltage (second harmonic) whose amplitude is proportional to the time-independent component U_0 of the actual

A. Polyethylene glycol

$$\mathrm{H[-O-CH_2-CH_2-]_nOH}$$

B. Dibucaine

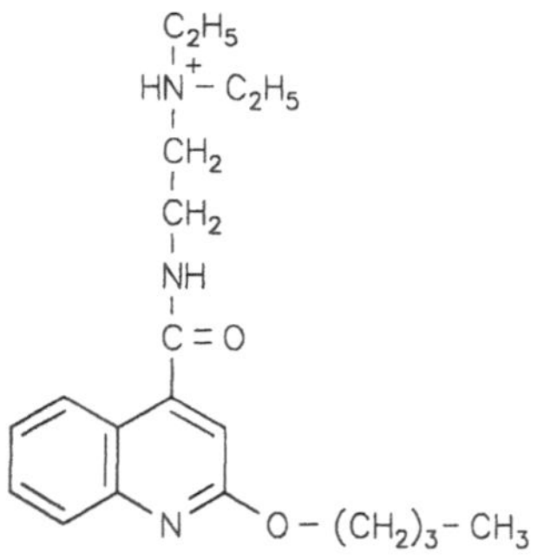

C. Salmeterol

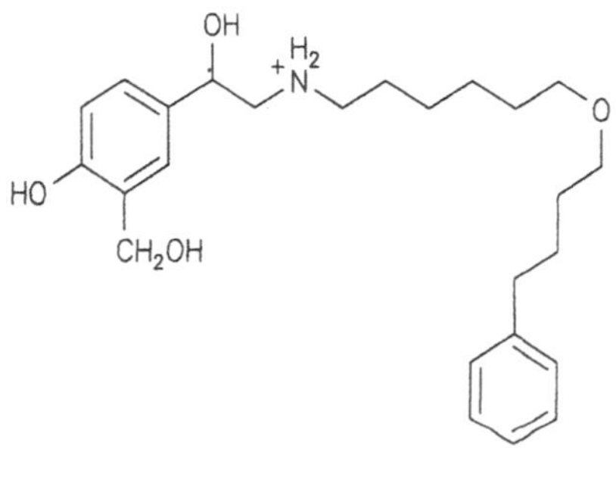

Fig. 1 Chemical structure of polyethylene glycol (A), Dibucaine (B), and Salmeterol (C)

transmembrane field [26]. When the applied DC voltage at the value U_{min} compensates the intrinsic transmembrane voltage, U_0 vanishes and the amplitude of the second harmonic signal monitored on an oscilloscope reaches a minimum (in theory it reaches zero, but due to noise, one only can see a minimum). The difference ΔU_{min} in the value of the DC component for the minimal second harmonic signal before and after addition of the substance is a measure for the change of the transmembrane electric potential difference caused by the substance. To measure this difference we used a set-up which allows the second harmonic signal to be observed directly on the oscilloscope. The corresponding block diagram is shown in Fig. 2. It includes 1) a sinewave voltage generator (Hewlett Packard, Model 200 CD, wide range generator), 2) a reject filter of the second harmonic, 3) a variable DC voltage source, 4) the cell with the bilayer lipid membrane (BLM), 5) a current to voltage converter (Keithley, 427 Current Amplifier), 6) a reject filter of the first harmonic, 7) a selective amplifier tuned to the second harmonic, 8) an oscilloscope (Tektronix 7403 N) and 9) a voltmeter (Fluke 77 Multimeter). Not shown in the scheme of Fig. 2 is a low pass filter (3kHz) and an amplifier which amplified the

signal five times after the selective amplifier of the second harmonic. The adsorption of amphiphilic ions was judged by the change ΔU_{min} of the electric potential difference across the membrane after introduction of a sample into one of the cell compartments. The change of the signal over a longer period without adding substance is less than 7 mV, as we checked with several measurements up to 100 min duration. Part of this change could be due to electrode effects. Before measuring we waited for a stable value of the DC voltage needed for the minimum of the second harmonic signal with fluctuations less than 3 mV for at least 5 min. The most points in the figures are averages of up to five measurements in less than 1 min. The standard deviation of this series is in most cases less than 1 mV.

The bilayer was made in the hole of a wall separating two compartments of a Teflon cell. The volume of the two compartments was 5 ml each and the diameter of the hole approximately 0.4 mm. The hole was prepainted with 5 μl of a 1% (w/v) lipid solution (POPC:POPS 9:1 in isooctane) on each side. We waited 10 min for evaporation of the solvent. Then the compartments were filled with the electrolyte (50 mM KCl). On a Teflon loop 2 μl of the 1% (w/v) lipid solution were applied. With this loop a bilayer could be easily drawn at the trans side of the cell. Via the capacitance of the membrane and a microscope for optical inspection, the development of the bilayer could be observed.

Results

Effect of polyethylene glycol

In a first series of measurements we added polyethylene glycol (PEG-35,000) to the aqueous subphase of a monolayer trough. The pure air-water interface was taken as a reference. At a bulk concentration of about 10^{-6}% (w/v) the surface potentials starts to raise with increasing PEG concentration. At about 10^{-3}% (w/v) a plateau value of 440 mV is reached. At higher bulk concentrations than 0.1% (w/v) the potential started to decrease. At 1% (w/v) the potential decreased to 360 mV and at 10% (w/v) 200 mV were reached. This measurement was repeated in presence of a dense lipid monolayer of DPhPC. A dense lipid monolayer (surface pressure of 32 mN/m) of DPhPC showed a surface potential of about 450 mV in agreement with previous measurements. In this case injection of PEG-35,000 into the subphase at first did not show any variation on the surface potential. Only at about 0.1% (w/v) again a similar decay as in absence of the lipid monolayer was observed. This general feature was observed also with lipid monolayers made up from different lipids. Depending on the lipid composition a dense layer

Experimental setup for the measurement of the "Second Harmonic"

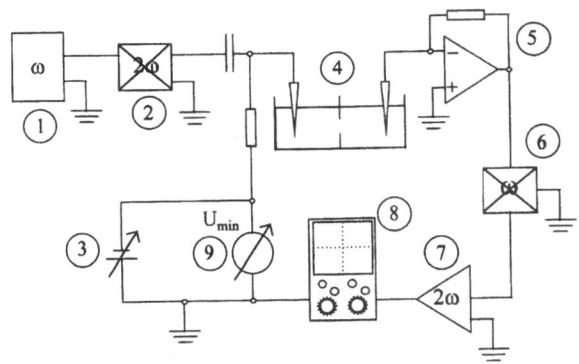

1) sine voltage generator   2) reject filter of the second harmonic

3) tunable DC voltage source   4) cell with BLM

5) current-voltage converter   6) reject filter of the first harmonic

7) selective amplifier of the second harmonic   8) oscilloscope

9) volt meter

Fig. 2 Scheme of the inner field compensation method as described in material and methods

shows surface potentials at the air-water interface between 250–600 mV [12, 13, 22]. The decay at high polymer concentrations in the bulk was observed for a great variety of PEGs of different polymerization degree ranging from 2000 to 100 000. We concluded from this measurement that PEG does not penetrate the lipid phase. At low concentrations it forms random coils in the water phase. At higher concentrations, corresponding to densities at which the polymer coils start to touch each other, it apparently forms partly higher ordered structures. This conformational change seems to cause a reduction of the dipole potential of the boundary layer between bulk phase and air, which comprises not only the lipid monolayer, but also the region where concentration, orientation or conformation of the molecules deviates from the bulk. One explanation could be a depletion of the water layer next to the membrane or a reorientation of the water molecules which have a relatively high dipole moment. Another explanation could be the appearance of a net dipole potential of the PEG molecule itself, caused by the higher order of the polymer after the conformational change.

These results motivated us to investigate whether such an effect can be observed on a bilayer. Surprisingly, no significant influence on the electrical properties such as conductance, capacitance and electrical transmembrane gradient of the bilayer could be observed. A possible explanation might be that in the monolayer experiments we are measuring voltage by a device with very high internal resistance (voltmeter configuration) what leads to a basically static (i.e. without an electrical net current) measurement. But in bilayer experiences we recorded the electrical current by a device having very low internal resistance (amperemeter configuration), which implies flow of ions. Likely the ordered structure of PEG at high concentrations is located at some distance from the lipids, modifying only the static potential. The PEG there is surrounded by mobile water and does not modify the electrical current.

Effect of Dibucaine

In a second series we investigated the effect of Dibucaine, a local anaesthetic of MW 379. The chemical structure is shown in Fig. 1B. The amino group has a pK of about 8.5; this causes at pH 7.4 about 93% of the drug to be positively charged [27]. The driving force for incorporation in the lipid phase is likely the hydrophobic force stemming from the aliphatic tail so that the molecule will penetrate the hydrophobic core of the lipid. At pH 7 the partition coefficient between the bulk water phase and the air-water interface is about 80 000 M^{-1} [28] whereas that for the lipid-water interface is about 11 000 M^{-1} at a surface pressure of 32 mN/m (calculated from [27]). Careful studies

at the air-lipid interface suggested an area per incorporated drug molecule of about 0.55 nm^2. It was proposed that the charged amino group of Dibucaine is located in the headgroup region of the lipids.

In the first part a lipid monolayer of POPC/POPS 9 : 1 was spread as described in Materials and methods. The lipid monolayer was compressed until a surface pressure of 32mN/m was reached. Recent investigations on lipid monolayers showed that at this pressure many physical properties of Dibucaine on monolayers are comparable to those of Dibucaine interacting with bilayers [27]. We first recorded the surface potential of the dense lipid monolayer without Dibucaine and used it as reference. In the following we injected Dibucaine into the subphase. After a few minutes of gentle stirring we observed an increase in the electric surface potential. In Table 1, we show the concentration-dependent plateau values typically reached after 15–60 min. Simultaneously we monitored the successive increase in area due to the insertion of the drug at constant surface pressure. The increase in area per drug agreed with previously published values [27].

In the second part we performed incorporation measurements on bilayers made of the same lipid composition as the monolayers. Here the drug was added asymmetrically on one side of the bilayer. Incorporation of the positively charged Dibucaine then caused a concentration dependent change of the transmembrane potential difference. In Fig. 3A we show a typical development of the

Table 1 Effect of Dibucaine and Salmeterol on the surface potential of lipid monolayers and on the potential difference across bilayers. ΔU_{min} denotes the change of the respective quantity (absolute value) after adding the substance. The conditions for subphase and electrolyte, respectively and the used lipids are given in the text

Bilayer Conc. [µM]	ΔU_{min} [mV]	Monolayer Conc. [µM]	ΔU_{min} [mV]
Dibucaine			
4	6	1000	20
16	14	2000	37
40	24	2300	30
60	14	4000	88
64	30		
80	29		
144	43		
200	39		
300	52		
304	59		
Salmeterol			
3.4	16	1.4	27
3.4	11	2.8	39
3.4	12	4.2	40
3.4	13	50	95
3.4	15	226	136
Average:			
3.4	13		

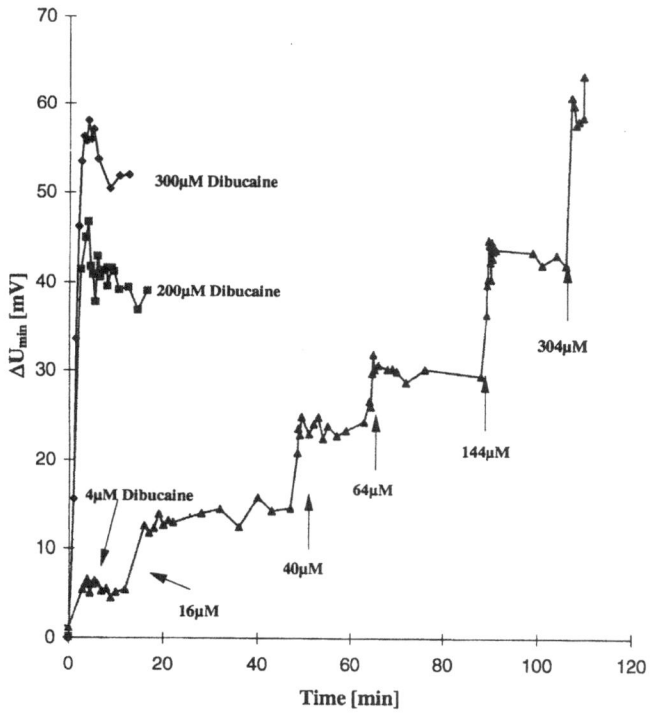

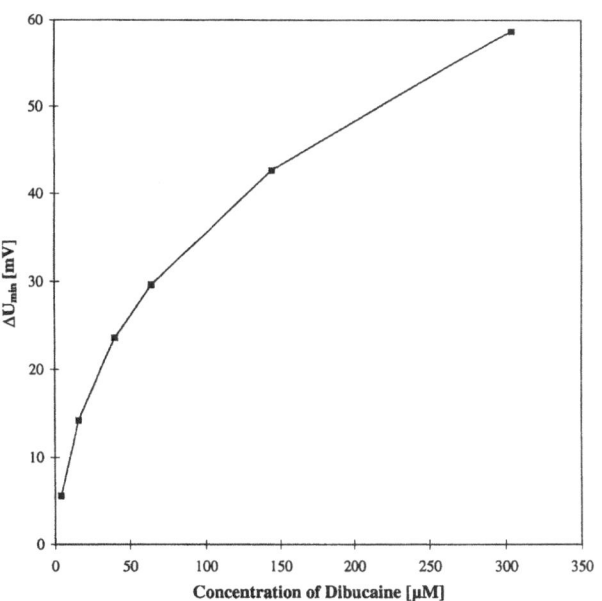

Fig. 3 A) Change of transmembrane potential difference ΔU_{min} vs. time after repeated addition of Dibucaine on right side [cf. Fig. 2] of the lipid bilayer. The potential was measured with the inner field compensation method as described in the text. B) Change of transmembrane potential difference ΔU_{min} vs. concentration of Dibucaine on right side of the lipid bilayer. The potential was measured with the inner field compensation method as described in the text

potential difference with time when Dibucaine is added repeatedly. Always within a few minutes a plateau value was reached. A summary is shown in Fig. 3B demonstrating

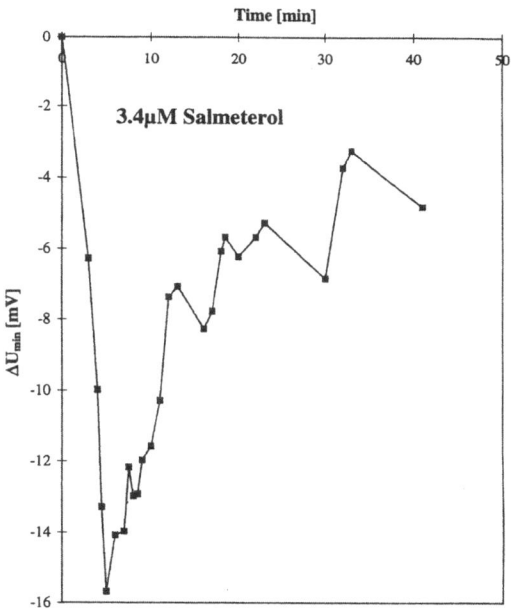

Fig. 4 Change of transmembrane potential difference ΔU_{min} vs. time after adding Salmeterol on left side [cf. Fig. 2] of the lipid bilayer. The potential was measured with the inner field compensation method as described in the text

the surface potential versus bulk concentration of the drug. It should be noted that the slope in Fig. 3B only apparently leads to a plateau for concentrations about one order higher than in figure 3B. According to Eq. (1) the potential up to a transmembrane potential of 25 mV is proportional to the charge density and hereby to the amount of adsorbed drug. Larger potentials require the application of the full Poisson-Boltzmann equation and a logarithmic relation between surface potential and surface charge density is obtained (Eq. (2)).

Effect of Salmeterol

In a third series, we investigated the effect of the cationic drug Salmeterol of MW 415. The partition coefficient was found to be about 12 500M^{-1}. Inspection of the chemical structure (Fig. 1C) shows that again a positive charge stems from the protonated amino group at pH 7. In contrast to Dibucaine, here the amino group is located nearer the hydrophobic part of the molecule. Therefore we expect the charge plane to be deeper inside the lipid layer.

In the first part, we spread a lipid monolayer of POPC/DOPS 9:1 as described in Materials and Methods. Again the lipid monolayer was used as reference. Injection of Salmeterol into the subphase caused an increase in the surface potential (see Table 1) within a few minutes. In contrast to the previous experiment, here we used a mono-

336
G. Bähr et al.
Electric potential across mono and bilayers

layer trough with fixed area. Instead of an increase in area the insertion of the drug therefore caused an increase of the surface pressure.

In the second part we performed incorporation studies on lipid bilayers. In Fig. 4 we show a typical increase in the transmembrane potential after addition of Salmeterol on one side. As expected, the potential increases during the first 5 min. But instead of increasing monotonically further as in the case of Dibucaine, the surface potential suddenly starts to decrease. We attributed this finding to a translocation or permeation of the drug to the other side.

Conclusion

We presented here our preliminary studies on the action of various surface active substances on the electric potential difference across membranes.

The first molecule in our investigation was polyethylene glycol (PEG). From previous studies it is known that PEG in high concentrations changes the water structure [29, 30]. On the other hand, to date no direct interaction between PEG and dense lipid membranes could be demonstrated. Our monolayer results suggest that at high concentrations part of the PEG in the aqueous phase has a higher ordered structure causing a decrease in the monolayer dipole potential. This change in potential was not visible in our bilayer measurements. A possible explanation could be that the two measurements are recording different physical properties. Monolayer in contrast to bilayer dipole potential measurements require no flow of ions across the membrane. In bilayer experiments, the potential difference caused by a small imbalance of the charge concentrations on both sides and partially oriented dipoles, is prone to collapse when a certain amount of charges begins to flow.

In Fig. 5, we summarized the effect of two membrane binding drugs having a positively charged amino group. Interestingly, Salmeterol with the charged group located nearer to the hydrophobic part of the molecule showed lager effects in the mono-as well as in the bilayer measurements. On the other hand, this could partially be also due to the slightly higher partition coefficient of Salmeterol

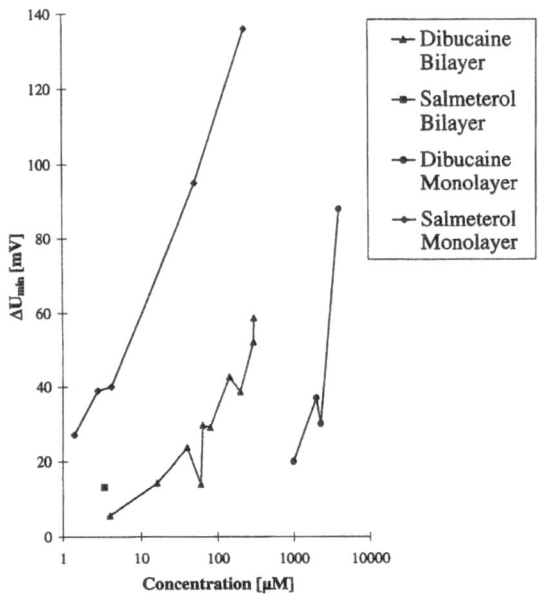

Fig. 5 Comparison between mono- and bilayer measurements of Dibucaine and Salmeterol. Data are obtained from Table 1

against Dibucaine ($12\,500\,\mathrm{M}^{-1}$ vs. $11\,000\,\mathrm{M}^{-1}$ at the lipid-water interface). Our studies on the time-dependent increase in the transmembrane potentials upon addition of a drug allow to conclude on the kinetics of the drug binding.

The method of second harmonic generation allowed us to measure the binding of charged molecules to a planar bilayer and, in the case of Salmeterol, even the permeation of the membrane could be determined with high precision. Both topics are of particular interest in the research of drug delivery.

Further studies remain to be done. Especially the kinetics of the binding and permeation, the dependence of pH and salt concentration needs more experiments. Also of interest is the comparison of monolayer and bilayer data at the same concentrations, where more data is needed (compare Table 1 and Fig. 5).

Acknowledgement We wish to thank Rolf Jiricek, Gerhard Wackerbauer and Ingrid Weiss for helpful discussions. This work was supported by grant No. 31.042045.94 from the Swiss National Science Foundation.

References

1. Chang DC et al (eds) (1992) Guide to Electroporation and electrofusion. Academic press, New York
2. Winterhalter M (1995) In: Lasic DD, Barenholz Y (eds) Nonmedical applications of liposomes. CRC Press, Boca Raton
3. Orlowski S, Mir LM (1993) Biochim Biophys Acta 1154:51
4. Cafiso, DS (1995) In: Disalvo EA, Simon SA (eds) Permeability and Stability of lipid bilayers CRC Press, Boca Raton
5. McLaughlin S (1989) Ann Rev Biophys Chem 18:113
6. Wininski AP et al (1986) Biochem 25:8206
7. Hiemenz PC (1986) Principles of Colloid and Surface Chemistry, Marcel Dekker Inc., New York and Basel
8. Winterhalter M, Helfrich W (1992) J Phys Chem 96:327

9. Nelson AP, McQuarrie DA (1975) J Theor Biol 55:13
10. Brockman H (1994) Chem Phys Lipids 73:57
11. Cevc G (1995) Biophys Chem 55:43
12. Vogel V, Möbius DJ (1988) Coll Interface Sci 126:408
13. Beitinger H et al (1989) Biochem Biophys Acta 984:293
14. Smaby JM, Brockman HL (1990) Biophys J 58:195
15. Gawrisch K et al (1992) Biophys J 61:1213
16. Glaser R (1982) J Membr Biol 66:79
17. Farge E, Devaux PF (1992) Biophys J 61:347
18. Hauser H, Gains N, Müller M (1983) Biochem 22:4775
19. Marcelja S, Radic N (1976) Chem Phys Lett 42:129
20. Cevc G, Marsh D (1985) Biophys J 47:21
21. Simon SA, McIntosh TJ (1989) Proc Nat Acad Sci (USA) 86:9263
22. Pickar AD, Benz R (1978) J Membr Biol 44:353
23. Seelig A, Allegrini PR, Seelig J (1988) Biochim. Biophys Acta 939:267
24. Winterhalter M et al (1995) Biophys J 69:1372
25. Sokolov VS, Kuzmin VG (1980) Biophysics 25:174
26. Carius W (1976) K Colloid Interface Sci 57:301
27. Seelig A (1987) Biochim Biophys Acta 899:196
28. Wackerbauer G. PhD thesis.
29. Herrmann A et al (1983) Biochim Biophys Acta 733:87
30. Harris JM (1993) Poly(ethylene glycol) Chemistry, Plenum Press, New York

Progr Colloid Polym Sci (1996) 100:338–344
© Steinkopff Verlag 1996

INTERFACES, FILMS AND MEMBRANES

Evidence of entropic contribution to "hydration" forces between membranes

V.I. Gordeliy
V.G. Cherezov
A.V. Anikin
M.V. Anikin
V.V. Chupin
J. Teixeira

Dr. V.I. Gordeliy (✉)
Structural Biology
Forschungszentrum Jülich
52425 Jülich, FRG

V.I. Gordeliy · V.G. Cherezov
Frank Laboratory of Neutron Physics
Joint Institute for Nuclear Research
Dubna, Russia

A.V. Anikin · M.V. Anikin · V.V. Chupin
Moscow Institute for Fine Chemical
Technology
Moscow, Russia

J. Teixeira
Leon Brillouin Laboratory
CE de Saclay
France

Abstract The main purpose of this work was to measure the forces between phospholipid membranes, where out-of-plane motions of membrane surface (undulations as well as "protrusions") are suppressed.

With the help of small-angle and wide-angle x-ray diffraction out-of-plane and in-plane structure of polymeric phospholipid membranes of 1,2-bis(11, 13) tetradecadienoyl-sn-glycero-3-phosphocholine (DTDPC) was studied.

Small-angle x-ray diffraction study confirms the previous conclusions that polymeric DTDPC membranes are rigid due to high level of polymerization in the central part of hydrophobic region.

The measured via osmotic stress method forces between membranes show a dramatic decrease of membrane hydration in comparison with that of usual phospholipid membranes. It does not exceed two water layers between membranes. The repulsive forces decay quickly with the decay length of 0.7 Å.

Wide-angle x-ray diffraction showed the in-plane packing of lipid molecules similar to that of the usual phospholipid membranes in liquid phase. It means that hydration of polymeric DTDPC membranes cannot be explained in the framework of "hydration" (polarization) hypothesis of the origin of short-range repulsive forces between membranes.

Thus, if out-of-plane motions of membrane surface are "switched off", the repulsive forces dramatically decrease. This, alongside with the recent neutron scattering study of out-of-plane and in-plane membrane structure, shows that entropic repulsive forces between phospholipid membranes are dominating, at least, at the intermembrane distances larger than a few angstroms.

Key words Hydration force – polymeric membranes – x-ray diffraction – structure – interactions

Introduction

It seems as though I. Langmuir was the first to recognize that short-range repulsive forces at small distances between colloid particles cannot be described in the framework of attractive van der Waals and repulsive double-layer electrostatic forces: "The fact that particles in the bipolar coacervates remain separated by considerable distances and do not come into contact proves the presence of some kind of repulsive force .... Hydration seems to be most reasonable explanations" [1].

However, systematic study of short-range repulsive forces was started in 1976, when D.M. Le Neveu R.P.

Rand and V.A. Parsegian proposed the osmotic stress method to measure the forces between membranes [2]. For the first time, they applied the method to the measurement of forces between lecithin bilayers [2]. The pressure-distance (P-d_w) experimental dependence was quite well fitted with the equation $P = P_0 \exp[-d_w/\lambda]$, where the decay length λ is equal to 1.9 Å [2].

It was clear that the measured forces could not be described by the DLVO theory [3]. A short time after this work ([2]) S. Marcelja and N. Radic proposed a phenomenological theory of short-range forces between membranes in aqueous solutions [4]. The basis of the theory is a hypothesis that "near the surface with lecithin, water molecules have preferred orientation" [4]. The second hypothesis is that the order imposed by the interfaces is described by the order parameter $\eta(x)$ [4]. The Landau–Guinsburg expansion of the free energy density with correspondent uniform boundary conditions for $\eta(x)$ lead to an exponential law for repulsive forces $P = P_0 \exp[-d_w/\lambda]$. It was in good agreement with the experiment [2, 4].

This experimental and theoretical success motivated intensive experimental [5] and theoretical studies of short-range forces between membranes [6–12]. Interpretation of the experiments as well as the theories [6–12] were mainly based on the same "hydration" hypothesis about the origin of these forces as in [2].

However, the order parameter which deals with water polarization near membrane surface has not been measured or calculated. This means that these theories could not predict a fundamental thing: the magnitude of forces. Moreover, experiments show that the decay length λ of hydration forces varies in a wide interval (0.8 ÷ 2.4 Å) being dependent on the kind of membrane and even on its phase state [5]. The existing phenomenological theories were not able either to calculate the decay length of the forces or to explain its variation [5–12].

Only in 1989 was an unconventional explanation of the variation of the decay length value of the force proposed by Kornyshev and Leikin [13]. They used the same order parameter formalism as in [2]. However, inhomogeneous boundary conditions for the order parameter on membrane surface were assumed; i.e., it was considered that mean solvation "force fields" were produced by discrete polar groups. It was shown that the hydration force parameters P_0 and λ are connected with the surface distribution (packing) of lipid polar groups [13].

An important consequence of this approach is: the more orderly the in-plane packing of polar heads the shorter is the decay length of hydration forces [13]. The authors tried to explain the experimentally observed variation of λ. However, a drawback of such analysis was the use of the assumptions about the in-plane structure of membrane surface but not of the experimental data [13].

Thus, the hydration mechanism of the origin of short-range repulsive forces between membranes is based on certain assumptions about the structure and properties of membrane interface [2–13].

However, inspite of a widely spread belief in "hydration paradigm" at that time the basic hypotheses of the conception were not directly proved (or rejected).

In 1990, Israelachvili and Wennerström proposed an alternative and completely different explanation of the origin of short-range repulsive forces between amphiphilic surfaces. They concluded that these forces "originate from the entropic (osmotic) repulsion of molecular groups that are thermally excited to protrude from these fluid-like surfaces" [14].

Similar ideas were proposed by G. Cevc in [15]. The dependence of the propagation length of intermembrane interaction on the interface width of the polar membrane was demonstrated theoretically [15].

Thus, the alternative approach is based on the hypothesis of "short-range" out-of-plane fluctuations of amphiphilic molecules [15, 16]. In fact, the basis of this approach is not directly proved by an experiment either.

The discussion which followed showed credible pros and contras for both approaches [16–21], stressing once again the ambiguities of interpretation of experimental data at the present level of knowledge.

The main reason for this controversy is that there are few reliable experimental data about the in-plane and out-of-plane structure of membrane interface, while both, "hydration" and "entropic", hypotheses appeal to structural properties of interfaces.

Investigation of the fine details of in-plane and out-of-plane membrane structure is, therefore, a key to the problem of hydration forces.

The second approach to the problem could be the investigation of the forces between membranes with the known properties of their surface. In particular, a comparison of the forces between membranes, with chemically the same surface and in-plane structure but with the "switched on" and "off" out-of-plane motions, could be of crucial importance. It could allow to isolate the contribution of different kinds of sources to the total repulsive force.

Neutron diffraction with the use of highly oriented deuterium labelled lipid membranes has been recently applied to investigate the in-plane and out-of-plane structure of membrane surface [22].

Investigation of the temperature dependence of "hydration" forces was done via complementary use of neutron diffraction and small-angle scattering with lipid membranes [22, 24] and earlier via x-ray diffraction [23].

The "hydration paradigm" [2, 3] is not consistent with these experimental data [22, 24]. On the other hand, the data show the existence of thermal out-of-plane fluctuations of lipid molecules, considerable enough to result in the experimentally observed repulsive forces between membranes.

In this paper, we discuss the application of the "second" approach to the solution of the hydration force problem.

A study of the structure and interactions of highly oriented polymeric membranes of the polymerizable phospholipids with conjugated diene groups at the hydrocarbon chain ends: 1,2-bis(11,13 tetradecadienoyl)-sn-glycero-3-phosphocholine (DTDPC) [25] has been done.

The previous investigation of polymeric DTDPC membranes [25] and the present x-ray diffraction study show that DTDPC membranes are rigid. It is due to high level (> 95%) polymerization network between monolayers and along each monolayer of the lipid membrane. This means that out-of-plane motion of membrane surface is strongly suppressed. One can expect to observe in such a system only the hydration force plus steric component, which results from the steric interaction of mobile polar head groups [26].

If entropic forces between usual phospholipid membranes are dominating, then the forces between polymeric DTDPC membranes should be considerably smaller and/or shorter. The measurements of repulsive forces between DTDPC polymeric membranes via osmotic stress method, discussed in the following paragraphs of the paper, directly confirm this conclusion.

Materials and methods

1,2-bis(11,13) tetradecadienoyl-sn-glycero-3-phosphocholine (DTDPC) (Fig. 1) was synthesized in M.V. Lomonosov Institute of Fine Chemical Technology as is described in [25]. Thin-layer chromatography (TLC) was carried out on silufol UV-254 plates (Kavalier). Chromatographic purification was achieved on open columns packed with silica gel L-40/100 (Chemapol).

Highly oriented multilamellar membranes were prepared from ethanol solution of monomeric DTDPC by spreading it on quartz slides with the subsequent slow evaporation of the solvent. The rest of the ethanol was removed under vorevacuum. The samples were polymerized under x-ray radiation.

The control over the process of membrane polymerization was exerted with UV-spectroscopy and x-ray diffraction.

The samples were highly oriented with the mosaic spread $\sim 1°$.

The forces between membranes were determined via osmotic stress method [2].

X-ray diffraction, small-angle and wide-angle measurements were performed at x-ray diffractometer with Ni-filtered CuK_α radiation with the wavelength $\lambda_{x-ray} = 1.54$ Å.

The temperature was $T = 20.0 \pm 0.1$ °C. The repeat distance d of multilamellar DTDPC membranes was determined from the positions of diffraction peaks ($2d \sin \theta = \lambda_{x-ray}$, where θ is half the scattering angle).

From integral intensities the modules of structure factors were calculated by the equation

$$|F(h)| = I^{1/2}(h) L^{-1/2} P^{-1/2}, \tag{1}$$

h is the order of diffraction reflection, L-the Lorentz factor, proportional for oriented specimens (as was the case) at small angles to h, and P is the polarization factor equal to $1/2(1 + \cos^2\theta)$ [27]. The signs of structure factors were determined by the application of Shannon's sampling theorem to the structure factor data [28]. The signs of centrosymmetric membrane structure, such as DTDPC bilayers, can be equal to ± 1. Figure 2 demonstrates this approach. The electron density profile along the normal of membrane plane (the x-axis) were calculated by the equation

$$p\rho(x) = \sum_h F(h)\cos\frac{2\pi hx}{d}. \tag{2}$$

Results

A typical lamellar x-ray diffraction pattern from DTDPC membranes is shown in Fig. 3. At all hydrations six diffraction orders have been collected. The measurements were performed at 10 different values of osmotic stress in the range from 0 (the relative humidity $\psi = 100\%$) up to 2.9×10^9 dyn/cm^2 ($\psi = 12\%$).

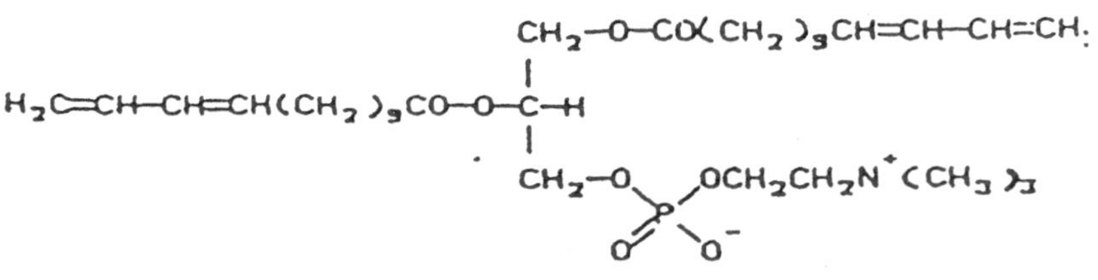

Fig. 1 Chemical structure of polymerizable 1,2-bis(11,13)tetradecadienoyl-sn-glycero-3-phosphocholine (DTDPC)

Structure factors of membranes at different humidities

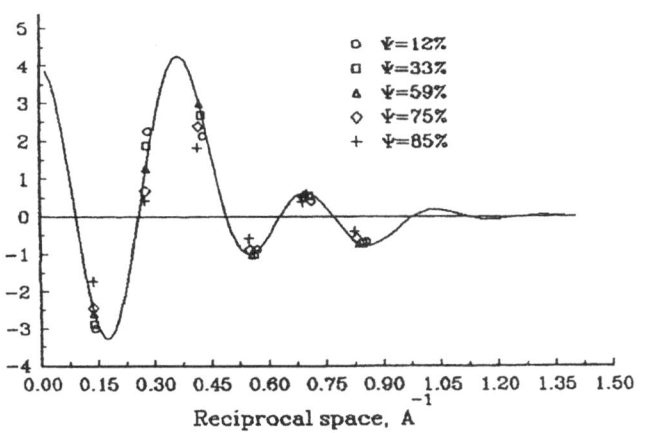

Fig. 2 Continuous transform, calculated by application of Shannon's sampling theorem (solid line) to the structure factor data for DTDPC membranes at different relative humidities Ψ

membranes from DTDPC

T=20 C, Ψ=60%

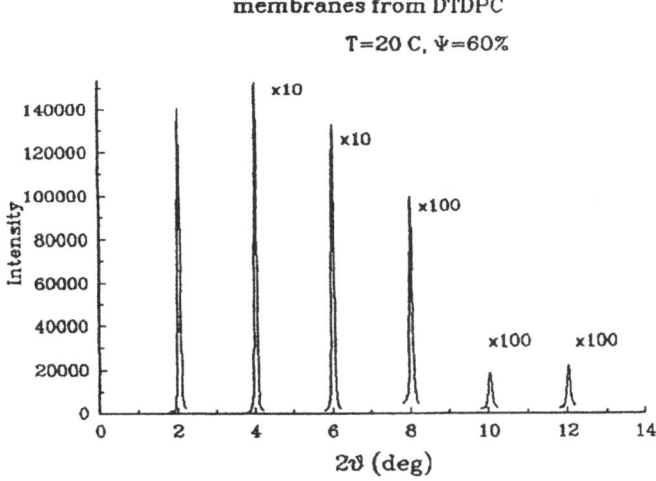

Fig. 3 X-ray lamellar diffraction pattern from DTDPC membranes at $T = 20\,°C$ and the relative humidity $\Psi = 60\%$ H_2O

Electron density profiles of membranes from DTDPC at different humidities

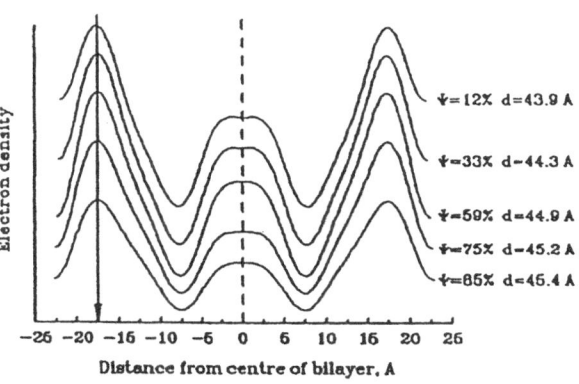

Fig. 4 Electron density profiles of polymeric DTDPC membranes at different humidities

Strip function model of bilayer from DTDPC

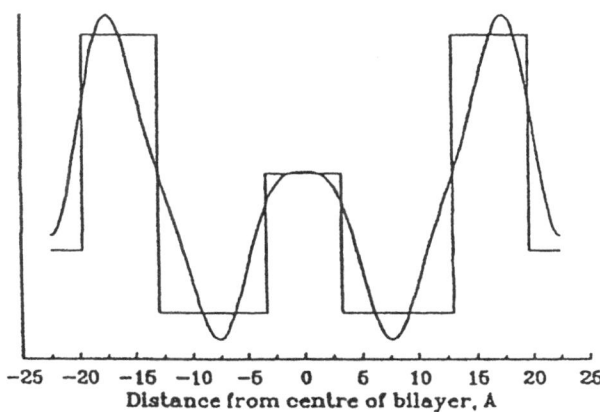

Fig. 5 Electron density profile and its strip function model for DTDPC membranes at 20 °C temperature (and relative humidity $\Psi = 75\%$ H_2O). Fit by the strip function model confirms the interdigitation of the ends of hadrocarbon chains of the opposite lipid monolayers in the lipid bilayer

The electron density profiles of polymeric DTDPC membranes calculated by Eq. (2) are shown for some membrane hydrations in Fig. 4.

The strip function model [29] was used to determine the thickness of the central part of electron density profiles (Fig. 5). This thickness is equal to 6.5 Å. Taking into account the geometry and chemical structure of the DTDPC molecule, we came to the conclusion that the ends of hydrocarbon chains of opposite monolayers are partially interdigitated. This means that there are covalent bonds between monolayers in polymeric DTDPC membranes. This is consistent with previous conclusions [25].

Another very important fact is that the positions of maxima of electron density do not change with hydration. The thickness of a DTDPC bilayer is about 43 Å. The thickness values of the central part of the bilayer (6.5 Å) and the bilayer itself (~43 Å) as well as the geometry of the DTDPC molecule lead us to the following conclusion. Hydrocarbon chains of polymeric DTDPC membranes are straight and oriented along the membrane normal. This means that membranes almost do not swell along their plane. This is consistent with the earlier discussed fact of the independence of membrane thickness from relative humidity, which is not the case with monomeric phospholipid membranes.

342
V.I. Gordeliy et al.
Evidence of entropic contribution to 'hydration' forces

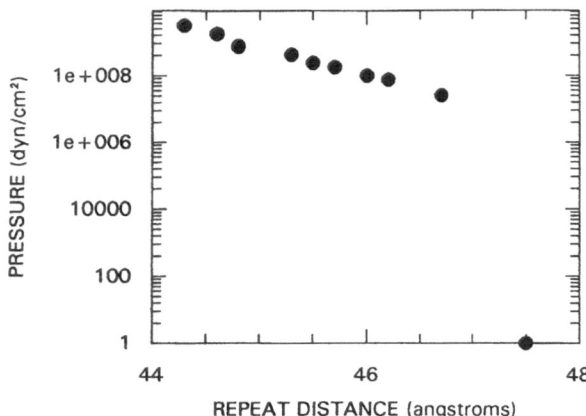

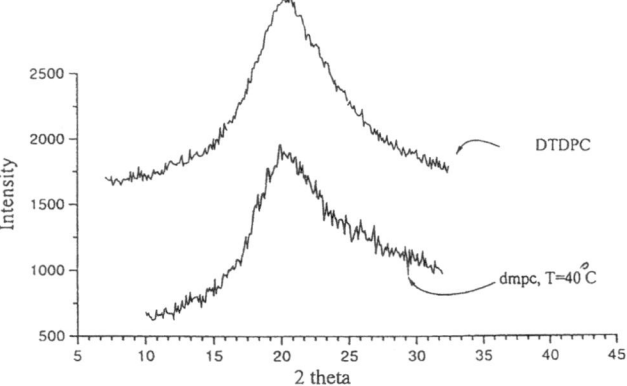

Fig. 6 The dependence of short-range forces between polymerized membranes of DTDPC on repeat distance (= intermembrane distance). The dependence can be described by equation $P = P_0 \exp[-d_w/\lambda]$, where $P_0 = 4 \times 10^9$ dyn/cm^2 and the decay length of the forces $\lambda = 0.7$ Å. The membrane repulsion decays very quickly in comparison with the monomeric state of the lipid (i.e., DMPC)

Fig. 7 Wide-angle diffraction from polymeric DTDPC and DMPC (usual, in liquid phase) membranes in excess water

The most striking phenomenon is that intermembrane distance changes only by 3.6 Å (d changes from 47.5 Å to 43.9 Å) with the increase of osmotic pressure from 0 up to 2.9 10^9 dyn/cm^2. It is much less than in the case of monomeric phospholipid membranes in liquid phase [2, 5, 19, 26].

The measured by osmotic stress method forces between DTDPC membranes are shown in Fig. 6. The dependence can be described by equation $P = P_0 \exp[-d_w/\lambda]$, where $P_0 = 4 \times 10^9$ dyn/cm^2 and $\lambda = 0.7$ Å.

Repulsion between membranes decays much more rapidly with the increase of intermembrane distance d than in the usual phospholipid membranes [5].

Discussion

The study of hydration and lipid DTDPC membrane interaction revealed a considerable suppression of short- and long-range repulsive forces (apparently resulting from the suppression of out-of-plane thermal fluctuations of membrane surface) (Fig. 6).

One should note that it is because of the fact that membrane thickness does not depend on hydration that we have only to plot the repeat period d dependence of P to be able to determine the decay length λ of the forces. As the accuracy of estimating d is much higher than that for intermembrane distance, we can assert that $\lambda = 0.7$ Å was indeed determined with good accuracy.

It is noteworthy that the intermembrane forces measurements were repeated with other membranes of

DTDPC and they have given the same result: $\lambda = 0.7$ Å, and the maximum modification of intermembrane distance (with the change in humidity from 12% to 100%) is equal only to 3.5 Å. If one looks at the adsorption isotherms of lipid membranes in different phases, the maximum hydration, which was observed in the case of phospholipid membranes at $\psi = 12\%$, is set equal only to two water molecules per lipid molecule. Taking into account the size of water molecule and the in-plane area per lipid molecule, we get that the membrane at $\psi = 12\%$ has maximum two water molecules per lipid [30] (i.e., there is one water layer between the two neighboring membranes). Higher hydration ($\psi = 100\%$) adds approximately one water molecule layer more. Altogether at $\psi = 100\%$ we have no more than a two water molecule layer between membranes!

The NMR study [25] showed that intrinsic dynamics of polar groups of DTDPC molecules in polymeric membranes is closer to that of dimiristoyl phosphocholine (DMPC) membranes in liquid rather than in gel phase. The wide-angle x-ray diffraction (Fig. 7) shows the absence of any ordering in in-plan packing of polar heads of polymeric DTDPC membranes. The packing and dynamics of polar heads in these membranes are very much like those of the usual phospholipid membranes in liquid phase. For comparison, we also show wide-angle diffraction pattern from DMPC membranes in excess water in Fig. 7.

These results mean that it is not possible to explain the dramatic changes in intermembrane interactions in DTDPC polymeric membranes in the framework of polarization (hydration) mechanism of the origin of intermembrane forces, appealing to the change of in-plane packing of polar heads to explain the difference in the decay length of hydration forces [13]. Here we have to remark that the typical values of the decay length in liquid and gel phases of phospholipid membranes are 2.1 and 1.2 Å respectively [5].

Can we or do we understand more about (very) short-range forces which remain after the "switching off" of out-of-plane motions of lipid molecules? It was shown experimentally that upward breaks in the pressure-distance dependences occur at intermembrane distances of ~ 5 Å for phosphocholine bilayer in both liquid and gel phases [19, 26, 31–32]. The calculated decay length of intermembrane forces at these short distances is estimated at $\sim 0.6 \div 0.8$ Å [19, 26, 31–32], which is very close to the decay length $\lambda = 0.7$ Å for polymeric DTDPC membranes. In [26] such kind of a force was "attributed to steric repulsion between mobile lipid head groups that extend 2–3 Å into the fluid space between bilayers". We should add to this only one remark: It seems that the nearest water molecules are strongly bounded to polar heads. Therefore one can speak about hydrated "mobile polar groups".

And finally, there is no theory of steric repulsion of membranes due to local motions of polar head groups. The phospholipid polar head groups are too short. Therefore, the application of the Alexander-de Gennes theory of the interaction of two brush layers of phospholipid membranes is problematic [33].

Nevertheless, the formal fit of the DTDPC data by the Alexander-de Gennes equation

$$P = \frac{kT}{s^3}\left[\left(\frac{2L}{d - d_L}\right)^{9/4} - \left(\frac{d - d_L}{2L}\right)^{3/4}\right] \qquad (3)$$

gives the following values of fitting parameters (L is the thickness of polar head group layer; d_L is the thickness of lipid bilayer; s is the mean distance between "grafting sites", i.e., between polar heads): $d_L = 43.4 \pm 0.1$ Å, $L = 1.9$ Å, $s = 8$ Å the regression coefficient r^2 is equal to 0.977 (Fig. 8).

It cannot be excluded that the reasonable values of s, d_L and L [19] are in favor of the application of the

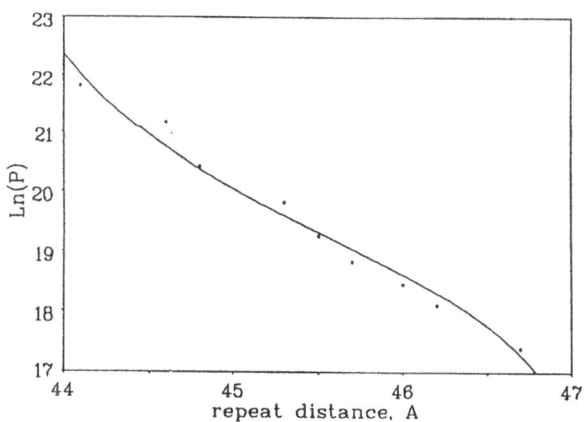

Fig. 8 The fit of the experimental pressure-distance P–d dependence by the Alexander-de Gennes equation (L is the thickness of polar head group layer; d_L is the thickness of lipid bilayer; s is the mean distance between "grafting sites", i.e., between polar heads): $d_L = 43.4$ Å, $L = 1.9$ Å, $s = 8.0$ Å, the regression coefficient r^2 is equal to 0.977

Alexander-de Gennes *shape* equation to membranes with short polar head groups [16].

Conclusions

i) the "simple" hydration force does not reach further than one water molecule layer (~ 3 Å) on phospholipid membrane surface;

ii) at short intermembrane distances ($d_w \le 5$ Å) the repulsive forces seem to be due to steric interactions of hydrated polar heads of opposite membranes;

iii) entropic forces due to out-of-plane motions (structure) of membrane surface [14–15] are dominant at the distances $d_w \ge 5$ Å.

Acknowledgments V.I. Gordeliy is very grateful to Prof. G. Büldt and the Alexander von Humboldt Foundation for support of this work.

References

1. Langmuir I (1938) J Chem Phys 6:873–896
2. Le Neveu DM, Rand RP, Parsegian VA (1976) Nature 259:601–603
3. Derjaguin BV, Landau LD (1941) Acta Physiochim URSS 14:633–662
4. Marcelja S, Radic N (1976) Chem Phys Lett 42(1):129–130
5. Rand RP, Parsegian VA (1989) Biochim et Biophys Acta 988:351–379
6. Cevc G, Podgornik R, Zeks B (1982) Chem Phys Letters 91:193–196
7. Marcelja S, Gruen DWR (1983) J Chem Soc Faraday Trans II 79:211–223
8. Jönson B, Wennerström H (1983) J Chem Soc Faraday Trans II 79:19–35
9. Shiby D, Ruckenstein E (1983) Chem Phys Letters 95:435–438
10. Cevc G, Marsch D (1985) Biophys J 47:21–31
11. Belaya ML, Feigal'man MV, Levadny VG (1986) Chem Phys Letters 126:361–364
12. Kornyshev AA (1986) J Electroanal Chem 204:79–84
13. Kornyshev AA, Leikin S (1989) Phys Rev A 40:6431–6437
14. Israelachvili JN, Wennerström H (1990) Langmuir 6:873–876

15. Cevc G (1991) J Chem Soc Faraday Trans 87(17):2733–2739
16. Israelachvili JN, Wennerström H (1992) J Phys Chem 96:520–531
17. Rand RP, Parsegian VA (1991) Langmuir 7:1299–1301
18. Leikin S, Parsegian VA, Rau DC, Rand RP (1993) Ann Rev Phys Chem
19. McIntosh TJ, Simon SA (1993) Biochemistry 32:8374–8384
20. Lipowsky R, Grotehaus S (1994) Biophys Chem 49:27–37
21. Cevc G (1995) Biophys Chem 55:43–53
22. Gordeliy VI, Bartels K, Bellet-Amalric E, Cherezov VG, Islamov AH, Hauß T, Teixeira J in preparation
23. Kirchner S, Cevc G (1993) Europhys Lett 23(3):229–235
24. Gordeliy VI, Golubchikova LV, Teixeira J (1993) Report of Leon Brillouin Laboratory, p 47, Saclay, France
25. Anikin AV, Chupin VV, Anikin MV, Serebrennikova G (1993) Makromol Chem 194:2663–2673
26. McIntosh TJ, Magid AD, Simon SA (1987) Biochemistry 26:7325–7332
27. Blaurock AE, Worthington CR (1966) Biophys J 6:305–312
28. Shannon CE (1949) Proc Inst Radio Eng n 4 37:10–21
29. Worthington CR (1969) Biophys J 9:222–234
30. Jendrasiak GL, Hasty JH (1974) Biochim et Biophys Acta 337:79–91
31. McIntosh TJ, Magid AD, Simon SA (1989) Biochemistry 28:7904–7912
32. McIntosh TJ, Simon SA, Needhan D, Shany GH (1992) Biochemistry 31:2020–2024
33. de Gennes PG (1987) Adv Colloid Interface Sci 27:189

Progr Colloid Polym Sci (1996) 100:345–350
© Steinkopff Verlag 1996

INTERFACES, FILMS AND MEMBRANES

Pore kinetics of mastoparan peptides in large unilamellar lipid vesicles

A. Arbuzova
G. Schwarz

A. Arbuzova · Dr. G. Schwarz (✉)
Department of Biophysical Chemistry
Biocenter of the University
4056 Basel, Switzerland

Abstract The apparent pore formation induced by polistes Mastoparan (MPP) and Mastoparan-X (MPX) in the membrane of large unilamellar POPC vesicles has been investigated. The time-course of the process was monitored by the fluorescence signal $F(t)$ reflecting the release of the marker substance carboxyfluorescein, which is entrapped in the vesicles at a self-quenching concentration. The data were analyzed according to a recently proposed theory allowing a quantitative evaluation of the pore kinetics and the mode of dye release. A mode of graded release was found for both peptides. The average dye retention factor of a single pore turned out to be $\rho \sim 0.70 \pm 0.05$ for MPP and $\rho \sim 0.55 \pm 0.05$ for MPX. The measured fluorescence signal $F(t)$ has been converted into the average retention function $R(t)$, i.e., the fraction of marker retained inside the liposomes after a given time of efflux. Then the pore formation rate per liposome could be determined and fitted to a pertinent time function. The relevant kinetic parameters are discussed in relation to the concentration of actually bound peptide.

Key words Mastoparan-X – polistes mastoparan – POPC – association isotherm – peptide-lipid interaction – carboxyfluorescein

Introduction

Cell membrane channels are supposed to be primarily made up of bundles of transmembrane α-helical peptide sequences. Hence a pore formation by α-helical peptides may be considered as a good model in order to investigate basic physicochemical features of channel structure and function.

Mastoparans are toxic tetradecapeptides isolated from wasp venom that cause degranulation of mast cells [1–3]. They also induce non-specific exocytosis of some other cells, stimulate the phospholipase A_2 activity [4] and activate G proteins in vivo and in vitro as they mimic a G-protein receptor [5]. The peptides associate with lipid bilayers altering the membrane structure so that the permeability for small molecules is substantially increased [4, 6–9]. The mechanisms of the interaction are not yet clear. Therefore it is important to investigate quantitatively the peptide-lipid binding and interaction. Pore mediated permeability is known to induce a leakage of a hydrophilic dye from liposomes [9].

Mastoparans adopt an amphipathic α-helical structure in their membrane bound state with a length about 2.1 nm

Abbreviations MPX, Mastoparan-X; MPP, polistes Mastoparan; POPC, 1-Palmitoyl-2-Oleoyl-sn-Glycero-3-Phosphocholine; LUV, Large Unilamellar Vesicle; HEPES, N-2-Hydroxyethyl-piperazine-N'-2-ethanesulfonic acid; EDTA, Ethylenediaminetetraacetic acid; CF, Carboxyfluorescein;

that is not enough to span the bilayer [10, 11]. As all the charged side chains are gathered on one side of the helix, mastoparans belong to the class L peptides that disturb a membrane [12]. It is not known yet if mastoparan acts just on the surface of a bilayer or spans it and forms a so-called barrel-stave channel.

For our studies, we have chosen two analogues of the original mastoparan: polistes mastoparan from the venom of *Polistes Jadwigae* [2] and mastoparan-X from *Vespa Xanthoptera* [3].

The primary structures of these peptides are:

ValAsp$^-$TrpLys$^+$Lys$^+$IleGlyGlnHis

   IleLeuSerValLeuNH$_2$ (for MPP)

IleAsnTrpLys$^+$GlyIleAlaAlaMetAlaLys$^+$Lys$^+$

   LeuLeuNH$_2$ (for MPX)

Here, we present an analysis of the pore formation kinetics induced by MPX and MPP in POPC large uniamellar liposomes exploiting an efflux assay approach. The data were evaluated according to a recently proposed theory [9, 13].

Theoretical proceeding

Efflux of dye, entrapped in liposomes at a self-quenching concentration, gives rise to an increase of the fluorescence signal $F(t)$ due to dilution of the dye, resulting in a final value F_∞ when the efflux is completed. The measurements are expressed in terms of a normalized efflux function $E(t)$:

$$E(t) = \frac{F_\infty - F(t)}{F_\infty - F_0}, \tag{1}$$

where F_0 is the signal at time $t = 0$ (when peptide is added). $E(t)$ must be converted into the retention function $R(t)$, which is equal to the fraction of dye still entrapped in vesicles at time t using the equation:

$$R(t) = E(t)\frac{1 - Q_0}{1 - Q_t} \tag{2}$$

This involves an initial static quenching factor Q_0 and an apparent transient quenching factor Q_t being determined in the course of efflux (see Materials and Methods).

In the case of a graded release using measured Q_0 and Q_t (where $Q_t > Q_0$) the experimental $E(t)$ may be transformed into $R(t)$. According to [9, 13], one can then calculate $p(t)$, i.e., the average number of pore openings per liposome up to the time t, by means of the relation:

$$p(t) = -\ln(R(t))/(1 - \rho), \tag{3}$$

with ρ being the average dye retention factor of a vesicle if only a single pore has been activated:

$$\rho = \tau_0/(\tau_0 + \tau_p), \tag{4}$$

where τ_0 is the relaxation time of dye release through a single pore (depending on the entrapped volume and pore properties) and τ_p is the average pore lifetime.

Materials and methods

The synthesized peptides polistes mastoparan and mastoparan-X were purchased from Bachem Feinchemikalien (Bubendorf BL, Switzerland). The concentration of peptide was determined by measuring tryptophan absorption with an UV spectrometer UVIKON 860 ($\lambda = 280$ nm, $\varepsilon = 5570$ cm^{-1}M^{-1}).

5(6)-carboxyfluorescein (CF) (mixed isomers, MW 376.3, 99% pure by HPLC) was bought from Sigma (St. Louis, MO, USA). HEPES was from Bioprobe. Triton X 100, NaCl and EDTA were supplied by Merck (Darmstadt). In order to dissolve the fluorescent dye and adjust the pH to 7.4 the necessary amount of 1 M NaOH was added (50 mM CF, 10 mM HEPES, 10 mM NaCl, 1 mM EDTA, approx. 135 mM NaOH). For measurements of the quenching curve this stock solution was diluted to lower concentrations with buffer.

HEPES buffer was used consisting of 100 mM NaCl, 10 mM HEPES, 1 mM EDTA, approx. 5.6 mM NaOH, pH 7.4.

POPC in chloroform was obtained from Avanti Polar Lipids Inc. (Birmingham, AL, USA) and used without further purification. The lipid concentration was determined by phosphate analysis [14]. A phospholipid film (10 mg) was dried by rotary evaporation and then placed overnight under oil pump vacuum.

The dry lipid film was dispersed in 2 ml CF solution (or in 2 ml buffer saline for binding measurements) and was vortexed for 5 min. A freeze-thaw cycle was repeated five times. The suspension was extruded 10 times through a 100 nm polycarbonate membrane (Nucleopore Corp., Pleasanton, CA, USA) at 15 bar N$_2$ pressure. The stock solution was stored at 4 °C. External dye was removed by gel filtration through a Sephadex G50 column (1 × 30 cm). Concentration of CF was checked after the preparation of vesicles by measuring absorption of CF at $\lambda = 490$ nm, $\varepsilon = 72\,000$ cm^{-1}M^{-1}.

Dynamic light-scattering measurements were performed on a commercial supplier ALV5000 at $\lambda = 632$ nm, detection at 90°.

Fluorescence measurements were done on a Jasco FP 777 spectrofluorometer at 20 °C in 1 cm (2 ml) quartz cuvettes for CF efflux measurements. Vesicles (lipid concentration 15–200 µM in a cuvette) were added to 2 ml buffer samples. Fluorescence of CF was measured at 500 nm (slit 10 nm); excitation at 490 nm (slit 1.5 nm), while stirring the solution. For a quenching curve we have measured the static quenching factor $Q_0 = F_0/F_\infty$ as a function of an entrapped dye concentration $C_{[CF]}$ immediately after removing the external dye. F_0 is the signal of the entrapped dye in self-quenching concentration and F_∞ is the final signal after addition of Triton X 100 (concentration in the cuvette 0.1%). The quenching curve for CF in LUVs was fitted to the polynomial function: $Q_0 = 1 - 3.17y + 3.84y^2 - 1.63y^3$, $y = C_{[CF]}/53$ mM, the parameters were used to evaluate $R(t)$, by means of Eqs. (1, 2) along the lines of the detailed theory [9].

After addition of vesicles into a cuvette with 2–2.5 ml buffer and measuring F_0, 0.1–4 µM of peptide was added and an increase of fluorescence intensity due to the efflux of CF was monitored continuously. 30–35 min later (after the efflux rate slowed down to not more than 3% fluorescence increase in 5 min) 1 ml of the sample was eluted with the same buffer on a Sephadex G50 column and approx. 2 ml were gathered directly into another cuvette (the whole procedure needing approx. 5 min). Then the appropriate F'_0 and F'_∞ for this sample were determined, which leads to $Q_t = F'_\infty/F'_0$, the quenching coefficient for vesicles after efflux for t minutes. The continuous measurement of the original sample was stopped by addition of Triton X 100. Then F_∞ was estimated. In some measurements efflux was stopped at a chosen time by adding an excess of empty POPC vesicles. Control measurements were made without adding peptide. Q_0 of the vesicle stock solution was practically constant over the whole time of experiments of approx. 8 h.

According to Eq. (3) the number of pore openings per vesicle was evaluated and fitted to a two-term exponential function, reflecting a very pronounced slowing down of the pore formation:

$$p(t) = v_i t + a_1(1 - e^{-k_1 t}) + a_2(1 - e^{-k_2 t}). \tag{5}$$

This involves some characteristic kinetic parameters, namely v_i: slow intermediate rate of pore formation; k_1, k_2-time constants, a_1, a_2-amplitude factors. The initial slope v_0 is:

$$v_0 = (dp(t)/dt)_{t \to 0} = v_i + a_1 k_1 + a_2 k_2. \tag{6}$$

The association isotherm of MPP with POPC LUVs of 100 nm diameter was measured using intrinsic tryptophan fluorescence: excitation 280 nm (slit 1.5 nm), emission 320 nm (slit 3 nm) in 0.5 cm cuvettes. The results were corrected for dilution and light scattering [6]. The partition coefficient K_p and the effective charge z were cal-

culated according to a Gouy-Chapman model approach as described elsewhere [6, 15].

The amount of bound peptide per lipid in recursive form is:

$$r = \frac{K_p C_f}{\alpha}, \tag{7}$$

where C_f stands for the peptide concentration in the aqueous phase, $\alpha = \exp(2z \sinh^{-1}(zbr))$ is the pertinent thermodynamic activity coefficient describing possible electrostatic interactions of bound molecules, b is a dimensionless quantity, depending on the salt concentration in solution. Knowing K_p and z, we can evaluate the amount of bound peptide for each given peptide/lipid concentration.

Results

The average radius of liposomes was 50 nm (measured by DLS) and did not change after addition of peptide. Hence there were no fusion or aggregation of vesicles. To destroy the vesicles Triton X 100 was added. It induces formation of mixed detergent micelles with outer radius of 3 nm.

Fitting of MPP association isotherms (Fig. 1) to the functional course of Eq. (7) yields an apparent partition coefficient $Kp = 4000$ M^{-1} (30% Std. Dev.) and effective charge $z_{eff} = 0.2$ (in other words, there is practically no

Fig. 1 Association isotherm for MPP including the data of three independent experiments. The calculated binding curve for MPP with $K_p = 4000$ [M^{-1}], $z_{eff} = 0.2$ is shown as a dashed line; $b = 11.5$. The solid line describes the binding isotherm for MPX calcualted with $K_p = 4000$ [M^{-1}], $z_{eff} = 2.4$ (the data for MPX are taken from [6])

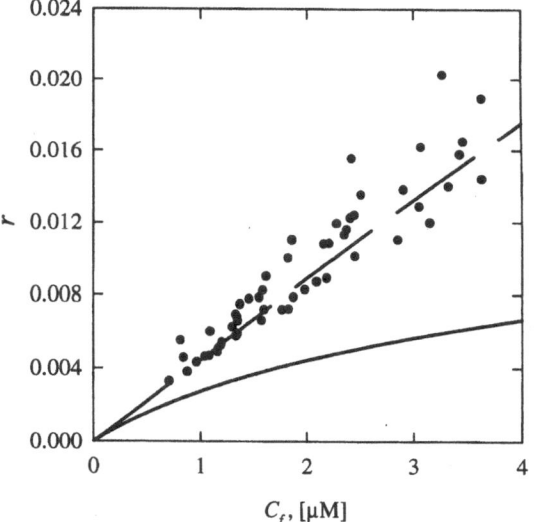

C_f, [µM]

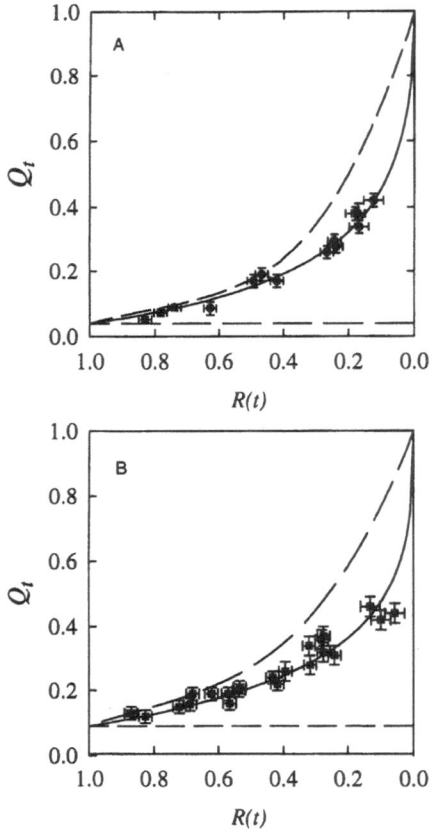

Fig. 2A, B Apparent quenching factor Q_t during efflux induced by MPP (A, ●) and MPX (B, ■). The initial CF concentration was 53 mM ($Q_0 = 0.04$) for experiments with MPP and 37 mM ($Q_0 = 0.1$) for MPX. The solid curve shows the fit with $\rho = 0.70 \pm 0.05$ (A) and that with $\rho = 0.55 \pm 0.05$ (B). Two extreme cases are shown by dashed lines: the "all-or-none" leads to $Q_t = Q_0$ at any $R(t)$, $\rho = 0.$, and very slow efflux with $\rho \rightarrow 1$

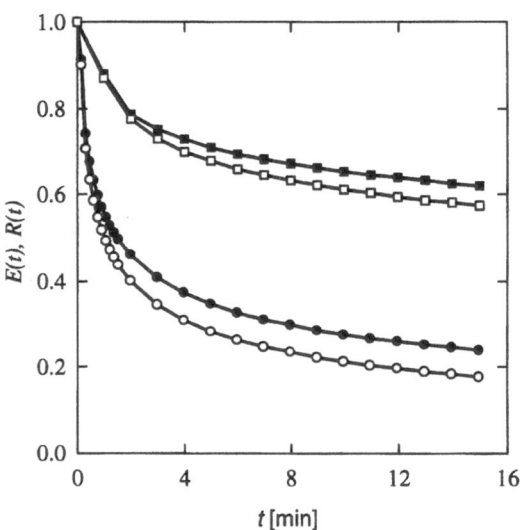

Fig. 3 Time dependence of the normalized efflux function $E(t)$ (open symbols) and retention function $R(t)$ (closed symbols) for MPP (○, ●) and MPX (□, ■) for the same amount of bound peptide ($r = 0.0017$). Cuvette contained 2 ml buffer, 34 µM POPC and 0.4 µM MPP or 0.06 µM MPX

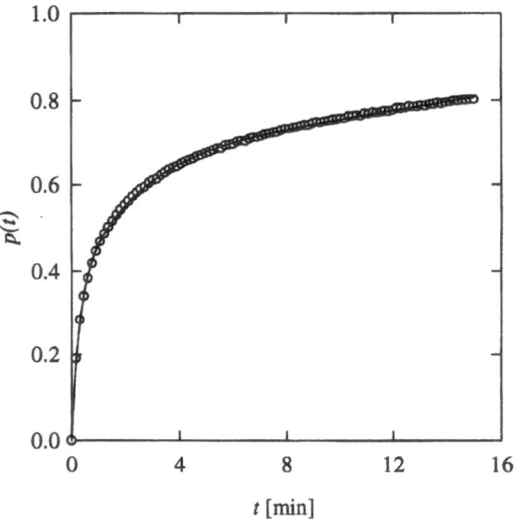

Fig. 4 An example of fitting the pore number per vesicle $p(t)$ to the exponential function $p(t) = v_i t + a_1(1 - e^{-k_{11}t}) + a_2(1 - e^{-k_2 t})$. The curve was measured with 0.3 µM MPP and 52 µM lipid in 2 ml buffer. The fitting of the experimentally obtained $p(t)$ curves is equally good in te beginning and at the end of the course ($\chi^2 = 0.06$)

electrostatic interaction). The binding of monomers is reversible; we can separate a peptide from vesicles by gel filtration and also stop efflux by changing the partition equilibrium by adding lipid in excess.

Association of peptide leads to pore formation and release of dye. The measurements of the apparent quenching factors reveal a graded release for both peptides with an average retention factor $\rho = 0.70 \pm 0.50$ for MPP and $\rho = 0.55 \pm 0.05$ for MPX (see Fig. 2A, 2B).

Figure 3 illustrates the difference in $E(t)$ and $R(t)$. For the same amount of bound peptide and lipid concentration MPP causes higher efflux: less dye is retained in the vesicles. Thus the rate of pore formation for MPP is higher than for MPX. This can be clearly seen in the Figs. 5A, B.

Figure 4 presents one example of the evaluated pore number function $p(t)$ and a fit according to Eq. (6) (see Materials and Methods).

Discussion

When we compare the association isotherms of MPP and MPX (the data for MPX are taken from [6]), we see that the partition coefficient K_p for MPP and MPX is the same.

Fig. 5 Double logarithmic diagram of the kinetics parameters dependent on the amount of bound peptide per lipid. A) The dependence of the initial pore formation rate on r. linear regression: solid line for MPP (○) and dashed line for MPX (■). B) The intermedite rate of pore formation for MPP and MPX, C) Time constant k_2

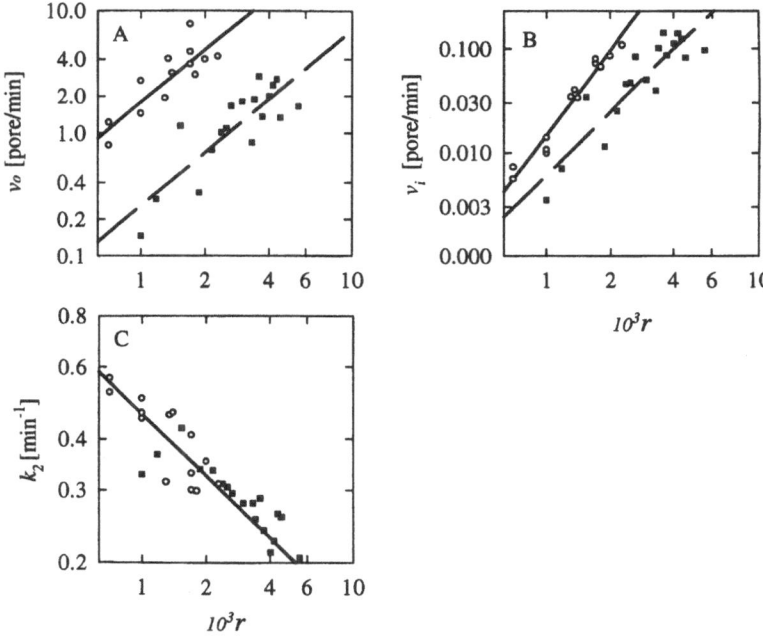

MPP has one negative charge in addition to three positive residues and therefore a net charge of $+1$ (for MPX: $+3$), therefore the effective charge of MPP is expected to be lower. Electrostatic effects play nearly no role in the partitioning of MPP as the effective charge is practically zero. This leads to a lower intermolecular electrostatic repulsion and therefore to the higher amount of bound molecules. Binding kinetics is very fast: partitioning reaches equilibrium in milliseconds [N. Hellmann, unpublished results]. Therefore, the pore formation kinetics is limited by some other processes, for instance: aggregation, incorporation into the bilayer.

From the average dye retention factors, one can calculate using Eq. (5) the ratio between the average pore lifetime τ_p and the relaxation time of dye release through a single pore τ_0 (in our case it could be estimated to be equal to some milliseconds). This ratio τ_p/τ_0 is equal to 0.43 ± 0.05 for MPP and 0.82 ± 0.05 for MPX, respectively.

The kinetic parameters are examined regarding their dependence on r, the amount of bound peptide per lipid (Fig. 5). From the double logarithmic plot, we can see that the initial rate of pore formation for MPX and MPP is nearly proportional to r. The intermediate rates have also similar dependence for both peptides (Fig. 5B). The first

time constant, k_1, does not depend on the amount of associated peptide and was equal for both peptides: $k_1 = 4.5 \pm 0.9$ [$\min^{-1}$]. On the other hand, Fig. 5C reveals that k_2 decreases with increasing r.

We envisage that the actual mechanism of pore formation should be the same for both peptides as they have a very similar amphipathic structure. They have nearly the same biological response, for instance, they are equally effective in facilitating phospholipase A_2 activity [4]. The fact that MPP is a weaker hemolytic agent [4] can be explained by some non-specific binding to proteins on the surface of red blood cells, as our experiments show that MPP includes higher leakage. As can be seen in Fig. 5 the dependence of the time constants on the amount of bound peptide per lipid is approximately the same for both peptides, also indicating some common mechanism. Further investigations towards a more detailed analysis of the underlying scheme of reactions are in progress.

Acknowledgments This work was supported No 31-32188.91 and No 31-042045.94 from Swiss National Science Foundation to G.S. The authors are grateful to Dr. M. Winterhalter for fruitful discussions and for reading the manuscript. We also thank Dr. F. Stieber (Institut für Physikalische Chemie, Universität Basel) for using his DLS apparatus.

References

1. Hirai Y, Yasuhura T, Yoshida H, Nakajima T, Fujino M, Kitada C (1979) Chem Pharm Bulletin 27:1942–1944
2. Hirai Y, Ueno Y, Yasuhura T, Yoshida H, Nakajima T (1980) Biomed Research 1:185–187
3. Hirai Y, Kuwada M, Yasuhura T, Yoshida H, Nakajima T (1979) Chem Pharm Bulletin 27:1945–1946
4. Argiolas, A, Pisano JJ (1983) J Biol Chem 258:13697–13702.
5. Higashijima T, Uzu S, Nakajima T, Ross EM (1988) J Biol Chem 263:6491–6494
6. Hellmann N (1995) PhD Thesis, University of Basel
7. Schwarz G, Blochmann U (1993) FEBS 318:172–176
8. Mellor TR, Sansom MSP (1990) Proc R Soc Lond B 239:383–400
9. Schwarz G, Arbuzova A (1995) BBA, 1239:51–57
10. Higashijima T, Wakamatsu K, Takemitsu M, Fujino M, Nakajima T, Miyazawa T (1983) FEBS lett 152:227–230
11. Wakamatsu K, Okada A, Miyazawa T, Ohya M, Higashijima T (1992) Biochem 31:5654–5660
12. Tytler EM, Segrest JP, Epand RM, Nie S-Q, Epand RF, Mishra VK, Venkatachalapathi YV, Anantharamaiah GM (1993) J Biol Chem 268:22112–22118
13. Schwarz G, Robert CH (1992) Biophys Chem 42:291–296
14. Böttcher CJF, Van Gent CM, Fries C (1961) Anal Chim Acta 24:203–204
15. Schwarz G, Beschiaschvili G (1989) BBA 970:82–90

Progr Colloid Polym Sci (1996) 100:351–355
© Steinkopff Verlag 1996

INTERFACES, FILMS AND MEMBRANES

Head-group variations and monolayer structures of diol derivatives

G. Brezesinski
K. de Meijere
E. Scalas
W.G. Bouwman
K. Kjaer
H. Möhwald

Dr. G. Brezesinski (✉) · K. de Meijere
H. Möhwald
Max-Planck-Institut für Kolloid-und
Grenzflächenforschung
Rudower Chaussee 5
12489 Berlin, FRG

E. Scalas
Universität Mainz Institut für Physikalische
Chemie Welder-Weg 11
55099 Mainz, FRG

W.G. Bouwman K. Kjaer
Physics Department
Risø National Laboratory
4000 Roskilde, Denmark

Abstract Monolayers of 5 chemically modified diols varying the head-group (nonadecane-1,2-diol (C1), hexadecyl-propane-1,3-diol (C2), hexadecyl-oxy-propane-1,2-diol (C3), hexadecyl-oxy-butane-1,2-diol (C4), hexadecanoyl-oxy-propane-1,2-diol (C5)) have been investigated by grazing incidence x-ray diffraction at 20 °C and at different lateral pressures. C1 and C5 exhibit a centred-rectangular lattice with NN (nearest neighbour) tilt and NN distortion directions. In the case of C1 on increasing the lateral pressure the distortion changes to NNN (next-nearest neighbour direction) without a change in tilt direction (NN). This behaviour could not be observed for the other compounds. C3 and C4 display a phase transition from a phase of NN tilt and NN distortion to a phase of NNN tilt and NNN distortion. At the transition pressure, which depends on the head group structure, both the tilt and distortion azimuths are changed. C2 forms very unstable films and is packed in an oblique lattice. The dependence of the tilt angle on the lateral pressure and the dependence of the signed distortion on the tilt magnitude, which shows a grouping behaviour of the diols, seem to imply that the NN-tilted rectangular phases of C3–C5 differ from that of C1.

Key words Amphiphiles – monolayers – grazing incidence x-ray diffraction – structure

Introduction

Monolayers at the air/water interface are appropriate systems to study the influence of chemical variations on the microscopic quasi-two-dimensional order. To achieve a better comprehension of these effects it is helpful to make use of simple surfactant molecules. Among these, derivatives of 1,2-diols have received interest because they can form thermotropic and lyotropic mesophases in the bulk and stable monolayers at the air/water interface [1–11].

One of the important influences on the packing of the monolayer is certainly the head-group, as a result of steric hindrance as well as of its specific interactions with the monolayer environment. Three molecular elements have been varied to investigate the role of the structure of the hydrophilic moiety. The first one is the introduction of heteroatoms. Starting from a simple n-nonadecane-1,2-diol (C1), an oxygen atom was introduced at two different positions in n-alkane-1,2-diol molecules (C3 and C4). The second element is the replacement of the ether by an ester linkage (C5). Such a replacement has a strong influence on the monolayer behaviour of double-chain

Table 1 The chemical names, molecular formulae and abbreviations of the substances investigated

Name	Formula	Abbreviation
nonadecane-1,2-diol	$C_{17}H_{35}$-CHOH-CH_2OH	C1
2-hexadecyl-propane-1,3-diol	$C_{16}H_{33}$-CH-$(CH_2OH)_2$	C2
3-hexadecyl-oxy-propane-1,2-diol (1-hexadecyl-glycerol)	$C_{16}H_{33}$-O-CH_2-CHOH-CH_2OH	C3
4-hexadecyl-oxy-butane-1,2-diol	$C_{16}H_{33}$-O-CH_2-CH_2-CHOH-CH_2OH	C4
3-hexadecanoyl-oxy-propane-1,2-diol (1-palmitoyl-gylcerol)	$C_{15}H_{31}$-CO-O-CH_2-CHOH-CH_2OH	C5

phosphatidylcholines [12]. The last structural element is the bulky 1,3-diol group (C2). In bulk systems the 1,2- and 1,3-diols with the same tail length behave in a completely different way [1, 2].

Table 1 shows the chemical formulae of the substances investigated. The grazing incidence x-ray diffraction measurements on monolayers of C1 and C2 at the air/water interface will be described in the results section. The results for the compounds C3–C5 were already partially published [7, 8, 10, 11], but will be compared with those of C1 and C2 in the discussion section.

Materials and methods

The compounds C1–C4 were kindly synthesized by colleagues in the group of Prof. C. Tschierske at the Institute of Organic Chemistry, University of Halle, Germany. Compound C5 was obtained chromatographically pure from Sigma and used as purchased. The compounds were spread from a 1 mM chloroform solution (C1–C4) or a 1 mM heptane/ethanol (9:1) solution (C5). In all cases the subphase was ultrapure water with a specific resistance of 18.2 MΩcm purified using a Millipore desktop unit. The surface pressure was detected using a Wilhelmy balance.

The Synchrotron grazing incidence x-ray diffraction (GID) experiments were performed at 20 °C using the liquid-surface diffractometer at the undulator beamline BW1 at HASYLAB, DESY, Hamburg, Germany. The experimental setup has been described in detail elsewhere [13, 14]. The diffracted radiation was detected by a linear position-sensitive detector (PSD) (OED-100-M, Braun, Garching, Germany), as a function of the vertical scattering angle. A Soller collimator in front of the PSD provided resolution of the horizontal scattering angle. The corrected scattering intensities were least-square fitted to model peaks which were taken as the product of a Lorentzian parallel to the water surface by a Gaussian normal to it. From the in-plane diffraction data it is possible to obtain

the lattice spacings, and from these the lattice parameters can be calculated [12, 14, 15, 16]. The out-of-plane data gives additional information on the molecular tilt and tilt direction in the monolayer plane. For a centred-rectangular structure the tilt azimuth can be in one of two symmetry directions which are conventionally called *nearest-neighbour* (NN) direction and *next-nearest-neighbour* (NNN) direction. In the case of a NN direction there is a two-fold degenerate peak Q_{xy}^d at $Q_z^d > 0$ Å^{-1} and a non-degenerate peak Q_{xy}^n at $Q_z^n = 0$ Å^{-1}, where Q_{xy}^n is the maximum position of the non-degenerate peak and Q_{xy}^d that of the two-fold degenerate peak [17]. The tilt angle t is then given by

$$t = \arctan\left(Q_z^d / \sqrt{(Q_{xy}^d)^2 - (Q_{xy}^n/2)^2} \right). \quad (1)$$

In the case of a NNN tilt both the two-fold degenerate and the non-degenerate peaks occur at Q_z greater than 0 Å^{-1} and the tilt angle can be calculated from

$$t = \arctan (Q_z^n / Q_{xy}^n). \quad (2)$$

We can introduce a convenient set of order parameters for the description of the monolayer structure. There are essentially two observable order parameters, each one with magnitude and direction. The first describes the tilt, whose magnitude is given by $\eta = \sin (t)$ and whose azimuth we shall call β. The second order parameter gives information on the orientation of the planes formed by the aliphatic zig-zig backbones and is essentially the distortion of the unit cell. Its magnitude is:

$$\xi = (l_1^2 - l_2^2)/(l_1^2 + l_2^2) \quad (3)$$

where l_1, l_2 are the major and minor axes of the ellipse passing through all six nearest neighbours of a molecule, respectively. The distortion azimuth, ω, is the azimuth of the major axis of the ellipse. Within a Landau theory of monolayer phase transitions the unit cell distortion is a secondary effect due to molecular tilt and ordering of the backbone planes [17, 18]. To separate the contribution of

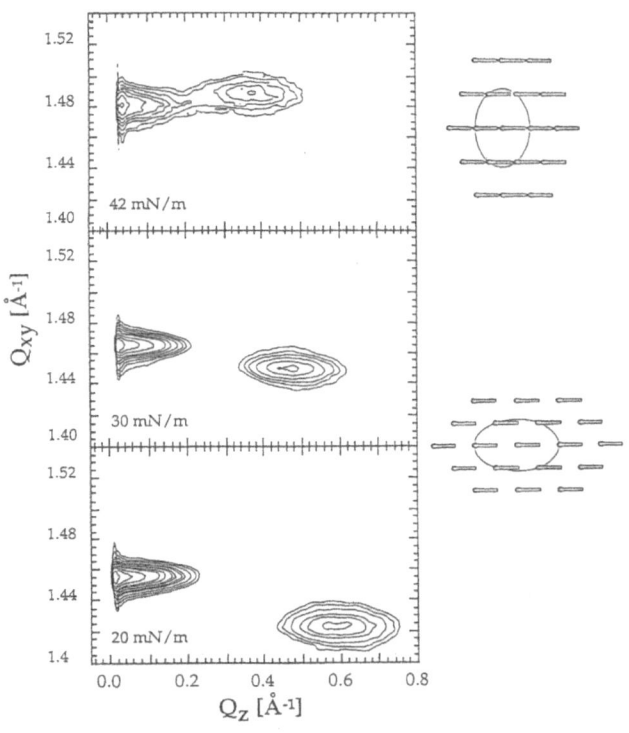

Fig. 1 Contour plots of the corrected intensity of nonadecane-1,2-diol (C1) versus in-plane and out-of-plane scattering vector components Q_{xy} and Q_z at different surface pressures (indicated) at 20 °C

tilt from that of backbone ordering, one can plot the signed distortion d:

$$d = \xi \cos 2(\omega - \beta) \tag{4}$$

as a function of η^2, as the distortion due to the tilt is proportional to η^2.

Results

The nonadecane-1,2-diol C1 forms stable monolayers on water over a wide range of temperatures. The pressure-area-isotherms indicate that up to 40 °C this compound forms only a condensed film. At 20 °C the first increase of pressure occurs at a molecular area of about 24 Å^2. At the pressure of film rupture the molecular area amounts to ~ 20 Å^2. Figure 1 shows contour plots of the corrected intensities as a function of in-plane scattering vector component Q_{xy} and of out-of-plane scattering vector component Q_z at different surface pressures (indicated). At all pressures investigated only two diffraction maxima were observed for the racemic C1. This indicates a centred-rectangular unit cell. One peak is located at the horizon ($Q_z = 0$ Å^{-1}) and the other peak has maximum intensity at $Q_z > 0$ Å^{-1}. With increasing lateral pressure the two-fold

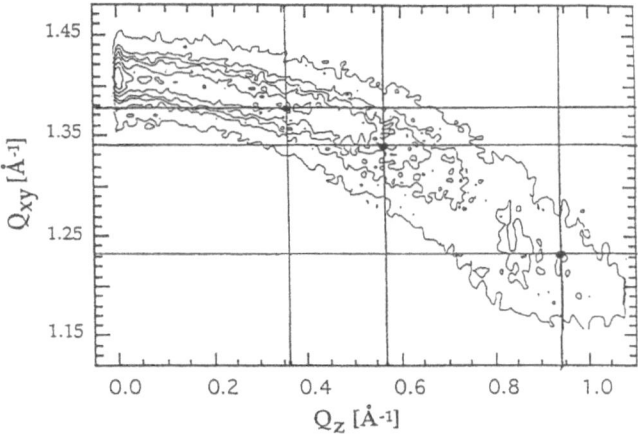

Fig. 2 Contour plot of the corrected intensity of 2-hexadecyl-propane-1,3-diol (C2) as a function of the in-plane and out-of-plane scattering vector components Q_{xy} and Q_z at 15 mN/m and 20 °C

degenerate peak moves to lower values of Q_z indicating a decrease of the tilt angle. Additionally, the two peaks change their positions in Q_{xy} on increasing the pressure. At 30 mN/m $Q_{xy}^n > Q_{xy}^d$, whereas at 42 mN/m $Q_{xy}^n < Q_{xy}^d$, indicating a change in distortion azimuth.

The diol C2 forms only a condensed film as well, but only up to temperatures around 32 °C. At 20 °C the monolayer is not very stable and collapses already at about 35 mN/m and a molecular area of about 25 Å^2. The larger area at dense packing compared to C1 must be caused by the larger 1,3-diol head-group. Because of the instability of the film only one x-ray measurement at 15 mN/m was performed. Figure 2 shows the contour plot of the corrected intensities as a function of Q_{xy} and Q_z. This plot has a pronounced "banana"-shape resulting from the overlap of different diffraction peaks. The averaged maximum positions of three diffraction peaks (oblique lattice structure) are indicated by points. But a complete description of such a smooth banana shape would need more than three peaks. Therefore, it seems to be possible that this monolayer is in a metastable state and consists of different coexisting phases.

Discussion

The diols investigated have almost the same hydrocarbon chain length (number of carbon atoms) but differ in head-group structure. As a consequence, the phase behaviour at 20 °C is rather different. C1 and C2 exhibit only condensed phases, whereas C3–C5 show a first-order phase transition between a liquid-expanded and a condensed phase. The pressure of this transition is only slightly influenced by the

354

G. Brezesinski et al.
Monolayer structures of diol derivatives

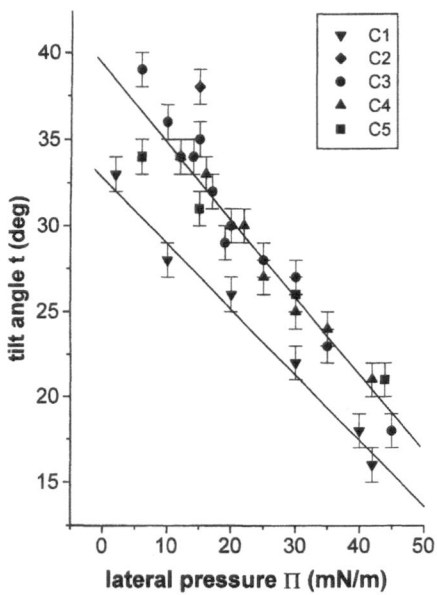

Fig. 3 Tilt angle t as a function of the lateral pressure Π for the monolayers of C1–C5

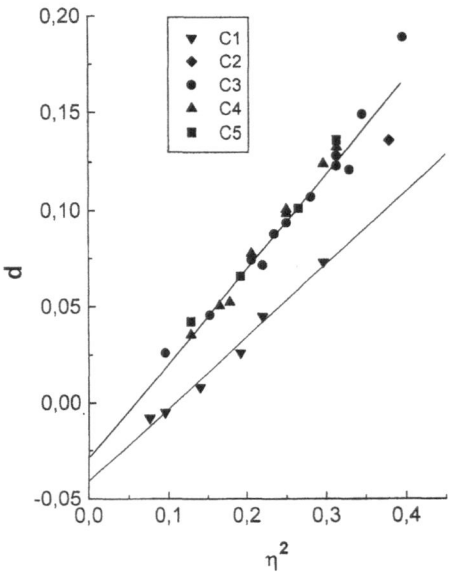

Fig. 4 Signed distortion d as a function of the parameter $\eta^2 = \sin^2(t)$ (see the second section) for the monolayers of C1–C5

head-group structure and amounts to values between 3 mN/m and 6 mN/m. The simple nonadecane-1,2-diol C1 is arranged in a rectangular lattice with NN tilt of the molecules. On increasing the lateral pressure the tilt angle decreases continuously. The chain cross-section amounts to (19.8 ± 0.2) Å^2. At lower pressures the distortion azimuth is directed towards NN. With increasing pressure the distortion magnitude ξ decreases continuously, reaches zero and increases again. Accordingly, the distortion azimuth ω changes from NN to NNN when passing through the undistorted state.

The compound C5 exhibits at all pressures investigated a centred-rectangular phase with NN tilt and NN distortion azimuth. Replacing the ester linkage by an ether one leads to a drastic change in the monolayer structure. At low surface pressures C3 and C4 are also arranged in a NN tilted rectangular phase, but on increasing the pressure there is a phase transition to a NNN tilted rectangular phase. At this transition pressure (18 mN/m for C3 and 12 mN/m for C4) also the distortion azimuth changes from NN to NNN.

The 1,3-diol C2 is not packed in a rectangular lattice. The observed metastable monolayer state must be caused by steric packing problems due to the swallow-tail head group structure.

Figure 3 shows the tilt angle as a function of lateral pressure. For all the diols investigated we observe a decrease of tilt upon increasing pressure. It seems that we can divide the diols into two groups. The compounds C2–C5 have almost the same pressure dependence of the tilt angle.

C1 exhibits always the smallest tilt angle but the slope of the curve is similar to that of the other diols. This grouping behaviour is supported by the plot of the distortion d as a function of $\sin^2(t)$ (Fig. 4). The distortion decreases linearly with decreasing $\sin^2(t)$. At a tilt angle of about 19° ($\Pi \cong 36$ mN/m) d amounts to zero, and changes the sign on further increase of pressure. As discussed above, this behaviour indicates a transition from a NN to a NNN distortion of the lattice without changing the tilt direction. At around 36 mN/m the lattice is undistorted, this means that a hexagonal packing of tilted molecules is observed. The extrapolation of d to zero tilt gives $d = -0.041$.

The observed distortion of the lattice of C2–C5 is larger than that of C1 at the same tilt angle. The extrapolation to zero distortion leads to an undistorted lattice of tilted molecules at a hypothetical surface pressure of about 57 mN/m ($t \approx 14°$). After extrapolating to zero tilt angle the group of C2–C5 has a distortion of -0.028. The greater absolute value of the extrapolated zero-tilt distortion in the case of C1 points to an ordering of the chain backbones. [17, 18]. This is not reflected by the chain cross-sections which are around 20 Å^2 in all the investigated cases except for C2, apparently suggesting that the chains are freely rotating. For C2 we find a chain cross-section of nearly 20.7 Å^2 probably due to the packing constraints induced by the 1,3-diol head group. In this case the larger head-group determines the large tilt angle of 38° at 15 mN/m. However, it is possible that the shape of such molecules cannot be approximated by a cylinder model [15, 16]. The transitions between the two rectangular phases of C3 and C4 cannot be seen in Fig. 4 because the change in distortion azimuth is connected with the change in tilt direction.

Inspection of the lattice parameters shows that the NN–NNN transition of C3 and C4 is connected to a drastic decrease of the short side of the rectangular unit cell and an increase of the long side (for example, C3 at 17 mN/m: $a = 5.52$ Å, $b = 8.58$ Å and at 19 m/N/m: $a = 4.93$ Å, $b = 9.31$ Å). Starting from the NN phase at lower pressure, increasing the lateral pressure leads to a decrease of the tilt angle and simultaneously to a decrease of the short-side length (direction of tilt) of the unit cell, whereas there is almost no change of the long side (C3 at 6 mN/m: $a = 6.01$ Å, $b = 8.58$ Å). At the transition from NN to NNN phase the short-side length decreases and the long-side length increases by almost 10%. Further increase of pressure yields a continuous decrease of the long-side length (the new tilt direction) whereas the short side is now not influenced (C3 at 45 mN/m: $a = 4.86$ Å, $b = 8.62$ Å).

Conclusion

Excluding C2, all the diols investigated exhibit at lower surface pressure a NN tilted and NN distorted centred-rectangular phase. From the dependence of the tilt angle as a function of pressure and the dependence of the distortion on the tilt magnitude the NN tilted phases of C3–C5 seem to differ from that of C1. Moreover, in the case of C1 we observed an undistorted lattice with tilted molecules which is not present for the other compounds. Another feature which distinguishes C1 is the change of distortion azimuth from NN to NNN at fixed NN-tilt azimuth. The behaviour of C1 is similar to that of the L_{2h} phase in fatty acids [17, 18]. The compounds with the heteroatom (C3–C5) have different phase sequences in dependence of the head-group structure. The ester compound C5 remains in the NN phase upon compression up to 44 mN/m. On the contrary the ether compounds C3 and C4 exhibit a phase transition to a NNN tilted and NNN distorted phase. The pressure range where the NN phase exists is sensitive to the head-group structure: for C4 it amounts to 6 mN/m and for C3 to 13 mN/m. The purpose of this work was to shed light on the importance of head-group interactions and we would like to discuss these by means of Fig. 4. The linear dependence of d on η^2 reflects the tail/tail interactions and their contribution is expected to vanish for $\eta \to 0$. Hence the ordinate for $\eta = 0$ reflects contributions from the heads. A negative value then physically means that the lattice is less compressed more normal to the chain tilt than perpendicular to it. This then suggests that there is a head-group interaction fixing this direction.

Acknowledgments This work was supported by the Bundesministerium für Forschung und Technologie, the Deutsche Forschungsgemeinschaft, the Danish Natural Science Research Council, and the EC through the Human Capital and Mobility program. We thank HASYLAB for providing excellent facilities and support. One of the authors (E.S.) could perform this work thanks to a Borsa di studio per il perfezionamento all'estero provided by Genoa University Italy. The contributions of B. Struth (University of Mainz), R. Rietz, F. Bringezu (University of Halle) and U. Gehlert, G. Weidemann (MPI for Colloids and Surfaces, Berlin) to the x-ray experiments are gratefully acknowledged. Useful discussion with V.M. Kaganer (Institute of Crystallography, Moscow) is also acknowledged.

References

1. Tschierske C, Brezesinski G, Kuschel F, Zaschke H (1989) Mol Cryst Liq Cryst Lett 6:139–144
2. Pietschmann N, Brezesinski G, Kuschel F, Tschierske C, Zaschke H (1990) Mol Cryst Liq Cryst Lett 7:39–46
3. Tschierske C, Brezesinski G, Wolgast S, Kuschel F, Zaschke H (1990) Mol Cryst Liq Cryst Lett 7:131–138
4. Rettig W, Brezesinski G, Mädicke A, Tschierske C, Zaschke H, Kuschel F (1990) Mol Cryst Liq Cryst 193:115–120
5. Moenke-Wedler T, Förster G, Brezesinski G, Steitz R, Peterson IR (1993) Langmuir 9:2133–2140
6. Rietz R, Brezesinski G, Möhwald H (1993) Ber Bunsenges Phys Chem 97:1394–1399
7. Brezesinski G, Rietz R, Kjaer K, Bouwman WG, Möhwald H (1994) Il Nuovo Cimento 16:1487–1492
8. Brezesinski G, Scalas E, Struth B, Möhwald H, Bringezu F, Gehlert U, Weidemann G, Vollhardt D (1995) J Phys Chem 99:8758–8762
9. Gehlert U, Vollhardt, D (1994) Progr Colloid Polym Sci 97:302–306
10. Rietz R, Rettig W, Brezesinski G, Bouwman WG, Kjaer K, Möhwald H (1995) Thin Solid Films, submitted
11. Scalas E, Brezesinski G, Möhwald H, Kaganer VM, Bouwman WG, Kjaer K (1995) Thin Solid Films, submitted
12. Brezesinski G, Dietrich A, Struth B, Böhm C, Bouwman WG, Kjaer K, Möhwald H (1995) Chem Phys Lipids 76:145–157
13. Kjaer K, Majewski J, Schulte-Schrepping H, Weigelt J (1992) HASYLAB Annu Rep 589–590
14. Als-Nielsen J, Jaquemain D, Kjaer K, Lahav M, Leveiller F, Leiserowitz L (1994) Phys Rep 246:251–321
15. Als-Nielsen J, Kjaer K (1989) In:Riste T, Sherrington D (eds) Phase Transitions in Soft Condensed Matter, NATO series B, 211. Plenum Press, New York and London, pp 113–137
16. Kjaer K (1993) Physica B 198:100–109
17. Kaganer VM, Peterson IR, Kenn RM, Shih MC, Durbin M, Dutta P (1995) J Chem Phys 102:9412–9422
18. Kaganer VM, Loginov EB (1995) Phys Rev E 51:2237–2249

Progr Colloid Polym Sci (1996) 100:356–361
© Steinkopff Verlag 1996

INTERFACES, FILMS AND MEMBRANES

I. Porcar
R.M. García
C.M. Gómez
V. Soria
A. Campos

Macromolecules in ordered media III. A fluorescence study on the association of poly-2-vinylpyridine with a phospholipid bilayer

I. Porcar · R.M. García
C.M. Gómez · V. Soria
Dr. A. Campos (✉)
Departament de Química Física
Universitat de València
46100 Burjassot, València, Spain

Abstract Association of low molecular weight poly-2-vinylpyridine taken as a polycation model with unilamellar vesicles based on dimirystoylphosphatidic acid has been experimentally investigated by means of fluorescence energy transfer experiments. The conventional profiles for the binding isotherms were obtained, confirming strong deviations from the ideality of these complex fluid composites. Two approaches, named binding and partitioning equilibrium, respectively, were used to obtain the association constant, K_A; the number of phospholipids that form a binding site, N; the partition coefficient, Γ; the activity coefficient, γ; and the effective charge, v. Analysis of these parameters as a function of polymer concentration and ionic strength was carried out.

Key words Fluorescence – phospholipid – poly-2-vinylpyridine – binding isotherms

Introduction

The study of polyion interaction at lipid interfaces is of indubitable interest from the viewpoints of biophysics and physical chemistry [1]. In particular, the interaction of polyions with liposomes of opposite charge is a current issue for its importance both in commercial processes and in biological systems [2–5]. On the other hand, the behavior of small macromolecules such as peptides in the presence of neutral of negatively charged lipid vesicles has been in-depth investigated for a long while [6–8]. In this context, theoretical treatments have been recently used to interpret the binding isotherms [9, 10].

In general, those models are of two types, i.e., a binding model in which the vesicles are considered as macromolecules to which a certain number of probe molecules can bind, and a partition model in which one considers the membrane of the vesicles as a separate lipid phase in which the probe can dissolve. The parameters characterizing the binding data are the association constant, K_A, and the number of charged lipids that form a binding membrane site, N, for the first approach, and the partition coefficient and the effective polymer charge per polymer chain at the interface, v, for the second one. Although these models have been intensively applied for interpreting peptide/lipid vesicle interaction, little work has been carried out considering charged macromolecules that, owing to their high molar mass and number of positive charges per marcromolecular chain, will show a more complex interaction with respect to peptide/liposomes systems. In this respect, and since detailed molecular information about the interaction of polymers with lipid vesicles in scarce, we have been studying the binding behavior of a commercial polycation such as poly(vinylpyridine) (PVPy) with lipid vesicles of Dimyristoylphosphatidic acid (DMPA). PVPy

Progr Colloid Polym Sci (1996) 100:356–361
© Steinkopff Verlag 1996

(used as model for a polyelectrolyte) is selected for this study because, firstly, it is fluorescent which allows for handy determination of its relative concentration in each phase, and secondly, it can be obtained as a commercial sample at distinct molar masses allowing a detailed study at different concentration conditions.

In previous papers we have studied the binding of poly-4-vinylpyridine (P4VPy) to DMPA vesicles as a function of temperature, ionic strength and total polymer concentration [11, 12] by fluorescence and viscosity measurements. These binding curves for P4VPy/DMPA have been interpreted firstly in terms of a partition equilibrium [11] of the polymer between the aqueous and lipid phases in conjunction with the Gouy-Chapman formalism, and secondly by comparing the effective polymer charge and the number of phospholipid heads in a binding site as a function of temperature and ionic strength [12].

In the present study, we report experimental results for the binding of poly-2-vinylpyridine (P2VPy) as a function of ionic strength and total polymer concentration. Binding isotherm data have been obtained by steady-fluorescence. A pH = 3.5 has been employed, being much below the $pK_a = 5.19$ [13] of vinylpyridine to ensure that the data are for binding the mono-cationic form of the polymer. These binding curves are analyzed in terms of binding model and the partition equilibrium. The Gouy-Chapman relationship will be employed to describe electrostatic effects that are considered to dominate the interaction.

Experimental

Materials

Dimyristoylphosphatidic acid (DMPA) was purchased from Sigma Chemical Co. (St. Louis, MO, USA) and P2VPy with a molecular weight (Mw) of 2900, from Polysciences Inc. (Warrington, PA, USA). Both materials were used without further purification. For fluorescence measurements the following buffers were used: 23.5 mM sodium acetate, 0.43 M acetic acid, 1 mM EDTA (pH 3.5; ionic strength $c_s = 0.027$ M) and 47 mM sodium acetate, 0.853 M acetic acid, 1 mM EDTA (pH 3.5, $c_s = 0.050$ M).

Preparation of small unilamellar vesicles (SUV)

Lipid powder dissolved in benzene/methanol (2:1, by volume) was evaporated in a round-bottomed flask by the use of a rotary evaporator to deposit a lipid film on the wall of a flask. An appropriate amount of acetate buffer solution was added to the dried lipid film to give the required final concentration of lipid suspension. The sample was

hydrated for 5 min at 60 °C, above the gel to liquid crystalline phase transition temperature of DMPA (55 °C), and it was then vigorously vortexed for 5 min followed by sonication (1 min/mL) in a bath sonicator at 60 °C to promote the formation of unilamellar liposomes. After this process, the lipid suspension was clarified by ultracentrifugation for 10 min at 12 000 rpm to remove large unilamellar vesicles (LUV) from the small unilamellar ones (SUV).

Fluorescence measurements

The association of P2VPy to the DMPA vesicles was studied by intrinsic fluorescence experiments. Fluorescence emission spectra of the pyridine group of P2VPy in buffer solution were obtained at 20 °C on a Perkin-Elmer Fluorescence Spectrophotometer (LS-5B) equipped with a 3700 Data station with excitation wavelength set at 263 nm and 5-nm slits. The fluorescence value at 262 nm was also recorded as a base-line check and the lipid contribution to the signal was subtracted after titration of vesicles without P2VPy, under otherwise identical conditions.

In lipid/polymer mixtures the changes in fluorescence intensity, I, at 407 nm were analyzed as a function of $R_i = [L]_T/[P]_T$ (being $[L]_T$ and $[P]_T$ the total lipid and polymer concentration and from the fluorescence intensity increase, the fraction of bound polymer α defined by $\alpha = (I - I_o)/(I_m - I_o)$ was estimated as I_o and I_m the fluorescence intensities of pure polymer and of polymer at lipid saturation, respectively.

Results and discussion

Because single molecules, oligomers and macromolecules exhibiting cationic behavior appears to bind primarily in the interface, it is reasonable to treat the binding of P2VPy to phospholipid bilayer as a simple adsorption equilibrium. In order to gain insight on this equilibrium, we use a methodology recently reported, based on fluorescence spectroscopy, to evaluate partition constants of fluorophores. Primary data consist of $\Delta I/I_o$ values, I_o being the fluorescence intensity of the P2VPy in a buffer solution free of phospholipids and $\Delta I(= I - I_o)$ the increment at lipid saturation. Figure 1A shows typical fluorescence spectra of P2VPy covering a whole lipid/polymer molar ratio, R_i^*, Parts B and C display the relative increment of the fluorescence intensity $\Delta I/I_o$ and the shift of the wavelength of the maximum emission, λ, as a function of R_i^*, respectively. Values recorded for either the polymer-buffer solution and polymer in the presence of DMPA vesicles show more energy transfer compared to solutions of the polymer in pure buffered environment.

Fig. 1 Binding of poly-2-vinylpyridine to DMPA vesicles. (a) Effect of addition of dimyristoylphosphatidic acid vesicles on the emission spectrum of poly-2-vinylpyridine at pH 3.5, c_s 0.027M and 20 °C for polymer concentration 2.50 μM; the successive lipid to polymer ratio (R_i^*) is shown in the figure. (b) Relative decrease of the fluorescence intensity at 402 nm versus R_i^* at different total polymer concentration: (○) 1.25; (□) 2.50; (◇) 3.75 and (△) 5.00 μM. (c) Apparent wavelength shift

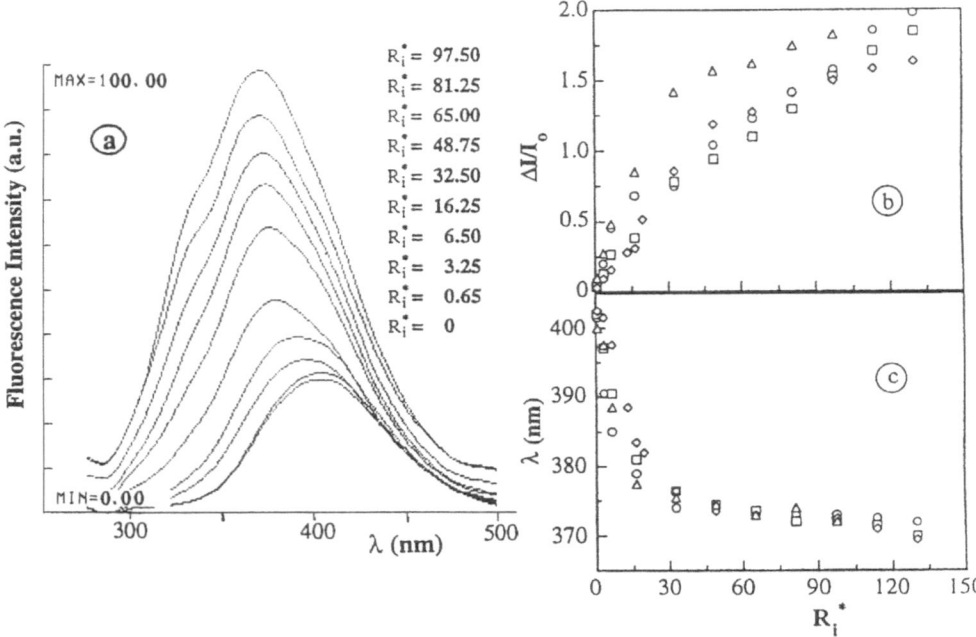

The increase in energy transfer efficiency clearly denotes strong interaction between the pyridine group and the phosphate heads of DMPA, which induces a reorganization of the polymer chains on the surface of the liposomes.

Since partitioning of a compound between a membrane and an aqueous phase is difficult to determine, and many vary with pH, ionic strength, temperature and composition of the membrane, attempts have been made to find a parameter for polyelectrolyte partitioning. Because the purpose of the present work is to examine and define experimental conditions for the study of the interfacial binding of polyelectrolytes onto the surface of phospholipid bilayers, we have developed protocols and obtained the binding isotherms for the system P2VPy in a buffered solution ($c_s = 0.027$ and 0.050 and pH = 3.5) containing DMPA based liposomes. In this context, Fig. 2 shows the binding isotherms in terms of the amount of associated polymer per accessible lipid, α/R_i^*, against the free polymer concentration, $[P]$, for P2VPy in acetate buffer (pH 3.5) at 20 °C and ionic strength of 0.027 M (panel A) and 0.050 M (panel B), respectively. Each binding isotherm shows an initial linear relationship between α/R_i^* and $[P]$ from which slope the partition coefficient is usually obtained [14, 15]. Nevertheless, as the concentration of free polymer increases a pronounced deviation from linearity is observed probably due to repulsive interactions among the polymer chains adsorbed at the outer lipid bilayer and those remaining in the bulk (close to the

interface) that hinder more polymer chains from approaching. These secondary effects are currently taken into account through the activity coefficient γ for describing a real partition coefficient. Under certain experimental conditions this deviation from the ideal behavior is accompanied by a sharp increase in α/R_i^*, as for instance in Fig. 2 panel A. Such behavior could indicate aggregate formation in the bilayer phase due to membrane saturation of adsorbed polycation, in a similar way as has been reported for melittin-Dimyristoylphosphatidicholine (DMPC) interaction [16]. It can also be observed that the total polymer concentration has no effect on the initial slope, and that there is an increase in the vesicle-associated polymer as the total polymer concentration increases, that is, the association is enhanced upon increasing the initial concentration of the macromolecule in the aqueous solution in the presence of DMPA liposomes. By comparing the binding results for both ionic strengths it is evidenced that increasing c_s slightly enhances the polymer association.

From data reported in Fig. 2, one can obtain some insight concerning the polymer-liposome association, mainly the association constant, K_A, and the number of membranes sites interacting with a polymer chain, N. Previously [11, 12], we developed two different in-depth approaches to interpret the binding isotherms, hence here we only present the fundamental equations needed for the discussion.

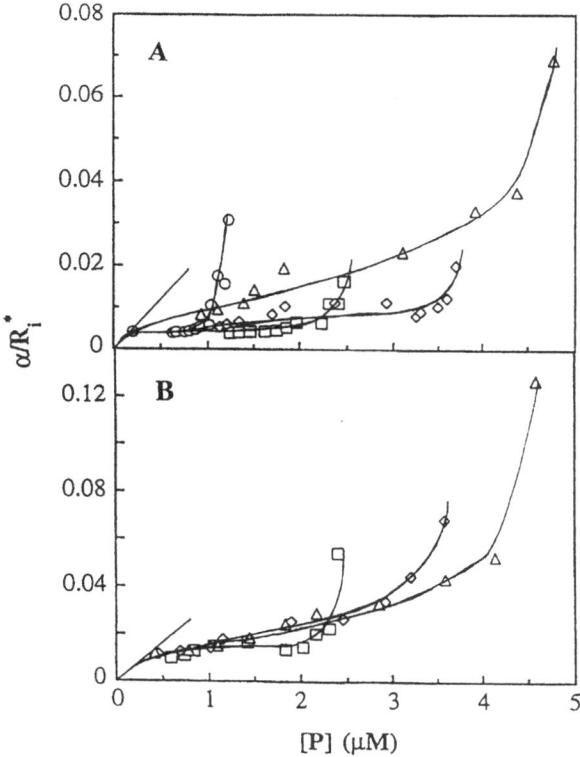

Fig. 2 Binding isotherms of P2VPy to DMPA lipid vesicles obtained at 20 °C and different total polymer concentration, $[P]_T$: (○) 1.25; (□) 2.50; (◇) 3.75 and (△) 5.00 μM. In panel A the ionic strength is 0.027 M, and in panel B 0.050 M. Solid lines are the best fit to the experimental data

The first treatment considers a binding equilibrium between the free polymer, P and a membrane site formed by N charged lipids, using the experimental data of the aforementioned isotherm falling into the linear part of the plot (see Fig. 2). In contrast, the second treatment refers to a partition equilibrium which can be explained as a buffered microphase and a liposome mesophase where the polyelectrolyte can be found.

The parameters of interest, K_A and N, are closely related with the binding isotherms by means of the α/R_i^* values at different free polymer concentration, $[P]$, as indicated from the following equation (see Eq. (4) from ref. [12]):

$$K_A = \frac{[PS_N]}{[P][S_N]} = \frac{\alpha[P]_T}{[P]\left(\frac{[L]_T}{N} - \alpha[P]_T\right)} = \frac{\alpha/R_i^*}{[P]\left(\frac{1}{N} - \frac{\alpha}{R_i^*}\right)}$$

$$(1)$$

where $[PS_N]$ is the concentration of polymer bound to N membrane sites; $[S_N]$ is the concentration of free membrane sites and $[P]$ is the concentration of the free poly-

mer. The fraction of bound polymer $\alpha = [PS_N]/[P]_T$; and $R_i^* = \beta R_i$, being $R_i = [L]_T/[P]_T$ the molar ratio between the total lipid concentration $[L]_T$ and the total polymer concentration, $[P]_T$. The prefactor $\beta = 0.65$, introduced to generate R_i^*, denotes the correction of R_i by assuming that two-thirds of the lipids stay in the outer shell of the lipid bilayer [17].

For charged substrate a more complex treatment was proposed [18]. In this context, the polymer concentration $[P]$ has two contributions: a fraction of polymer placed near to the liposome surface, $[P]_0$, and another fraction in the bulk solvent at infinite distance from the bilayer surface, $[P]_\infty$. Moreover, when both entities involved in the interaction are oppositely charged, which is the system used here, almost all polymer chains are accommodated near the liposome surface hence $[P]_\infty \ll [P]_0$ and the so-called Gouy-Chapman-Stern (GCS) equation [18] can be attained:

$$\left(\frac{\alpha}{R_i^*}\right)^3 = K_A x_L^2 \left(\frac{1}{N} - \frac{\alpha}{R_i^*}\right)[P]_\infty ,$$

$$(2)$$

where x_L denotes the mole fraction of a kind of phospholipid. In the present work $x_L = 1$ because the subunits forming the vesicle are DMPA exclusively. The value of $[P]_\infty$ comes from the expression:

$$[P]_\infty = [P]\left(\frac{\alpha}{R_i^*}\right)^2 \bigg/ x_L .$$

$$(3)$$

Values of K_A and N obtained from Eq. (2), and those obtained from equations explained in ref. [12] are compiled in Table 1 for the P2VPy in the presence of DMPA-based liposomes at diverse ionic strength and polymer concentration.

A careful inspection of these data reveals that the K_A and N values decrease when polymer concentration increases, this trend being more pronounced at lower ionic strength. This behavior is consistent with the argument based on the excluded volume effects. A portion of polymer chains is adsorbed onto the liposome surface an the next neighbor chain are electrostatically repelled from the liposome surface itself. Due to the approximation introduced in the present formalism the K_A and N values obtained have a qualitative meaning which is congruent with the previously reported for P4VPy, which exhibits the same trend. From the comparison of data obtained with both equation, it seems to be that Eq. (1) gives greater values for both K_A and N at any ionic strength and polymer concentration.

Another approach currently used to study the association of fluorescent probes to lipid vesicles is the partition

Table 1 Values of K_A and N obtained for P2VPy in aqueous solution at 20 °C and pH of 3.5 for different ionic strengths, c_s, and total polymer concentrations, $[P]_T$, in the presence of DMPA-liposomes

c_s(M)	$10^{-5} K_A(M^{-1})$				N			
	0.027[a]	0.050[a]	0.027[b]	0.050[b]	0.027[a]	0.050[a]	0.027[b]	0.050[b]
$[P]_T(\mu M)$								
1.25	21.0		20		60		60	
2.50	8.5	5.8	10	6	65	50	55	23
3.75	2.3	4.8	8	5	55	40	50	26
5.00	7.1	3.0	7	4	40	29	25	16

[a] data obtained from direct model (Eqs. (6) and (7) in ref. [12])
[b] data obtained from Eq. (3)

Table 2 Values of γ, Γ and ν calculated from Eqs. (5), (4) and (6), respectively, for the association isotherm of P2VPy on DMPA-liposomes at $[P]_T = 3.75$ μM

R_i^*	γ^a	$\Gamma \times 10^{-3}$ (M^{-1}) [a]	γ^b	$\Gamma \times 10^{-3b}$ (M^{-1}) [b]	$\nu^{a,b}$
$c_s = 0.027$ M					
84.5	0.740	12.7	1.308	22.4	100.0
74.8	0.653	12.6	1.143	22.0	83.4
68.3	0.662	11.6	1.254	22.0	83.4
65.0	0.567	16.3	0.832	23.9	74.3
48.8	0.494	9.3	0.969	18.2	62.9
$c_s = 0.050$ M					
78.0	1.238	16.7	1.594	21.5	97.7
65.0	1.409	16.5	2.031	23.8	87.7
52.0	1.776	17.8	3.360	33.6	78.6
39.0	1.435	16.5	3.321	38.3	62.1
26.0	0.775	11.3	2.336	34.1	44.3

[a] K_A and N data from Eqs. (6) and (7) in ref. [12] and compiled in Table 1
[b] K_A and N data from Eqs. (3) and compiled in Table 1

model [11]. In the frame of this model, the master equation is expressed as:

$$\frac{\alpha/R_i^*}{[P]} = \frac{\Gamma}{\gamma}, \tag{4}$$

where Γ represents the ratio of polymer concentration between the lipid phase and the buffered solution free of vesicles. γ denotes the conventional activity coefficient which takes into account the deviations of the ideality of the complex polymer-liposome buffered solution mainly due to the interpolyelectrolyte interactions near to the DMPA-based vesicle surface. Obviously the pH, ionic strength and temperature can also affect the γ values. As can be seen from Eq. (4), the ratio Γ/γ can be easily obtained from fluorescence measurements starting from the binding isotherms (see Fig. 2), however the evaluation of the individual values of this parameter is a difficult task. As can be observed from the functionality of Eq. (4) for each experimental $\alpha/R_i^*/[P]$ data, one can obtain infinite Γ, γ pairs of values. In order to select the more plausible pair of

values, we have recently derived an expression for γ based on data on the preceding model [12] given by:

$$\gamma = \frac{\beta\left(N\beta\dfrac{\alpha}{R_i^*} - 1\right)}{K_A[P]\left(1 - N\beta\dfrac{\alpha}{R_i^*}\right) - \beta} \tag{5}$$

This equation allows us to evaluate γ for every experimental point of the binding isotherm (see Fig. 2), under certain conditions. Insertion of values from Eq. (5) into Eq. (4) allows to obtain the respective Γ values for each system. An alternative way to evaluate γ was proposed by Schwarz et al. [9] based on the Gouy-Chapman model approach for charged surfaces. The activity coefficient is found to be expressed according to:

$$\ln\gamma = 2\nu \sinh^{-1}\left(\frac{\alpha}{R_i^*}\nu b + x_L z_L b\right), \tag{6}$$

where the parameter ν stands for the effective number of charges per polymer chain, P2VPy in the present work, located at the interface, z_L is the valency of the lipid head group, and b is a dimensionless quantity depending on the ionic strength of the buffered solution ($b = 3.126\,c_s^{-1/2}$ at 20 °C in our conditions). Data from Γ, γ obtained from Eqs. (4) and (5) as well as the v parameter from Eq. (6) are compiled in Table 2. Focusing our attention on Γ values, we observe that Γ increase with the ionic strength denoting that the charges of the P2VPy interact more favorably with their ionic atmospheres when located close to the DMPA bilayers than they can do in the buffered surroundings. The above argument is supported by the γ values greater than unity, as reported in Table 2 for $c_s = 0.050$ M. In contrast, for $c_s = 0.027$ M, the Γ values are around 25% lower than the preceding one and the activity coefficient, γ, is always lower than unity, denoting that the mutual electrostatic repulsion between P2VPy chains eventually become more effective. Moreover, looking at the v data of Table 2, we see a similar dependence on the R_i^*, so at both c_s values v decreases as R_i^* decrease and, as expected, when c_s decreases v increases. Nevertheless, we believe that the values obtained here are very close and must be regarded as a qualitative contribution to the description of the interphase in the polymer-liposome association phenomena.

Conclusions

Fluorescence intensity measurements is a powerful tool for establishing binding isotherms of charged polyelectrolytes onto the surface of oppositely charged liposomes. Non-ideal effects, such as electrostatic repulsion between the protonated P2VPy molecules, mainly at low ionic strength, lead to mutual electrostatic repulsion evidenced by strong deviation from the linearity of the adsorption isotherm (see Fig. 2) and cause a continuous flattening out which is, at least qualitatively, well described by the coefficient γ.

Acknowledgments Financial support from CICYT (Spain) under Grant No. PM95-0149 is gratefully acknowledged. I. Porcar is indebted to the Ministerio de Educación y Ciencia (Spain) for a fellowship grant.

References

1. Mc Quigg DW, Kaplan JI, Dubin PL (1992) J Phys Chem 96:1973–1978
2. Margolin A, Sherstynk SF, Izumrudov VA, Zezin AB, Kabanov VA (1985) Eur J Biochem 146:625–631
3. Shauer SL, Melancon P, Lee KS, Burgess MT, Record MT (1983) Cold Spring Harbor Symp Quant Biol 47:463–470
4. von Hippel PH, Bear DG, Morgan WD, McSwiggen JA (1984) Annu Rev Biochim 53:389–398
5. Yaroslavov A (1993) Polymer Preprints Vol 34 N1.
6. Alouf JE, Dufourcq J, Siffert O, Thiaudiere E, Geoffroy C (1989) Eur J Biochim 183:381–390
7. Dufourcq J, Faucon JF (1977) Biochim. Biophys. Acta 467:1–11
8. Bernard E, Faucon JF, Dufourcq J (1982) Biochim. Biophys. Acta 688:152–162
9. Schwarz G, Beschiaschvili G (1989) Biochim. Biophys. Acta 979:82–90
10. Stankowski S (1991) Biophys J 60:341–351
11. Porcar I, Gómez CM, Pérez-Payá E, Soria V, Campos A (1994) Polymer 35:4627–4637
12. Porcar I, Gómez CM, Codoñer A, Soria V, Campos A (1995) Makromol. Chem Makromol. Symp 94:171–180
13. The *Merck Index*, ed. M. Windholz, Merck & Co., Inc., tenth edition, 1983, p 1149
14. Schwarz G, Blochmann U (1993) FEBS Lett. 318:172–176
15. Stankowski S, Schwarz G (1990) Biochim Biophys Acta 1025:164–172
16. (a) Vogel H (1981) FEBS Lett 134:37–42; (b) Thiaudiere E, Siffert O, Talbot JC, Bolard J, Alouf JE, Dufourcq J (1991) Eur J Biochem 195:203–213
17. Kuchinka E, Seelig J (1989) Biochemistry 28:4216–4221
18. Barghouthi SA, Puri RK, Eftink MR (1993) Biophys Chem 46:1–11

Progr Colloid Polym Sci (1996) 100:362–367
© Steinkopff Verlag 1996

P. Jauregi
J. Varley

Lysozyme separation
by colloidal gas aphrons

P. Jauregi · Dr. J. Varley (✉)
Biotechnology and Biochemical Engineering
Group, University of Reading
P.O. Box 226
Reading RG6 6AP,
United Kingdom

Abstract Colloidal gas aphrons (CGAs) are micro bubbles (10–100 μm) created by intense stirring of a surfactant solution and are composed of a gaseous inner core surrounded by a thin surfactant film. Their size and structure confers to them colloidal behaviour, thus the main interaction forces governing these types of dispersions are surface forces and electrostatic interactions.

These dispersions have been used for several applications including separation e.g. ion flotation, dye separation. There have been no reports of their direct application to protein separation. Properties deriving from their size and structure (high interfacial area, high stability due to the thin surfactant film and the possibility of different mechanisms for protein adsorption such as hydrophobic and electrostatic interactions) suggest CGAs may be useful as a technique for protein recovery.

In the present work CGAs have been characterised for an anionic surfactant, sodium bis(2-ethyl hexyl) sulfosuccinate (AOT), and applied to lysozyme recovery. Preliminary results are presented for an AOT-lysozyme system. The protein was contacted with the aphrons. The aphron phase was separated from the bulk due to its greater buoyancy and recoveries of lysozyme in the aphron phase were determined. The effect of different factors such as initial concentration of surfactant and protein, ionic strength and pH, on the protein recovery was studied in order to optimise the recovery of the protein into the aphron phase. Initial results show that good separation can be achieved.

Key words Colloidal gas aphrons – lysozyme – AOT (sodium bis(2-ethyl hexyl)sullfosuccinate) – protein separation

Introduction

Colloidal gas aphrons (CGAs) are microbubbles composed of a gas core and surrounded by a surfactant film. They are created by intense stirring (5000–10000 rpm) of a surfactant solution. Their size range (10–100 μm) and structure (see Fig. 1 for structure proposed by Sebba [1]) confers to them colloidal behaviour and they can be considered as gas-liquid colloidal dispersions. CGAs possess several attractive features. Firstly, they have large interfacial area, and secondly, they are relatively stable because of the soapy shell surrounding the gas bubble. Electrostatic interactions (in the case of ionic surfactants) and surface forces contribute towards the stability of these types of dispersions.

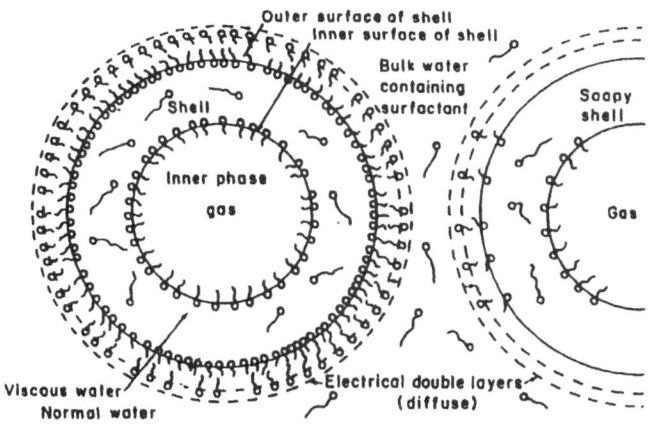

Fig. 1 Proposed structure of CGAs [1]

Recovery steps within downstream processing generally represent a large part of the total capital investment in a fermentation plant. There is a need for novel separation techniques in downstream separation. Bioseparation techniques, such as reverse micelles, aqueous two-phases and foam separation are finding application in protein separation. Selective separation has been achieved for some systems [2–8] but still these techniques need to be optimized in order to use them on an industrial scale. CGAs may provide an attractive alternative technique for protein separation.

Once the CGAs have been created and stirring is stopped, CGAs begin to cream because of the much lower density of the encapsulated gas. The CGA layer can then be separated from the bulk solution. If species can be contacted with CGAs prior to creaming this buoyancy effect is useful in the application of CGAs as a separation technique. Creaming can be delayed by stirring CGAs so that enough energy is input to overcome buoyancy forces and keep CGAs in suspension. In this state, due to their low viscosity (similar to that of water), CGAs are very easy to transport and can be pumped from one location to another. This is also a very interesting characteristic for their application as a separation technique.

In the work presented here, the potential of CGAs for protein separation has been investigated for the sodium bis(2-ethyl hexyl) sulfosuccinate (AOT)-lysozyme system. The direct application of CGAs for protein recovery has not been reported yet, but CGAs have been used as a separation technique for some other applications such as: intensification of mass transfer in aqueous two-phase systems for enzyme extraction [9]; clarification of suspensions [10]; predispersed solvent extraction of dilute products [11]; removal of sulphur crystals [12]; removal of heavy metals from aqueous solutions [13]; coflotation and

solvent sublation processes [14]; separation of organic dyes from wastewater [15] and harvesting of bacteria [16].

In the first stage of the work presented here, CGAs are characterised for an anionic surfactant (AOT) by studying the factors affecting their stability and gas hold-up. Subsequently, the CGAs are contacted with the protein in solution from where protein molecules adsorb to the aphron phase. It is expected that the protein interacts with the surfactant in the aphrons mainly due to electrostatic and/or hydrophobic forces. The aphron phase separates easily from the solution due to its buoyancy. This aphron phase is enriched in protein and effective recovery is therefore achieved. Protein recovery requires aphrons to be relatively stable: stability should be high enough to allow protein adsorption, but since protein-surfactant interaction may cause protein denaturation [17–19], aphrons need to collapse in a relatively short time period to minimise deleterious effects. Small bubbles and a uniform bubble size distribution are preferable to maximise interfacial area for the protein adsorption.

Experimental

Materials

AOT (sodium bis-(2-ethyl hexyl) sulfosuccinate) was obtained from Fisons plc. The aphronic solutions were made from buffered solutions of AOT (acetate buffer, CH_3COOH/CH_3COONa 0.1 M, for pH = 4; phosphate buffer, H_2PO_4Na/HPO_4Na_2 0.025 M and 0.01 M, for pH = 6 and 8 respectively). Acetic acid, sodium acetate (anhydrous), sodium dihydrogen orthophosphate monohydrated, di-sodium hydrogen orthophosphate (anhydrous), sodium chloride and urea, were supplied by BDH (AnalaR grade). The laboratory mixer (SL2T model) fitted with a four-bladed impeller (D = 30 mm) surrounded by a high shear screen and with speed digital read-out was supplied by Silverson Ltd. Lysozyme from chicken egg white (approx. 95% pure), BCA (Bicinchoninic acid) and copper (II) sulphate pentahydrate (4%, w/v) solutions were supplied by Sigma Chemical Company.

CGA formation

A fully baffled beaker (volume = 1 L, diameter = 105 mm) containing a buffered solution (0.4 L) of AOT was stirred at very high speed (8000 rpm) using a laboratory mixer (see Fig. 2). For each experiment, the stability of the aphrons was measured in terms of τ, which is defined here as the time required for half the amount of original liquid to drain. The gas hold-up (ε) which is the gas volumetric ratio

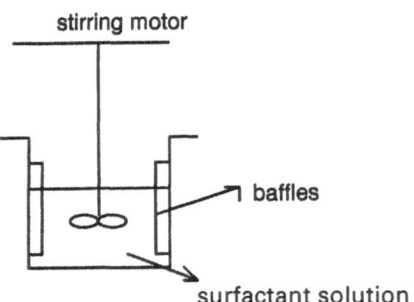

Fig. 2 Stirring system for the formation of the CGAs

i.e., the ratio between the gas volume and the dispersion final volume (after stirring has ceased), was also measured for each of the experiments.

CGA characterisation

In order to determine the effect of different parameters on the stability and gas hold-up of the aphrons a statistical design (central composite) was developed. Five factors were chosen and studied at three different levels (low, medium and high), except for the temperature factor which was studied at two levels. This combination resulted in 38 experiments. The factors studied and levels chosen were as follows: concentration of surfactant (0.1, 2.5, 61 mM), concentration of salt (0, 0.07, 0.14 M), pH (4, 6, 8), time of stirring (4, 10 and 16 mins), and temperature (controlled, by using a water jacket, and non controlled, by allowing the temperature to rise while stirring).

Lysozyme recovery

In these experiments, the protein solution was stirred using a magnetic stirrer and a volume of CGAs was added by pipette. The mixture was stirred for 15 min using a magnetic stirrer and then the aphronic phase (foam phase) and the liquid phase (retentate phase) were separated out by pipetting the total retentate phase (bottom phase). The amount of protein present in each phase was determined by using the BCA method [20] as indicated below. In some cases, urea was added to solubilize the precipitated protein. For each experiment the following parameters were calculated: percentage of protein separated into the foam (% recovery), percentage of protein which precipitated out (% lost), enrichment ratio (EnR), which is the ratio between the protein concentration in the foam phase, $(Cp)_f$, and the initial protein concentration, $(Cp)_o$, and separation ratio (SpR), which is the ratio between the

protein concentration in the foam phase and the protein concentration in the retentate phase, $(Cp)_r$.

Protein determination

The determination of protein was carried out by following the Sigma protocol. The BCA method was chosen for protein determination as it determines protein accurately even in the presence of surfactant [20].

Results and discussion

Characterization of CGA

From the statistical study, it can be concluded that the concentration of surfactant and salt are the main factors affecting the stability of CGAs. Effects of temperature, time and pH follow in decreasing order. The effect of concentration of surfactant (C_{sf}) shows that higher concentrations of surfactant give higher stability. This can be explained on the basis of Sebba's proposed structure for CGAs (see Fig. 1). Increasing concentration of surfactant will favour this structure, as additional surfactant molecules will contribute towards the surfactant shell, which might be composed of several surfactant layers. This shell, surrounding the gas core, will delay the coalescence of the aphrons resulting in a high stability dispersion. Another explanation for this effect could be that a larger amount of aphrons is formed at higher surfactant concentration. The effect of concentration of salt (C_{sl}) shows that stability decreases with increasing concentration of salt. This can be explained on the basis of electrostatic interactions among the aphrons. When the aphrons are formed, the resulting dispersion behaves as a colloidal dispersion. The repulsive electrostatic interactions between the double layers of the negatively charged aphrons will stabilise the system. Increasing the concentration of salt (NaCl) will cause these interactions to be suppressed leading to the formation of a less stable dispersion. Concentration of salt interacts with concentration of surfactant as shown in Fig. 3. At low concentrations of surfactant there is no significant effect of salt concentration on the stability, but as surfactant concentration increases, the effect of the salt concentration increases. Therefore the maximum stability is obtained at the highest concentration of surfactant and lowest concentration of salt. pH does not show a significant effect on CGA stability. A possible reason for this is that for the pH range in which the experiments were performed, the surfactant in solution was present mainly as a salt (pKa = 2.9. The ionic form of the surfactant will contribute to a larger extent towards the stabilisation of the aphrons.

Progr Colloid Polym Sci (1996) 100:362–367
© Steinkopff Verlag 1996

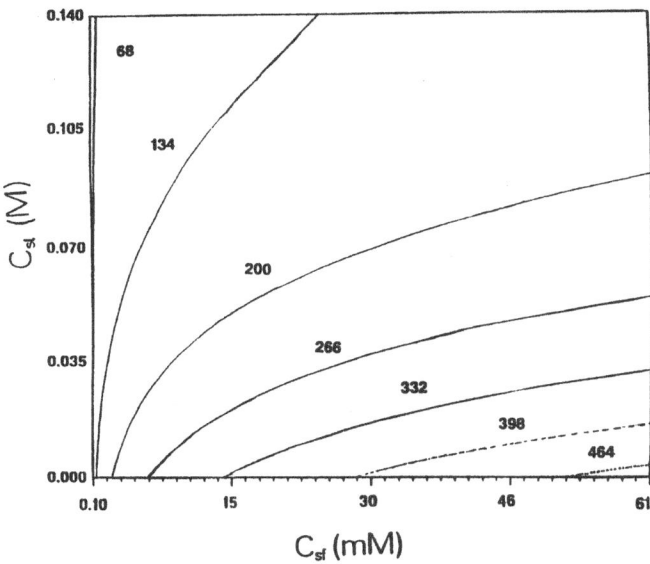

Fig. 3 Contour plot for the interaction of concentration of surfactant (C_{sf}) and salt (C_{sf}) at pH = 6, t = 4 min, under controlled temperature. The numbers on the curves correspond to the predicted stability values (τ) in seconds

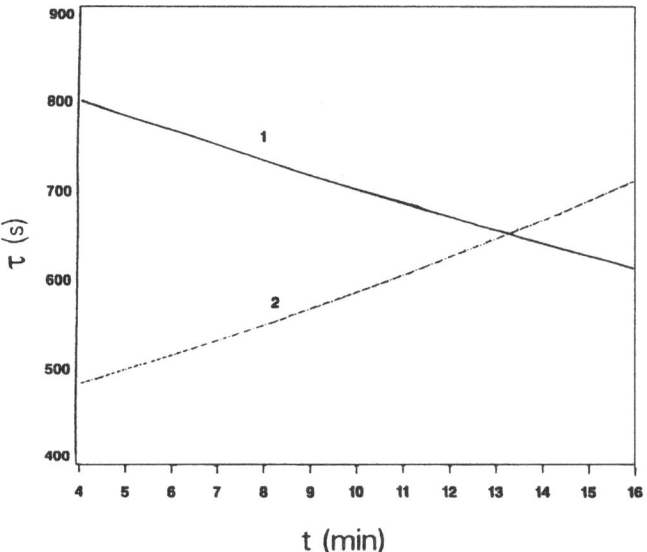

Fig. 4 Plot of the predicted values of stability (τ) and time (t) for the two different temperatures studied (1 and 2, non controlled and controlled respectively) at C_{sf} = 61 mM, C_{sf} = 0 and pH = 6

The interaction between time and temperature shows a significant effect on stability whereas time and temperature individually do not. The effect of interaction of time with temperature shows that the highest stability is obtained at a longer time of stirring under controlled temperature (below or equal to 25°C) and at shorter time of stirring under non controlled temperatures (see Fig. 4).

Thus the effect of temperature is negative at long times of stirring so it needs to be controlled.

Gas hold-up (ε) shows no significant dependence on pH and temperature in the range of studied values. As mentioned above the AOT is present in the ionic form which will favour formation of the aphrons and high gas hold-ups. Gas hold-up depends mainly on concentration of surfactant followed by concentration of salt and time. There are no interactions shown in this case which means that surfactant concentration, salt concentration and time affect independently the response i.e. gas hold-up. Gas hold-up tends to a maximum ($\varepsilon = 0.6$) at about 12 mM surfactant concentration. Concentration of salt has a negative effect on the gas hold-up, the higher the concentration of salt, the lower the gas hold-up. Time has an effect to a smaller extent, but increasing time of stirring leads to higher gas hold-up.

Lysozyme recovery using CGAs

Initial experiments were carried out to study the separation of lysozyme by using CGAs from a solution of AOT. The main variables studied were: initial concentration of surfactant and protein present in the mixture, $(Cs)_o$ and $(Cp)_o$ respectively, initial surfactant to protein ratio $(ms/mp)_o$, pH and ionic strength of the protein and surfactant solutions, $(is)_p$ and $(is)_s$ respectively (see Table 1). Ionic strength in the protein solution was varied by using different concentrations of the buffer salt, i.e.: for pH = 8, phosphate buffer 0.01 M (is = 0.029 M) and 0.1 M (is = 0.29 M); for pH = 6, phosphate buffer 0.025 M (is = 0.031 M) and 0.1 M (is = 0.12 M); for pH = 4, acetate buffer 0.1 M (is = 0.018 M) and 0.1 M NaCl in acetate buffer 0.1 M (is = 0.12 M).

When high concentrations of surfactant (above the critical micelle concentration value, 2.4 mM) were used no significant separation was observed, therefore following experiments were carried out at lower concentrations of surfactant. First five experiments were carried out at fixed pH and ionic strength and only the initial amount of protein and surfactant were varied. Results obtained (see Table 2) show that as the ratio $(ms/mp)_o$ decreases the recovery of the lysozyme increases for experiments 1–3. On the contrary, if experiments 4 and 5 are compared, it is observed that the highest recovery is obtained at the highest $(ms/mp)_o$ ratio. In most of these experiments a considerable amount of protein precipitates out in both phases. Only experiment 5 shows no significant precipitation in the retentate phase. These results indicate that there is an optimum surfactant to protein ratio for which the recovery can be maximized.

Table 1 Initial conditions for the lysozyme recovery experiments

Exper.	$(ms/mp)_o$	pH	$(is)_p$ (M)	$(is)_s$ (M)	$(Cp)_o$ (mg/ml)
1	0.78	8	0.29	0.29	0.23
2	0.52	8	0.29	0.29	0.31
3	0.41	8	0.29	0.29	0.37
4	0.43	8	0.29	0.29	0.33
5	0.32	8	0.29	0.29	0.39
6	0.43	8	0.29	0.29	0.34
7	0.43	8	0.029	0.29	0.34
8	0.43	6	0.031	0.031	0.34
9	0.43	6	0.12	0.031	0.34
10	0.43	4	0.018	0.018	0.34
11	0.43	4	0.12	0.018	0.34

In experiments 6–11 the effect of pH and ionic strength on the recovery was studied by fixing the initial surfactant and protein amount in the mixture and varying pH and ionic strength. Results show the highest recoveries at pH = 6 (75%) but recoveries at pH = 8 are not significantly lower (69%). The lowest recoveries were measured at pH = 4 (54%). The retentate and foam phase of experiments carried out at pH = 4, show a significant amount of precipitate which could not be redissolved using urea. Only the retentate of experiment 11 shows a clear solution after adding urea. Differences in results could be explained on the basis of the variation in the density of charge of the lysozyme with the pH. A titration curve of the lysozyme showed that at pH = 8 lysozyme is not significantly positively charged. At pH = 6 there is not a significant difference in the density of charge whereas at pH = 4 lysozyme is mainly positively charged. Electrostatic interactions will be favoured at this pH which could cause formation of protein-surfactant insoluble aggregates. Ionic strength seems to have an effect only in the experiments carried out at pH = 4; the experiment carried out in the presence of

0.1 M NaCl results in a clear retentate with a low amount of protein remaining (see experiment 11 in Table 2).

Conclusions

From the characterisation study of CGAs it can be concluded that stability is optimised at a higher concentration of surfactant (above the CMC value) at any of the studied pHs. The latter factor does not show a significant effect; this is consistent with observations reported by Subramaniam et al. [10] and Save and Pangarkar [9, 16]. Addition of salt i.e. high ionic strength, causes a decrease of stability of CGAs which implies that electrostatic forces play an important role in the formation and stability of CGAs. This effect has not been studied in previous reports. Amiri and Woodburn [12] reported a significant effect of pH on the stability but this was due to changes in ionic strength. Thus stability of CGAs does not depend significantly on pH but it does depend on ionic strength. Gas hold-up depends mainly on time, concentration of surfactant and concentration of salt, each of which has an independent effect. Gas hold-up increases with time but this effect is not very significant which means that most of the gas is entrained before 4 min stirring.

From the study of the recovery of lysozyme with CGAs, it can be concluded that the protein can be separated from aqueous solutions using CGAs. Maximum recoveries (75%), enrichment (25.3) and separation ratios (143) obtained in these experiments, indicate that CGAs have potential as a separation method. Initial surfactant to protein ratio, pH and ionic strength were proved to be important parameters for the optimisation of the recovery. High surfactant to protein ratios lead to low recovery. This could be because the formation of surfactant and protein aggregates are favoured at such conditions. Conversely, if

Table 2 Results for the lysozyme recovery experiments; the amount of protein measured in the retentate phase (R) and in the foam phase (F) are indicated

Exps.	R(mg)	F(mg)	% rec.	% lost	$(Cp)_r$ (mg/ml)	$(Cp)_f$ (mg/ml)	SpR	EnR
1	1.125	0.08	4	35	0.15	0.16	01.1	0.7
2	1.125	0.38	14	45	0.15	0.25	01.7	0.8
3	0.58	1.11	30	54	0.08	0.44	05.7	1.2
4	0.08	0.72	20	46	0.01	0.80	100	2.4
5	1.57	0.39	8	16	0.14	0.43	3.2	1.1
6	0.61	2.47	69	14	0.06	8.23	143	24.2
7	0.48	2.29	64	23	0.05	1.76	35.2	5.2
8	0.91	2.69	75	0[†]	0.10	2.45	25.3	7.2
9	1.47	2.14	59	0[†]	0.15	3.06	20.4	9
10	1.15	1.62	45	23	0.12	2.31	20.1	6.8
11	0.61	1.96	54	29	0.06	4.90	82.7	14.4

[†] The accounted losses in experiments 8 and 9 were 0.14% and -0.08% respectively, which could be due to experimental error

Progr Colloid Polym Sci (1996) 100:362–367
© Steinkopff Verlag 1996

surfactant to protein ratio is too low, the amount of protein present is higher than the amount of protein per aphron that can be adsorbed, which leads to poor recoveries. The pH of the protein solution only plays an important role when pH = 4. Further experiments need to be carried out in order to fully understand protein adsorption to CGAs and consequently to optimize protein recovery.

Acknowledgments The authors would like to thank the Basque Government and BBSRC for their financial support and Dr. Steven Gilmour from the department of Applied Statistics of Reading University, for his assistance in the statistical design. We extend our gratitude to Dr. Bob Rustall from the Biotechnology and Biochemical Engineering Group of Reading University, for his advice concerning the protein study.

References

1. Sebba F (1987) In: *Foams and Biliquid Foams-Aphrons.* John Wiley & Sons Ltd Chichester
2. Goklen KE, Hatton TA (1987) Sep Sci Technol 22: 831–841
3. JA, Andrews BA (1990) In: Asenjo JA (ed) Separation Processes in Biotechnology. Marcel Dekker, New York, pp 143–175
4. Kula MR (1979) Applied Biochem Bioeng 2:71–95
5. Charm SE (1972) In: Lemlich R (ed) Adsortive bubble separation techniques. Academic press, New York and London, pp 157–174
6. Ahmad SI (1975) Sep Sci 10:6731-pt–1pt688
7. Sarkar P, Bhattacharya P, Mukherjea RN, Mukherjea M (1987) Biotechnol Bioeng 24:9341pt–940

8. Brown L, Narsimhan G, Wankat PC (1990) Biotechnol Bioeng 36:947–959
9. Save SV, Pangarkar VG, Kumar SV (1993) Biotechnol Bioeng 41:72–78
10. Subramaniam MB, Blakebrough N, Hashim MA (1990) J Chem Technol Biotechnol 48:41–60
11. Matsushita K, Mollah AH, Stuckey DC, Del Cerro C, Bailey AI (1992) Colloids Surf 69:65–72
12. Amiri MC, Woodburn ET (1990) Trans Ins Chem Eng 68:154–160
13. Ciriello S, Barnett SM, Deluise FJ (1982) Sep Sci Technol 17:521–534
14. Caballero M, Cela R, Perez-Bustamante JA (1989) Sep Sci Technol 24:629–640
15. Roy D, Valsaraj KT, Kottai SA (1992) Sep Sci Technol 25:573–588

16. Save SV, Pangarkar VG, Kumar SV (1995) Technol Biotechnol 62:192–199
17. Helenius A, Simons K (1975) Biochim Biophys Acta 415:29–79
18. Lapanje S (1978) In: Survey of denaturation studies. Denaturation by detergents. John Wiley, New York
19. Attwood DA, Florence AT (1983) In: Surfactant Systems: their chemistry, pharmacy and biology. Chapman and Hall, London, pp 629–638
20. Smith PK, Krohn RI, Hermanson GT, Mallia AK, Gartner FH, Provenzano MD, Jujimoto EK, Goeke NM, Olson BJ, Klenk DC (1985) Anal Biochem 150:76–85

Progr Colloid Polym Sci (1996) 100:368–369
© Steinkopff Verlag 1996

Amokrane S   186
Andrade SM   195
Anikin MV   338
Anikin AV   338
Arbuzova A   345
Avramiotis S   286
Azemar N   132

Babič-Ivanćič V   24
Bähr G   330
Banić K   221
Bartsch E   127, 151
Bastos-González D   217
Bault P   43
Bee A   212
Belloni L   54
Benabdeljalil K   143
Berger K   9
Berjano M   246
Bikanga R   43
Blondel JAK   311
Bobola P   186
Boerboom JFG   321
Bonet Avalos J   117
Bordi F   170
Bouwman WG   351
Branda VD   156
Brasselet S   68
Brezesinski G   351
Bru R   276
Buess-Herman C   112
Burns NL   271
Butt HJ   91
Bydén M   6

Cabrerizo-Vílchez M   73
Calderó G   132
Camardo M   177
Cametti C   170
Campos A   356
Carlà M   182
Carrera I   132
Castle J   259
Cebers A   101
Cerpa A   266
Chantres JR   29
Cherezov VG   338
Chittofrati A   182
Chupin VV   338
Cinteza O   156
Codastefano P   170
Coderch L   230
Cooke DJ   311
Costa SMB   195

D'Angelo M   177
De Cuyper M   306
de la Maza A   230
de las Nieves FJ   217
de Meijere K   351
Degiorgio V   201
del Val MI   296
Dhont JKG   81

Di Pietro G   201
Dokic P   235
Dumont F   112

Edlund H   6
Egea MA   107
Elorza B   29
Elorza MA   29
Erdeljan A   235

Fainerman VB   316
Farago B   162
Farid A   259
Fauconnier N   212
Fernández-Barbero A   73
Filipović-Vinceković N   24
Fioretto D   177
Fischer-Palković I   36
Forcada J   87

Gallegos C   246
Gambi CMC   182
Garcell LR   266
García MJ   132
García ML   107
García RM   356
García-Carmona F   276
García-González MT   266
Gerharz B   91
Giri MG   182
Godé P   43
Gómez CM   356
Gordeliy VI   338
Gradzielski M   162
Grigoriev P   330
Guerrero A   246

Hianik T   301
Hidalgo-Alvarez R   217
Hiltrop K   9
Hoffmann H   64
Holmberg K   271, 281
Hu Y   19

Imae T   1
Infante M-R   290

Jauregi P   362
Joanny JF   117
Jochems AMP   321
John E   330
Johner A   117
Joos P   316

Kasper A   127, 151
Kato T   15
Khan A   6
Kirsch S   127, 151
Kizling J   281
Kjaer K   351
Kochurova NN   328
Köhler W   127
Kostarelos K   206

Krivánek R   301
Kunieda H   1

Lachaise J   143
Langevin D   162
Ledoux I   68
Lianos P   286
Lin T   19
Lindström B   6
Liu W-J   19
Llorens J   252
Lope O   230
López-Nicolás JM   276
López-Quintela MA   191
Lu RJ   311
Luckham PF   206
Lutterbach N   54

Mahía J   191
Manresa M-A   290
Mans C   252
Marion G   143
Martínez-García R   73
Martinis M   36
Maugras M   290
Miguel MC   96
Miller R   316
Mira J   191
Möhwald H   351
Molina L   290
Momper B   91
Mortensen K   19
Müller J   121
Mutz M   330

Nakamura K   1

Oh S-G   281
Olteanu M   156
Onori G   177
Otero C   296
Ottewill RH   60

Pagés X   107
Paillette M   68
Palberg T   121, 241
Palmieri L   177
Parra JL   230
Partal P   246
Peikov V   64
Penfold J   311
Perani A   290
Pertz S   156
Piazza R   201
Pons JM   212
Pons R   132
Popović S   36
Porcar I   356
Prins A   321
Punčec S   36

Radeva Ts   64
Raicu V   156

Regnaut C   186
Rennie AR   60
Renth F   127, 151
Requena J   266
Reus V   54
Rivas J   191
Robledo L   296
Rodado P   29
Roger J   212
Ronco G   43
Rouch J   170
Rubí JM   96
Rudé E   252
Rusanov AI   328

Salager J-L   137
Salgueiro MA   107
Samseth J   19
Sánchez-Ferrer A   276
Sanguineti A   182
Santos RM   87
Santucci A   177
Sargent DF   301
Scalas E   351
Schäfer R   127
Schwarz G   345
Sciortino F   170
Selve C   290

Semenov AN   117
Senatra D   182
Serna CJ   266
Shchipunov YA   39
Shigeta K   1
Shulepov SYu   148
Shulepov YuV   148
Shumilina EV   39
Sillescu H   127, 151
Škrtíc D   24
Sokoliková L   301
Solans C   132
Soria V   356
Sovilj M   235
Sovilj V   235
Stébé M-J   290
Stölken S   127
Stoylov SP   64
Stubičar M   221
Stubičar N   221

Tadros ThF   206
Tartaglia P   170
Tartaj P   266
Teixeira J   338
Težak Đ   36
Thomas RK   311

Valero J   107
Valls O   107
Valls R   107
van der Linden CC   117
Van Helden AK   48
van Kalsbeek HKAI   321
Varley J   362
Vázques-Vázquez C   191
Verbeiren P   112
Verduin H   81
Versmold H   54
Villa P   43
Vinceová K   301
von Hünerbein S   241

Wijnen ME   321
Williams A   321
Winterhalter M   330
Woodcock LV   259
Würth M   241

Xenakis A   286

Zemb T   54
Zyss J   68

Progr Colloid Polym Sci (1996) 100:370–371
© Steinkopff Verlag 1996

SUBJECT INDEX

5-fluorouracil 29
5-lipoxygenase 276

acetal group 87
ACTH 301
activation 276
adhesive spheres 186
adsorption 39, 212
AFM-MFFT 91
aggregation 36
alkylcyanoacrylates 107
amphiphiles 351
antibiotic 290
AOT (sodium bis(2-ethyl hexyl) sullfosuccinate) 177, 362
apoptosis 290
aqueous solutions 328
association 39
– isotherm 345

bacterial adherence 271
bending energy 162
bile salt 24
binding isotherms 356
biocompatibility 290
bromopolystyrene particles 54

carboxyfluorescein 345
cations 206
ceramics 252
change of surface charge 64
citric acid 212
CMC 43
coarsening 101
colloidal aggregation 73
– crystal 54
– dispersion 81, 241
– gas aphrons 362
– systems 96
colloids 127, 201, 259
complex fluids 60
concentrated emulsion 132
– dispersions 148
– solutions 15
– suspensions 266
concentration domains 101
conductivity 170
core-shell particles 151
critical behavior 81
crosslink density 127
crown ether 281
cryo-electron microscopy 1
cyclodextrin 276
cytochrome c 306
– oxidase 306

D-glucose derivatives 43
deflocculants 252
deoxycholate 29
detergents 48
diclofenac 107
dielectric relaxation 177
– study 182

diffusion 132
dioctadecyldimethyl-ammoniumbromide (DODAB) 306
dipole potential 330
dispersion of electric conductivity with frequency 156
dodecyl pyridinium bromide 1
DPH 29
drainage 143
drug delivery systems 29
dynamic light scattering 121, 151
— tension 328
– scaling 73
– surface tension 316

Eicke-Hall theory 170
electric field 330
– light scattering 64
electroosmosis 271
electrostriction 301
emulsifier 246
emulsion 246
– polymerization 87, 91
ensemble averaging 121
enzymes (membrane-bound) 306
extrusion (prep. meth.) 29

ferrofluids 96
fluorescence 195, 356
– anisotropy 29
fluorinated compound 182
foam 143
forced Rayleigh scattering 127
formulation 137
fractal structure 73
frequency dependence 182
functionalized polymer colloids 217

gas transfer 143
Gd_2CuO_4 191
gel layer 321
gluconic acid 212
goethite 266
graphite suspensions 156
grazing incidence x-ray diffraction 351

heat treatment 217
heterogeneous latexes 91
hexagonal phase 101
HIPRE 132
HLB 137
HPLC 212
hydration force 338

immunoassays 87
inclusion complex 276
inner field compensation 330
interactions 338
interfaces 117
interfacial tension 162
ion-selective electrode 39
IR spectra 117
isoelectric point 266

kinetics of crystal growth 221

lamellar mesophase 36
– phase 9
lead fluoride 221
lecithin 286
light microscopy 151
– scattering 73, 143
linoleic acid 276
lipid bilayers 301
– membrane 330
liposome interactions 29
liposomes 206
liquid crystal 6, 9, 24, 281
long-range interactions 101
loops and tails 117
lubricants 48
luminescence decay 286
lyotropic 9
– liquid crystals 15
lysozyme 362

magnetic fluids 212
– particles 96
– properties 191
magneto(proteo)liposomes 306
magnetophoresis (high-gradient) 306
Mastoparan-X 345
maximum bubble pressure method 316
mean squared displacements 151
membranes 306
micelles 15, 43, 48
micellization 39
– of Trito X-100 221
microemulsion 68, 137, 162, 170, 177, 182, 186, 191, 195, 281, 286
micronetwork spheres 151
microparticles 191
mixed layers 311
– micelles 19
molecular, organized system 290
monolayers 48, 351
morphology 91
multiscaling 36

nanoparticles 107
neutron reflection 311
– scattering 60
non-equilibrium phase transitions 241
— phenomena 96
— ionic surfactant 221
nonequilibrium surface electric potential 328
nonergodic systems 121
nonionic surfactant 15, 43, 201
nonlinear rheology 241
– scattering 68

octyl glucoside 230
organic synthesis 281
osmotic pressure 54
overlap potential 186

peptide-lipid interaction 345
percolation 182
– threshold 170
permeability 235
pF-stat measurements 221
phase diagram 6
– separation 201
– transfer agent 281
– transitions 301
phospholipids 306, 356
photon correlation spectroscopy 121
Piroxicam 195
PIT 137
polistes mastoparan 345
poly-2-vinylpyridine 356
polyacrylamide adsorption 156
polydispersity 162
polymer-adsorption 107, 117
— desorption 107
polymeric membranes 338
polystyrene latices 60
POPC 345
precipitation 24
protein adsorption 271
– separation 362
proteins 321
PTFE particles 64

Q salt 281
quadratic hyperpolarizability
   coefficients 68

reflectometry 143
refractive index 112
relaxation dynamics 96
release 132
– kinetic 235
reverse micelles 195, 286
– vesicles 1
rheology 246, 252, 259, 266

SANS 162
scaling laws 117
SDS-gelatin interaction 235
self-diffusion 5, 127
— similarity 36
serpentine 266
shear flow 81, 24
short-chain lecithins 19
simulation 48
single particle detection 73
skin formation 321
SLS 9
small-angle light scattering 81
— neutron scattering 19
sol–gel 191
spectroturbidimetry 112
β-lactam 290
stability range 212
steric stabilization 206
stratum corneum dissociation 230
— solubilization 230
stripe phase 101

structure 60, 338, 351
– factor 148
substitution reaction 281
sucrose alkanoates 1
– ester 246
surface aggregation 39
– charge 271
– friction 259
– potential 330
– rheology 321
surfactant 6, 29, 39, 48, 137, 281, 290, 311
– solutions 316

tellurium 112
temperature effects 316
thermodynamic data 43
– theory 19
thin films 112
thixotropy 252
trypsin 286
turbidity 81

ultrasonic velocimetry 301
USAXS 54

vesicles 206
video enhanced microscopy 1
viscoelasticity 246

x-ray diffraction 9, 338
— light or neutron scattering 148

zeta-potential 217